Improving Homeland Security Decisions

What are the risks of terrorism and what are their consequences and economic impacts? Are we safer from terrorism today than before 9/11? Does the government spend our homeland security funds well? These questions motivated a 14-year research program of the National Center for Risk and Economic Analysis of Terrorism Events (CREATE) at the University of Southern California, funded by the Department of Homeland Security. This book showcases some of the most important results of this research and offers key insights on how to address the most important security problems of our time.

Written for homeland security researchers and practitioners, this book covers a wide range of methodologies and real-world examples of how to reduce terrorism risks, increase the efficient use of homeland security resources, and thereby make better decisions overall.

Ali E. Abbas is Professor of Industrial and Systems Engineering and Professor of Public Policy and Management at the University of Southern California. He also serves as the director of the Center for Interdisciplinary Decisions and Ethics (DECIDE). His research interests focus on all aspects of decision making under uncertainty broadly defined. He is the author of two books and two edited volumes, and has more than 100 refereed publications. He has organized numerous workshops on decision making in large-scale systems with NASA Langley, the Transportation Security Administration (TSA), the Department of Homeland Security, and the National Science Foundation.

Milind Tambe is Professor of Computer Science and Professor of Industrial and Systems Engineering at the University of Southern California. He is a founding co-director of the USC Center on Artificial Intelligence in Society. Among his pioneering contributions in Artificial Intelligence is the security games framework that has been deployed by agencies such as the U.S. Coast Guard, Federal Air Marshals Service, and various nongovernmental organizations, for which he has received several commendations. He is a fellow of the Association for Advancement of Artificial Intelligence (AAAI) and the Association for Computing Machinery (ACM).

Detlof von Winterfeldt is Professor of Industrial and Systems Engineering and Professor of Public Policy and Management at the University of Southern California. His research interests are in the foundation and practice of decision and risk analysis. In 2004, he cofounded the National Center for Risk and Economic Analysis of Terrorism Events (CREATE) and serves as its current director. He is the author of two books, three edited volumes, and more than 100 refereed articles and chapters. He is a fellow of the Institute for Operations Research and the Management Sciences (INFORMS) and of the Society for Risk Analysis.

Improving Homeland Security Decisions

Edited by

ALI E. ABBAS

University of Southern California, Los Angeles

MILIND TAMBE

University of Southern California, Los Angeles

DETLOF VON WINTERFELDT

University of Southern California, Los Angeles

CAMBRIDGE
UNIVERSITY PRESS

CAMBRIDGE
UNIVERSITY PRESS

University Printing House, Cambridge CB2 8BS, United Kingdom

One Liberty Plaza, 20th Floor, New York, NY 10006, USA

477 Williamstown Road, Port Melbourne, VIC 3207, Australia

314–321, 3rd Floor, Plot 3, Splendor Forum, Jasola District Centre, New Delhi – 110025, India

79 Anson Road, #06-04/06, Singapore 079906

Cambridge University Press is part of the University of Cambridge.

It furthers the University's mission by disseminating knowledge in the pursuit of education, learning, and research at the highest international levels of excellence.

www.cambridge.org
Information on this title: www.cambridge.org/9781107161887
DOI: 10.1017/9781316676714

First published 2017

Printed in the United Kingdom by Clays Ltd.

A catalogue record for this publication is available from the British Library.

Library of Congress Cataloging-in-Publication Data
Names: Abbas, Ali E. (Ali El-Sayed), editor. |
Tambe, Milind, 1965– editor. | Von Winterfeldt, Detlof, editor.
Title: Improving homeland security decisions / edited by Ali E. Abbas, University of Southern California, Los Angeles, Milind Tambe, University of Southern California, Los Angeles, Detlof von Winterfeldt, University of Southern California, Los Angeles.
Description: Cambridge, United Kingdom; New York, NY:
Cambridge University Press, 2017. |
Includes bibliographical references and index.
Identifiers: LCCN 2017001016 | ISBN 9781107161887 (hardback)
Subjects: LCSH: National security – United States. | Internal security – United States. |
Civil defense – United States.
Classification: LCC UA23.I338 2017 | DDC 353.3/2330973–dc23
LC record available at https://lccn.loc.gov/2017001016

ISBN 978-1-107-16188-7 Hardback

Contents

Contributors

Ali E. Abbas, University of Southern California

Bo An, Nanyang Technological University

Misak Avetisyan, Texas Tech University

Charles Baschnagel, University of Maryland, College Park

Steven P. Bennett, United States Department of Homeland Security

Vicki Bier, University of Wisconsin, Madison

Avrim Blum, Carnegie Mellon University

Andrew J. Collins, Old Dominion University

Tony Cox, Cox Associates

Jinshu Cui, University of Southern California

Francesco Delle Fave, Disney Research Boston

Maged Dessouky, University of Southern California

Robin L. Dillon, Georgetown University

Peter B. Dixon, Victoria University, Melbourne

Barry Ezell, Old Dominion University, Virginia

Scott Farrow, University of Maryland Baltimore County

Ryan J. B. Garcia, Rochester Institute of Technology

Peter Gordon, University of Southern California

Nika Haghtalab, Carnegie Mellon University

William Haskell, University of Southern California

Nathaniel Heatwole, University of Southern California

Willis H. Henry, RAND Corporation

Manish Jain, Armorway Inc.

Hongzhong Jia, University of Southern California

Richard S. John, University of Southern California

Edward H. Kaplan, Yale

Greg Keeney, University of Southern California

Ralph L. Keeney, Duke University

Christopher Kiekintveld, University of Texas, El Paso

Yunkyung Kim, University of Southern California

Howard Kunreuther, University of Pennsylvania

John Lathrop, Decision Strategies, LLC

Philip Leclerc, Virginia Commonwealth University

Russell Lundberg, RAND Corporation

Isaac Maya, University of Southern California

Jason R. W. Merrick, Virginia Commonwealth University

Erwann Michel-Kerjan, University of Pennsylvania

James E. Moore, II, University of Southern California

Thanh H. Nguyen, University of Southern California

Fernando Ordóñez, Universidad de Chile

Jiyoung Park, University at Buffalo, The State University of New York

Gregory S. Parnell, University of Arkansas

James Pita, Armorway, Inc.

Fynnwin Prager, University of Southern California

Ariel D. Procaccia, Carnegie Mellon University

Harry W. Richardson, Autonomous University, State of Mexico, Toluca

Maureen T. Rimmer, Victoria University, Melbourne

Bryan Roberts, Econometrica, Inc.

Adam Rose, University of Southern California

Heather Rosoff, University of Southern California

Xiaojun Shan, University at Buffalo

Zhihong Shen, University of Southern California

Eric Shieh, University of Southern California

Johannes Siebert, University of Bayreuth

John Sokolowski, Old Dominion University

Dušan M. Stipanović, University of Illinois at Urbana-Champaign

Milind Tambe, University of Southern California

Matthew E. Taylor, Washington State University

Catherine H. Tinsley, Georgetown University

Dan Wei, University of Southern California

Lawrence M. Wein, Stanford University

Detlof von Winterfeldt, University of Southern California

Glyn Wittwer, Victoria University, Melbourne

Jun Zhuang, University at Buffalo

1

Improving Homeland Security Decisions

Ali E. Abbas, Milind Tambe, and Detlof von Winterfeldt

1.1 Introduction

The Second World War marked an important era of significant advances in the history of science. Major developments in the fields of operations research, control theory, dynamic systems, game theory, utility theory, and information theory were established. These developments were motivated by the problems encountered such as the need to better assign resources, to schedule routes, to keep inventory, to send coded messages, and to handle the complexities of these challenges in the face of great uncertainty.

The attack on the World Trade Center on September 11, 2001, also spurred a significant effort to increase our nation's scientific capabilities to deal with adversaries. The worldwide threats of terrorism, drug smuggling, and crime have led to a large increase in new models and tools for homeland security research. The new struggles we encounter today are ubiquitous, across the globe. Yet, given limited security resources, a key question is how to best utilize our limited resources and make sound decisions related to homeland security.

The Center for Risk and Economic Analysis of Terrorism Events (CREATE) has been at the forefront of this research since it began operations in 2004. As the nation's first university-based center of excellence, CREATE provides research and deployed applications of risk analysis, economic analysis,

Research for many chapters in this book was supported by the U.S. Department of Homeland Security (DHS) through the National Center for Risk and Economic Analysis of Terrorism Events (CREATE) at the University of Southern California (USC) under award number 2010-ST-061-RE00010. However, any opinions, findings, and conclusions or recommendations in this book are those of the authors and do not necessarily reflect views of the DHS. We would like to thank Dr. Matthew Clark, Director of the Office of University Programs (OUP) and Georgia Harrigan, CREATE's Program Manager at OUP for their long-term support of CREATE.

decision analysis, game theory, and behavioral research in many areas of the homeland security enterprise.

The purpose of this book is to provide an overview of the models and analysis tools that have emerged from CREATE's research on improving homeland security decisions. The focus of this book will be on models that use risk analysis, economic consequence analysis, decision analysis, and game theory, among other models, to help determine optimal strategies for security.

Risk Analysis combines data analytics, mathematical modeling, and expert judgment to estimate and manage risks from many sources, including industrial accidents, environmental hazards, natural disasters, and terrorism. Risk analysis has been applied successfully to natural and technical phenomena for many years, but its application to terrorism is relatively new. Chapters 2–6 cover various aspects of risk analysis in the context of terrorism.

Economic Consequence Analysis is at the heart of the problem of allocating scarce resources, both efficiently and equitably. Economic consequence analysis, formerly referred to as *hazard loss estimation*, translates the physical damage wrought by terrorist attacks and natural disasters into direct economic terms and then extends their measurement into broader multiplier (supply-chain), general equilibrium (adding market interactions), or macroeconomic (adding the influence of aggregate behavior and fiscal and monetary policy) impacts on gross domestic product and employment. It also incorporates our resilience in responding to these threats. Benefit-cost analysis is more comprehensive in that it examines both sides of the ledger. Avoided consequences are the benefit side, while the cost side involves the evaluation of the price, including broader social costs, of interdiction, mitigation, and resilience. Other areas of economic consequence analysis involve the analysis of insurance and the design of various counterterrorism policy instruments. Chapters 7–13 describe the approach CREATE uses to assess economic consequences and Chapters 14 and 15 cover cost-benefit analysis and insurance methods.

Decision Analysis provides a normative method for thinking about decisions in the face of uncertainty. This approach relies on six fundamental elements to help guide the decision maker to the right decision. These elements include: (i) the frame, making sure you are working on the right problem; (ii) creating a set of meaningful and doable alternatives; (iii) capturing information and quantifying uncertainty using probability distributions; (iv) incorporating preferences and attitude toward risk using value and utility functions; (v) identifying the decision maker and the stakeholders affected by the decision; and, most important, (vi) using a sound decision-making system to combine these elements. We shall discuss various aspects of these elements throughout this book together with numerous decision analytic methods with applications in several areas. Chapter 16 shows how terrorists' objectives can be identified and structured using decision analysis methods. Chapter 17 describes both the process of decision analysis and tools for multi-objective decision making. Chapter 18 combines utility and theory and control theory to illustrate their use in multi-objective decision making with limited resources. Chapter 19 provides an example of a decision analysis for evaluating approaches to protect commercial airplanes against surface-to-air missile attacks by terrorists.

Game Theory provides a sound mathematical approach to improve defensive decisions in the face of an adaptive adversary. CREATE is at the forefront of this research, providing the first, long-term, deployed applications of game theory for security at important sites, including ports such as New York, Boston, Los Angeles, and Houston, transportation systems such as international flights of U.S. air carriers, and many others. Indeed, fueled by these applications, an entirely new area of research called "security games" has been established, with dozens of research groups worldwide focused on this topic. This book will present basic materials in key aspects of research in security games, starting with examination of major applications, computational approaches for solving security games, human behavioral aspects of security games, use of machine learning in security games, and others. Chapters 21 and 22 introduce the central concept of the Stackelberg Security Games and their solution methods. Chapters 23–25 discuss various aspects of security games and provide examples of deployed security systems based on game theory.

Behavioral Research over the past decades has shown that people generally do not make rational judgments and decisions when left to their own intuition. This research is relevant for homeland security in two ways: First, we should not necessarily assume that terrorists are rational when trying to pursue their goals and understanding what they want, and how they may fall short in achieving this is an important ingredient in fighting terrorism. Second, people react to risks in general and to terrorism in particular in often biased and irrational ways. This can lead to an amplification of risks through an overreaction by decision makers in government and industry and through behavioral responses by the public. Chapters 26–30 cover these topics.

Other models and techniques will also be covered in this book. For example, the use of control theory combined with new models of preferences enables a mechanism for the real-time control of unmanned aerial systems in pursuit of multiple objectives. The number of unmanned aerial systems might be limited and their objectives are to maximize the surveillance of an area, while maintaining close proximity to each other and also avoiding collision. Dynamic systems and control theoretic methods will be used with preference models to make this widely used application feasible.

Improving Homeland Security Decisions provides an overview of state-of-the-art research in homeland security, a description of the various applications where analytical methods were incorporated, as well as an in-depth perspective of methods, techniques, and algorithms that were developed.

This book is designed to be of interest to:

(i) Researchers and graduate students who are interested in pursuing research in homeland security applications and those who wish to understand some of the basic principles of this research topic in more depth;
(ii) Security analysts and practitioners interested in obtaining an overview of the various applications of analytical methods in homeland security research (even if they skip exact details);
(iii) Other researchers, generally familiar with analytical methods, who wish to jump into this area of research.

Although most chapters are written in simple expository form, this book will not provide a full introduction to some of the topics presented, where we assume readers have some general background in each of these areas.

2

Probabilistic Risk Analysis and Terrorism Risk

Barry Ezell, Steven P. Bennett, Detlof von Winterfeldt, John Sokolowski, and Andrew J. Collins

2.1 Introduction

"Probability is the guide to life."

–Cicero (107 BC)

"We have to identify and prioritize risks – understanding the threat, the vulnerability and the consequence. And then we have to apply our resources in a cost-effective manner."

(Michael Chertoff, Former Secretary of the Department of Homeland Security, 2006)

For more than 30 years, probabilistic risk analysis (PRA) has been a major tool for assessing risks and informing risk management decisions by government and industry, in areas as diverse as environmental protection, industrial safety, and medical decision making. Applications of PRA to terrorism risks are new, however, and not uncontroversial. Here, we take a broad view of PRA, including any probabilistic approach involving tools like event trees, fault trees, and decision trees. We also introduce other tools such as game theoretic approaches and system dynamics, which may prove to be useful in dealing with the intelligent adversary. A major challenge in risk analysis of terrorism is the fact that terrorists, unlike nature or engineered systems, are intelligent adversaries and may adapt to our defensive measures.

There has been recent criticism of PRA approaches to terrorism risk analyses, especially (but not only) by the National Research Council's Committee on Methodological Improvements to the Department of Homeland Security's (DHS) Biological Agent Risk Analysis (referred to hereafter as the NRC

Reprinted by permission, Ezell, B.C., Bennett, S.P., von Winterfeldt, D., Sokolowski, J. & Collins, A.J. (2010), Probabilistic risk analysis and terrorism risk. *Risk Analysis*, 30, pp. 575–589.

Address correspondence to Barry Ezell, 1030 University Blvd., Suffolk, VA 23435; bezell@odu.edu.

Committee). The NRC Committee has argued that because of this adaptive nature, it is problematic to assess probabilities of terrorism events or to use traditional PRA tools like event trees, suggesting alternative tools to assess the risks of terrorist events. One purpose of the article is to justify the use of PRA for terrorism risk analysis, while acknowledging its limitations. A secondary purpose of the article is to propose a pluralistic approach to terrorism risk analysis, which allows alternative approaches to be examined and tested. To this end, we examine some alternative approaches and discuss their contributions and limitations. While we do not take issue here with the possible value of these alternative approaches, we aim to make a case that (1) probabilities of terrorism events are useful to assess terrorism risks; (2) event trees can be used as part of a terrorism PRA to decompose the universe of terrorism scenarios; and (3) alternatives suggested by the NRC Committee like extended forms of games or decision trees constructed from the terrorists' perspective, like all approaches, have limitations.

This chapter is organized in the following way. Section 2.1 provides a short background on the Department of Homeland Security (DHS) Bioterrorism Risk Assessment (BTRA) methodology and the context that motivated this article. Following the BTRA background, it concludes with a summary of the NRC Committee's criticism of the use of probability to assess the likelihood of terrorism events and the use of event trees in favor of approaches that consider terrorist events as actions that can be derived from their objectives. Section 2.2 details the usefulness of probabilities in bioterrorism risk analysis. Section 2.3 provides an overview of several tools that have been used or might be used to account for the intelligent adversary in terrorism risk. The review of tools in Section 2.3 is not intended to be exhaustive, but rather to note that modeling tools have limitations in dealing with the intelligent adversary. For example, some tools and approaches, while promising, may require additional development before ready for use in real- world applications while others are more mature and established. The final section summarizes, and advances, the claim that no single model or approach is sufficient to cover the entire landscape of terrorism risk and support the difficult decisions that must be made by Homeland Security decision makers.

2.1.1 *2006 Bioterrorism Risk Assessment Background*

Signed in 2004, Homeland Security Presidential Directive[1] (HSPD) 10 focused on improving the nation's ability to prevent, prepare for, respond to, and recover from terrorism attacks that employed biological agents as their means. An important component of HSPD-10 was the president's requirement for DHS to develop "periodic assessments of the evolving biological weapons threat," explaining that "the United States requires a continuous, formal process for conducting routine capabilities assessments to guide prioritization of . . . on-going investments in biodefense-related research, development, planning, and preparedness." The first national Bioterrorism Risk Assessment would be required by January 2006. To meet this requirement, in early 2005, DHS investigated three methodologies varying in complexity, depth, and community familiarity. A Technical Expert Review Panel reviewed each methodology.[2] Based on resulting comments, and other factors, DHS determined that BTRA should primarily be a PRA-based methodology. BTRA has evolved over the years, incorporating new tools and techniques as science progresses and as program realities allow.

DHS requested the National Academy of Sciences' National Research Council (NRC) to review BTRA in 2006. The NRC's Draft Final Report, delivered in January 2008, recommended "to assess the probabilities of terrorist decisions, DHS should use elicitation techniques and decision-oriented models that explicitly recognize terrorists as intelligent adversaries who observe U.S. defensive preparations and seek to maximize achievement of their own objectives." Also, the committee chairman proposed a decision tree approach from the "terrorist point of view" (NRC, 2008).

DHS identified several concerns with the NRC Committee's report (U.S. Department of Homeland Security, 2008). In particular, the conclusion that probability of terrorism events and event trees are not suitable for bioterrorism risk analysis appeared to be controversial and not shared by many in the risk analysis community. This article challenges the NRC Committee's conclusion.

2.1.2 *Intelligent Adversary Analysis*

An essential aspect of any terrorism risk assessment is the approach used to represent and model terrorist adversaries. It is arguable that one of the best sources of information on the nature and intelligence of our adversaries, although limited, uncertain, and incomplete, is the intelligence community (IC). The IC persistently observes, collects, fuses, and assesses terrorist activities, motivations, intent, and capabilities. The ongoing challenge for DHS risk analysts, then, is how best to consult, incorporate, and transform relevant intelligence information into meaningful inputs for terrorism risk analysis, in conjunction with other models of terrorists' behavior outside of the IC.

Intelligence products exist in a range of forms, from opinions based on anecdotal information, to assessments based on tradecraft, and in other cases, technical methods and models. How, then, might DHS and the IC transform intelligence information into meaningful inputs for bioterrorism risk analysis? The NRC Committee advised DHS to model potential bioterrorists as "intelligent adversaries" as a part of its risk assessment – assuming that at each decision point in the planning of an attack, the adversary will always make the choice that maximizes his or her objectives, thus making terrorism attack probabilities *outputs* of decision models, rather than incorporating intelligence information as input (NRC, 2008).

In decision analysis terminology, the NRC Committee proposed to conceptualize the interaction between defenders and attackers in an evolving terrorist attack as a decision tree, in which the attacker's choices are modeled as decisions that maximize expected utility and the defender's choices are modeled as uncertain events, related to the relative effectiveness of the defenses. Three other possibilities are (1) a decision tree in which the defender's choices are modeled as decisions that maximize expected utility and the attacker's choices are modeled as uncertain events that are influenced by the defender's decision; (2) a decision tree in which both the attacker's and the defender's choices are modeled by decisions that maximize expected utility, e.g., an extended form of a game; and (3) an event tree that models both the attacker's choices and the defender's responses as uncertain events.

Clearly, there are advantages and disadvantages to these ways of representing attacker–defender interactions and there is no "correct" answer.

2.2 Probabilities Are Useful to Quantify the Risk of Terrorist Attacks

In the first issue of the journal *Risk Analysis*, Kaplan and Garrick published an important paper that defined risk as the triplet of scenario, likelihood, and consequence (Kaplan & Garrick, 1981). For the following three decades, the risk and decision analysis communities have cited this seminal paper and used many of the concepts and tools developed in it. More recently, Garrick and colleagues (2004) advocate the use of PRA for assessing terrorism risk, specifically for assessing the probabilities of terrorist attacks. Work based on Garcia, McGill and colleagues, Paté-Cornell and Guikema, Rosoff and von Winterfeldt, Willis, and von Winterfeldt and O'Sullivan is an example of risk analyses that use PRA, and that externally estimate probabilities of terrorist attacks as inputs (Garrick et al., 2004; Garcia, 2006; McGill, Ayyub, & Kaminskiy, 2007; Pate-Cornell & Guikema, 2007; Willis, Morral, Kelly & Medby, 2003; von Winterfeldt & O'Sullivan, 2006).

Willis, McGill and colleagues, and other terrorism risk researchers operationalize terrorism risk as the product of threat, vulnerability, and consequences. More specifically, threat is usually defined as the probability of an attack (weapon, delivery mode, target, etc.), vulnerability as the probability of an attack's success given that it occurs, and consequences are the losses that occur (fatalities, injuries, direct and indirect economic impacts, among others) given a successful attack. Equation (2.1), then, is a common expression of homeland security risk (McGill et al., 2007; Wilson, 2003)

$$Risk = P(A) \times P(S \mid A) \times C \tag{2.1}$$

Hence, a useful first-order indicator of terrorism risk is the expected consequences (loss of lives, economic losses, psychological impacts, etc.) against which the benefit of existing or potential terrorism strategies, policies, and countermeasures can be evaluated and estimated. In this probabilistic framework, the attack probabilities ($P(A)$ in Equation (2.1)) are for the most part agreed to be the most challenging to estimate. Quantifying $P(A)$ requires knowledge, data, or modeling about the motivations, intent, and capabilities of terrorists (largely the domain of the intelligence community), in addition to or instead of knowledge about historical attacks and their relevance to current risk.

It is very difficult to elicit absolute probability (or frequency) judgments that permit this kind of output. However, relative judgments in terms of rank orders or ratios are easier to acquire from intelligence or other experts. For example, while it may be difficult to assess the absolute probability that a particular terrorist group will engage in a terrorism attack using nuclear materials in the United States in the next 10 years, experts can more easily reason comparatively, and might judge a "dirty bomb" attack using radiological material from a medical facility is more likely or less likely than an attack using an improvised nuclear device by considering the relative technical difficulties of executing these attacks. There is extensive literature regarding methods for eliciting uncertain probability judgments (often as probability distributions) from experts, which suggests how one might elicit probabilities in the face of intelligence complexities and uncertainties inherent in terrorism risk analysis (for a recent summary, see Bedford and Cooke, and Hora (Bedford & Cooke, 2001; Hora, 2007).

When intelligence analysts estimate a probability of attack, they are making a statement of belief about what a terrorist might do, based on available intelligence information as well as their personal experience and judgment. Apostolakis (1990) makes this crystal clear: "there is only one kind of uncertainty stemming from our lack of knowledge concerning the truth of a proposition. Distinctions between probabilities are merely for our convenience in investigating complex phenomena. Probability is always a measure of degree of belief."

There are two common arguments against the use of expert-estimated attack probabilities for terrorism risk analysis: (1) that the level of uncertainty and incompleteness associated with intelligence data prevents reasonable probability estimates from being made, even when using expert elicitation approaches that are designed to capture and represent uncertainty, and (2) that these probabilities are not static – i.e., the adversary is intelligent, observing U.S. defensive actions and shifting attack preferences accordingly. Regarding the first argument, it is important to note that intelligence information is already in use for decision support at the highest levels in government; uncertainty and incompleteness are managed and communicating by representing judgments verbally with associated caveats. This approach, however, has historically led to some significant misunderstandings of intelligence information, a notable example being a (now declassified) 1951 National Intelligence Estimate (NIE 29–51), entitled "Probability of an Invasion of Yugoslavia in 1951." In this intelligence document appeared the statement: "Although it is impossible to determine which course the Kremlin is likely to adopt, we believe that an attack on Yugoslavia in 1951 should be considered a serious possibility." When asked by State Department staff what odds the authors of the assessment placed on an attack in 1951, Sherman Kent (2007) of the National Board of Estimates replied "65 to 35 in favor of an attack." The State Department had interpreted "serious possibility" as being "very considerably lower" than Kent's 65/35 reply. Kent then polled the other authors of the document to determine the odds they had in mind when they agreed to the wording, observing that the odds in the minds of the authors ranged from 80/20 to 20/80 in favor of an attack. The example above is not intended to criticize the production and communication of intelligence information; rather, it highlights an opportunity for improved clarity and understanding of uncertainty when a mathematical language for capturing and expressing degree of belief – probability theory – is used. Expression of intelligence information in a consistent manner that reflects uncertainty and is able to be incorporated into other models is helpful and arguably can improve the interpretation and utility of the information, particularly as it informs risk analysis.

Regarding the second argument against using expert-elicited attack probabilities, the adaptive nature of the adversary is certainly an important consideration. Nevertheless, it is reasonable to start with a baseline of defensive actions, current terrorist motivations, intent, and capabilities (based on data, intelligence, and other expertise), and then assess probabilities conditional on this baseline. We take it for granted that all probabilities are conditional on our current state of knowledge. While it is perhaps more difficult to spell out these conditions precisely in terrorism risk analysis, there is no fundamental difference in this type of conditioning compared to conditioning probability judgments in the case of natural or engineered systems.

Once we introduce new defensive actions, it is, of course, important and necessary to reassess these probabilities in light of the preventative, protective, or deterrence effects of the defensive actions. For example, as von Winterfeldt and O'Sullivan (2006) pointed out, the use of countermeasures to Man-Portable Air-Defense Systems (MANPADS) is assessed to have a strong deterrence effect on terrorists who may contemplate the use of MANPADS weapons (such as shoulder-fired missiles, etc.) to attack commercial airplanes. It is important to note, however, that re evaluation of probabilities following defensive action is not necessarily always a best estimate of risk, since for a terrorist adversary, it is next to impossible to determine whether or not, or the degree to which, the adversary is in fact (1) aware of particular defensive actions and their subsequent implications, and (2) adjusting the adversary's decisions and preferences based on awareness of defensive actions. Additional intelligence information can assist in determining the "penetrance" of U.S. defensive adjustments into the adversary's decision-making process, but any newly determined "postdefensive adjustment" risks may well be best presented to decision makers alongside, or in addition to, baseline risks rather than instead of them.

2.3 Tools for Terrorism Risk Analysis

Probabilities associated with complex events are difficult to assess directly, and it is therefore often useful to decompose these events into components and to determine the overall event probability by assembling the components' probabilities using standard probability calculus. There are many alternative decomposition tools, including event trees, fault trees, decision trees, influence diagrams, and belief nets. When the intention is to divide a very large universe of events into a structured set, event trees are useful as part of a baseline assessment of terrorism risk, beginning with an initial choice of weapon and target, and following through the path from attack, through success or failure, to eventual consequences. Event trees have been used to decompose terrorism scenarios in a number of efforts (Ezell, Haimes, & Lambert, 2001; Koller, 2000; Viscusi, 2003). Rosoff and von Winterfeldt use event trees to track the paths to failure or success of a dirty bomb attack and von Winterfeldt and O'Sullivan use a combination of decision and event tees to quantify the costs and benefits of countermeasures to MANPADS (Rosoff & von Winterfeldt, 2007; von Winterfeldt & O'Sullivan, 2006).

We present three categories of tools for use in PRA as it applies to terrorism risk, beginning with an introduction of logic trees under which we group forward logic trees and fault trees. Next, we briefly review additional methods – influence diagrams, systems dynamics models, and Bayesian networks (BN) – as potentially useful in transforming conceptual terrorist actions into computational models. For the final category, we discuss game theoretic approaches. For each we discuss the potential advantages and limitations.

2.3.1 *Logic Trees*

Logic trees are important tools for exploring the scenario space, analyzing uncertain events, defining scenarios, and assessing risk (Dillon-Merrill, Parnell, & Buckshaw, 2008). The use of logic trees in probabilistic seismic hazard analysis has a long history, ranging from weighting of a few alternative assumptions to full uncertainty treatment for all of the inputs to a probabilistic assessment. Logic tree analysis consists of specifying a sequence of assessments that must be made in order to perform an analysis and then addressing the uncertainties in each of these assessments in a sequential manner. Thus, it provides a convenient approach for breaking a large, complex assessment into a sequence of smaller, simpler components that can be more easily addressed (Cornell & Merz, 1975). In this next section, we divide logic trees into two categories: (1) probability, event, and decision trees, and (2) fault, attack, and success trees. Where some may draw a serious distinction between probability, event, and decision trees, they fundamentally all use forward logic in their design. Parnell *et al.* arrived at a similar conclusion in a report by the Homeland Security Institute for DHS that represented "consensus among the authors" and detailed 20 risk assessment frameworks (Parnell et al., 2005). In this chapter, we do not attempt to recount what has already been published. Instead, we narrow our focus to trees and game theory; two areas (minus PRA event tress) that the NRC Committee strongly recommended as the appropriate way to do bioterrorism risk analysis.

2.3.1.1 Probability, Event, and Decision Trees

Probability trees model a sequence of uncertain events in order to calculate the probabilities of events in the outcome space (Figure 2.1). A probability tree is a succession of circular nodes (uncertain state variables) with branches. The branches emanating from each node represent the different possible values of the uncertain variables associated with the node. Probability trees have the following properties: (1) event nodes and branches; (2) forward logic; and (3) downstream events conditioned on previous nodes. Probability trees have many uses such as (1) to graphically represent the fundamentals of probability theory; (2) to describe probabilistic relationships between two or more

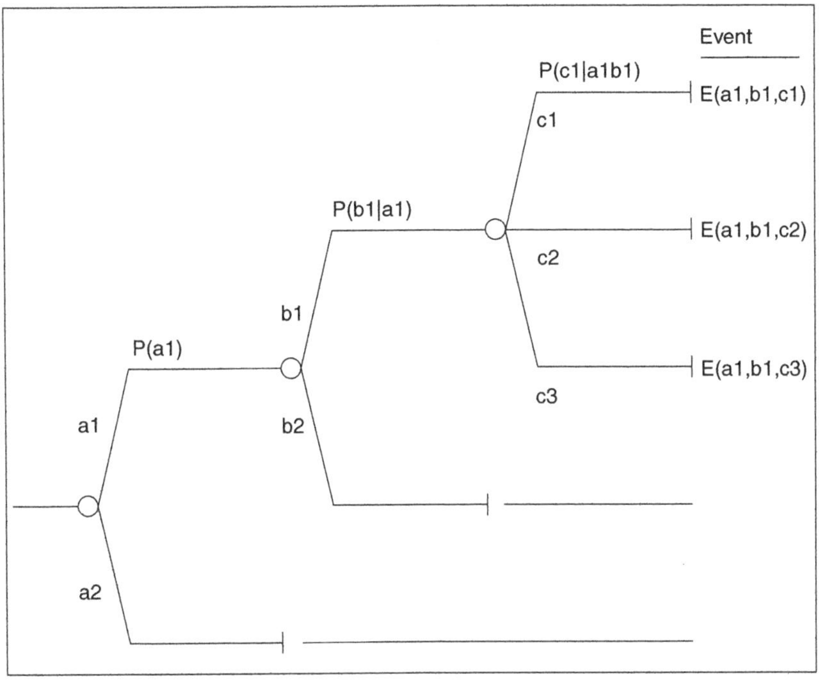

Figure 2.1. Probability tree.

events; and (3) to serve as the mathematical foundation for more advanced tree structures such as event trees or decision trees.

Event trees inductively model the sequences of events that lead to consequences (Kumamoto & Henley, 1996). Event trees have the following properties: (1) events (nodes) and branches; (2) forward logic; and (3) downstream events conditioned on previous events. Event trees are an extension of probability trees by adding: initiating event, mitigating events, and consequences. Consequences are added for each probability path. Event trees have been used in many fields. For large systems, event trees have been used in nuclear reactor safety studies. Ezell and colleagues employed event trees to understand cyber risk to supervisory control and data acquisition systems for water supply (Ezell et al., 2011). In PRA, event trees operate by identifying the likelihood of any given probability path (from initiating event through the leaves of the all tree branches).

Probabilities are assigned to event tree branches to represent the relative likelihood or degree of belief about the outcome of each branch. Probabilities at a given node are assessed conditionally on the assumption that all the branches

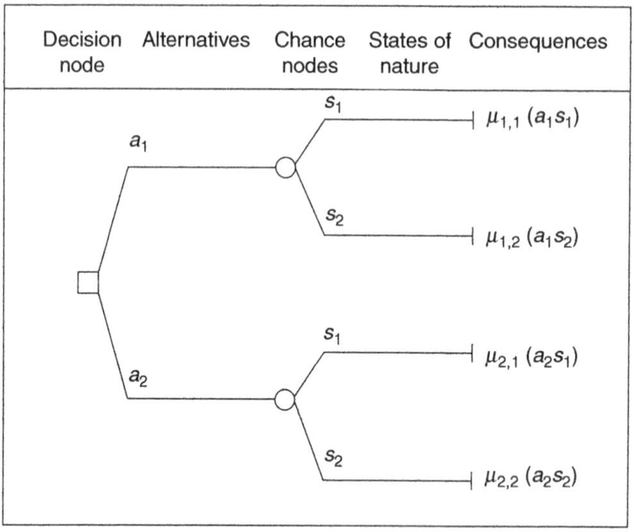

Figure 2.2. Decision tree (Haimes, 2009).

leading to that node represent the true states of the preceding events. Because they are conditional probabilities for assumed mutually exclusive and collectively exhaustive events, the sum of the conditional probabilities at each node is unity (Bommer, Scherbaum, Bungum, Cotton, Sabetta, & Abrahamson, 2005).

Event trees have been used for very large complex systems. For example, NASA has a mature program using PRA for decision making and managing project risk – Mars missions, Space Station construction, Space Shuttle flights, etc. Compared to the BTRA event tree, NASA PRAs involve extremely large sets of unknowns. While it is desirable to create small and compact event trees that are simply described, this is often inadequate for the representation of real uncertainties. Consider as an example, a comparison between the BTRA and NASA PRA. The BTRA is comprised of one event tree, 16 events, and 74 branches. A NASA Space Shuttle PRA has 5,000 event trees, 6,000 events, and 2,000,000 branches, and approximately 100 off-line supporting models (Vesely, 2005).

Decision trees are logic trees that include decision nodes in addition to events. A decision tree is effectively a diagram of a decision, read left to right (Kirkwood, 2002). The leftmost node in a decision tree is the root node and is usually a decision node (represented by a square). Branches emanating to the right from a decision node represent the set of decision alternatives that are available. Small circles in the tree are chance nodes that represent uncertainty in outcomes (Figure 2.2). In the same fashion as probability trees and event trees, probabilities are assigned to the branches, referred to as states of nature, emanating from the chance nodes in the tree. The leaves of each path through the tree are called endpoints. Each endpoint represents the final outcome of a path from the root node of the decision tree to that endpoint (Haimes, 2009). In decision analysis, decision trees are used as a decision support tool to find the alternative with the best expected value. In the terrorism context, decision trees can structure the attacker's actions as decisions and the defender's as chance nodes vs. structuring the defender's actions as decisions and the attacker's actions as chance nodes. The NRC report emphasizes the use of decision trees that structure the attacker's actions as choices and the defender's responses as chance nodes. Other studies do the reverse (von Winterfeldt & O'Sullivan, 2006). When both the attacker's and the defender's choices are modeled as decision nodes, such approaches are considered to be in the domain of game theory.

Logic trees have limitations. First, probability assignments are usually based significantly on subjective judgments (1) because of the often limited availability of numeric data, and (2) because human judgment is often needed to weigh alternative interpretations of whatever data are available. In the context of the BTRA, intelligent adversary decisions are modeled as probability estimates, which are structured to incorporate the intelligence community expert knowledge about the intelligent adversary's preferences and capabilities. Another limitation is that terrorist choices are not random events such as failure events in engineered systems or natural hazard events. At best, the probabilities assigned by the intelligence community represent its belief as to the choices terrorist will make. Terrorists are not static and change their preferences based on what we choose to do so probability judgments must be updated as the situation changes. Event tree PRA is therefore only a "snapshot" of threat, vulnerability, and consequences for a given time period.

In BTRA, DHS used an event tree and modeled terrorist decisions as probability estimates elicited from the intelligence community – estimates that were elicited specifically to incorporate the adversary's level of sophistication, or "intelligence" from the intelligence community's perspective. The NRC Committee, however, rendered an opinion that the government should use decision trees from the terrorist's perspective, where at each decision point in the tree, an adversary will always choose the "branch" that maximizes the consequences of an attack. Decision trees (as the Committee points out) have the advantage that they are a simpler form of analysis, and more easily explainable.

Using decision trees to model terrorism from the perspective of terrorists (as proposed by the Committee) requires two important assumptions. First, it assumes ideal adversary intelligence and rationality – that the adversary *knows* which branch choice at a particular decision node best maximizes consequences at tree endpoints. For most terrorist attack paths, and certainly for technically complex attack paths requiring some degree of sophisticated knowledge and skills such as those for bioterrorism, this can be a very unsafe assumption for the U.S. government (Hees & Roy, 2007). This assumption is problematic for a few reasons – first, successfully executing a WMD terrorism attack such as bioterrorism is a technically challenging enterprise, often with counterintuitive relationships between technical planning decisions and resulting consequences (e.g., terrorists could assume that a particular mode of production for "virus A" might be optimal in terms of producing casualties when used in attack since that mode is known to be optimal for "virus B"; such assumptions in biology are often false due to the complexity of biological systems). As such, many decisions that may appear to be the consequence-maximizing ones may in fact be the exact opposite, and the level of technical proficiency – an aspect of terrorist groups that is studied by the intelligence community – can be the determining factor in whether or not the adversary knows this. Further, the assumption in proposed decision tree approaches that terrorist adversaries have a perfect working knowledge of all defensive countermeasures arrayed against them is potentially problematic for two reasons; first that the adversary really does know all the details of our defenses, and second that even if the adversary did, he or she would act rationally in possession of that information. Were intelligence information to indicate that either of these were not the case, the objective functions for the terrorist would be undermined to the point that the utility of the analysis for estimating risk from overidealized terrorists would be greatly reduced.

The second critical assumption, also related to the adversary's objective function in the decision mode, is that the intelligence community (or anyone else for that matter) knows the objectives the adversary is trying to maximize in the first place. While consequence maximization may be a good guess, it is not easy to determine which types of consequences are most desirable to the adversary, and in what proportions. This is an important aspect of the analysis in that decision points for the adversary may be consequence maximizing for some consequence measures (such as fatalities), but not necessarily for others (such as economic or psychological consequences). For example, consider the 2001 anthrax attacks in the United States, in which four people were killed and 18 (known) infected. It has been noted that the widespread panic, closure of government and postal facilities, and massive public expenditures and preventive actions were of higher impact than the four fatalities, and that psychological and economic impacts may be more important adversary objectives than killing Americans (Cordesman, 2005). However, one way to manage this challenge is to examine data, and elicit experts that seek to assess the *outputs* of terrorists' objective functions – that is, what adversaries are actually doing or planning. Specifically, while intelligence information describing what the adversary is actually *doing* is uncertain and incomplete, data about the specifics of the objective function a terrorist may be trying to maximize – that is, what the terrorist is *thinking* is at least as uncertain and incomplete, and probably more so. Specifically, this assumption requires knowledge of the adversary's preferences and capabilities relating to

- representational and computational capacities (Hees & Roy, 2007)
- the nature of the adversary's objective function (i.e., whether the goal is to maximize his or her consequences);
- adversary knowledge and reasoning about the problem commensurate with the United States; and
- adversary capability of representing a model and computing its solution (Myerson, 1991).

To summarize, probability trees, event trees, and decision trees share common elements of forward logic and appear very similar. Event trees have a long history in PRA and in the past 10 years many applications to terrorist risk have been published. Applications using decision trees in terrorism risk, however, are not well developed in the risk analysis literature. In comparing Figures 2.3 (event tree) and 2.4 (decision tree from an attacker's perspective), the distinction between the two formulations becomes clear. In Figure 2.3, the event tree formulation expresses *the intelligence community's uncertainty in the intelligent adversary's true utility*. This uncertainty is represented by a probability assignment for a terrorist choice based on the state of information available to the intelligence community. In Figure 2.4, and as described above, the decision tree assumes (1) *we know the intelligent adversary's true utility for the choices in the tree*, and (2) the adversary has the knowledge and rationality to actually maximize that utility; we argue that this is not the case.

2.3.1.2 Fault, Attack, and Success Tree

Fault trees deductively model the combinations of system failures and human errors that could lead to an accident (Figure 2.5). Fault trees are a useful tool for analyzing, visually displaying, and evaluating failure paths in a system to evaluate system level risk (Ericson, 1999). Similarly, success trees model the combinations of events that lead to success. Attack trees model the actions of an intelligent adversary to defeat a defensive system (Mauw & Oostdijk, 2006). Fault, attack, and success trees all use Boolean algebra, reliability theory, and probability theory. Fault trees provide insight into why mitigating events in an

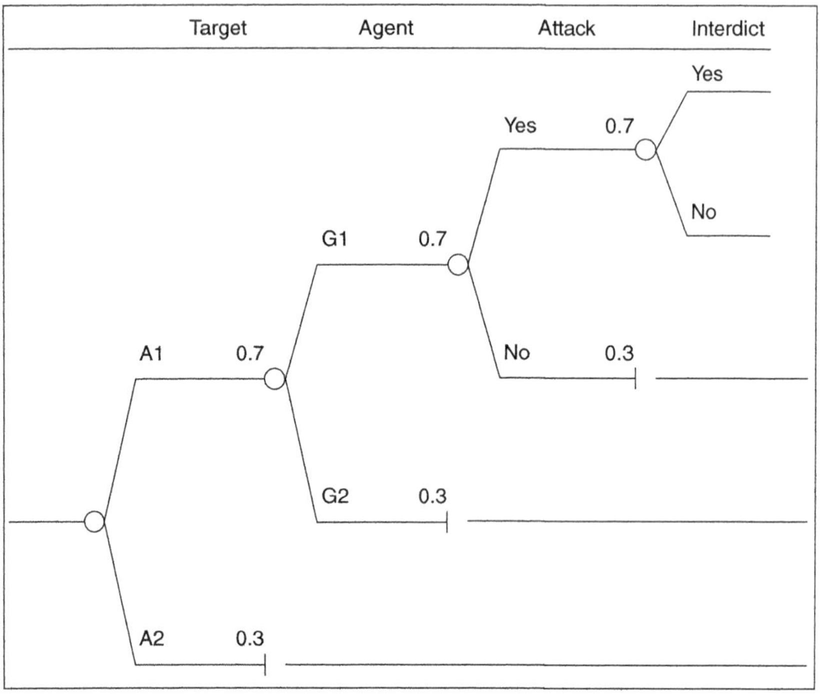

Figure 2.3. Notional bioterrorism event tree.

event tree may fail. Attack trees categorize the different ways in which a system can be attacked. And, success (defense) trees are the complement of fault trees. Fault trees have been used for years in concert with event trees. For example: 1961 – Minute Man Missile, aircraft design, and nuclear power plant safety analyses have all used fault trees to support planning and decisions. Similarly, attack trees are special case of fault trees that have been used to represent an adversary's successful defeat of a countermeasure or system, such as a fire- wall system, a mitigating event for an event tree (Ezell, 1998). Along this same line, other techniques may prove to be useful with logic trees. For example, in the next section, a combination of systems dynamic modeling and Bayesian networks is explored to provide more structure to how one may underpin estimates from the intelligence community for the behavior of the intelligent adversary.

2.3.2 *Influence Diagrams*

Influence diagrams depict the relationships between decisions, events, and outcomes by depicting them as nodes and arcs in a directed acyclic graph (Figure 2.6).

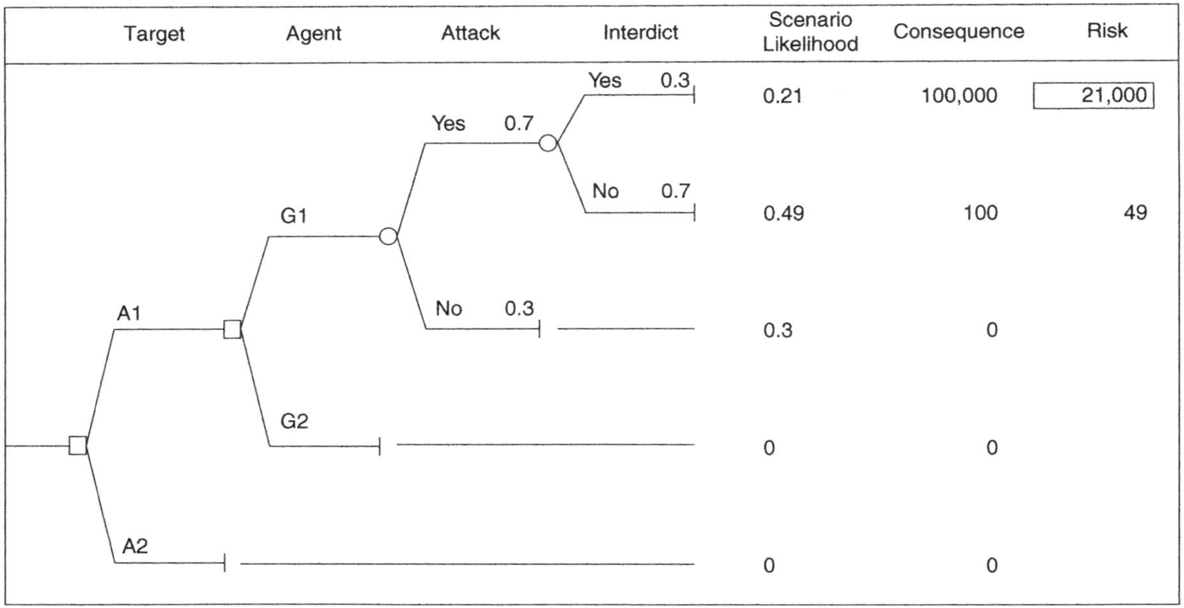

Figure 2.4. Notional bioterrorism decision tree.

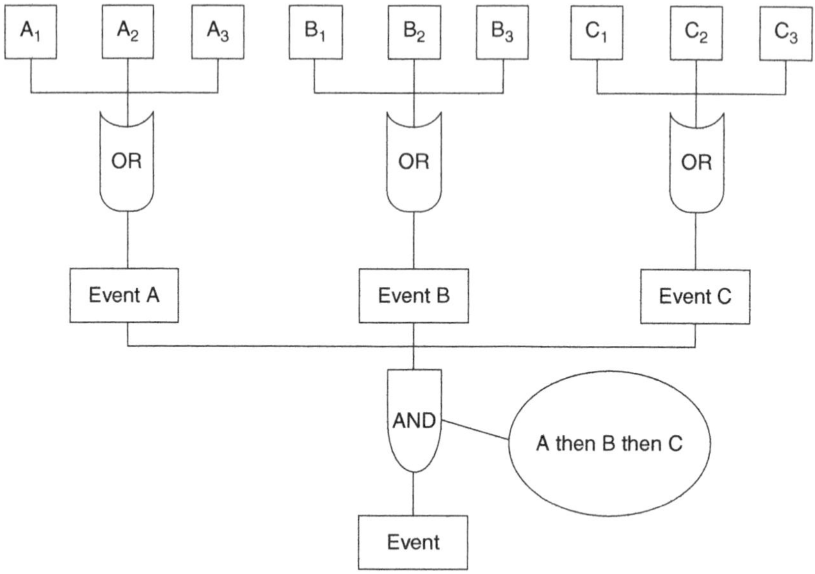

Figure 2.5. Generic fault tree.

Following influence diagram conventions, uncertain variables and events are typically shown as ellipses, variables calculated from predecessors are shown as double ellipses, decision nodes are shown as rectangles, and value nodes are shown as diamonds (or hexagons). Influence diagrams are directed acyclic graphs, in which an arrow connecting node A to node B is interpreted as follows:

1. If node A is an event node and node B is a decision node, it means that "the event in A will be known prior to making the decision in D."
2. If both nodes A and B are both chance nodes, it means that "knowing the event in node A affects the probabilities of events in node B."
3. If node A is a chance or decision node and node B is an outcome node, it means that "the outcome depends on the predecessor nodes."
4. If nodes A and AN are chance nodes and node B is a node characterizing a calculated variable, it means that the variable is node B is calculated from the numbers representing the uncertain variable in A and AN.

Influence diagrams are also compact representations of decision trees, which hide the branches of the tree and display only its nodes. Every influence diagram can be represented by a symmetric decision tree, in which all possible

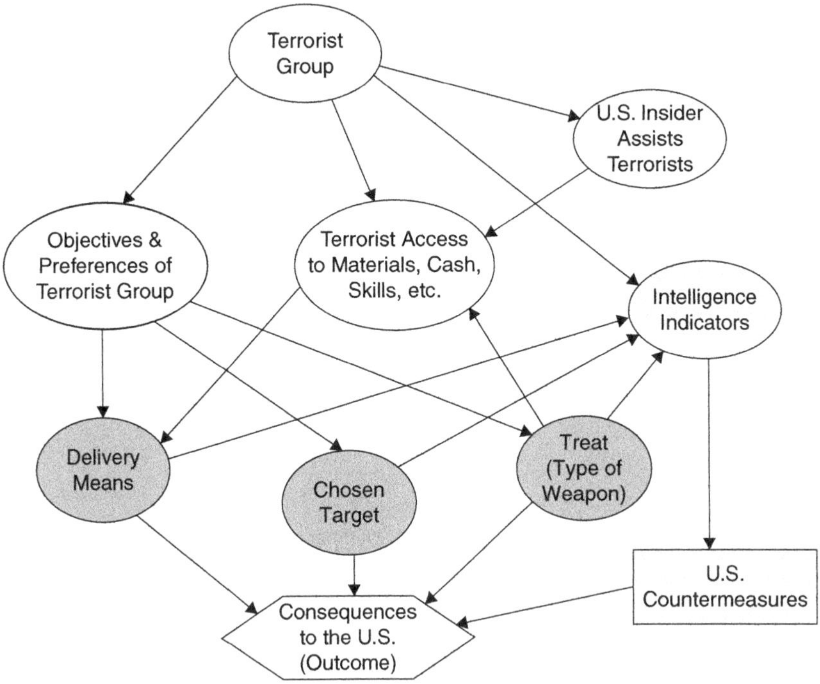

Figure 2.6. Example terrorist influence diagram.

combinations of decision, events, and outcomes are represented. Conversely, all decision trees can be enriched to symmetric trees and converted to an influence diagram.

Figure 2.6 shows an influence diagrams for a terrorist group's decision to choose a mode and a target of attack (Pate-Cornell, 2007). In this influence diagram, there is only one decision node, characterizing the possible countermeasures, and one outcome node, the consequences of a combination of the type of weapon, target, and delivery means (shaded event nodes in Figure 2.6). The other nodes are event nodes, characterizing terrorist states and information available to the U.S. government.

This influence diagram analysis was applied in an illustrative analysis to determine the relative likelihood of alternative types of terrorist attacks on the United States. The results suggested that IED attacks, "dirty bombs," and biological attacks are most likely.

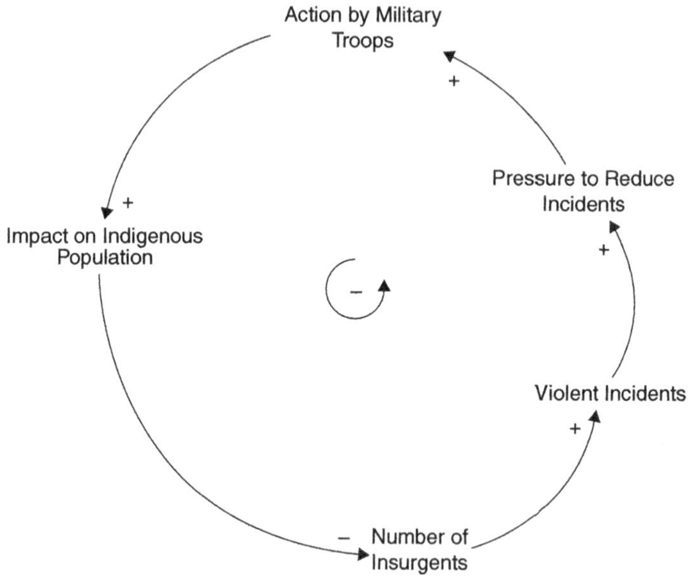

Figure 2.7. Insurgency causal loop diagram.

2.3.3 *Causal Loop Diagrams and Systems Dynamic Models*

Representing terrorist behavior in a simulation is a unique challenge. When simulating physical phenomena one can utilize detailed mathematical models to capture the behavior of those systems. Models such as those representing flight dynamics or fluid structures are well understood and have been developed to a point where they closely represent reality. Some representations have become good enough to substitute the models and simulations for real-world systems and still attain the same or nearly the same training and testing benefits as the actual system. The same cannot be said for the terrorist and social systems. Human systems are far more complex than physical systems and therefore much more difficult to represent in computational models. However, strides have been made in this area and a few points are worth mentioning.

A modeling paradigm that lends itself to capturing the many factors that influence complex system behavior is system dynamics (SD). SD is a way to model and depict factors contributing to the behavior of a system and the causal relationships that exist among those factors. To employ this modeling method one begins by developing a causal loop diagram that shows these variables and their causal linkages. Figure 2.7 represents one such causal loop diagram.

This figure is a simple example of factors that are causing a change in an insurgent population. The words in the diagram represent the factors linked

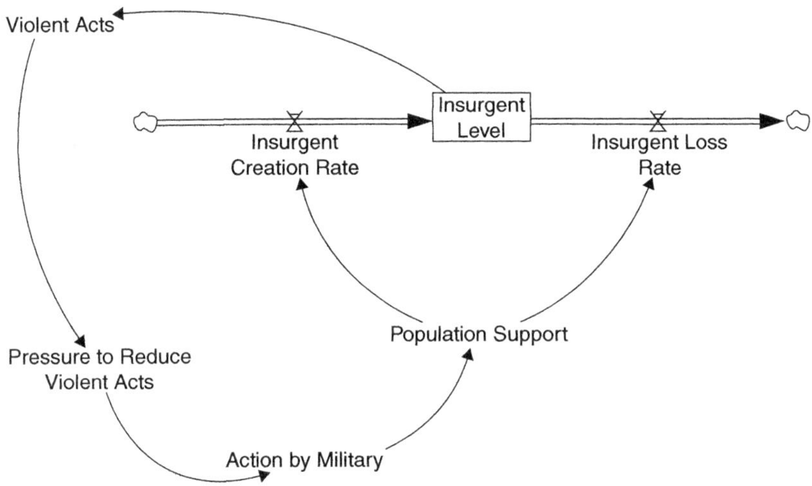

Figure 2.8. Insurgent stock and flow diagram.

to the insurgent level and the arrows show the causal linkages. A plus sign indicates a change in the same direction as the source factor change. A minus sign indicates a change in the opposite direction. One can interpret this diagram as follows. As the number of insurgents change, this causes a change in the number of violent incidents. As those change, pressure on the government to reduce incidents changes. This change causes a subsequent response by military troops, which leads to a change in how the indigenous population responds. Their response then contributes to the number of people participating in the insurgency.

Once the causal loop diagram is developed the modeler can proceed to the next step in SD, which is the development of a stock and flow diagram. This type of diagram is derived from the causal loop relationships and represents actual levels and rates of change of the system variables. Figure 2.8 is the stock and flow representation for this insurgent model.

The rectangular box is the level variable of interest that is governed by creation and loss rates. These rates are in turn influenced by the factors noted in the causal loop diagram. With an SD model one has a graphical representation of the system that clearly shows the cause and effect relationships and provides a sense of the interdependencies that exist in complex systems.

Although a simple example, one can see that this approach lends itself to modeling many types of systems. By including variables that affect the risk level of a particular problem, a sense of how changes in those variables will affect the risk level may be gleaned.

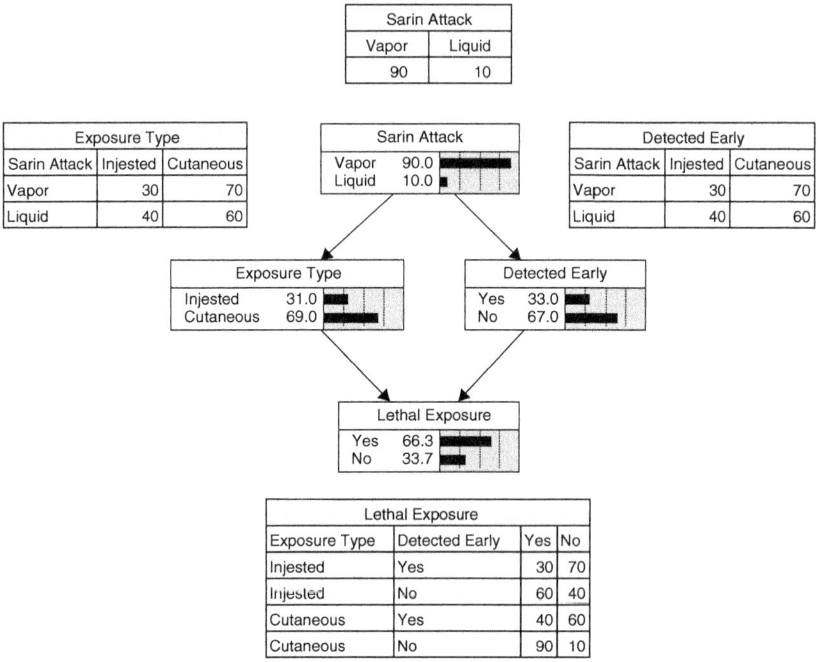

Figure 2.9. Notional Sarin nerve agent Bayesian network.

2.3.4 *Bayesian Network Analysis*

A Bayesian network is a directed acyclic graph or belief network where nodes represent random variables and the directed arcs indicate probabilistic dependence. The arcs in the network define probabilistic dependence between pairs of variables; the direction of the arc indicates which of two possible conditional probability distributions has been captured (Spirtes, Glymour, Scheines, Kauffman, Aimale, & Wimberly, 2000). In describing a network, modelers use tree and family metaphors to describe the relationships between nodes. Parents are nodes that point to children. Ancestors of a node include all nodes that point to it directly or indirectly through its parents. Descendants are those nodes to which a node points directly or indirectly (Hudson, Ware, Laskey, & Mahoney, 2005).

Bayesian networks have been used in the development of anti-terrorism modeling. BN have also been used to predict distribution for lethal exposure to chemical nerve agents such as Sarin. For example, in Figure 2.9, conditional probability tables are shown for each node: Sarin Attack, Exposure Type, Detected Early, and Lethal Exposure. In this notional model, Sarin Attack is the root node and parent to Exposure Type and Detected Early. By changing the evidence for a given variable, the inference algorithm applies Bayes's rule and updates the probability distribution for Lethal Exposure. BN are useful as well because as the situation changes, they are easy to update; as the evidence changes, the posterior probability changes. BN software such as Netica™ or Genie™ is well known and can interact and exchange data with other models.

2.3.5 *Game Theoretic Models*

Game theory (GT) is the study of multi-agent decision problems. The classic references include von Neumann and Morgenstern (1944), and Nash (1951), though it was first considered in the nineteenth century by Cournot (1838). Most of the current research and applications are conducted by micro-economists, though the subject is not confined to that area as there have also been several successful applications of the technique in areas as diverse as computer science and evolutionary biology (Dash & Jennings, 2003; Smith, 2004).

An important assumption in game-theoretic models is that all the possible utilities of the different consequences for each player must be derivable and usable within the model. This implies knowledge of the possible goals and aspirations of the different players. Classical game theory only considers one set of utilities for each player; however, if there is uncertainty about the other players' intentions, multiple utilities can be modeled using games of incomplete information (Harsanyi, 1967, 1968a, 1968b). In addition, another well-discussed assumption of game theory is the need for the players to be *rational* and *intelligent* enough to work out the consequence of their actions (Binmore, 1990). This assumption can be quite limiting especially when players turn out to be less sophisticated than first thought (e.g., they have misinterpreted the consequences of their actions). It has also been questioned whether any sensible decisions can be made about terrorist behavior for analytical application (Plaut, 2004). However, these limiting assumptions have not prevented game theory from contributing to analysis in numerous related areas (Borm, Hamers, & Hendrick, 2001).

One response that is often offered in defense of the assumption above, that adversary objectives are strictly consequence maximizers, is that it defaults to the worst case and is therefore "erring" in the right direction. The thinking is that planning or defending against the worst an intelligent agent can do is not necessarily a bad thing. However, there are some limitations to this argument. For example, if resources have to be removed from defending against a small-scale attack to help defend against a large-scale attack this could make attacking on the small scale beneficial to the terrorist *especially* if that was the intention anyway. Further, the worst-case consequences for terrorism scenarios (particularly in the case of WMD attacks) can be many orders of magnitude larger than the assessed expected consequences given less-than-optimal choices made by the adversary. This can greatly bias assessment conclusions and have strong impacts on planning and defense strategies.

There have been several successful applications of game theory within the nuclear, biological, and chemical (NBC) arena. For example, Robert Aumann, Nobel Memorial Prize in Economics 1995, and Michael Maschler developed game theoretic models during the Cold War for dealing with nuclear disarmament problem (Aumann & Maschler, 1995). Most of the work using game theory within an NBC arena focuses on the human element of the problem, to include negotiating with terrorists, formation of terrorists groups, and willingness of the domestic population to accept solutions to a terrorism problem (Anderton & Carter, 2006; Bauch, Galvani, & Earn, 2003; Douglas & Mars, 2003; Lapan & Sandler, 1988). More recent applications of game theory can be seen in Kott and McEneaney, who discuss several applications of the game theory to Department of Defense problems and cover a wide variety of possible implementations (Kott & McEneaney, 2006).

It is important to realize that unlike other analytical methods (where finding the maximum or minimum of some value is the sole goal), there is no "one size fits all" solution method to a game. For example, possible solution methods include min-max (Nash, 1951; von Neumann & Morgenstern, 1944). Also, these solution methods can produce multiple solutions to the same problem (due to the nonlinear nature of games). This has led to much research into the selection of the solution when multiple solutions are presented (Harsanyi & Selten, 1998; Herings, Mauleon, & Vannetelbosch, 2004).

Recent useful applications of game theory have been used in analyses supporting the protection of multiple targets and to interdependent security (Hausken, Bier, Zhuang, 2009; Heal & Kunreuther, 2005). A very important application of game theory was developed by Milind Tambe and his colleagues, who used the Stackelberg game to develop optimal randomizations of inspections and patrols (Jain, Pita, Tambe, Ordonez, Paruchuri, & Kraus, 2008). Applications are spreading rapidly to include randomizations of patrols and inspections at the Los Angeles Airport to randomization of assignments of Federal Air Marshalls, among other areas.

Game theory is a normative technique as opposed to a descriptive or positive one (Binmore, 1990). This means that given a game, game theory will tell you how the game should be played as opposed how it will actually be played. This may lead an analyst to gain some unexpected and interesting insight into the terrorism problem, which other techniques fail to provide. However, given the problems of determining an opponent's rationale and intentions, a good solution might be given by game theory but it could be for the wrong game in the first place.

2.4 Conclusion

There remains significant debate in the risk analysis, decision analysis, security, and intelligence communities about the validity of the NRC Committee's underlying assumption that terrorists have the understanding to always make optimal choices that maximize consequences in the wide array of challenging technical disciplines required for successful execution of a bioterrorism attack. Despite decades of advances in consequence modeling in the U.S. government, there may be significant uncertainty and inaccuracy in our own estimates of consequences from a WMD attack given a set of initial conditions – Can we base our strategic biodefense planning on the assumption that terrorists do better? That they have the information at their disposal to always make consequence-maximizing choices? Given this, eliciting intelligence experts that have a level of understanding of what our adversaries are capable of accomplishing is a quite reasonable and responsible option; representing this information probabilistically on an event tree is reasonable as well, noting the relevant limitations.

Professor Yacov Haimes (2009) suggests the use of multiple techniques for assessing terrorist actions as probabilities. He reminds us that "no single model or methodology can effectively meet all the challenges of tracking terrorism through scenario generation and structuring, updating and quantifying the value of intelligence, assigning priorities to the scenarios in a well-established risk-based methodology, or track terrorists' attack plans." We agree that multiple approaches, perhaps in combination are needed to address the complex issue of terrorism, including event trees, decision trees, fault trees, Bayesian belief networks, game theory, and agent-based models, among others. In this context, however, PRA and event trees have been shown to be useful approaches for assessing terrorism risks, especially for creating a baseline comparison of these risks. Decision trees, like PRA and all approaches, have limitations and are not on their own a complete solution. In the case of applications for terrorism risk analysis, the NRC's decision tree approach has limitations that may be difficult to surmount upon implementation to include the fact that adversaries' objective functions and level of ability to predict tree outcomes are unknown and difficult to estimate.

Acknowledgments

The authors would like to thank the contributions of the Area Editor Professor Yacov Haimes and the reviewers for this journal article. Dr. Detlof von Winterfeldt's participation in writing this article was supported by Decision Research, Oregon and the U.S. Department of Homeland Security under Grant 2008-GA-T8-K004 (Federal Emergency Management Agency, National Preparedness Directorate) to the University of Southern California. Points of view and opinions expressed in this article are solely those of the authors and do not necessarily represent the official position or policies of Decision Research or FEMA's National Preparedness Directorate or the Department of Homeland Security.

NOTES

1. Homeland Security Presidential Directive 10: Biodefense for the 21st Century, 2004. Available at www.fas.org/irp/offdocs/ nspd/hspd-10.html.
2. Detlof von Winterfeldt was a member of the Technical Expert Review Panel in 2005.

REFERENCES

Anderton, C. H., & Carter, J. R. (2006). Applying intermediate microeconomics to terrorism. *Journal of Economic Education*, *37*(4), 442–458.

Apostolakis, G. (1990, December). The concept of probability in safety assessments of technological systems. *Science*, *250*(4986), 1359–1364.

Aumann, R. J., & Maschler, M. B. (1995). *Repeated games with incomplete information*. MA: MIT Press.

Bauch, C. T., Galvani, A. P., & Earn, D. J. (2003, August). Group interest versus self-interest in smallpox vaccination policy. *Proceedings of the National Academy of Sciences of the USA*, *100*(18), 10564–10567.

Bedford, T., & Cooke, R. M. (2001). *Probabilistic risk analysis foundations and methods*. Cambridge, UK: Cambridge University Press.

Binmore, K. (1990). *Essays on the foundations of game theory*. Basil Blackwell.

Bommer, J., Scherbaum, F., Bungum, H., Cotton, F., Sabetta, F., & Abrahamson, N. (2005, April). On the use of logic trees for ground-motion prediction equations in seismic-hazard analysis. *Bulletin of the Seismological Society of America*, *95*(2), 377–389.

Borm, P., Hamers, H., & Hendrick, R. (2001, December). Operations research games: A survey. *TOP*, *9*(2), 139–199.

Committee on Methodological Improvements to the Department of Homeland Security's Biological Agent Risk Analysis, National Research Council (NRC). (2008). Department of Homeland Security Bioterrorism risk assessment: A call for change. Available at: http://books.nap.edu/openbook. php?record id=12206.

Cordesman, A. (2005). *The challenge of bioterrorism*. MA: CSIS Press.

Cornell, C. A., & Merz, H. (1975, October). Seismic risk analysis of Boston. *Journal of the Structural Engineering Division*, *101*(10), 2027–2043.

Cournot, A. (1838). *Recherches sur les principes math ematiques de la thorie des richesses* (*Researches into the mathematical principles of the theory of wealth*). Hachette.

Dash, R. K., & Jennings, N. R. (2003, December). Parkes DC. Computational-mechanism design: A call to arms. *IEEE Intelligent Systems*, *18*(6), 40–47.

Dillon-Merrill, R., Parnell, G., & Buckshaw, D. (2008). Logic trees: Fault, success, attack, event, probability, and decision trees. In *Wiley Handbook of Science and Technology for Homeland Security*.

Douglas, M., & Mars, G. (2003). Terrorism: A positive feedback game. *Human Relations*, *56*(7), 763–786.

Ericson, C. (1999, August). Fault tree analysis: A history. Proceedings of the 17th International System Safety Conference.

Ezell, B. (1998, May). *Risks of cyber attack to supervisory control and data acquisition for water supply*. Thesis, University of Virginia.

Ezell, B., Haimes, Y., & Lambert, J. (2001). Risks of cyber attack to water utility supervisory control and data acquisition systems. *Military Operations Research*, *6*(2), 30–46.

Garcia, M. (2006). *Vulnerability assessment of physical protection systems*. Sandia National Laboratories. US: Elsevier.

Garrick, B., Hall, J., Kilger, M., McDonald, J., O'Toole, T., Probst, P., Parker, E., Rosenthal, R., Trivelpiece, A., Arsdale, L., & Zebroski, E. (2004). Confronting the risks of terrorism: Making the right decisions. *Reliability Engineering and System Safety*, *86*(2), 129–176.

Goodwin, D. (2005). *The military and negotiation: The role of the soldier-diplomat*. Taylor & Francis.

Haimes, Y. (2009). *Risk modeling, assessment, and management*. New York: John Wiley and Sons.

Harsanyi, J. (1967). Games with incomplete information played by Bayesian players I. *Management Science*, *14*(20), 159–182.

Harsanyi, J. (1968a). Games with incomplete information played by Bayesian players II. *Management Science*, *14*(20), 320–334.

Harsanyi, J. (1968b). Games with incomplete information played by Bayesian players III. *Management Science*, *14*(20), 486–502.

Harsanyi, J., & Selten, R. (1998). *A general theory of equilibrium selection in games*. Cambridge, MA: MIT Press.

Hausken, K., Bier, V. M., & Zhuang, J. (2009). *Defending against terrorism, natural disaster, and all hazards, game theoretic risk analysis of security threats*. Springer.

Heal, G., & Kunreuther, H. (2005). You only die once: Interdependent security in an uncertain world. In H. W. Richardson, P. Gordon, & J. E. Moore, II (Eds.), *The economic impacts of terrorist attacks*. Cheltenham, UK: Edward Elgar.

Hees, M., & Roy, O. (2007). Intentions, decisions and rationality. Available at: www.illc.uva.nl/Publications/ResearchReports/ PP-2007–21.text.pdf.

Herings, P. J., Mauleon, J. A., & Vannetelbosch, V. J. (2004). Fuzzy play, matching devices and coordination failures. *International Journal of Game Theory*, *32*(4), 519–513.

Hora, S. (2007). Eliciting probabilities from experts. In W. Edwards, R. Miles Jr., & D. von Winterfeldt (Eds.) *Advances in decision analysis*. Cambridge, UK: Cambridge University Press.

Hudson, L., Ware, B., Laskey, K., & Mahoney, S. (2005). An application of Bayesian networks to antiterrorism risk management for military planners, C4ISR Papers Mason Archival Repository. Available at: hdl.handle.net/1920/268.

Jain, M., Pita, J., Tambe, M., Ordonez, F., Paruchuri, P., & Kraus S. (2008). Bayesian Stackelberg games and their application for security at Los Angeles International Airport. Available at: http://www.sigecom.org/exchanges/volume7/2/jain.pdf.

Kaplan, S., & Garrick, B. J. (1981). On the quantitative definition of risk. *Risk Analysis*, *1*(1), 11–27.

Kent, S. (2007). The Board of National Estimates: Collected Essays, Center for the Study of Intelligence. Available at: www.cia.gov.

Kirkwood, C. (2002). A decision tree primer. Available at: www.public.asu.edu/k˜irkwood/.

Koller, G. (2000). *Risk modeling for determining value and decision making*. FL: Chapman and Hall.

Kott, A., & McEneaney, W. M. (Eds). (2006). *Adversarial reasoning: Computational approaches to reading the opponent's mind*. FL: Chapman and Hall.

Kumamoto, H., & Henley, E. (1996). *Probabilistic risk assessment and management for engineers and scientists*. US: IEEE Press.

Lapan, H., & Sandler, T. (1988). To bargain or not to bargain? That is the question. *American Economic Review*, *78*(2), 16–21.

Mauw, S., & Oostdijk, M. (2006). Foundations of attack trees. Available at: http://www.cs.ru.nl/~martijno/publications/ papers/attacktrees.pdf.

McGill, W., Ayyub, B., & Kaminskiy, M. (2007). Risk analysis for critical asset protection. *Risk Analysis*, *27*(5), 1265–1281.

Myerson, R. (1991). *Game theory: Analysis of conflict*. Washington, DC: Harvard University Press.

Nash, J. (1951). Non-cooperative games. *Annals of Mathematics*, *54*(2), 286–295.

Parnell, G., Liebe, R., Dillon-Merrill, R., Buede, D., Scouras, J., Colletti, B., Cummings, M., McGarvey.,D., Newport, R., Vinch, P., Ayyub, B., Kaminskiy, M., & Scouras, J. (2005). *Homeland security risk assessment: Volume 1 – An illustrative framework and Volume 2 – Appendices of methods*. Washington, DC: Homeland Security Institute.

Pate-Cornell, M. E. (2007). The engineering risk analysis method and some applications. In W. Edwards, R. Miles Jr., & D. von Winterfeldt (Eds). *Advances in decision analysis*. Cambridge, UK: Cambridge University Press.

Pate-Cornell, M. E, & Guikema, S. D. (2002, December). Probabilistic modeling of terrorist threats: A systems analysis approach to setting priorities among countermeasures. *Military Operations Research*, *7*(4), 5–23.

Plaut, S. (2004). Misplaced applications of economic theory to the Middle East. *Public Choice*, *118*(1), 11–24.

Rosoff, H., & von Winterfeldt, D. (2007). A risk and economic analysis of dirty bomb attacks on the ports of Los Angeles and Long Beach. *Risk Analysis, 27*(3), 533–546.

Smith, M. (2004). The theory of games and the evolution of animal conflicts. *Journal of Theoretical Biology*, *47*(1), 209–221.

Spirtes, P., Glymour, C., Scheines, R., Kauffman, S., Aimale, V., & Wimberly F. (2000). Constructing Bayesian network models of gene expression networks from microarray data. Proceedings of the Atlantic Symposium on Computational Biology.

Vesely, W. (2005). Quantitative risk analysis: Challenges and opportunities at NASA, Manager Risk Assessment for NASA Headquarters. Available at: ti.arc.nasa.gov/projects/ishem/Papers/Vesely PRA.doc.

Viscusi, W. K. (2003). *The risk of terrorism*. US: Kluwer Academic Publisher.

von Neumann, J., & Morgenstern, O. (1944). *Theory of games and economic behavior*. Princeton, NJ: Princeton University Press.

von Winterfeldt, D., & O'Sullivan, T. (2006). Should we protect commercial airplanes against surface to air missile attacks from terrorists? *Decision Analysis*, *3*(2), 63–75.

Willis, H., Morral, A., Kelly, T., & Medby, J. (2003). Estimating terrorism risk. *RAND Center for Terrorism Risk Management Policy*.

Wilson, R. (2003). Combating terrorism: An event tree approach. *Proceedings of the International Seminar on Nuclear War and Planetary Emergencies 27th Session*, 122–143.

U.S. Department of Homeland Security. Response to National Academies of Science. (2008). Available at: http://www.dhs.gov/xabout/structure/gc1222807486868.shtm.

3

Integrating Stakeholder Values into Strategic Planning through Comparative Risk Analysis

Willis H. Henry and Russell Lundberg

The need for deliberative risk analysis in homeland security studies is increasing. Following the attacks of Sept. 11, 2001, expansion of the role of security has taken place at all levels of the U.S. government. While the Department of Homeland Security (DHS) is the lead agency for addressing homeland security risks, other entities have responsibilities in that domain, including the military, the Federal Bureau of Investigation, the Department of Health and Human Services and the Centers for Disease Control and Prevention, and others. Even DHS is not a unitary actor, but contains multiple agencies with multiple responsibilities. Outside of the executive branch, Congress is involved in the allocation of security resources and has priorities of its own, taking into consideration the views of the public. Additionally, a large portion of the nation's critical infrastructure is in the hands of private business, and securing that infrastructure will need to involve private engagement.

These multiple stakeholders are facing multiple threats and trying to achieve multiple goals. Communities are threatened by natural disasters, aging infrastructure, terrorism, criminal networks, and large-scale industrial accidents, and need to make themselves more resilient to the repercussions of these types of hazards. While the final goal of policy would be to select an optimal set of risk management activities, the large number of such activities and limited budgets make it necessary to set priorities. An initial consideration taken by many is identification of current and emerging hazards and threats of greatest concern (Hagel, 2014; Rademaker, 2008; DHS, 2010).

Risks in the homeland security domain vary greatly in kind and consequence. Comparing the economic consequences of a potential cyber-attack to the health consequences of a potential influenza pandemic, for example, is a challenging exercise in comparative risk analysis (CRA), for not only are the risks plagued with uncertainty, but they are driven by multiple differing consequences. Weighing the importance of differing consequences (such as deciding how many acres of wetlands lost in an oil spill are worth one life) necessarily involves subjectivity (Willis & LaTourrette, 2008).

Similar conditions exist in other domains, including areas of environmental health and safety, national security, and others. These issues first came to structured consideration in the area of environmental policy. Since the creation of the Environmental Protection Agency in the 1970s, the agency had the challenge of allocating funds across very dissimilar environmental risks. Comparative risk assessments in that environment were too often ad hoc and not useful for standardized comparisons of risk (see DeKay et al. for an overview of studies examining earlier CRAs) (DeKay et al., 2001; EPA 1987).

More recently, national and global risk assessments have emerged that include security concerns. At the global level, the World Economic Forum Global Risks Report has recently released its tenth annual report in 2015, with other reports including the World Health Organization and the United Nations (Alliance Development Works, 2012; World Economic Forum, 2015; World Health Organization).

National risk assessments are widespread across Europe (including the United Kingdom, Ireland, the Netherlands, and more) and in more developed nations of Asia (Department of the Taoiseach, 2014; OECD, 2009; Willis et al., 2012; Vlek, 2013). In the United States, there are a number of risk rankings that relate to security, including national-level considerations of risk in the Chairman's Risk Assessment in the Department of Defense and risk assessments supporting the Quadrennial Homeland Security Review in the Department of Homeland Security, and widespread consideration of jurisdictional risk through the THIRA process (Meinhart, 2006; DHS, 2010, 2013).

Each of these processes aims to identify and prioritize hazards that are being managed. However, the methods used in each process vary. Accordingly, the results of these processes vary both in validity and reliability.

The goal of a comparative risk assessment, like those used to prioritize national security threats, should be to capture stakeholder values, informed by the best available science, in a reliable manner. Achieving this requires two conditions be met. First, those who are considering the risks must have awareness of accurate assessments of any relevant information, using quantitative risk information where possible and acknowledging the uncertainty about the current state of knowledge. Second, the subjectivity of the value judgments must be captured in a valid and reliable way, using the best methods for expert elicitation. There is a well-established literature that identifies the biases associated with the perception of risks (including availability bias, anchoring, and others) and that develops elicitation methods to address them (Slovic, 1992; Slovic et al., 2004; Wildavsky & Dake, 1990).

Methods to incorporate diverse judgment into risk rankings have been developed in other domains that similarly face a diverse set of risk. To address the need to integrate judgment into informed rankings of risk in environmental policy (EPA, 1987), researchers at Carnegie Mellon University developed an approach that would elicit informed perceptions of risk in a structured and analytical fashion (Morgan et al., 1996). Their methodology, which we call the Deliberative Method for Ranking Risks, has been applied in a number of domains, for laypeople and experts, in the United States and abroad. This deliberative method was initially tested in a ranking of the health and safety risks at a school and was validated for both laypeople and risk experts (Florig, et al., 2001; Morgan, 1999; Morgan et al., 2001). Later papers expanded the set of risks to consider environmental hazards (Willis et al., 2004; Willis et al., 2005). Researchers also applied the risk-ranking method in high-level governmental decision-making contexts (Willis et al., 2010) and examined the utility of the method in different contexts of the United Arab Emirates (Willis et al. 2010) and China (Xu, Florig, & DeKay, 2011). More recently, the Deliberative Method for Ranking Risks has been applied to the homeland security domain, examining risks in a strategic perspective as hazards at the national level (Lundberg, 2013). While this latest ranking considered only one perspective of homeland security risk ranking, it did provide some evidence that the method can be used to elicit informed rankings of homeland security risks from a lay population.

This chapter serves as a practitioner's guide to implementing this Deliberative Method for Ranking Risks, particularly for different kinds of terrorist and homeland security risks. The method will be described in detail, along with the materials used to support the ranking and the resulting outcomes.

3.1 Overview of Five Steps on Process

The Deliberative Method for Ranking Risks produces informed, judgment-based rankings of risks by applying the range of knowledge of how people learn about risks. The method has five steps (see Figure 3.1). The first two steps involve conceptualizing the risks, deciding the risks to be compared and the attributes needed to compare them. Then the selected risks are described using the selected attributes as a foundation. These descriptions are used to inform the risk-ranking session in step four, the results of which are analyzed in step five. This section will describe each of these steps individually.

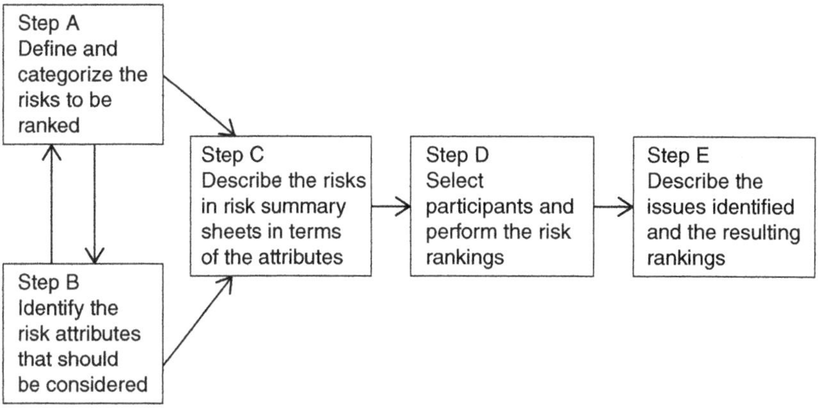

Figure 3.1. Steps of the Deliberative Method for Ranking Risks (adapted from Florig et al., 2001).

3.2 Categorizing the Risks

A first step in ranking risks is deciding what risks you want to compare. In the whole universe of risks, there is a need to create a manageable number of clear, mutually exclusive, discrete risks to compare. A first consideration is bounding the problem. In the context of terrorism, there are certain choices to be made: Do you consider only terrorism in a certain area, such as within the nation or within a city? Do you distinguish between terrorism and crime? Do you consider only terrorism, or do you consider it along with natural disasters and accidents in an all-hazard approach?

A second consideration is how discrete risks will be constructed. In environmental policy, Willis and colleagues (2004) found that risks were categorized in many ways, including by "specific agent or stressor (e.g., lead), the activity giving rise to the hazard (e.g., mining), the location (e.g., specific streams and rivers), the physical medium (land, water, or air), the endpoint of concern (e.g., biological diversity)." Similarly, there is the potential for multiple conceptualizations of risk in homeland security; homeland security risks can be categorized by hazard (e.g., hurricanes, terrorist bombings), area (e.g., Washington, DC), sector (e.g., power plants), cause of risk (e.g., terrorists), or others. There is no objectively correct way to categorize risks, as any of these approaches can be useful in some circumstances. Instead, the categorizations should be matched to the purpose for which they are being used and for the organization that is using them. For example, DHS may want to consider risks by hazard when working on a strategic level (such as under the Quadrennial Homeland Security Review), but risks by area when allocating funds (such as under the

Urban Area Security Initiative). Similar choices have to be made in deciding how to conceptualize risk to fit the problem in any risk ranking.

In addition to being administratively compatible, Morgan and colleagues (2000) identify three other attributes desired in ideal risk-ranking systems. First, they should be logically consistent: exhaustive, so that no relevant risks are overlooked; mutually exclusive such that none is double counted; and homogeneous so that all the risks should be compared using the same attributes. Second, they should be equitable, fairly drawn from the range of shareholder interests. And third, they should be compatible with cognitive constraints and biases. One important aspect of that is the number of hazards to consider. People can only consider a limited number of risks at one time, and previous studies have limited the number of risks to around a dozen. If a greater number of risks must be considered, they can be divided into smaller groups that are considered in different sessions with a few risks that are considered across sections to calibrate the rankings. Other biases to be aware of include the "lamppost effect" (only considering easy-to-understand risks) and aspects of dread and the unknown (and to divide similar-seeming risks with different perceptions of dread and the unknown into different categories).

A final consideration is the specific risks to consider. Risk categorization requires substantial judgment, as grouping the thousands of possible hazards into a manageable number of categories requires understanding and weighing the importance of different causes and attributes (Morgan et al., 2000). Additionally, the process should be iterative (Kemshall, 2003), and for that reason risk categorization in this method proceeds in parallel with and is informed by the selection of attributes in step two, as the process of considering consequences may identify additional risk types that are important or recommend the removal or consolidation of other risk types (Florig et al., 2001). Even after placing risks into categories, there may be more risks than can be handled in a risk-ranking session. Previous studies have used between 9 and 14 risks in a single risk-ranking session; when larger numbers of risks are considered, it is standard to break those risks into smaller subsamples and run multiple sessions with around 10 risks per session. Additionally, while it is preferable to use the complete set of all risks, using a select subset of risks that covers the larger set in an interesting way may be useful on some occasions.

3.3 Selecting Attributes

While deciding the risks to compare, it is also important to identify the attributes needed to describe those risks. This step should be conducted concurrently with the categorization and selection of risks. The attributes describing the risks should be both comprehensive and parsimonious.

There are several ways that attributes can be selected, each with its benefits and drawbacks. One way to identify attributes is through an empirically based approach, directly identifying the factors that concern people for your particular risk set. The initial application of the Deliberative Method for Ranking Risks applied an empirically based approach undertaken by Jenni (Florig et al., 2001). Jenni (1997) outlines two empirical approaches to identify risk attributes. The first relies on examining attributes that result in different risk standards – for example, the identification that voluntary risks are held to a looser standard than involuntary ones identified by Starr (1969). The second approach is to draw on the "psychometric paradigm," using psychological scaling techniques to represent people's perceptions of risks in a particular domain (Slovic, 1987, 1992). These techniques often find perceptions of risk explained in terms of three factors: dread, the unknown, and societal and personal exposure. For each of the nine qualitative risk variables from Fischhoff and colleagues (1978) (voluntariness, immediacy, known to exposed, known to science, controllability, familiarity, chronic/catastrophic, dread, and severity), Jenni identified attributes or proxies on the basis of being justifiable, clearly defined, and measurable. This approach has been used to identify risk attributes from scores of attributes used across risk assessments in the fields of health and safety (Morgan, 1999) and ecological risk management (Laso, Kinnell, & Fisher, 2000; McDaniel, Axelrod, & Slovic, 1995; Willis et al., 2004).

An alternative approach is to draw on the literature in selecting attributes. Rather than reinventing the wheel, researchers conducting a risk ranking can draw from this literature when identifying their attributes. As there are thousands of papers that examine the consequences of disasters from the point of a single attribute, it is useful to focus on papers that use a comprehensive framework. Some of these frameworks involve theoretical consideration of the risks (including Mileti, 1999; Lindell and Prater, 2003; and Keeney and von Winderfeldt, 2011), while others represent applied frameworks (including Homeland Security Presidential Directive 7, the National Planning Scenarios, the National Infrastructure Protection Plan, the Quadrennial Homeland Security Review, and multiple methodologies as identified in the Committee to Review DHS's Approach to Risk Analysis, 2010) (Committee to Review the DHS's Approach to Risk Analysis, 2010; HSC/DHS, 2005; Keeney & von Winterfeldt, 2011; Lindell & Prater, 2003; Mileti, 1999; DHS, 2009, 2010; White House, 2003). As any one framework can omit important variables, it is important to examine many such frameworks. These frameworks, however, omit the important perceptual characteristics associated with dread and the

The narrative introduction is followed by a table presenting what is known about the risk in terms of the attributes selected to describe it. The purpose of the table is to summarize the state of scientific knowledge in terms of all risk attributes. This summary table should reflect the range of values for quantitative attributes, not only the best estimate for each attribute but lower and upper bounds as well, each rounded to the proper significant digit.

The risk tables should be followed by up to three additional pages that describe what is known about the risks in a structured format. These pages should answer three questions: 1) What is known about the risk, its damage, and how it creates that damage? 2) What is the exposure to the risk, including how likely it is, if the risk is probabilistic or reflects intelligent adversaries, and if there are certain areas or times that are at greater exposure to the risk? and 3) What has already been done about the risk? The final question reinforces that the risk summary sheets reflect the residual risk that currently remains after risk mitigation measures have been implemented. These summaries should include not only what is known about the risks, but also recognition of what is not known about the risks. These interior pages should be readable, with readability scores lower than a college level. Additionally, while the statements used in the risk summary sheets should be well documented, the risk summary sheets actually used in the risk-ranking sessions should leave out the references to make it more readable.

Production of the risk summary sheets should be undertaken following best practices of risk assessment. Each risk should be examined using a multidisciplinary team to cover the intricacies of each risk. For example, the experts who understand cyber-attacks will be different from those who understand pandemics, so both should be included in the process. Alternatively, multidisciplinarity can be accomplished through peer review. But regardless of whether it adds multidisciplinarity, peer review should be incorporated into the risk assessments as part of recognized scientific practices to make sure processes are valid and all relevant research is covered. The risk summary sheets developed for the risk assessment should be well documented with citations for each claim. While these citations will be left off of the risk summary sheets used in the risk ranking, there should also always be a most current version with citations to substantiate all claims.

3.5 Unique Aspects of Assessing Comparative Risk in the Homeland Security Domain

Several characteristics of homeland security risks make their assessment unique from other risk management domains. While some security risks have a long history of data to support empirical estimates (the average number of terrorist bombings per year averaged over the past 40 years, for example), many security risks are so uncommon that averaging is not useful for decision making. In those cases, expected value should be decomposed into likelihood and consequence, each of which can be estimated separately and then combined. If historical events are not available, either for combined averages or for decomposed likelihood or consequences, there are alternative approaches: projecting from analogous data, either the same event but in different times or different places, or a different but similar event (such as applying an accidental blackout for a blackout resulting from a cyber-attack); estimates from models; and expert opinion.

Additionally, estimates of risk should also account for the uncertainty in how much we know. While we may not know how many tornadoes will strike in a given year, we can calculate a reasonable average for how many tornadoes will strike per year in a decade; our estimates of the average for a terrorist nuclear detonation are far less certain – is a nuclear detonation in the next decade nearly certain to happen or nearly certain not to? For this reason, estimates should reflect the imprecision of those estimates. One way this imprecision should be incorporated is through bounding the estimates, not only presenting a best estimate for each attribute, but lower and upper bounds for those estimates as well.

Another way that this imprecision should be addressed is through the significant digits of the estimates. There is no hard and fast rule for the proper number of significant digits, but they should reflect the data used in the assessment. In Lundberg, for example, some estimates had an upper bound more than 1,000 times higher than the lower bound, and chose to present all attributes rounded to one significant digit. The proper number of significant digits for your ranking should reflect an empirical examination of the data.

3.6 Conducting Risk Rankings

Once the support materials for the risk ranking have been developed, the risk ranking itself can be conducted. This section will discuss the risk-ranking sessions, including what needs to be done during the sessions and what needs to be done to prepare for them.

3.6.1 *Recruiting Participants*

Soliciting individuals' perceptions of risks requires having a group of participants. As the risk-ranking sessions include group activities, they are conducted in small groups. Previous applications of the Deliberative Method for Ranking Risks have conducted rankings with between 8 and 20 people. The small size of the groups limits the ability of the rankings to be representative of the nation as a whole. Instead, participants have been selected using two general approaches.

One possible group of participants is the general public. Volunteers have been solicited online and through interested groups reflecting the general population (such as parent-teacher organizations) (Lundberg, 2013; Willis et al., 2004; Xu et al., 2011). As the ranking sessions can be lengthy (5–6 hours), participants may be paid for their time; the amount of payment varies depending on the context, but in the United States in 2012 this was around $20 per hour. Given the challenges of developing public samples representative of the nation, it may be useful to employ purposeful sampling. Lundberg used a sample of participants that covered a range of interesting attributes of the population, selecting participants with a variety of genders, ages, races, and educational levels. Other purposeful groups can be constructed, including identifying samples of typical cases, extreme cases defining the maximum variation, cases of highly engaged individuals, or others. Patton describes a variety of other purposeful sampling techniques that can be used to achieve diverse participant groups (Patton, 1990).

Another approach is to select samples of expert populations. While it may not be possible to generate a representative sample of the public generally, it may be possible to generate results that are representative of certain groups. Morgan and colleagues, for example, conducted their risk ranking using a sampling of risk experts enrolled in a risk course at the Harvard School of Public Health (Morgan et al., 2001). These expert populations can also be drawn from the actual decision makers on an issue. Willis, for example, conducted a risk ranking using high-level government officials that covered the range of issues of environmental policy in the United Arab Emirates (Willis et al., 2010). The results of that ranking session are not necessarily representative of the views of the country's population, but are representative of the views of high-level decision makers in that area. An analogous situation in security would be in the Quadrennial Homeland Security Review, where selected security experts identify the risks of greatest concern in an unstructured fashion; the results of a risk ranking with this group would not necessarily be representative of the nation, but could be representative of homeland security experts.

There are trade-offs between these two approaches. The rankings that result from expert groups may be more representative, representing either expert concerns or governmental priorities. Expert rankings may also be more likely to be used as the expert populations may be more aware of and supportive of

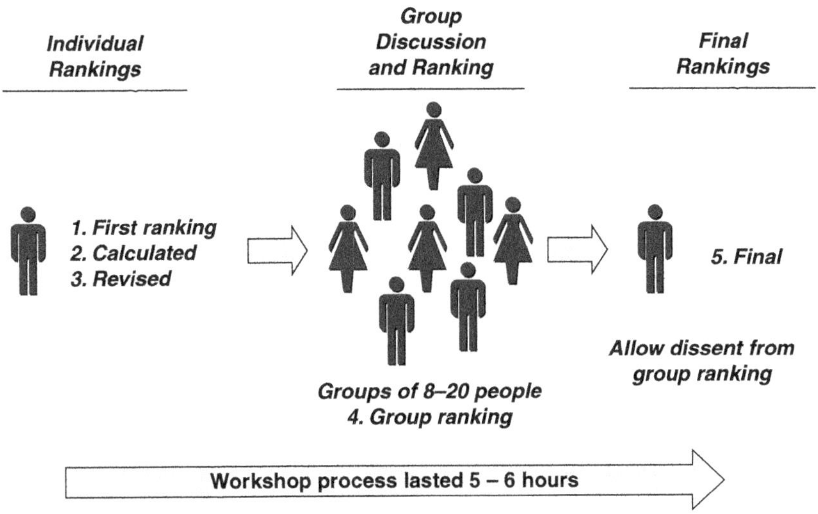

Figure 3.3. Steps to the risk-ranking session (adapted from Willis et al., 2010).

the resulting rankings. However, engaging expert participation is more challenging, and identifying a time that all relevant experts can come together can be difficult. Additionally, expert opinions may not reflect public concerns. As Andrews notes, risk rankings done "behind closed doors without a public mandate" have been ignored (1998). Additionally, risk rankings that include public opinion may be more defensible after the fact should an event occur.

3.6.2 *Carrying Out Sessions*

The process for ranking risks builds on validated decision-sciences methods of eliciting preferences from individuals and groups and involves several steps (see Figure 3.3). The initial rankings focus on the risks on an individual level. The risks are then approached through the perspective of the attributes, with individuals drawing on a calculated risk to develop a more considered revised ranking. Next, because people can learn from the perspectives of others, a group ranking is performed. But as groups can lead to forced consensus, the rankings conclude with an opportunity to dissent from the group. The following pages describe this process in greater detail, identifying how to execute a risk-ranking session as well as the materials required to support those rankings (summarized in Table 3.1).

There are several steps to the risk-ranking sessions. Before any consideration of the risks occurs, participants are orientated to the process. Reading from a script, participants are given an overview of the goals, process, and

Table 3.1. *Materials to support risk-ranking sessions*

<table>
<tr><th>Initial Ranking</th><th>Calculated Ranking</th><th>Revised Ranking</th><th>Group Ranking</th><th>Final Ranking</th></tr>
<tr><td colspan="5">Script</td></tr>
<tr><td colspan="5">Agenda</td></tr>
<tr><td colspan="5">Notes on the numbers</td></tr>
<tr><td colspan="5">Risk summary sheets</td></tr>
<tr><td colspan="5">Big sheet of risks</td></tr>
<tr><td>Risk ranking response form</td><td></td><td></td><td></td><td></td></tr>
<tr><td colspan="5">Spreadsheet to record rankings</td></tr>
<tr><td></td><td>Risk characteristics forms</td><td></td><td></td><td></td></tr>
<tr><td></td><td>Attribute cards</td><td></td><td></td><td></td></tr>
<tr><td></td><td>Spreadsheet to calculate risks</td><td></td><td></td><td></td></tr>
<tr><td></td><td>Examples of calculated risks</td><td></td><td></td><td></td></tr>
<tr><td></td><td></td><td>Comparison of first and calculated rankings</td><td></td><td></td></tr>
<tr><td></td><td></td><td></td><td></td><td>Session evaluation sheet</td></tr>
</table>

agenda for the day. Then the items needed to understand the risks are described. The risks are described in several documents, not only the risk summary sheets (described earlier), but definitions of the attributes of the risks in a document called "Notes on the Numbers" and a compilation of the summary tables in a document called the "Big Sheet of Risks." The Big Sheet of Risks should present the risks in two ways: first, providing the estimates for each of the risks and each of the attributes with the columns representing the risks and the rows representing the attributes; and second, the rankings of each risk by attribute (such as the ranking based solely on lives lost, the ranking based solely on economic damage, etc.). In this way, the Big Sheet of Risks allows a participant to easily scan the full set of risks to understand how they compare to each other in terms of each attribute and across attributes. Depending on the dedication and compliance of the participating group, review of these materials can be conducted individually in advance of a ranking session. In either case, the ranking session should begin with a review of these materials.

After the introductory materials have been reviewed, the initial ranking is undertaken. For the first step of the ranking process, participants are asked to consider the risks in a structured process. First, they sort the risks into high, moderate, and low risks. Then, they compare a risk from the high category and a risk from the moderate category, to ensure that the high risk is a greater concern. A third risk is then considered and ordered as higher, lower, or in between the two risks already considered. This process is continued until all risks are ordered from highest to lowest. Usually participants are allowed to reflect equal concern, i.e., ties, in this ranking. The results of the initial rankings are recorded on a risk-ranking response form.

Following the initial ranking, participants are guided through a calculated ranking based on a ranking of the attributes of concern. Some description of this process is useful for the participants so that they can more easily understand the ranking that results from the calculation. Our script described the calculated ranking as follows:

> To understand how this might work, let's start with an analogy. Cars have lots of different characteristics, including power, fuel efficiency, size, safety features, styling, and price, among others. Different people may have different opinions about which of these characteristics are most important. ... However, if I knew how each of these people thought about the characteristics of cars, and if I had information about how the various makes and models of cars scored on each of these characteristics, I think I could do a decent job of predicting how these people would rank order a list of cars in terms of their desirability.

The first step in the calculated ranking is to develop an understanding of the participants' individual perceptions of the attributes. First, it is important to determine the directionality for the attributes. While some attributes have a clearly better endpoint and a clearly worse one (for example, fewer lives lost is better than more lives lost), other attributes are less clear (for example, whether you are more concerned about things caused by people or things that are natural in origin is a personal judgment) (Willis et al., 2005). While research indicates that people are more concerned about particular non-consequence aspects of dread and uncertainty in the aggregate, it is not certain which aspects are of more or less concern on an individual level. For this reason, it is important to identify the endpoint of concern for each of the non-consequence attributes for each individual, and a form to solicit those characteristics is needed.

After the endpoints of concern are identified, the individuals must be guided through the process of ranking the attributes from the one that affects their concern the most to the one that affects their concern the least. A set of cards should be printed, each with one attribute on it and the range of values for the attributes for that set, to support the physical sorting of attributes. The ranking of attributes has typically followed an approach based on swing weights (Florig et al., 2001; Lundberg, 2013; Morgan et al., 2001; Willis et al., 2004; Willis et al., 2010; Xu et al., 2011). In this process, participants are first asked to discard any attributes that do not affect their concern at all. Then they are presented with a set of attribute cards, one for each attribute and each showing the lowest and highest value in the risk set for that attribute. For example, our attribute card describing the average lives lost shows the average number of deaths ranging from zero (from cyber-attacks on critical infrastructure) to 4,000 (from pandemic influenza). At this point, the swing weight method should be described to the participants. The participants are asked to imagine a risk that had all of the attributes at the high endpoint of concern, and further to imagine that they could make the risk less of a concern by changing its value on one and only one of the attributes from the high end to the low end (for example, changing the greatest number of deaths in a single event from 3 million down to one or changing the average economic damage per year from $10 billion to $7 million). Using this guidance, participants identify which of these changes would reduce their concern the greatest amount, then the second greatest amount, etc., until all characteristics have been ranked (ties can be allowed). These rankings of concern over the attributes should be recorded on a sheet prepared for this purpose, recording for each participant whether they care about the attribute at all and, if so, the rank order of the relative concern.

Each individual's relative concern for the attributes must then be used to create a calculated ranking. Weights should be applied such that the attribute of greatest concern receives the most weight and the attribute of least concern the least weight. The weighting method used to convert ranks to weights is an analytic choice; common options include inverse rank and Rank-Order Centroid approaches. Criteria for choosing these include both the validity of the resulting ranking and the simplicity in explaining the results to participants. Willis compared these approaches and demonstrated that inverse rank best supports the Deliberative Method for Ranking Risk (Willis, 2002). In our model, this was implemented as follows: an initial calculation was made for raw weights, with the attribute of least concern given one point, the attribute of next least concern given two points, etc., while all attributes about which they expressed no concern received zero points (because not all respondents will be concerned about the same number of attributes, the number of points given to the attribute of most concern will vary from person to person). The relative weight was then calculated from the raw weight by dividing the raw weight for that attribute by the sum of the raw weights for all of the attributes. These weights were then applied to the attributes, but because the scales for the attributes varied (for example, economic damages were in the millions or billions while lives lost only ranged to the hundreds), the scales for each of the attributes should be normalized (in our case, with the lowest value given a 0 and the highest value given a 100 for each attribute). The normalized values for the attribute for each risk were multiplied by the weight for each attribute to give a weighted value for that attribute, which was summed to give a concern score for each risk. These concern scores were then used to rank each of the risks from greatest calculated concern to least calculated concern. As these calculations are complicated, a spreadsheet prepared to execute the calculations automatically is important; integrating this calculation into a spreadsheet recording all the rankings may be convenient.

These calculated rankings are then used to further inform participants' consideration of the risks. Each individual gets his own calculated ranking based on his own reported concern for the attributes. But before he considers his own ranking, each individual should be given some examples to help him understand the weighting process: a first, using only one attribute of concern; a second, using a different single attribute of concern that results in a different ranking; and third, a ranking using both of those two attributes of concern. After gaining context, participants are given their own weighted risks, printed from the calculating spreadsheet. This should include the rankings for the attributes (which they can confirm) and the calculated rankings of the risks, as well as the participants' initial rankings.

The initial rankings and the calculated rankings may differ, and it is important to assure the participants that neither the initial nor the calculated rankings are objectively better than the other. Instead, participants should be encouraged to consider those risks where the difference between the initial and calculated risks are the greatest and attempt to understand the reason for those differences. Participants are then asked to consider what they have learned and revise their rankings as they see fit.

As a fourth step, participants are led through a facilitated group process to produce a consensus ranking of the hazards. As people can learn about hazards from each other, the participants are asked to consider the risks together. The process for the group ranking is similar to the initial ranking, first ordering the risks into groups, then developing an overall ranking. Throughout the process, it is important to identify the reasons behind their concerns. The diversity of opinions should be explored; for each risk, at least one argument for it being a low concern and one argument for it being a high concern should be encouraged. Additionally, the process should be inclusive, and each participant should be encouraged to comment at least once in the group discussion.

The group rankings are useful for considering the hazards from other points of view. But as groups can lead to forced consensus, it is important to give participants an opportunity to dissent from the risks and present their own perspective while still incorporating the learning of the group discussion. To adjust for these effects, participants are asked to provide a final ranking at the individual level. Participants should be encouraged to form their own final rankings, dissenting or agreeing with the group ranking as they think fit.

After the final ranking, it is important to understand the participants' perception of the process and results; especially when the method is used in a research setting. Typically, this survey is done using an evaluation sheet with several questions. First, participants are asked where their knowledge of the risks came from (prior knowledge, risk-ranking sheets, calculated ranking, or group discussion). Second, they are asked the extent to which their final rankings were based on prior steps (first ranking, calculated ranking, and group ranking). Finally, they are asked if they found the process inclusive, if they were satisfied with the group ranking, if the group ranking represented their ranking, and if they approve of the ranking for making decisions. Soliciting answers for these questions on a six-point scale is standard and will make them comparable to previous rankings done using the method.

The entire process can be lengthy, typically lasting four to six hours, but potentially longer. For greater details on the process and materials used, see Lundberg (2013).

Attempts to reduce the time commitment of the rankings have produced limited results. Internet surveys that provide information similar to the initial steps have proven of limited use, and grappling with the risks as in this structured process is required (Lundberg & Willis, n.d.). However, there are ways to reduce the time commitment. Some of the initial consideration of the risks can be conducted before meeting in order to save time. Similarly, it could be possible (although perhaps difficult) to perform the calculated and revised rankings remotely. The group ranking, however, presents significant challenges for remote rankings. The group process involves a synchronous aspect of give and take, and that kind of engagement could be difficult in an asynchronous environment.

3.7 Interpreting Results

The final step of the Deliberative Method for Ranking Risks is to interpret the results of the risk-ranking sessions. The methods discussed in this section were developed to evaluate the validity and reliability of the ranking results and are reported in several prior studies (Florig et al., 2001; Lundberg, 2013; Morgan et al., 2001; Willis et al., 2004; Willis et al., 2010; Xu et al., 2011).

While the most immediate outcomes of interest are the rankings themselves (with several approaches to consider those rankings), there are several other outcomes that are typically explored.

The primary outcome of the method is the rankings, both the group ranking and the final individual rankings. The average of the final individual rankings should be considered as an initial step, either in terms of the mean or the median values. This can be useful to identify which risks are greater than others. But it is also important to compare these averages with consideration of the disagreement about these rankings that exists among participants. Disagreement among individuals can be captured through the confidence intervals or quartile boundaries across participant rankings, as it may not be possible to actually differentiate between some similar risks. Additionally, it may be useful to examine the shape of the distributions of the individual rankings; for example, while most of our rankings had distributions that were unimodal (suggesting some consensus around a central value), the distribution for terrorist nuclear detonations was nearly flat (suggesting little to no consensus as to the risk).

A second outcome of interest is the rankings of how much the attributes influence their concern. The average ranking of the influence can be useful, similar to the ranking of the risks, as well as the number of people who considered the attribute important at all. Additionally, it may be useful to consider how the attributes correlate with the rankings of risks. This can be done by correlating the rankings based on each attribute of risk individually with the final rankings (in our case, correlating the final rankings with the rankings based solely on lives lost, then based on economic damage, etc.). This correlation provides an indicator of whether and to what extent participants' judgments were influenced by each attribute.

It may also be useful to understand how well the process worked. One process issue is how well the rankings built consensus. There are several ways to identify this. The first is to examine the standard deviation of the rankings at each stage in the process; the standard deviation should decrease (reflecting greater consensus) as the process unfolds. This can also be examined more quantitatively, comparing the correlation of the initial rankings to the correlation of final rankings, which should increase.

As this increased consensus may reflect forced consensus, a second process issue is how much the final ranking represented informed consideration rather than forced consensus. The final survey can be useful for this, examining the extent to which the first ranking, the calculated ranking, and the group ranking each contributed to the final ranking. Alternatively, this can be tested quantitatively using a regression, testing both the initial ranking and the group ranking as predictors of the final ranking; both the initial ranking and final ranking should be significant. Additionally, the survey can be useful to examine forced consensus, asking people the extent to which the group considered and discussed different points of view and encouraged each member to express his or her opinion.

A final issue to test is the extent to which the group supports the results. In the final survey, participants should be asked how satisfied or dissatisfied they were with the group's risk ranking, the extent to which the group's risk ranking is representative of their own concerns, and how strongly they would approve of the rankings being used in making decisions in a real organization. These three questions can tell a story of how people support the process even more than they support the results.

3.8 Extensions

While the Deliberative Method for Ranking Risks has only been applied to the homeland security domain at a national, strategic level using public participants, studies in other domains suggest that the method can be usefully applied to a range of problems and a range of participants. It can be useful at the national level comparing risks as hazards by experts in the Quadrennial Homeland Security Review or comparing the risks to states in the budget process. At the level of critical infrastructure sectors, the method can be useful for supporting prioritizations of one facility over another, for example, comparing one chemical plant to another. At the jurisdictional level, the method can be useful to support comparing the risks to one asset or another alongside of the Threat and Hazard Identification and Risk Assessment (THIRA) process. Given the challenges involved in quantitative comparative analyses of homeland security risk (Committee to Review the DHS's Approach to Risk Analysis, 2010), the Deliberative Method for Ranking Risks may be a useful tool for structuring previously ad hoc qualitative risk assessments across the homeland security domain.

REFERENCES

Alliance Development Works. (2012). *World risk report*. Alliance Development Works, in cooperation with the United Nations University – Institute for Environment and Human Security and the Nature Conservancy.

Andrews, C. J. (1998). *Substance, process, and participation: Evaluating a decade of comparative risk*. Paper presented at the Annual Meeting of the American Collegiate Schools of Planning. Pasadena, CA.

Committee to Review the DHS's Approach to Risk Analysis. (2010). *Review of the Department of Homeland Security's approach to risk analysis*. National Research Council of the National Academies, Editor, Washington, DC: National Academies Press. 148.

DeKay, M. L., et al. (2001). The use of public risk ranking in regulatory development. In P. Fischbeck & R. S. Farrow (Eds.), *Improving regulation: Cases in environment, health, and safety* (pp. 208–230). Washington, DC: Resources for the Future.

Department of the Taoiseach. (2014). *Draft national risk Assessment*. R.o.I. Department of the Taoiseach, Editor. Dublin, Ireland.

Fischhoff, B., et al. (1978). How safe is safe enough? A psychometric study of attitudes towards technological risks and benefits. *Policy Sciences*, *9*(2), 127–152.

Florig, H. K., et al. (2001). A deliberative method for ranking risks (I): Overview and test bed development. *Risk Analysis*, *21*(5), 913–921.

Hagel, C. T. (2014). *Quadrennial defense review*. Washington, DC: Department of Defense.

Homeland Security Council (HSC/DHS). (2005). *National planning scenarios* (2005). Homeland Security Council in partnership with the U.S. Department of Homeland Security, Editor. Washington, DC.

Jenni, K. E. (1997). *Attributes for risk evaluation*. Pittsburgh, PA: Carnegie Mellon University.

Keeney, R. L., & von Winterfeldt, D. (2011). A value model for evaluating homeland security decisions. *Risk Analysis* *31*(9), 1470–1487.

Kemshall, H. (2003). *Understanding risk in criminal justice*. McGraw-Hill Education (UK).

Lazo, J. K., Kinnell, J. C., & Fisher, A. (2000). Expert and layperson perceptions of ecosystem risk. *Risk Analysis*, *20*(2), 179–194.

Lindell, M. K., & Prater, C. S. (2003). Assessing community impacts of natural disasters. *Natural Hazards Review*, *4*, 176.

Lundberg, R. (2013). *Comparing homeland security risks using a deliberative risk ranking methodology*. Santa Monica, CA: Pardee RAND Graduate School. 408.

Lundberg, R., & Willis, H. H. (n.d.) Perceptions of homeland security risk: Comparing responses of a deliberative risk ranking to a nationally representative survey. Working paper.

McDaniels, T., Axelrod, L. J., & Slovic, P. (1995). Characterizing perception of ecological risk. *Risk Analysis*, *15*(5), 575–588.

Meinhart, R. (2006). *Strategic planning by the Chairmen, Joint Chiefs of Staff, 1990 to 2005*. DTIC Document.

Mileti, D. S. (1999). *Disasters by design: A reassessment of natural hazards in the United States*. Washington, DC: National Academies Press.

Morgan, K. M. (1999). *Development and evaluation of a method for risk ranking*. Pittsburgh, PA: Carnegie Mellon University.

Morgan, K. M., et al. (2001). A deliberative method for ranking risks (II): Evaluation of validity and agreement among risk managers. *Risk Analysis*, *21*(5), 923–937.

Morgan, M. G., et al. (1996). *A Proposal for Ranking Risk within Federal Agencies. Comparing Environmental Risks: Tools for Setting Government Priorities*, 111–147.

Morgan, M. G., et al. (2000). Categorizing risks for risk ranking. *Risk Analysis*, *20*(1), 49–58.

Organization for Economic Co-operation and Development (OECD) (2009). *Innovation in country risk management*. Paris, France: Author.

Patton, M. Q. (1990). *Qualitative evaluation and research methods*. SAGE Publications.

Rademaker, M. (2008). National security strategy of the Netherlands: An innovative approach. *Information & Security. An International Journal*, *23*(1).

Slovic, P. (1987). Perception of risk. *Science*, *236*(4799), 280–285.

Slovic, P. (1992). Perceptions of risk: Reflections on the psychometric paradigm. In S. Krimsky & D. Golding (Eds.), *Social theories of risk* (pp. 117–152). New York: Praeger.

Slovic, P., et al. (2004). Risk as analysis and risk as feelings: Some thoughts about affect, reason, risk, and rationality. *Risk Analysis*, *24*(2), 311–322.

Starr, C. (1969). Social benefit versus technological risk. *Science*, *165*, 1232–1238.

U.S. Department of Homeland Security (DHS). (2009). *National infrastructure protection plan*. Department of Homeland Security, Editor. Washington, DC.

U.S. Department of Homeland Security (DHS). (2010). *Quadrennial homeland security review report: A strategic framework for a secure homeland*. Department of Homeland Security, Editor. Washington, DC.

U.S. Department of Homeland Security (DHS). (2013). *Comprehensive preparedness guide 201: Threat and hazard identification and risk assessment guide*. Department of Homeland Security, Editor. Washington, DC.

U.S. Environmental Protection Agency (EPA). (1987). *Unfinished business: A comparative assessment of environmental problems*, U.S. Environmental Protection Agency (EPA), Editor. 1987, National Technical Information Service Report: Alexandria, VA.

Vlek, C. (2013). How solid is the Dutch (and the British) national risk assessment? Overview and decision-theoretic evaluation. *Risk Analysis*, *33*(6), 948–971.

White House. (2003). *Homeland Security Presidential Directive (HSPD-7): Critical infrastructure identification, prioritization, and protection*. Washington, DC.

Wildavsky, A., & Dake, K. (1990). *Theories of risk perception: Who fears what and why? Daedalus*, *119*(4), 41–60.

Willis, H. H. (2002). *Ecological risk perception and ranking: Towards a method for improving the quality of public participation in environmental policy*. August. Pittsburgh, PA: Carnegie Mellon University, Carnegie Institute of Technology.

Willis, H. H., et al. (2004). Ecological risk ranking: Development and evaluation of a method for improving public participation in environmental decision making. *Risk Analysis*, *24*(2), 363–378.

Willis, H. H., et al. (2005). Aggregate, disaggregate, and hybrid analyses of ecological risk perceptions. *Risk Analysis*, *25*(2), 405–428.

Willis, H. H., et al. (2010). Prioritizing environmental health risks in the UAE. *Risk Analysis*, *30*(12), 1842–1856.

Willis, H. H., et al. (2012). *The validity of the preference profiles used for evaluating impacts in the Dutch National Risk Assessment.* Santa Monica, CA: RAND Corporation.

Willis, H. H., & LaTourrette, T. (2008). Using probabilistic terrorism risk modeling for regulatory benefit-cost analysis: Application to the Western Hemisphere Travel Initiative in the Land Environment. *Risk Analysis*, *28*(2): 325–339.

World Economic Forum (EFF). (2015). *Global risks report: 2015*. in *World Economic Forum*.

World Health Organization (WHO). (2009). *Global health risks*. Author.

Xu, J., Florig, H. K., & DeKay, M. L. (2011). Evaluating an analytic–deliberative risk-ranking process in a Chinese context. *Journal of Risk Research*, *14*(7), 899–918.

4

Validating Terrorism Risk Assessment Models – Lessons Learned from 11 Models

John Lathrop and Barry Ezell

4.1 Introduction

A defining risk of our time is the possibly growing capability of terrorist groups to fabricate and deliver weapons of mass destruction (WMDs). That risk is characterized by extreme possible consequences, including tens of thousands of fatalities and initiation of global conflict. Yet by some definitions of WMDs, we have not, as of this writing, observed even a single full-scenario event. There are three other interrelated aspects of that risk: (1) the essential terrorist-defender game aspect of the risk, where the terrorist may be intelligent and adaptive to defensive actions, and may make decisions based on poorly understood processes of radicalization and poorly understood foreign and domestic subcultures; (2) terrorist incentives to develop and launch WMD attacks may be changing due to "The Great Unraveling" of international processes (Cohen, 2014; Haas, 2014); (3) terrorist capabilities can include step function increases due to Internet information, random meetings of individuals, and random opportunities.

These considerations combine to create an almost overwhelming risk management challenge and an almost overwhelming risk assessment challenge for risk analysts. We pose that latter challenge as: How, in this context, do analysts apply all available data and analysis tools to generate the most effective risk management advice? That challenge has many facets.

In this chapter, we address one particular facet: How should we validate WMD terrorism risk assessment models (here abbreviated as TRAMs, always as applied to WMD terrorism risk) in the absence of observed events? We define WMD as not including IEDs and IEDs with effects multiplied through, e.g., attacks on infrastructure. We began to address the challenge in "Validation in the Absence of Observed Events," recently published in *Risk Analysis* (Lathrop & Ezell, 2015). There we create a special definition of validation of models concerning not-yet-happened events. The basis of that definition rests on considering why we validate models. Typical validation involves some form of testing how well the model results correlate with observed events. Those tests are not based on a desire simply to correlate with observed events, but more fundamentally are based on a desire to test how well the model can advise decision makers in their decisions.

We find support for that reasoning in three sources: (1) ISO Standard 15288, on "Systems and software engineering – System life cycle processes," which defines validation as "confirmation, through ... objective evidence, that the requirements for a specified intended use ... have been fulfilled" (ISO, 2015); (2) MIL-STD-3022 on Verification, Validation, and Accreditation (VV&A) for Models and Simulations, which states that "Validation [is the] process of determining the degree to which a model ... and [its] associated data are accurate representations of the real world from the perspective of the intended use(s)" (Department of Defense, 2008). Then finally, we cite Sargent, who lists four key entities in his paradigm for model verification and validation (Sargent, 2013). Two entities apply to model validation. (1) "Conceptual model validation is defined as determining that the theories and assumptions underlying the conceptual model are correct and that the model representation of the problem entity is 'reasonable' for the intended purpose of the model"; and (2) "Operational validation is defined as determining that the model's output behaviour has a satisfactory range of accuracy for the model's intended purpose over the domain of the model's intended applicability."

Note that all three definitions are based on intended use(s) and purpose. What is the intended use and purpose of TRAMs? To improve WMD terrorism risk management. Applying that reasoning to the challenge we present here leads to a one particular definition of validation we will focus on in this chapter: *Validation in the absence of observed events should test how well the model, including the process by which the model is built and operated and its results presented to decision makers, can improve risk management decisions.*

Underlying that broader definition is a broader conceptualization of what a risk assessment is. In this context, a risk assessment is typically not like a voltmeter – it is not a "riskmeter." In many cases we have studied, the results format issues involved are ones of representation of a complex, uncertain situation – that is, analytically reducing a large amount of information into summary renditions targeted to improving risk management decisions. That perspective frames the reasoning of this chapter.

We apply these points to specify four concepts of validation:

First: Validation should test how well the model performs the modeling, model quality, and engagement functions called for by the use and purpose of TRAMs to improve risk management decisions. In Sections 4.3 through 4.6, we present 13 tests of how well the model fulfills these functions.

Second: Validation should include tests of full disclosure. The 13 tests we specify are quite challenging. In fact, none of the 11 TRAMs in our experience would perform well on many of the tests. Again invoking our central principle of support of decision makers, validation should include an evaluation of the degree to which a TRAM achieves full disclosure, defining that in the following terms: full disclosure of *how* and *how much* the TRAM falls short on the tests we specify here, and the *risk management implications* of each of those shortfalls.

Third: Validation is not a matter of valid/not valid. It is not a matter of being above some threshold of minimum validity. Petty, in his work on verification, validation, and accreditation, stipulates that it is largely incorrect to refer to complex models as "valid" or "validated." More appropriately, validation is a qualified description of validity that is interpreted by its users as accurate enough to be useful within the parameters and context (bounds of validity) set forth in its design (Petty, 2010).

Fourth: Validation is not a matter of generating a "validity score." It is a matter of testing the model to determine how well it performs the modeling, model quality, and engagement functions called for by the use and purpose of TRAMs to improve risk management decisions, then full disclosure as defined earlier. Others have suggested that we develop constructed scales for each of our 13 tests, to measure the degree of performance of a model on each of the 13 tests, as if they are attributes in a multiattribute utility score. We respectfully disagree. Constructed scales are quite useful in many contexts, but in this context would involve the collapse of several different considerations onto a single N-level scale within each of the consideration sets represented by each test. That would represent an unnecessary and unhelpful loss of information, unnecessary since no decision maker is typically in the position of evaluating alternative TRAMs based on a validation "score." Rather, a decision maker is in the position of wanting all applicable risk management advice available from a TRAM project. Again citing our central principle of support for the decision maker, that decision maker would not be well served by a score of performance on each of the 13 tests. He or she would be well served by a report on how well the model performs on each test, its shortfalls with respect to each test, and the risk management implications of those shortfalls.

That last point bears restating, in terms summarizing the insights of the previous paragraphs: Once we decide that validation in this context should be based on how well the model can improve risk management decisions, that leads to defining validation on two levels:

(1) Performance: How well the model performs the modeling, model quality, and engagement functions called for by the use and purpose of TRAMs to improve risk management decisions.
(2) Disclosure: How well any shortfalls in that performance, and the risk management implications of those shortfalls, are communicated to risk management decision makers.

As mentioned, in this chapter, we present 13 validation tests. The first four are from the *Risk Analysis* article (Lathrop & Ezell, 2015). Those tests are based on insights we have gained over 10 years of work with the 11 TRAMs we have developed, augmented, analyzed, compared, or reviewed, and on a systematic examination of the strong and weak points of each TRAM, how those strong and weak points matter for the usefulness of the decision advice available from each TRAM, and differences among the TRAMs. Table 4.1 lists those 11 TRAMs. It also lists a twelfth analysis, a survey, which we do not count as a model. All of the concepts presented in this chapter are based on a combined consideration of those 11 models. That combined consideration concept is central to this chapter, in that it allows us to make statements based on the strong and weak points of the models that could not be made specifically about any one of the models, due to confidentiality and classification issues.

This chapter is written to be useful to four groups of people: those who may be commissioning TRAMs, who may be using TRAMs, who may be validating TRAMs, and who may be performing TRAMs. As should become clear in the course of this chapter, all four groups have a vested interest in maintaining a level of TRAM performance and disclosure specified by the concepts of validation presented here. This chapter ranges over a wide variety of validation issues, but those issues are not chosen arbitrarily. They are based on our systematic study of the 11 TRAMs listed in Table 4.1.

The model validation issues addressed here are, frankly, quite challenging. We address them by introducing a conceptual framework in Section 4.2, then applying each of four perspectives, presented in Sections 4.3 through 4.6, to develop 13 validation tests. Section 4.7 applies those tests to BTRA as it is reported in the 2008 NAS study. In Section 4.8, we present a combined look at the third and fourth perspectives, evaluating how well the model engages with the decision-making process and with the terrorist-defender game. Section 4.9 recaps the findings of this chapter, then Section 4.10 presents paths forward, a discussion of the real-world practical considerations of implementing the validation processes presented here.

Table 4.1. *The 11 terrorism risk assessment models in the authors' experience*

WMD Terrorism Risk Assessment Model (TRAM)	Customer	Description of Work
Bioterrorism Risk Assessment (BTRA) (Ezell & Collins, 2010; Ezell & von Winterfeldt, 2009; National Research Council, 2008; Pillai, 2011)	DHS Science and Technology	Third Party Review (both)[a] Project Manager: Lathrop
Radiological Nuclear Terrorism Risk Assessment (RNTRA) (Pillai, 2011)	DHS Domestic Nuclear Detection Office	Third Party Review[b] (both) Project Manager: Ezell
Risk Assessment Process for Informed Decision Making (RAPID) (Ezell et al., 2011)	DHS Strategy, Planning, Analysis and Risk	Reviewed, then recommended methodological enhancements. (both) Project Manager: Ezell
Global Biological Threat Priority Model (GBTP) (Caskey, Ezell, & Dillon-Merrill, 2013)	Dept. of State, through Sandia National Laboratories	Verified and validated this multiple objective decision model. (Ezell only)
Maritime Security Risk Analysis Model (MSRAM) (U.S. Coast Guard, Port Security Evaluation Division, n.d.)	U.S. Coast Guard	Third Party Review (Lathrop) Recommending enhancements (both)
National Laboratory Experience[c]		
Modeling the Adversary for Responsive Strategy (MARS): Evaluates defensive measures vs. WMDs and IEDs on a "Common Yardstick."	Department of Energy, through Lawrence Livermore National Laboratory (LLNL)	Developed model, including Adaptive Adversary function, Dueling Portfolio framework.
Make, Buy, Steal (MBS): Compares IND risks by types, sources, to U.S. targets. Compares defensive strategies.	" " "	Developed model.
Matthew Bunn: A mathematical model of the risk of nuclear terrorism (2006)	" " "	Compared four Nuclear TRAMS: two listed earlier, RNTRA and MBS, then Bunn, Mueller, and the Lugar Survey (Lugar, 2005)

Table 4.1 *(cont.)*

WMD Terrorism Risk Assessment Model (TRAM)	Customer	Description of Work
John Mueller: IND risk model presented in his book. *Atomic Obsession* (2010)	" " "	
Mosaic Effect: Evaluating IND disclosures	" " "	Developing model
Unnamed model	DOE, through three national laboratories	Supported ongoing development, including adversary modeling

[a] That is, both authors participated.
[b] This project began as an independent verification and validation test, but changed to a third party review.
[c] Lead author has performed or is performing this work as part of his support to national laboratories.

4.2 Introducing a TRAM Validation Conceptual Framework

This chapter takes an unconventional approach to validation. We maintain that the most fundamental aspect of the problem we are addressing here is: how to validate a model whose use is to improve the management of the risk of a set of high-consequence probabilistic events that are 1) in part the result of terrorist choices and actions; and 2) outside of our experience. Taleb, in his book *The Black Swan* (2010), touches on some of the issues involved. His basic thesis, as applied to this topic, is that people do not perform well at anticipating events from outside of their experience. In more typical modeling situations, a model organizes information within experience, calculating results helpful for decision advice. That experience includes data, past results, conventionally accepted assumptions and algorithms, and subject matter expert (SME) opinions. With the TRAMs considered here, in general and in all 11 cases in Table 4.1, important steps of the modeling rely on SME judgments, or assumptions that are effectively SME judgments. But Tetlock (2005) tells us that SME judgments are typically importantly and systematically flawed, biased in the direction of overconfidence and, in particular for the cases such as those considered here, biased by an inability to adequately consider events outside of the SMEs' experience. Crowther (2014) cites a case where in 100 calibrated judgments in a calibration course, 99% were overconfident in a first calibration session, 72% in a second session. Hubbard (2009) recommends four calibration sessions to improve overconfidence in experts. Consider Augustine's Law XLV (1986) where he writes that one should expect that the expected can be prevented, but the unexpected should have been expected. Crowther (2014) points out that whereas Black Swans are less than underestimated epistemic uncertainty – overconfidence in models, parameters, and decision relevance, Perfect Storms are less then underestimated aleatory uncertainty – underestimates occur in variability, dependencies, and combinatory impacts. Then add to that the particular problem in this context that we should expect that terrorists are deliberately planning attacks that are not on the defender's list of attacks planned for. We can think of these attacks as "Deliberate Black Swans." So we are confronting a critical risk management problem that our normal modeling processes seem intrinsically inadequate to address. How *does* a society manage such a risk?

This chapter presents ways to approach that problem. One key concept is Best Use of Available Data, where those "Available Data" include the Tetlock and related data on shortfalls in SME judgments. Combining that with our definition of validation from Section 4.1, we include in our validation tests the degree to which model results are presented with prominent caveats concerning how the model falls short of veridicality in the ways characterized here, and with advice as to how to take these shortfalls into account in risk management decisions. The validation tests listed later specifically test for these caveats and that advice. The word *veridicality* is carefully chosen. Typical definitions include the words *truthful* and *veracious* (Merriam-Webster, 2017). Those meanings correspond to words in MIL-STD-3022, cited earlier: "accurate representations of the real world" (Department of Defense, 2008).

To expand on the caveat side of that: a modeler, here specifically a TRAM modeler, has a responsibility broader than building the best feasible model. He has the responsibility to advise the decision maker in risk management, including advice that specifically takes into account the known shortfalls of his TRAM, such as those we list in this chapter. As part of that, the modeler has an obligation to make best efforts to avoid having the decision maker interpret the model results as being more complete in describing the risk space than they actually are. The most direct way to address that obligation is to build the appropriate caveats directly into the model/results-results/delivery process.

The concepts presented here expand on the definition of validation presented in Section 4.1. As such, they represent an important shift from model validation as concerning only the model itself, to testing how well the model advises the decision maker in risk management. The rest of this chapter implements this line of thinking in terms of 13 tests, based on our experience with the 11 TRAMs listed in Table 4.1. Each test is worded as a question to be answered, that question designed to test the model, and its use, as to how well the concerns listed here are addressed. Each question is to be answered in text form, answering at the different levels discussed earlier: how well the model performs on that test, then how well that performance, shortfalls with respect to that test, and the risk management implications of those shortfalls are disclosed.

4.3 Perspective I: Capturing the Risk-Generating Process

The first perspective asks the question: Does the TRAM capture the essential elements of the risk-generating process? By that term we mean the process that generates the WMD terrorism risk, involving terrorist decisions and actions. The elements of that process will become clear in the course of the four tests we specify in this section. This first perspective was covered in the *Risk Analysis* article, and is presented here in summary form as four tests to be posed and answered as validation tests (Lathrop & Ezell, 2015):

Test 1: Does the TRAM capture the initiation process, if that is in scope?

By initiation process we mean the process by which a terrorist or terrorist group moves from inaction to action and launches the sequence of decisions and actions intended to ultimately result in a WMD terrorist attack. To clarify: in the rest of this chapter, when we refer to "terrorist," we mean either an individual terrorist or a terrorist group. The question posed as Test 1 seems a simple question, but based on our experience, it is not at all simple. It contains within it three issues that must be explicitly addressed in determining the validity of a TRAM: What external events and uncertainties affect initiation, and how well are they accounted for? Are all applicable information bases and SME elicitations brought to bear, including bases and elicitations available from previous studies? Where SME elicitations are employed, are the best available SMEs elicited, are they elicited with elicitation best practices, and what was done to explicitly address the problem of SME overconfidence? At this point, a range of different TRAMs have been performed. Many of them involve modeling the same or at least similar processes and datasets. Given that, it is a clear shortfall in model quality if any one TRAM does not at least seek to cross-reference with other applicable data sets associated with other TRAMS.

Test 2: Does the TRAM capture the scenario-shaping process?

By scenario shaping process we mean the sequence of probabilistic events that describe the step-by-step mechanics of the scenario as it unfolds after initiation. Those steps include terrorist decisions and actions, where the decisions are probabilistic (to the defender and in general) and the actions involve a probability distribution over outcomes. Typically those steps, in various orders, include deciding among WMDs; deciding how and where to acquire materials or devices; acquiring materials, devices, and expertise; fabrication; deciding among route-target pairs; transportation to the target; and detonation or release. Based on our experience, this Test 2 contains within it the same three issues listed in Test 1 (with the first issue slightly reworded), and again, those issues must be explicitly addressed in determining the validity of a TRAM.

Test 3: Does the TRAM consider, capture, and account for unanticipated scenarios?

In particular, does the TRAM include in its risk management advice, advice specifically to address the risk management implications of unanticipated scenarios? Is that advice developed through every possible means, including, where appropriate, Red Team SME panels? This test addresses the Achilles heel of probabilistic risk assessment: no matter how well structured, data-populated, elicited, and calculated, a probabilistic risk assessment can only be as good as its completeness, i.e., the degree to which its scenarios cover the scenario space of the risk-generating process. While it would seem obvious to have as a validation test some test of completeness, in fact, the issue of unanticipated scenarios cannot be well addressed by simply trying harder to anticipate all scenarios. When challenged along the lines of Test 3, the modelers can simply claim that they systematically examined all possible ways that the risk can be generated, and concluded that any unanticipated scenarios or alternative probabilistic causal chains will not contribute significantly to the assessed risk. But of course, that simply begs the question of unknown unknowns. So it follows that unanticipated scenarios will always be a concern, no matter how competently built and operated the TRAM. It follows that this validation issue is best addressed by asking how well a TRAM accounts for the unanticipated scenario issue in its risk management advice.

Some risk analysts may object to Test 3, on the grounds that it pushes the concept of risk assessment outside of its typical boundaries. But recall the broader concept of validation presented in Section 4.1, based on quality of risk management advice. Working from that concept, we have no choice but to push validation tests in the direction of how well the TRAM process supports the risk management process, even if that pushes the concept of risk assessment outside its typical boundaries.

Test 4: Does the TRAM consider alternative probabilistic causal chains?

This test is a different form of Test 3, bringing a more fundamental perspective to bear on the concept of unanticipated scenarios. All the comments raised concerning Test 3 apply to Test 4 as well. To repeat/modify those: Does the TRAM include in its risk management advice, advice specifically to address the risk management implications of alternative probabilistic causal chains? Is that advice developed through every possible means, including, where appropriate, Red Team SME panels? As with Test 3, the same concerns arise about pushing the concept of risk assessment outside its typical boundaries. And again, all the concepts raised in response to those concerns with Test 3 apply to Test 4 also.

The four tests are broadly structured. That foreshadows the tone of the rest of this chapter. WMD terrorism must be a process based on investigation and disclosure. It cannot be accomplished well by performing correlation tests between model results and observed events since full-scenario events have not yet been observed and intelligence on component events is generally not adequate for correlation tests. That investigation and disclosure should be guided by a set of questions to be investigated, that set designed to test the model on the degree to which it performs the modeling, model quality, and engagement functions called for by the use and purpose of TRAMs to improve risk management decision making. This chapter develops one such set of questions.

4.4 Perspective II: Model Quality

The last paragraph leads to the second perspective, model quality. Rather than list general tests of general model quality, we here focus on those aspects of model quality that we have found to be critical in particular for TRAMs. We compiled the following list of questions based on the strong and weak points of the 11 TRAMs we have studied, and differences among those TRAMs. As mentioned earlier, we could provide detailed examples from particular models of our set of 11 TRAMs, but to do so would raise problems of confidentiality and classification.

Test 5: Does the TRAM perform all of its steps from data and assumptions to results in ways that can be effectively checked by a third party?

A central part of that: Does the model have a mathematical formulation that can be peer reviewed? Some TRAMs are so complex, it may seem like the modelers feel that the internal workings of the model are simply too complex to be communicated effectively to a third party. But the net effect of that is for the modelers to take a "trust us" position, which is simply unacceptable.

Test 6: Are the assumptions implicit in the setting of scope boundaries explicitly questioned and analyzed for their implications for risk management?

We found one case in particular where two TRAMs, both assessing effectively the same WMD risk, made different assumptions in the setting of scope boundaries that quite substantially affected the results. The TRAM with broader scope found scenarios with significant risk in the scope region between its scope and the narrower scope of the other TRAM, which missed those scenarios.

We have found that scope boundaries may be set for either of two reasons. In the first case, a judgment is made that scenarios outside of the scope boundaries do not contribute significantly to the risk. In the second case, institutional scope is reflected in TRAM scope. For example, if the commissioning agency is charged with detection of WMD of a particular type, it is reasonable and appropriate for a TRAM commissioned by that agency to be scoped to reflect that agency's mission. We maintain, however, that in those cases or any others, the risk management implications of those scope boundaries should be analyzed and communicated to the decision makers. For example, if a TRAM scope is limited to evaluating alternative detection systems, those implications include the limitations of feasibility of completeness in covering all routes to the specified list of targets and all transportation detection defeat strategies, and the leverage implications of a defend-N-targets strategy.

Test 7: Are all results framed and presented within the context of the key assumptions made?

"Framed within the context" includes disclosure of how veridicality may be impacted by those assumptions. That framing should be immediately adjacent to the results, to counter the natural tendency for users to interpret a graphic or table of numbers without the context of the assumptions and their implications.

Perspectives I and II concern capturing the risk-generating process and model quality. In our investigation of the 11 TRAMs, we have found two other perspectives are quite important to validation that extend significantly beyond the concepts of perspectives I and II.

4.5 Perspective III: Engaging With the Risk Management Decision-Making Process

Test 7 relates to how well the TRAM communicates into the risk management decision-making process it is to serve. That leads to a third perspective on validation. Does the TRAM do the best feasible job of supporting the terrorism risk management decision-making process (DMP), considering that TRAM and that DMP working together as a system? This third perspective represents a quite significant step. It moves validation beyond the model itself, to how well the model engages with the DMP it is to serve. That is, the modeler should not view himself as working in isolation to get the analytics of the model right. Likewise, the model validator should not view himself as validating the model in isolation from the DMP the model is to serve.

That perspective gives us a broader-scope basis for a minimal standard for validation: Validation should ask, viewing the TRAM-DMP system, does the TRAM do more good than harm? This perspective can be supported from a quote from a book by Box and Draper that is in fact a better quote than the typical one attributed to Box: "all models are wrong, some are useful" (Box & Draper, 1987). The better quote, from the same book: "Remember that all models are wrong; the practical question is how wrong they have to be to not be useful." Such a minimal standard for validation may seem too minimal, but it applies quite directly regarding a 2008 National Academy of Sciences (NAS) review of one of the 11 TRAMs, the Bioterrorism Risk Assessment (BTRA) model (National Research Council, 2008).

The 2008 NAS review of BTRA found many flaws. The one flaw we will discuss here is that BTRA lacked an adaptive adversary module. That review recommended that BTRA not be used until it is extended to account for the adaptive adversary. Setting aside considerations of how the model has been used, from a model validation perspective, we need to ask: Would the WMD terrorism risk management process have been better served without the BTRA model, versus with that model lacking an adaptive adversary module? The point here is, recalling Box and Draper that every model has veridicality shortfalls, validation should consider whether a veridicality shortfall is troubling but also, looking at the TRAM-DMP system as a system, does the model, with its veridicality shortfall, do more harm than good? This extended concept of validation is challenging. It calls for validating the model in its DMP context, not as an analysis abstraction.

We use the BTRA 2008 NAS review example to illustrate one part of the TRAM-DMP system concept. But that concept should be applied much more broadly. The modeler and the validator must take into account how the model, in this case the TRAM, is *actually* to be used. In our experience, a TRAM is never used in isolation, with its results directly mapped into a decision. Rather, the TRAM is only one input into a much broader process typically, if not always, involving: (1) broad and non-analytically structured discussion among users of the model results; and (2) other decision factors. That reality has direct significance for the last test of Perspective II, Test 7, and provides a foundation for this Perspective III. We will explore this concept with two examples:

Example 1: Putting uncertainty into terms relevant to the risk management decision-making process (DMP)

When a TRAM explicitly calculates error bars on its results metrics, it typically displays those results with those error bars, with error bars of two or more alternatives often overlapping. Yet often what matters to the DMP cannot be inferred simply by observing overlapping error bars. A typical decision maker will combine the TRAM results with other decision factors. If alternative A is evaluated as more desirable than B, with overlapping error bars, he will want to know if he picks A, if he will be found to be wrong, i.e., that B will later be found to be more desirable. That rank flip could be due to another at least equally credible analysis showing B preferred to A, or it could be due to new data arriving that show B preferred to A. But the probability of rank flip cannot be inferred from overlapping error bars. If the errors are completely correlated, there will be no rank flip. If the errors are uncorrelated, the probability of rank flip cannot be calculated without knowing the probability distributions along the error bars. A TRAM based on an appropriately structured probabilistic risk assessment event tree could calculate probability of rank flip with feasible additional analysis. Yet we have yet to see any TRAM presented with probability of rank flip. We view that lack of results format as evidence of a lack of orientation toward the DMP by the risk analysts of those TRAMs.

Example 2: Considering how the DMP will address risk aspects left out of a TRAM

TRAMs sometimes evaluate the relative effectiveness of various suites of risk reduction measures. In some cases, those risk reduction measures involve technical devices, which we will call "Gadgets," that operate and are effective in ways largely independent of staffing, while other defensive measures fall under the category "Guns, Guards, and Gates," which we will abbreviate "Guards," that operate and are effective in ways largely independent of Gadgets. Those terms may seem glib, but they are in fact often useful summary descriptors. A TRAM may evaluate defensive suites and layers consisting of combinations of Gadgets and Guards. The defensive processes associated with Gadgets and Guards are qualitatively different. It may be challenging to calibrate assessments of the effectiveness of the two types of defensive measures. TRAM modelers may consider "decoupling" the two sets of evaluations, evaluating Gadget risk reduction measures separately from Guard risk reduction measures, and so deliver two evaluations into the DMP. Yet the DMP, somewhere, needs to consider Gadget and Guard defensive measures as parts of an overall defensive system. So if the TRAM modelers choose not to address the calibration of the two types of measures, the calibration problem does not go away, but rather is deferred up into the DMP, where it will be treated in a less analytic way.

So the question the TRAM modelers and model validators must ask is: Is that calibration best addressed by the modelers within the TRAM, or non-analytically within the DMP? Yes, that calibration can only be imperfectly performed within the TRAM, can be criticized by an NAS review, and may push the modelers out of their comfort zone. But the real question to be addressed is: How is the WMD terrorism risk management process better served? By Gadgets – Guards calibration performed within the TRAM, or by that calibration performed within the DMP, unaided by analysis-based calibration?

Looking carefully at the BTRA example and Example 2, they present two completely different TRAM–DMP interactions. In the NAS 2008 BTRA review, the question is: Is the risk management process better served by using BTRA lacking an adaptive adversary module, vs. by not using BTRA at all? In Example 2, the question is: Is the risk management process better served by Gadgets-Guards calibration within the TRAM, vs. within the non-analytic decision process? These two questions involve completely different considerations, but they represent the same concept: validation of a model, viewed in isolation, may miss the point of how the risk management process is better served.

The concept of this Perspective III, that validation should involve looking at the TRAM-DMP system viewed as a system, leads us to view the validation of a model as asking which way the risk management process is better served. Asking that question sometimes has to address a possible difference of interests between TRAM modelers and the risk management process. There may be a C. P. Snow "Two Cultures" problem here: the culture of TRAM modelers and NAS, arguing out qualities of TRAMs in isolation, vs. the culture of the risk management process, asking how it can be best served by TRAMs (Snow, 1959).

This discussion of Perspective III has covered a number of concepts and examples. Time now to pull together this section into three validation questions:

Test 8: Does the TRAM do the best feasible job of supporting the risk management decision-making process (DMP)?

Test 8 can be broken down into two questions: (1) Does the TRAM do the best feasible job supporting the terrorism DMP, considering the TRAM and that DMP as a system? (2) In particular, does the TRAM communicate results in formats that support risk management decisions as clearly as possible, making minimum demands on the part of the users to understand the significance of the results for the decisions they face? We have found cases where results formats seemed selected from the modeler perspective, not the user perspective. In one case, the risks of different WMD variants were simply ranked, with no clear further analysis supporting how that risk ranking might be used for actual risk management decisions. On the positive side, in another case currently under development, results formats are being developed in what we call a "high bandwidth" format, by which we mean risk results are presented in multiple forms as required to best support the risk management process: "Geography of Risk" (where risks are highest), locations in the scenario space and magnitudes of risk reductions due to new defensive measures, and which terrorist groups are precluded from successful attacks due to particular defensive measures. Repeating our earlier discussion: in many cases, the results format issues involved are ones of representation of a complex uncertain situation – that is, analytically reducing a large amount of information into summary renditions targeted to best support risk management decisions.

It may be surprising that this test does not explicitly and separately evaluate the quality of the insights provided. That is because the phrases "best feasible job" and "as clearly as possible" cover that part of the evaluation with due recognition that limitations in scope and available data may limit the quality of insights provided in ways beyond the control of the modelers.

An additional issue relates to Test 8. A key concern is how the risk management process integrates the TRAM results into its decisions. While that is a serious concern, we maintain that there is only so much a TRAM process can control, and that process cannot, in general, be expected to control the decision-making process side of the TRAM-DMP system, other than to make best efforts to integrate its results into that TRAM-DMP system and the corresponding risk management process. That is best tested by the two questions of Test 8.

Test 9: Given the central role of uncertainty in the problems addressed by TRAMs, does the TRAM do the best feasible job of capturing the uncertainties and communicating those into the DMP in formats most helpful for the effective incorporation of those uncertainties into risk management decisions?

Test 10: Does the TRAM explicitly address the question of its scope boundary with that DMP, in particular whether challenging issues are best addressed within the TRAM, vs. addressed in the DMP outside of the analytic structure of the TRAM?

And then a related issue: whether challenging issues are best addressed using the TRAM in spite of its veridicality flaws, vs. denying the DMP the use of the TRAM. That is, does the TRAM, with its veridicality shortfalls, do more good than harm?

4.6 Perspective IV: Engaging with the Terrorist-Defender Game

The tests of Perspectives I, II, and III cover many key aspects of TRAM validation, yet those perspectives miss what may be the single most driving challenge of TRAMs. Once it is understood that adversary choice is or could be a key element, then a key fact must be addressed. The defender is in a terrorist-defender game, that is, a game not against some well-behaved abstract process or agent, but against perhaps many highly motivated and creative human adversaries. As such, the game is not defined by the defender, but is defined by terrorist–defender interaction. From the defender's point of view, it is defined by terrorist action and reaction. The rest of this section presents six tests along those lines, again based on strengths and weaknesses we find among the 11 TRAMs in our experience, and a systematic examination of the differences between them.

The first test is based on the concept of a Tactical-Strategic Stack (TSS). The example of Table 4.2 illustrates the general concepts involved. For any given situation, there could be a different TSS. Note from Table 4.2 that a TSS need not have a fully ordinal progression from the lowest level to the highest level, but rather different concepts may come into play at different levels. A key point of a TSS and Table 4.2 is that a game may be played at any of several levels, and those different levels should be explicitly considered in designing the defender's strategy for how best to play that game.

Test 11: Does the TRAM play the terrorist-defender game at the most effective level of the Tactical-Strategic Stack (TSS)?

At what level on the TSS does the terrorist play the game, and then does the TRAM play that game at that level or higher? If the TRAM does not play the terrorist-defender game at the most effective level, are the implications of that fully reported along with all other TRAM results?

Here is an example of the defender playing at less than the most effective level on a TSS: suppose the defender is allocating detectors (radiation, biological, chemical) among N cities, allocating M detectors among N cities to minimize defender casualties, where the N cities are selected according to a combination of size and iconic significance. That is a defensive strategy of defending each of N cities. On Table 4.2, that would be playing on Level #2. That assumes the terrorist is (probabilistically to the defender) choosing among those N cities as a target, with the goal of maximizing defender casualties. But what if the terrorist chooses some city outside of that set of N cities? The terrorist may not know those N cities exactly, but he may have a good enough idea of that to select a city outside of that set. The terrorist may select that city for either or both of two reasons: (1) to avoid the detector network; (2) because the

Table 4.2. *Example Tactical-Strategic Stack (from previous work). ("Blue" is the defender, "Red" is a terrorist.)*

Level #	Level of Blue Decision	Adversary/Game Modeling Considerations
12	Overall terrorism risk management, "Meta Game"	When you find yourself playing an unfavorable game, change the game. Example: tactically defending one target at a time.
11	Overall terrorism risk management	What are high-level Blue strategies, policies that take a big-picture look at the game?
10	Overall game strategies that consider Red vs. Blue leverage	What Blue, Red strategies exploit their own leverage, counter adversary leverage?
9	Policies, other types of alternatives affecting motivation, deterrence	What makes Red go proactive? Or stand down? Can we model Red Line behavior helpfully?
8	Strategies re resilience, robustness	How model adv choice to challenge resilience, robustness, seek scenarios not on Blue list, "Sinister Black Swans"
7	Intelligence resource allocation	What are most-high-valued Red signatures? Value of information analyses? Network-level signatures? Network-disrupting-opportunity signatures?
6	Countermeasure, CM R&D Portfolio	Broad, Dynamic Dueling Portfolios: Blue (dynamic) vs. Red (dynamic) Portfolios
5	Countermeasure Portfolio	Narrower Dueling Portfolios: Blue (static) vs. Red (dynamic) Portfolios
4	Countermeasure R&D funding	Red future capabilities, weapon-target choice
3	Which CMs to deploy where? How many?	Red weapon-target choice
2	Allocation of CMs among targets	Red target choice
1	Allocation of detection, medical CMs among agents	Red bio agent choice

terrorist may be seeking to maximize some other goal than defender casualties, e.g., terror or embarrassment of the defender's government, and may feel that that goal is best attained by attacking a city outside of the N city set.

If the terrorist is playing any other game than choosing among those N cities to maximize defender casualties, and/or if the terrorist is playing the game at any level in a TSS effectively above Level #2 in Table 4.2, then the defender is playing a suboptimal game, even if the defender's defensive goal is to minimize defender casualties. The detector allocation problem is like any other probabilistic allocation problem. If the actual probability distribution of terrorist attacks among the N cities is importantly different from the probability distribution the defender assumes, in any way that the defender could realize by a more strategic consideration of a TSS, then the defender is playing a suboptimal game.

There are subtleties involved in this point. This is not a criticism of a defender strategy of allocating M detectors among N cities. This is a call to explicitly, with deliberate separate effort, challenge any such strategy scope of a TRAM by considering the different levels in a TSS that the terrorist could be playing the game, then to report any risk management implications of the findings of that consideration as an integral part of results reporting. An example of a result of such an investigation of a TRAM working at Level #2 of Table 4.2: it could find that somewhere, perhaps not within the scope of the TRAM being validated, the defender should consider the allocation of defensive resources between allocation of detectors among cities versus higher-TSS-level measures, e.g., resilience to unexpected attacks, fly-in response resources, disincentivization of the terrorist, etc. Then the risk management implications of those considerations should be reported along with other TRAM results. Validation should check to see that that has happened.

There are, of course, institutional reasons for a given defender government agency to focus on, e.g., allocating M detectors among N cities. This Test 11 is not a call for any agency to work outside of its scope. It is a call for the reporting of TRAM results to include any risk management implications of any terrorist working at any level of a TSS other than the level implicitly or explicitly assumed by the TRAM. That, in turn, raises a more general concept of this chapter: validation should not demand that any given TRAM analytically work outside of its desired scope, but validation should include testing to see if the TRAM considered the questions listed in this chapter, and to see that the risk management implications of those considerations are explicitly made clear in any reports of TRAM results. These considerations are part of the full disclosure considerations discussed earlier. This is related to Perspective III: Engaging With the Risk Management Decision Making-Process (DMP). The TRAM should be validated in terms of its role in the DMP, and that role includes reporting the risk management implications of its findings within its scope, and the risk management implications of any limitations of its scope relative to a complete look at the terrorist-defender game being played at all levels of a TSS.

This discussion can be broadened with an example of a very high strategic interaction: Can the terrorist play against defender TRAM practices? One way in which that could happen: the terrorist may be able to use his knowledge of the defender's TRAM practices to determine the attacks "on the defender's list," i.e., the list of attacks on which the defender bases his defense strategy. With that knowledge, the terrorist may be able to design "Black Swan attacks," attacks deliberately designed to not be on the defender's list, and so attacks that may be particularly effective due to their avoidance of defender defensive measures. We can refer to those attacks as "Deliberate Black Swans." We can't think of a reason that this high strategic interaction could *not* take place, therefore it should be taken into account in a TRAM's risk management advice

This aspect of validation could be seen as beyond the scope of TRAM validation, on the premise that the validation of a TRAM should be limited to validating how well that TRAM performs within its desired scope. But we argue otherwise. We are forced to argue otherwise by the central concept of validation put forth in this chapter. To repeat that central concept from Section 4.1: Validation in the absence of observed events should test how well the model, including the process by which the model is built and operated and its results presented to decision makers, advises those decision makers in their decisions. A TRAM cannot best advise its decision makers if it does not report the risk management implications of limitations of its scope. Just as any analysis report should begin by listing its assumptions, its scope, and the implications of those assumptions and scope, so too a TRAM report should make the reader aware of the implications of what it does not address in terms of any terrorist-defender game further up in a Tactical-Strategic Stack.

Test 12: Does the TRAM posit terrorist groups with a consistent set of capabilities?

A TRAM often involves the modeling of terrorist groups engaging in several actions including, e.g., material acquisition, fabrication of a WMD, then transporting that WMD while evading detection. Terrorist groups should be posited with a consistent set of intelligence and skills over those three different tasks. That is, if a terrorist group can acquire the material and fabricate the WMD, such a group is apt to be able to apply considerable intelligence and skill to transporting the WMD while evading detection. It is less plausible, and should be criticized in validation, if terrorist groups are posited with adequate skills in one or two of these three tasks, then less competent in any remaining tasks.

Test 13: Does the TRAM adequately treat the problem of defender uncertainty about terrorist knowledge and ability to use that knowledge?

This test is a key aspect of engaging the TRAM in the actual terrorist-defender game being played, in particular how that game playing creates this particular special and large uncertainty. Two examples, from our experience:

Example 1. If a TRAM depends on detectors covering anything less than all routes to targets, and/or depends on detectors with defeat strategies, its results crucially depend on assumptions of how much the terrorist may know about those detectors' locations and defeat strategies, and the terrorist's ability to use that information to evade or defeat those detectors. Clearly, that TRAM should include sensitivity analyses on terrorist knowledge and ability to use that knowledge.

Example 2. (This example presented in a deliberately abstract form.) If a TRAM includes a target that can be attacked N equivalent ways, and a new defender defensive measure suite is able to eliminate M of those ways, the risk from an uninformed or unintelligent terrorist can be considered reduced by M/N, while the risk from an informed and intelligent terrorist may be reduced by zero. Clearly again, that TRAM should include sensitivity analyses on terrorist knowledge and ability to use that knowledge.

Note how both examples involve interaction with Test 12. That test's "consistent set of capabilities" includes capabilities regarding knowledge of detector location and defeat strategies, and regarding which attacks are effectively defended.

An overview of the tests of Perspective IV: those three tests overlap, and deliberately so, to create a framework of validation that addresses the TRAM strengths and weaknesses we have found in our experience, and in particular to prompt insight-building validation investigations. Together, the three tests test to see how well the TRAM advises the defender to play the terrorist-defender game in the most effective way.

4.7 An Example Application of the 13 Tests

One way to develop an understanding of the usefulness of the 13 tests is to apply them to a TRAM. Unfortunately, of the 11 models listed in Table 4.1, only three are described well in publicly available references, and two of those, Bunn and Mueller, are quite simple mathematical models that do not exercise the 13 tests in an insight-building way. The remaining of the three models, BTRA, is only extensively publicly reviewed in the 2008 National Academy of Sciences review of that model, "Department of Homeland Security Bioterrorism Risk

Assessment: A Call for Change" (National Research Council, 2008). That review was limited by limited access to the model, but is the best publicly available example of a description of a TRAM. The next paragraphs consider each test in turn, covering the issues of performance on that test, disclosure of any shortfalls, and the risk management implications of any shortfalls. The 13 tests vary in nature, so what is presented here under "Disclosure" and "Implications" varies accordingly. In all of these considerations, the presentation here is limited to what can be determined from the NAS report. Past tense is used since later versions of BTRA involved modifications to the model reviewed in the NAS report.

Test 1: Does the TRAM capture the initiation process, if that is in scope?

Performance: BTRA reporting did not meet the three conditions presented with this test. But more than that, it effectively defined overall initiation as out of scope. While BTRA divided frequency of initiation among four terrorist group types, it normalized its results in such a way that the overall initiation rate was taken out of calculations, and so limited its results to assessing the relative risks among biological agents. The NAS criticized that normalization, arguing that non-normalized results were called for to provide risk management advice, and those results could not be calculated from BTRA results as reported. This is an example of a case where the TRAM did not perform well on one of the 13 tests, not due to some lack of thoroughness, but through a deliberate modeling decision, in this case to report its results only in relative terms.

Disclosure: While BTRA reporting presented the normalization, it did not present that as a shortfall.

Implications: Failed to inform decision makers of the magnitudes of the overall risks, and so to provide that data for them to use in risk management decisions of larger scope than allocations based on the relative risks among the biological agents.

Test 2: Does the TRAM capture the scenario-shaping process?

Performance: Yes. BTRA was built with an event tree structure that captures events and dependencies.

Disclosure: Does not apply.

Implications: Does not apply.

Test 3: Does the TRAM consider, capture, and account for unanticipated scenarios?

Performance: No. BTRA implicitly assumed its tree structure covered all scenarios. Modelers could argue that the tree structure was developed to parse the scenario space in such a way that completeness follows from the tree structure, but a critical review of that issue was not presented. NAS touched on this issue with a recommendation that BTRA should allow for changing situations and new information, and that DHS should develop a strategy for the rapid assessment of newly recognized and poorly characterized threats. (National Research Council, 2008)

Disclosure: None.

Implications: Failed to inform decision makers of the risks of unanticipated scenarios.

Test 4: Does the TRAM consider alternative probabilistic causal chains?

Performance: No. Same answer as for Test 3.

Disclosure: None.

Implications: Failed to inform decision makers of the risks of alternative causal chains.

Test 5: Does the TRAM perform all of its steps from data and assumptions to results in ways that can be effectively checked by a third party?

Performance: No. The NAS review found this to be a central concern.

Disclosure: None.

Implications: Significantly impacts decision maker confidence in the efficacy of the model and underlying results. Effectively prevents independent validation and impedes third party review.

Test 6: Are the assumptions implicit in the setting of scope boundaries explicitly questioned and analyzed for their implications for risk management?

Performance: No.

Disclosure: None.

Implications: Decision makers informed by this model may invest in defensive measures based on an overly narrow scope of the risks.

Test 7: Are all results framed and presented within the context of the key assumptions made?

Performance: Although assumptions are discussed in the BTRA report, the enabling presidential directive required only a summary of key findings (White House, 2004). But key assumptions were made that could seriously impact the relative ranking of the biological agents by risk, in such a way that decision makers could be unaware of that issue.

Disclosure: Key assumptions were disclosed, though typically not in ways such that the implications of those assumptions were made clear in close proximity to key results displays.

Implications: Decision makers may be unaware of the caveats that should be applied. This is a good example of the issue that the TRAM may perform in a way that could be argued to be in compliance with its enabling text, yet fall short in terms of the tests specified here.

Test 8: Does the TRAM do the best feasible job of supporting the risk management decision-making process (DMP)?

... considering the TRAM and the DMP as a system? ... communicating results in formats that support risk management decisions as clearly as possible, making minimal demands on the part of the users to understand the significance of the results for the decisions they face?

Performance: No. The NAS documented many errors in the conceptual and computational models that impacted the efficacy of the results. The NAS called out the need for an increased focus on risk management (as opposed to risk assessment) in a specific recommendation (National Research Council 2008). BTRA was not developed with interaction with the DMP in scope. Formats for results were not helpful for key risk management decisions concerning biological terrorism risk.

Disclosure: None. Not in the scope of BTRA.

Implications: The usefulness of BTRA could be markedly impacted by this shortfall. This is a good example of a key issue: The scope of BTRA did not include support of the risk management decision-making process, other than assessing the relative risks of a list of biological agents. As with Test 7, this is a case of the TRAM process working effectively within its specified scope, yet falling short in terms of the tests specified here.

Test 9: Given the central role of uncertainty in the problems addressed by TRAMs, does the TRAM do the best feasible job of capturing the uncertainties and communicating those into the DMP in formats most helpful for the effective incorporation of those uncertainties into risk management decisions?

Performance: No. Typically, biological agents were ranked by risk with overlapping error bars, yet no guidance to the decision makers was provided concerning the likelihood or significance of rank reversals, other than to recommend that agents be grouped into cohorts.

Disclosure: None, other than as suggested by recommendations that agents be grouped into cohorts.

Implications: Decision makers could make risk management decisions that could be found to be suboptimal upon more complete analysis of the implications of the uncertainties for those decisions.

Test 10: Does the TRAM explicitly address the question of its scope boundary with that DMP, in particular whether challenging issues are best addressed within the TRAM, vs. addressed in the DMP outside of the analytic structure of the TRAM?

And then a related issue: Whether challenging issues are best addressed using the TRAM in spite of its veridicality flaws, vs. denying the DMP the use of the TRAM. That is, does the TRAM, with its veridicality shortfalls, do more good than harm?

Performance: Not addressed. As with Tests 7 and 8, this is an example of a case where the TRAM could perform adequately within the bounds of its dictated scope, yet fall short with respect to one of the tests specified here.

Disclosure: None.

Implications: Decision makers could make risk management decisions that could be found to be suboptimal upon more complete analysis of the implications of the scope boundaries between BTRA and the DMP.

Test 11: Does the TRAM play the terrorist-defender game at the most effective level of the Tactical-Strategic Stack (TSS)?

Performance: Not in the model environment. BTRA was structured at an understood strategic level for choices each terrorist group would make at each terrorist decision node in the BTRA event tree. The NAS raised this issue in a specific discussion of the need to consider enhanced, engineered, emerging, or suspected biological agents (National Research Council, 2008).

Disclosure: None.

Implications: Decision makers could make risk management decisions that could be found to be suboptimal upon more complete analysis of the implications of the assumptions made concerning at what level on the Tactical-Strategic Stack the terrorist group(s) may be playing.

Test 12: Does the TRAM posit terrorist groups with a consistent set of capabilities?

Performance: Terrorist groups were grouped by resources, but BTRA does not make statements broader than the connection between resources, generally characterized as financial, and what is capable given those resources. One linkage that could be considered: a terrorist group that may generate an enhanced or engineered biological agent may also be more apt to choose advanced modes of agent production, processing, and/or dissemination.

Disclosure: None.

Implications: Limits the defender's ability to consider the threats of different terrorist groups considering a more detailed analysis of different sets of capabilities.

Test 13: Does the TRAM adequately treat the problem of defender uncertainty about terrorist knowledge and ability to use that knowledge?

Performance: Limits that consideration to very basic terrorist decisions at each of several nodes in the event tree. The NAS made one of its central criticisms of BTRA its lack of modeling an adaptive/intelligent adversary, i.e., modeling terrorist knowledge and ability to use that knowledge.

Disclosure: None.

Implications: Limits the defender's ability to take into account the risk management implications of his uncertainty of terrorist knowledge and ability to use that knowledge. For example, in this context, the terrorist may be able to exploit his knowledge of defender actions concerning which biological agents to develop detection and response systems for. The terrorist may also be able to exploit his estimation of defender assumptions regarding the relative risks of contagious vs. non-contagious agents, and agents differing in treatment requirements.

Reviewing these results, one overall finding stands out: several of the ways in which BTRA fell short in terms of the 13 tests were not due to a modeling shortfall within the modeling effort, but in fact resulted from how the model was scoped and the interaction between the TRAM and the DMP, where that scope and interaction were both outside of the control of the modeling effort. This issue was discussed in the sections on Tests 7, 8, and 10. More broadly, the issues of validation identified in this chapter involve more than the model itself. A model can fulfill all conditions of its specified scope and enabling text, and still fall short in terms of the 13 tests specified here. The concepts of validation presented in this chapter involve more than the model itself – they include aspects of the model scope and enabling text.

But those considerations only extend to performance. The BTRA modeling effort could still have done better in terms of disclosure of the shortfalls and disclosure of the risk management implications of those shortfalls. For example, while the presidential directive enabling BTRA required only a summary of key findings, that requirement did not prevent more complete disclosure of shortfalls and of the risk management implications of those shortfalls (White House, 2004).

The BTRA model, as reviewed by the NAS, did not perform well on the 13 tests. Without being specific, we make the general statement that none of the 11 models in our experience would perform well on the 13 tests. This makes a point that we deliberately did not make more prominently in this chapter: while this chapter specifies a definition of validation, that definition is one on which no current TRAM, in our experience, performs well. So a second message of this chapter would be that current TRAMs should be expanded beyond their current forms and practices. While that is an important conclusion, we only mention it in passing here, so as not to distract from the central point of this chapter, which is to define validation in meaningful terms directly targeted to WMD terrorism risk assessment models.

4.8 Combining Perspectives III and IV

Reviewing Sections 4.5 and 4.6, Perspectives III and IV, we see that both perspectives involve testing how well the TRAM engages with the overall

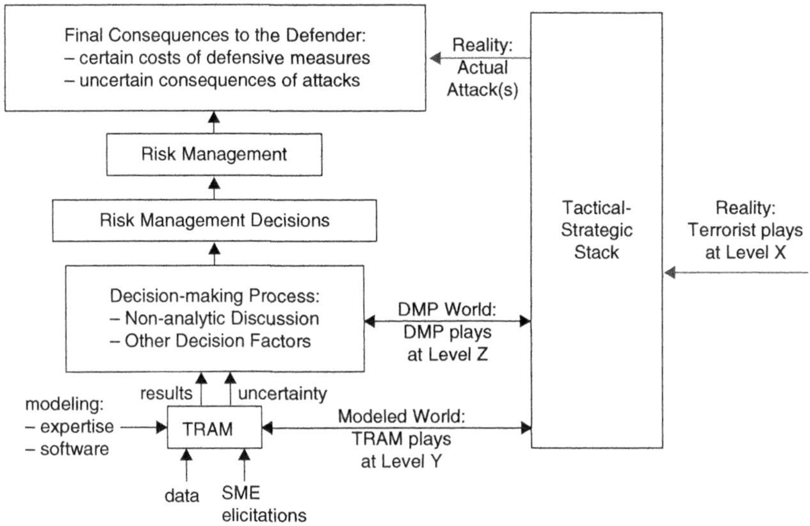

Figure 4.1. Essential elements between a TRAM and final consequences to the defender.

risk-generating process. Perspective III tests how well the TRAM engages with the risk management decision-making process. Perspective IV tests how well the TRAM engages with the Tactical-Strategic Stack of the terrorist-defender game being played. Here we take one more step, and relate those two validation concepts to a higher-order goal: we want the validation to test the TRAM to see how well it supports the utility maximization of the defender. While that concept may seem overly vague, we can use it to gain an overview that provides insight into TRAMs and their validation. That overview is presented graphically in the following two figures. Figure 4.1 presents that overview in terms of six entities. On the bottom is the TRAM, on the right side is the Tactical-Strategic Stack, and on the top left is the final consequences to the defender, the inputs to his utility maximization: the set of certain costs (i.e., of defensive measures) and uncertain consequences of terrorist WMD attacks. Up the left column is the sequence of defender processes between the TRAM and final consequences to the defender: the TRAM advises the decision-making process, which generates risk management decisions, which generates risk management, i.e., the set of defensive measures. Then that set of defensive measures counters uncertain terrorist WMD attacks at each of many stages. To present one example listing of those stages: disincentivization, deterrence, prevention, monitoring-intelligence, detection, interdiction, hardening,

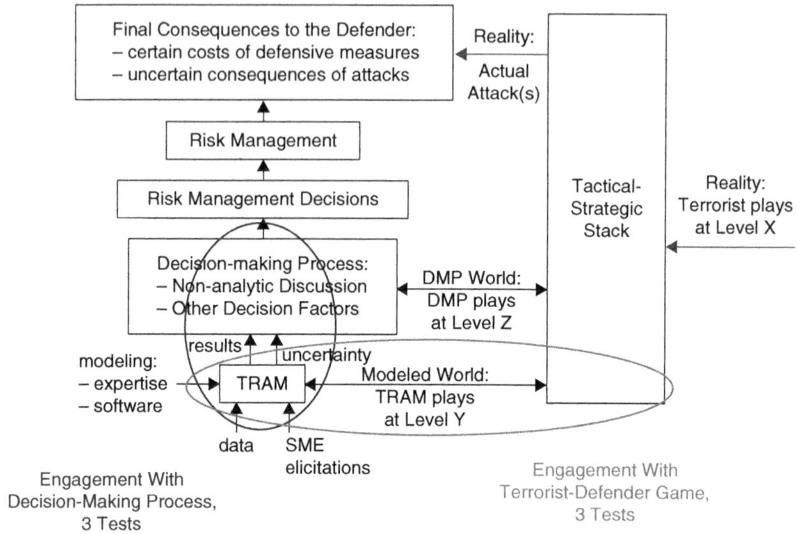

Figure 4.2. How a TRAM engages with the elements of Figure 4.1.

robustness, resilience, response, and recovery. The right side of Figure 4.1 summarizes the terrorist-defender game aspect of the risk management process, based on the Tactical-Strategic Stack, the terrorist playing at Level X in that stack, the TRAM modeling at Level Y in that stack, and the defender's decision making process playing at Level Z in that stack. The modeled world and the DMP world are effectively abstractions of the real world. "Reality" is marked in the figure by "Terrorist plays at Level X," and "Actual Attack(s)."

The bottom line of Figure 4.1 is that many processes and veils of uncertainty exist between a TRAM and final consequences to the defender. So then, if validation is to test how well a TRAM supports defender utility maximization, in the framework of Figure 4.1, what can validation test? Validation of a TRAM can only test that TRAM support in terms of the ways in which the TRAM interfaces with the overall process, as illustrated in Figure 4.2. Quite reasonably, the TRAM can only interface with the overall process via the two elements directly adjacent to it: (1) how well the TRAM interfaces with the defender decision-making process; and (2) how well the TRAM plays at what level in the Tactical-Strategic Stack, vs. the level at which the terrorist plays in that stack. Both of those aspects can be considered engaging: how well the TRAM engages with the decision-making process, Perspective III, and how well the TRAM engages with the terrorist-defender game, Perspective IV.

4.9 Summary of Main Points

In the absence of observed events, we redefine WMD terrorism risk assessment model (TRAM) validation as testing how well the model can improve risk management decision making. We characterize risk assessments in this context as typically not simply "riskmeters," but in fact representations of a complex, uncertain situation, those representations analytically reducing a large amount of information into summary renditions targeted to improving risk management decisions. Based on that reasoning, we then develop four concepts of validation:

(1) Validation should test how well the model performs the modeling, model quality, and engagement functions called for by the use and purpose of TRAMs to improve risk management decisions. We develop and present 13 validation tests of that model performance.
(2) Validation should include tests of full disclosure: TRAMs should include in their results *how* and *how much* they fall short on each of the validation tests we specify here, and the *risk management implications* of those shortfalls.
(3) Validation is not a matter of valid/not valid. Rather, it is a qualified description of validity.
(4) Validation is not a matter of a validity "score." It is a matter of investigation and disclosure.

The 13 validation tests of model performance we present are based on our examination of the strong and weak points of and the differences among the 11 TRAMs we have developed, augmented, analyzed, compared, or reviewed. These tests are each worded as a question to be posed and answered in a validation process. They arise from four perspectives on TRAM validation:

Perspective I, Section 4.3: Capturing the Risk-Generating Process: four tests.
Perspective II, Section 4.4: Model Quality: three tests.
Perspective III, Section 4.5: Engaging With the Risk Management Decision-Making Process: three tests.
Perspective IV, Section 4.6: Engaging With the Terrorist-Defender Game: three tests.

4.10 Paths Forward

This chapter specifies a validation of TRAMs in the absence of observed events, based on a validation process based on 13 performance tests, and tests of full disclosure. What is now called for is to specify a procedure for implementing such a validation process. Based on the experience we have gained in working with the 11 TRAMs listed in Table 4.1, we here specify such a procedure, in terms of four requirements. Sadly, we believe meeting these four requirements represents greater challenges than conducting the 13-test/disclosure validation process presented here. The four requirements:

Requirement 1: Establish a general acceptance, of the general TRAM using/commissioning community, that TRAMs must perform to a level of quality specified by the 13 performance tests and related disclosure tests listed here, or their equivalent. That is, a general acceptance of the principle that any TRAM that does not perform acceptably as measured by the validation process presented here, or its equivalent, shall not be used to advise decisions. One existing example of enforcing this type of principle is U.S. Coast Guard (USCG) Commandant Instruction 5200.40, which establishes a rigorous process to be followed such that modeling assumptions are accurate and well documented; results produced are stable, consistent, and repeatable; and the correlation between the model behavior and real-world behavior is clearly understood (U.S. Coast Guard, 2006). A second existing example is DoD Instruction 5000.61, which requires that models and associated data used to support processes, products, and decisions must undergo validation throughout their life-cycles (Department of Defense, 2009). In the course of implementing these two official instructions, the USCG and each military service have established regulations and policy directives that test that models correctly represent the real world; test to what degree the model results are comparable to the real world; and specify the range of inputs where the model is useful (Petty, 2010). More broadly, these two official instructions and their implementation demonstrate that the USCG and the DoD have each established a culture of enforced model quality. Other federal departments and agencies can do the same.

Requirement 2: Establish an institutional arrangement to conduct reasonably standardized TRAM validations, based on the 13 performance tests and related disclosure tests listed here, or their equivalent, then to report out the results of those tests, as best as can be reported consistent with sensitivity and classification issues. Those issues could be addressed by reporting limited to methodological considerations, though as mentioned in the discussion of high strategic interaction in Test 11, even reporting limited to methodological considerations would have to be carefully screened. That arrangement could be based on a pattern of third party reviews, such as the three cited in Table 4.1.

Requirement 3: Establish a willingness of TRAM commissioning agencies to fund those validations.

Requirement 4: Establish a willingness of those commissioning agencies to fund investments in TRAM model quality to bring TRAMs up to the performance levels specified by the 13 performance tests and related disclosure tests listed here, or their equivalent.

REFERENCES

Augustine, N. (1986, May 24). Augustine laws. United Press International.

Box, G., & Draper, N. (1987). *Empirical model-building and response surfaces.* New York: Wiley.

Bunn, M. (2006). A mathematical model of the risk of nuclear terrorism. *The Annals of the American Academy of Political and Social Science, 607.*

Caskey, S., Ezell, B., & Dillon-Merrill, R. (2013). Global, chemical, biological, and nuclear threat potential prioritization model (G-CBN-TP). *Bioterrorism & Biodefense, 4*(1).

Cohen, R. (2014, September 16). *The great unraveling. New York Times.*

Crowther, K. (2014). *Risk calibration.* Security Analysis Risk Management Association.

Department of Defense. (2008). Documentations of Verification, Validation, and Accreditation (VV&A) for Models and Simulations, MIL-STD-3022.

Department of Defense. (2009, December 9). Instruction 5000.61, DoD Modeling and Simulation (M&S) Verification, Validation, and Accreditation (VV&A).

Ezell, B., & Collins, A. (2010). Response to Parnell, Smith, and Moxley, Intelligent adversary risk analysis: A bioterrorism risk management model. *Risk Analysis, 30*(6).

Ezell, B., Dillon-Merrill, R., Parnell, G., Lathrop, J., Barrett, A., & Krempley, M. (2011). Methodology Enhancements to Risk Assessment Process for Informed Decision Making (RAPID), Innovative Decisions, Inc. Report, Contract # HSHQDC-10-P-00253.

Ezell, B., & von Winterfeldt, D. (2009). Probabilistic risk analysis and bioterrorism risk. *Biosecurity and Bioterrorism: Biodefense Strategy, Practice, and Science, 7*(1).

Haas, R. (2014). The unraveling. *Foreign Affairs, 93*(6).

Hubbard, D. (2009). *The failure of risk management: Why it's broken and how to fix it.* Hoboken, NJ: Wiley.

International Organization for Standardization (ISO). 2015. ISO/IEC/IEEE 15288. Systems and software engineering – System life cycle processes. www.iso.org/iso/home/store/catalogue_tc/catalogue_ detail.htm?csnumber=63711 (last accessed April 30, 2017).

Lathrop, J., & Ezell, B. (2015). Validation in the absence of observed events. *Risk Analysis.* 36(4). doi: 10.1111/risa.12442.

Merriam-Webster Dictionary (2017).

Mueller, J. (2010). *Atomic obsession.* New York: Oxford University Press.

National Research Council. (2008). *Department of Homeland Security bioterrorism risk assessment: A call for change.* Washington, DC: National Academies Press.

Petty, M. (2010). Verification, validation, and accreditation. In J. Sokolowski & C. Banks (Eds.), *Modeling and simulation fundamentals: Theoretical underpinnings and practical domains*. New York: Wiley.

Pillai, S. (2011). Taking Measure of Countermeasures (Part 1). Department of Homeland Security website, www.dhs.gov/news/2011/04/14/written-testimony-st-house-homeland-security-subcommittee-emergency-preparedness (last accessed April 30, 2017). Department of Homeland Security.

Sargent, R. (2013). Verification and validation of simulation models. *Journal of Simulation, 7*.

Snow, C. (1959). *The two cultures*. Cambridge: Canto Imprint, Cambridge University Press.

Taleb, N. (2010). *The black swan*. New York: Random House.

Tetlock, P. (2005). *Expert political judgment*. Princeton, NJ: Princeton University Press.

U.S. Coast Guard. (2006, December 22). Commandant Instruction 5200.40, Verification, Validation and Accreditation (VV&A) of Models and Simulations (M&S).

U.S. Coast Guard, Port Security Evaluation Division. (n.d.). Maritime security risk analysis model. http://aapa.files.cms-plus.com/PDFs/MSRAMBrochureTrifold.pdf (last accessed April 30, 2017).

White House. (2004). Biodefense for the 21st century. *Homeland Security Presidential Directive* 10.

5

Coping with Uncertainty in Adversarial Risk Analysis

Vicki Bier and Tony Cox

Morgan, Henrion, and Small (1990) present a useful taxonomy of types of uncertainty. They note that uncertainties can arise from: (1) statistical variations (e.g., "random error in direct measurements of a quantity"); (2) systematic errors (e.g., "biases in the measuring apparatus and experimental procedure"); (3) subjective judgments (e.g., for quantities where empirical data are largely unavailable); (4) linguistic imprecision (e.g., translation of verbal phrases into numerical probabilities); (5) variability (e.g., quantities that vary over time or space, or from one person to another); (6) inherent randomness or unpredictability (which cannot be reduced by further research); (7) disagreement (e.g., among multiple experts); (8) approximation (e.g., due to limits in the spatial resolution of a model); and (9) model uncertainty (i.e., uncertainty about the most appropriate model to represent some phenomenon).

In the context of terrorism risk, statistical variation might arise from the sparseness of data on actual terrorist attacks; systematic error could reflect the fact that many terrorist plots are likely to go unobserved; and uncertainty about the most appropriate model (e.g., game theory vs. decision theory vs. risk analysis) is likely to be widespread. Even within a single model, experts may disagree about the relative importance of different terrorist objectives (e.g., fatalities vs. economic damage). These importance weights might also vary over time, and from one terrorist group to another. Thus, good strategies for dealing with uncertainty – both statistical methods for drawing reliable conclusions from sparse data sets (e.g., about whether the frequency or severity of terrorist events is increasing over time; Mohtadi and Murshid 2009), and also methods for dealing with "deep uncertainty" (e.g., about whether terrorist groups are optimizing vs. satisficing; Shan and Zhuang 2013) – are crucially important.

Predicting where and when a terrorist or other adversary is likely to strike invites speculation about a host of unknowns. What do adversaries care about, and how do they assess their chances of success? What are their priorities? What deters them? Are attackers who perceive their cause as righteous, just, or holy also more likely to believe that their attacks will succeed? How can defenders best prepare to defend against adversarial attacks when the answers to some or all of these questions are unknown? (In terms of the taxonomy by Morgan and Henrion, such questions are concerned with disagreement and model uncertainty.)

This chapter discusses methods for making defensive decisions in the presence of realistic uncertainties about the answers to these and other questions. We consider methods ranging from preparing for the worst case (as assessed by the defender), to defending against a wide range of plausible attacks, to accepting the fact that we must learn and adapt as we go.

5.1 What Types of Uncertainties Can Be Important in Terrorism Risk Analysis?

Adversarial risk models face challenges involving numerous real-world uncertainties that arise in anticipating and attempting to reduce risks of terrorist attacks. For example, attacker goals and capabilities, as well as defense costs and cost-effectiveness, must usually be estimated subjectively; e.g., using a Bayesian Network (BN), influence diagram (ID), or nested attacker and defender decision trees.

Our ability to deal with and assess uncertainties – whether quantitatively (e.g., through formal expert-elicitation processes) or qualitatively, as is common in many parts of the intelligence community – is subject to numerous heuristics and biases, such as the planning fallacy, hindsight bias, and overconfidence (Hora, 2007; Montibeller & von Winterfeldt, 2015; Tversky & Kahneman, 1974; von Winterfeldt, 1999). There is no reason to believe that such biases will be less prevalent in terrorism risk analysis than in other domains, and, in fact, there is good reason to believe that they may be more prevalent. After all, empirical data on terrorism risk are likely to be sparse (Bier & Lin, 2008), and experts may be subject to extensive political pressures to provide estimates that support the preferred policies of key decision makers (see, e.g., *Morningside Post*, 2007; Tetlock, 2009). This makes the availability of good strategies for dealing with uncertainty (as discussed in Section 5.2) more important in terrorism than in other domains.

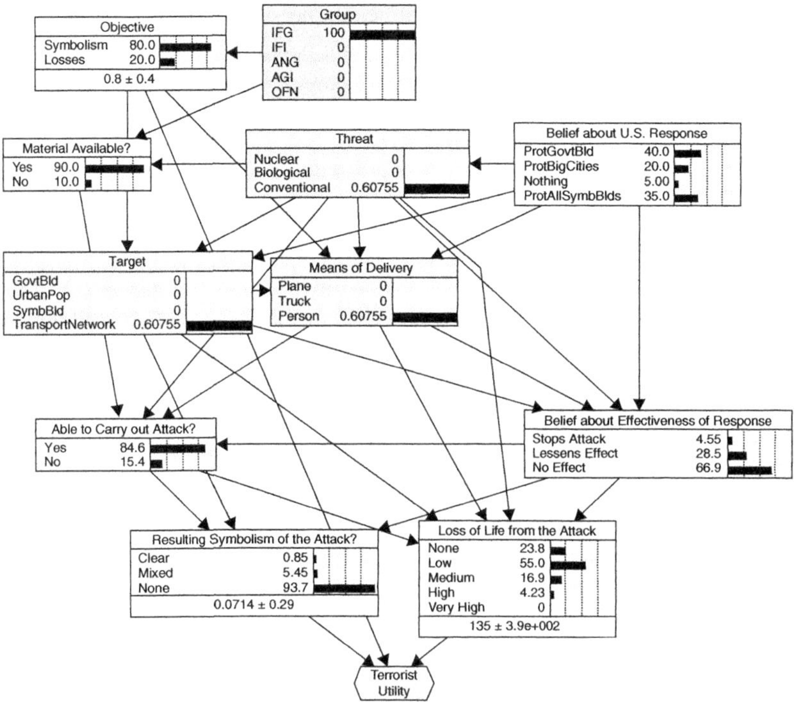

Figure 5.1. An ID model for terrorist decisions made by an Islamic fundamentalist group (IFG; Pate-Cornell & Guikema, 2002).

5.1.1 *Terrorists' Uncertainty about the Consequences of Various Possible Acts*

Clearly, terrorists will typically be uncertain about the consequences of their attack efforts, for numerous reasons. Figure 5.1 illustrates some of these, such as uncertainty about the symbolic value of an attack, as well as the number of casualties that may result; the notation and meaning of such influence diagram (ID) models are explained in Section 5.2. As shown in Figure 5.1, attacks are not guaranteed to succeed, and may, for example, have some probability of not resulting in casualties. Moreover, attackers generally will not have full knowledge of the defenses arrayed against them. Even in cases where the existence of a particular defense is known (e.g., air marshals on airplanes), attackers will generally not know which assets are likely to be defended at any given point of time (Dighe et al., 2009).

If a desired consequence of an attack is to damage assets that are highly valued by the defender, the attacker could also be uncertain about the defender's valuations of particular targets. Thus, for example, disclosure of defenses could have the undesired consequence of informing potential attackers about which targets are most valued by the defender (Zhuang & Bier, 2010).

5.1.2 *Defender Uncertainty about Terrorists*

Defenders will also generally be deeply uncertain about the possible terrorist threats they are facing. To begin with, defenders must simultaneously protect against multiple different terrorist groups (Hausken & Bier, 2011a), with different goals, beliefs, and capabilities (as shown, e.g., in the upper-right node of Figure 5.1). They will thus face uncertainty about which groups are most likely to launch an attack, and/or which group will attack first.

In addition, unlike national governments (which typically make many types of basic information public), terrorist organizations are generally small and highly secretive. Thus, defenders will typically be uncertain about current terrorist goals and objectives (Rosoff & John, 2009; Wang & Bier, 2011). Attacker groups will typically differ in their underlying values, and defender experts may have differing opinions about those values. For example, in a sample analysis in which "proxy experts" were asked to evaluate the attractiveness of different attack scenarios (Maya et al., 2012), the rankings given by one expert were consistent with a high weight on fatalities (reflecting a preference for highly lethal attacks), while other experts gave much lower rankings to scenarios with large numbers of fatalities. As a result of such differences in opinions, decision makers may be uncertain even about whether a particular group wishes to maximize or minimize fatalities (assuming all else equal – i.e., equal levels of success in achieving their other objectives).

Even if defenders had comprehensive knowledge of current adversary goals and objectives (e.g., due to massive intelligence gathering, or successful infiltration of one or more terrorist groups), they would still face uncertainty about how those goals and objectives may evolve in the future, due to factors such as changing political climate, leader propensity to violence, or leader decapitation (Miller, 2012; Price, 2012; Sticha et al., 2009). This is important, since many defensive strategies are capital-intensive, and are thus likely to be in effect for a period of years or even decades, during which attacker goals may change and adversary groups may decline in power or even cease to exist. ID models can in principle represent such effects, but the needed probabilities may be difficult or impossible to obtain, forcing the use of more general models that allow for such uncertainty.

Defenders also face uncertainty about terrorist resources and capabilities (which intelligence experts generally believe are at least as important as goals; Wang & Bier, 2015), about how those goals and capabilities may translate into specific intentions and plans (e.g., bombs in printer cartridges), and about the likely timing of possible attacks (since, e.g., large attacks may need to be timed far enough apart for security protocols to have relaxed after earlier attacks; Woo, 2002).

5.1.3 *Uncertainty about One's Own Values (Present and Future)*

Finally, both defenders and attackers may face considerable uncertainty about their own values. Thus, for example, terrorists may face uncertainty about which attacks will be impressive enough to lead to improved terrorist recruitment (Woo, 2003), and which will be so extreme as to inspire revulsion even among those who share the terrorists' goals (Bier et al., 2012). Likewise, democratic governments must satisfy multiple stakeholders with differing values, and may be unsure about the public's tolerance for particular interventions (e.g., violations of privacy), and/or the public's views about particular levels of consequences due to terrorism. For example, there is evidence that events due to terrorism may be perceived as more severe than other events with similar levels of consequences (Breckenridge & Zimbardo, 2006; Burns & Slovic, 2010; Covello et al., 2001), in part because of the outrage they provoke. Those values may also evolve over time (McArdle et al., 2012).

Most past work (e.g., Bier et al., 2007; Major, 2002) assumes that defenders know both the costs and the cost-effectiveness of the available defensive strategies. However, in reality, even the costs may be uncertain for defenses that have not yet been implemented on a broad scale. On the one hand, large projects are notorious for extensive cost/schedule overruns, suggesting that the costs of some defenses are likely to be underestimated. On the other hand, Ruth Ruttenberg and Associates (2004) have pointed out that widespread adoption of new requirements typically leads to dramatic cost decreases, since the requirements create an automatic incentive to develop new and less costly ways of coming into compliance or new technologies for achieving the same goals; see also Finkel (2013), Goodstein and Hodges (1997), Harrington and colleagues (2000), Hazilla and Kopp (1990), Office of Technology Assessment (1995), Pizer and Kopp (2003), and Putnam, Hayes, and Bartlett, Inc. (1980).

Even if the cost of a particular defensive strategy is well known, the defender may be deeply uncertain about the effectiveness of that strategy – even after it has been implemented. In the absence of frequent attacks, it will typically be unclear whether a particular defense would be effective at reducing the consequences and/or success probabilities of attacks that have not occurred (Jamshidi & Bier, 2009). For example, more than a decade after September 11, we may still be unclear about whether a particular investment has reduced risk by 20%, by 2%, or by 0.002%.

Finally, even if the costs and effectiveness of a particular defense in reducing the consequences or success probabilities of an attack are well understood, there will typically be uncertainty about whether a given reduction in risk will be sufficient to deter attacker(s) (Bier & Kosanoglu, 2015), or whether retaliation will inspire hatred (Hausken & Bier, 2011b).

5.2 Strategies for Dealing with Statistical Uncertainty and "Deep Uncertainty"

Key concepts and methods of statistical uncertainty quantification are well developed for the case where probabilistic risk models and plentiful data are available. The following sections first review these methods, and then present methods for making effective risk management decisions even in situations where not enough is known to predict attacker behaviors with useful accuracy and confidence.

5.2.1 *Are We Being Attacked Yet? Computational Methods for Detecting Changes and Quantifying Uncertainty*

Since 2001, when letters containing anthrax led to five deaths and 17 other infections from which the victims recovered, the U.S. Environmental Protection Agency, the Centers for Disease Control and Prevention, and the Department of Health and Human Services have invested more than a billion dollars to develop surveillance methods and prevention and preparedness measures to help reduce or mitigate the consequences of bioterrorism attacks should they occur again in the future (Grundmann, 2014). In practice, detecting a significant upsurge in hospital admissions with similar symptoms may often be the most practical and reliable way to identify that a bioterrorism attack is in progress (ibid.). The statistical challenge of detecting such changes against the background of normal variability in hospital admissions has motivated the development of computational intelligence methods that seek to reduce the time to detect attacks when they occur, while keeping the rates of false positives acceptably small (Cami et al., 2009; Shen & Cooper, 2010).

Such statistical data analysis and pattern detection – carried out in settings for which the patterns for which one is searching are well understood (e.g., a jump in hospitalization rates for patients with similar symptoms that could be caused by a biological agent), and where enough surveillance data are available to quantify background rates and to monitor changes over time – illustrate

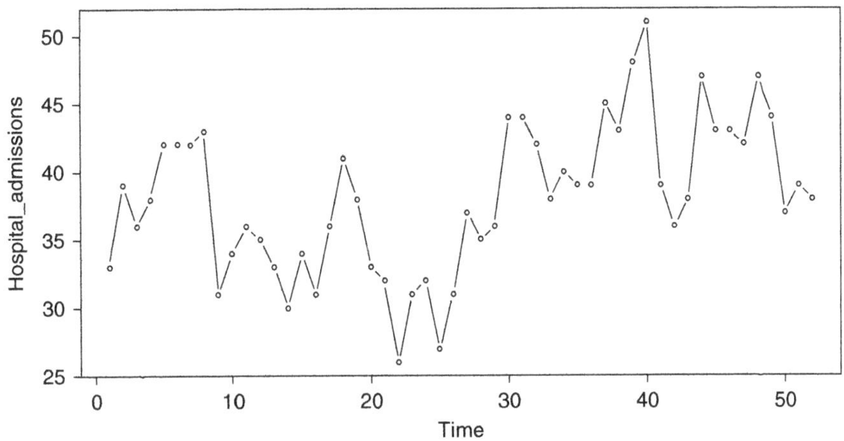

Figure 5.2. A hypothetical example of hospital admissions data over 52 weeks. Did an attack or other adverse event occur? If so, when, and how large was its effect?

the types of uncertainty for which sophisticated quantitative techniques are already available. We briefly recall some of the key concepts of uncertainty analysis for such applications. Figure 5.2 presents a hypothetical example showing weekly counts of hospital admissions with a specified symptomology for a city. Given such surveillance data, the inference task is to determine whether the hospitalization rate increased at some point in time (suggestive of an attack), and, if so, when and by how much. Intuitively, it appears that hospitalization counts are somewhat greater on the right side of Figure 5.2 than the left, but might this just be due to chance, or is it evidence of an underlying problem?

Figure 5.3 illustrates the types of results that can be provided by applying current statistical (or computational intelligence, computational Bayesian, machine learning, pattern recognition, data mining, etc.) technology to such questions, by combining statistical evidence with risk models to draw inferences about what is probably happening in the real world. The main idea is simple: the peaks in Figure 5.3 indicate the times that are estimated to be most likely for when an attack occurred, based on the data in Figure 5.2. The likelihood-based algorithm that produces this information, described later in this chapter, also estimates how admission rates changed from before to after the attack. Figure 5.3 shows that the algorithm is unable to make a strong guess about the attack time (if any) from the somewhat ambiguous data in Figure 5.2 – namely, between weeks 25 and 29.

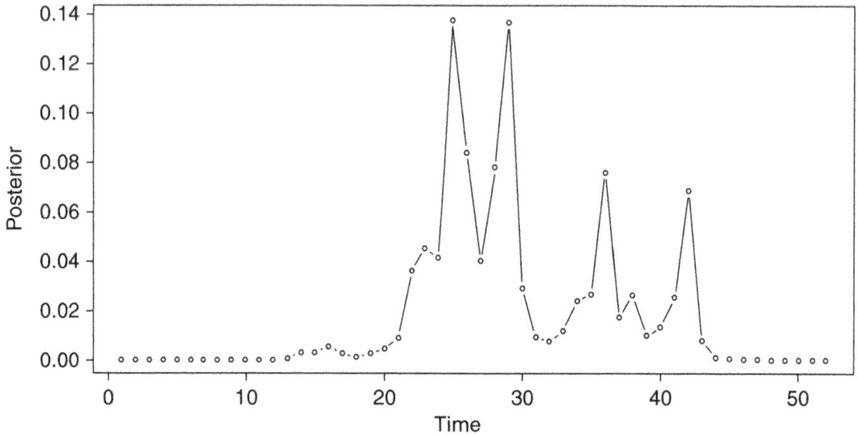

Figure 5.3. A Bayesian posterior distribution for the timing of an attack, if one has occurred.

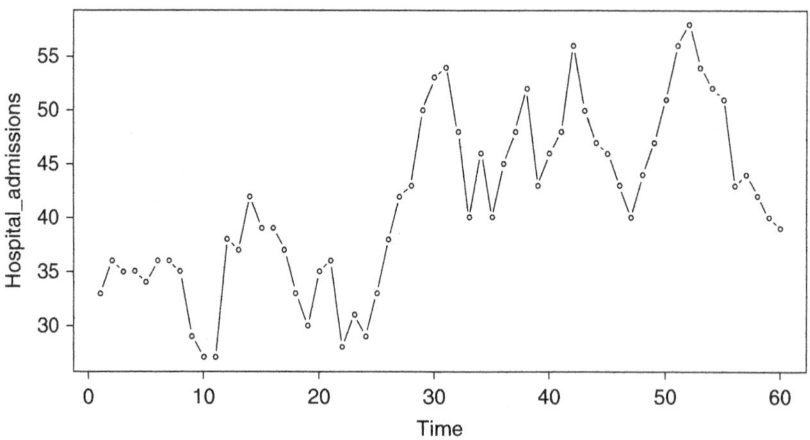

Figure 5.4. A different time series with much clearer evidence of an increase in hospitalization rates.

By contrast, Figures 5.4 and 5.5 repeat the same types of data and analysis results, using the same algorithm, for a case in which the effect of an attack on increasing hospitalization rates is twice as great. Now, the signal shows more clearly through the noise. The algorithm correctly identifies the time of the attack (week 25), and estimates the magnitude of its effect (not shown in Figure 5.5).

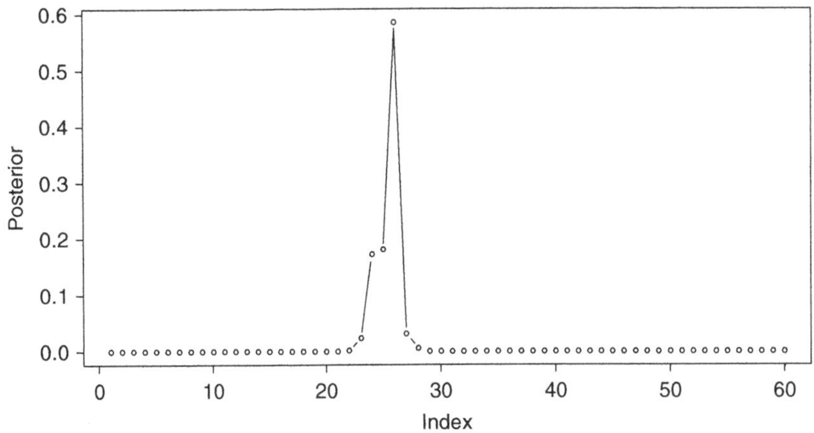

Figure 5.5. Bayesian posterior distribution for the timing of the attack in Figure 5.4, if one has occurred.

Thus, existing computational methods do a good job of estimating important unknown quantities (such as the likely time of an attack) from data when:

(1) A risk model is known (e.g., in this example, the effect of an attack is known to be a one-time increase in admission rates, so the pattern to look for is known);
(2) Plentiful surveillance data are available, to allow accurate maximum-likelihood estimates (MLE) of admission rates to be formed when the correct week of attack, w, is assumed; and
(3) The effect size is large enough to show clearly through the random noise in the surveillance count time series (as indicated by the high peak in Figure 5.5, which can be shown via simulation to be many orders of magnitude greater than the peaks that occur by chance under the null hypothesis of no real change in admission rates).

Numerical experimentation shows that these algorithms detect attacks quite quickly (within one to two weeks with high confidence) under these conditions.

However, the success of such algorithms depends on conditions 1–3 being satisfied. Under these conditions, likelihood-based methods and computational Bayesian methods can be used to quantify uncertainties and display useful results, as in Figures 5.2–5.5. Moreover, since they are concerned with detection and estimation rather than prediction, these methods are not undercut by uncertainties about attacker motivation, capabilities, intentions, plans, resource constraints, or behaviors. They simply address the question of whether an attack has yet occurred, constantly updating the answer and providing likelihood-based estimates as surveillance data are obtained and processed.

This simple, useful focus and well-developed technology contrast with the much harder problems of predicting probable attacks and optimizing defensive resource allocations to deter them or blunt their impacts. We consider methods for answering these more difficult questions next.

5.2.2 *Predicting and Quantifying Uncertainties about Adversarial Risks Using BN, ID, or Nested Decision-Tree Models*

Figures 5.2–5.5 illustrate uncertainty analysis under the best-case conditions of a known risk model and plentiful data. These favorable conditions make it possible to develop likelihood functions relating uncertain quantities to observations, and this in turn makes it practical to apply MLE and computational Bayesian methods to quantify the probability distributions of the unknown quantities, such as the time of attack, *w*, and the magnitude of the increase in hospital admission rates caused by the attack. We now discuss how risk can be assessed and managed when there is no clear basis for developing a probabilistic risk model, the form of the risk increase over time caused by an attack is uncertain (e.g., there may be an incubation period, then a period of increased risk, then a period of recovery as the effects of the attack die away, all complicated by contagion dynamics), and the assessment is directed to possible future attacks, rather than to recent or currently ongoing ones. Several technical options are available.

The essence of Bayesian methods is to provide a way to calculate the probabilities of the values of unobserved, uncertain quantities from the values of observed quantities (i.e., data) and a risk model. Conceptually, this may be written as

$$P(\mathit{unknown\,quantities}\,|\,\mathit{observed\ data;risk\ model}),$$

or simply *P*(*unknowns* | *datag*; *model*), where the conditional sign "|" is read as "given" or "conditioned on." When the risk model is known and fixed throughout a discussion, it is common to omit mentioning it, and the key relation computed by Bayesian methods can be written as *P*(*unknowns*|*data*). The risk model, in turn, specifies the prior probabilities for values of the unknown quantities – that is, their joint probability distribution before taking into account any observed data values, *P*(*unknowns*) – as well as the likelihood model relating *attacker_action* ⟶ *infection_rate* ⟶ *hospital_admissions* ⟵ *non_hospitalized_population*

Figure 5.6. A simple BN model.

observations to the unknown quantities, *P(data|unknowns)*. Bayes' Rule is the following relation:

$$P(unknowns = x \mid data = d) = \frac{P(data = d \mid unknowns = x)\,P(unknowns = x)}{\sum_u P(data = d \mid unknowns = u)\,P(unknowns = u)}$$

where x denotes a possible joint value (i.e., a specific combination of values) for all the unknowns. In Figures 5.2–5.5, that unknown variable is simply the attack time w, but more typically it would be a vector of several unknowns. Similarly, d denotes all the observed data. The denominator,

$$\sum_u P(data = d \mid unknowns = u)\,P(unknowns = u)$$

is a formula for calculating the probability of the observed data, $P(data = d)$, from the risk model given by prior $P(unknowns = u)$ and likelihood $P(data = d \mid unknowns = u)$. It is based on the idea that the probability of any event, including $P(data = d)$, can be found by summing up its probabilities for each of the events in any mutually exclusive, collectively exhaustive set of events (here, the events that the unknowns have joint values u, where u ranges over all possible joint values for the unknowns). If some or all of the unknowns are continuous, then summation must be replaced by integration, and computational tractability may then require evaluating the integral using rapid approximate techniques such as Monte Carlo sampling, but the basic idea is the same.

In practice, the enormous number of possibilities implied by considering all possible combinations of values for all of the unknowns, as required by the right side of $P(data = d) = \Sigma_u P(data = d \mid unknowns = u)\,P(unknowns = u)$, can usually be tamed by exploiting the fact that the values of some of the unknowns (modeled as random variables) typically depend directly on the values of only a few other variables. To illustrate, consider the model shown in Figure 5.6.

In this model, the four italicized variables are interpreted as random variables, each with several possible values. For example, *attacker_action* might have the possible values {*yes*, *no*}, indicating whether an attack has occurred yet; *infection_rate* might have possible values of {*base_value*, *new_value*}; *non_hospitalized_population* is a count variable indicating the number of non-hospitalized people in the surveillance population at the start of a week; and *hospital_admissions* is a count variable giving the number of patients admitted with a specified symptomology during the course of the week. This last variable has a binomial distribution with parameters *non_hospitalized_population* for the population size (number of trials) and *infection_rate* for the probability.

In Figure 5.6, the probability distribution for the value of each variable depends only on the values of the variable(s) that point into it. For example, *hospital_admissions* depends on *non_hospitalized_population* and on *infection_rate*, but not on *attacker_action* (once *infection_rate* is known). This allows a reduction in the total number of possibilities that must be considered. Each node (representing a variable) in the directed acyclic graph (DAG) shown in Figure 5.6 can be thought of as having an associated probability table that gives the probabilities for each possible value of that node, for each combination of possible values of the variables that point into it (if any). This is called the conditional probability table (CPT) for the node. The probability tables for the input nodes (those with only outward-directed arrows, such as *attacker_action*) are sometimes called marginal distributions or marginal probability tables. The full joint probability distribution for all of the values of all variables can be represented by specifying the probability table for each node (i.e., the marginal probability tables for input nodes and the CPTs for nodes with incoming arrows). This can greatly reduce the total number of combinations of variable values that need to be explicitly considered, while still permitting Bayesian inference.

This DAG-based representation of the dependency relations (and conditional independence relations) among random variables is called a Bayesian Network (BN). Algorithms and commercial software products such as *Hugin®*, *Bayesia®*, and *Netica®*, as well as free routines in R, Python, and various statistics packages, are available that allow Bayes' Rule to be applied to such networks, so that observed values (or "findings") entered at any nodes can be used to automatically create posterior distributions for the values of all other nodes. Several of these also support learning the structure of BN models from data (although there is still no perfect approach for this), statistically testing hypothesized model structures to see how well they fit or explain the available data, and quantifying probability tables at the nodes directly from underlying data. Multiple BN models for single periods (here, weeks), such as the one in Figure 5.6, can be linked together to form a dynamic BN (DBN) model, to reflect how changes in one period affect the probable values of variables in subsequent periods. The technology of building, testing, and using BN models to draw inferences is mature and sophisticated (Koller & Friedman, 2009).

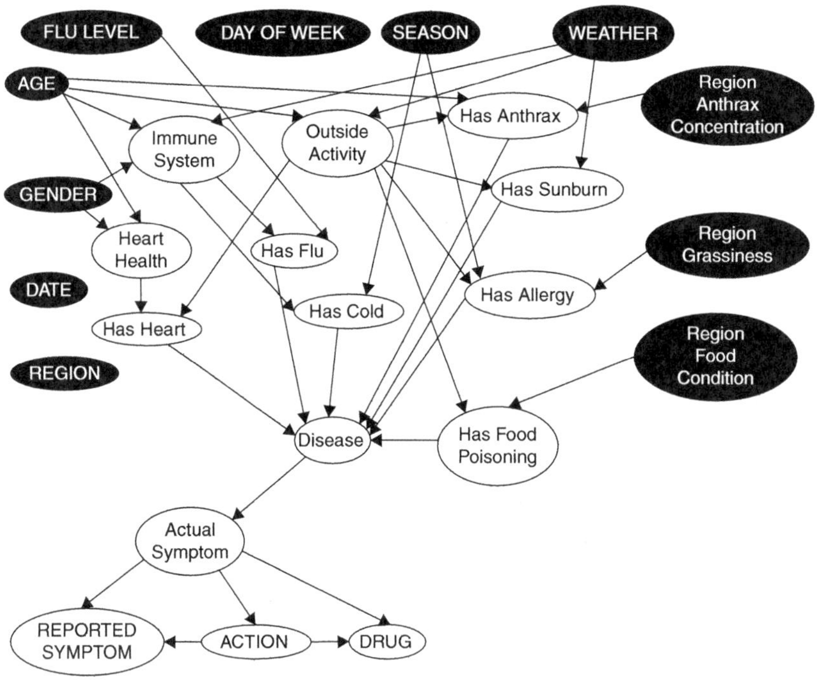

Figure 5.7. A BN for patient health status (Wong et al., 2003).

Figure 5.7 shows the DAG structure of a BN model for part of the "What's strange about recent events" (WSARE) 3.0 Bayesian inference system (Wong et al., 2003) for early detection of disease outbreaks. (Here, the variables "Date" and "Region" on the left-hand side of Figure 5.7 are assumed to affect all nodes, so the arrows from these global variables to the other nodes are not shown, for simplicity.) WSARE 3.0 integrates monitoring of time series for health care end points (such as hospital admissions with specific symptoms, as in Figures 5.2 and 5.4) with monitoring of other related variables (e.g., medication sales, temperature and weather, day of week, flu season, allergy season) to help detect possible anomalies caused by bioterror attacks (e.g., anthrax) as quickly as possible. The BN in this model specifies the probabilistic relations among multiple variables so that anomalies (i.e., significant departures from the modeled background joint probability distribution of the variables in the model) can be detected quickly. The authors conclude that WSARE 3.0 can successfully identify outbreaks in simulated data extremely quickly, while keeping false positives counts low.

Finally, Figure 5.1 shows a model (developed using the *Netica®* software) for predicting attack threats and perceived probable consequences from the perspective of a particular terrorist group. This BN is augmented with two types of nodes that do not represent random variables. There is a *value node* (the hexagon labeled "Terrorist Utility" at the bottom of the DAG) and three *choice nodes* (the "Threat," "Target," and "Means of Delivery" nodes) that reflect decisions or choices made by the terrorist group. BNs with choice and value nodes are also called ID models, and BN algorithms can be adapted to solve for the best decisions in an ID (Zhang, 1998); i.e., the best values to choose at the choice nodes in order to maximize the expected value of Terrorist Utility. The number displayed in the choice nodes in Figure 5.1 (0.60755) is the expected utility created by the terrorist group's best choices; in this example, a conventional weapon delivered by a person (rather than a vehicle) against a transport network. The posterior probability distributions for loss of life and resulting symbolism are displayed in the chance nodes (the usual BN type nodes, representing random variables) near the bottom of the model.

ID models support individual decision makers by computing the combinations of available actions that maximize expected utility. Older formalisms for probabilistic risk analysis (such as event trees, fault trees, and decision trees) can be automatically converted to equivalent BN or ID representations. IDs, in turn, can be generalized to allow multiple agents, such as an attacker and a defender, who seek to maximize their own expected utilities, taking into account the actions of others (Koller & Milch, 2001).

BNs (including DBNs and IDs) represent a substantial contribution to the current state of the art in adversarial risk analysis. They automatically quantify uncertainties, in the form of posterior distributions for uncertain quantities (and expected utilities for different possible decisions). They also provide a practical technology for integrating many diverse sources of evidence (as illustrated in Figure 5.7), and for integrating different types of knowledge about adversaries and what matters to them, to anticipate how adversary perceptions and values are likely to affect their attack decisions, as in Figure 5.1.

However, a fully specified BN requires a probability table at every chance node. When these cannot realistically be estimated with useful confidence and precision from available data, it is common practice for experts to supply their best guesses. BNs can then be used to encode expert knowledge or speculations in both the DAG structure of the BN itself, and the prior probabilities and conditional probabilities represented in the probability tables for the nodes in the model.

For completeness, we note that some analysts use nested attacker and defender decision trees to represent and quantify uncertainties about adversarial risks.

In this approach, a decision tree is constructed "in which both the attacker's and the defender's choices are modeled by decisions that maximize expected utility" (Ezell et al., 2010), analogous to an extensive-form game. This approach is helpful in small examples for displaying the chronological order of events and making conditional probabilities explicit (e.g., showing how the outcomes of possible terrorist actions depend on defenders' past actions, and/or the reverse), and therefore is useful in communicating the assumptions of the model (especially to stakeholders who are not technically trained) when there are only a few discrete choices. Similarly, project-risk analysis (i.e., modeling a terrorist attack as a risky project) has been used as a way of quantifying the success or failure probabilities of particular attack strategies, and identifying the attack stages at which terrorists are most vulnerable to detection or interdiction (Rosoff & von Winterfeldt, 2007). However, these modeling approaches can be limiting, especially in representing situations where the attacker or defender actions may be continuous rather than discrete, as in many resource-allocation problems.

As noted in Section 5.1, experts (like other people) tend to be overconfident, to use narrow framing (focusing on scenarios of interest, but not on the wider world that might change outcomes), and to offer opinions or judgments that frequently prove to be dramatically wrong – especially about political and military topics for which there is little chance to learn through repeated trials with clear, prompt feedback (Kahneman, 2011; Montibeller & von Winterfeldt, 2015; Tetlock, 2009). Fortunately, there are useful alternatives to using unassisted expert opinion to cope with uncertainties about adversarial risks.

5.2.3 *Predicting Adversarial Risks with Unknown Probabilities and Scarce Data*

This section presents techniques for modeling adversarial risks using BN models (including DBNs or IDs) or decision trees when insufficient knowledge and data are available to support trustworthy expert or data-driven (e.g., MLE) estimates of precise probabilities. Several different approaches are discussed next.

Imprecise Probabilities, Second-Order Probabilities, and Bounding Approaches

When probabilities are unknown, one possibility is to use interval estimates of the unknown probabilities. For example, one might have high confidence that the prevalence of a certain health condition, such as flu, is between 0.05 and 0.15, and yet be unable or unwilling to specify a particular value within that range. In this case, algorithms for propagating imprecise probabilities can be used (e.g., Ferson & Donald, 1998; Ferson & Hajagos, 2006); the resulting probabilities of different outcomes (or the values at particular output nodes) will then themselves be intervals, representing imprecise probabilities. More generally, instead of specifying marginal and conditional probability tables at the nodes of a BN or decision tree, uncertainty about the probabilities can be modeled by providing a (usually convex) *set* of probability distributions at each node.

BNs that have been generalized in this way are called *credal networks*, and algorithms for propagating sets of probabilities through such networks have been developed (Corani et al., 2012). These algorithms can be extended to allow optimization of risk management decisions in credal ID models (de Campos & Ji, 2008).

Alternatively, *second-order probability distributions* ("probabilities of probabilities") for the uncertain probabilities at BN or decision-tree nodes can be specified. If these uncertainties about probabilities are well approximated by Dirichlet or beta probability distributions (as happens naturally when probabilities or proportions are estimated from small samples using Bayesian methods with uniform or Dirichlet priors), then uncertainty analysis can be used to propagate these uncertain probabilities efficiently through the BN model or decision tree, leading to uncertainty distributions for the posterior probabilities of the variable values (Kleiter, 1996). Imprecise Dirichlet models have also been used to learn credal sets from data, resulting in upper and lower bounds for the probability that each variable takes on a given value (Corani et al., 2012).

Robust Inference and Optimization

Henrion and colleagues (1996) note that, in many situations, the key inferences and insights from BN models are quite insensitive (or "robust") to variations in the estimated values in the probability tables for the nodes. When this is the case, best guesses (e.g., MLE point estimates) of probability values may be adequate, even if the data and expertise used to form those estimates are scarce, and the resulting point estimates are quite uncertain. In practice, randomly sampling different values to enter into the node probability tables from a distribution, and then seeing how close the resulting inferences are, can help to build confidence in the recommendations from an analysis, if the results are relatively insensitive to the specific probability values used as inputs.

In many such models with imprecise probabilities, *robust optimization* identifies actions that maximize the minimum expected utility for all probability distributions in the uncertainty set of probability distributions for uncertain quantities. For decision problems with multiple interacting decision makers (e.g., a defender and an adversary), McLay and colleagues (2012) consider robust optimization of "level-k" adversarial games. In such games, players face "deep" uncertainties that can be bounded, but not easily described using probability distributions; in addition, they are assumed to have bounded rationality, and thus consider only a limited number of moves and countermoves in modeling adversarial behavior.

Adaptive Learning, Bayesian Model Averaging (BMA), Model Ensembles, and "Plural Analysis"

At the forefront of current methods for quantifying uncertainty about BN models are techniques for learning BNs directly from data, while characterizing the uncertainties about the resulting models. Often, more than one BN model (or credal BN, when imprecise probabilities are used) will fit the available knowledge and data approximately equally well, and it may be far from clear which DAG model structure best describes the real world. In such cases, Bayesian logic can be extended to include uncertainty about the "best" model as one more uncertainty about which inferences can be drawn – although of course no model may be perfectly "correct" (Bier, 1995; Box, 1979). To do so, likelihood functions are created that treat alternative models as another "parameter" to be selected. The posterior probability that a random variable of interest, X, has a particular value, x (e.g., that an attack will occur next July) is then given by

$$P(X = x) = P(X = x \mid M_1)P(M_1) + P(X = x \mid M_2)P(M_2) + \cdots + P(X = x \mid M_k)P(M_k)$$

where $M_1, \ldots, M_k$ denote alternative models (e.g., different hypothesized or fitted BN or ID models, all of which are plausible for the data as assessed by the likelihood function). The posterior probabilities of different models are obtained by Bayes' Rule (or by rescaling the likelihood function, if a flat prior is assumed, as in Figures 5.3 and 5.5).

Bayesian Model Averaging (BMA) refers to the use of an ensemble of multiple models to develop final estimates or risk predictions by weighting the predictions from each model based on its posterior probability or likelihood; typically, such BMA predictions are more accurate than those from any single model (Hoeting et al., 1999). (A closely analogous process called credal model averaging is available for models with imprecise probabilities; Corani et al., 2012.) As more data are collected, the model weights change to reflect the new data, with the new weights producing new predictions; thus, for example, the weighted-average modeling of Shan and Zhuang (2013) could be extended to update the weights on the different models in a Bayesian manner over time, as more information is collected. BMA also has methods to avoid the pitfall of giving too much weight to models that are excessively complex, and generally performs reasonably well in practice (Raftery and Zheng, 2003).

Such multiple-model analysis with adaptive updates (an automated version of the "plural analysis" encouraged by Brown & Lindley, 1986) allows best available predictions and recommended decisions based on current information to be generated at any time, while learning effectively from new data when they become available. These tools cope with uncertainty in the adversarial risk model by using multiple models to express uncertainty about how the world works. They use data-driven weighting of many different models to improve predictive accuracy for any given set of information, hedge bets about which model is best, learn from experience, and adaptively improve predictive performance over time.

5.3 How Has Uncertainty Analysis Been Used in Adversarial Risk Analysis?

Risk analysis for terrorism and counterterrorism has to deal with numerous uncertainties, as described previously; in fact, uncertainties about terrorism risk are greater than those about many other applications of risk analysis. However, the field has not always dealt with this effectively. Sometimes, practitioners have invested too much effort in quantifying uncertainties that are of little practical importance for decision making; other times, they have made assumptions reflect more confidence than is justified.

Emphasis on the importance of uncertainty quantification in risk analysis has sometimes led to slavish (and potentially costly) quantification of uncertainties (e.g., so-called second-order uncertainties, or uncertainty about probabilities), even when that quantification effort is unlikely to lead to any significant benefit in decision making. The *Bioterrorism Risk Assessment* by the U.S. Department of Homeland Security (2006) was a (much-maligned) example of this approach. Moreover, this study quantified uncertainty using a computationally demanding simulation procedure, even though the same quantification could have been done analytically, at less cost and with greater accuracy.

Cox and colleagues (2008) criticize this approach as uninformative, stating that "Uncertainties about consequence probabilities are not worth characterizing when final actions must be taken based on information available now." Strictly speaking, that is not entirely true; as shown by Bier (1994) and Mosleh and Bier (1996), it may also be important to quantify uncertainties about probabilities when the consequences of interest are a nonlinear function of the uncertain probability (e.g., when decision makers are concerned about the likelihood of multiple consecutive failures, or about the time to failure). However, the observation is still widely valid.

In the quest to transform itself from a hodge-podge collection of agencies into an effective bureaucracy, the Department of Homeland Security sometimes went overboard on routine quantification of uncertainties, without a sufficiently deep (e.g., decision-theoretic) view of when that quantification effort was likely to be fruitful. As a result, the National Research Council (2008) recommended that the department spend more time and effort quantifying those uncertainties that are likely to have an impact on decisions, such as uncertainties about terrorist behavior, or about indirect attack consequences (rather than just direct fatalities).

At the simplest level, many applications of terrorism risk analysis use qualitative risk-ranking matrices. These typically require that each hazard be given a unique rating (e.g., "high," "medium," or "low") in both frequency and severity. Cox and colleagues (2005) and Cox (2008a) have indicated that neglecting the ranges of possible outcomes for particular hazards (e.g., the fact that the consequences of a particular attack type could range from negligible to catastrophic, depending on unpredictable features such as attacker skill or even the weather at the time of the attack) can lead to significant errors and misinterpretations, including uninformative rankings (where numerous hazards all receive the same overall ranking), failure to achieve the maximum possible risk reduction for a given expenditure (Cox et al., 2005), and even ranking reversals (where a lesser threat mistakenly receives a higher ranking). As a result, Ball and Watt (2013) conclude that "risk matrices … may be creating no more than an artificial and even untrustworthy picture of the relative importance of hazards." Challenges in the application of risk matrices to terrorism in particular are detailed by Cox (2009). By contrast, slightly more sophisticated graphical portrays of risk (such as Figure 5.8), which are also commonly used in homeland security, provide much more information, and can even be used to portray correlations between frequency and severity (e.g., the fact that severe earthquakes may be much less common than minor earthquakes).

At a somewhat more sophisticated level, Brown and Cox (2011a) point out that both estimates of terrorism risk and the optimal risk-reduction strategy depend critically on the options that an adversary has available to reduce uncertainties before launching an attack. More to the point, they identify cases in which reducing the level of adversary uncertainty may actually reduce risk; for example, ensuring that adversaries know about the implementation of highly effective countermeasures may act as a deterrent to attacks. A focus on adversary uncertainty and how it affects counterterrorism decision making is rare in practice, although it has been addressed at a more theoretical level – e.g., by Powell (2007) and Zhuang and colleagues (2010). Moreover, Cox

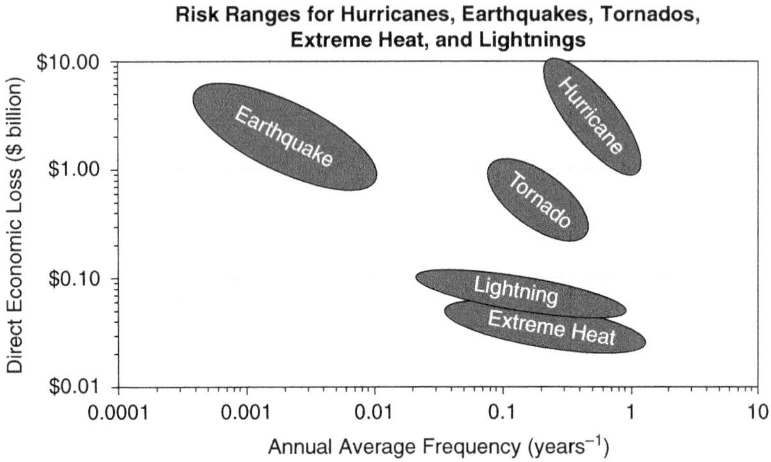

Figure 5.8. Scatter plot of economic loss versus frequency (Argonne National Laboratory 2009).

(2008b) emphasizes that differing interpretations of even such a seemingly well-defined parameter as "success probability of an attack" can lead to widely differing implications for risk reduction. In particular, Bier and Lin (2013) point out that an event that may with small probability affect a vast number of people should be treated differently than an event that is guaranteed to occur, but will affect any given person with small probability; for example, in the face of risk aversion about severe consequences, it may be rational to spend much more to prevent the first type of event than the second. In fact, it is difficult to assess the importance of any given attack scenario without considering possible defender risk aversion, which requires development of an entire probability distribution for consequences (Cox 2008b), rather than just a point estimate or an expected value. Asbeck and Haimes (1984) attempt to address concerns about catastrophic risk through calculation of a "conditional expected value" (e.g., conditional on consequences exceeding some threshold of severity), to supplement the use of traditional expected value. However, this approach lacks the appeal of providing a single summary statistic (such as expected value), without fully addressing concerns regarding risk aversion (Bier et al., 2004).

As noted earlier, the cost-effectiveness of proposed defenses may be highly uncertain in many practical situations. This is important, since Bier and colleagues (2007) show that cost-effectiveness should ideally be a major determinant of any resource-allocation decision. However, in practice, there has been little if any effort to systematically quantify the cost-effectiveness of our defenses. Moreover, despite the (possibly excessive) emphasis on risk as the product of threat, vulnerability, and consequences, at least as recently as 2008, the Department of Homeland Security was still using a value of 1.0 for vulnerability at all possible locations (Government Accountability Office, 2008), rather than attempting to assess vulnerability (and the potential reductions in vulnerability from defensive investment) realistically, and characterizing the extensive uncertainties affecting these estimates.

As a result of such challenges, Brown and Cox (2011b) conclude that "Rather than continuing to devote 'considerable resources' to creating simplistic ... and low-performing 'new theories and approaches' ... the United States would be served far better by having competent risk analysts apply well-established techniques from operations research and risk analysis to model uncertainty and to robustly improve our infrastructure resilience and defensive resource allocations."

5.4 Making Prudent Decisions Despite Realistic Uncertainties

We hope that this chapter has made two points: (1) that uncertainties in adversarial risk assessment are serious and pervasive; but (2) that they can be represented and usefully addressed via modeling methods such as those illustrated in Figures 5.1–5.7 (e.g., BN or decision-tree models, and extensions that allow for imprecise or uncertain probabilities). The structures of these network (DAG) models display what is known or hypothesized about the relations among variables that affect attacker decisions, as in Figure 5.1.

When there is deep uncertainty about what the diagram of important causal relations should look like and how its probability tables should be quantified, viable options for representing and calculating such uncertainties include methods for using multiple plausible models (as in BMA), and robust inference or credal network models with imprecise or uncertain probabilities. On the other hand, if probabilistic risk models are available and data are plentiful, then more straightforward likelihood-based methods can be used (as illustrated in Figures 5.2–5.5, to swiftly detect anomalies in surveillance data that might indicate a bioterror attack). This array of techniques offers a substantial repertoire for modeling realistic uncertainties and making choices about the best combinations of defensive actions in the presence of realistic (usually large) uncertainties about attacker beliefs, capabilities, and intents.

REFERENCES

Asbeck, E., & Haimes, Y. Y. (1984). The partitioned multiobjective risk method. *Large Scale Systems 6*, 13–38.

Ball, D. J., & Watt, J. (2013). Further thoughts on the utility of risk matrices. *Risk Analysis 33*, 2068–2078.

Bier, V. M. (1993, October 20–22). Some illustrative examples of model uncertainty. *Proceedings of Workshop I in Advanced Topics in Risk and Reliability Analysis, Model Uncertainty: Its Characterization and Quantification*, pp. 91–98, Annapolis, Maryland. Reprinted in Mosleh, A., Siu, N. Smidts, C, & Lui, C. (Eds.). (1995). *Model uncertainty: Its characterization and quantification* (pp. 93–100). University of Maryland.

Bier, V. M. (1994). Uncertainty irrelevance and preferences over ambiguity in expected utility decision theory. In B. M. Ayyub & M. M. Gupta (Eds.), *Uncertainty modeling and analysis: Theory and applications* (pp. 211–220). North-Holland.

Bier, V. M., Ferson, S., Haimes, Y. Y., Lambert, J. H., & Small, M. J. (2004). Risk of extreme and rare events: Lessons from a selection of approaches. In T. McDaniels & M. J. Small (Eds.), *Risk analysis and society: Interdisciplinary perspectives* (pp. 74–118). Cambridge University Press.

Bier, V. M., & Kosanoglu, F. (2015). Target-oriented utility theory for modeling the deterrent effects of counterterrorism. *Reliability Engineering and System Safety, 136*, 35–46.

Bier, V. M., & Lin, S.-W. (2008). A study of expert overconfidence. *Reliability Engineering and System Safety, 93*, 711–721.

Bier, V. M., & Lin, S.-W. (2013). On the treatment of uncertainty and variability in making decisions about risk. *Risk Analysis, 33*, 1899–1907.

Bier, V. M., Menoyo, J., & Wang, C. (2012). Achieving realistic levels of defensive hedging based on non-monotonic and multi-attribute terrorist utility functions. In J. W. Herrmann (Ed.), *Handbook of operations research for homeland security* (pp. 125–139). Springer.

Bier, V. M., Oliveros, S., & Samuelson, L. (2007). Choosing what to protect: Strategic defensive allocation against an unknown attacker. *Journal of Public Economic Theory 9*, 563–587.

Box, G. E. (1979). Robustness in the strategy of scientific model building. In R. Launer & G. Wilkinson (Eds.), *Robustness in statistics* (p. 202). Academic Press.

Breckenridge, J. N., & Zimbardo, P. G. (2006). The strategy of terrorism and the psychology of mass-mediated fear. In B. Bongar (Ed.), *Psychology of terrorism* (pp. 116–133). Oxford University Press.

Brown, G. G., & Cox Jr., L. A. (2011a). How probabilistic risk assessment can mislead terrorism risk analysts. *Risk Analysis, 31*, 196–204.

Brown, G. G., & Cox Jr., L. A. & Cox Jr., L. A. (2011b). Making terrorism risk analysis less harmful and more useful: another try. *Risk Analysis, 31*, 193–195.

Brown, R. V., & Lindley, D. V. (1986). Plural analysis: Multiple approaches to quantitative research. *Theory and Decision, 20*, 133–154.

Burns, W. J., & Slovic, P. (2010). Predicting and modeling public response to a terrorist strike. In P. Slovic (Ed.), *The feeling of risk: New perspectives on risk perception*. Earthscan.

Cami, A., Wallstrom, G. L., & Hogan, W. R. (2009). Measuring the effect of commuting on the performance of the Bayesian Aerosol Release Detector. *BMC Medical Informatics and Decision Making, 9* (Supplement 1), S7.

de Campos, C. P., & Ji, Q. (2008). Strategy selection in influence diagrams using imprecise probabilities. In *Proceedings of the 24th Conference in Uncertainty in Artificial Intelligence (UAI)* (pp. 121–128).

Corani, G., Antonucci, A., & Zaffalon, M. (2012). Bayesian networks with imprecise probabilities: Theory and application to classification. In D. E. Holmes & C. Jaim (Eds.), *Data mining: Foundations and intelligent paradigms* (pp. 49–93). Intelligent Systems Reference Library, Volume 23.

Covello, V. T., Peters, R. G., Wojtecki, J. G., & Hyde, R. C. (2001). Risk communication, the West Nile virus epidemic, and bioterrorism: Responding to the communication challenges posed by the intentional or unintentional release of a pathogen in an urban setting. *Journal of Urban Health: Bulletin of the New York Academy of Medicine, 78*, 382–391.

Cox Jr., L. A. (2008a). What's wrong with risk matrices? *Risk Analysis, 28*, 497–512.

Cox Jr., L. A. (2008b). Some limitations of "risk = threat × vulnerability × consequence" for risk analysis of terrorist attacks. *Risk Analysis, 28*, 1749–1761.

Cox Jr., L. A. (2009). Improving risk-based decision making for terrorism applications. *Risk Analysis, 29*, 336–341.

Cox Jr., L. A., Babayev, D., & Huber, W. (2005). Some limitations of qualitative risk rating systems. *Risk Analysis, 25*, 651–662.

Cox Jr., L. A., Brown, G. G., & Pollock, S. M. (2008).When is uncertainty about uncertainty worth characterizing? *Interfaces, 38*, 465–468.

Dighe, N. S., Zhuang, J., & Bier, V. M. (2009). Secrecy in defensive allocations as a strategy for achieving more cost-effective attacker deterrence. *International Journal of Performability Engineering, 5*, 31–43.

Ezell, B. C., Bennett, S. P., von Winterfeldt, D. Sokolowski, J., & Collins, A. J. (2010). Probabilistic risk analysis and terrorism risk. *Risk Analysis, 30*, 575–589.

Ferson, S., & Donald, S. (1998). Probability bounds analysis. In A. Mosleh & R. A. Bari (Eds.), *Probabilistic safety assessment and management* (pp. 1203–1208). Springer-Verlag.

Ferson S., & Hajagos, J. G. (2006). Varying correlation coefficients can underestimate uncertainty in probabilistic models. *Reliability Engineering and System Safety, 91*, 1461–1467.

Finkel, A. M. (2014). Emitting more light than heat: Lessons from risk assessment controversies for the "job-killing regulations" debate. In C. Coglianese, A. M. Finkel, & C. Carrigan (Eds.), *Does regulation kill jobs?* (pp. 128–149). University of Pennsylvania Press.

Goodstein, E., & Hart Hodges, H. (1997). Polluted data: Overestimating environmental costs. *American Prospect, 8*(35), 64–69.

Government Accountability Office. (2008, November 14 and December 15). Homeland security grant program risk-based distribution methods. Presentation to Congressional Committees.

Grundmann, O. (2014). The current state of bioterrorist attack surveillance and preparedness in the US. *Risk Management and Healthcare Policy*, 177–187.

Harrington, W., Morgenstern, R. D., & Nelson, P. (2000). On the accuracy of regulatory cost estimates. *Journal of Policy Analysis and Management, 19*, 297–322.

Hausken, K., & Bier, V. M. (2011a). Defending against multiple different attackers. *European Journal of Operational Research, 211*, 370–384.

Hausken, K., & Bier, V. M. (2011b). Endogenizing the sticks and carrots: Modeling terrorism and the emergence of hatred. *Annals of Operations Research, 186*, 39–59.

Hazilla, M., & Kopp, R. J. (1990). Social cost of environmental quality regulations: A general equilibrium analysis. *Journal of Political Economy, 98*, 853–873.

Henrion, M. Pradhan, M., Del Favero, B., Huang, K., Provan, G., & O'Rorke. P. (1996). Why is diagnosis using belief networks insensitive to imprecision in probabilities? In *Proceedings of the Twelfth Annual Conference on Uncertainty in Artificial Intelligence (UAI96)* (pp. 307–314). Morgan Kaufmann Publishers.

Hoeting, J., Madigan, D., Raftery, A., & Volinsky, C. (1999). Bayesian model averaging. *Statistical Science, 14*, 382–401.

Hora, S. (2007). Eliciting probabilities from experts. In W. Edwards, R. F. Miles, & D. von Winterfeldt (Eds.), *Advances in decision analysis: From foundations to applications* (pp. 129–153). Cambridge University Press.

Jamshidi, T., & Bier, V. M. (2009, December). Cost-effectiveness of investments in defense of critical infrastructure. Annual Meeting, Society for Risk Analysis, Baltimore, Maryland.

Kahneman, D. (2011). *Thinking, fast and slow*. Macmillan.

Kleiter, G. D. (1996). Propagating imprecise probabilities in Bayesian networks. *Artificial Intelligence, 88*, 143–161.

Koller, D., & Friedman, N. (2009). *Probabilistic graphical models: Principles and techniques*. MIT Press.

Koller D., & Milch, B. (2001). Multi-agent influence diagrams for representing and solving games. *Proceedings of the 17th International Joint Conference on Artificial Intelligence*.

Major, J. A. (2002). Advanced techniques for modelling terrorism risk. *Journal of Risk, Finance, 4*, 15–24.

Maya, I., John, R. S., Rosoff, H., Bier, V. M., & Barrett, A. (2012, July). Adaptive adversary decision modeling for global terrorism risk. World Congress on Risk, Society for Risk Analysis, Sydney, Australia.

McArdle, S. C., Rosoff, H., & John, R. S. (2012). The dynamics of evolving beliefs, concerns, emotions, and behavioral avoidance following 9/11: A longitudinal analysis of representative archival samples. *Risk Analysis, 32*, 744–761.

McLay, L., Rothschild, C., & Guikema, S. (2012). Robust adversarial risk analysis: A level-k approach. *Decision Analysis, 9*, 41–54.

Miller, E. (2012). Patterns of onset and decline among terrorist organizations. *Journal of Quantitative Criminology, 28*, 77–101.

Mohtadi, H., & Murshid, A. P. (2009). Risk of catastrophic terrorism: An extreme value approach. *Journal of Applied Econometrics, 24*, 537–559.

Montibeller, G., & von Winterfeldt, D. (2015). Cognitive and motivational biases in decision and risk analysis. *Risk Analysis, 35*, 1230–1251.

Morgan, M. G., Henrion, M., & Small, M. (1990). *Uncertainty: A guide to dealing with uncertainty in quantitative risk and policy analysis*. Cambridge University Press.

Morningside Post. (2007). The bipartisan politicization of the national intelligence estimate. www.themorningsidepost.com/2007/12/08/the-bipartisan-politicization-of-the-nie/.

Mosleh, A., & Bier, V. M. (1996). Uncertainty about probability: A reconciliation with the subjectivist viewpoint. *IEEE Transactions on Systems, Man, and Cybernetics (Part A: Systems and Humans), 26*, 303–310.

National Research Council. (2008). *Department of Homeland Security bioterrorism risk assessment: A call for change*. National Academies Press.

Office of Technology Assessment. (1995). *Gauging control technology and regulatory impacts in occupational safety and health: An appraisal of OSHA's analytic approach.*

Pate-Cornell, E., & Guikema, S. (2002). Probabilistic modeling of terrorist threats: A systems analysis approach to setting priorities among countermeasures. *Military Operations Research, 7*, 5–20.

Pizer, W. A., & Kopp, R. (2003). Calculating the costs of environmental regulation. Resources for the Future Discussion Paper 03-06.

Powell, R. (2007). Allocating defensive resources with private information about vulnerability. *American Political Science Review, 101*, 799–809.

Price, B. C. (2012). Targeting top terrorists: How leadership decapitation contributes to counterterrorism. *International Security, 4*, 9–46.

Putnam, H., & Bartlett, Inc. (1980). *Comparisons of estimated and actual pollution control capital expenditures for selected industries*. U.S. Environmental Protection Agency.

Raftery, A. E., & Zheng, Y. (2003). Discussion: Performance of Bayesian model averaging. *Journal of the American Statistical Association, 98*, 931–938.

Rosoff H., & John, R. S. (2009). Decision analysis by proxy for the rational terrorist. *Proceedings of the International Joint Conference on Artificial Intelligence (IJCAI-09)*. Workshop on Quantitative Risk Analysis for Security Applications.

Rosoff, H., & von Winterfeldt, D. (2007). A risk and economic analysis of dirty bomb attacks on the ports of Los Angeles and Long Beach. *Risk Analysis, 27*, 533–546.

Ruth Ruttenberg and Associates. (2004). *Not too costly, after all: An examination of the inflated cost-estimates of health, safety and environmental protections.* Prepared for the Public Citizen Foundation, Inc.

Shan, X., & Zhuang, J. (2013). Hybrid defensive resource allocations in the face of partially strategic attackers in a sequential defender-attacker game. *European Journal of Operational Research*, 228, 262–272.

Shen Y., & Cooper, G. F. (2010). A new prior for Bayesian anomaly detection: Application to biosurveillance. *Methods of Information in Medicine, 49*, 44–53.

Sticha, P. J., Weaver, E. A., Tatman, J. A., Mahoney, S. M., & Buede, D. M. (2009). Reading the behavior signature: Predicting leader personality from individual and group actions. *Association for the Advancement of Artificial Intelligence Spring Symposium: Technosocial Predictive Analytics*, pp. 130–136.

Tetlock, P. E. (2009). *Expert political judgment: How good is it? How can we know?* Princeton University Press.

Tversky, A., & Kahneman, D. (1974). Judgment under uncertainty: Heuristics and biases. *Science, 185(4157)*, 1124–1131.

U.S. Department of Homeland Security. (2006). *Bioterrorism risk assessment.* Biological Threat Characterization Center, National Biodefense Analysis and Countermeasures Center.

Von Winterfeldt, D. (1999). On the relevance of behavioral decision research for decision analysis. In J. Shanteau, B. A. Mellers, & D. A. Schum (Eds.), *Decision science and technology: Reflections on the contributions of Ward Edwards* (pp. 133–154). Springer.

Wang C., & Bier, V. M. (2011). Target-hardening decisions based on uncertain multi-attribute terrorist utility,. *Decision Analysis, 8*, 286–302.

Wang C., & Bier, V. M. & Bier, V. M. (2015). Quantifying adversary capabilities to inform defensive resource allocation. *Risk Analysis*, http://onlinelibrary.wiley.com/doi/10.1111/risa.12399/.

Wong, W.-K., Moore, A., Cooper, G. & Wagner, M. (2003). Bayesian network anomaly pattern detection for disease outbreaks. In T. Fawcett & N. Mishra (Eds.), *Proceedings of the Twentieth International Conference on Machine Learning* (pp. 808–815). AAAI Press.

Woo, G. (2002). Quantitative terrorism risk assessment. *Journal of Risk Finance, 4*, 7–14.

Woo, G. (2003). Insuring against Al Qaeda. NBER Insurance Project Workshop.

Zhang, N. L. (1998). Probabilistic inference in influence diagrams. *Computational Intelligence, 14*, 475–497.

Zhuang J., & Bier, V. M. (2010). Reasons for secrecy and deception in homeland-security resource allocation. *Risk Analysis, 30*, 409–418.

Zhuang, J., Bier, V. M., & Alagoz, O. (2010). Modeling secrecy and deception in a multiple-period attacker-defender signaling game. *European Journal of Operational Research, 203*, 409–418.

6

A Risk and Economic Analysis of Dirty Bomb Attacks on the Ports of Los Angeles and Long Beach

Heather Rosoff and Detlof von Winterfeldt

6.1 Introduction

6.1.1 *The Dirty Bomb Threat*

Since the events on September 11, 2001, the prospect of a terrorist attack using a radiological dispersal device (dirty bomb) is cited as among one the most serious terrorist threats (Blair, 2002). Several recently reported incidents confirm the concerns of security officials. In June 2002, the United States arrested Jose Padilla for his involvement with Al Qaeda in planning a dirty bomb attack on the United States (Johnson & Locy, 2002), and in January 2003, British officials found documents in the Afghan city of Herat indicating Al Qaeda successfully built a small dirty bomb as well as possessed training manuals on using the explosive device (Garner, 2003).

There are several reasons why terrorists may consider dirty bombs to be an attractive weapon. Radioactive materials are relatively easy to obtain and building a dirty bomb is a fairly simple process, requiring little more than the skills required for assembling a conventional bomb (King, 2004). Furthermore, dirty bombs can create large radioactive plumes, cause health and psychological effects, and have major economic impacts due to the need for decontaminating large areas.

The primary challenge faced by terrorists is procuring the radioactive material. The International Atomic Energy Agency (IAEA) (2002) states that nearly every country has devices containing radioactive material useful for the creation of dirty bombs and questions whether security in many of these locations is adequate. Significant quantities of radioactive material have been lost, stolen, or abandoned – referred to as "orphan sources" – from U.S. and international facilities. According to an August 2003 General Accounting Office (GAO) report (GAO, 2003), since

1998 more than 1,300 radioactive sources have become orphaned in the United States. A primary concern of U.S. and international security experts is the number of orphan sources scattered throughout the former states of the Soviet Union and the security of nuclear facilities in Pakistan, India, and other developing countries.

A dirty bomb consists of radioactive material packaged in conventional explosives. When detonated, the radioactive material scatters into the environment, some forming a radioactive plume, and the remaining quantity falling in clumps or large particulate matter near the location of the explosion. No nuclear-fission and/or fusion reaction takes places as in a nuclear weapon. However, a dirty bomb can result in both death and injuries from the initial blast of the conventional explosives as well as radiation sickness and cancer from exposure to the radioactive material. Furthermore, the dirty bomb is widely recognized as having psychological and long-term economic impacts that could outweigh its health consequences. More specifically, depending on the amount of radioactive material released and dispersed, the contaminated area could require complete evacuation, followed by decontamination efforts that could take months or even years. Locally, evacuations and decontamination efforts impact the economy and instill public fear about returning to the contaminated area. Nationally, this could result in dirty bomb scares, both real and hoaxes, and instigate residual repercussions throughout the economy.

This chapter presents a risk and economic analysis of a dirty bomb attack on the ports of Los Angeles and Long Beach. We attempt to answer the following three questions:

1. What are the threats and vulnerabilities of a dirty bomb attack upon the ports?
2. If a dirty bomb attack was successfully carried out at the ports, what might be the health and economic impacts?
3. Given our risk and economic analysis, what are potential policy recommendations for more effective countermeasures?

The next section of this chapter describes the sources of radioactive material in the United States and abroad that could be used to construct a dirty bomb. Section 6.3 summarizes an analysis of 36 attack scenarios and describes a methodology and some preliminary findings for estimating the relative likelihood of a successful attack. Section 6.4 presents an analysis of the consequences of the most likely attack scenarios in terms of the health effects and economic impact of a port shutdown. Section 6.5 examines possible countermeasures and their cost effectiveness.

6.2 Sources of Radioactive Material

Millions of radioactive sources are distributed worldwide, with hundreds of thousands in varying quantities and sizes currently being used, stored, and produced. In the United States alone, approximately 2 million licensed sealed sources are in use (GAO, 2003). Among the 15 member states of the European Union, the European Commission reported that about 500,000 sealed sources have been located (GAO, 2003). As seen in Table 6.1, there are multiple sources of radioactive material that pose different levels of security risk given the amount of curies (unit of measurement of radioactivity) they could generate. Spent fuel rods from nuclear reactors and waste facilities, industrial and blood irradiators, and radiography equipment are among some of the primary sources that contain radioactive material. For a terrorist to build a dirty bomb, any of the radioactive material necessary for these applications could be employed. Most reports of trafficking incidents or unauthorized movement of radioactive material involve sealed sources, with a few incidents involving unsealed sources such as contaminated scrap metal.

Table 6.1. *Sources of Radioactive Material*

Radioactivity Source	Radioisotope	Level (Curies)
Spent fuel assembly	Multiple sources	300,000–2,000,000
Industrial irradiator (sterilization and food preservation)	Cobalt 60 (Co 60) Cesium 137 (Cs 137)	Up to 3,000,000 Up to 4,000,000
Blood irradiator	Co 60 Cs 137	2,400–25,000 50–15,000
Radiotherapy (single and multibeam)	Co 60 Cs 137	4,000–27,000 500–13,500
Medical radiography	Co 60 Iridium 19 (Ir 192)	1,000 1–200
Industrial radiography	Co 60 Ir 192	3–250 3–250
Calibration	Co 60 Cs 137 Americium 241	20 60 10

Sources: Modified (1) Center for Nonproliferation Studies (CNS), *The Four Faces of Nuclear Terrorism*, 2005; (2) CNS, *Commercial Radioactive Sources: Surveying the Security Risks*, 2003; (3) IAEA, *Categorization of Radioactive Sources*, 2003; (4) personal communication with nuclear safety expert, Pacific Northwest National Laboratory, August 2004.

6.2.1 *Nuclear Reactor and Waste Facilities*

In the United States, nuclear power and waste facilities contain millions of curies of radioactive material that is the mostly deadly in nature, but also extremely difficult to obtain and handle. Special licenses are issued by the U.S. Nuclear Regulatory Commission (NRC) to ensure the facilities are designed, constructed, and operated in accordance with safety standards. In addition, security surrounding nuclear power and waste sites is extremely high. While the large inventories of radioactive material may be appealing to terrorists, such precautions present a formidable challenge to acquiring the material. However, material from the nuclear fuel cycle are less protected in some countries and may be available from "rogue" countries that are developing a nuclear power capability (see below).

6.2.2 *Medical, Research, and Industrial Facilities*

The NRC also issues licenses for medical, research, and industrial applications requiring radioactive material. Medical and research institutions use radioactive material in medical diagnosis, sterilization of medical equipment, radiotherapy (both internal and external), and for research in nuclear medicine. Radiotherapy, the treatment of disease with radiation, employs radioisotopes that are susceptible to security risk (Ferguson, Kazi, & Perera, 2003). In contrast, the material used for sterilizing equipment and medical diagnosis present a smaller security concern, since they require relatively low amounts of radioactive material with short half-lives.

Industrial facilities use radioactive material to operate machinery such as food irradiators, gauging devices, well-logging devices, and industrial radiography systems. Irradiators pose the greatest security risk because they typically contain thousands to millions of curies (Ferguson et al., 2003). Industrial radiography contains lower quantities of radioactive material, but they are placed in portable devices that present a security risk (Ferguson et al., 2003). Gauging and well-logging devices typically contain inconsequential amounts of radioactive material (Ferguson et al., 2003). While the NRC is responsible for issuing licenses and monitoring such facilities, security requirements are less stringent than those found at nuclear reactor and waste facilities.

6.2.3 *Foreign Sources of Radioactive Material*

Internationally, experts are concerned about the security risk associated with spent fuel assemblies and reprocessed material abandoned, lost, or poorly guarded in the former states of the Soviet Union. There are also approximately

1,000 radioisotope thermoelectric generators (RTGs) that have exhausted their design life and are in need of dismantlement. The amount of radioactivity generated by these sources can be in the millions of curies. Surplus radioactive material coupled with a large number of sites with inadequate protection present opportunities for illegal stealing, selling, and trafficking. Compared to the United States, acquiring material of this quantity in some foreign countries may be less challenging mostly because of less stringent accountability and security standards.

The former Soviet Union also houses weapons-grade plutonium and highly enriched uranium produced in excess during the Cold War. If a terrorist were to acquire plutonium or highly enriched uranium, he or she would most likely save these materials for the construction of a nuclear weapon. However, experts have noted that of all known cases of attempted trafficking of weapons-grade nuclear materials, the total acquired material is not enough to build a single nuclear bomb (Council on Foreign Relations, 2004).

6.3 Scenarios and Probabilities

Ports are attractive terrorist targets because of the potential for a successful attack to result in lives lost and economic damage to local businesses, harbor operations, and the flow of trade worldwide. Overall, ports are major trade nodes, have complex business infrastructures, and are difficult to secure due to their extensive size and accessibility by water and land. Most ports are located near major metropolitan regions that rely heavily on the resources and jobs provided by the businesses within the harbors. Also, ports are connected through several transportation modes (e.g., road, ship, and rail), and often industries, businesses, and tourist attractions are close by, presenting terrorists with several options for deception and attack scenarios.

In this analysis, we examined possible dirty bomb attacks on the Los Angeles and Long Beach harbors as an example. Combined, they are the third busiest port in the world, which handles 14.2 million 20-foot unit equivalent containers annually with a value of about $295 billion (Port of Long Beach, 2007; Port of Los Angeles, 2007). In addition, 44% of U.S. imports enter into the country through these two ports (Port of Long Beach, 2007; Port of Los Angeles, 2007). Dispersed across the harbors are oil refineries, business offices, storage facilities for hazardous materials and cargo, container terminals, and more. Cargo is transported to the ports via land, ship, or rail, increasing the challenge of securing the region. And whether coming to the ports for work or to make a delivery, many people enter the Los Angeles and Long Beach harbors daily.

Immediately surrounding the ports are parks and various roads leading to fishing wharfs and tourist attractions such as the Queen Mary and cruise line terminals. Also in the proximity are downtown Long Beach and San Pedro. Major highways, roads, and bridges pass through or alongside the ports. The activity in the nearby metropolis and recreational areas makes a terrorist attack on the ports of significant consequence both to the local livelihood as well as to the regional and national economy.

To analyze the dirty bomb threat to the ports of Los Angeles and Long Beach, we explored the danger of varying sources and quantities of radioactive material (measured in curies – Ci), as well as the differences in such attacks when the material originates from domestic versus international locations. We considered three source scenarios, each depicting either a small, medium, or large-scale attack:

1. Low radioactivity scenario: Theft of radioactive material from a radiotherapy device in a U.S. hospital.
2. Medium radioactivity scenario: Theft of radioactive material from an industrial irradiator in a U.S. facility.
3. High radioactivity scenario: Purchase of a spent fuel assembly from a former Soviet Union nuclear power or reprocessing plant.

In collaboration with a counterintelligence expert, we examined these three scenarios in more detail. In particular, we studied the motivations and capabilities of terrorists to engage in any of the three scenarios to attack the ports of Los Angeles and Long Beach and conducted a qualitative "red teaming" exercise for each. In a red teaming exercise, the scenario is played out from the perspective of the terrorists to better understand the opponents' thinking and plans. For each scenario, we considered four transportation modes (truck, ship, train, and plane or helicopter) and three locations (bridge, harbor-elevated, harbor-ground). We examined a total of 36 possible terrorist attack scenarios.

For illustrative purposes, this chapter will focus primarily on the medium radioactivity scenario and its transportation and location subscenarios. We assumed that a moderate quantity of radioactive material (100,000 curies) is stolen from a U.S. blood or industrial irradiator. Once the radioactive material is acquired, we assumed it would be transported to a warehouse near the port for dirty bomb construction. A separate terrorist cell, equipped with technical expertise, would be accountable for building the dirty bomb during this phase of preparation for the attack. Finally, upon construction completion, a third cell would drive the dirty bomb to the selected site and remote detonate the device at a safe distance from the explosion.

6.3.1 *Type of Bomb Constructed*

The type of dirty bomb constructed can vary in sophistication depending on the quantity and type of radioactive material used and the amount of time provided to assemble the device. Furthermore, the level of the terrorist's expertise in balancing the use of explosives with the nature and quantity of radioactive material determine the severity of the blast effect and plume formation. A successfully built dirty bomb might result in very minor consequences (dispersing a few clumps of radioactive material over a fairly small area) or significant consequences (dispersing a large fraction of radioactive material as aerosols or fine particulates into the air).

Also, the time allocated for bomb construction is sensitive to the possibility of detection following material theft or black market purchase. If detected, only limited time may be provided for building the bomb. Under time constraints, the terrorists might simply use the vehicle carrying the radioactive material as the detonation device.

6.3.2 *Delivery Modes*

Terrorists are likely to select a delivery mode that has a low probability of detection by port security, yet maximizes the potential for damage to the ports. As such, the vehicle of choice is based upon what is the ideal means of dirty bomb transport to the detonation site. The ports of Los Angeles and Long Beach are accessible by land, air, and sea. A truck, car, or train might be the best mode of transport if entering the port by one of the surrounding access roads or as a package on a cargo train. With respect to arriving through the ports' waterways, a cargo ship or recreational boat most likely provide the most flexibility. Nearby helicopter landing pads and airports make planes and helicopters alternative modes of transport, although less likely because of additional security barriers associated with gaining access to their launch sites.

In addition, the vehicle selected depends on the size and weight of the dirty bomb. A bomb's dimensions vary based on the amount of conventional explosive and radioactive material used in construction. Typically, radioactive material tends to be easily packaged because it comes in either a powder or pellet form. However, the shielding material can be bulky and heavy. The bomb's surface area is altered most significantly when explosives are packaged around the radioactive material. Ultimately, the bomb can be designed to fit into something as small as a suitcase or as large as a van.

6.3.3 *Detonation Site*

To increase the effects of the dirty bomb, the detonation site is carefully selected based upon factors such as the ease with which it can be accessed, and its compatibility with the weather conditions surrounding the ports. Detonation site access is evaluated based on variables such as population density, location within or outside of the ports, and the selected mode of transport for executing the attack. Finally, weather conditions, as well as wind direction and velocity, are considered as they affect the size and directional flow of the radioactive plume. Overall, to a terrorist, the optimal detonation site causes damage resulting in lives lost and economic consequences. A location that is less visible and susceptible to suspicious behavior is critical to enhancing the probability of attack success. However, too few people in the surrounding vicinity, winds blowing out to sea, and a detonation site located miles from the harbors might deem the attack insignificant.

6.3.4 *Pruning Scenarios and Assessing Relative Likelihoods*

An important step was to determine which combinations of radioactive source × transportation mode × delivery location were implausible, not likely, or likely, given the intent to attack the ports of Los Angeles and Long Beach. We worked with a counterintelligence expert to help make these qualitative judgments. Plausibility was judged simply by considering the logical combination of the scenario factor (amount and type of bomb, delivery mode, location of detonation). For example, we considered it implausible that a terrorist group would use a small amount of radioactive material to attack the ports. If terrorists were to obtain a small quantity of radioactive materials, they probably would plan for its release within an enclosed facility or building where the dispersal effects would have a greater impact. Another example is the combination of a truck as a delivery mode and an elevated harbor location. This would require getting the truck into the harbor, waiting for the container to be placed on a ship, and exploding it in mid air. This all seems very complicated compared to simply detonating the truck either within or close to the perimeter of the harbor.

Table 6.2 shows the four transportation scenarios and the three detonations site scenarios considered for the medium radioactivity source scenario. Using qualitative judgments, we were able to narrow the 12 scenarios down to two likely ones. These two transportation/location scenarios were not significantly different in judged probability or consequences, so only one was analyzed for this scenario (a similar process was conducted for the high radioactivity scenario, though different cells were judged to be likely). Because of the sensitivity of information, the plausibility and qualitative likelihood judgments of this portion of the project are not included in Table 6.2.

Table 6.2. *Transportation and Location Scenarios*

		Transportation			
		Truck	Ship	Train	Plane/Helicopter
Location	Bridge				
	Harbor – ground				
	Harbor – elevated				

6.3.5 *Probabilities of Success*

We used Microsoft Project © to lay out the details for the selected scenarios. This software originally was created to provide businesses with a computer tool that tracks a project's progress by task, timeline, and resources. A terrorist attack operates much like any other complex business project, starting with an attack planning phase, followed by the actual preparations for the attack, and culminating with the attack execution. Microsoft Project © was used to outline planning, preparing, and execution tasks, and defined each in terms of task duration and number of resources (people) required. For example, in the medium radioactivity scenario, the project starts with tasks such as planning how and where the attack will take place, determining who will be involved in the attack scenario, and establishing a means of communication among the operatives. Next, preparations begin, which include tasks such as traveling into the United States and purchasing explosives for the dirty bomb. Ultimately, the planning and preparation tasks come together with the execution of the dirty bomb attack on the ports of Los Angeles and Long Beach.

Each task was entered into Microsoft Project © through a table format known as a Gantt chart. Tasks were inserted chronologically and described by relevant details, such as predecessor information, task duration, and resources needed. Once the Gantt chart was completed, the tasks were grouped together to form what is termed a network diagram, also known as a PERT chart. The network diagram is a graphic layout of the entire attack scenario from start to finish. Figures 6.1 and 6.2 are snapshots taken from the medium radioactivity scenario network diagram. They illustrate the steps involved for two separate tasks, building the dirty bomb and transporting the dirty bomb into the harbors. For example, Figure 6.1 shows how building the dirty bomb involves obtaining the explosive and radioactive material prior to assembling the device.

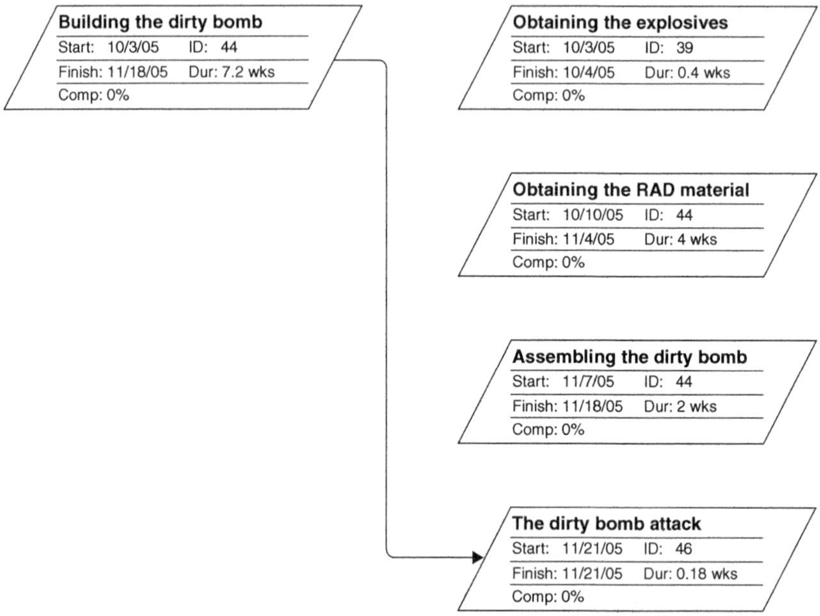

Figure 6.1. Microsoft Project © tasks – building the dirty bomb.

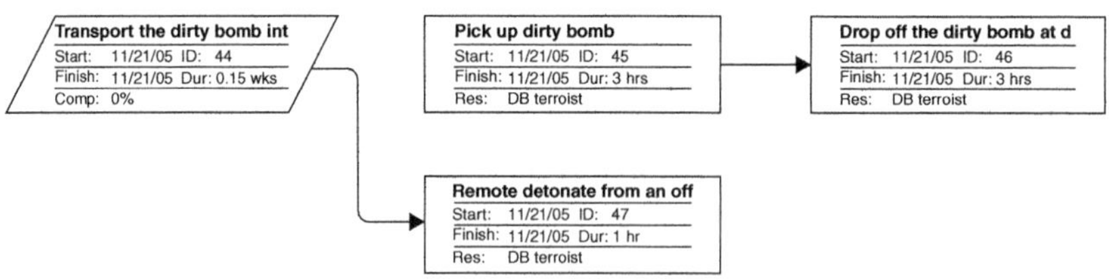

Figure 6.2. Microsoft Project © tasks – transporting the dirty bomb.

Figure 6.2 shows how detonating the dirty bomb involves transporting and dropping off the bomb at the selected site and then using a remote detonation device to generate the explosion. Figure 6.3 depicts how all the individual tasks come together to form the network diagram. For security reasons, we do not provide the details of each of the boxes, but only show the overall schematic. The upper-left parallelogram represents the start of the initial planning for the dirty bomb attack. The box on the far right signifies project completion with dirty bomb detonation.

Each of the planning, preparing, and execution tasks was associated with a probability of detection and disruption of the project. To determine how the

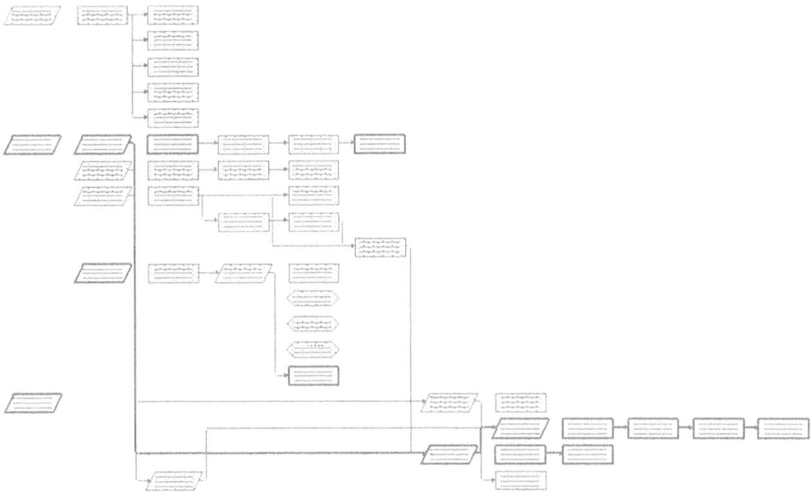

Figure 6.3. Schematic view of the complete project.

probability of detection affects overall attack success, we collaborated with a counterintelligence analyst with whom we identified the most vulnerable tasks and assigned a probability of success to each. Table 6.3 lists some of these tasks for the medium radioactivity scenario. For example, the theft of radioactive material is clearly a very vulnerable task from the perspective of the terrorists.

The probability of success for each of these tasks depends upon the complexity of the task, the number of people involved, and the time required to perform the task. Preliminary assessments of success probabilities were

Table 6.3. *Vulnerable Tasks of the Medium Scenario*

Tasks
Travel into the U.S. – the coordinator
Obtain a job at the selected facility (for stealing the radioactive material)
Steal radioactive material from research hospital
Transport radioactive material from the research hospital
Casing of the Los Angeles Port
Travel into the U.S. – attack executioners
Assemble the dirty bomb
Transport the dirty bomb into the LA Port
Dirty bomb detonation – first explosion
Second explosion

made for a given estimate of the number of people involved and task duration. A logit model was used to estimate variations in these probabilities as a function of changes in the number of people and time to task completion. We then developed probability distributions over the number of people and time for each task and used a probabilistic simulation model (@Risk by Palisades, Inc.) to simulate the uncertainty around the overall success probability of each task.

The research team, with the help of a counterintelligence analyst, used only publicly available, open-source data to make all assessments. The data represent very preliminary estimates and are largely illustrative of the methodology used. Refinements of these probability estimates would require access to classified data as well as the use of established procedures for formal elicitation of probabilities from personnel currently working in counterintelligence and counterterrorism operations.

An example of the results from the medium radioactivity scenario probabilistic simulation is shown in Figure 6.4. Interestingly, the probabilities of success are relatively small (less than 20–40%). This is because for the overall project to be successful, all individual tasks must be successful. As the uncertainty and risk affecting the success of the vulnerable tasks listed in Table 6.3 varies, this in turn affects the overall probability of project success. Of course, terrorists may engage in multiple, independent projects, thus increasing the probability that at least one of them succeeds.

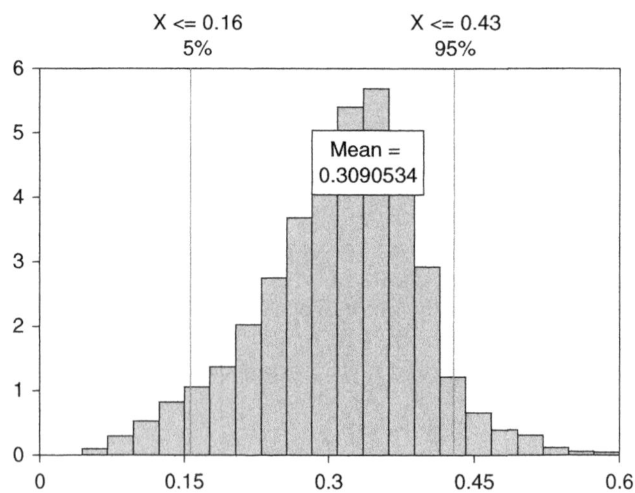

Figure 6.4. Distribution over the probability of a successful attack (medium radioactivity scenario).

6.4 Consequences

The consequences of a dirty bomb attack fall into three categories: (1) immediate fatalities and injuries due to blast effects and acute radiation exposure, (2) medium-and long-term health effects caused by airborne dispersal of radioactive material, and (3) economic impacts resulting from shutting down port operations – including evacuations, business losses, property losses, and decontamination costs. In the medium radioactivity scenario, we assumed that 5–30% of the material contained in the bomb was released into the air as aerosols or fine particulates. This results in a plume carrying roughly 500–3,000 Ci. The ranges of various damage estimates are shown in Table 6.4. Explanations for these ranges are provided below. We tried to be conservative on the upper end of the ranges, using information, model assumptions, and existing estimates that are at the high end. The low end of these ranges is usually self-explanatory, resulting from a failure of a successful dispersal of radioactive materials into the air.

Table 6.4. *Ranges of Consequence Estimates*

Consequences	Medium Scenario	High Scenario	Measure
Blast and acute radiation effects	0–10	0–10	Fatalities
Latent cancers	0–20	0–1,000	Fatalities
Port shutdown and related business losses	0–200 million	30–100 billion	Dollars
Evacuation cost (plume)	Negligible	10–100 million	Dollars
Business loss (plume)	Negligible	1–3 billion	Dollars
Property values (plume)	Negligible	100–200 million	Dollars
Decontamination cost (plume)	10–100 million	10–100 billion	Dollars

Note: The lower end of the health effect ranges include cases in our simulation where one or more of the following might occur: (1) unsuccessful airborne releases due to faulty construction of the dirty bomb; (2) wind direction toward populated areas; and (3) low radioactive doses (10 mrem or less) that produce no health effects. The higher end of the health effects were based on (1) 20% release of the source term; (2) wind direction flowing toward populated areas; and (3) NARAC estimates of doses (up to 100 mrem) and a linear dose–response function. The ranges in Table 6.4 should be considered preliminary for the purpose of an illustration of the analysis' capabilities.

6.4.1 *Blast Effects and Acute Radiation*

The immediate fatalities and injuries following the explosion of the dirty bomb depend on the amount of explosives used and the population density in the area near the detonation site. To explode a dirty bomb, only a limited amount of explosive material is needed and, therefore, the blast effects are limited to an area within 100 feet of the detonation point (National Council on Radiation Protection and Measurements, 2004). Unless the bomb is set off in a very densely populated area, the effects are likely to cause only a few fatalities and several injuries. Acute radiation sickness might occur if bystanders or emergency workers who rush to assist blast victims suffer from prolonged exposure to highly radioactive material. For example, during a 2004 dirty bomb exercise held in Long Beach, emergency workers rushed to the blast site, unaware of the radioactive material and without protective clothing. Had this been a real attack, they probably would have suffered from some level of radiation exposure, though most likely not in a range that produced acute radiation effects. Overall, the severity of radiation sickness depends on the dose and duration of exposure. For example, total body exposure of about 100 rem can result in radiation sickness, where 400 rem causes radiation sickness and death in half of the exposed individuals (Medline Plus, 2005).

6.4.2 *Health Effects Due to Airborne Releases*

The incidence of health effects following the detonation of a dirty bomb depend largely on the source and amount of radioactive material used, and the sophistication of the detonation device. If successfully detonated, a respirable fraction of the material will be released into the air that varies from about 1% to 80% of the original source (Krikorian, 2005). The remaining material will fall in clumps or larger particles within hundreds of feet of the detonation site. In addition, weather conditions, wind direction, and wind velocity exacerbate the situation as they predicate the formation of the radioactive plume.

Figure 6.5 shows the medium radioactivity scenario Gaussian plume. This plume is hypothetical and not based on specific models. However, we have obtained similar plumes from the National Atmospheric Release Advisory Center (NARAC) to verify that these examples are realistic. While the following calculations were conducted with the NARAC plumes (not included), the results would be very similar when applied to the plumes shown (see Bradley, 2005, for a summary of the plume and dose modeling capabilities of NARAC).

The plume in Figure 6.5 defines an inner ellipse with more than 1 mrem exposure per hour and an outer ellipse covering an area exposed to more than 0.1 mrem/hour. NARAC model calculations for a similar plume suggest that

Figure 6.5. Hypothetical plume due to a release in a medium radioactivity scenario.

the total four-day effective dose equivalent exceeds 1,000 mrem or 1 rem in the inner ellipse and 100 mrem in the outer ellipse. To put these numbers into perspective:

- Public background radiation exposure is about 300 mrem/year
- A single CAT scan (for medical diagnostic purposes) creates an exposure of 1.3 rem
- Worker radiation standards are set at 5 rem/year
- Radiation effects occur around 400 rem or higher.

While these numbers may not be comforting to those exposed to 100 mrem or more, it is clear that the health impacts will be relatively small.

Initial exposure to radioactivity occurs through inhalation of contaminated material as the plume passes over an area. Typical calculations assess the amount of exposure during the first four days following the event. To get a rough first-order approximation of the four-day exposure, the analysis assumed median exposure values (500 mrem) in the outer ellipse of the plume and higher exposures in the inner ellipse (2 rem) to calculate and integrate population doses. All persons located in the area covered by the radioactive plume are susceptible to radiation exposure and contamination (both internal and external). Using this crude approximation and a linear dose-response function for the population estimates provided by NARAC, results indicated there would be no more than 10 latent cancers for the medium scenario and no more than 500 latent cancers for the high scenario. All assumptions made in these calculations were conservative, so the actual latent cancers are likely to be much lower.

While Figure 6.5 identifies the area in which short-and medium-term exposure to radioactive material could occur, there also might be a significant level of ground deposition resulting in long-term exposure consequences. Radiation from deposition is usually referred to as "ground shine." The process by which deposed material is resuspended, inhaled, or gets into the food chain is complicated. Only a fraction of this radioactive material eventually is absorbed by people, thus creating the same effect as the inhalation of material transported through the plume. This process will occur continually until decontamination procedures are effective.

According to the NARAC models, the ground shine contours are similar to those shown in Figure 6.5 with the outer ellipse defining areas above 100 mrem/year and the inner ellipse defining areas exceeding 1 rem/year. To get a first-order approximation of the health effects, we assumed all the ground shine would be absorbed by people living in the plume area during the first year following the attack. This assumption is clearly conservative, since only a fraction of the ground contamination would be resuspended or get into the food chain during this time. Furthermore, we assumed that workers and the public would return to the contaminated areas and not take any particular precautions. Next, we assumed decontamination would be successful within a year following the attack and that no additional ground shine occurs thereafter. Together, these assumptions imply that the health effects due to ground shine are approximately the same as those due to the first four days of plume exposure. Both estimates are included in the health effect ranges shown in Table 6.4.

6.4.3 *Economic Consequences*

One of the major concerns about the dirty bomb threat to the ports of Los Angeles and Long Beach is the potential for an extended shutdown of the region's operations. While it is very hard to predict how long the ports would be inoperable following the medium radioactivity scenario, it is understood that large areas of the ports would be subjected to short-, medium-, or even long-term closures because of:

- Concerns of dock workers about returning to work
- Concerns of shippers about delivering goods to the harbors
- Extensive procedures related to decontamination activities

Several shut-down scenarios were analyzed, ranging from short (15 days) to medium (120 days) to long (one year). To assess the economic impacts, the Southern California Planning Model (SCPM) was used. This is a highly disaggregated regional input-output model of the southern California economy that was previously used to estimate the impacts of earthquakes and other disasters in southern California (Gordon, Richardson, & Moore, 2004).

The results are shown in Table 6.4. The 15-day shutdown has a small impact (about $300 million) because most ships would simply wait out the port closures and businesses would be supplied through other ports. The 120-day and one-year shutdowns, in contrast, have significant impacts ($63 and $252 billion, respectively) because they account for the economic impacts of a delay of delivering goods as well as all ripple effects throughout the nation's economy that such long-term delays involve. This includes costs ranging from the loss of local dock worker jobs to the reduced income and possible forced closure of nationwide businesses not receiving necessary parts or retail products.

Additional analysis focused on the costs associated with the evacuation of the plume area, reductions of property values, and business losses resulting from stigmatization of businesses in the contaminated region. We assumed that all residents and businesses would evacuate for one week from a plume with higher than 100 mrem activity (Figure 6.5). In addition, property values in the plume area were estimated to drop by 25% during the first year following the attack and then recover to previous levels (Gordon, Moore, Pan, & Richardson, 2005). Finally, we assumed business activity would be reduced by 10% for the first year following the attack and then return to former levels (Gordon et al., 2005).

The results in Table 6.4 show that the economic impacts of the evacuation are small. This occurs because the evacuees would likely continue their business as usual, albeit from shelters, homes of family or friends, or hotels. The cost of the (temporary) reduction in property values can be in the hundreds of millions, but not nearly of the same magnitude as the cost of shutting down the ports. The costs of business disruptions could be fairly large, certainly in the billions of dollars, but only if one assumes the majority of businesses relocate outside of the region or cease to exist.

In addition to the social costs inflicted upon the contaminated region, there are extensive costs associated with decontaminating surfaces with depositions of radioactive material. More specifically, the cost of decontamination depends on the required clean-up level and the cost of disposing low-level radioactive material. One study estimated extremely large costs (in the trillion dollars) even for the high radioactivity scenario plume (Reichmuth, 2005). This was based on the assumption that the clean-up standards would be those promulgated by the Environmental Protection Agency (15 mrem/year) and the cost of disposal would be similar to that imposed by the current low-level radioactive waste sites at Barnwell in North Carolina or at Envirocare in Utah. Using less stringent clean-up standards (e.g., 100 mrem/year) and disposal costs closer to those of a landfill, these cost estimates can be reduced by a factor of 1,000. Nevertheless, the clean-up costs are still in the billions (Table 6.4).

6.5 Countermeasures

Current efforts to counter the threat of a dirty bomb attack involve plans to check all cargo for radiological materials – both dirty bombs and actual nuclear devices (Martonosi, Ortiz, & Willis, 2005). For example, on June 4, 2005, Secretary Chertoff announced that the Los Angeles and Long Beach ports will be equipped with sensitive radiological detection devices in the form of portals to screen all international cargo entering the harbor (Krikorian, 2005). This is certainly a step in the right direction, as radiation portals are very effective and relatively unobtrusive measures to detect even very low levels of radiation (Board on Energy and Environmental Systems, Division of Engineering and Physical Sciences, and National Research Council, 2002). However, the following discussion shows that significant threats remain, even within the specific set of scenarios analyzed in this chapter.

In addition to radiation portals at the entry and exit points of the harbor, it would also be useful to install radiation detection devices in the outer perimeter of the harbor, especially in areas where an radiological dispersal device device could do damage. Furthermore, one of the complicated aspects of countering terrorism is that terrorists shift their attack modes in response to our defensive actions. In the case of radiological detection devices, it seems likely that terrorists would attempt to develop attack scenarios that avoid any newly installed radiation detection devices. Thus, trucks or cars having to go through screening checkpoints would be a less likely method of attack. Instead, terrorists might opt for delivery vehicles that completely bypass detection measures.

Another problem with radiological detection devices is the anticipated rate of false alarms. These devices can detect very low radioactivity levels. They have the potential to pick up radiation from many sources other than weapons-grade material or radioactive material used in dirty bombs. For example, some naturally occurring material, such as granite, give off low levels of radioactivity that might be detected. People who recently received medical procedures involving radiography also are likely to set off alarms. It is very important to define the optimal sensitivity level of the detection devices (balancing the costs of missing a threatening device against the cost of too many false alarms). Significant research exists in this area, known as "signal detection theory," that can guide the operators of these systems on how to set the optimal level of sensitivity (see Green & Swets, 1966, for a general introduction and Heasler & Woods, 2005, for a specific example).

When optimizing the sensitivity of the detection devices, the costs and benefits of false alarms, hits, misses, and correct rejections (using the signal detection terminology) have to be considered carefully together with the probability that a piece of cargo might contain a radiological device. The initial inspection at the radiation portal is a relatively efficient process. However, if the alarm is set off, the truck or container must go into a special inspection cue. Such secondary inspections create shipment delays, require significant amounts of manpower, and incur large operational costs (GAO, 2006).

In addition to highlighting ways of modifying current countermeasure efforts at the ports of Los Angeles and Long Beach, our research demonstrated how a terrorist attack can be interrupted at many stages. The project risk analysis identifies the attack tasks most susceptible (from the terrorists' point of view) to disruption and thus defines the terrorists' vulnerabilities (see Table 6.2 for an example). In the dirty bomb scenarios discussed in this chapter, the findings suggest that the most cost-effective solution is to prevent or interdict the purchase or theft of radiological material. Radioactive material in the United States is highly regulated by the NRC and thefts are difficult to carry out successfully. In our attack scenario involving theft from a research or industrial facility, we hypothesized that an employee would assist in attempting to bypass NRC barriers. As such, one implication of focusing on this phase of the attack would be the benefit associated with improving security of the facility, particularly management of employees with access to radioactive sources. Similarly, in the scenario involving theft or purchase of significant material in the former Soviet Union and other foreign countries, we recognize the importance of improving safeguards and security at these facilities.

6.6 Conclusions

A terrorist attack using a dirty bomb in the United States is possible, perhaps even moderately likely, but would not kill many people. Instead, such an attack primarily would result in economic and psychological consequences. Moreover, it would not be easy to carry out a dirty bomb attack. Considering the difficulties associated with obtaining and transporting radioactive material, building the dirty bomb, and detonating the device successfully, our preliminary analyses suggest that the chances of a successful attempt are no better than 15–40% for the medium radioactivity scenario, and less likely for the high radioactivity scenario. Of course, multiple independent attempts would increase these chances. While our probability estimates are mostly illustrative, the chances of terrorists succeeding with an attack that involves relatively low-level radioactive material from a U.S. facility are larger than their chances of succeeding with the import of a large quantity of foreign sources. This is because transporting foreign source material through a number of international ports increases susceptibility to detection.

If a dirty bomb attack is successful, the consequences depend primarily on the amount of radioactive material in the detonated source term, the amount released into the air, weather conditions, and the population density in the impacted region. The medium radioactivity scenario analyzed in detail suggests there would be some, but fairly limited, health effects and possibly significant economic impacts.

The most costly economic impact would result from a lengthy shutdown of the ports and decontamination efforts. The length of the harbor shutdown would in part depend on the decision to declare access to the ports as safe. In a national emergency, standards of safety different from those promulgated by the EPA may be appropriate. For example, worker safety standards may be more appropriate than public safety standards. The same also holds true for clean-up standards. Because we don't know how policymakers and harbor workers will react in such an emergency, we have parameterized the length of the harbor shutdown, from 15 days to one year, corresponding to roughly \$130 million to \$100 billion in costs.

The economic consequences of evacuations, property value impacts, and business losses due to stigmatization in the plume area are in the billions, but not in the tens or hundreds of billions. People and the economy are likely to respond in a resilient way. Many people would relocate for some time out of the areas with relatively high levels of radioactivity (100 mrem or more), but they would not stop working. Also, businesses may relocate and later return to their original location. Similarly, effects on property values may be severe in the short term but, like in many other disasters, return back to normal in a year or so.

Regarding countermeasures, our analysis clearly supports ongoing programs to install radiation detection technology around the harbor. In addition, the analysis raises concerns regarding the security risks associated with cargo material as it is offloaded from ships but not yet transported through the portals, incoming containers from the U.S. mainland (by truck, small boat, or air), and harbor perimeter control. Finally, the analysis suggests preventing terrorism by interdicting vulnerable activities during the planning and preparing stages of an attack scenario. Such action might include being more proactive in controlling and protecting the original sources of radioactive material.

Acknowledgments

This research was supported by the U.S. Department of Homeland Security through the National Center for Risk and Economic Analysis of Terrorist Events (CREATE) under Grants EMW-2004-GR0112 (FEMA) and N0014-05-0630 (ONR). However, any opinions, findings, and conclusions or recommendations in this document are those of the authors and do not necessarily reflect views of the U.S. Department of Homeland Security. We would like to thank Richard John for major contributions to the probabilistic modeling. We would also like to thank Elsa Lee for major contributions to developing scenarios and estimating probabilities. The staff of the National Atmospheric Release Advisory Center (NARAC) at the Lawrence Livermore National Laboratory provided us with example plumes and exposure estimates. The CREATE economics team headed by Peter Gordon, Jim Moore, and Harry Richardson provided the economic impact estimates. We also benefited from many discussions with teams at national laboratories working on dirty bomb issues, including Battelle Pacific Northwest Laboratories, Sandia National Laboratory, and Lawrence Livermore National Laboratory.

REFERENCES

Blair, B. (2002). *What if terrorists go nuclear?* Available at www.cdi.org/terrorism/nuclear-pr.cfm.

Board on Energy and Environmental Systems, Division of Engineering and Physical Sciences, and National Research Council. (2002). *Committee on Alternatives for Controlling the Release of Solid Materials from Nuclear Regulatory Commissioned-Licensed Facilities.* Washington, DC: National Academy Press.

Bradley, M. (2005). NARAC: *An emergency response resource for predicting the atmospheric dispersion and assessing the consequences of airborne radionuclides.* Available at www.llnl.gov/tid/idf/documents/pdf/324174.pdf.

Council on Foreign Relations. (2004). *Terrorism Q&A: Making a bomb.* Available at http://cfrterrorism.org/weapons/makingprint.html.

Ferguson, C. D., Kazi, T., & Perera, J. (2003). *Commercial radioactive sources: Surveying the security risks.* Monterey Institute of International Studies, Center for Nonproliferation Studies.

Gardner, F. (2003). Al Qaeda "was making a dirty bomb." *BBC News Online* on the web January 31, 2003. Available at http://news.bbc.co.uk/2/hi/uknews/2711645.stm.

General Accounting Office (GAO). (2003). *Nuclear security: Federal and state action needed to improve security of sealed radioactive sources*. Available at www.gao.gov/new.items/d03804.pdf.

Gordon, P., Moore, J. E., Pan, Q., & Richardson, H. (2005). Los Angeles–Long Beach seaports radioactive plume economic impact. Draft Report, Center for Risk and Economic Analysis of Terrorist Events, University of Southern California, 2005.

Gordon, P., Richardson, H., & Moore, J. (2004). The economic impacts of a terrorist attack on the twin ports of Los Angeles-Long Beach. Presentation at the CREATE Economics Symposium: Economic Costs and Consequences of a Terrorist Attack. Los Angeles, University of Southern California, August 2004.

Green, D. M., & Swets, J. A. (1966). *Signal detection theory and psychophysics*. New York: John Wiley and Sons Inc.

Harper, F. (2005, April 26). Presentation at the Department of Homeland Security Conference on Research and Development, Boston. Presentation at the Department of Homeland Security Conference "Working Together: R&D Partnerships in Homeland Security," Boston, April 27–28, 2005.

Heasler, P., & Woods, T. (2005). Game theoretic modeling of detection and deterrence. Presentation at the Department of Homeland Security Conference on Research and Development, April 26, 2005.

International Atomic Energy Agency. (2002). *Inadequate control of world's radioactive sources*. Available at www.iaea.org/search97cgi/s97 cgi?action=View&VdkVgw Key=http%3A%2F%2Fwww%2Eiaea%2Eorg%2FN ews Center%2FPressReleases%2F2002%2Fprn0209%2Eshtml &QueryZip=dirty+ bomb&&viewTemplate=cvw smpl.hts &collection=PressR.

Johnson, K., & Locy, T. (2002). Threat of dirty bomb softened. *USA Today* on the web June 12, 2002. Available at www.usatoday.com/news/nation/2002/06/11/ bushothers.htm.

King, G. (2004). *Dirty bomb: Weapons of mass disruption* (p. 25). New York: Penguin Group.

Krikorian, G. (2005, June 4). Port complex to get radiation detection devices to forestall terrorism. *Los Angeles Times*. Available at www.latimes.com/news/local/la-mechertoff4jun04,1,7330940.story?ctrack=1&cset=tru.

Martonosi, S. E., Ortiz, D. S., & Willis, H. (2005). *Evaluating the viability of 100 per cent container inspection at America's ports*. RAND Corporation.

Medline Plus (2005). *Medical Encyclopedia Radiation Sickness*. Accessed February 19, 2005. Available at www.nlm.nih.gov/medlineplus/ency/article/ 000026.htm.

National Council on Radiation Protection and Measurements (NCRP). (2004). *Management of terrorist events involving radiological material.* National Academy Report, NCRP Report No. 138, NCRP, Bethesda, MD, October 24, 2001.

The Port of Long Beach. (2007). *About the port – port statistics index*. Viewed on January 2007. Available at www.polb.com/html/1 about/portStatsMain.html.

The Port of Los Angeles. (2007). *About the port – FAQs*. Viewed on January 2007. Available at www.portoflosangeles.org/aboutfaq.htm.

Reichmuth, B. (2005). Economic consequences of a rad/nuc attack: Cleanup standards significantly affect cost. Presentation at the Department of Homeland Security Conference "Working Together: R&D Partnerships in Homeland Security," Boston, April 27–28, 2005.

U.S. Government Accountability Office. (2006). *Combat nuclear smuggling: DHS has made progress deploying radiation detection equipment at U.S. ports-of-entry, but concerns remain.* GAO Report No: GAO-06-389. Washington, DC.

7

Regional Transportation and Supply Chain Modeling for Large-Scale Emergencies

Maged Dessouky, Fernando Ordóñez, Zhihong Shen, and Hongzhong Jia

7.1 Introduction

Large-scale emergencies (or major emergencies) are defined as any event or occurrence that overwhelms local emergency responders, that severely impacts the operation of normal life, and that has the potential to cause substantial casualties and property damage. Examples are natural disasters (earthquake, hurricane, flooding, etc.) and terrorist attacks, like the one that took place on September 11, 2001. To manage the risks and consequences of large-scale emergencies and to mitigate their impacts to the population, wide-scale distribution plans of medical supplies must be developed.

Careful and systematic preplanning, as well as efficient and professional execution in responding to a large-scale emergency, can save many lives. Rational policies and procedures applied to emergency response could maximize the effectiveness of the scarce resources available in relation to the overwhelming demands. A key ingredient in an effective response to an emergency is the prompt availability of necessary supplies at emergency sites. It is challenging for practitioners to manage the huge volume of medical supplies to guarantee the readiness of the delivery and the freshness of the stock, as well as to deliver these massive supplies in a short time period to dispersed demand areas. Operations research models can play an important role in addressing and optimizing the logistical problems in this complex distribution process. Larson and colleagues (2005, 2006) conducted a detailed analysis based on well-known and recent large-scale emergencies. They emphasized the need for quantitative, model-oriented methods provided in the operations research field to evaluate and guide the operational strategies and actions in response to major emergencies.

In certain emergency settings, medication or antidotes must be applied within a specified time limit from the occurrence of the event to maximize their

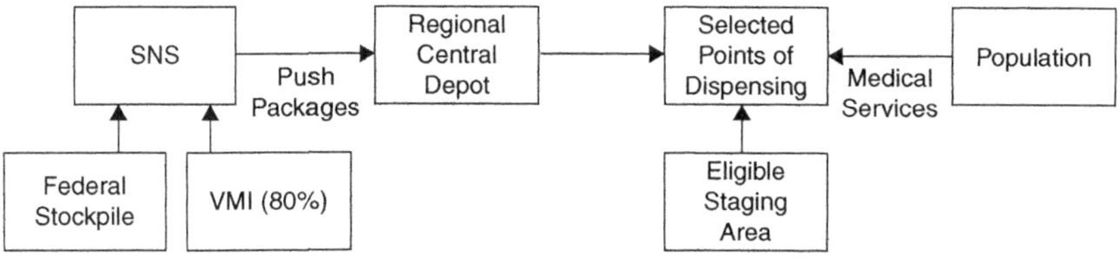

Figure 7.1. Distribution process of large-scale emergency response.

effectiveness and, in some situations, to save lives. Traditional pharmaceutical supply chains are not adequate to push a huge volume of medical supplies to the affected population within a short time frame. Usually the local caches are only enough to serve the first responders. The majority of the medical supplies used would be from the Strategic National Stockpile (SNS). The Strategic National Stockpile is part of the federal response to a bioterrorist attack on the United States and it is composed of a combination of vaccines, prophylaxis, and medical supplies aimed at minimizing the damage of an attack. This stockpile is separated into prepositioned push packages and a larger inventory managed by manufacturers (known as Vendor-Managed Inventories – or VMI – that correspond to about 80% of the SNS). The VMI provide additional flexibility and cost efficiency as they can deploy the specific medicine to combat a bio threat. Given the possible biological and chemical threats facing the nation and the limited budget to set up the SNS, the effective allocation of these resources in designing the SNS is key in managing the risk posed by these threats.

The SNS is administrated by the Centers for Disease Control and Prevention in the United States. For example, to address emergencies of infectious diseases, the SNS contains 300 million doses of smallpox vaccines and enough antibiotic to treat 10 million people for anthrax during a large-scale bioterror attack. We illustrate the distribution process in Figure 7.1. If there is a public health emergency severe enough to cause local supplies to run out, the SNS supplies can be accessed. Initial medical supplies from the push packages will be delivered to any state in the United States within 12 hours of federal and local authorization. This initial delivery can be followed by supplies from the VMI if needed. After the stockpile reaches the regional central depot, it is each state's responsibility to have plans to receive and distribute SNS medicine and medical supplies to local communities as quickly as possible. The state investigates the efficient strategies and best practices in identifying potential points of dispensing (PODs) sites where the population in the nearby neighborhood comes to seek medical services, as well as in constructing suitable routes to distribute the supplies from the regional central depot to the selected PODs.

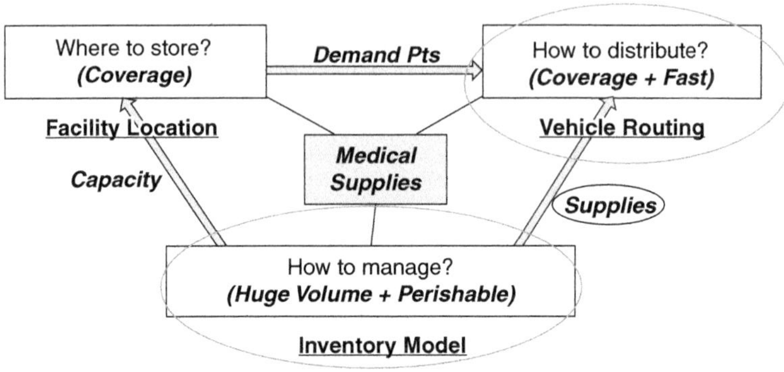

Figure 7.2. Problem decomposition for emergency response.

As the primary goal is to establish an effective and rapid distribution system for the medical supplies to respond to a large-scale emergency, it is natural to decompose the whole problem into three separate but interconnected key decisions, which can be integrated to provide a complete medical supply chain management solution to large-scale emergencies (see Figure 7.2). The three interconnected key decisions in the distribution process are:

- How to manage these medical supply inventories?
- Where to store them?
- How to distribute them?

A perishable inventory management model (Section 7.2), a facility location model (Section 7.3), and a routing model (Section 7.4) are formulated to address these three questions separately to respond to the challenges presented earlier. The models and solutions presented in this chapter are based on our previous work (Jia et al., 2007a, 2007b; Shen et al., 2009a, 2009b, 2011), and we refer the reader to these articles for in-depth explanations and additional results. The solutions to each problem can be coordinated into an overall efficient management strategy for the distribution process in the emergency response context. The minimum amount that must be kept in the inventory system is enforced by the SNS policy, which is the key constraint in the inventory model and also provides the capacity constraint to the location model as well as the total supply constraint to the routing model. The results of the location model are the selected PODs, which will feed into the routing model as the demand points to service. The goal is to establish efficient inventory management policies and to dispatch plans to help save lives through a timely delivery plan.

In a large-scale emergency setting, the initial prepositioned supplies will typically not be sufficient to cover all demands in the recommended times. The unmet demand in an emergency situation can directly result in loss of life, an impact that outweighs any operational costs. Another particular feature of large-scale emergencies is that the massive inventory in stock (medical supplies) is made up of perishable items. Therefore, the overriding objective in our work is to mitigate the impact of an emergency by minimizing unmet demand. It requires both an efficient inventory management scheme for the massive supplies to guarantee the readiness and freshness of the stockpile for continuous service and an effective plan to distribute the huge quantity of supplies as much as possible within the preferred timeline. We use a mix of optimization solvers and heuristics to solve the three models presented in this chapter. In this chapter, for the inventory management model, we developed a search procedure to find the optimal production policy. For the facility location model, we developed a standard locate-allocate heuristic to solve the problem while the routing model is optimally solved using the CPLEX commercial solver.

7.2 Inventory Management Problem

The readiness and freshness of the continuous massive supply from the VMI, managed by private vendors, is crucial to have the ability to satisfy the sudden demand that might arise in a large-scale emergency. It is imperative to enable massive stockpiles of medical supplies in an economical manner that would not put undue stress on vendors, who would otherwise require a higher compensation to participate. Consider, for instance, the requirement of the stockpile of Cipro, a common antibiotic used for anthrax infections and other illnesses. The federal government aims to disperse antibiotics to treat 10 million people in the case of a massive anthrax attack. This stockpile represents enough Cipro to meet regular market demand for several years. Current SNS policy allows firms to partially recoup costs by selling the stockpile in the private market when supplies are six months from expiration.

Since the size of the stockpile is large compared with the expected market demand, which leads to waste, an opportunity exists to implement a combined inventory policy that addresses both the regular market demand and the SNS minimum inventory in an efficient manner. From the manufacture's perspective, a more sophisticated inventory holding policy that could enable using the stockpile to meet the regular market demand and production to maintain the stockpile inventory has the potential to reduce costs and thus reduce the price charged to the government. In this section, we explore such a combined inventory model and estimate its economic impact. For a general review of inventory models for emergency management, the reader is referred to the paper by Ozguven and Ozbay (2014).

7.2.1 *Inventory Model*

We begin with a brief summary of the modified EMQ model proposed by Shen and colleagues (2011), which is used to obtain the optimal ordering policy for a perishable inventory system with a large minimum volume constraint. We assume a single fixed-life perishable item is produced, consumed, and stored for an infinite continuous time horizon. There are two types of demand. One is a known regular market demand with a constant rate of D items per period. The production can start at any time at a constant maximum allowable rate of P, which is greater than D. The other demand is that the inventory system must maintain at every point in time at least I_{min} of non-spoiled inventory.

Excess supplies could be disposed to some secondary market (e.g., overseas market) as long as it is before expiration of the item. The model assumes that inventory is consumed using the natural first-in-first-out (FIFO) policy and prices for both the primary and secondary markets are uniform. In this inventory model, the inventory cycle T_{inv} is defined as the minimum length of time that an inventory pattern repeats and it is a given parameter. It is assumed that excess supplies above the I_{min} level will be disposed (salvaged) once at the end of each T_{inv} so that exactly I_{min} items are present at the beginning of the next inventory cycle. To maintain a fresh inventory, every T_{inv}, the I_{min} amount produced in the previous inventory cycle is consumed and another I_{min} amount is produced. The production batch size Q is the sole decision variable and we can determine all the relevant quantities to describe this model as a function of Q. For any given Q, we initially run a regular EMQ cycle (underlying regular EMQ cycle with cycle length T = QD), and we may need to make some adjustment near the end of the inventory cycle to satisfy the minimum inventory requirement (last production cycle).

Given different possible combinations of T_{inv} and I_{min}, there is a total of five different possible scenarios. Figure 7.3 provides illustrative inventory plots for all five cases. The classification is based on three criteria: 1) if a last production cycle is needed; 2) where the inventory cycle ends relative to a regular underlying EMQ cycle; 3) where the last production cycle starts. Detailed explanation can be found in Shen and colleagues (2011). The total cost of the modified EMQ system is a non-continuous, non-differentiable function of Q.

Let TRC(Q) be the total cost function. Shen and colleagues (2011) show that TRC(Q) for each case among the five cases is a quadratic function of the order quantity Q as well as boundary conditions for transitions between the different cases. To illustrate we present TRC(Q) for case 1. The total costs within

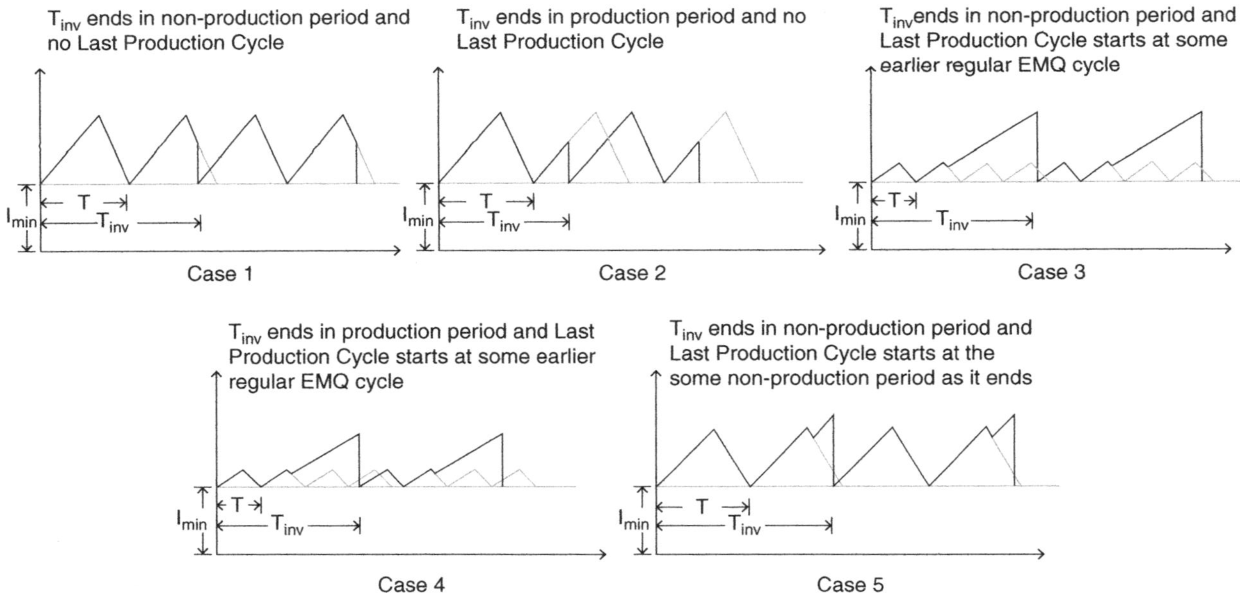

Figure 7.3. Five cases for combined inventory policy: perishable inventory, I_{min} minimal inventory requirement, T_{inv} inventory cycle.

a single inventory cycle are composed of three parts: inventory holding cost h, fixed setup cost A, and unit purchasing cost v. After some simplifying steps, Shen and colleagues (2011) show that

$$TRC(Q) = A(N+1) + \left(\frac{Q}{D} - T_{inv} + N\frac{Q}{D}\right)Dv + \frac{1}{2}h\left[(N+1)\frac{Q^2}{D}\left(1-\frac{D}{P}\right) - D\left(\frac{Q}{D} - T_{inv} + N\frac{Q}{D}\right)^2\right]$$

where N is the number of complete regular production cycles in one T_{inv} cycle. For a given segment, N is fixed so TRC(Q) is a quadratic function of Q for case 1. By proving that when N is greater than a threshold value the total cost will monotonically increase as N increases, the exact optimal Q can be obtained with pseudo-polynomial complexity. Shen and colleagues (2011) show that this model only holds when $I_{min} \le T_s D$ and $T_{inv} \le T/2$, where T_s is the shelf life of the inventory.

7.2.2 *Example*

Computational experiments have compared the cost of maintaining a massive I_{min} inventory in an integrated system to the cost of separate inventories for regular demand and for the SNS stockpile. Given the following problem parameters

Parameter	Unit	Estimation
I_{min}	million pills	1200 (= 10 × 60 × 2)
T_{inv}	month	36
Y	month	36 (= 108–2 × T_{inv})
P_{gov}	mil \$/mil pill	0.95
P_{market}	mil \$/mil pill	4.67
v	mil \$/mil pill	0.2
w	mil \$/mil pill	–0.3 (3 years)/ –0.075(6 years)
D	mil pill/year	300
P	mil pill/year	600
A	mil \$/time	2 (= 100 × h)
h	mil \$/mil pill/year	0.02

we obtain that the integrated system has a total cost of \$78 million per year, which is less than the \$111.45 million per year obtained for a standard model. This standard model splits the production capacity of 600 million pills per year into two parts: an I_{min} inventory renewed every T_{inv} and a separate inventory for the regular demand. We use 400 million pills per year for the regular market

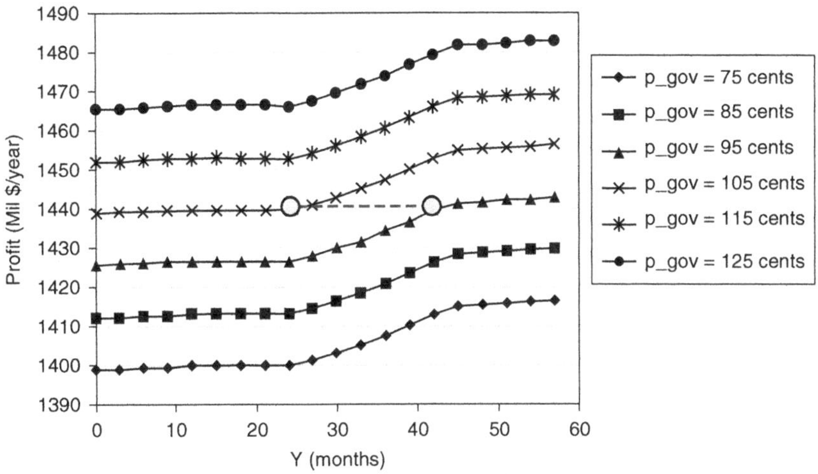

Figure 7.4. Optimal vendor profit for combined inventory as function of remaining time of salvaged inventory (Y).

demand with the regular EMQ model and the remaining 200 million pills per year constantly running to build the 1.2 billion pills I_{min}, which is completely refreshed every six years. The disposed I_{min} inventory is sold at 7.5 cents per pill on the secondary market.

Shen and colleagues (2011) study the sensitivity of the optimal solution with respect to various parameters. For instance, we observe that the industry profits per year increase with the price per pill and also with the amount of time remaining in salvaged inventory (increasing Y).

The different slopes observed on the profit versus. flexibility (Y) plot in Figure 7.4 have to do with the different impacts of increasing flexibility (Y) the amount of time remaining in the salvaged inventory. The flat slopes are due to cost savings in the regular EMQ part of the inventory for changes in Y, while the sharp profit improvement is due to the additional gain from the extra salvage amount.

7.3 Storage Problem

In the event of a large-scale emergency such as a terrorist attack (e.g., September 11) or a major natural disaster (e.g., hurricane), tremendous demands for medical supplies will occur at the incident site(s) in a short time period. Emergency preparedness requirements for such events need pharmaceutical caches to be prepositioned or staging areas to be determined so as to enable a rapid disbursement of supplies from the national stockpiles (CDC, 2005). An important consideration in selecting the locations of facilities (i.e., the pharmaceutical caches or staging areas) is the coverage of the demand areas. A sufficient coverage of the demand areas by the facilities can ensure prompt medical services to the population, hence minimizing the loss of life caused by the emergencies.

Most facility location models, including the ones in a context of emergency services, consider providing a single facility (possibly including a few backup facilities) to cover a demand point (Church & ReVelle, 1974; Hogan & ReVelle, 1986; Huang et al., 2010; Marianov & ReVelle, 1996; Paluzzi, 2004; Schilling et al., 1979). Such emergency models, however, typically have not considered conditions where a tremendous demand and low frequency combine to create situations with insufficient supplies and large uncertain demands. These conditions, which may occur in large-scale emergencies, require a modification in the definition of facility coverage to allow for redundant facility placements and tiered facility services to ensure an acceptable form of coverage of all demand areas. Given the occurrence of a large-scale emergency, the resources of a number of facilities need to be applied to quell the impact of the emergency. This implies that, to service a demand point, a number (multiple quantity) of facilities should be used, which can be classified in terms of the distance/time to the demand point. A facility that is close to a demand point provides a better quality of coverage to that demand point than a facility located far from that demand point (Dessouky et al., 2006).

Jia and colleagues (2007a, 2007b) introduced a new facility location model for the medical supply distribution for large-scale emergencies. This model aims to maximize the population coverage, and addresses the uncertainty of demand and lack of supplies at each facility by requiring each demand point to receive a multiple facility quantity and quality of coverage. Specifically, the multiple facility quantity of coverage means that each demand point is covered by a multiple quantity of facilities that can be decided based on the attributes of the demand point such as the population density and the likelihood to be impacted by an emergency. The multiple facility quality of coverage means that the demand points are to be serviced by the facilities at different and tiered distance levels. We note that facility location problems for routine demand services with uncertainties have been previously investigated; for example, see Chen and Lin (1998) and Snyder and Daskin (2006). For a complete review of facility location problems with uncertainties, see Snyder (2006).

7.3.1 *Facility Location Model*

The facility deployment problem considers a given geographical territory and assumes that the requests for medical services are concentrated among various (finite) demand areas within the territory. To properly allocate the medical resources, the eligible facility sites need to be identified and the potential demand areas need to be categorized. Each demand area has distinct attributes such as population density, economic importance, geographical feature, weather pattern, etc. Therefore, the likelihood for a certain type of large-scale emergency to occur in one area, as well as the corresponding impact level, are different from the other areas. As such, the quantity of facilities that needs to be allocated to a demand area should consider the attributes of the demand point and hence may be different from the number of facilities allocated to other areas. The selection of eligible facility sites must also consider a set of criteria that are suited for large-scale emergencies. For instance, the facilities should have easy access to more than one major road/highway. The sites should be secured and insusceptible from damages caused by the emergencies.

To formulate the covering facility location model for large-scale emergencies, we consider a set I of demand points and a set J of eligible facility sites. Indexed on these sets the following types of integer variables are defined, requiring each demand point to be serviced at quality levels $r \in \{1,...q\}$. Readers interested in details of the model are referred to Dessouky and colleagues (2006) and Jia and colleagues (2007a, 2007b).

$x_j = \begin{cases} 1 \\ 0 \end{cases}$ if a facility is placed at eligible site j;
otherwise.

$u_i^r = \begin{cases} 1 \\ 0 \end{cases}$ if demand point i is covered at quality level r;
otherwise.

Furthermore, the following input parameters are defined:

Inputs:

M_i = the population of demand point i;
Q_i^r = the minimum number of facilities that must be allocated to demand point i so that i can be considered as covered at quality level r;
P = the maximal number of facilities that can be placed in J;
d_{ij} = the distance from eligible facility site j to demand point i;
D_i^r = the distance requirement for demand point i to be serviced at quality level r;
$a_{ij}^r = \begin{cases} 1 \\ 0 \end{cases}$ if eligible facility site j can cover demand point i at quality level r, i.e. $d_{ij} \leq D_i^r$
otherwise
c^r = the importance weighting factor of the facilities that have quality level r.

In what follows, we present a maximal covering model that is suitable to the location problems of large-scale emergencies. In this model, each demand point is considered covered only if it can be serviced by a multiple quantity of facilities that are located at different quality levels (distances). The model has an objective of maximizing the demands covered by sufficient quantity of facilities at different quality levels.

$$v(\text{MCLP}) = \max \sum_r c^r \left(\sum_i M_i u_i^r \right) = \sum_r \sum_i c^r M_i u_i^r \tag{7.1}$$

Subject to:

$$\sum_{j \in J} x_j \leq P \tag{7.2}$$

$$x_j = \{0,1\} \qquad \forall j \in J \tag{7.3}$$

$$\sum_{j \in J} a_{ij}^r x_j \geq Q_i^r u_i^r \qquad \forall i \in I, r = 1,...q \tag{7.4}$$

$$u_i^r = \{0,1\} \qquad \forall i \in I, r = 1,...q. \tag{7.5}$$

In the objective function, the weight parameter, c^r, is used to prioritize the importance of the facilities at each different quality level. For any demand point, in general, a high-quality-level facility coverage is more crucial than a low-quality-level facility coverage. Therefore, we define $c^1 \geq c^2 \geq \cdots \geq c^q$. Constraints (7.2) and (7.3) are used to represent that there are at most P facilities to be located in a set J of possible locations. Constraint (7.4) states that demand point i is considered as covered at quality level r only if there are more than a required quantity (Q_i^r) of facilities located within the corresponding distance constraint servicing it. Note that the number of quality levels (q) and the quantity of the facilities at each quality level (Q_i^r) for each demand point can be determined by considering the population, the importance, and the emergency occurrence likelihood at each demand point. In general, a demand point that is of a greater economic/political importance, has more population, and is more likely to receive an emergency should be covered by more facilities located at different quality levels. Also note that we consider that a facility of quality r is also of quality $r + 1,...,q$ since it is closer than required for larger quality of coverage. Thus we must satisfy the requirements: $Q_i^r \leq Q_i^{r+1}$ for all qualities $r = 1,...,q-1$. A facility that services a demand point at a high quality level (e.g., the first quality level) is also considered able to service this demand point at lower quality levels. Finally, constraint (7.5) enforces the integrality of variables u_i^r. This problem is similar to the maximal covering location problem (MCLP) considered by Church and ReVelle (1974), Meggido and colleagues (1983), and Galvão and ReVelle (1996), but with different conditions to consider a point covered. Note that any solution x that locates P facilities is a feasible solution to the problem; simply set all $u_i^r = 0$ to satisfy constraint (7.4). Variables u are auxiliary variables that quantify the number of points that are covered for any location x of up to P facilities.

For different large-scale emergencies, the demand for medical supplies at any point i may vary, and it depends on various factors such as the impact of the emergency and the likelihood for the emergency to affect a demand point. In the model, we use the population parameter, M_i, to represent the demand for medical supplies at point i.

7.3.2 *Illustrative Example*

The emergency instance we consider is an anthrax emergency in Los Angeles County, which has a total population of 9.5 million. Anthrax is an acute infectious disease caused by the spore-forming bacterium. There are several types of anthrax infections (cutaneous, inhalation, and gastrointestinal), and each requires vaccines and antibiotics to cure infected people and immunize the high-risk population. During an anthrax emergency, the Strategic National Stockpile (SNS) is mobilized for local emergency medical services. The problem is then to select a given number of eligible facility sites as staging areas for mass distribution of the medical supplies from the SNS to cover as much of the population as possible.

To formulate the facility location problem, we first identify the demand points. We use the population density pattern that is available for Los Angeles County (ESRI, 2005), and consider the centroid of each census tract as a demand point to represent the aggregated population in this tract. Thus we obtain 2,054 discrete demand points that have different population densities. In addition, we select the set of eligible facility sites by deviating the facility sites that were identified by the emergency management of Los Angeles County by a small distance. The county has identified a little more than 200 eligible facility sites. In this chapter, we use 200 as the number of eligible facility sites.

To ensure adequate facility service, we consider multiple quantity coverage requirements for the demand points. The quantity requirements for a single quality level are dependent on the population at each demand point and they are defined as follows:

(1) $Q_i = 1$, if the population of demand point i is less than 3,000;
(2) $Q_i = 2$, if the population of demand point i is between 3,000 and 6,000;
(3) $Q_i = 3$, if the population of demand point i is between 6,000 and 9,000;
(4) $Q_i = 4$; if the population of demand point i is greater than 9,000.

Furthermore, we use double-quality coverage to provide medical services for each demand point. The distance requirements for the first and second quality levels are defined respectively as 8 km and 16 km. We consider the quantity of facilities required to service a demand point at the second quality level double the quantity of facilities required at the first quality level. For example, if a demand point has a population of 5,392, then two facilities will be required at quality level 1 (within 8 km of the demand point), and four facilities will be required at quality level 2 (within 16 km of the demand point). In addition, for simplicity, we define the two different quality level coverages as having the same importance, i.e., $c^1 = c^2$.

We consider the location problem of incrementally opening the complete set of the eligible facility sites in Los Angeles County. In a large-scale emergency, it will most likely be impractical to open all the sites at the same time due to the limited availability of the emergency responders and medical supplies (i.e., they do not arrive all at one time). Thus, it is likely that the sites will have to be opened sequentially. In this set of experiments, we gradually increase the set of facilities by 20 each time. Thus the numbers of opened facilities after each time interval are: 20, 40, 60, 80, 100, 120, 140, 160, 180, 200, and so on. Furthermore, we consider that once a facility is opened, it cannot be closed.

We use the LocAlloc heuristic (see Jia et al., 2007a) to solve the series of location problems. To evaluate the solutions, we also solve these problems by uniformly selecting facilities throughout Los Angeles County, which is a plausible solution strategy during an emergency. We refer to this solution as the uniform solution. Figure 7.5 depicts the coverage results obtained from the heuristic solutions and the uniform solutions, including the primary (first facility) coverage, first quality level coverage, and both first and second quality level coverage for the sequential location problems.

From these results, it can be seen that the heuristic is able to generate much better location solutions than the uniform selection approach for all the location problems (except for the last one, in which the results are equal). Based on the solutions obtained from the heuristic, a significantly larger percentage of the population can receive the primary (first facility) coverage, first quality level coverage, and both first and second quality level coverage.

In addition, it is shown from the figure that most of the population is covered by the first 100 opened facilities and the later opened facilities provide only limited marginal coverage. This implies that during the emergency, the resources for the first 100 facilities should be deployed as early as possible so that the impact of the emergency can be mitigated and the loss of life can be minimized.

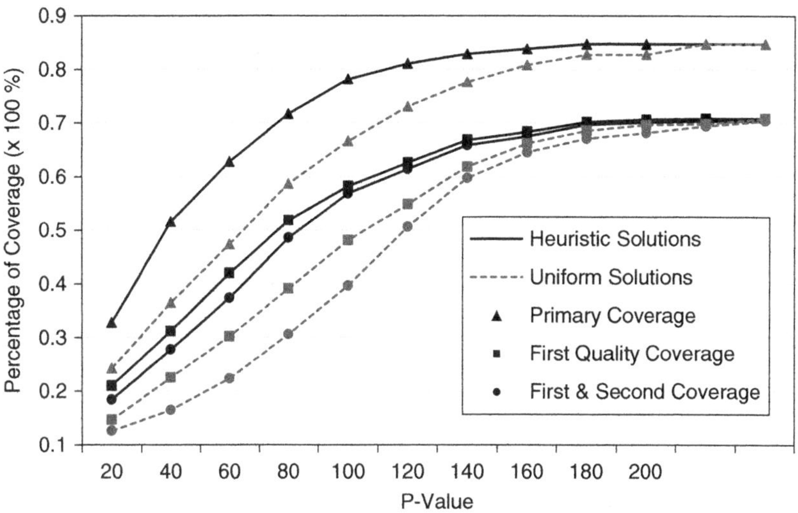

Figure 7.5. The coverage results with sequentially available resources.

7.4 Distribution Problem

The objective of the routing problem we consider in a large-scale emergency situation is to minimize the unsatisfied demand that can occur because resources and first responders are overwhelmed. In fact, unmet demand in an emergency situation can result in loss of life, an impact that outweighs other commonly used vehicle routing problem (VRP) objectives such as the travel time or number of vehicles used. We note some researchers have combined the location and routing problems (see, e.g., Oran et al., 2012). In this chapter, we treat these two problems separately since the combined problem becomes computationally tractable when considering all the uncertainties associated with a large-scale emergency.

Unmet demand in these emergency situations can be due to having insufficient resources or delays in reaching the affected population. In particular, the administration of medication to the population within a time frame makes an appreciable health difference in a large-scale emergency. It is therefore important to associate each demand with a time window, to address situations where late delivery directly leads to loss of life.

The unpredictable nature of large-scale emergencies leads to many stochastic elements in the problem, including stochastic demand and travel times. First, at a given demand point (e.g., a neighborhood block), the quantity of required emergency supplies (antidotes, protective equipment, medication, etc.) is often proportional to the unknown size of the population and/or numbers of casualties.

Second, the casualties' exposures, or demands among "worried well" are hard to accurately predict. In addition, in an emergency situation, the traffic condition can become highly uncertain, due to unpredictability in people's behavior. At the same time, the emergency event itself may directly affect road conditions. For example, an earthquake may destroy a segment of the freeway.

When we consider the problem of routing inventory in response to large-scale emergencies, both routes and delivery quantities are important in creating an effective dispatching plan. These two aspects of the solution are related to the following uses of efficient inventory routing plans for a large-scale emergency: First, from a planning perspective, agencies can use efficient routes in mock trial runs of an emergency event to predesign the routing policies and provide training opportunities. Second, from an operational perspective, the models can be used to update the predesigned routes and promptly determine exact delivery quantities to dispensing or treatment sites to generate real-time, near-optimal dispatching solutions after an emergency situation has happened. In this chapter, we focus on the planning aspect of the routing problem, which appeared in Shen and colleagues (2009a). The recourse operational aspect appears in Shen and colleagues (2009b).

7.4.1 *Vehicle Routing Model*

As mentioned earlier, in the planning aspect, we aim to minimize the unmet demand, taking into account the uncertainty in demand and travel time. Such a problem is related to the routing problem with profits, since minimizing the unmet demand maximizes the collected profits when the profit of each node is its demand level. The route generation is also constrained by deadlines associated with every demand node. However, in our problem, the unmet demand is not only caused by the deadlines, but also potentially by the vehicle capacity and the insufficient supply at the depot.

In our problem setting, the large-scale emergency makes both the demand level and the traffic condition highly unpredictable; hence uncertainty exists both in demand and in travel time. The overwhelming demand with limited resources and the urgency of the timely delivery make the problem constrained by both the vehicle capacity (in demand dimension) and the service deadline (in temporal dimension). We formulate the route-planning stage as a stochastic programming problem where we quantify all unmet demand, which can be due to insufficient capacity, lack of supply, or late delivery. We use chance constraints to represent the uncertainty in demand and travel times.

In addition to minimizing the unmet demand, the planning problem considers a secondary objective to guide the construction of a *complete route*. Instead of ignoring the nodes not selected by the routing problem with profits due to tight deadline, insufficient supply, or limited vehicle capacity, we include them at the end of each route according to this secondary objective. These complete routes that visit all demand nodes, even after the deadline or with an empty truck, provide a blueprint for recourse actions in the operational stage. The secondary objective is the arrival time at each node. We use a significantly small coefficient k to weigh this secondary objective so it will not interfere with the route generation guided by the minimum unmet demand primary objective.

Next, we define the notation and present the mathematical model for the vehicle routing problem with uncertainty in demand and travel time used to minimize unmet demand while visiting all nodes. Readers interested in details of the model are referred to Dessouky and colleagues (2006) and Shen and colleagues (2009a, 2009b). We consider a set K of vehicles and a set D of demand nodes. We identify an additional node, node 0, as the supply node (depot) and let $C = D\text{U}\{\text{node } 0\}$ represent the set of all nodes. Indexed on sets K and C, we define the following decision variables

---

$X_{i,j,k} = \begin{cases} 1 \\ 0 \end{cases}$ if vehicle k goes from node i to node j
otherwise.

$S_{i,k} = \begin{cases} 1 \\ 0 \end{cases}$ if node i can be serviced by vehicle k;
otherwise.

$Y_{i,j,k}$ = amount of commodity traveling from node i to node j using vehicle k
U_i = amount of unsatisfied demand of commodity at node i
$T_{i,k}$ = visit time at node i of vehicle k
$\delta_{i,k}$ = delay incurred by vehicle k in servicing i.

We also define the following input parameters:

Inputs:

s = amount of supplies at the supply node (depot)
c_k = load capacity of vehicle k
dl_i = service deadline at demand node i.
$\tau_{i,j,k}$ = uncertain time required to traverse from node i to node j for vehicle k
ζ_i = uncertain amount of demand at node i.
α_D = confidence level for chance constraint that defines the unmet demand
α_T = confidence level for chance constraint that defines vehicle arrival time

---

We also let M be a large constant used to express nonlinear relationships through linear constraints. With this notation we express the minimum unmet demand routing problem as the problem below. We notice that the first three constraints in the formulation plus cycle canceling constraints enforced by the chance constraints define feasible vehicle routes. The fourth constraint defines the delays at nodes and measures arrival time only for the nodes visited by a vehicle. The next three constraints identify the nodes serviced by vehicle k and ensure no flow is left at i by k if the node is not serviced by k. The eighth constraint enforces the initial supply of resources at the depot and the ninth constraint allows the transfer of supplies only where vehicles have moved. The last two constraints are chance constraints that require the solution be such that the demand satisfaction constraints and the definition of arrival time are satisfied with high probability.

$$\min \sum_{i \in D} U_i + \kappa \sum_{i \in D, k \in K} T_{i,k}$$

$$s.t \quad \sum_{j \in C} X_{0,j,k} = \sum_{j \in C} X_{j,0,k} = 1 \qquad k \in K$$

$$\sum_{j \in D, k \in K} X_{i,j,k} = \sum_{j \in D, k \in K} X_{j,i,k} = 1 \qquad i \in D$$

$$\sum_{j \in D} X_{i,j,k} = \sum_{j \in D} X_{j,i,k} \qquad i \in D, k \in K$$

$$0 \le T_{i,k} \le \min \left\{ M \sum_{j \in C} X_{i,j,k}, \delta_{i,k} + dl_i \sum_{j \in C} X_{i,j,k} \right\} \qquad i \in D, k \in K$$

$$\delta_{i,k} \le M(1 - S_{i,k}) \qquad i \in D, k \in K$$

$$S_{i,k} \le \sum_{j \in C} X_{i,j,k} \qquad i \in D, k \in K$$

$$MS_{i,k} \ge \left(\sum_{j \in C} Y_{j,i,k} - \sum_{j \in C} Y_{i,j,k} \right) \qquad i \in D, k \in K$$

$$s \ge \sum_{k \in K} \left(\sum_{j \in C} Y_{0,j,k} - \sum_{j \in C} Y_{j,0,k} \right)$$

$$Y_{i,j,k} \le c_k X_{i,j,k} \qquad i, j \in C, k \in K$$

$$X_{i,j,k}, S_{i,j} \in \{0,1\} \qquad i, j \in C, k \in K$$

$$U_i \ge 0, T_{i,k} \ge 0, T_{0,k} = 0, \delta_{i,k} \ge 0 \qquad i \in D, k \in K$$

$$\mathrm{P}\left\{ \sum_{k \in K} \left(\sum_{j \in C} Y_{j,i,k} - \sum_{j \in C} Y_{i,j,k} \right) + U_i \ge \zeta_i \right\} \ge 1 - \alpha_D \qquad i \in D$$

$$\mathrm{P}\left\{ \left(T_{i,k} + \tau_{i,j,k} - T_{j,k} \right) \le M \left(1 - X_{i,j,k} \right) \right\} \ge 1 - \alpha_T \qquad i \in D, j \in C, k \in K$$

Since the chance constraints in this formulation have only one random parameter on each probability statement that appears by itself, if the uncertain parameters have known distributions, the chance constraints can be simplified by using the 1-α critical level of the distribution. That is, given an univariate random variable Z with cumulative distribution F_Z, the statement $P\{Z \leq f(x)\} \geq 1-\alpha$ is equivalent to $F_Z^{-1}(1-\alpha) \leq f(x)$. We can therefore simplify the chance constraints in the above formulation if we assume that travel times between nodes of the graph and the demand at a node follow lognormal distributions. We assume that $\log(\zeta_i)$ follows $N(\mu_{\zeta i}, \sigma_{\zeta i})$ and that $\log(\tau_{ijk})$ follows a $N(\mu_{\tau ijk}, \sigma_{\tau ijk})$. This means we can solve the problem with chance constraints using the following deterministic constraints, where κ_T and κ_D correspond to the inverse of the standard normal at the 1-α critical level:

$$\left(T_{i,k} + \exp\left(\mu_{\tau_{i,j,k}} + \kappa_T \sigma_{\tau_{i,j,k}}\right) - T_{j,k}\right) \leq M\left(1 - X_{i,j,k}\right) \quad i \in D, j \in C, k \in K$$

$$\sum_{k \in K}\left(\sum_{j \in C} Y_{j,i,k} - \sum_{j \in C} Y_{i,j,k}\right) + U_i \geq \exp\left(\mu_{\zeta_i} + \kappa_D \sigma_{\zeta_i}\right) \qquad i \in D$$

7.4.2 *Example*

Our computational experiments are primarily focused on comparing the alternatives between short-and-risky routes and long-and-safe routes as well as exploring how the chance constraints safety factors influence the quality of the routes. We test our model on 10 different problem scenarios. For each scenario, we uniformly generate: 50 demand nodes and one depot node in a 200 by 200 square domain; and a mean demand quantity for each demand node ranging from 5 to 15. We service this demand with a fleet of 10 incapacitated vehicles. We use the Euclidean distances between any two of the nodes and assume a symmetric complete graph topology. The mean travel time between any two nodes is proportional to the distance. In the deterministic model (DP), we use the mean value of the demand quantities and travel times as its parameters. In the chance constrained model (CCP), we use a lognormal distribution with the same mean value as was used in the DP model; the standard deviation is set to be proportional to the mean value for the demand (20% of the mean value) and inversely proportional to the mean value for the travel time ($\sigma = \mu\ (UB - \mu)/100$), *UB* is the upper bound for the inversely proportional transformation, whose value is dependent on the graph topology, which must be greater than the longest arc in the graph. We restrict our analysis to this type of travel time distribution because we

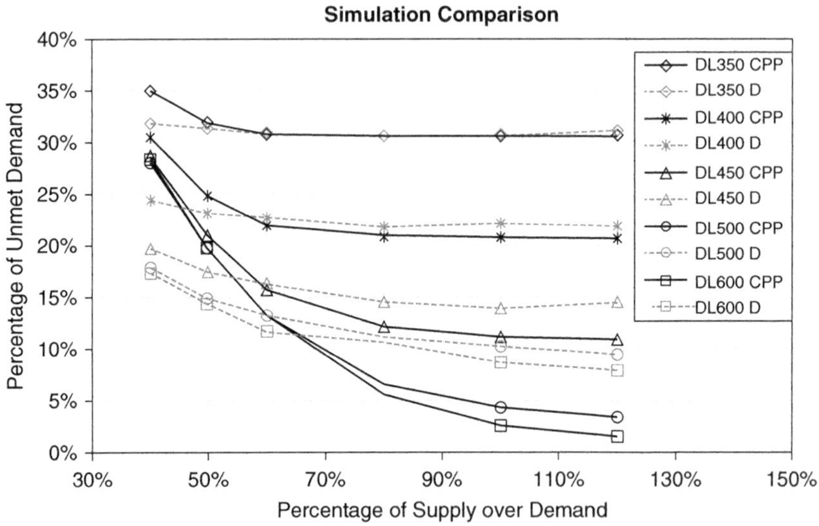

Figure 7.6. Percentage of unmet demand as a function of available supply and deadlines for chance constrained (CPP) and deterministic (D).

aim to compare the trade-offs between short-and-risky routes and long-and-safe routes. For the chance constrained model, we set the confidence level as 95%, thus setting the "safety factor" values κ_D and κ_T in the deterministic equivalent of the chance constraints to 1.65.

We conduct a simulation to compare the DP and CCP routes. For 30 *test cases* over each one of 10 problem scenarios, we solve the problem formulations using the same tabu heuristic. We then generate 100 *instances* of emergencies to evaluate how well these preplanned routes behave. We allow adjusting the supplies on the route after observing the realization of the demand and travel time uncertainty. Our results appear in Figure 7.6.

The results suggest that, when the deadline is very tight (between 40% to 60% of the *base route length*), the deterministic routes outperform the CCP routes. There is no benefit for utilizing a conservative routing strategy under tight deadlines since the longer, less-risky trips (arcs) in most generated instances are longer than the deadlines. In this case, the short-and-risky routes generated by the DP model at least have a chance in some of the generated instances of arriving on time, especially for the demand points in the beginning of the route. In the situation where there is limited supply (e.g., when the total supply is only 70% of the total demand), the DP and CCP routes yield roughly the same amount of unmet demand even under a more relaxed deadline. Because of the insufficient supplies at the depot, the extended deadline cannot provide a better coverage.

Under moderate deadlines and supply quantities (when the percentage of the total supply over the total demand and the percentage of the deadline over the *base route length* both are between 80% and 120%), we can conclude that CCP routes outperform DP routes 2% to 6% in terms of the percentage of the unmet demand over the total demand, according to our simulation experiments. As the deadline and the total supply are relaxed, the percentage of the unmet demand is decreasing for both DP and CCP routes; at the same time, the advantage of CCP routes over DP routes gradually increases. The simulation results on the percentage of the unmet demand from the CCP routes are approaching to its lower bound due to the shortage of the supply, with an increasing deadline. We believe that it is because of, under the relaxed enough deadline and supply quantities, its conservative nature that the CCP model shows its merits. It leads to similar number of nodes in different routes in the CCP model. In contrast, the DP routes are more prone to have an uneven number of nodes between different paths, leading to a higher chance of having unmet demand when they are tested over different realization *instances*.

Another experiment we conduct is to fix the deadline and the total supply at the depot and vary the confidence level, which has a one-to-one correspondence to the κ_T and κ_D values in the chance constraints. We use the same value for both κ_T and κ_D. For one specific network topology and demand distribution (a *problem scenario*), we fix the deadline at 80% of the *base route length* and the total supply at 90% of the total demand. We generate different sets of CCP routes with different safety factors, and run the same simulation as described previously. We plot the percentage of the unmet demand over the total demand in Figure 7.7. For the specific problem scenario we are testing, the estimated safety factor is 0.75 by averaging the variance value of all the edges in the graph in a statistical sense. It is shown as a dot on the plot. The point marked by a square gives the best average unmet demand ratio, which is also the value we used for the CCP model for the comparison of Figure 7.6. It corresponds to the confidence level at 95% and the safety factor as 1.65.

Figure 7.7 shows that with an increasing safety factor starting from the deterministic point coverage is improved. However, if we are too conservative as further increasing the safety factor after the "best CCP point," the quality of the result deteriorates. Under a given deadline, if the safety factor is too big such that all the short and risky arcs have been "stretched" to exceed the

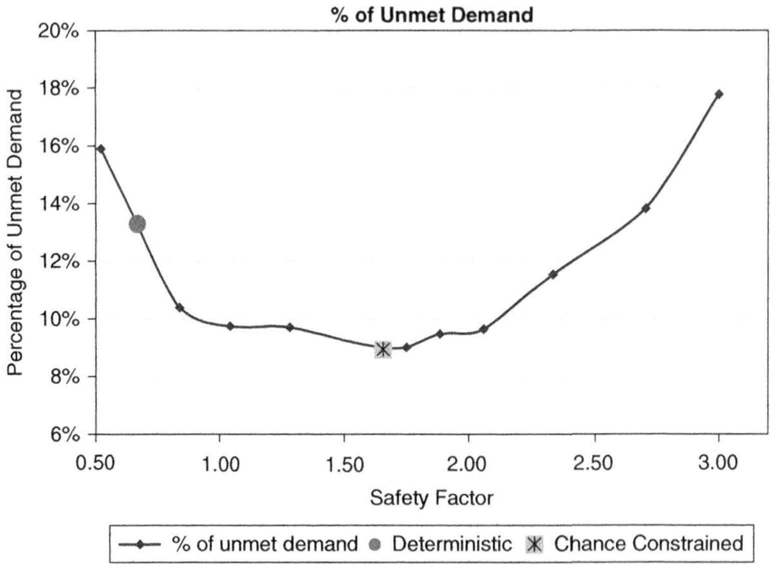

Figure 7.7. Percentage of unmet demand as varying safety factors for chance constrained (CPP) formulations.

deadline, then the route planning based on those "over-stretched" arcs is not very meaningful in providing any guidance since every node would miss its deadline. Hence it is crucial to plan using an "optimal" safety factor, which is neither too opportunistic nor too conservative. Robust optimization, which aims to optimize the worst-case scenario, also shares the same problem structure and computational complexity as the deterministic and the CCP formulations under a bounded uncertainty set or a given set of scenarios. Robust optimization seeks a solution that satisfies all possible constraint instances. It achieves this by enforcing that the strictest version of every constraint is satisfied by substituting the largest uncertainty parameter value in each constraint. This maximal value for the uncertainty parameter has a one-to-one correspondence with a safety factor value in the chance-constrained formulation and lies somewhere on the plot in Figure 7.7.

7.5 Conclusion

In this chapter, we decompose the overall transportation and distribution of supplies in response to a large-scale emergency into three interrelated models: (1) Perishable Inventory Model; (2) Facility Location Model; and (3) Vehicle Routing Model. For the perishable inventory management system, a trivial extension of a regular EMQ model is adequate when the required minimum inventory is not significant compared with the amount consumed by the regular market demand rate within the shelf life since it can be timely and completely depleted and refreshed by the regular market demand. However, when we consider the VMI in the SNS for the large-scale emergency setting, the minimum inventory requirement is in a scale that is comparable with the total market consumption within the shelf life. A more sophisticated inventory management strategy is required to provide the fresh and massive stockpile throughout the time horizon in the system. Hence in this chapter, we modeled the perishable inventory management problem with a minimum inventory volume constraint as a modified economic manufacturing quantity (EMQ) model. We observed that at a given profitability level of the firm, there are trade-offs between the lower amount paid by the government to firms (by reducing the unit price *pgov*) with the higher flexibility the government allows to firms (longer time before the expiration to salvage the pills). For the facility location problem, we show how the traditional models can be extended to incorporate redundancy in the system to provide better coverage. In experimental settings, we demonstrate how this modified facility location provides more redundant coverage than traditional models on a representative large-scale emergency scenario in Los Angeles County. For the routing problem, we proposed a chance constrained model to minimize the unmet demand for the planning stage. The influence of the safety factor for the planning stage model at different values is also illustrated by simulation results. We demonstrated that it is crucial to plan with an "optimal" safety factor.

Acknowledgment

This work was supported by the U.S. Department of Homeland Security through the Center for Risk and Economic Analysis of Terrorism Events (CREATE). However, any opinions, findings, and conclusions or recommendations in this document are those of the authors and do not necessarily reflect the views of the U.S. Department of Homeland Security.

REFERENCES

CDC website (2005). www.cdc.gov/

Chen, B., & Lin, C. S. (1998). Minmax-regret robust 1-median location on a tree. *Networks, 31*, 93–103.

Church, R., & ReVelle, C. (1974). The maximal covering location problem. *Papers of the Regional Science Association, 32*, 101–118.

ESRI website (2006). http://reports.esribis.com/

Dessouky, M. M., Ordóñez, F., Jia, H. Z., & Shen, Z. H. (2006). Rapid distribution of medical supplies. In R. Hall (Ed.), *Delay management in health care systems.* Springer.

Galvão, R. D., & ReVelle, C. (1996). A lagrangean heuristic for the maximal covering location problem. *European Journal of Operational Research, 88*, 114–123.

Hogan, K., & ReVelle, C. (1986). Concepts and applications of backup coverage. *Management Science, 32*, 1434–1444.

Huang, R., Kim, S., & Menezes, M. (2010). Facility location for large-scale emergencies. *Annals of Operations Research, 181*, 271–286.

Jia, H. Z., Ordóñez, F., & Dessouky, M. M. (2007a). A modeling framework for facility location of medical services for large-scale emergencies. *IIE Transactions, 39*, 41–55.

Jia, H. Z., Ordóñez, F., & Dessouky, M. M. & Dessouky, M. M. (2007b). Solution approaches for facility location of medical supplies for large-scale emergencies. *Computers and Industrial Engineering, 52*, 257–276.

Larson, R. C. (2005). Book Chapter, Decision models for emergency response planning, *The McGraw-Hill handbook of homeland security*. The McGraw-Hill Companies.

Larson, R. C., Metzger, M., & Cahn, M. (2006). Responding to emergencies: Lessons learned and the need for analysis. *Interfaces, 36*, 486–501.

Marianov, V., & ReVelle, C. (1996). The queueing maximal availability location problem: A model for the siting of emergency vehicles. *European Journal of Operational Research, 93*, 110–120.

Megiddo, N., Zemel, E., & Hakimi, S. L. (1983). The maximum coverage location problem. *SIAM Journal of Algebraic and Discrete Methods, 4*, 253–261.

Oran, A., Chuan Tan, K., Hooi Ooi, B., Sim, M., & Jaillet, P. (2012). Location and routing models for emergency response plans with priorities. *Future Security, Communications in Computer and Information Science, 318*, 129–140.

Ozguven, E., & Ozbay, K. (2014). Emergency inventory management for disasters – a review. *Journal of Emergency Management, 12*, 269–286.

Paluzzi, M. (2004). Testing a heuristic *P*-median location allocation model for siting emergency service facilities. Paper presented at the Annual Meeting of Association of American Geographers, Philadelphia, PA.

Schilling, D., Elzinga, D., Cohon, J., Church, R., & ReVelle, C. (1979). The TEAM/FLEET models for simultaneous facility and equipment siting. *Transportation Science, 13*, 163–175.

Shen, Z., Dessouky, M. M., & Ordóñez, F. (2009b). A two-stage vehicle routing problem for large-scale bioterrorism emergencies. *Networks, 54*, 255–269.

Shen, Z., Dessouky, M. M., & Ordóñez, F. (2011). Perishable inventory management system with a minimum volume constraint. *Journal of the Operational Research Society, 62*, 2063–2082.

Shen, Z., Ordóñez, F., & Dessouky, M. M. (2009a). The Stochastic Vehicle Routing Problem for Minimum Unmet Demand. *Optimization and Logistics Challenges in the Enterprise, Springer Series on Optimization and Its Applications.*

Snyder, L. V. (2006). Facility location under uncertainty: A review. *IIE Transactions*, *38*, 537–554.

Snyder, L. V., & Daskin, M. S. (2006). Stochastic *p*-robust location problems. *IIE Transactions*, *38*, 971–985.

8

Economic Consequences of Terrorism and Natural Disasters: The Computable General Equilibrium Approach

Peter B. Dixon, Maureen T. Rimmer, Glyn Wittwer, Adam Rose, and Nathaniel Heatwole

8.1 Introduction

This chapter is an introduction to computable general equilibrium (CGE) modeling and its application to the analysis of terrorism and other disruptive events.

CGE belongs to the economy-wide class of models, that is, those that provide industry disaggregation in a quantitative description of the whole economy. Economy-wide models emphasize the links between different parts of the economy, and, in effect model it as a set of integrated supply chains. The first economy-wide model was Leontief's (1936, 1941) input-output (I-O) model, which quantified links between industries as suppliers and customers for each other's products. In Leontief's world, an increase in household demand for cars stimulates the motor vehicle industry, which in turn stimulates the steel industry, which in turn stimulates the iron ore and coal industries, etc.

Following Leontief, the next stage of economy-wide modeling was the multi-sector programming models of Chenery and Clark (1959), Sandee (1960), Manne (1963), Bruno et al. (1970), Evans (1972), and others. Input-output and programming models lacked clear descriptions of the behavior of individual agents. In input-output modeling, the *economy* organizes production of each commodity (the vector X) to satisfy a vector of final demands (the vector Y) with given technology specified by the input-output coefficient matrix (A). In programming models, the *economy* organizes production to maximize a welfare function subject to Leontief's technology specification and subject to constraints on the availability of primary factors.

CGE modeling started with Johansen (1960). By contrast with the earlier economy-wide models, Johansen's CGE model identified behavior by *individual agents*. Households in CGE models offer their labor services to industries and maximize utility subject to their budget constraints.

The authors thank Scott Farrow for comments on an earlier draft.

Industries choose inputs to minimize costs subject to production-function constraints and the need to satisfy demands for their outputs. Capitalists allocate capital between industries to maximize their returns. The overall outcome for the economy is determined by the actions of individual agents coordinated through price adjustments that equalize demand and supply in product and factor markets.

Relative to I-O and linear programming models, CGE models are an effective framework for understanding and measuring how different parts of the economy are linked. CGE models go much further than the traditional I-O model in which the principal focus is on links through sales of goods and services between industries and from industries to final users. CGE models emphasize links provided by competition for scarce resources: labor in different skill categories, capital, and land, as well as institutional accounts relating to trade, households, and government. As originally recognized by Johansen, an industry may be harmed by the expansion of other industries through shortages of labor and capital. Johansen's CGE model balanced the benefits of Leontief's input-output links against the inhibiting effects of increases in the costs of labor and capital. Modern CGE models emphasize links occurring through the exchange rate, which connects import-competing and export-oriented industries. This is important in analyzing tariff policy. With cuts in protection, import-competing industries contract, lowering the exchange rate and allowing expansion of export-oriented industries. Another link encapsulated by modern CGE models is through the public-sector budget and competition for scarce public funds (see, e.g., Shoven & Whalley, 1984). For a further discussion of similarities and differences between I-O and CGE models, the reader is referred to Rose (1995). More modern versions of I-O approaches, applied to natural disasters and terrorism, have been developed, but are more cumbersome in their application than are CGE models (see Park et al., 2013; Richardson et al., this volume).

Because of their flexibility and realism, CGE models have gradually become the dominant form of economy-wide modeling.[1] Over the past 50 years, they have been used in the analysis of an enormous variety of questions. These include:

the effects on

- macro variables, including measures of nationwide or even global economic welfare;
- industry variables;
- regional variables;
- labor market variables;
- distributional variables; and
- environmental variables

of

- taxes, tariffs, and public expenditures;
- environmental policies;
- technology developments;
- changes in international commodity prices and interest rates;
- labor market policies and union behavior;
- exploitation of mineral deposits (the Dutch disease); and
- terrorism and other disruptive events.

While most of these questions have been analyzed in single-country, single-period models, there are now numerous CGE models that are either multi-regional or multi-period (dynamic) or both. By going multi-regional, CGE modeling has thrown light on both intra-country and intercountry regional questions. By going dynamic, CGE modeling has the potential to broaden and deepen its answers to all the questions with which it has been confronted. It has also entered the baseline forecasting area. CGE models are now used to generate forecasts for industries, labor force groups, and regions.

CGE models have been applied to the analysis of natural disasters for more than two decades (see Boisvert, 1992; Brookshire & McKee, 1992) and to terrorism for more than one (see, e.g., Rose, 2015). The major modification needed is to be able to "shock" the model to represent the direct property damage, death and injury, and/or shortage of critical inputs caused by a disruption. This has typically been accomplished in two ways. One is to reduce the capital and labor stocks in the model when property damage and death/injury take place. The second is to explicitly incorporate constraint on the availability of outputs or inputs. The first is relatively straightforward, while the second implicitly involves a form of disequilibrium (see, e.g., Rose et al., 2009). It is pertinent in cases such as earthquakes, for example, where water service disruptions differ across economic sectors because the damage to the pipeline distribution system, as well as the industrial makeup of its customers, is not uniform throughout the affected region (see, e.g., Rose et al., 2011). The shortfalls in critical input supplies are not the only disequilibria that can be introduced into CGE models (imbalances in trade, government treasuries, and labor markets have been modeled for many years).

Early CGE models applied to disasters employed Cobb-Douglas production functions, which reflected a rather flexible substitution capability between inputs, and were thus found unsatisfactory for this purpose (Boisvert, 1992; Rose & Guha, 2004). It is now the standard practice to utilize nested (multi-decision-level) constant elasticity of substitution (CES) production functions, which allow for a wide range of elasticity of substitution values. In addition, some models have replaced standard elasticity estimates with those that reflect relatively short periods of analysis applicable to all but major disasters.

Other advances in the CGE modeling of natural disasters and terrorism include combining them with GIS to reveal sub-regional distinctions in damages and commodity flows (Rose et al., 2007, 2011), dynamic analysis to reveal the time-path of impacts and adjustments (Dixon et al., 2010), incorporation of recovery stimuli (Sue Wing et al., 2016), incorporation of resilience (Rose et al., 2005), and incorporation of behavioral factors such as fear (Giesecke et al., 2012; Rose et al., 2009). The latter two are discussed at greater length later in this chapter.

Although CGE modeling has proved valuable in policy analysis, it has met considerable resistance from parts of the economics profession. There is a feeling that CGE models are black boxes. This is understandable because the theory, data, and computational requirements for CGE models are daunting relative to those for I-O models. In an effort to increase the accessibility of CGE modeling, its practitioners have been active in providing textbooks[2] and training not only for economists, but also for policy advisors. CoPS and Purdue University's Global Trade Analysis Project (GTAP) are prominent training providers with courses in many parts of the world, including Washington, DC. CGE modelers have also worked hard to increase the transparency of their results through a variety of means, including back-of-the-envelope (BOTE) justifications that can be readily understood by people without CGE backgrounds (see, for example, Dixon & Rimmer, 2013). Reduced-form analysis (discussed later in this chapter) is another approach to reducing the cost for decision makers of achieving a practical understanding of what CGE modeling can contribute.

The remainder of this chapter is organized as follows. Section 8.2 is an introduction to the basic design of CGE modeling. Section 8.3 is a discussion of the phenomena that must be captured by the CGE model so that it can provide insights on the economic implications of terrorism and other disruptive events. Section 8.4 is an introduction to USAGE and USAGE-TERM.[3] USAGE is a detailed national CGE model of the United States developed at the Centre of Policy Studies (CoPS), Victoria University, with support from the U.S. International Trade Commission and other federal government agencies. USAGE-TERM is a multi-region version of USAGE developed at CoPS with support from CREATE. Section 8.5 gives an illustrative application of USAGE-TERM demonstrating the relevance of the model in terrorism analysis. Section 8.6 contains concluding remarks, including comments on how the modeling capacity of USAGE-TERM can be made available in a timely fashion to decision makers.

8.2 An Introduction to Basic Design of a CGE Model

Most CGE models start with an input-output database that gives a quantitative picture of the economy for a given year (the base year or year 0). To a large extent, a CGE model can be thought of as a system of equations that describes the movements in the components of the input-output database away from their year 0 values. This interpretation puts us in the CGE modeling tradition of Johansen descending from Leontief. The alternative tradition is associated with Arrow and Hahn (1971), Debreu (1959), and Scarf (1967, 1973). Contributors in this tradition were concerned with the existence of equilibria. In what turned out to be the more practical Johansen tradition, existence is not an issue: the input-output table gives an initial solution. The emphasis is on movements away from the initial solution.

In comparative-static analysis, these movements take the economy from the initial picture of year 0 to an alternative picture of year 0. In dynamic analysis, the movements take the economy from year 0 to a picture of year 1; then further computations can show how the economy moves from year 1 to year 2, etc.

The equations describing the movements in the components of the input-output database are derived from neoclassical optimizing assumptions (utility maximizing, profit maximizing, cost minimizing) tied together by demand and supply equations and zero-pure-profit equations. Because of our emphasis on movements from one picture of an economy to another, it is natural for us to present the equations in percentage change form, that is, as connecting percentage changes in variables. The percentage change form can also be interpreted as a linear approximation to the true equations (those consistent with the model's theory and data). The linear approximation is the basis for an effective and simple computational method.[4]

In the remainder of this section, we describe an input-output database for a CGE model and the relationship between the input-output data and the national accounts. Understanding this relationship is important in checking and interpreting CGE results and in understanding how CGE modeling is related to macroeconomics. Then we set out some illustrative CGE equations and show how their implementation uses the input-output data.[5] Next we discuss the percentage-change solution technique for CGE models, originally implemented in linear approximation form by Johansen, but now implemented in a multi-step form that eliminates linearization errors. Finally, we comment on the relationship between CGE modeling and more traditional econometric techniques.

		Absorption Matrix					
		1	2	3	4	5	6
		Prod-ucers	Invest-ors	House-holds	Exports	Govern-ment	Invent-ories
	Size	← *I* →	← *I* →	← 1 →	← 1 →	← 1 →	← 1 →
Basic Flows	↑ *C*×*S* ↓	V1BAS	V2BAS	V3BAS	V4BAS	V5BAS	V6BAS
Margins	↑ *C*×*S*×*N* ↓	V1MAR	V2MAR	V3MAR	V4MAR	V5MAR	V6MAR
Sales Taxes	↑ *C*×*S* ↓	V1TAX	V2TAX	V3TAX	V4TAX	V5TAX	V6TAX
Labor	↑ *M* ↓	LABOR					
Capital	↑ 1 ↓	CAPITAL					
Output	↑ 1 ↓	VOUTPUT					

C = Number of commodities
I = Number of industries
S = Supply sources for commodities
M = Number of occupations
N = Number of commodities used as margins

	Import Duty
Size	← 1 →
↑ *C* ↓	TARIFF

Figure 8.1. Input-output database for a typical single-country CGE model.

8.2.1 *The Starting Point: The Input-Output Database*

For understanding how a CGE model is constructed and encapsulates linkages, it is useful to study Figure 8.1. This is a slightly simplified representation of the input-output database used in single-country CGE models such as the USAGE model of the United States (Dixon et al., 2013).

Figure 8.1 has two parts, an absorption matrix and a vector of tariff collections on imports. The first row of the absorption matrix, V1BAS, …, V6BAS, shows flows in the database year (year 0) of commodities to producers, investors, households, exports, public consumption, and inventory accumulation. Each of these matrices has CS rows, one for each of C commodities from S sources. Here we will assume that S is 2, domestic and imported.[6] C can be large, several hundred in some models.

V1BAS and V2BAS each have *I* columns where *I* is the number of industries. The typical component of V1BAS is the value of good c from source s [good (c,s)] used by industry i as an input to current production, and the typical component of V2BAS is the value of (c,s) used to create capital for industry i. As shown in Figure 8.1, V3BAS, V4BAS, V5BAS, and V6BAS each have one column. Thus, Figure 8.1 refers to a model that recognizes one household, one category of public demand, one destination for exports, and one category of inventory demand. These dimensions can be extended in work concerned with income distribution, multiple levels of government, and trade agreements in which the essence is special access for exports to particular markets.

All of the flows in V1BAS, ..., V6BAS are valued at basic prices. The basic price of a domestically produced good is the price received by the producer (that is, the price at the factory door or the farm gate, or equivalently, the price paid by users excluding sales taxes, transport costs, and other margin costs). The basic price of an imported good is the landed-duty-paid price, i.e., the price at the port of entry just after the commodity has cleared customs.

Costs separating producers or ports of entry from users appear in the input-output data in the margin matrices and in the row of sales tax matrices. The margin matrices, V1MAR, ..., V6MAR, show the values of N margin commodities used in facilitating the flows identified in V1BAS, ..., V6BAS. The margin commodities in USAGE are the domestic varieties of: wholesale trade, retail trade, road transport, rail transport, water transport internal, air transport internal, pipelines, water transport international, and air transport international. Each of the matrices V1MAR, ..., V6MAR has CSN rows corresponding to the use of N margin commodities in facilitating flows of C commodities from S sources. The sales tax matrices V1TAX, ..., V6TAX show collections of sales taxes (positive) or payments of subsidies (negative) associated with each of the flows in the BAS matrices.

Payments by industries for M occupational groups are recorded in Figure 8.1 in the matrix LABOR. In models and applications focusing on labor market issues, such as training needs and immigration, M can be large. For example, some versions of the USAGE model distinguish 350 occupations.

In most CGE models, payments by industries for the use of capital are recorded in the input-output data as a vector: CAPITAL in Figure 8.1. There can be additional rows for the costs of using land and other miscellaneous costs.

As an accounting truism, the value of inputs to an industry equals the value of output. This can be insured by calculating the returns to capital as a residual after deducting the cost of other inputs from the value of output. Then under the assumption that each commodity is produced by just one industry and each industry produces just one commodity, the absorption matrix satisfies a balance condition for each industry/commodity. The c-th column sum of V1BAS,

V1MAR, V1TAX, LABOR, and CAPITAL, that is, the value of inputs to the c-th industry, equals the basic value of sales of the c-th commodity (the product of the c-th industry). If c is a non-margin commodity, then the basic value of sales of the domestic variety of the c-th commodity is the sum across the (c,"dom")-rows of V1BAS1 to V6BAS. If c is a margin commodity, then the basic value of sales of the domestic variety of c is the direct uses, i.e., the sum across the (c,"dom")-rows of V1BAS to V6BAS, plus the margins use of commodity c. The margins use of c is the sum of the components in the (k,s,c)-rows of V1MAR to V6MAR for all commodities k and sources.

By forming the sum of all inputs to domestic production and the sum of the basic value of sales of all domestic commodities, and then applying the input-output balance conditions, we see that:

$$\begin{aligned}&\mathrm{Sum(V1BAS)+Sum(V1MAR)+Sum(V1TAX)+Sum(LABOR)+Sum(CAPITAL)}\\&\quad=\sum_{k=1}^{6}\mathrm{Sum(V\,k\,BAS)}+\sum_{k=1}^{6}\mathrm{Sum(V\,kMAR)}-\sum_{\substack{k=1\\k\neq 4}}^{6}\mathrm{Sum(V\,k\,BAS(imp))}\end{aligned}\tag{8.1}$$

where

> the notation Sum(matrix) means the sum of all components of matrix,
> VkBAS(imp) refers to the import components of VkBAS, and
> we assume that there are no imported commodities that are simply re-exported, that is, we rule out V4BAS(imp).

Cancelling common terms from the left- and right-hand sides of (8.1) and bringing in additional tax terms to both sides leads to:

$$\begin{aligned}&\mathrm{Sum(LABOR)+Sum(CAPITAL)}+\sum_{k=1}^{6}\mathrm{Sum(VkTAX)+Sum(TARIFF)}\\&\quad=\ \mathrm{Sum(V\,2\,BAS)+Sum(V\,2TAX)+Sum(V\,2MAR)}\\&\qquad+\mathrm{Sum(V\,3\,BAS)+Sum(V\,3TAX)+Sum(V\,3MAR)}\\&\qquad+\mathrm{Sum(V\,4\,BAS)+Sum(V\,4TAX)+Sum(V\,4MAR)}\\&\qquad+\mathrm{Sum(V\,5\,BAS)+Sum(V\,5TAX)+Sum(V\,5MAR)}\\&\qquad+\mathrm{Sum(V\,6\,BAS)+Sum(V\,6TAX)+Sum(V\,6MAR)}\\&\qquad-\left[\sum_{\substack{k=1\\k\neq 4}}^{6}\mathrm{Sum(V\,k\,BAS(imp))-Sum(TARIFF)}\right]\end{aligned}\tag{8.2}$$

which is the GDP identity. The left-hand side is GDPINC, that is, GDP as income, consisting of payments to factors of production *plus* indirect taxes. The right-hand side is GDPEXP, that is, GDP as expenditure, consisting of investment, private consumption, exports, public consumption, and inventory accumulation all valued at purchasers prices, *less* imports valued at c.i.f prices (landed-duty-paid less tariffs). Equation (8.2) is not only of interest as the foundation of national income accounting, but it also plays an important checking role in CGE modeling. If equation (8.2) is violated by results from a simulation, then we are alerted to the existence of an error.

8.2.2 *Moving Off the Starting Point*

The components of Figure 8.1 move in response to shocks to exogenous variables such as changes in technologies, preferences, world commodity prices, and taxes. As the components move, demand and supply equations and zero-profit equations maintain the balance for each commodity between the basic values of costs and sales. The movements in the individual components of Figure 8.1 are combinations of price and quantity movements. Price movements are driven by costs reflecting production technologies and scarcities of primary factors. Quantity movements are governed by optimizing behavior: industries choose input and output quantities to maximize profits; investors choose input quantities to minimize the costs of creating units of capital; and households choose consumption quantities to maximize utility.

CGE equations are often most easily understood in percentage change form. For example, many CGE models handle the demand for primary factors via equations of the form:

$$\begin{aligned} \mathrm{xfac}(n,j) - \mathrm{afac}(n,j) = {} & \mathrm{x0output}(j) \\ & - \sigma\mathrm{fac}(j) * \left(\mathrm{pfac}(n,j) - \sum_k \mathrm{SFAC}(k,j) * \mathrm{pfac}(k,j) \right) \\ & - \sigma\mathrm{fac}(j) * \left(\mathrm{afac}(n,j) - \sum_k \mathrm{SFAC}(k,j) * \mathrm{afac}(k,j) \right) \end{aligned} \tag{8.3}$$

where

σfac(j) is a positive parameter denoting the elasticity of substitution between primary factors in industry j;

SFAC(k,j) is the share of factor k in the total cost of primary factors to industry j that can be calculated from input-output data; and xfac(n,j), x0output(j), pfac(k,j) and afac(k,j) are all percentage changes in variables away from their values in an initial situation represented by a database such as Figure 8.1.

xfac(n,j) is the percentage change in the demand for factor n to be used in industry j;

x0output(j) is the percentage change in the output of industry j;

pfac(k,j) is the percentage change in the price of factor k to industry j; and

afac(k,j) is factor-k-using technical change in industry j, expressed as a percentage.[7]

By looking at (8.3), we can immediately identify three assumptions that have been made about the production function for industry j. First, the absence of intermediate input prices on the right-hand side of (8.3) implies that primary factors and intermediate inputs are being treated as non-substitutes. Second, the implicit coefficient of one on x0output(j) implies constant returns to scale: if there are no changes in factor prices or technologies, then demands for factors move in line with output. Third, the absence of an n argument on σfac(j) implies that the elasticity of substitution between all pairs of primary factor has the same value, implying that the underlying production function must be of the CES form (see Arrow et al., 1961). Of course, these assumptions can be varied, but whatever assumptions are adopted, their implications are often transparent in percentage change equations.

Other typical CGE equations include

$$\begin{aligned} \text{p0dom}(c) = {} & \sum_{v,s} \text{R1}(v,s,c) * \text{p1}(v,s,c) + \sum_{k} \text{RFAC}(k,c) * \text{pfac}(k,c) \\ & + \sum_{v,s} \text{R1}(v,s,c) * \text{a1}(v,s,c) + \sum_{k} \text{RFAC}(k,c) * \text{afac}(k,c) \end{aligned} \tag{8.4}$$

and

$$\begin{aligned} \text{x0output}(c) = {} & \sum_{j} \text{T1}(c,\text{dom},j) * \text{x1}(c,\text{dom},j) + \sum_{j} \text{T2}(c,\text{dom},j) * \text{x2}(c,\text{dom},j) \\ & + \text{T3}(c,\text{dom}) * \text{x3}(c,\text{dom}) + \text{T4}(c) * \text{x4}(c) + \text{T5}(c,\text{dom}) * \text{x5}(c,\text{dom}) \\ & + \text{T6}(c,\text{dom}) * \text{x6}(c,\text{dom}) + \sum_{j}\sum_{s}\sum_{k} \text{TMAR}_j(k,s,c) * \text{xmar}_j(k,s,c) \end{aligned} \tag{8.5}$$

In (8.4), the percentage change in the basic price of domestically produced commodity c [p0dom(c)] is determined by the percentage changes in input prices [p1(v,s,c) and pfac(k,c)] to the production of c and by technical changes [a1(v,s,c) and afac(k,c)] affecting inputs per unit of output of c. In (8.5), the percentage change in the output of c [x0output(c)] is determined by percentage changes in direct demands and margin demands. Direct demands are made: by industries that use domestic c as an input to production and capital creation [x1(c,dom,j) and x2(c,dom,j)], by households [x3(c,dom)], by foreigners [x4(c)], by government [x5(c,dom)], and by inventory accumulators [x6(c,dom)]. If c is a margin commodity, then further demands arise from the use of c to facilitate the flow of commodity k from source s to user j [$xmar_j$(k,s,c)]. The R coefficients in (8.4) are shares of each input in total costs of production in industry c, and the T coefficients in equation (8.5) are shares of each demander in the total sales of domestic commodity c. Both the R and T coefficients are calculated from the input-output data in Figure 8.1.

A detailed CGE model such as USAGE has millions of equations. In addition to equations such as (8.3) to (8.5), there must be equations to define: demands for intermediate inputs to current production and capital creation, demands by final users (households, exports, government, inventory accumulation), and margin demands. There must be equations to link purchasers' prices [such as the p1's appearing on the right-hand side of (8.4)] to sales taxes, the costs of margins, and basic prices [such as p0dom on the left-hand side of (8.4)]. There must be market-clearing equations not only for commodities [illustrated by (8.5)], but also for primary factors. So that results can be reported conveniently, there must be defining equations for a wide range of macro variables such as the expenditure components of GDP (C, I, G, X, and M), the income components of GDP (factor payments and indirect taxes), the price level, the average wage rate, the balance of trade, and the public-sector deficit. While CGE models have huge numbers of equations, when expressed in linear percentage change form, these are easily interpreted.

The linear percentage change form also facilitates computation. The linear equations can be written as

$$A * v = 0 \tag{8.6}$$

where v is the vector of percentage changes in variables [the x's, a's, and p's in equations such as (8.3) to (8.5)] and A is a matrix of coefficients containing shares [e.g., the SFACs in (8.3),the R's in (8.4), and the T's in (8.5)], parameters [e.g., σfac in (8.3)], and many zeros and ones. The dimensions of A are mxn where m is the number of equations and n is the number of variables, which is always greater than m. To solve (8.6), we must choose m variables to be endogenous and n-m variables to be exogenous. CGE modelers gain flexibility by varying this choice between applications. For example, in some applications, it is appropriate to fix aggregate employment exogenously and determine market-clearing wage rates endogenously. In other applications, it is appropriate to fix wage rates exogenously and determine employment endogenously. Once the endogenous/exogenous choice (closure) is made, then CGE software such as GEMPACK[8] rearranges (8.6) as

$$A_1 * v_1 + A_2 * v_2 = 0 \tag{8.7}$$

where

v_1 is the vector of percentage changes in the m variables chosen to be endogenous,
v_2 is the vector of percentage changes in the remaining n-m variables, which are exogenous, and
A_1 and A_2 are the mxm and mx(n-m) matrices formed from the columns of A corresponding to the endogenous and exogenous variables.

From (8.7), GEMPACK computes percentage changes in endogenous variables for given percentage changes in exogenous variable according to[9]:

$$v_1 = -A_1^{-1} * A_2 * v_2 \tag{8.8}$$

Linearization errors can be avoided by imposing the changes in the exogenous variables in a series of steps.[10] For example, if we want to know the effects of a 50% increase in a tax rate, then we could proceed by: (a) computing the effects of a 25% increase; (b) updating all the flows in Figure 8.1 to reflect the situation after the 25% increase; (c) reevaluating share coefficients such as those in (8.3) to (8.5); (d) reassembling the A_1 and A_2 matrices; and (e) reapplying (8.8) to compute the effects of the remainder of the 50% tax increase. In practice, the effects of quite large changes in exogenous variables can often be computed without noticeable linearization error in just a few steps.

Increased dimensionality in multi-country and multi-regional models makes highly efficient computational techniques such as multi-step versions of (8.8) implemented in well-developed software, such as GEMPACK, almost mandatory. For a multi-country or multi-regional[11] model, we need a starting point such as Figure 8.1 and associated equations for each country/region with consistency between exports from source s to destination d and imports in destination d from source s. For a dynamic model, we need data such as those in Figure 8.1 plus data for stock variables such as capital in each industry, public-sector debt, and foreign assets and liabilities. We also need equations that link the values of stock variables through time and allow for lagged processes. The USAGE model, for example, has dynamic equations ensuring: (a) that an industry's capital stock at the start of year t+1 is its capital at the start of year t plus its investment during year t minus depreciation; (b) that public-sector debt at the start of t+1 is public-sector debt at the start of t plus the public-sector deficit incurred during t; (c) that net foreign liabilities at the start of t+1 are net foreign liabilities at the start of t plus the current account deficit in t plus the effects of revaluations of assets and liabilities caused by changes in price levels and the exchange rate; and (d) that wages adjust sluggishly to eliminate gaps between supply and demand for labor. With dynamic equations in place, computations such as (8.8) generate percentage changes that move variables from their values in year t to those for year t+1 to t+2 and so on.

8.2.3 *How Does CGE Modeling Relate to Regression-Based Econometric?*

The most popular tool in applied economics is regression analysis implemented with time-series or cross-section data. Consequently, we are often asked about how regression analysis relates to CGE modeling. The main use of regression analysis in the CGE field is in the estimation of price-response parameters [e.g., σfac(j) in (8.3)], including: substitution elasticities between capital, labor, energy, and materials; substitution elasticities between imported and domestic products in the same commodity classification; substitution elasticities in household demand between different commodities; and export-demand elasticities. Regression-based econometrics plays a particularly prominent role in the family of CGE models pioneered by Dale Jorgenson (for an overview, see Jorgenson et al., 2013).

CGE modeling complements regression analysis by providing a strong theoretical framework into which regression results can be embedded. Where no regression results are available, the CGE framework forces modelers to be explicit about parameter assumptions and directs their attention to sensitivity analysis. This facilitates assessments of the reliability of policy conclusions based on model results.

8.3 Extensions of CGE to Incorporate Resilience and Behavioral Linkages

8.3.1 *Resilience*

Two major refinements have been made in CGE modeling over the past decade to account for unique aspects of the consequences of terrorism and natural disasters. Resilience has become a popular buzzword, but it has been given specificity in the economic context. Rose (2004, 2009) distinguishes two major types of this concept. *Static Resilience* refers to the ability of a system to maintain function after the initial shock. This is attained by using remaining resources as efficiently as possible. He defines *Dynamic Resilience* as the ability to accelerate the pace of recovery. Efficient reallocation of resources is not sufficient to do so, but rather investment in repair and reconstruction is needed. Rose also distinguishes between *inherent* and *adaptive* resilience. The former refers to resilience that exists at the time of the shock and can be implemented straightaway, while the latter requires some change in a resilience tactic before it can be implemented in response to the shock. Finally, resilience is operative at the micro (individual firm, household, institution), meso (industry and market), and macro (economy-wide) levels.

Some examples are:

<u>micro</u>

inherent (existing inventories, ability to substitute gas for coal in steam-fired electricity boilers)
adaptive (conservation not previously thought possible, business relocation)

<u>meso</u>

inherent (workings of the price system in reallocating resources to highest value use)
adaptive (rearranging supply chains)

<u>macro</u>

inherent (monetary and fiscal policy)
adaptive (importing scarce resources from the suppliers outside the region)

While the definition of resilience focuses on the post-disaster stage, economic resilience is a process. For example, it can be enhanced prior to a disaster, but implementation takes place following the shock (e.g., building up inventories, backing up data on clouds).

In addition, Rose (2004, 2009) has developed an operational metric of resilience as the avoided losses divided by the maximum potential losses. The metric has been applied extensively (see Kajitani & Tatano, 2009 Rose et al., 2007, 2009, 2011). Studies of actual and simulated disasters indicate that as much as 90% of business interruption losses, as measured by GDP, can be reduced through the right combination of resilience tactics (see also the chapter on resilience in this volume).

Here we note how resilience can be incorporated into CGE models:

- input substitution: inherent is already embodied; adaptive can be modeled by changing substitution elasticities (Rose and Liao, 2005)
- conservation: inherent is considered to already be in effect through the optimizing nature of CGE models; adaptive can be modeled by changes in the production function productivity term (Rose and Liao, 2005).
- relocation: adaptive (within the region) can be modeled by reducing the constraint on labor, capital, or intermediate inputs (Rose et al., 2009)
- inventories (inherent and adaptive by reducing the constraint on intermediate inputs (Sue Wing et al., 2016)
- production rescheduling (recapture): adaptive by reducing the constraint on capital, labor, or materials (Rose et al., 2011)
- price system: already embodied in the CGE model
- import substitution: inherent already embodied; adaptive can be modeled by changes in import elasticities
- fiscal and monetary policy: adaptive through interest rate changes and fiscal stimuli.

8.3.2 *Behavioral Linkages*

The second major extension of CGE modeling is to capture the economic consequences of the unique behavioral changes that are engendered by terrorism and to a lesser extent natural disasters. The majority of these are associated fear. Psychologists, risk analysts, and decision analysis often refer to them as stemming from the social amplification of risk in the short run (Burns & Slovic, 2010; Kasperson et al., 1988; Slovic, 1987) and stigma affects in the long run (McClelland et al., 1990). For example, Rose et al. (2009) found that more than 80% of the net economic consequences of the September 11, 2001, attacks on the World Trade Center stemmed from a nearly two-year decline in airline travel–related tourism (even after substitution effects are taken into account).

Fear effects translate into three types of behavior. First is *avoidance*, often exemplified by staying away from public gatherings after a terrorist attack on that or similar sites or due to communicable diseases. This can be modeled by either a reduction in the direct demand for the economic activity or through a "Phantom tax" penalty on it (see Rose, 2009).

The second subcategory is *aversion* effects, where changes at the margin can be affected by prices. Examples include the wage and investor premia that these two categories of factor suppliers require to return to the affected area. Similar to these would be discounts for shoppers and restaurant and hotel patrons to return. Giesecke et al. (2012) model these effects in terms of the response to a hypothetical radiological dispersion device ("dirty bomb") attack on the Los Angeles financial district over a 10-year span (with decaying fear over time), and found the behavioral effects to be 15 times greater than the ordinary partial and general equilibrium effects resulting from damage, injury, and downtime (due to decontamination repair).

A third category of behavioral effects relates to the many types of behavioral responses to terrorism and disasters that are associated with the concept of *bounded rationality*, which encompasses inability to process information, misperceptions, myopia, and non-optimizing behavior such as "minimizing regret" (Girgenzer & Selten, 2002). Kunreuther and others have often pointed to these as primarily affecting mitigation decisions, such as the failure of people to adequately insure themselves against disaster risks for reasons noted previously, as well as outright denial of the threats. Closely related are *contagion* effects or interdependence, effects whose prevalence of magnitude are heightened in densely populated areas (Heal & Kunreuther, 2007). The major ways to incorporate this subcategory are to alter utility and production functions in various ways and to explicitly incorporate uncertainty. Contagion effects are much more complicated, because they involve interdependence of utility functions, which are difficult to model in general, but especially in a CGE framework.

8.4 USAGE and USAGE-TERM

This section describes USAGE and USAGE-TERM. USAGE is a widely used national CGE model. USAGE-TERM is a regional version of USAGE that is being developed to facilitate the effects of localized regional shocks, particularly terrorism and other disruptive events. Following the description of the two models, we present an illustrative USAGE-TERM application in Section 8.5.

8.4.1 *USAGE*

USAGE is a 500-industry, dynamic, CGE model of the U.S. economy. The model has been used by and on behalf of the U.S. International Trade Commission, the U.S. Departments of Commerce, Agriculture, Energy, Transportation, and Homeland Security, as well as private-sector organizations such as the Cato Institute and the Mitre Corporation. Applications of the model involve preparation of baseline forecasts and analyses of a variety of issues, including the effects of: trade policies, environmental regulations, carbon taxes, energy security, illegal immigration, road infrastructure, Next-Gen aviation infrastructure expenditures, the Obama stimulus package, the National Export Initiative, an H1N1 epidemic, security-related port closures, and a Chlorine attack in Los Angeles.[12]

USAGE is essentially a national model, although it does have a facility for disaggregating national results in a top-down fashion to the 50 states and the District of Columbia.[13] This facility is effective for working out the regional implications of national policies that are unlikely to have a significantly different effect on costs of production in one state compared with other states. A legitimate application of the top-down facility is the U.S. International Trade Commission's analysis of the effects on the state economies of changes in import restraints (tariffs and quotas).[14] However, a limitation of the top-down facility is that it is unsuitable for projecting the effects of policies and other shocks (including terrorism events) that are initiated at the regional level and affect costs in one region relative to those in other regions.

8.4.2 *USAGE-TERM*

To overcome this limitation, the CoPS team, with considerable support from colleagues at CREATE, has developed a series of bottom-up regional versions of USAGE. All of these are in the family of TERM models developed initially for Australia.

The first USAGE-TERM model was USAGE-R51. This version, created in 2011, treats the 50 states plus the District of Columbia as 51 highly integrated economies connected by: trade, factor movements, and a common currency. In USAGE-R51, policies such as carbon taxes levied at the state level cause changes in production costs in one state relative to those in others, and lead to changes in trade and factor flows. Thus USAGE-R51 allows assessments of the costs and benefits to states of state-initiated policies.

The initial version of USAGE-R51 was comparative static. In 2012–2013, the model was given a dynamic dimension similar to that in the national USAGE model, covering capital and debt accumulation together with lagged processes (Subsection 8.2.2). Thus it became capable of tracing out effects of a shock over a number of years.

In 2013–2014, we extended the regional detail from the state to the county level to create USAGE-County. This work was motivated by wanting to improve the capabilities of USAGE for modeling terrorism shocks and other disruptive events. These events occur at a localized level, often well below the state level. Thus, for analyzing such events, extending the USAGE capability to the county level is an important enhancement. We have also created a version, USAGE-CD, in which the identified regions are the 436 congressional districts.

The key data requirement for these regional versions of USAGE is job shares matrices with components, S(j,r), being the share of region r in U.S. employment in industry j. For USAGE-County, the main source for these estimates is county data for about 400 industries from the 2010 Census.[15] For USAGE-CD, we also use the county data. Most congressional districts are an aggregation of counties. For these districts, we derived jobs data by addition. However in some western states, counties are large and encompass several congressional districts. For these congressional districts, we estimated job shares using county data with their detailed industry coverage supplemented by congressional district data available in the Census for 13 broad industries. The 13-industry coverage for congressional districts is in www2.census.gov/acs2011_1yr/CD113/. For congressional districts that are contained within a county but are only part of it, we converted the 13 industries into the USAGE 500 industries by assuming that the detailed industry breakdown within each of the 13 industries is the same as that in the county to which the congressional district belongs. For a congressional district that lies across county borders, we allocated the 13-industry data for the congressional district to the relevant counties, taking account of the locations of cities and rural areas. Then we split the allocated 13-industry data for each of the relevant counties to the 500 level as earlier. Finally, we combined the 500-level data from the relevant counties to form the 500-industry breakdown for the cross-county congressional district.

Another important data requirement for regional versions of USAGE is interregional trade flows. For each region and each commodity, we can estimate net trade flows from data on output and absorption (use of the commodity within the region). Then applying a modified gravity formula, initially devised by Mark Horridge (see Dixon et al., 2007a), we estimate interregional trade flows that are consistent with our estimates of net trade flows. These interregional trade estimates take into account: the tradability of commodities, home bias (the tendency to buy the local variety), and distance between supplying and consuming regions.

USAGE-County has potentially huge dimensions: 500 industries in 3,000 counties supplying their products to 500 industries and final users in 3,000 counties. The dimensionality problem is reduced for USAGE-CD, where the regional dimension is 436. Nevertheless, USAGE-CD computations at full dimension are impractical, and even if they could be carried out, the interpretation of the results would be unnecessarily time-consuming. To address this problem, CoPS has developed a flexible aggregation program that allows model users to specify the regions and industries of interest.[16] The program then aggregates the full-dimension master database and creates a version of USAGE-TERM in which only the regions and industries of interest are identified.

8.5 An Illustrative Application of CGE Modeling: The Economic Effects of a One-Kiloton Nuclear Detonation in Downtown Los Angeles

Here we describe an illustrative application of USAGE-CD, concerned with a hypothetical terrorism event in downtown Los Angeles that takes place at the start of 2012. First, we discuss how USAGE-TERM was set up. This requires decisions concerning the levels of regional and industry disaggregation. On the one hand, we need disaggregation that is sufficient to capture the main features of the terrorism event. On the other hand, we need to avoid unmanageable levels of detail with associated slow computational turnaround. Second, we list the shocks that were applied. USAGE-CD is able to cope with detailed scenarios implemented through a long list of shocks covering: length of shutdown period, immediate and subsequent deaths, medical expenditures, clean-up expenditures, capital destruction, labor force participation effects, and aversion and resilience behavior. We are also able to introduce explicit assumptions concerning the financing of event-related, public-sector expenditures (in this case, clean-up, medical, and evacuation expenditures). Third, we present results covering a 20-year simulation period.

8.5.1 *Setting up USAGE-TERM and Specifying the Shocks*

For the terrorism event in our illustrative simulation, we assume that detonation of a one-kiloton nuclear device takes place in California Congressional District 34 (CA-34), downtown Los Angeles. For the simulation, we formed a four-region version of USAGE-TERM with the regions being:

- CA-34;
- Rest of Los Angeles County (RoLA);
- Rest of California (RoCalif); and
- Rest of the United States (RoUS).

We chose an 18-industry disaggregation, which can be seen in Figure 8.9.

The shocks and financing assumptions applied in USAGE-CD are listed in the following nine points.

1. Impact Area, Immediate Deaths, Affected Surviving Population

We assume that property damage and fallout are confined to CA-34 and affect only part of this region, which we refer to as the **impact area**. The residential population of CA-34 is about 700,000. The impact area has a daytime population of about 315,000 and a night-time population of around 125,000 (Heatwole et al., 2014). We assume that the detonation occurs in the daytime. Thus the **affected population** is 315,000.

The immediate deaths number in the tens of thousands (DNDO, 2013).[17] The **affected surviving population** represents the affected population (315,000) less the number of immediate deaths. We assume that the percentage of the population killed immediately is equal across age groups. This is relevant for working out labor-supply effects (point 3).

2. Shutdown Period

We assume that the impact area is shut down for two years with clean-up expenditures and repairs spread evenly between year 1 and year 2. During the shutdown period, no business is undertaken in the impact area.

3. Evacuations and Labor Supply

We assume that 87,250 people are evacuated from the impact area (Heatwole et al., 2014), and stay away for the full shutdown period. In calculating the number of evacuees, we assumed: (a) that the surviving night-time population of the impact area requires evacuation; and (b) the night-time population suffers the immediate death rate that applies to the daytime population.[18]

We need to specify where the evacuated people go and how much it costs to accommodate them during the shutdown period. We assume that 25% of the evacuees from the impact area spend the shutdown period elsewhere in CA-34, 25% go to the rest of LA, 25% go to the rest of California, and 25% go to the rest of the United States.

We assume that evacuation and subsequent accommodation expenses total $3.97 billion over the evacuation period of two years (Heatwole et al., 2014). This allows for an evacuation expense of $9,000 per person plus accommodation expenses of $50 per day per person for two years [3.97b = 87,250*(9,000 + 50*2*365)]. Beyond the shutdown period, we assume that the surviving evacuees who do not return to CA-34 stay in the regions to which they were evacuated. However, beyond the shutdown period, we do not count their accommodation expenses as an additional cost associated with the event.

Evacuations and immediate deaths have an impact effect in the event year that reduces labor supply in CA-34. We arrive at the specific number in several steps. First, a portion of the night-time population of the impact area (125,000) suffers immediate death (DNDO, 2013). On the assumption that 60% of these people were participating in the labor force, the labor supply in CA-34 is accordingly reduced. Second, a portion of the daytime workers who come into the impact area suffer immediate death. All of these people were participants in the workforce. Thus, immediate deaths in this group reduce the labor supply in CA-34 (Heatwole et al., 2014). Third, 65,438 (=0.75*87,250) night-time residents of the impact area evacuate from CA-34. We assume that the workers in this group cease offering labor to CA-34, further reducing CA-34's labor supply by 39,263 (= 0.6*65,438). Fourth, the surviving non-evacuated workers reduce their participation rate by 25% (see point 5).

The 65,438 people who are evacuated to locations outside CA-34 add to labor supply in those locations. Their impact effect on labor supply in each of RoLA, RoCalif, and RoUS is 9,816 (= 65,438 *0.6*0.75/3). This takes account of the baseline participation rate (0.6), the participation effect of the event (0.75), and our assumption that equal numbers of evacuees go to each of the three regions outside CA-34.

With regard to national labor supply, the event-related reduction in 2012 is the first, second, and fourth effects plus 0.25 times the third effect. In percentage terms, the reduction in labor supply is 0.070. Over time, the national labor supply effect gradually disappears: the people in the impact area at the time of the event eventually would have left the labor force even without the event.

4. Later Deaths, Morbidity, and Labor Supply

There are also thousands of later deaths caused by cancer related to exposure to radiation, and the number of cancer deaths is equal to about 17% of the number of immediate deaths (DNDO, 2013). In our simulation, we take account of later deaths in the labor supply. For working out the time path of labor supply effects, it is necessary to make assumptions concerning: (a) the time path of later deaths; (b) the age profile of the affected population at the time of the event; (c) the life expectancy of people in different age groups in the affected population in the absence of the event; and (d) the labor force participation behavior of the affected population particularly in the years immediately prior to cancer-related death. Assumptions with regard to (a)–(c) are informed by the experience of Hiroshima and by Census and life expectancy data. On (d) we assume: (i) labor force participation rates for members of the affected surviving population are reduced in year 1 to 75% of their baseline participation rates, namely, the rates that would have applied in the absence of the event; (ii) labor force participation rates for the affected surviving people recover over succeeding years, nearing 100% of their baseline participation rate by year 20; and (iii) labor force participation rates for the affected surviving people who are facing delayed death within three years fall to 75% of their baseline participation rate three years before death, 50% two years before death, and 25% one year before death (Heatwole et al., 2014).

Most of the later deaths occur many years after the event, and many of those affected by cancer will live beyond working age. Consequently, the effect on labor supply of later deaths and morbidity is minor, especially in our simulation period of 20 years.

5. Medical Expenditures

The cost of medical treatment in the year of the event is hundreds of millions of dollars (DNDO, 2013). We assume that treatment takes place in the regions to which the evacuees have relocated. Non-cancer medical costs arising from the event are assumed to gradually diminish over the next 20 years, reaching an accumulated total of $6 billion (Heatwole et al., 2014).

Beyond the shutdown period, we assume that additional medical expenditures are required to deal with cancer costs. The cumulative total is $0.33 billion (2012 dollars) after 20 years. This may seem a low cost, but recall that most of the cancer deaths (and therefore associated expense) occur well after 20 years.

We assume that health expenditures are half covered by public spending and half by private spending. In a national emergency, there may be a case for increasing the public contribution. However, those who can afford private health spending may prefer to pay their own expenses rather than wait for public provision.

6. Radiation Clean-Up and Decontamination Cost

We treat these costs as additional government spending, occurring for two years and amounting to tens of billions of dollars in each year (DNDO, 2013). We assume that the bulk of this expenditure, about 86%, is on labor drawn from the rest of LA. The remaining expenditure is attributed to CA-34.

7. Treatment of Capital

We assume that property damage (destruction of capital) caused by the detonation is $16 billion (Heatwole et al., 2014). This is 18% of CA-34's capital. As we will see, whether the destroyed capital is replaced is determined endogenously in the model simulation.

8. Aversion Behavior and the Return of the Evacuated Population

Following a nuclear attack, it is reasonable to suppose that workers will require elevated wages to induce them to take jobs in the region (CA-34) in which the event took place, even after the measurable danger from radiation has passed. Similarly, some goods and services produced in CA-34 may suffer stigmatization in the marketplace.

The labor market issue is handled via the equation specifying labor supply to CA-34. We set a shift variable in this equation to give a degree of aversion to working in CA-34. We assume the labor supply at any given wage will return to only 80% of its pre-event level following the two-year shutdown. For subsequent years, we maintain the initial aversion level. In sensitivity analysis not reported here (see Heatwole et al., 2014), we assumed a gradual lessening of aversion. As we will see, the path of the labor market aversion shift variable is important in determining the extent of long-run recovery in CA-34.

With regard to goods and services, we assume that international demand for exports exiting the United States from CA-34 moves 25% below baseline in 2012 and 2013.[19] In 2014, CA-34's international exports recover to approximately 90% of their baseline level. For exports from CA-34 to other U.S. regions, we assume no aversion beyond the period of the shutdown. However, these exports are very subdued because costs are high in CA-34, making the region uncompetitive.

After the shutdown period, business activity returns to the impact area, that is, the region exhibits resilience. This effect is endogenous in USAGE-TERM. Owners of surviving capital in the impact area restart businesses and offer jobs. Despite the aversion behavior described previously, workers flow back into the region to accept these jobs, although only at higher wages.

9. Financing: The Public-Sector Deficit and Foreign Debt

We assume that the government finances its event-related expenses in 2012 and 2013 by allowing a blowout in the public-sector deficit. This causes inflationary pressure in the United States and a loss of international competitiveness. Consequently, there is a deterioration in the trade accounts and a sharp increase in net foreign liabilities. In effect, the U.S. cleanup and other event-related expenditures are initially financed by foreigners. Beyond 2013, we assume that the United States gradually pays off the event-related foreign debt. It does this by adopting tighter fiscal policy (with lower public and private consumption) than would have applied without the event.

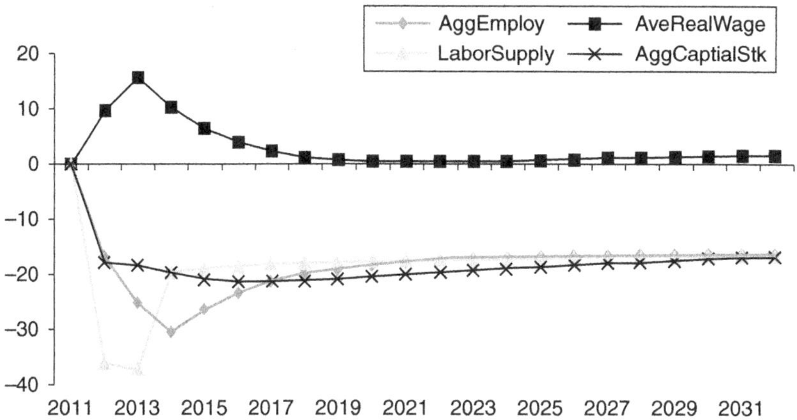

Figure 8.2. Impacts on labor market and capital in CA-34 (% deviation from baseline).

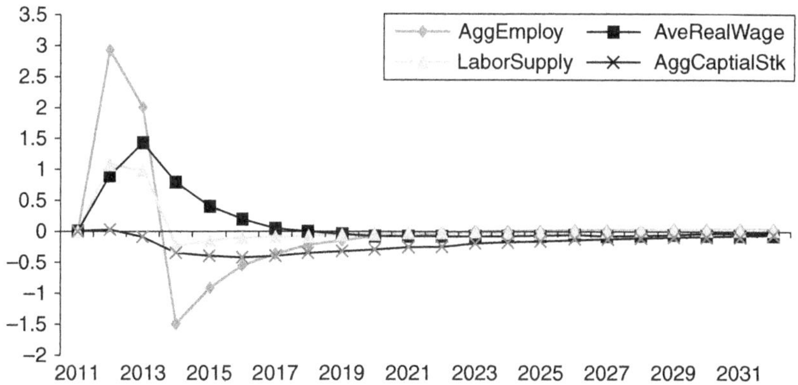

Figure 8.3. Impacts on labor market and capital in rest of LA (% deviation from baseline).

8.5.2 *Results*

Figures 8.2 to 8.9 show results for the four regions (CA-34, rest of Los Angeles, rest of California, and rest of the United States) and for the nation. All of the results are deviations from a no-event baseline. For example, Figure 8.2 shows that the event (nuclear attack in downtown Los Angeles) causes labor supply in CA-34 to be 36% less in 2012 than in the baseline.

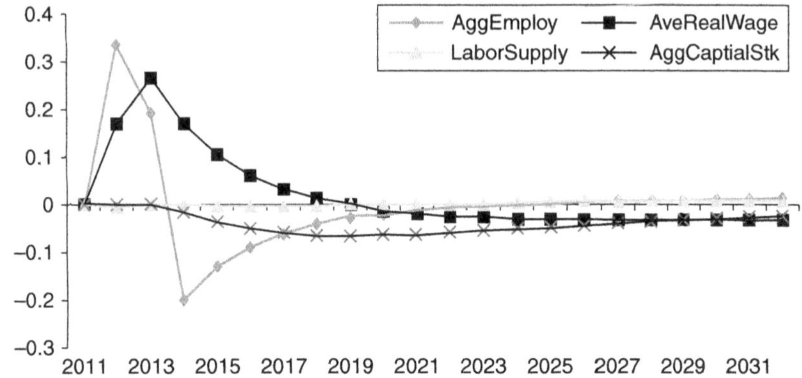

Figure 8.4. Impacts on labor market and capital in rest of California (% deviation from baseline).

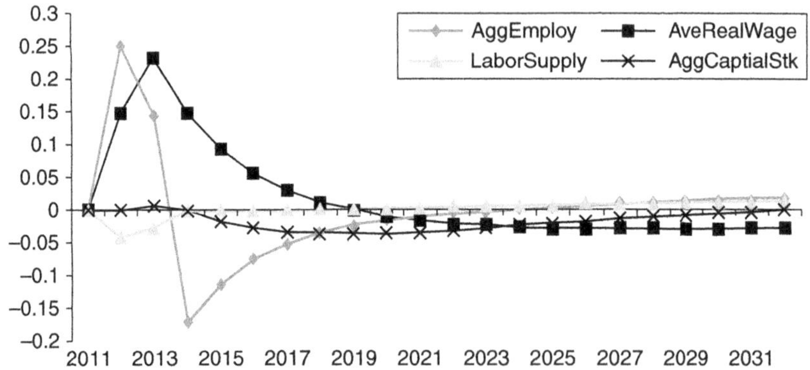

Figure 8.5. Impacts on labor market and capital in the rest of the United States (% deviation from baseline).

Apart from the short-run reduction in labor supply (imposed as a shock), Figure 8.2 shows that the main short-run macroeconomic effects in CA-34 of the event are:

- a reduction in available capital of 18% (point 7 in Subsection 8.5.1);
- a sharp increase in wage rates to overcome labor shortages caused by the evacuation (labor supply reducing) and the clean-up (labor demand increasing); and
- a reduction in employment of about 17%, implying a considerable tightening of the CA-34 labor market.

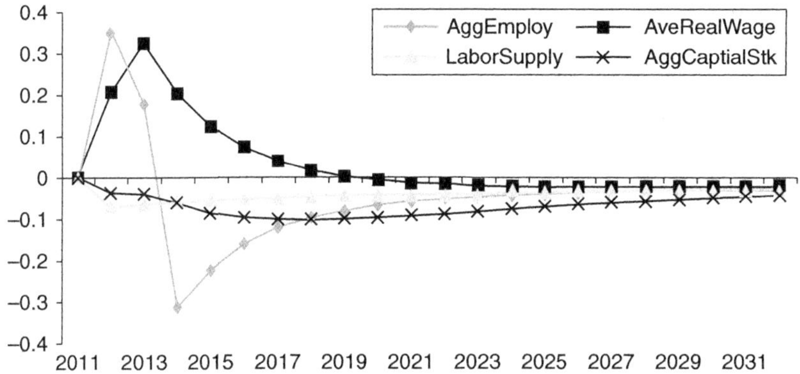

Figure 8.6. Impacts on national labor market and capital (% deviation from baseline).

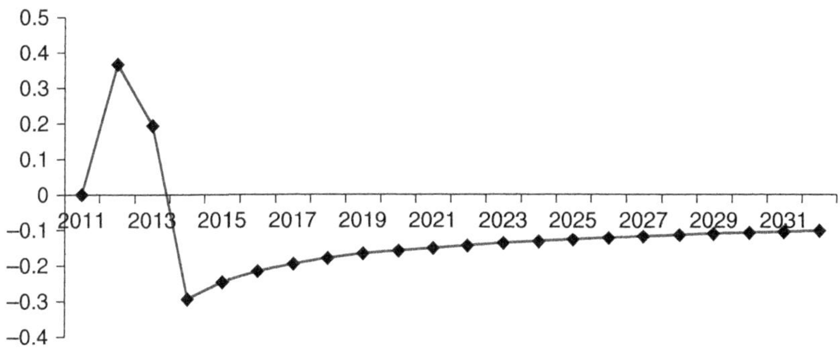

Figure 8.7. Impacts on national aggregate private consumption (% deviation from baseline).

In the long run, CA-34 stays permanently depressed in the event simulation, with employment, capital stocks, and labor supply all remaining about 20% below their baseline levels. Why 20%? In the long run, CA-34 must produce goods and services that are competitive with those produced in the rest of the United States. Consequently, CA-34 cannot sustain high wage rates relative to those in the rest of the nation. With CA-34 wage rates eventually returning to near their baseline levels (reflecting those in the rest of the United States), the aversion assumption outlined in point 8 in Subsection 8.5.1 means that labor supply to CA-34 is reduced in the event simulation by 20% relative to the baseline. With labor supply reduced by 20%, the rest of the CA-34 economy is also reduced by 20%.

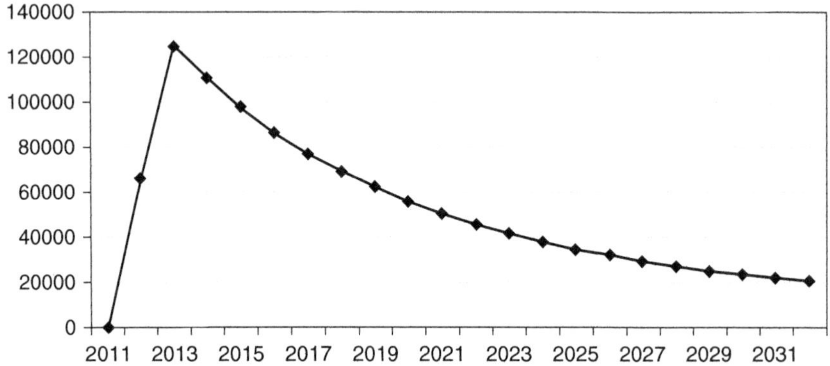

Figure 8.8. Impacts on net foreign liabilities ($m 2012 deviation from baseline).

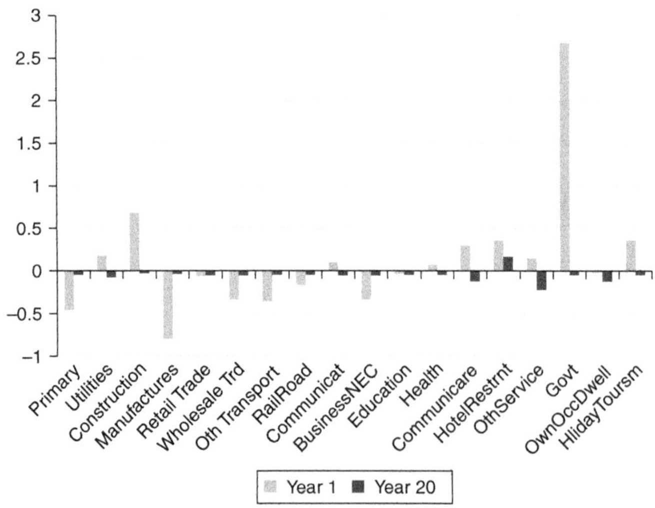

Figure 8.9. Impacts on national industry output (% deviation from baseline).

In the rest of Los Angeles, increased clean-up expenditures raise employment above the baseline in 2012 and 2013 (Figure 8.3). This is because the majority of clean-up services is supplied in each of these years from this region (point 6). Increased demands for health care services and accommodation also contribute to employment growth (points 5 and 3). In 2013, employment is subdued by rising wages, which are 0.9% above baseline in 2012 and 1.3%

above baseline in 2013. Since real wages persist above baseline once the clean-up has finished, employment falls to 1.5% below baseline in 2014. Increased demands by government and households in 2012 and 2013 crowd out other final demand expenditures, including investment. Therefore, for several years, aggregate capital stocks fall slightly below baseline before eventually recovering to near baseline levels.

The proportional impact of the event on the rest of California (see Figure 8.4) in 2012 and 2013 is much smaller than that on RoLA. For example, aggregate employment in 2012 rises 0.34% above baseline in RoCalif, compared with 2.9% in RoLA. This is because additional demands in RoCalif are limited to the health and accommodation needs of CA-34 evacuees. Unlike RoLA, RoCalif does not make any direct contribution to the clean-up effort in CA-34.

Real wage increases in CA-34 and RoLA during the clean-up force up wages in RoCalif and RoUS, though by smaller percentages. Since real wages adjust sluggishly, they persist above baseline in RoCalif and the RoUS (Figure 8.5) when the clean-up ends.

In RoLA, RoCalif and RoUS, employment moves below baseline in 2014 and stays noticeably below for several years until real wages have fallen sufficiently to enable labor demand to match labor supply. Once full wage adjustment has taken place, each of these regions has slightly larger employment than in the baseline, due to persistent labor market aversion in CA-34 and a consequent permanent exodus of workers.

Figures 8.6 to 8.8 refer to the national macro effects of the event. In the short run, the event boosts national employment, wage rates (Figure 8.6), and private consumption (Figure 8.7). However, in the long run, the event leaves the United States economically poorer as it pays off event-related debt (Figure 8.8 and point 9). Consumption, a reasonable indicator of economic welfare, is still 0.1% below baseline 20 years after the event (Figure 8.7).

Figure 8.9 shows industry effects at the national level. The event causes a sharp short-run increase in output of government activities. The construction industry also shows a significant national gain. By contrast, trade-related industries (primary and manufactures) contract, reflecting higher costs as real wages rise. In the long run, the event causes all industries other than health care and hotels/restaurants to be smaller than in the baseline. The long-run result for the health care industry is related to continuing treatment of the surviving affected population. The long-run result for hotels and restaurants is related to foreign visitors: tourism in the United States by foreigners is encouraged in the long run by a lower real exchange rate associated with paying off event-related foreign debt.

8.6 Concluding Remarks

Economy-wide modeling has had an important role in formulating policy regarding major disruptive events for many years. As explained by McGuire (1987), the development of economy-wide techniques, including input-output analysis and linear programming, received a major boost in the U.S. Defense Department as tools for managing resource allocation in World War II. Emerging since the 1960s, CGE modeling is now the most flexible and comprehensive framework for dealing with economy-wide issues. The CGE framework encompasses many links between different parts of the economy: input-output links, links via competition for labor and capital, links through public-sector budgets, and links through foreign exchange markets. CGE has also proved capable of incorporating factors specific to disaster analysis: resilience, fear, and aversion behavior.

To improve the accessibility of their work to policy makers, CGE modelers have adopted several strategies, including provision of training, textbook explanations, and back-of-the-envelope result justifications. Another technique that is currently under development is CGE reduced-form models. The simplest version is based on equation (8.8) in Subsection 8.2.2.[20] The matrix $(-A_1^{-1} * A_2)$ on the right-hand side gives the elasticities of endogenous variables (e.g., GDP, welfare, employment, etc.) with respect to exogenous variables (characteristics of a disruption scenario such as deaths, medical expenditures, clean-up expenditures, etc.). Key rows and columns (corresponding to key endogenous and exogenous variables) can be stored at the back of a spreadsheet program. Then users of the spreadsheet program, without any knowledge of CGE modeling or any need to rerun the CGE model, can vary the characteristics of the scenario [the v_2 vector in (8.8)] and, at least to a linear approximation, instantaneously compute the implications for endogenous variables. However, this simple approach has an important limitation: the elasticities matrix is too specific. It refers to a particular type of event in a particular region, e.g., a nuclear attack in downtown Los Angeles (see Section 8.5). Users of the spreadsheet might also apply the elasticity matrix worked out for Los Angeles in an application to a similar downtown location, say, in Chicago. But this would lack credibility.

Current research by CoPS and CREATE is aimed at improving the generality of quick spreadsheet computations that can be carried out without inputs from specialist CGE modelers. Our approach is to conduct a large number of USAGE-TERM simulations covering a variety of events occurring in diverse regions. These events will have different assumed fatality, injury, and capital-destruction characteristics. For each simulation, we will generate results for macro variables and for variables describing employment by industry in the region in which the incident takes place, in nearby regions, in the rest of the state, and in the rest of the United States. We will then regress these USAGE-TERM results against the assumed characteristics of the incident. This will give us regression equations into which DHS officials and other policy advisors will be able to insert characteristics of actual events or hypothetical events of interest on the right-hand side and compute fitted outcomes for macro and regional variables on the left-hand side. These fitted outcomes will approximate the results that would be obtained from a full USAGE-TERM simulation.

NOTES

1. The GTAP network alone contains 11,000 CGE models from government departments, academia, and private organizations in about 150 countries (see www.gtap.agecon.purdue.edu/).
2. See, for example, Burfisher (2011).
3. USAGE is an acronym for U.S. Applied General Equilibrium. TERM is an acronym for The Enormous Regional Model. Thus USAGE-TERM is a version of the USAGE model with enhanced regional detail. USAGE evolved from Australia's MONASH model (Dixon and Rimmer 2002). The original TERM model was created by Mark Horridge for Australia; see Horridge et al. (2005).
4. See Horridge et al. (2013) for a comparison of GEMPACK software, which relies on the percentage change representations of models, and GAMS software in which models are presented in level form. The comparison is highly favorable to GEMPACK.
5. Readers can find complete expositions of the CGE equation system in numerous books, including Johansen (1960), Dixon et al. (1982), Dervis et al. (1982), and Hertel (1997).
6. Some CGE models, even single-country models, identify many sources of supply for imports, making S large.
7. Equation (8.3) can be derived from a cost-minimizing problem of the form: choose XFAC(k,j) for all k to minimize $\sum_k PFAC(k,j) * XFAC(k,j)$ subject to

   $XOUTPUT(j) = CES_k\left(\dfrac{XFAC(k,j)}{AFAC(k,j)}\right)$ where the uppercase symbols, PFAC, XFAC, etc., refer to the levels of the variables denoted by the corresponding lowercase symbols in (8.3).
8. See Horridge et al. (2013) and Harrison et al. (2014).
9. This method was introduced in the seminal CGE work of Johansen (1960).
10. This method was introduced by Dixon et al. (1982) and later embedded in GEMPACK; see Pearson and Codsi (1991).
11. The regions can be subnational areas, nations, or groups of nations.
12. Published applications of USAGE include: Dixon and Rimmer (2004, 2010, 2011, 2013); Dixon et al. (2007a, 2007b, 2010, 2011a, 2011b, 2013, 2014); Fox et al. (2008); Gehlhar et al. (2010); Giesecke (2011); Giesecke et al. (2012); USITC (2004, 2007, 2009, 2011, 2013); and Zahniser et al. (2012).
13. See Dixon et al. (2007a).

14. See U.S. International Trade Commission (2004).
15. See www2.census.gov/econ2010/CBP_CSV/.
16. See, for example, Wittwer (2012).
17. The actual number of immediate deaths is sensitive or For Official Use Only (FOUO) information. No FOUO information appears in this chapter. Instead, all FOUO data are replaced with qualitative statements that give the reader a sense of the magnitude of the numbers, without revealing any FOUO information in the process.
18. We ignore the possibility that some of the night-time population was out of the impact area at the time of the event.
19. These exports are mainly services (e.g., financial services), CA-34 has no ports.
20. This version is implemented in Heatwole et al. (2014).

REFERENCES

Arrow, K., Chenery, H., Minhas, B., & Solow, R. (1961). Capital-labor substitution and economic efficiency. *Review of Economics and Statistics, 45*, 225–250.

Arrow, K., & Hahn, F. (1971). *General competitive analysis*. San Francisco, CA: Oliver and Boyd, Edinburgh; Holden-Day.

Boisvert, R. (1992). Indirect losses from a catastrophic earthquake and the local, regional, and national interest. In *Indirect economic consequences of a catastrophic earthquake* (pp. 204–266). Washington, DC: FEMA.

Brookshire, D., & McKee, M. (1992). Other indirect costs and losses from earthquakes: issues and estimation. In *Indirect economic consequences of a catastrophic earthquake* (pp. 235–267). Washington, DC: FEMA.

Bruno, M., Dougherty, C., & Fraenkel, M. (1970). Dynamic input-output, trade and development. In A. P. Carter & A. Brody (Eds.), *Applications of input-output analysis* (pp. 48–69). Amsterdam: North-Holland.

Burfisher, M. (2011). *Introduction to computable general equilibrium models*. Cambridge: Cambridge University Press.

Burns, W., & Slovic, P. (2010). Predicting and modeling public response to a terrorist strike. In P. Slovic (Ed.), *The feeling of risk: New directions in the perception of risk* (pp. 285–306). London: Earthscan Publishing.

Chenery, H., & Clark, P. (1959). *Interindustry economics*. New York: Wiley.

Debreu, G. (1959). *Theory of value: An axiomatic analysis of economic equilibrium*. New Haven, CT: Cowles Foundation, Yale University Press, pp. xi+114.

Dervis, K., de Melo, J., & Robinson, S. (1982). *General equilibrium models for development policy*. Cambridge: Cambridge University Press.

Dixon, P. B., Giesecke, J. A., Rimmer, M. T., & Rose, A. (2011b). The economic costs to the U.S. of closing its borders: A computable general equilibrium analysis. *Defence and Peace Economics, 22*(1), 85–97.

Dixon, P. B., Johnson, M., & Rimmer, M. T. (2011a). Economy-wide effects of reducing illegal immigrants in U.S. employment. *Contemporary Economic Policy, 29*(1), 14–30.

Dixon, P. B., Koopman, R. B., & Rimmer, M. T. (2013). The MONASH style of CGE modeling: A framework for practical policy analysis. In P. B. Dixon & D. W. Jorgenson (Eds.), *Handbook of computable general equilibrium modeling* (chapter 2). Amsterdam: Elsevier.

Dixon, P. B., Lee, B., Muehlenbeck, T., Rimmer, M. T., Rose, A. Z., & Verikios, G. (2010). Effects on the U.S. of an H1N1 epidemic: Analysis with a quarterly CGE model. *Journal of Homeland Security and Emergency Management, 7*(1), Article 75.

Dixon, P. B., Osborne, S., & Rimmer M. T. (2007b). The economy-wide effects in the United States of replacing crude petroleum with biomass. *Energy and Environment, 18*(6), 709–722.

Dixon, P. B., Parmenter, B. R., Sutton, J., & Vincent, D. P. (1982). *ORANI: A multisectoral model of the Australian economy*. Contributions to Economic Analysis 142. Amsterdam: North-Holland.

Dixon, P. B., & Rimmer, M. T. (2002). *Dynamic general equilibrium modelling for forecasting and policy: A practical guide and documentation of MONASH.* Contributions to Economic Analysis 256. Amsterdam: North-Holland Publishing Company, pp. xiv+338.

Dixon, P. B., & Rimmer, M. T. (2004). The US economy from 1992 to 1998: Results from a detailed CGE model. *Economic Record, 80* (Special Issue), pp. S13–S23.

Dixon, P. B., & Rimmer, M. T. (2010). Validating a detailed, dynamic CGE model of the U.S. *Economic Record, 86* (Special issue), pp. 22–34.

Dixon, P. B., & Rimmer, M. T. (2011). You can't have a CGE recession without excess capacity. *Economic Modelling, 28*(1–2), 602–613.

Dixon, P. B., & Rimmer, M. T. (2013). Validation in CGE modeling. In P. B. Dixon & D. W. Jorgenson (Eds.), *Handbook of Computable General Equilibrium Modeling* (chapter 19) Amsterdam: Elsevier.

Dixon, P. B., Rimmer, M. T., & Roberts, B. W. (2014). Restricting employment of low-paid immigrants: A general equilibrium assessment of the social welfare implications for legal U.S. wage-earners. *Contemporary Economic Policy, 32*(3), 639–652.

Dixon, P. B., Rimmer, M. T., & Tsigas, M. E. (2007a). Regionalizing results from a detailed CGE model: macro, industry and state effects in the U.S. of removing major tariffs and quotas. *Papers in Regional Science, 86*(1), 31–55.

Domestic Nuclear Detection Office (DNDO). (2013). Spreadsheet of consequence information for attacks using improvised nuclear devices, sent to the National Center for Risk and Economic Analysis of Terrorism Events (CREATE), University of Southern California. *For Official Use Only/FOUO.*

Evans, H. D. (1972). *A general equilibrium analysis of protection: The effects of protection in Australia.* Contributions to Economic Analysis 76. Amsterdam: North-Holland Publishing Company.

Fox, A., Powers, W., & Winston, A. (2008). Textile and apparel barriers and rules of origin: What's left to gain after the agreement on textiles and clothing? *Journal of Economic Integration, 23*(3), 656–684.

Gehlhar, M., Somwaru, A., Dixon, P. B., Rimmer, M. T., & Winston, A. R. (2010). Economywide implications from US bioenergy expansion. *American Economic Review: Papers & Proceedings, 100*(3), 172–177.

Giergenzer, G., & Selten, R. (Eds.) (2002). *Bounded rationality: The adaptive toolbox.* Cambridge, MA: MIT Press.

Giesecke, J. A. (2011). Development of a large-scale single U.S. region CGE model using IMPLAN data: A Los Angeles County example with a productivity shock application. *Spatial Economic Analysis* 6(3).

Giesecke, J. A., Burns, W., Barret, A., Bayrak, E., Rose, A., Slovic, P., & Suher, M. (2012). Assessment of the regional economic impacts of catastrophic events: CGE analysis of resource loss and behavioral effects of an RDD attack scenario. *Risk Analysis, 32*(4), 583–600.

Harrison, J., Horridge, J. M., Jerie, M., & Pearson, K. R. (2014). *GEMPACK manual*, GEMPACK Software, ISBN 978-1-921654-34-3, available at www.copsmodels.com/gpmanual.htm.

Heal, G., & Kunreuther, H. (2007). Modeling interdependent risks. *Risk Analysis, 27*(3), 621–634.

Heatwole, N., Rose, A., Dixon, P., Rimmer, M., Wittwer, G., & Wei, D. (2014). *Modeling the temporal and spatial consequences of nuclear terrorism events.* Final report to the U.S. Domestic Nuclear Detection Office (DNDO) by the National Center for Risk and Economic Analysis of Terrorism Events (CREATE), University of Southern California.

Hertel, T. (Ed.) (1997). *Global trade analysis: Modeling and applications.* Cambridge: Cambridge University Press.

Horridge, J. M., Madden, J. R., & Wittwer, G. (2005). Impact of the 2002–03 drought on Australia. *Journal of Policy Modeling, 27*(3), 285–308.

Horridge, M., Meeraus, A., Pearson, K., & Rutherford, T. (2013). Software platforms: GAMS and GEMPACK. In P. B. Dixon & D. W. Jorgenson (Eds.), *Handbook of computable general equilibrium modeling* (chapter 20, pp. 1331–1382). Amsterdam: Elsevier.

Johansen, L. (1960). *A multisectoral Study of economic growth.* Contributions to Economic Analysis 21. Amsterdam: North-Holland Publishing Company, pp. x+177.

Jorgenson, D., Jin, H., Slesnick, D., & Wilcoxen, P. (2013). An econometric approach to general equilibrium modeling. In P. B. Dixon & D. W. Jorgenson (Eds.), *Handbook of computable general equilibrium modeling* (chapter 17). Amsterdam: Elsevier.

Kajitani, Y., & Tatano, H. (2009). Estimation of lifeline resilience factors based on empirical surveys of Japanese industries. *Earthquake Spectra, 25*(4), 755–776.

Kasperson, R., Renn, O., Slovic, P., Brown, H., Emel, J., Goble, R., Kasperson, J., & Ratick, S. (1988). The social amplification of risk: a conceptual framework. *Risk Analysis, 8*(2), 177–187.

Leontief, W. W. (1936). Quantitative input-output relations in the economic system of the United States. *The Review of Economics and Statistics, 18*(3), 105–125.

Leontief, W. W. (1941). *The structure of the American economy 1919–1929.* Cambridge, MA: Harvard University Press.

Manne, A. S. (1963). Key sectors of the Mexican economy 1960–1970. In A. S. Manne & H. M. Markowitz (Eds.), *Studies in process analysis* (pp. 379–400). New York: Wiley.

McClelland, G., Schulze, W., & Hurd, B. (1990). The effect of risk beliefs on property values: A case study of a hazardous waste site. *Risk Analysis,* 10, 485–497.

McGuire, M. G. (1987). Defence economics. In J. Eatwell, M. Milgate, & P. Newman (Eds.), *The new Palgrave dictionary of economics* (pp. 760–762). London: McMillan.

Park, J. Y., Moore, J. E., & Richardson, H. W. (2013). The Gulf oil spill and economic impacts: Extending the National Interstate Economic Model (NIEMO) to account for induced impacts. *Journal of Homeland Security and Emergency Management, 10*(1), 231–244.

Pearson, K. R., & Codsi, G. (1991). The update and multi-step version of TABLO: User guidelines. GEMPACK document No 30, Impact Project, ISSN 1030–2514. University of Melbourne. Available from the Centre of Policy Studies, Victoria University, Melbourne.

Rose, A. (1995). Input-output economics and computable general equilibrium models. *Structural Change and Economic Dynamics, 6*(3), 295–304.

Rose, A. (2004). Defining and measuring economic resilience to disasters. *Disaster Prevention and Management, 13*(4), 307–314.

Rose, A. (2009). *Economic resilience to disasters.* CARRI Research Report #8, ORNL.

Rose, A. (2015). Macroeconomic consequences of terrorist attacks: Estimation for the analysis of policies and rules. In C. Mansfield & V. K. Smith (Eds.), *Benefit transfer for the analysis of DHS policies and rules* (pp. 172–200). Cheltenham, UK: Edward Elgar.

Rose, A., & Guha, G. (2004). Computable general equilibrium modeling of electric utility lifeline losses from earthquakes. In Y. Okuyama & S. Chang (eds.), *Modeling spatial economic impacts of natural hazards* (pp. 119–142). Heidelberg: Springer.

Rose, A., & Guha, G. & S. Liao (2005). Modeling regional economic resilience to disasters: A computable general equilibrium analysis of water service disruptions. *Journal of Regional Science, 45*(1), 75–112.

Rose, A., Liao, S., & Bonneau, A. (2011). Regional economic impacts of a Verdugo earthquake disruption of Los Angeles water supplies: A computable general equilibrium analysis. *Earthquake Spectra, 27*(3), 881–906.

Rose, A., Oladosu, G., Lee, B., & Beeler Asay, G. (2009). The economic impacts of the 2001 terrorist attacks on the World Trade Center: A computable general equilibrium analysis. *Peace Economics, Peace Science, and Public Policy 15,* Article 6.

Rose, A., Oladosu, G., & Liao, S. (2007). Business interruption impacts of a terrorist attack on the electric power system of Los Angeles: Customer resilience to a total blackout. *Risk Analysis, 27*, 13–31.

Sandee, J. (1960). *A long-term planning model for India.* New York: Asia Publishing House and Calcutta: Statistical Publishing Company.

Scarf, H. E. (1967). On the computation of equilibrium prices. In W. Fellner (Ed.), *Ten essays in honor of Irving Fisher.* New York: Wiley.

Scarf, H. E. (1973). *The computation of economic equilibria.* New Haven, CT/London: Yale University Press, pp. x + 249.

Shoven, J. B., & Whalley, J. (1984). Applied general-equilibrium models of taxation and international trade: An introduction and survey. *Journal of Economic Literature, 22*(3), 1007–1051.

Slovic, P. (1987). Perception of risk. *Science,* 236(4799), 280–285.

Sue Wing, I., Rose, A., & Wein, A. (2016). Economic consequences of the ARkStorm scenario. *Natural Hazards Review, 17*(4), 1–10.

U.S. International Trade Commission. (2004, 2007, 2009, 2011, 2013). *The economic effects of significant U.S. import restraints: Fourth, fifth, sixth, seventh & eighth update*. Investigation no. 332-325, Publications 3701, 3906, 4094, 4253, 4440.

Wittwer G. (Ed.) (2012). *Economic modeling of water: The Australian CGE experience*. Dordrecht, Heidelberg, Germany: Springer.

Zahniser, S., Hertz, T., Dixon, P. B., & Rimmer, M. T. (2012). Immigration policy and its possible effects on U.S. agriculture and the market for hired farm labor: A simulation analysis. *American Journal of Agricultural Economics*, *94*(2), 477–482.

9

Economic Resilience to Terrorism and Natural Hazards

Adam Rose

9.1 Introduction

The use of the term *resilience* in the context of natural disasters and terrorism has increased exponentially over the past 20 years. In 2005, it reached a major milestone with the publication of two prominent volumes. Economist Howard Chernick (2005) edited a collection of papers on the resilience of New York City's economy in the aftermath of 9/11, and planners Lawrence Vale and Thomas Campanella (2005) edited a volume of essays on a range of disasters throughout world history, including earthquakes, floods, epidemics, and wars. However, only one of the nearly 30 articles in these volumes provided an explicit definition of *resilience*. Over the course of the past 10 years, the pendulum has swung the other direction with the offering of dozens of alternative definitions of *resilience*. Moreover, the term is in such widespread use and has been applied so broadly that it is in danger of merely becoming a popular buzzword, thereby undercutting its substance and usefulness.

Over the past 10 years, researchers at the Center for Risk and Economic Analysis of Terrorism Events (CREATE) have answered the following important questions (see, e.g., Rose, 2009a; Rose & Krausmann, 2013; Rose, 2017):

- Is resilience a meaningful concept? Yes, it is, especially from an economic perspective.
- Can resilience be rigorously defined? Yes, though it is complex and has multiple dimensions.
- Can resilience be empirically measured? Yes, and an operational metric has been established and successfully applied in several contexts.
- How effective has resilience been today? It has been very effective, and moreover has been relatively low-cost and hence cost-effective.

- Can a meaningful resilience index be established? Yes, if it has solid conceptual grounding, is based on actionable variables, and if appropriate weights for individual indicators can be specified.

Advances in conceptualizing and measuring resilience have contributed significantly to the establishment of CREATE's Economic Consequence Analysis (ECA) framework (Rose, 2009b, 2015; Rose, Prager, Chen, & Chatterjee, 2017). This approach has significantly advanced the traditional hazard loss estimation methodology by broadening its scope through the inclusion of critical features previously neglected or given short shrift. In addition to resilience, this includes behavioral linkages, which reflect responses such as fear arising from the social amplification of risk or stigma effects (see, e.g., Giesecke et al., 2012; Rose et al., 2015). While studies of the effects of resilience indicate that it can reduce economic consequences of disasters, in terms of losses in GDP and employment, by as much as 90%, these behavioral effects relating to fear can potentially increase the negative impacts by one or two orders of magnitude.

Other factors incorporated into the framework include the impacts of remediation or decontamination, which are especially large in the case of insidious threats stemming from chemical, biological, radiological, or nuclear (CBRN) threats. Unlike benefit-cost analysis, which automatically considers them a cost, the ECA framework evaluates them in terms of their bottom-line impact on macroeconomic indicators, often depending on whether the economy is operating at or near full employment. Also, moving it toward further consideration of the cost side of the ledger, the CREATE ECA framework has included mitigation and spillover effects stemming from its implementation, such as delays, invasion of privacy, and the impact on the business environment, associated with countermeasures like traffic stops, bag and parcel checks, and closed-circuit television (Rose et al., 2014).

This chapter explains how the aforementioned critical questions surrounding economic resilience have been answered, modeled, and applied over the past decade. The focus is on the economic dimensions, but we begin with a discussion of the use of the term *resilience* in other disciplines, and conclude that there are many more commonalities than differences across them. We then define economic resilience in detail, including its various subtleties such as static vs. dynamic and inherent vs. adaptive resilience. We also discuss the role of resilience at the micro, meso, and macroeconomic levels. We illustrate many of these differences in exploring the time path of resilience. Next, we present an operational metric of resilience that is relatively simple and straightforward, but has not been supplanted by more sophisticated definitions and metrics. We then discuss the application of the metric by CREATE researchers and others in measuring resilience in case studies of simulated and actual events. We then turn to the discussion of the recent flurry of activity on construction of resilience indices, pointing out that most of them are misguided because they simply view resilience as the flip side of vulnerability. We emphasize the importance of actionable variables that can lead to recovery and how to weight these variables in the overall index.

At the outset, we note that research on resilience is split into two camps. About half of the researchers view resilience as any action that can reduce losses from disaster, ranging from pre-disaster mitigation to post-disaster recovery. Not surprisingly, this group is dominated by engineers, whose work is primarily in the area of mitigation (see, e.g., Bruneau et al., 2003; Haimes, 2006). The other camp focuses on resilience following a disaster. Steps can be taken to enhance resilience, noting that it is very much a process, but such measures are usually not implemented until afterward (e.g., stockpiling of critical materials, development of emergency plans). Another aspect of the difference is that the former group typically measures resilience in terms of reduction in property damage, while the latter focuses on the reduction of the *flow* of goods and services stemming from the property *stock*, often referred to as business interruption (BI). While property damage takes place at a point, or short period of time, BI begins at that point and continues until the economy has recovered or reached a new normal (Rose, 2004). As such, it is more complex because it involves behavioral and policy responses. Recent events indicate that BI can be as large as or larger than property damage following a disaster. For example, Blomberg and Hess (2009), Roberts (2009), Rose and colleagues (2009), and Werling and Horst (2009), utilizing a variety of methodologies, all found that the reduction in U.S. GDP following the World Trade Center attacks exceeded property damage by fourfold. The BI associated with Hurricane Katrina began to exceed the property damage several years ago, in part because New Orleans has not recovered according to most definitions. We focus on the second approach, noting that much of our analysis is applicable to mitigation as well.

9.2 Defining *Resilience* across Disciplines

The etymology of *resilience* is the Latin term *resilio*, meaning to rebound. Although published accounts of its use date back to ancient Rome in Cicero's *Orations* (Alexander, 2013), and to some early twentieth-century physicists and psychologists (Manyena, 2006), ecologists were the first to embrace, formulate, and make extensive use of the general concept of resilience more than 30 years ago (see, e.g., Holling, 1973). Since then, it has been adapted or reinvented for the case of short-term disasters (see, e.g., Bruneau et al., 2003;

Rose, 2004, 2007; Tierney, 1997) and long-term phenomena, such as climate change (see, e.g., Dovers & Handmer, 1992; IPCC, 2007). The analysis of economic resilience benefits from a comparison of its definitions in ecology, engineering, organizational behavior, planning, psychology, and sociology over the past 40 years. We focus here on points of agreement. This is the basis for establishing criteria for operational metrics that are consistent with fundamental principles, the needs of potential users, and the practical matters of data availability and computational manageability.

9.2.1 *Ecological Origins*

Ecologists have pioneered a useful, broad definition of resilience relating to the survival of complex systems. Holling's definition is "the ability of systems to absorb changes … and still persist" (1973, 17). He sometimes refers to it as "buffer capacity," and resilience is measured in this paradigm in relation to the size of the shock that is absorbed. Pimm (1991) is usually cited as the source of an alternative ecological emphasis in the definition of resilience in terms of the speed at which the system returns to equilibrium. In most disciplines, the term *resilience* is more in line with the buffer concept, as the ability to mute the influence of the external shock. It is not just the decrease in activity, but rather the decrease relative to the potential decrease from the external shock. Perrings also defines resilience in a relative manner: "As a first approximation, this may be measured by an index of the level of pollution or depletion relative to the assimilative or carrying capacity of the ecological system concerned" (2001, 322). Subsequently, Perrings defines it in terms of the "gap between current and critical loads" (p. 323) to the ecosystem and even to the ecological-economic system.

9.2.2 *Individual Resilience*

At the most fundamental level, resilience pertains to how individuals cope with crises. People everywhere are very adept at both self-preservation and extending help to others in a crisis.

Another source of individual resilience stems from various economic roles, including producer, consumer, and provider of labor and capital services. Economic incentives help promote resilience, though this is affected by other key considerations. First, workers and managers only focus on the enterprise once they know their families are safe and receiving the proper care. Second, disasters instill fear in people, which affects some of their behavior. This fear can erode individual resilience. Recent research has focused on ways to reduce this fear through improved risk communications, which is yet another way of strengthening resilience (Rosoff et al., 2013).

Flynn (2008) has emphasized the key role of individuals in resilience, and sees resilience as empowering. It provides people the opportunity to reach their full potential in a crisis. It also provides cohesion to the community and nation. Flynn points out that "a terrorist chooses battlegrounds that are likely to be occupied by civilians, not soldiers." He notes the importance of resilience as a weapon against the spread of fear, which is one of the terrorist's greatest objectives. One of the dividends of empowering individuals is that doing so releases an enormous amount of energy and skills to cope with disasters. Flynn also notes that it lessens the paternalistic role of government in this area. The actions of many governments that consistently bail out disaster victims, even if they have engaged in moral hazard (e.g., continuous rebuilding in flood plains) is a classic example (see, e.g., Mileti, 1999).

Another strong role for the individual stems from a major theme of development – public participation in decisions and processes. A related key theme is fairness or equity, one version of which is the basis for promoting equality in both the participation and in the outcomes of resilient activities. It includes special consideration for the aged, the infirm, women, and racial/ethnic minorities.

Synergies and economies of scale and scope arise when individuals band together to address a crisis. Specialization, organizational memory, and official sanctioning are some of the many reasons for the formation of institutions in this area. Similar motivations, as well as those relating to social cohesion, are the basis for community resilience at the neighborhood, town/city, province/state, and national levels.

9.2.3 *Community Resilience*

Adger (2000) was one of the first to extend the ecological definition of resilience to human communities as a whole. He measured *social resilience* as related to social capital and in terms of economic factors (e.g., resource dependence), institutions (e.g., property rights), and demographics (e.g., migration). Norris and colleagues (2008) have approached the matter in a similar fashion for *community resilience*. They developed a framework for it that encompasses stress, adaptation, wellness, and resource dynamics. They state that "community resilience is a process linking a network of adaptive capacities (resources with dynamic attributes to adapt after a disturbance or adversity)." Adaptive capacities include economic development, social capital, information and communication, and community confidence. Community is defined in a broad sense to include both the built and natural environments and the economy in addition to the social structure. Still, the major focus is less on community organization than on preventing injury, both physical and mental.

Norris and colleagues (2008) defined population wellness as "a community-level outcome indicative of a successful adaptation, defined as high and non-disparate levels of mental and behavioral health, role functioning, and quality of life in constituent populations." They use "wellness" as an indicator of the success of adaptive resilience at the individual level. They acknowledge that the community as a whole is greater than the sum of the parts. Community adaptation is then defined as population wellness, linked to a high prevalence of individual wellness in the community.

In the Norris framework, resilience stems from a set of networked adaptive capacities. The adaptive aspect stems from a combination of resources themselves and community responses to crises. Regarding network resources, Norris and colleagues (2002) emphasize the importance of the resource base in economic development and note that socioeconomic status is a main indicator of vulnerability, especially in developing countries. Resilience is a function of the size and diversity of the resource base and of resource equity. Social capital is a second important type of adaptive capital and stems from people using social networks primarily for personal gain. Support networks are especially important in disaster communication and recovery. For example, those linked into networks are more likely to evacuate and those connected to networks are more likely to engage in mutual support and cohesive behavior. Public participation is also a key feature of community resilience.

9.2.4 *Engineering-Based Definitions*

Bruneau and colleagues (2003) provide a comprehensive analysis of the many aspects of earthquake loss reduction, all under the heading of resilience. The authors apply the concept at four levels: technical, organizational, social, and economic. They contend that resilience has four dimensions: robustness (ability to withstand a shock), redundancy (e.g., parallel or back-up systems), resourcefulness (stabilizing measures), and rapidity (with respect to rebuilding and recovery). Bruneau and colleagues also stipulate that the resilience of a system has three aspects: reduced probability of failures, reduced consequences from failures, and reduced time to recovery. These pertain more to the general risk equation, but have obvious overlaps with the four dimensions.

A major criticism of the definition of Bruneau and colleagues (2003) is that it lumps all aspects of hazard loss reduction under the banner of resilience, including mitigation. This is in contrast to the position of Klein and colleagues (2003), and many others, to keep the definition of resilience from becoming too broad. They propose the concept of "adaptive capacity" as the more appropriate umbrella concept that covers many of the features identified by Bruneau and colleagues. This is also more consistent with defining resilience as an outcome or system attribute, rather than as a tactic like mitigation. Chang and Shinozuka state that: "It is useful to view robustness and rapidity as the desired *ends* of resilience-enhancing measures. Redundancy and resourcefulness are some of the *means* to these ends" (2004, 741).

It appears that Bruneau and colleagues, as well as non-engineers such as Mileti (1999), have envisioned a goal of a community that is able to take many steps to minimize its vulnerability to hazards. *Resilience* has become a convenient term to characterize all of these possibilities. However, this broad usage is inconsistent with the etymology of the term in general and its use in ecology, economics, and other areas of research. Ideally, another term can be found to characterize this ideal community response, so that the term *resilience* can be applied to the subset of characteristics to which it is best suited.

9.2.5 *Organizational Behavior*

Organizational behavior, and the closely related area of institutional behavior, focus on resilience as a process (Hill & Paton, 2005). As such, it is a strategy in risk management under the subheading of crisis and continuity management. Paton and Johnston (2001) define resilience in this dimension as "a capacity of people and systems that facilitate organizational performance to maintain functional relationships in the presence of significant disturbances as a result of a capability to draw upon their resources and competencies to manage the demands, challenges and changes encountered." This viewpoint extends even more fundamentally to natural ecosystems, whereby The Resilience Alliance (2005) includes as one of its three dimensions of resilience "the degree to which the system is capable of reorganization." Adger and colleagues (2005) extend this to the social-ecological nexus.

Comfort (1994) confines resilience to actions and processes after the event occurs, or to the *consequences* of failure. This also relates to process-oriented counterparts of the concept of dynamic resilience, where the focus is not on attaining a target level of output, but rather a target level of "functioning." However, the trajectory of this functioning is clear from the major themes of nonlinear and adaptive dynamics (Comfort, 1999). It also leaves no doubt that the dynamic version of resilience (the rapidity to bounce back) is uniquely applicable to the post-disaster stages. Moreover, the recovery process this characterizes is another way of reducing the consequences of the hazard ensuing from structural or system damage. Manyena (2006) contends that resilience has evolved from an emphasis on outcomes to an emphasis on process in holistic terms (see also Pfefferbaum et al., 2005).

Klein and colleagues (2003) have taken this even further to suggest that resilience goes beyond the Holling definition to include the functioning and interaction of interlinked systems (see also UN/ISDR, 2002). However, this still does not go as far as suggesting that resilience includes all aspects of adaptation or mitigation.

In contrast to resilience activities emphasized in the economics literature, the focus of organizational theory is on "competencies and systems" (Hill & Paton, 2005; see also an extension of this theme to the community as a set of networked adaptive capacities). The relationship between the two approaches can be viewed as follows: most standard treatments of resilience in engineering and economics identify a set of options and assume that managers can optimize among their choices (see, e.g., Rose & Liao, 2005). Organizational analysis identifies vulnerabilities and limitations in managerial abilities and how they can be overcome through resilience. The economics approach to reconciling these two views would be to assume some form of "bounded rationality" (see, e.g., Gigerenzer & Selten, 2002; Kunreuther et al., 2013) and to view managerial resilience as an improvement over the basic outcome. Hill and Paton (2005) analyze several aspects of the theory and practice of business continuity management and how it relates to resilience. They emphasize that a major prerequisite of success in this area is the willingness of an organization to adapt to its new environment.

9.2.6 *Planning*

Sustainable communities and the supporting theme of smart growth emanate from the collaborative visions of ecologists, sociologists, geographers, economists, and planners. Planners have been adept at developing practical approaches to the broader design, while the other disciplines have been more niche-oriented, including the nexus of ecological economics, in reorienting individual business operations to principles of industrial metabolism (see, e.g., Ayres & Simonis, 1989; Daly & Farley, 2004).

The planning profession has as a goal the creation of hazard-resilient communities (Burby et al., 2000; Godschalk, 2003), primarily through comprehensive land-use strategies. This holistic approach is superior to the piecemeal way that ordinary hazard mitigation is usually promulgated, which has actually enticed development in hazardous areas. For example, the presence of dikes and levees in New Orleans gave residents a feeling of false security. Many similar examples have led to the general trend of fewer disaster events, but the ones now taking place have much larger damages. Smart growth has tended to avoid such outcomes. Mileti (1999) has stated that "no single approach to bringing sustainable hazard mitigation into existence shows more promise at this time than increased use of sound and equitable land use management."

Burby and colleagues (2000) identify four major themes related to integrating mitigation in land-use planning in to promoting community resilience, but only one of them, and only in part, pertains to the post-disaster period. This is an example of the tension in the planning field about terminology, similar to the discussion in other fields. Godschalk concludes that "Traditional hazard mitigation programs have focused on making physical systems resistant to disaster forces" (2003, 137). He goes on to state, however, that "future mitigation programs must also focus on teaching the city's social communities and institutions to reduce hazard risk and respond effectively to disasters, because they will be the ones most responsible for building ultimate urban resilience." In fact, Geis (2000) has explicitly stated a preference for the term "disaster-resistance" with respect to planning themes and practices in this area, concluding it is more appropriate and attractive than is "disaster resilient." Godschalk (2003) makes the point, however, that "Resilient cities are constructed to be strong and flexible, rather than brittle and fragile." It is this flexibility (adaptability) that is the key to resilience as interpreted by others (e.g., Comfort, 1999; Rose, 2007; Zolli & Healy, 2012). Godschalk summarizes the work of several researchers to identify eight categories of resilience responses, seven of which have been emphasized by Rose (2004, 2007): redundant, diverse, efficient, autonomous, strong, adaptable, and collaborative.

9.3 Defining Economic Resilience

9.3.1 *Basic Attributes*

Resilience has four roles in the economics literature. Most generally, it is noted in vague terms, as an attribute of the economy in studies of economic shocks (see, e.g., Dhawan & Jeske, 2006). In ecological economics, it is a major focus of analysis as a key attribute necessary for sustainability (see, e.g., Folke, 2006). Some initial attempts have been made to extend this research to the socioeconomic arena and have it overlap with the study of institutions (see, e.g., Levin, 1998). In the disaster literature, it has been an important dimension of hazard economic loss estimation and terrorist consequence analysis (Rose, 2009b). Following Rose (2004, 2009a), we offer the following definitions of resilience to disasters:

Here we provide specific definitions of *economic resilience* by linking it to the common essence of resilience definitions in the previous section.

In general, *static resilience* refers to the ability of the system to maintain a high level of functioning when shocked. *Static economic resilience* is: the efficient use of remaining resources at a given point in time. It refers to the core economic concept of coping with resource scarcity, which is exacerbated under disaster conditions.

Dynamic resilience refers to the ability and speed of the system to recover. *Dynamic economic resilience* is: the efficient use of resources over time for investment in repair and reconstruction. Investment is a time-related phenomenon – the act of setting aside resources that could potentially be used for current consumption in order to reestablish productivity to be used in the future. Static economic resilience does not completely restore damaged capacity and is therefore not likely to lead to complete recovery.

Note that the definitions are couched in terms of function, typically measured in economics as the *flow* of goods and services, such as gross domestic product (GDP) or broader measures of human well-being, as opposed to property damage. It is not the property (capital *stock*) that directly contributes to economic welfare, but rather the flows that emanate from these stocks either for businesses or for households. Two things should be kept in mind. First, while property damage takes place at a point in time, the reduced flow, often referred to on the production side as business interruption (BI), just begins at the time of the disaster, but continues until the system has recovered or has attained a "new normal." Second, the recovery process, and hence the application of resilience, depends heavily on the behavior of economic decision makers and on public policy. Of course, recovery is a multifaceted activity. It is not as simple as, for example, rebuilding a school destroyed by an earthquake, hurricane, or armed attack.

Resilience ability implies a level of attainment will be achieved. Hence, the definitions of economic resilience are contextual – the level of function has to be compared to the level that would have existed had the ability been absent. This means a reference point must be established. In the case of static economic resilience ability, this reference point is the worst-case outcome. In the case of dynamic resilience ability, this reference point is the normal recovery path.

Another important distinction exists between *inherent* and *adaptive* resilience. The former refers to aspects of resilience already built into the system, such as the availability of inventories, excess capacity, substitutability between inputs, and contingent contractual arrangements accessing suppliers of goods from outside the affected area (imports). Resilience capacity can be built up through these means and is then accessed after the disaster. Adaptive resilience arises out of improvisation under stress, such as Draconian conservation otherwise not thought possible (e.g., working many weeks without heat or air conditioning), changes in the way goods and services are produced, and new contracting arrangements that match customers who have lost their suppliers with suppliers who have lost their customers.

We analyze resilience pertaining to the economy at three levels:

- Microeconomic (individual business or household)
- Mesoeconomic (individual industry or market)
- Macroeconomic (combination of all economic entities, including their interactions)

Underlying each of the levels of analysis is an extensive body of economic principles, such as consumer and producer theory, the theory of markets, industrial organization, macroeconomic theory, and growth theory. Over the years these have been infused with the complexities of uncertainty, expectations, and bounded rationality that make them even more applicable to resilience to disasters.

At the micro level, on the business supplier side, static economic resilience includes redundant systems, improved delivery logistics, and planning exercises. Several options also exist on the business customer side. Broadening the supply chain (see, e.g., Sheffi, 2005) by expanding the range of suppliers in place or on a contingency basis is an increasingly popular option. Another is conservation of resources made scarcer by the disaster, such as automated controls to monitor the flow of water to help make sure they are used only in times when they are needed and to reduce nonessential uses. Conservation is only minimally inherent because economists typically assume that most available efficiencies in resource use are currently being utilized; thus, most resilient conservation options pertain to adaptive applications. All inputs (capital, labor, infrastructure services, and materials) can be conserved. The major obstacle is the necessity of the input in the production process. Other resilience tactics include input and import substitution, use of inventories and excess capacity, cross-training workers, relocation, and production recapture (working overtime and extra shifts when functionality is restored to make up lost production). Most of the same resilience tactics associated with businesses are applicable to government and household operations as well.

At the mesoeconomic level, resilience can bolster an industry or market and include, for instance, industry pooling of resources and information and innovative pricing mechanisms. What is often less appreciated is the inherent resilience of market prices that act as the "invisible hand" to guide resources to their best allocation in the aftermath of a disaster (see, e.g., Horwich, 1995). Some pricing mechanisms have been established expressly to deal with such a situation, as in the case of non-interruptible service premia that enable customers to estimate the value of a continuous supply of electricity and to pay in advance for receiving priority service during an outage. The price mechanism is a relatively costless guide to redirecting goods and services. Price increases, to the extent that they do not reflect "gouging," serve a useful purpose of reflecting highest-value use, even in the broader social setting. Moreover, if the reallocation violates principles of equity (fairness), the outcomes can be adjusted by income or material transfers to the needy. Of course, markets are likely to be damaged by a major disaster, in an analogous manner to buildings and humans. In this case, we have two alternatives for some or all of the economy: 1) substitute centralized decree or planning, though at a significantly higher cost of administration; 2) bolster the market, such as by improving information flows (e.g., the creation of a clearinghouse to match customers without suppliers to suppliers without customers).

At the macroeconomic level, resilience is very much influenced by interdependencies between sectors. Consequently, macroeconomic resilience is not only a function of resilience measures implemented by single businesses, but it also is determined by the actions taken by all individual companies and markets including their interaction. Examples of resilience options at the macro level would be economic diversity to buffer impacts on individual sectors or geographic proximity to economies not affected by disaster to facilitate access to goods or aid. Others include fiscal strategies (e.g., infrastructure spending to boost the affected economy) and monetary policy (e.g., keeping interest rates low to stimulate private-sector reinvestment). The macro level overlaps with the popular focus on "community resilience" and represents a more holistic picture. However, economists have long appreciated the importance of microeconomic foundations of macroeconomic analysis for several reasons. First, the macroeconomy is composed of individual building blocks of producer and consumer behavior as underpinnings for macroeconomic considerations stemming from group interactions. Second, behavioral considerations are best addressed first at the most elemental level because of the prominence of individual motivations for survival and coping mechanisms in anticipation of and in response to disasters.

The previous examples relate primarily to *static economic resilience*. *Dynamic economic resilience* is applicable at all three levels as well as in terms of expediting the recovery process and enhancing its outcome. At the micro level, this can be promoted through rapid processing of insurance claims and arranging financing to facilitate repair and reconstruction. At the meso and macro levels, it includes hastening and improving the economic effectiveness of the recovery process by improving logistics and coordinating recovery across sectors. Cross-cutting all three levels is the need for adapting to changing conditions by promoting flexibility and translating short-run practices in to sustainable ones through a continuous learning process (see, e.g., Chang & Rose, 2012; Rose, 2014; Zolli & Healy, 2012).

Overall, there are more commonalities than differences in the various economic and other definitions of resilience. All definitions relate to reducing losses from disasters. All emanate from a survival motivation. Nearly all emphasize the importance of adaptive behavior. Most view resilience as a process and emphasize the need to expand resilience capacity at multiple levels. Most emphasize interactions within a broader community. Most stress the importance of natural, societal, and/or produced resources and their wise utilization in a crisis. Moreover, there are few inconsistencies between the definitions. They differ mainly in terms of some unique terrain they stake out from their own discipline.

Note, however, that resilience is sometimes conflated or confused with related terms such as *vulnerability* and *sustainability*. See Rose (2013, 2014, 2017) for a more detailed discussion.

9.3.2 *Time-Path of Resilience*

Figure 9.1 illustrates some of the temporal features of resilience and aids in illustrating some of the distinctions between various dimensions of resilience made in the previous subsection. It depicts the normal time-path of a system like the economy, the initial ability to limit the consequences of a shock, and rebounding and learning aspects, such that the system might even be able to attain a stronger growth path if recovery is undertaken properly.

Following Rose (2009b), Chang and colleagues (2011), and Petersen and Rose (2013), in Figure 9.1, the vertical axis represents system performance, and the horizontal axis represents time. The performance increases starting at Point A until the system is hit by a shock at Point B. The shock causes it to degrade to a lower level at Point D. *Robustness* is a measure of the difference between the drop to the level at Point D and a maximum drop for the specific disaster, which we can, for the purpose of illustration, say is a zero performance level. Resilience is of two types depicted in relation to Points F and G. *Static economic resilience* begins at Point D and refers to using the *remaining resources* to move to a higher level of performance during recovery. *Dynamic economic resilience* begins at Point E and refers to the ability to recover more quickly by more effective *repair and reconstruction*. As a result of static resilience, the

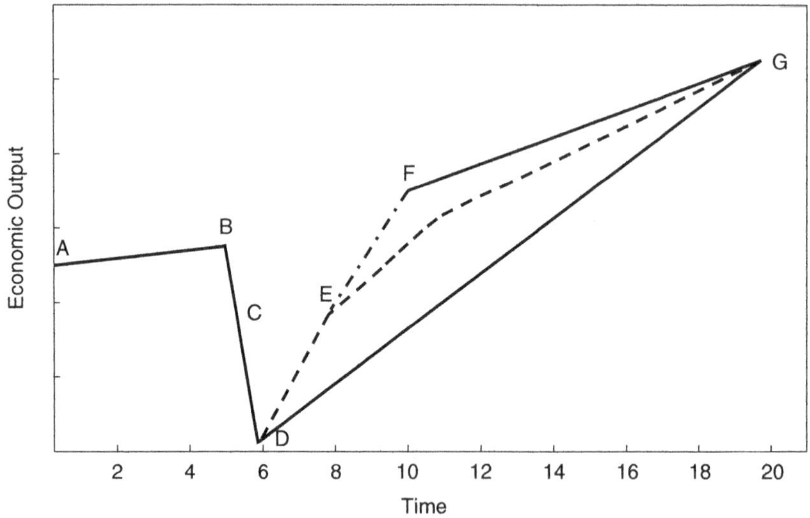

Figure 9.1. The time-path of resilience.

system recovers to its pre-disaster trajectory level at Point G at best. It may not be able to return to the original path at all if investment in repair and reconstruction is not undertaken. Dynamic resilience, however, enables the system to recover earlier than it otherwise would have at Point F. The magnitude of static resilience is the reduction in losses corresponding to the area (difference) between the base recovery path and the dashed arc (DEG). The magnitude of dynamic resilience corresponds to the area (difference) between the dashed arc and the dot/dash and then solid line (EFG). An alternative measure of dynamic resilience would be the difference between a normal repair/reconstruction path and an accelerated one.

Resilience need not be monotonically increasing, meaning that the recovery path need not always be increasing. One reason is that resilience can be degraded or overwhelmed, causing dips in the recovery path, as in the case of Hurricane Katrina. The preceding discussion is most appropriate in relation to engineering and economic systems. It does not, however, adequately convey many aspects of resilience as a process, which is much more important in organizational and community resilience.

While the entire time-path of resilience is key to the concept for many analysts, it is important to remember that this time-path is composed of a sequence of individual steps. Even if dynamics are the focal point, it is important to understand the underlying process at each stage, i.e., why a level of function is achieved and why that level differs from one time period to another. As presented here, static resilience helps explain the first aspect, and changes in static resilience, along with repair and reconstruction of the capital stock, help explain the second.

9.4 An Operational Metric

Following Rose (2004, 2009a), we provide an admittedly crude but operational metric of resilience. *Direct Static Economic Resilience* (*DSER*) refers to the level of the individual firm or industry (micro and meso levels) and corresponds to what economists refer to as "partial equilibrium" analysis, or the operation of a business or household entity itself. *Total Static Economic Resilience* (*TSER*) refers to the economy as a whole (macro level) and would ideally incorporate what is referred to as "general equilibrium" effects, which include all of the price and quantity interactions in the economy, macro-aggregate considerations, and the ramifications of fiscal, monetary, and security policy.

An operational measure of *DSER* is the extent to which the estimated direct output reduction deviates from the likely maximum potential reduction given an external shock, such as the curtailment of some or all of a critical input. **In essence, *DSER* is the percentage avoidance of the maximum economic disruption that a particular shock could bring about.**

A major measurement issue is what should be used as the maximum potential disruption. For ordinary disasters, a good starting point is a linear or proportional relationship, as, for example, between an input supply shortage and the direct disruption to the firm or industry. Note that while a linear reference point may appear to be arbitrary or a default choice, it does have an underlying rationale. A linear relationship connotes rigidity, the opposite of the "flexibility" connotation of static resilience. Analogously, the measure of *TSER* is the difference between a linear set of indirect effects, which implicitly omits resilience, and a nonlinear outcome, which incorporates the possibility of resilience.

We illustrate the application of the definition with the following case study by Rose and colleagues (2009), who estimated the national and regional economic impact of the September 11, 2001, terrorist attack on the World Trade Center. The researchers refined available data indicating that more than 95% of the businesses and government offices operating in the WTC area survived by relocating; the vast majority to midtown Manhattan or across the river in northern New Jersey. Had all of these firms gone out of business, the potential direct economic loss in terms of GDP would have been $43 billion. However, relocation was not immediate, taking anywhere from a few days to as long as eight months for the vast majority of firms. Rose and colleagues (2009) calculated this loss in GDP at $11 billion. They were then able to apply the resilience definition provided here to estimate that the effectiveness of relocation as a resilience tactic in the aftermath of the 9/11 attacks was 72% ($43 minus $11, divided by $43). This study highlights the importance of excess capacity as a resilience tactic (in general and office space in particular). This more intensive use of resources is also the theme of the recovery in the current great recession in the United States and other countries, as employment recovery significantly lags the recovery of output. The experience of New York thus signals a significant change in approaches to disaster recovery, which has typically emphasized prompt rebuilding. Coupled with stronger requirements for mitigation, and hopefully some general accumulated wisdom, we are recovering less by reflex action and more by intelligent planning (see also Chang & Rose, 2012; Vale & Campanella, 2005).

For the most part, *TSER* refers to off-site considerations. These would include infrastructure interdependence, supply chain interactions, community/institutional systems, and features of the economy of the entire region or nation, including the influence of fiscal and monetary policy. Clearly, Federal Reserve Bank policies of providing $100 billion for interbank loans in the immediate aftermath of 9/11 and keeping interest rates low for a much longer period greatly helped to reduce the potential losses of this disaster (Rose & Blomberg, 2010).

TSER pertains primarily to interdependencies that affect resilience. The interdependencies apply to relationships up and down each individual industry supply chain (meso level), but also the integration of all these chains (macro level). Even if two of these supply chains do not appear to be related, the fact that they utilize many of the same inputs means that they compete for their availability, and that affects their price. It also means each activity or unit of analysis affects macroeconomic variables, where the whole is not just the simple sum of the parts. Resilience can also stem from fiscal, monetary, and security policies. The production interdependencies are referred to as simple multiplier effects, the price–quality relationships are referred to as general equilibrium (GE) effects, and the aggregate relationships are referred to as macroeconomic effects. The size of *TSER* is influenced by the magnitude of the DSER but its proportion less so. A high *DSER* will have a high *TSER* associated with it because of the high base value. But the basic/GE/macro multiplier is more dependent on the structure of the economy. More technologically advanced economies and more self-sufficient ones (and the former significantly affect the latter) have higher multipliers. This applies not only to the more obvious aspects of the technological relationships of factory production, but also to less visible aspects of the economy such as the financial system.

Recall that the definition of resilience is contextual, i.e., resilience in comparison to a reference base of the absence of resilience, or, more specifically, the worst potential outcome from a given threat. This would best be modeled by a linear or fixed-coefficient production function at the micro and meso levels, and an input-output (I-O) model, composed of such functions, at the macro level. Also, while *DSER* can be directly empirically measured (e.g., every firm is aware of the amount of electricity it purchases directly), *TSER* cannot be, primarily because of its complexity arising from the many rounds of interdependencies (the firm does not know how much electricity its direct and indirect suppliers and customers use). A formal modeling approach is needed. I-O is ideal for the establishment of the reference or worst-case scenario (it does not allow for input or import substitution, conservation, or technological adaptation). In contrast, computable general equilibrium (CGE) models are able to incorporate several types of inherent resilience and can readily be modified to include adaptive ones (see, e.g., Rose, 2015; Rose & Liao, 2005). Macreconometric models are superior for capturing the resilience effects of monetary and fiscal policy. Multiplier values are somewhat dependent on the targets of the disaster in cases where they are not evenly distributed. Otherwise, the multipliers typically fit some well-known patterns depending on the model used. For example, I-O multipliers would be around 3.0 for Los Angeles and 2.0 for Portland. CGE multipliers would be on the order of about half of the indirect effects (i.e., they would have values of 2.0 and 1.5, respectively, for the aforementioned cities), because of the more flexible adjustments, including input substitution and price changes, accounted for in CGE models dampen negative impacts.

Our basic resilience metric can be applied to *Direct Dynamic Economic Resilience* (*DDER*) and *Total Dynamic Economic Resilience* (*TDER*) as well. The metric itself is unchanged. It is still expressed in terms of percentage losses avoided, but is influenced by a different set of factors than is the static counterpart. The primary considerations relate to expediting investment in repair and reconstruction. The reference base is normal recovery time in the presence of ordinary resilience activities or absence of resilience entirely (for the purists), which is not easy to measure precisely, but can be approximated by historical averages adjusted for estimates of resilience that were implemented. Rules of thumb again provide insights, in that cutting recovery time in half will likely lead to a reduction of losses by an equal percentage. The complication comes from the likelihood of enhanced productivity associated with the new capital stock, which can be factored in to further offset losses and hence increase the resilience score. Likewise, the effect of repaired or new capital stock that makes a business, industry, or overall economy less vulnerable to future disasters can be factored into the metric as well.

9.5 Measurement of Economic Resilience

Several studies have examined economic resilience in actual disasters or with the use of simulation studies. The major pioneer is Tierney (1997), who surveyed businesses in the aftermath of the Northridge earthquake and Midwest floods. Rose and Lim (2002) translated Tierney's findings into specific measures of resilience of the Los Angeles electricity system. They identified such factors as time-of-day-use, electricity "importance" (dependence), and production recapture as key to understanding why businesses that averaged an X% reduction of electricity were able to continue operation at much less than an X% reduction in their production of goods and services. In fact, they found that these micro-level tactics resulted in a reduction of business interruption losses by more than 90% of baseline estimates, a level consistent with Tierney's survey responses.

Rose and Liao (2005) also used the Tierney survey responses relating to water service disruptions to evaluate two major types of resilience: conservation of scarce inputs and enhanced ability to substitute other inputs for water. Overall, the aforementioned resilience strategies only yielded a reduction of losses by a few percentage points, however. In this study, macro impacts follow from micro impacts of resilience through various types of general equilibrium effects. The macro impacts of resilience doubled the micro-level impacts.

Other estimates of resilience have been less evidence-based, but are still prominent in the literature. Several resilience factors were incorporated by the author and colleagues into FEMA's loss estimation tool – HAZUS (FEMA 2013), which is being adapted for use in many other countries. The Direct Economic Loss Module (DELM) includes factors for individual businesses making up lost production at a later date by working overtime or extra shifts after their utility lifelines had been restored or after building damage had been repaired. These "recapture factors" were based on a synthesis of the literature, and the indication is that these factors are very high for short periods, ranging between 50% and 98% for most sectors. That is, customers are unlikely to cancel their orders for the output of disaster-stricken industries for short periods of time, because they have inventories on hand or long-standing supply chain relationships. On the other hand, this type of resilience is likely to decline over time, and is likely to fall to zero after one year, if not after several months. The HAZUS Indirect Loss Module (IELM) includes such resilience factors as inventories, excess capacity, and the ability of increased imports and exports. However, there are no definitive estimates of the effectiveness of these resilience tactics for all cases, so it is necessary for the user to access primary or secondary data to specify the effectiveness of individual types of resilience.

Several other simulation studies have been undertaken to estimate the effects of resilience on losses from disasters, using the metric presented in the previous section. Kajitani and Tatano (2009) used a survey to estimate the resilience of Japanese industries to various types of lifeline disruptions from disasters. Their findings are the most definitive to date on a broad spectrum of resilience tactics. Rose and colleagues (2007a, 2007b) estimated the resilience of the Los Angeles water and power systems to a two-week outage due to a terrorist attack. They found that resilience could be as high as 90%, primarily due to production recapture. Rose and Wei (2013) examined such resilience tactics as excess capacity, inventories, and export diversion to reduce potential losses from a 90-day shutdown of a major U.S. seaport complex. They also examined the effectiveness of major government policies such as accessing the Strategic Petroleum Reserve, but found it likely to be of limited help at the cost of considerable political capital. Overall, they found that the implementation of these resilience tactics could reduce GDP losses in the region by more than 70%.

Note, however, these results should be carefully applied to natural disasters. In the case of terrorism, a specific infrastructure provider is targeted, but the rest of the economy is unscathed. Hence, when the water or power is restored, firms can resume production immediately. This is not the case for natural disasters, which inflict widespread damage so that factories need to be repaired in addition to having their lifelines restored. Thus, the context needs to be factored into the potential effectiveness of some resilience tactics, such as production recapture, inventories, or conservation, where the effectiveness of these tactics wanes over time.

In addition, the simulation studies cited previously are biased to estimating resilience at its maximum effectiveness. This is not always the case due to the disarray accompanying most disasters, administrative obstacles, and personal failings. Moreover, Rose (2009a) has pointed out that resilience can be eroded during large disasters as inventories are depleted, Draconian conservation becomes onerous, and opportunities for production recapture decline as customers abandon their traditional suppliers unable to deliver.

Two other factors impinge on effectiveness. The first is cost, but it turns out that resilience tactics are inexpensive, especially compared to mitigation measures such as structural reinforcement of buildings or building of levees. Conservation more than pays for itself, input substitution requires a small penalty, the cost of inventories is just the carrying charge, emergency drills take little time, production rescheduling involves the payment of overtime wages. Still, even small costs can be serious obstacles in poor countries. Second, resilience may have negative side effects. In the rush to rebuild, requests are often made to suspend environmental and safety regulations. Such requests must be carefully considered in light of the ancillary damages that could occur.

Overall, more case studies are needed to contrast successes (e.g., recovery from Superstorm Sandy) and failures (e.g., recovery to Hurricane Katrina). Best-practice examples can be very helpful, especially at the small business and household levels, where expertise is limited. Cross-cultural studies are enlightening as well in both broadening horizons and illustrating limits of resilient practices.

9.6 Resilience Indices

Recently, interest has shifted to identifying individual resilience indicators that can be aggregated into an overall index. This has emanated in part from the successful compilations of vulnerability indices (see, e.g., Cutter et al. 2003). Several well-intentioned examples of resilience indices include Cutter and colleagues (2010) and Sherrieb and colleagues (2010). One major conclusion is that many of the components of resilience indicators to date are not in fact important to the resilience of individual businesses or the economy as a whole during the early stages of the recovery process. Specifically, resilience is not just the flip side of vulnerability. For example, the Index derived by Cutter and colleagues (2010) includes housing capital, equitable incomes, employment, business size, and position access. Subcomponent variables include percent employment, percent home ownership, business size, female labor force participation, and a proxy for single-sector employment dependence. Hardly any of these indicators match those derived from a conceptual economic framework. Also, few have much to do with the operation of an individual firm or the economy as a whole. Percent employment is a good initial measure of excess capacity in the labor force from which to draw; however, it does not take into account the fact that disasters are able to draw additional labor from neighboring communities either through market signals (higher wages) or for altruistic motivations.

Another approach to measuring and evaluating resilience is the establishment of a resilience threshold. This is analogous to the poverty threshold measured by incomes of $1.25/day per household. Some, for example, have proposed a resilience threshold of $5/day. This is, however, a narrow and limited approach. First, it appears arbitrary because there is no basis for the dollar value. Second, it does not help much in relation to the destruction of major assets, such as houses or businesses. Here, a better measure would be asset holdings or access to credit or other financial assistance. Third, it is weak from a purely personal standpoint. Aspects of connectedness to social support groups (families, neighborhood associations, religious or social organizations) should be included. Also important is access to social services and to information about recovery alternatives. Last, one would ideally proceed beyond the minimum that this threshold indicator implies to a functional relational relationship that tracks the degree of resilience attained by higher levels of income or assets. It would be valuable to determine whether the indicator is characterized by increasing and then diminishing returns.

A major reason to construct a resilience index is not only to study the recovery process, but also to improve it. This speaks to the importance of *actionable variables*. Several indicators included in prior resilience indices refer to background conditions and general trends that can hardly be improved in the near-term aftermath. Moreover, improvement of some of them is not necessarily consistent with other economic goals. For example, diversification of the economy may come at the cost of some economic activity. Other indicators need to be acknowledged as being immutable during the key period in which businesses take resilient actions. Examples include literacy rates, percent disabled, and percent minority population. More research is needed to replace these indicators with ones that really matter to economic resilience and that can be implemented in the short run.

Finally, it is important to identify resilience that is inherent in the survival mechanisms of businesses and households separate from resilience tactics that require government policy assistance. This way, future recovery efforts can better capitalize on existing capabilities and minimize duplication of government services. The focus of government can then be on facilitating this inherent resilience by removing obstacles to private enterprise, reducing wait times for assistance, and more effectively targeting its role.

Our analysis indicates that few resilience frameworks and no actual indices adequately focus on business operations in the aftermath of a disaster. We contend that business behavior, in relation to static and dynamic resilience, is the key to economic recovery, at least in the short run. Thus, while many prior attempts at developing a resilience index are useful for long-run analyses (more than a year after the event), hardly any are useful for the short run.

Following Rose and Krausmann (2013), we now outline the development of a short-run economic resilience index focusing on business behavior that is intended to help gauge recovery potential. The key issues are: 1) if and how to combine resilience actions at the micro, meso, and macro levels; and 2) how to weight the various components. Both of these considerations are applicable to household/community resilience as well.

The first challenge is whether a single index can best reflect the features of resilience at various levels of the economic aggregation hierarchy. It is best to separate the three levels first, and then explore ways to combine them. At the microeconomic level, i.e., the level of individual companies and organizations, a plethora of actionable measures to increase economic resilience exists. Business resilience consists of resilience options on both the customer and supplier sides. While the former relates to coping with disaster impacts on the delivery of inputs and making effective use of resources, the latter pertains to ways to guarantee the delivery of outputs to customers. Some of these options relate to static economic resilience by aiming at diminishing losses (excess capacity, input or import substitution, etc.), while others facilitate speedy business recovery (reduce operating impediments, management, etc.) and hence belong to the dynamic resilience category.

The second consideration is how to evaluate the relative contribution of each resilience tactic. This issue of weighting has been finessed in nearly all prior cases. For instance, weights can be derived empirically, e.g., using surveys, or, as a fallback, from a theoretical model of the indicator to be measured. An alternative way to assign weights is by consensus or relevance with respect to specific policy initiatives. We propose that the best way to develop weights is based on evidence of the relative effectiveness of each type of resilience tactic. The weights would then have to be adjusted for differences in the context in which the index is to be applied, if it differs, and it likely will, from the original case. There is no one-size-fits-all approach to deriving weights and the method of choice will depend on the particular problem at hand.

Note that economics is only one part of the recovery from disasters. We are not suggesting economics is the only thing that matters in recovery, even just economic recovery. In addition to the importance of noneconomic factors, we need to consider the existence of market failure, where the interests of business are not necessarily consistent with the interests of the community as a whole. In our context, this is exemplified by the possibility that the dispersal of toxic waste or ordinary pollutants might be accelerated by more rapid recovery, partly because environmental concerns may be given lower priority. One approach is to go beyond ordinary market-based economic indicators, such as gross domestic product (GDP), and utilize broader measures of economic well-being, including the value of environmental resources and the value of household services.

9.7 Conclusion

While it is desirable to reduce risk, this must be placed in the broader context of the several alternatives available. We can mitigate many threats and reduce vulnerability through various means. However, we can also reduce consequences via the static and dynamic economic resilience discussed earlier. It is important to balance the strategies among the most cost-effective alternatives. It is also important to keep in mind that it is impossible to eliminate all threats and that the goal should be to minimize vulnerability. Resilience then becomes our next best line of defense against major disruptions.

A broader strategy is to translate ingenuity in coping with disasters in the short run into long-run decisions and practices that continuously promote sustainability. Resilience tactics to address resource shortages, such as conservation, input substitution, and technology modification, can be further refined for long-run application. Disasters can also provide opportunities for transitions to more sustainable paths in the reconstruction process through revised land-use planning, downsizing, and industrial targeting, in addition to enhanced structural mitigation. Resilience offers many important lessons for enhancing sustainability. It places greater emphasis on flexibility and responding effectively to the realities of disequilibria, as opposed to unrealistically smooth equilibrium time-paths.

REFERENCES

Adger, W. N. (2000). Social and ecological resilience: Are they Related? *Progress in Human Geography, 24*(3), 247–364.

Adger, W. N., Hughes, T. P., Folke, C., Carpenter, S. R., & Rockstrom, J. (2005). Social-ecological resilience to coastal disasters. *Science, 309*, 1036–1039.

Alexander, D. (2013). Resilience and disaster risk reduction: An etymological journey. *Natural Hazards and Earth Systems Science Discussions, 1*, 1257–1284.

Ayres, R., & Simonis, U. (1989). *Industrial metabolism*. Tokyo: United Nations Press.

Blomberg, S. B., & Hess, G. (2009). Estimating the macroeconomic consequence of 9/11. *Peace Economics, Peace Science and Public Policy, 15*(2), Article 7.

Bruneau, M., Chang, S., Eguchi, R., Lee, G., O'Rourke, T., Reinhorn, A., Shinozuka, M., Tierney, K., Wallace, W., & von Winterfeldt, D. (2003). A framework to quantitatively assess and enhance seismic resilience of communities. *Earthquake Spectra, 19*, 733–752.

Burby, R., Deyle, R., Godschalk, D., & Olshansky, R. (2000). Creating hazard resistant communities through land-use planning. *Natural Hazards Review, 1*(2), 99–106.

Chang, S., & Rose, A. (2012). Economic recovery to disasters. *International Journal of Mass Emergencies and Disasters, 30*(2), 171–181.

Chang, S., A., Rose, A., & Johnson, L. (2011). National Research Council, "What is national earthquake resilience?" In *National earthquake resilience: Research, implementation, and outreach*. Washington, DC: The National Academies Press.

Chang, S., & Shinozuka, M. (2004). Measuring and improving the disaster resilience of communities. *Earthquake Spectra, 20*, 739–755.

Chernick, H. (ed.) (2005). *Resilient city: The economic impact of 9/11*. New York: Russell Sage Foundation.

Comfort, L. (1994). Risk and resilience: Inter-organizational learning following the Northridge earthquake of 17 January 1994. *Journal of Contingencies and Crisis Management* 2(3), 157–170.

Comfort, L. (1999). *Shared risk: Complex seismic response.* New York: Pergamon.

Cutter, S. et al. (2003). Social vulnerability to environmental hazards. *Social Science Quarterly, 84*(2), 241–261.

Cutter, S. L., Burton, C. G., & Emrich, C. T. (2010). Disaster resilience indicators for benchmarking baseline conditions. *Journal of Homeland Security and Emergency Management, 7*, Article 51.

Daly, H., & Farley, J. (2004). *Ecological economics*. Washington, DC: Island Press.

Dhawan, R., & Jeske, K. (2006). How resilient is the modern economy to energy price shocks? *Economic Review*, Federal Reserve Bank of Atlanta, Third Quarter: 21–32.

Dovers, R., & Handmer, J. (1992). Uncertainty, sustainability and change. *Global Environmental Change, 2*(4), 262–276.

Federal Emergency Management Agency (FEMA). (2013). *Multi-hazard Loss Estimation Methodology (HAZUS®MH MR4)* www.fema.gov/library/viewRecord .do?id=3726.

Flynn, S. (2008). America the resilient: Defying terrorism and mitigating natural disasters. *Foreign Affairs, 87*(2), 2.

Folke, C. (2006). Resilience: The emergence of a perspective for social-ecological systems analyses. *Global Environmental Change, 16*(3), 253–267.

Foster, H. (1997). *The Ozymandias principles: Thirty-one strategies for surviving change*. Victoria, Canada, University of British Columbia Press.

Geis, D. (2000). By design: The disaster-resistant and quality-of-life community. *Natural Hazards Review, 1*(3), 151–160.

Giesecke, J. A., Burns, W. J., Barrett, A., Bayrak, E., Rose, A., Slovic, P., & Suher, M. (2012). Assessment of the regional economic impacts of catastrophic events: CGE analysis of resource loss and behavioral effects of an RDD attack scenario. *Risk Analysis, 32*(4), 583–600.

Gigerenzer, G., & Selten, R. (eds.) (2002). *Bounded rationality: The adaptive toolbox.* Cambridge, MA: MIT Press.

Godschalk, D. (2003). Urban hazard mitigation: Creating resilient cities. *Natural Hazards Review, 4*(3), 136–143.

Haimes, Y. Y. (2006). On the definition of vulnerabilities in measuring risks to infrastructures. *Risk Analysis, 26*(2), 293–296.

Hill, R., & Paton, D. (2005). Managing company risk and resilience through business continuity management. In D. Paton & D. Johnston (Eds.) *Disaster resilience: An integrated approach* (pp. 249–264). Springfield, IL: Charles C. Thomas Publishers.

Holling, C. (1973). Resilience and stability of ecological systems. *Annual Review of Ecology and Systematics, 4*, 1–23.

Horwich G. (1995). Economic lessons of the Kobe earthquake. *Economic Development and Cultural Change, 48*(3), 521–542.

IPCC. (2007). *Climate change 2007: Mitigation of climate change. Working Group III contribution to the fourth assessment report of the IPCC.* Cambridge, UK: Cambridge University Press.

Kajitani, Y., & Tatano, H. (2009). Estimation of lifeline resilience factors based on empirical surveys of Japanese industries. *Earthquake Spectra, 25*(4), 755–776.

Klein, R., Nicholls, R., & Thomalla, F. (2003). Resilience to natural hazards: How useful is this concept? *Environmental Hazards, 5*, 35–45.

Kunreuther, H., Pauly, M. V., & McMorrow, S. (2013) *Insurance and behavioral economics*. New York: Cambridge University Press.

Levin, S. (1998). Resilience in natural and socioeconomic systems. *Environment and Development Economics. Special Issue on Resilience and Sustainability, 3*(2), 221–235.

Manyena, S. (2006). The concept of resilience revisited. *Disasters, 30*(4), 433–450.

Mileti, D. (1999). *Disasters by design: A reassessment of natural hazards in the United States*. Washington, DC: Joseph Henry Press.

Norris, F., Friedman, M., & Watson, P. (2002). 60,000 disaster victims speak: Part II, summary and implications of the disaster mental help research. *Psychiatry, 65*, 240–260.

Norris, F., Stevens, S., Pfefferbaum, V., Whyche, K., & Pfefferbaum, R. (2008). Community resilience as a metaphor, theory, set of capacities and strategy for disaster readiness. *American Journal of Community Psychology, 41*, 127–150.

Paton, D., & Johnston, D. (2001). Disasters and communities: Vulnerability, resilience and preparedness. *Disaster Prevention and Management, 10*, 270–277.

Perrings, C. (2001). Resilience and sustainability. In H. Folmer, H. L. Gabel, S. Gerking, & A. Rose (Eds.), *Frontiers of environmental economics*. Cheltenham, UK: Edward Elgar.

Petersen, K., & Rose, A. (2013). MSB/DHS Program on Community Resilience: An overview.

Pfefferbaum, B., Reissman, D., Pfefferbaum, R., Klomp, R., & Gurwitch, R. (2005). Building resilience to mass trauma events. In L. Doll, S. Bonzo, J. Mercy, & D. Sleet (Eds.), *Handbook on injury and violence prevention interventions*. New York: Kluwer Academic Publishers.

Pimm, S. L. (1991). *The balance of nature*. Chicago: University of Chicago Press.

Resilience Alliance. (2005). Research on social-ecological systems: A basis for sustainability, www.resilience.org.

Roberts, B. (2009). The macroeconomic impacts of the 9/11 attack: Evidence from real-time forecasting. *Peace Economics, Peace Science and Public Policy, 15*(2), Article 8.

Rose, A. (2004). Defining and measuring economic resilience to disasters. *Disaster Prevention and Management, 13*, 307–e14.

Rose, A. (2007). Economic resilience to natural and man-made disasters: Multidisciplinary origins and contextual dimensions. *Environmental Hazards, 7*(4), 383–395.

Rose, A. (2009a). *Economic resilience to disasters*. CARRI Research Report #8. ORNL.

Rose, A. (2009b). A framework for analyzing and estimating the total economic impacts of a terrorist attack and natural disaster. *Journal of Homeland Security and Emergency Management, 6*(1), Article 6.

Rose, A. (2013). *Defining and measuring societal resilience from an economic, environmental and personal security perspective*. Background paper for the United Nations Development Programme Human Development Report.

Rose, A. (2014). Economic resilience and the sustainability of cities in the face of climate change: An ecological economics framework. In H. Blanco & D. Mazmanian (Eds.), *Elgar companion to sustainable cities: Strategies, methods and outlook* (pp. 336–353). Cheltenham, UK: Edward Elgar.

Rose, A. (2015). Macroeconomic consequences of terrorist attacks: Estimation for the analysis of policies and rules. In V. K. Smith & C. Mansfield (Eds.), *Benefit transfer for the analysis of DHS policies and rules* (pp. 172–201). Cheltenham, UK: Edward Elgar.

Rose, A. (2017). *Defining and measuring economic resilience from a societal, environmental and security perspective*. Singapore: Springer.

Rose, A., Avetisyan, M., & Chatterjee, S. (2014). A framework for analyzing the economic tradeoffs between urban commerce and security. *Risk Analysis, 34*(5), 1554–1579.

Rose, A., Avetisyan, M., Chan, O., & Rosoff, H. (2015). The role of behavioral responses in the total economic consequences of terrorist attacks on U.S. air travel targets. CREATE, University of Southern California.

Rose, A., & Blomberg, S. B. (2010). Total economic impacts of a terrorist attack: Insights from 9/11. *Peace Economics, Peace Science, and Public Policy, 16*(1), Article 2.

Rose, A., & Blomberg, S. B., & Krausmann, E. (2013, October). An economic framework for the development of a resilience index for business recovery. *International Journal of Disaster Risk Reduction*, *5*, 73–83.

Rose, A., & Blomberg, S. B. & Liao, S. (2005). Modeling resilience to disasters: Computable general equilibrium analysis of a water service disruption. *Journal of Regional Science, 45*(1), 75–112.

Rose, A., & Blomberg, S. B. & Lim, D. (2002). Business interruption losses from natural hazards: Conceptual and methodology issues in the case of the Northridge earthquake. *Environmental Hazards: Human and Social Dimensions, 4*, 1–14.

Rose, A., Oladosu, G., & Liao, S. (2007a). Business interruption impacts of a terrorist attack on the electric power system of Los Angeles: Customer resilience to a total blackout. *Risk Analysis, 27*(3), 513–531.

Rose, A., Oladosu, G., & Liao, S. (2007b). Regional economic impacts of a terrorist attack on the water system of Los Angeles: A computable general disequilibrium analysis. In H. Richardson, P. Gordon, & J. Moore (Eds.), *The economic costs and consequences of a terrorist attack.* Cheltenham, UK: Edward Elgar Publishing Company.

Rose, A., Oladosu, G., Lee, B., & Beeler-Asay, G. (2009). The economic impacts of the 2001 terrorist attacks on the World Trade Center: A computable general equilibrium analysis. *Peace Economics, Peace Science, and Public Policy, 15*(2), Article 4.

Rose, A., Prager, F., Chen, Z., and Chatterjee, S. (2017). *Economic Consequence Analysis of Disasters: The E-CAT Software Tool.* Singapore: Springer.

Rose, A., & Wei, D. (2013). Estimating the economic consequences of a port shutdown: The special role of resilience. *Economic Systems Research, 25*(2), 212–232.

Rosoff, H., Siko, R., John, R., & Burns, W. (2013). Should I stay or should I go?: An experimental study of health and economic government policies following a severe biological agent release. *Environment, Systems and Decisions, 33*(1), 121–137.

Sheffi, Y. (2005). *The resilient enterprise*. Cambridge, MA: MIT Press.

Sherrieb, K., Norris, F., & Galea, S. (2010). Measuring capacities for community resilience. *Social Indicators Research, 99*, 227–247.

Tierney, K. (1997). Impacts of recent disasters on businesses: The 1993 Midwest floods and the 1994 Northridge earthquake. In B. Jones (Ed.), *Economic consequences of earthquakes: Preparing for the unexpected* (pp. 189–222). Buffalo, NY: National Center for Earthquake Engineering Research.

United Nations International Strategy for Disaster Reduction (UN/ISDR). (2002). *Living with risk: A global review of disaster reduction initiatives by the ISDR.* Geneva, Switzerland.

(UN/ISDR). (2006). Hyogo framework for action 2005–2015: Building the resilience of nations and communities to disasters. Extract from the final report of the World Conference on Disaster Reduction (A/CONF.206/6).

Vale, L., & Campanella, T. (2005). *The resilient city: How modern cities recover from disaster*. New York: Oxford.

Werling, J., & Horst, R. (2009). Macroeconomic and industry impacts of 9/11: An interindustry macroeconomic approach. *Peace Economics, Peace Science, and Public Policy, 15*(2), Article 5.

Zolli, A., & Healy, A. M. (2012). *Resilience: Why things bounce back.* New York: Free Press.

10

The Temporal Regional Economic Impacts of a Hurricane Disaster on Oil Refinery Operations: A FlexNIEMO Approach

Jiyoung Park, Peter Gordon, Yunkyung Kim, James E. Moore, II, and Harry W. Richardson

10.1 Introduction

Many economic impact estimates have relied on input-output (IO) models. Spatially disaggregated IO models permit decision makers and analysts to apply IO techniques to estimate local effects. Applied to the greater Los Angeles area, the Southern California Planning Model (SCPM) estimates impacts at the level of traffic analysis zones. The 2001 version of the National Interstate Economic Model (NIEMO) is a multiregional IO model that involves the 50 states and the District of Columbia. NIEMO has both supply- and demand-side capability. While there are some debates about using supply-side models, in principle, the loss of exports is best modeled via the demand-side version of NIEMO whereas the loss of imports is better modeled via the supply-side version.

Markets will eventually make substitutions via market price changes after major disruptions, but these adjustments are not captured by IO models, and these models may significantly overestimate long-term impacts. In response to this substantial limitation associated with IO methods, we suggest a transformative approach that internalizes the adjustment process in the IO system itself. Post-event information about concurrent demand and value-added changes contributes to identifying technological (production function) changes after a major disaster event. Comparing such results to the economic impact estimates from a baseline NIEMO exercise demonstrates how the detailed impacts of the many post-event substitutions and adaptations occur.

Table 10.1, based on the data of the U.S. Bureau of Economic Analysis, suggests real gross state product (GSP) for the United States, the Southeast, and Louisiana. Louisiana experienced an economic decline of 1.5% in 2004–2005, largely as a result of Hurricanes Katrina and Rita. With the exceptions of Louisiana and Alaska, the GSP of all the other states grew. The mining sector,

Table 10.1. *Contributions to percent change in real GSP, 2004–2005 by sector*

State and region	United States	Southeast	Louisiana
Agriculture, forestry, fishing, and hunting	–0.05	–0.01	–0.05
Mining	–0.04	–0.12	–1.66
Utilities	0.01	0.00	–0.07
Construction	0.13	0.27	–0.07
Durable goods manufacturing	0.40	0.37	–0.04
Non-durable goods manufacturing	0.08	0.08	0.48
Wholesale trade	0.07	0.13	–0.04
Retail trade	0.20	0.32	0.03
Transportation and warehousing	0.11	0.11	0.02
Information	0.34	0.40	0.14
Finance and insurance	0.54	0.54	0.22
Real estate, rental, and leasing	0.32	0.67	–0.41
Professional and technical services	0.48	0.47	–0.13
Management of companies	0.01	0.04	0.05
Administrative and waste services	0.21	0.32	0.19
Educational services	0.01	0.01	–0.02
Health care and social assistance	0.33	0.34	–0.08
Arts, entertainment, and recreation	0.02	0.03	–0.05
Accommodations and food services	0.13	0.17	0.00
Other services	0.06	0.06	–0.06
Government	0.18	0.41	0.07
Total	3.60	4.60	–1.50

1. Source: U.S. Bureau of Economic Analysis
2. Real GSP is adjusted based on 2000 dollars
3. Southeast region includes AL, AR, FL, GA, LA, MS, NC, SC, TN, VA, WV

including oil and gas production, was most negatively impacted with respect to GSP. The mining sector in Louisiana is relatively minor aside from oil and gas production. Three states (Texas, Louisiana, and New Mexico) in Petroleum Administration for Defense District III (PADD III), which includes the Gulf of Mexico, ranked first, second, and seventh with respect to the U.S. Oil and Gas Extraction industry (Park, 2009; U.S. Bureau of Economic Analysis, 2012). The aggregation of mining activities into this sector is not likely to affect the accuracy of the results.

Clear and systematic analytical results are needed for local and federal policy makers to prepare appropriate financial and mitigation support to the region. This study provides detailed examination of severe economic losses of oil-refinery industries following from Hurricanes Katrina and Rita. After introducing our previous research on the 2005 hurricanes, a new approach to MRIO modeling is described that builds on our previous research concerning these 2005 hurricanes. Our Flexible NIEMO (FlexNIEMO) simulation is provided, including estimates of input data for FlexNIEMO. The study concludes with a summary and implications of the research.

10.2 IO Operation and NIEMO

Leontief's first contributions to IO studies (1936, 1941) were to generalize interdependences between industries in an economy based on the general static equilibrium of the economy. This Leontief or demand-driven IO model is the classic IO system. The classic IO system accounts for technical inter-industrial relations throughout an economy via fixed-coefficient production functions.

In the standard Leontief IO model, the matrix form of an economy is identified in equation (10.1).

$$X = Z + D, \tag{10.1}$$

where

Z is the NxN matrix of intermediate interindustry flows presenting the deliveries in dollar values from industry sector i to j;

D is the NxK matrix showing various kinds of final demands of industry sector i to final users k, where k contains private households, governments, investments, and exports; and finally

X denotes the total outputs for each sector i, which is the column sum of intermediate flows and final demands associated with outputs from sector i.

The Leontief technical coefficients are interindustry coefficients used to produce total inputs corresponding to demand requirements. The Leontief technical coefficients matrix A, which identifies the backward effects of interindustry relationships, can be obtained from a flow matrix Z; but divided by the total outputs, that is, $A = Z(\widehat{X})^{-1}$, where the $\widehat{X}$ is the diagonal matrix version of X.

Equation (10.2) is redefined from equation (10.1) as

$$X = (I - A)^{-1}Y. \tag{10.2}$$

Final demands change exogenously only for specified sectors as a vector ΔY. New total inputs, ΔX needed to satisfy the required backward changes can be derived via equation (10.3),

$$\Delta X = (I - A)^{-1} \Delta Y. \tag{10.3}$$

The National Interstate Economic Model (NIEMO) is the first operational MRIO model of the United States since 1990 (Park et al., 2007). NIEMO was widely applied to suggest plausible public policy alternatives based on regional and national economic impact measures resulting from various disasters. In particular, NIEMO has addressed regional and national security problems, combined with various econometric methods that quantify security costs.

Chenery (1953) and Moses (1955) suggested a relatively simplified MRIO framework that could serve as an alternative to measuring interregional economic connections and avoid issues associated with excessive spatial aggregation. Sparse to nonexistent reliable interstate trade data is the main reason their proposed model was not operationalized for the United States. Matters were also complicated by the shift in the U.S. system of economic data sectors from the Standardized Industry Code system to the North American Industrial Sector Code (NAISC) system.

NIEMO was created by applying Commodity Flow Survey (CFS) data that the Bureau of Transportation Statistics (BTS) has released every five years since 1993 to estimate interstate trade flows. The details of the process can be found in Park and colleagues (2009a). Reconciling the different sector definitions and classifications of the commodities among multiple data sources, a new 29-commodity and 18-service sector system was developed. NIEMO's utility for impact analysis is demonstrated in Park and Gordon (2005).

Table 10.2 summarizes how NIEMO has been applied to various empirical studies and to produce total impact results. Applications include hypothetical terrorist attacks (Park, 2008; Park et al., 2008a; Richardson et al., 2007a); diverse natural disaster studies (Park et al., 2013, 2011a); and the hypothetical closure of all U.S. borders during the international avian influenza epidemic (Gordon et al., 2009).

NIEMO's evolution continues. While a temporal extension will be discussed in this study, the approach has been extended also to the sub-state level (Southern California Interregional Input–Output: SCI–IO) and to a world MRIO model. NIEMO has been combined with the North American transportation network system (TransNIEMO) and environmental models

Table 10.2. *Various economic impact studies using NIEMO*

Nature of Disruption	Targets	Type of Economic Impact	Total Economic Impacts ($M)	Base Year/Duration/ Model	Citations	Note
Explosives	LA/LB, Houston, and NY/NW	Ports shut down	23,258	2001/1month/demand-driven NIEMO	Park et al. 2007	Direct/Indirect state-by-state impacts
Dirty bomb	LA/LB	Ports shut down	26,905	2001/1 month/supply-driven NIEMO	Park 2008	Direct/Indirect state-by-state impacts
9/11	U.S. Airports	Loss of air passengers	214,347–420,455	Sep. 2001–Aug. 2003/ 2 years/IMPLAN IO model	Gordon et al. 2007	Direct/Indirect/ Induced U.S. impacts
Mad cow disease	U.S. Bovine animals	Cessation of foreign exports	13,681	Dec. 2003–Apr. 2005/ 15 months/demand-driven NIEMO	Park et al. 2006	Direct/Indirect state-by-state impacts
Explosives	13 U.S. theme parks	Consumer losses	20,747–24,921	2004/18 months/ demand-driven NIEMO	Richardson et al. 2007a	Direct/Indirect state-by-state impacts
International avian influenza epidemic	Loss of Air passengers, U.S. seaports closing, loss of cross-border shopping, loss of legal and illegal labors	U.S. border closures	1,734,075–5,408,796	2001/1 year/demand- and supply-driven USIO and NIEMO	Gordon et al. 2009	Direct/Indirect state-by-state or U.S. impacts
Hurricanes Katrina and Rita	Louisiana customs district	Seaports shut down	44,374	Aug. 2005–Mar. 2006/ 8 months/demand- and supply-driven NIEMO	Park et al. 2008b	Direct/Indirect state-by-state impacts

(EnviNIEMO). These extended NIEMO models provide a means to investigate the magnitudes and policy implications of negative economic impacts resulting from disruptive events, as well as from climate change.

10.3 Hurricanes in the Gulf of Mexico

A tropical depression formed on August 23, 2005, and moved toward the Gulf of Mexico, strengthening to a Category 5 hurricane over the central Gulf of Mexico (NCDC, 2005). On August 29, 2005, Hurricane Katrina made landfall on the Louisiana coast with Category 3 intensity. With winds reaching 130 mph, the New Orleans levees were breached, devastating the city due to flooding followed by the storm. According to the Louisiana Geographic Information Center (2005), the disaster was historically the costliest natural disaster in U.S. history as 80% of the City of New Orleans was flooded, and more than 1,800 casualties occurred.

A month later, Hurricane Rita hit the U.S. coast of the Gulf of Mexico on September 24, 2005. It strengthened to a Category 5 storm while at sea, but weakened to Category 3 when it landed near the border of Texas and Louisiana. However, Katrina had already devastated Louisiana. The total number of reported fatalities was 130, and Rita's damage was far less than that of Katrina (Knabb et al., 2006).

Prior to the two hurricanes, the three costliest natural disasters in terms of dollar damages recorded in U.S. history were the 1988 drought, with estimated losses of $39+ billion in business disruption; Hurricane Andrew in 1992, with $30 billion in property damage; and the Northridge earthquake in 1994, with $44+ billion in property damages (National Research Council, 1999). Several studies of the economic impacts of Hurricanes Katrina and Rita were completed relatively quickly after these catastrophic storms. Louisiana received federal reimbursements for losses from Katrina and Rita of about $105 billion and $10 billion, respectively (Kent, 2006). Referencing the economic losses from U.S. hurricanes since 1950, Nordhaus (2006) quoted $81 billion in damage from Katrina and $11.3 billion from Rita based on the Monthly Tropical Weather summary of the National Hurricane Center (2007).

However, the majority of the research measured the magnitude of direct losses or speculations about future impacts on the affected area. Considering the interrelations between economic sectors and household consumption, total economic losses may well exceed those estimates. For example, accounting for the New Orleans port's seven months of nonoperation following the hurricanes, Park and colleagues (2008b) estimated the direct and indirect economic losses from the storms as $62.1 billion, with the most severely affected sectors relying heavily on waterborne commerce.

However, it is apparent that port inoperability is only part of the story. Several oil and gas refineries were shut down for more than a week. For example, the Minerals Management Service of the U.S. Department of the Interior (MMS, 2006) reported that some 115 offshore oil platforms were missing, sunk, or had gone adrift. According to Katz and colleagues (2006), one-half of 1.3 million evacuees from the New Orleans metropolitan area could not return within a month after the storm.

Based on a report released by the Energy Information Administration (EIA, 2006a) that analyzed historical impacts of tropical cyclones on Gulf of Mexico crude oil and natural gas production (1960–2005) and refinery operations over the past 20 years, production foregone during precautionary periods stemming from severe weather threats (EIA, 2006a) reached 1.4% for crude oil and 1.3% for natural gas compared to normal annual production from wells on the Outer Continental Shelf. Another report released by the Government Accountability Office (GAO, 2006) found factors resulting in natural gas price increases, influences on consumers resulting from these higher prices, and the adequacy of the role federal government agencies played in ensuring competition in natural gas prices. Natural gas spot prices in September 2005 increased to roughly twice the average price of $7.60 per million BTUs in July 2005. This was associated with a substantial portion of domestic supply disruption and excessive demand due to the colder weather than expected in early December (GAO, 2006).

How did the hurricanes change employment? The Bureau of Labor Statistics (BLS, 2006) presented Katrina's impacts on the employment level in the Gulf Coast area as down nearly 40% in September 2005 compared to a year before. Additionally, Colgan and Adkins (2006) estimated the proportion of employment and wages for the ocean-related industries. These include oil and gas exploration, marine transportation, and related goods and services. They reported the region encompassing FL, AL, MS, LA, and TX produced about 292,000 jobs, paying nearly $7.7 billion in wages in 2004.

Crude oil and petroleum products in the area accounted for nearly 60% of the U.S. total in 2004 (EIA, 2006b). The paths of Hurricanes Katrina and Rita relative to oil and gas production platforms on the U.S. coast of the Gulf of Mexico are shown in Figure 10.1. Because offshore installations near the Gulf of Mexico play a substantial role in the U.S. oil and gas industries, domestic gasoline prices escalated significantly right after the two storms. Underlying production trends are shown in Figure 10.2. Total U.S. production parallels the Gulf Coast flow, while the rest of the United States shows a relatively steady output.

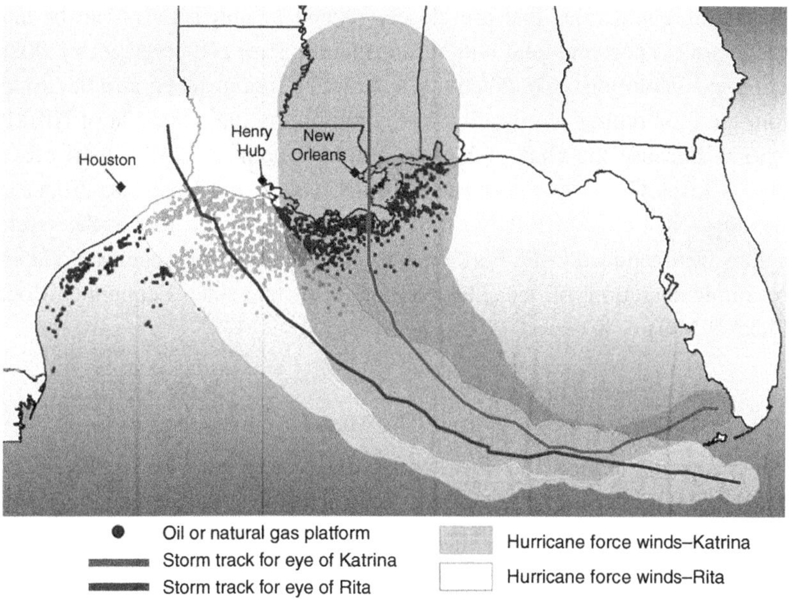

Figure 10.1. Paths of Hurricanes Katrina and Rita relative to oil and gas production platforms on the U.S. coast of the Gulf of Mexico.
Source: Government Accountability Office, 2006

Figure 10.2. Petroleum production pre- and post-Katrina and Rita.
Source: Authors' calculations of raw data from the EIA official website (http://tonto.eia.doe.gov/dnav/pet/pet_pnp_wiup_dcu_r30_w.htm)

Crude oil industries that are closely related to port activity can be analyzed using export and total output information. Park and colleagues (2008b) addressed disruptions of port activity, subtracting foreign and domestic exports from the total output of oil refineries by modifying the definition of NIEMO regions. Because the Gulf of Mexico region is defined as PADD III (Texas Inland, Texas Gulf Coast, Louisiana Gulf Coast, North Louisiana-Arkansas, and New Mexico), NIEMO's original 52 (including the rest-of-the world) regions were reduced to 47. Further, the 47-region NIEMO model was adapted for implementation of the Flexible National Interstate Economic Model (FlexNIEMO).

10.4 The Model

This research applied the Flexible National Interstate Economic Model (FlexNIEMO) to construct a month-to-month supply-side version of NIEMO. The Holt-Winters time-series approach is used to estimate normal economic trends assuming that the two hurricanes had not occurred. The time-series approach provides the direct impacts necessary for input data for FlexNIEMO. This approach used to estimate the normal economic trends is described in several previous articles (Gordon et al., 2007; Park et al., 2008b, 2009b, 2006; Richardson et al., 2007b). This time-series method allows the estimated coefficients to change gradually over time, based on data for previous periods and exponentially declining weight values. The forecasts are statistically reliable based on the R-Squared value of 0.87. Based on the estimated coefficients, the forecast oil-refinery industry values are obtained at the end of 2005 (August to December) and for the first three quarters of 2006. Direct impacts are calculated using the difference between the actual and predicted oil refinery production. Figure 10.3 gives the 13 months of forecasts using the Holt-Winters method.

FlexNIEMO is used to construct monthly versions of the supply-side NIEMO. A widely cited, key limitation of all IO-type models is that all coefficients in the models are fixed. Hence, these models necessarily ignore substitution opportunities that should be prompted by market signals. The approach developed recently by Park and colleagues (2011b) and Park and Richardson (2014) allows the fixed coefficients in input-output representations to be continuously modified, responding to previous economic events and interindustry relations. Flexible IO (Gordon et al., 2009) and FlexNIEMO approaches allow for the construction of new IO coefficients that capture the substitution effects experienced in any industrial sector. Built with both demand- and supply-driven

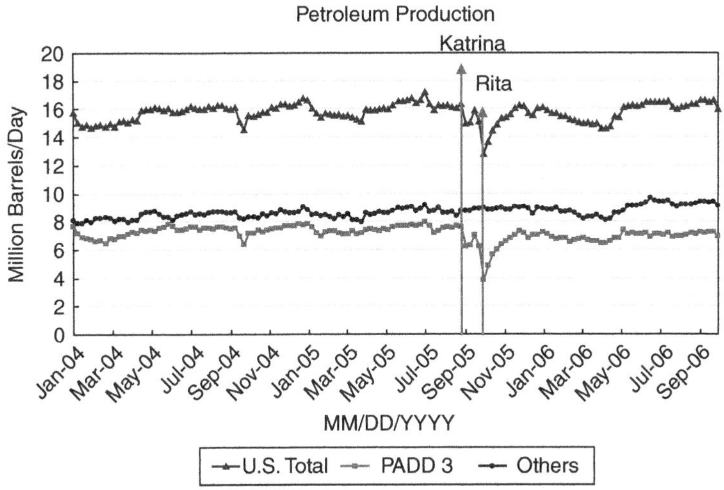

Figure 10.3. Changes of actual and estimated oil refinery values: PADD 3 (Gulf of Mexico).
Source: Authors recalculated and estimated, based on Refiner Petroleum Product Prices by Sales Type (http://tonto.eia.doe.gov/dnav/pet/pet_pri_refoth_dcu_R30_m.htm) Weekly Petroleum Production by Product http://tonto.eia.doe.gov/dnav/pet/pet_pnp_wiup_dcu_r30_w.htm

IO models, this flexible approach relaxes the assumption of fixed coefficients by adjusting coefficient matrices.

The FlexIO and FlexNIEMO approaches begin with empirical, substitution-driven changes in value-added and final-demand vectors. These approaches are useful for explaining economic resilience and dynamics. Usually, value-added and final-demand changes stemming from a natural disaster or a terrorist attack do not terminate in the target year. Instead, economic activities are affected in a more dynamic way. Combining a forecasting method with an IO approach provides only a limited capacity for updating IO coefficients. Instead, temporal changes in the final-demand and value-added elements are relatively easy to identify and collect, so new IO coefficient sets can be endogenously updated via FlexIO and FlexNIEMO approaches.

In this study, the supply-side version of NIEMO is applied because oil refinery production supports the economy in the Gulf of Mexico and the United States. A problem with using the supply-side version of the model is that it is unclear how to account for demand-side adjustments during the recovery. However, the supply-side version of FlexNIEMO overcomes some of the major shortcomings inherent in the IO model. Furthermore, the requirement that the six states in the Gulf region be aggregated into a single zone dictated development of a new aggregated version NIEMO that has (47x47)x(47x47) different coefficients for each month (August 2005 to September 2006). Therefore, a 47-region FlexNIEMO is applied.

Based on the flexible and spatially disaggregated approach suggested by Park and colleagues (2011b) and Park and Richardson (2014), the general equations for the demand-driven NIEMO with 47 regions are described, beginning with estimating total impacts. Let $X^s(t)$ and $X^d(t)$ be the total input row vector and the total output column vector respectively for various commodity and service sectors. In these cases, the total input vector is the sum of interindustry sales ($A(t)X^s(t)$) and total final demands ($F(t)$); and the total output vector is the sum of interindustry purchases ($B(t)X^d(t)$) and total value-added factors ($W(t)$). i.e.,

$$X^s(t) = A(t)X^s(t) + F(t), \tag{10.4}$$

$$X^d(t) = B(t)X^d(t) + W(t). \tag{10.5}$$

$A(t)$ is a $X^s(t)$-based requirements matrix composed of a technical flows matrix for industries within a region, $Z\ (t)$, and a block diagonal matrix of interregional trade flows, $C^s(t)$. $A(t)$ is defined as

$$A(t) = \Gamma_t\left[Z\ (t), C^s(t)\right]. \tag{10.6}$$

A column vector $F(t)$ represents region-specific final demand changes. Similarly, $B(t)$ is a $X^d(t)$-based requirements matrix composed of a technical flows matrix for industries, $Z\ (t)$, and a block diagonal matrix of interregional trade flows, $C^d(t)$, defined as

$$B(t) = \Theta_t\left[Z\ (t), C^d(t)\right] \tag{10.7}$$

In addition, $W(t)$ is a row vector of region specific changes in value added. Note that Γ_t and Θ_t are matrix functions that are used to update the requirement matrices based on the technical flows matrix and the trade flows matrix. Two vectors $F(t+k)$ and $W(t+k)$ can be acquired or computed for period $(t+k)$ to be applied in various scenarios. Following the detailed procedures suggested by Park and colleagues (2011b), new vectors $F(t+k)$ and $W(t+k)$ and a new technical flows matrix $Z(t+k)$ are derived for the next period.

However, a challenge occurs when applying zero-changes in sectors and states. For example, only one industry sector (USC Sector 10: coal and petroleum products) dataset is needed for this exercise. It would be inefficient to obtain all the other regions' total output changes data to update the technical coefficient matrices by period. To allow these unspecified sectors and states to be indifferent to total output changes between pre- and post-event circumstances, and to only reflect the targeted region effects, FlexNIEMO needs $X^s(t=0)$ and $X^d(t=0)$ to be unspecified. Instead of developing a difference-based approach for the FlexNIEMO process, a proportionality approach is adopted, constraining untargeted sectors and states to unity between the initial period (e.g., t = 0) and the next period (e.g., t = 1). This implies that there are no changes between the pre-event period and post-period for the untargeted sectors and regions.

The proportional final demand and value-added changes between the first two periods are defined via this approach, and the new technical coefficient matrix for the demand-side NIEMO and the new allocation coefficient matrix for the supply-side NIEMO are estimated. Because this study uses only the supply-side NIEMO, the exogenous direct losses of value added $\Delta\hat{w}(t=i)$ for period i are calculated using the difference between econometrically estimated forecasts and observed decreases in total output as

$$\Delta\hat{w}(t=i) = (\hat{x}^s(t=i) - x^s(t=i))\hat{r}_s, \tag{10.8}$$

where

$\hat{x}^s(t=i)$ is the total output vector estimated with a quasi-experimental time-series model, assuming a normal economic status for period i,
$x^s(t=i)$ is the observed total output vector occurring in period i, and
$\hat{r}_s$ is a constraint diagonal matrix enforcing no change in the proportionate relationship between value added and total input through all periods.

The supply-driven model estimates total economic impacts $\Delta\hat{EI}(t=i)$ for period i via equation (10.9).

$$\Delta\hat{EI}(t=i) = \Delta\hat{w}(t=i)ISNIEMO(t=i), \tag{10.9}$$

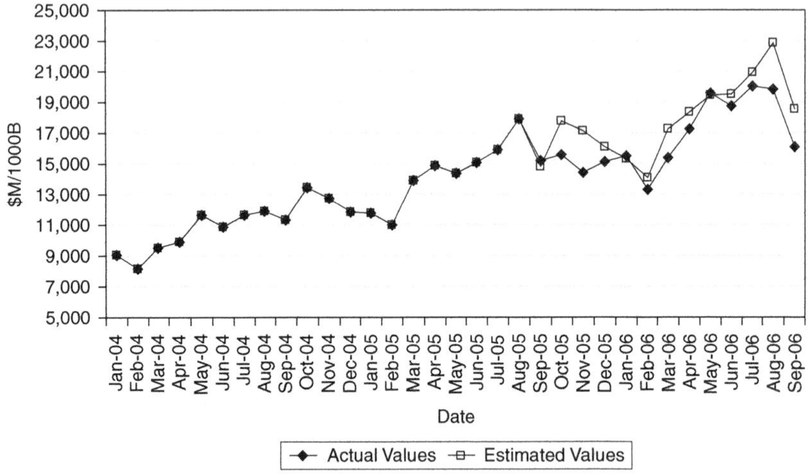

Figure 10.4. Total negative impacts of oil refinery losses from Hurricanes Katrina and Rita: FlexNIEMO vs. NIEMO results.

where *ISNIEMO*($t=i$) is the period i inverse matrix of the new allocation coefficient matrix for the supply-side NIEMO.

Note that the total sum of the impacts for each period ($\sum_i \hat{EI}(t=i)$) can be obtained for the 47 regions (indexed over m) and 47 USC sectors (indexed over n), and the total for the entire United States is

$$\sum_m \sum_n \sum_i \hat{EI}(t=i). \tag{10.10}$$

10.5 FlexNIEMO Results

Figure 10.4 compares the results estimated from an application of NIEMO with those from FlexNIEMO. We focus on the top five U.S. states as identified by the total impacts results of the original NIEMO, the Gulf of Mexico region, the rest of the United States, and the rest of the world. Given the $4.54 billion losses per 1,000 barrels as the direct losses for USC Sector 10, the original NIEMO estimated negative impacts of $8.32 billion in total. However, FlexNIEMO estimated $4.85 billion losses in total based on the same direct impacts. As shown in Figure 10.4, the estimated regional and industrial spillover effects predicted from the standard IO perspective are substantially diminished when regional and industrial adaptations are accounted for. Results for the Gulf of Mexico region are similar.

The month-to-month output changes for the top five states are presented in Table 10.3. Positive impacts appear because of the positive direction of the

Table 10.3. *Monthly changes of total impacts of oil-refinery losses from Hurricanes Katrina and Rita for the top five regions*

Period	CA	FL	GA	PA	Gulf_MX	U.S.Total	Rest of World	World Total	Direct Impact on Gulf_MX
Sep. 2005	0.54	0.99	0.86	0.31	122.11	124.80	0.32	125.12	120.42
Oct. 2005	–1.24	–2.17	–1.33	–0.90	–625.06	–630.69	–1.69	–632.38	–616.37
Nov. 2005	–4.75	–2.87	–2.16	–1.96	–784.12	–795.86	–1.41	–797.27	–767.52
Dec. 2005	–1.27	–0.78	–0.55	–0.56	–282.18	–285.33	–0.75	–286.08	–276.62
Jan. 2006	0.42	0.22	0.17	0.17	57.84	58.83	0.12	58.95	56.20
Feb. 2006	–1.02	–0.58	–0.41	–0.42	–228.22	–230.64	–0.55	–231.19	–224.00
Mar. 2006	–4.88	–2.56	–1.88	–1.96	–556.89	–568.17	–1.43	–569.60	–536.58
Apr. 2006	–2.10	–1.17	–0.83	–0.86	–335.05	–340.01	–1.10	–341.11	–325.40
May. 2006	0.41	0.22	0.16	0.17	44.91	45.87	0.12	45.99	43.15
Jun. 2006	–1.26	–0.70	–0.50	–0.51	–214.25	–217.23	–0.65	–217.88	–208.61
Jul. 2006	–2.40	–1.28	–0.93	–0.96	–260.02	–265.59	–0.71	–266.30	–249.79
Aug. 2006	–5.69	–3.16	–2.23	–2.30	–877.47	–890.85	–2.92	–893.77	–851.32
Sep. 2006	–5.58	–3.00	–2.16	–2.22	–729.66	–742.63	–1.65	–744.28	–707.04
Sum	–28.81	–16.85	–11.78	–12.00	–4,668.06	–4,737.50	–12.30	–4,749.81	–4,543.47

Unit: $million

Note: Quantities for direct impacts of the USC Sector 10 losses are identified in the last column of this table.

direct impacts obtained from the times-series model for the Gulf of Mexico. These indicate that there were slightly positive gains in September 2005, January 2006, and May 2006. These impacts would not be captured using traditional IO models. FlexNIEMO captures the adaptation processes in the form of economic structure changes associated with the occurrence of an extreme event. The essence of these adjustments involves considerable learning and adaptation, and this method is suitable for this type of analysis. While not presented here, results are available for each industry.

10.6 Conclusions

Impact modeling using widely available input-output approaches routinely includes the caveat that an assumption of fixed-coefficient technologies produces estimates that necessarily overstate results. A new and operational multiregional input-output model of the United States, FlexNIEMO is applied to identify the scale and scope of changes to technical coefficients in the case of oil and gas refinery losses in the Gulf of Mexico following Hurricanes Katrina and Rita in late 2005. Results show that the detailed study of adaptations is useful because overstated impacts from the application of conventional IO are indeed substantial. FlexNIEMO gives coefficient adaptations by month, by industry, and by region. Questions about the effects of time, distance, and industry linkage can be posed and studied using this approach.

NIEMO includes millions of empirical multipliers and the detailed examination of selected individual adaptations remains. These are in the tradition of input-output structural decomposition analysis (see Rose & Casler, 1996), and will be taken up in future research. NIEMO is currently being extended to address local land use modeling, transportation and multimodal systems; demand-price elasticity modeling; and environmental modeling for estimating greenhouse gas (GHG) emission effects. Another important extension is to place NIEMO in a game-theoretic context, which can be more directly dedicated to the dynamics of conflict activities.

REFERENCES

Bureau of Labor Statistics (2006). The labor market impact of Hurricane Katrina: an overview. *Monthly Labor Review*, 129(8): 3–10. Available at www.bls.gov/opub/mlr/2006/08/art1abs.htm.

Chenery, H. B. (1953). Regional analysis. In H. B. Chenery, P. G. Clark, & V. C. Pinna (Eds.), *The structure and growth of the Italian economy* (pp. 98–139). Rome: U.S. Mutual Security Agency.

Colgan, C. S., & Adkins, J. (2006). Hurricane damage to the ocean economy in the U.S. Gulf region in 2005. *Monthly Labor Review*, *129*(8), 76–78. Available at www.bls.gov/opub/mlr/2006/08/art7abs.htm.

Energy Information Administration (EIA). (2006a). *The impact of tropical cyclones on Gulf of Mexico crude oil and natural gas production* (released June 7, 2006). Available at http://tonto.eia.doe.gov/FTPROOT/features/hurricanes.pdf.

Energy Information Administration (EIA). (2006b). *Weekly petroleum status report.* Available at www.eia.doe.gov/oil_gas/petroleum/data_publications/weekly_petroleum_status_report/wpsr.html.

Gordon, P., Moore, II, J. E., Park, J. Y., & Richardson, H. W. (2007). The economic impacts of a terrorist attack on the US commercial aviation system. *Risk Analysis*, *27*(3), 505–512.

Gordon, P., Moore II, J. E., Park, J. Y., & Richardson, H. W. (2009). The economic costs of border closure: A state-by-state analysis. In H. W. Richardson, P. Gordon, and J. E. Moore, II (Eds.), *Global business and the terrorist threat* (pp. 341–374). Cheltenham: Edward Elgar.

Government Accountability Office (GAO). (2006). *Natural gas: Factors affecting prices and potential impacts on consumers (GAO-06-420T).*

Katz, B., Fellowes, M., & Mabanta, M. (2006). *Katrina index: Tracking variables of post-Katrina reconstruction (Updated March 2, 2006).* Washington, DC: The Brookings Institution.

Kent, J. D. (2006). *2005 Louisiana Hurricane Impact Atlas Volume 1, Louisiana Geographic Information Center*. Available at http://lagic.lsu.edu/lgisc/publications/2005/LGISC-PUB-20051116-00_2005_HURRICANE_ATLAS.pdf.

Knabb, R. D., Brown, D. P., & Rhom, J. R. (2006). *Tropical Cyclone Report Hurricane Rita: 18–26 September 2005. National Hurricane Center.* Available at www.nhc.noaa.gov/pdf/TCR-AL182005_Rita.pdf

Leontief, W. (1936). Quantitative input and output relations in the economic system of the United States. *Review of Economic Statistics*, *XVIII*(3), pp. 105–125.

Leontief, W. (1941). *The structure of American economy, 1919–1929: An empirical equilibrium analysis.* Cambridge, MA: Harvard University Press.

Louisiana Geographic Information Center (2005). *Louisiana Hurricane Impact Atlas, Vol. 1.* Available at http://lagic.lsu.edu/images/hurricanes/2005_LAGIC_HURRICANE_ATLAS.pdf.

MMS. (2006). *Deepwater Gulf of Mexico 2006: America's expanding frontier.* Available at www.gomr.mms.gov/homepg/whatsnew/techann/2006/2006–022.pdf.

Moses, L. N. (1955). The stability of interregional trading patterns and input-output analysis. *American Economic Review*, *45*, 803–832.

National Climate Data Center (NCDC). (2005). *Climate of 2005: Summary of Hurricane Katrina.* Available at www.ncdc.noaa.gov/oa/climate/research/2005/katrina.html#top.

National Hurricane Center of National Weather Service. (2007). *Tropical Weather Summary – 2005 Web Final.* Available at www.nhc.noaa.gov/archive/2005/tws/MIATWSAT_nov_final.shtml.

National Research Council. (1999). *The impacts of natural disasters: A framework for loss estimation.* Available at www.nap.edu/catalog/6425.htmlb.

Nordhaus, W. D. (2006). December, NBER Working Paper No. 12813, *The economics of hurricanes in the United States*. Cambridge, MA: National Bureau of Economic Research.
Park, J. Y. (2008). The economic impacts of a dirty-bomb attack on the Los Angeles and Long Beach port: Applying supply-driven NIEMO (National Interstate Economic Model). *Journal of Homeland Security and Emergency Management*, *5*(1).
Park, J. Y. (2009, March 26). Application of a price-sensitive supply-side input-output model to an examination of the economic impacts of Hurricanes Katrina and Rita's disruptions of the U.S. oil industry. Invited to present at the 2009 Upstate New York of Society Chapter for Risk Analysis Symposium, 408 Baker Hall, State University of New York College of Environmental Science and Forest.
Park, J. Y., Cho, J., & Rose, A. (2011a). Modeling a major source of economic resilience to disasters: Recapturing lost production. *Natural Hazards*, *58*(1), 163–182.
Park, J. Y., & Gordon, P. (2005, November 10–12). An evaluation of input-output aggregation error using a new MRIO model. Presented at the *North American Meetings of the Regional Science Association International 52nd Annual Conference*, Riviera Hotel & Casino, Las Vegas, NV.
Park, J. Y., & Richardson, H. W. (2014). Refining the Isard Multiregional Input-Output Model Theory. In P. Nijkamp, A. Rose, & K. Kourtit (Eds.), *Regional science matters: Studies dedicated to Walter Isard* (pp. 35–54). Springer-Verlag.
Park, J. Y., Gordon, P., Jun, E., Moore, II, J. E., & Richardson, H. W. (2009b). Identifying the regional economic impacts of 9/11. *Peace Economics, Peace Science and Public Policy*, *15*(2), Article 6.
Park, J. Y., Gordon, P., Moore, II, J. E., & Richardson, H. W. (2008a). The state-by-state economic impacts of the 2002 shutdown of the Los Angeles-Long Beach ports. *Growth and Change*, *39*(4), 548–572.
Park, J. Y., Gordon, P., Moore, II, J. E., & Richardson, H. W. (2009a). A two-step approach to estimating state-by-state commodity trade flows. *The Annals of Regional Science*, *43*(4), 1033–1072.
Park, J. Y., Gordon, P., Moore, II, J. E., & Richardson, H. W. (2011b, June 13–17). Constructing a flexible National Interstate Economic Model. Presented at the 19th International Input-Output Association Conference, Alexandria, VA.
Park, J. Y., Gordon, P., Moore, II, J. E., & Richardson, H. W. (2013). The interregional and interindustry impacts of the Gulf oil spill: Applying the National Interstate Economic Model (NIEMO). *Journal of Homeland Security and Emergency Management, 10*(1), 231–244.
Park, J. Y., Gordon, P., Moore, II, J. E., Richardson, H. W., Kim, S., & Kim, Y. (2008b). Estimating the state-by-state economic impacts of Hurricane Katrina. In H. W. Richardson, P. Gordon, & J. E. Moore, II (Eds.), *Natural disaster analysis after Hurricane Katrina* (pp. 147–186). Cheltenham: Edward Elgar.
Park, J. Y., Gordon, P., Moore, II, J. E., Richardson, H. W., & Wang, L. (2007). Simulating the state-by-state effects of terrorist attacks on three major U.S. ports: Applying NIEMO (National Interstate Economic Model). In H. W. Richardson, P. Gordon, & J. E. Moore, II (Eds.), *The economic costs and consequences of terrorism* (pp. 208–234). Cheltenham: Edward Elgar.

Park, J. Y., Park, C. K., & Nam, S. J. (2006, July 23–26). The state-by-state effects of mad cow disease using a new MRIO model. Presented at the 2006 *American Agricultural Economic Association (AAEA) Annual Meeting*, Long Beach, CA.

Richardson, H. W., Gordon, P., & Moore, II, J. E. (Eds.). (2007b) *The economic costs and consequences of terrorism.* Cheltenham: Edward Elgar.

Richardson, H. W., Gordon, P., Moore, II, J. E., Kim, S. J., Park, J. Y., & Pan, Q. (2007a). Tourism and terrorism: The national and interregional economic impacts of attacks on major U.S. theme parks. In H. W. Richardson, P. Gordon, & J. E. Moore, II (Eds.), *The economic costs and consequences of terrorism* (pp. 235–253). Cheltenham: Edward Elgar.

Rose, A., & Casler, S. (1996). Input-output structural decomposition analysis: A critical appraisal. *Economic Systems Research*, 8(1), 33–62.

U.S. Bureau of Economic Analysis. (2012). www.bea.gov, the BEA GDP data access on June 5, 2012.

11

Risk-Informed Benefit-Cost Analysis

Scott Farrow

11.1 Introduction

Many government organizations seek risk-informed decisions. Evaluating and managing risk is clearly central to the mission of some agencies such as the Environmental Protection Agency (EPA) or the U.S. Department of Homeland Security (DHS). Risks clearly exist behind numerous governmental issues such as the financial viability of Medicare or the management of offshore oil and gas resources. However, the risk assessment community has tended to separate itself from management issues by attempting to draw a line between the "scientific" aspects of risk assessment and the management of risk.

Managers, however, request analytical metrics to inform their decisions that often aggregate, weight, or transform metrics from their natural units such as probability, fatalities, or monetized business consequences. The choice of analytical metrics – the raw data, ranked risks, "rational" or "behavioral" utility metrics, monetized risks, pain indices, or other measures – are either explicitly or implicitly framed by a model. Such models may be "mental" and reside within a decision maker or explicitly incorporate a formal structure for decision making with risk. Each metric is thus based on explicit or implicit assumptions.

This chapter first compares alternative risk metrics on the basis of their assumptions. No criteria are used to select a "best" metric (although some biases may appear). The initial sections of this chapter extend the concept of "meta-choice" analysis (Vining & Boardman, 2006). Those authors argued that different types of analysis can be characterized along the two dimensions

The author thanks the Center for Risk and Economic Analysis of Terrorism Events at the University of Southern California for funding and participants at the benefit-cost workshop at the Toulouse School of Economics for comments. Ali Abbas and Debra Elkins were helpful in motivating this work.

of *goal choice* and *monetization*. Meta-choice analysis in their framework consisted of matching the desired type of analysis with the analytical purpose. This chapter focuses on the meta-analytic choices surrounding risk. Later sections focus on one type of analysis, benefit-cost analysis, and the more detailed analytic choices to be made about risk for that particular model. The decision-making context is made specific by focusing on the DHS.

11.2 Risk Principles and Enterprise Risk Management in the U.S. Government

Guidance or requirements for risk-informed decisions come from numerous sources for the U.S. government. Preeminent are laws passed by Congress. Some laws direct that risk in a particular context be managed to a specific level, for example, by requiring flood insurance for all within a 100-year flood plain for those with certain types of mortgages. Other laws may limit the application of some metrics to inform risk, such as directing the EPA to protect the health of sensitive populations with an adequate margin of safety and to ignore cost in the development of some regulations that have the force of law. Other legislation requires agencies to consider benefits and costs of risk management choices.

In the absence of clear and direct guidance from Congress, the Executive Office of the President provides guidelines for agencies managing risk. One recent guideline is the inclusion of enterprise risk management (ERM) elements in agency annual budget submissions (OMB, A-11, 2014). This new task in budget preparation can be carried out using methods tailored to each agency, but states that one of the purposes of such analysis is to "Facilitate identification and adoption of opportunities for improvement, including risk management" (OMB, A-11, 2014, section 270.9). In the context of environmental, health, and safety risks, the Office of Management and Budget (OMB) provides additional guidance while noting that the guidance may also be relevant to other fields (2007).

Is it useful to consider whether the DHS risk profile falls within the environmental, health, and safety (EHS) context? Might risks in the DHS domain such as flooding and biological or nuclear contaminants be considered environmental or health or safety? Might concern about terrorist attacks on people and the built environment be considered safety or environmental? Less obvious might be issues about immigration or customs, which have some element of health and safety, but other objectives as well. A restricted interpretation of DHS activities might result in only some activities lying specifically within the

EHS context. A broad interpretation, including the OMB suggestion that the guidance may be relevant to other fields, could permit the application of OMB risk guidance to DHS.

What do OMB's risk principles state about risk management? The two management principles are stated next[1]:

1. *In making significant risk management decisions, agencies should analyze the distribution of the risks and the benefits and costs (both direct and indirect, both quantifiable and non-quantifiable) associated with the selection or implementation of risk management strategies. Reasonably feasible risk management strategies, including regulation, positive and negative economic incentives, and other ways to encourage behavioral changes to reduce risks (e.g., information dissemination), should be evaluated. Agencies should employ the best available scientific, economic, and policy analysis, and such analyses should include explanations of significant assumptions, uncertainties, and methods of data development.*

Agencies should refer to Circular A-4 for updated guidance regarding agency best practices for regulatory analysis.

2. *In choosing among alternative approaches to reducing risk, agencies should seek to offer the greatest net improvement in total societal welfare, accounting for a broad range of relevant social and economic considerations such as equity, quality of life, individual preferences, and the magnitude and distribution of benefits and costs (both direct and indirect, both quantifiable and non-quantifiable).*

Agencies should refer to Circular A-4 for updated guidance regarding agency best practices for regulatory analysis. (OMB, 2007)

Risk management, as guided by OMB, focuses on risk, benefits, and costs (although not all interpret this to mean benefit-cost analysis as practiced by economists). Further, the objective is the greatest net improvement in total societal welfare considered in broad terms. Additional guidance on regulatory issues is provided in a supplemental publication (OMB, A-4) and separately on benefit-cost analysis and discounting (OMB, A-94), but the risk guidance is not uniquely focused on regulation; in fact, it specifically calls for non-regulatory approaches such as incentives and information.

11.3 The Department of Homeland Security Context

DHS's many components could be said to be focused on risk, and the agency itself was a creation of Congress shortly after the 9/11 terrorist attacks revealed a new magnitude of risk. DHS is charged with managing a wide range of natural- and human-caused risks, including those relevant to the Coast Guard, the Federal Emergency Management Agency, the Transportation Security Agency, Customs and Immigration, and more.

Table 11.1. *DHS threat and hazard types: National strategic risk assessment*

Threat/Hazard Type	
Animal Disease Outbreak	Aircraft as a Weapon
Earthquake	Armed Assault
Flood	Biological Terrorism Attack (non-food)
Human Pandemic Outbreak	Chemical/Biological Food Contamination Terrorism Attack
Hurricane	Chemical Terrorism Attack (non-food)
Space Weather	Cyber-Attack against Data
Tsunami	Cyber-Attack against Physical Infrastructure
Volcanic Eruption	Explosives Terrorism Attack
Wildfire	Dam Failure
Biological Food Contamination	Radiological Substance Release
Chemical Substance Spill or Release	Nuclear Terrorism Attack
	Radiological Terrorism Attack

Source: DHS, 2011.

DHS prepares risk assessments on an agency-wide basis every four years and numerous assessments on various combinations of risks. A strategic planning and risk office exists within the office of policy, as do other offices that support DHS-wide efforts at risk analysis.

A recent report summarized the task of risk management in the agency as:

> The safety, security, and resilience of the Nation are threatened by an array of hazards, including acts of terrorism, malicious activity in cyberspace, pandemics, manmade accidents, transnational crime, and natural disasters. At the same time, homeland security organizations must manage risks associated with workforce management, acquisitions operations, and project costs.
>
> Collectively, these external and internal risks have the potential to cause loss of life, injuries, negative psychosocial impact, environmental degradation, loss of economic activity, reduction of ability to perform mission essential functions, and loss of confidence in government capabilities. (DHS, 2011)

Consider the problem faced in trying to synthesize risks across the many components of DHS. In public documents, DHS has recognized 23 types of strategic risk it seeks to manage as presented in Table 11.1 (DHS, 2011).[2] DHS has also stated in the Strategic National Risk Assessment that it is concerned about numerous kinds of consequences falling into six categories of:

1. loss of life,
2. injuries and illnesses,

3. direct economic costs,
4. social displacement,
5. psychological distress, and
6. environmental impact.

DHS also recognizes other or more detailed consequences in various settings (DHS, 2011). Unsurprisingly, DHS's problem involves multiple risks with multiple consequences.

This context of a variety of risks, consequences, and decision makers will be used to make specific some of the general implications of choosing different risk management models and metrics.

11.4 Risk Management Models and Metrics

In this section, models are distinguished from metrics, although some models will be uniquely associated with some metrics. This approach complements that taken by Vining and Boardman (2006), who distinguished among models based on the goals identified (economic efficiency or not) and monetization (or not) of impacts. That two-way breakdown was determined to be too aggregate for the purposes of this chapter and importantly, Vining and Boardman did not explicitly address issues related to risk.

11.5 Meta Model Choice

Models as abstractions from reality are commonly used to inform decision makers concerned with risks. Such models occur at least in the political science, public policy, risk analysis, engineering, economic, and decision analysis communities.

The models chosen here are not from a comprehensive review of the literature, but based on experience in cross-agency practices and academic experience in engineering, public policy, and economics. Other researchers may wish to add models. The models considered here are:

- Impact analysis as exemplified by a National Environmental Policy Act type of analysis (CEQ, 2005), where impacts are identified, quantified to the extent possible, and discussed.
- "Standard" Decision Analysis as exemplified by expected utility or an expected outcome analysis hence requiring probabilities and some common unit of measurement (Clemen & Reilley, 2001; Haimes, 2005; Keeney & Raiffa 1976).

- "Advanced" Decision Analysis as exemplified by a multi-criteria/multi attribute analysis (Clemen & Reilley, 2001; Haimes, 2005; Keeney & Raiffa, 1976).
- "Standard" Benefit-Cost Analysis as exemplified by equal utility weighting and expected values whether monetary or utility (Boardman et al., 2011; OMB A-94).
- "Advanced" Benefit-Cost Analysis as exemplified by equity weighting and information on statistical outcomes (Boardman et al., 2011; OMB A-4).
- Economic consequence analysis as exemplified by input-output or computable general equilibrium analysis (Rose, 2014; Santos, 2006).
- Cost-Effectiveness Analysis as exemplified by the least cost to achieve a stated objective (Boardman et al., 2011; OMB A-94, A-4).

While the lines between some of the analyses can be blurred, the attempt here is to identify those assumptions, as standardly used, which distinguish these alternative risk management analytical framings.

Each model builds from conceptual assumptions, implemented with data and quantitative procedures, and results in one or more performance measures. The performance measures may integrate several elements, such as expected net present value, which aggregates outcomes over time and reports the probabilistically weighted average. The purpose of this section is to set out the assumptions that distinguish the several models and the standard performance measures. Characteristics of various models as identified by the author are set out in Appendix Table 11.1.

Areas in which the models were consistent as generally applied (exceptions may always exist), included:

1) Ability to define who has standing (and so models could be consistent);
2) Individual preferences generally modeled as being independent of one another.

The distinguishing characteristics among the models, some of which are bundled together in the discussion to follow for ease of presentation, are: a) whose preferences are being modeled; b) the goal and associated metric; c) the method of aggregation, including weights if relevant; and d) risk measures and preferences. Each is discussed briefly here with a focus on those models with the most divergent approaches.

1. Whose preferences are modeled? This assumption distinguishes most forms of decision analysis from benefit-cost analysis. In standard decision analysis, the analyst is typically modeling the preferences of one or a few people

generally considered to be taking the role of a decision maker, stakeholder, or representative consumer. In contrast, benefit-cost analysis models the preferences of those in society who have standing (although a representative consumer may frequently be invoked). The question of whose preferences are being modeled is ambiguous for an NEPA-type impact analysis.

2. The goal and associated metrics. The goal or objective function of some model types is clear while others are more ambiguous. A NEPA-type analysis has no unifying objective function and reports metrics in their natural units (illnesses, crimes, health outcomes, dollars). Impact analysis typically reports gross domestic product, but also dimensions such as unemployment without a unifying objective. Decision analysis typically uses monetary or utility measures for the outcomes with the goal being the maximization of the common metric. The outcomes may be decomposed into attributes. Benefit-cost analysis aspires to use surplus measures (measures of value taking into account different valuation for the intra-marginal units), but standard benefit-cost analysis typically applies an average value that may or may not be based on concepts of surplus. The goal is to maximize total surplus (net benefits), which economists define as (economic) efficiency. Advanced benefit-cost analyses may take into account measures of surplus possibly including option price or other variations involving risk preferences. Regarding descriptive or other statistics used, the expected value of the preferred measure is typically reported although increasingly, more information about the statistical distribution of the risk metric is presented.
3. Method of aggregation, including aggregation weights: There are often two dimensions of aggregation: 1) for one individual, aggregating across impacts if multiple impacts exist; or 2) aggregating across individuals. Here the assumption is that social impacts, broadly based, are the object of estimation in contrast to impacts on an identified group such as a select group of consumers, firms, or the government alone. In an impact analysis, individual impacts such as new jobs are typically aggregated equally across individuals for impacts in a particular category, but no attempt is made to aggregate across impact categories, as across jobs and environmental impacts. Economic impact, or consequence analyses, are defined in monetary terms and typically aggregate equally within and across economic impact categories. With decision analysis, numerous methods exist to develop weights to aggregate impacts for one individual, but the standard application is without aggregation across individuals. Sometimes, if a group of people is analyzed, then various aggregates such as mean, median, or minimum distance measure might be reported but are seldom interpreted to represent the aggregate of society. An exception is the aggregation of

expected monetary values, which may be little distinguished from a benefit-cost analysis based on expected values. Standard benefit-cost analysis uses monetary values to aggregate both for an individual and across individuals. The latter typically assumes an equal social and individual value per incremental dollar to each member of society. Advanced benefit-cost analyses will sometimes weight dollar impacts to different groups in the aggregation process, often though not necessarily based on income. (This difference can be thought of as adjusting simultaneously for societal preferences across the different groups and for individual differences in the incremental value of income.)

4. Risk preferences: Risk preferences are here taken to be any of risk-neutral, risk-loving, risk-averse, or some behavioral risk approach (Eeckhoudt et al., 2005). Such preferences can be revealed at two stages of an analysis. The first are the risk preferences of individuals and the second are the risk preferences in any social aggregation. Impact analysis typically ignores such risk preferences. Some advanced types of economic consequences analysis may include behavioral risk responses, but the majority of economic impact analyses assume a risk-neutral approach on the part of both individuals and in the aggregation process. Consequently, the standard risk metric is an expected value impact, often an expenditure or income. Basic decision analysis often assumes risk neutrality by analyzing the expected value of a decision. Advanced decision analyses can consider alternative risk preferences as measured by utility, and then may aggregate using expected or a non-expected utility approach. Standard benefit-cost analysis is carried out in the world of certainty for individual outcomes and the aggregation of individual values is based on the risk neutrality, or expected value, of both the individual and any governmental decision maker. Advanced benefit-cost analyses may take into account risk preferences in assessing the valuation of individuals, and then typically, although not necessarily, present the expected value of such valuations (Arrow & Lind, 1970; Haimes, 2005; Keeney & Raiffa, 1976).

An assumption that tends to vary across model types includes the treatment of impacts in different time periods, which is usually aggregated into a present value metric by discounting. Benefit-cost analysis routinely uses discounting of future values or annualization of values as do many decision analyses. Discounting may be used less commonly in other analyses.

These common and disparate dimensions among the models extend the two-way monetization and goal dimensions of Vining and Boardman (2006). The Vining and Boardman choice problem is graphed is Figure 11.1, where

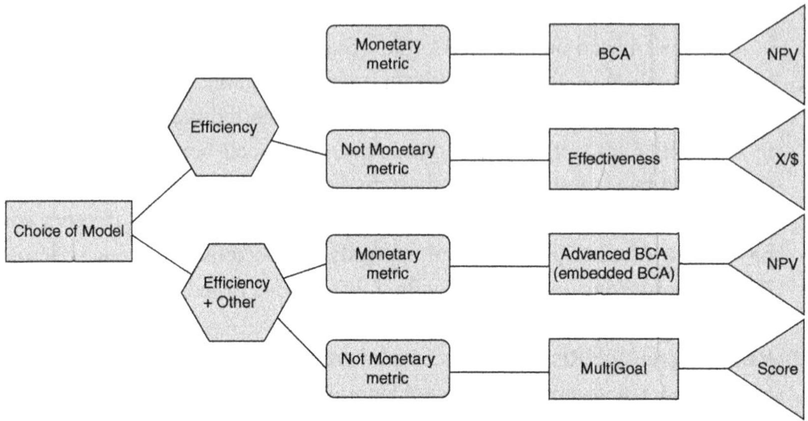

Figure 11.1. Meta-choice logic model of Vining and Boardman.

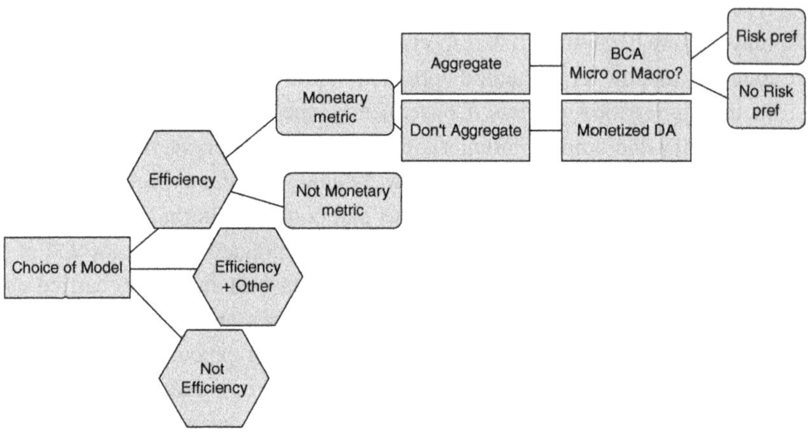

Figure 11.2. Extended meta-choice including aggregation and risk preferences.

various choices in the logic model, starting on the left, lead the analyst to a chosen model on the right-hand side and its associated metric. For example, the sequence of the goal of efficiency and monetization leads to benefit-cost and net present value.

Model choices involving aggregation and weighting refine the Vining and Boardman meta-analysis. As the left-hand side of Figure 11.1 remains the same, Figure 11.2 highlights additional choices beyond monetization that lead to finer characterization among the models.

The distinction among models can be investigated in greater detail should one desire. For instance, Adler (2012) has developed a related logical sequence for standard and nonstandard methods of aggregation based on which of several axioms are met by social welfare functions. The "tree" sequence, however, is not unique in the sense that some reordering of the axioms can lead to a different visual representations.

11.6 Benefit-Cost Analysis and the Implementation of Risk Concepts

This section investigates the specific ways in which risk and uncertainty affect the choice of what is to be measured – the conceptual metric – and the numerical estimate. The conceptual issues are highlighted by focusing on benefit-cost analysis and more detail is provided on empirical measurement.

11.7 Risk and BCA

Risk appears in numerous guises in the design and implementation of a benefit-cost analysis (Farrow & Viscusi, 2011). Risk is here interpreted broadly to include both known and unknown statistical distributions and the several ways in which these concepts reveal themselves in assumptions about metrics and model specification. For instance, an incorrectly applied constraint that cross-price elasticities are zero among some markets introduces a kind of degenerate or deterministic distribution on a value that may be incorrect.

This section separates levels of benefit-cost analysis into deterministic BCA (DBCA) as that standardly used in practice and BCA incorporating risk (RBCA) as a more advanced version.[3] As will be shown, numerous issues that may appear relevant only in an RBCA are in fact implicit in the assumptions leading to the restricted DBCA.

First, define a standard DBCA, with Consumer Surplus (CS) and Producer Surplus (PS) as measures net of payments to government, GR as government revenue, and EX as externalities and other symbols as standardly used. The DBCA objective function, which is to be maximized by the choice among alternatives or scale, is then

$$\textit{Present Value Net Benefit} = \sum_{t=0}^{t=} \frac{B_t - C_t}{(1+r)^t} = \sum_{t=0}^{t=} \frac{CS_t + PS_t + GR_t + EX_t}{(1+r)^t} \qquad (11.1)$$

A standard RBCA objective function is defined as below where π_i is probability (here assumed constant over time, although this is not necessary), D(t) is a discount function, CSR, PSR, and ER are measures that may include risk-adjusted values, and ε_{it} is an additive error that may vary by outcome or time:

$$Expected\ Present\ Value\ Net\ Benefit = \sum_{i=1}^{n}\sum_{t=0}^{t=}\pi_i\left(\frac{CSR_{it} + PSR_{it} + GR_{it} + ERx_{it}}{D(t)}\right) + \epsilon_{it} \qquad (11.2)$$

Each of these terms and more that are implicit will be discussed for the way in which risk affects their definition or measurement. The discussions will be arbitrarily divided into the following categories:

- Meta-choice assumptions within BCA, addressing specification issues related to means of aggregation and the extent of markets,
- Metrics, addressing issues of decision criteria and risk-based preferences including discounting,
- Randomness, addressing issues of the probability of events, variability in parameters, random error.

11.8 Meta-choice and Specification within Benefit-Cost Analysis

The first part of this chapter discussed the meta-choices between major classes of decision-support models. Within a given analytical choice, such as benefit-cost analysis, several large specification or framing issues remain for the analysis. A "wrong" assumption on these issues can create modeling error, or in statistical terms, impose a constraint or degenerate distribution on the analysis. While assumptions are central to abstraction, they also introduce specification risk to the extent they are incorrect. Standard macro-level framing specification assumptions for benefit-cost analysis are discussed next:

- General versus partial equilibrium: This choices defines the scope of the terms in the numerator of equations (11.1) or (11.2). Benefit-cost analysis may be carried out in either a general or partial equilibrium setting (Florio, 2014; Goulder & Williams, 2003). Choosing a partial equilibrium specification imposes zero cross-price elasticity constraints on economic interactions while simplifying the analysis. Alternatively, if one carries out a general equilibrium analysis, the functional specifications and parameters may not be accurate; the difficult question is which is more accurate compared to the "true" model. A specification risk occurs with either choice.

- Aggregation across economic actors: This choice defines the weighting of impacts in the numerator of equations (11.1) and (11.2). The default aggregation behind equations (11.1) and (11.2) is equal and independent social weighting for each individual. This can be explained as equal social weighting and equal marginal utility of income for all participants; or if one assumes diminishing marginal utility of income, then equal weighting assumes society more heavily weights impacts on the wealthy (Nyborg, 2012). The most frequent alternative to equal weighting subjectively applies larger weights to impacts on lower-income or other identified populations (e.g., Loomis, 2013). Other weighting approaches can but seldom do include interdependent weights. A specification risk occurs with any aggregation choice for the social welfare function.

11.9 Metrics

11.9.1 *Decision Criteria Metrics*

Performance metrics for benefit-cost analysis are generally inferred from criteria for optimization. The standard optimization problems involve a single social welfare optimizer and implicitly or explicitly includes the aggregation function (social welfare function) and the risk perspective of the decision maker.

- Deterministic benefit-cost analysis seeks to maximize equation (11.1) with the resulting optimal solution being the largest present value of net benefits among mutually exclusive alternative considered. Several variations of this problem exist although often ignored in practice (Bellinger, 2007). One variation is when the scale of the project can be chosen such that some benefits and costs vary with scale. In that case, the decision criteria equates the marginal (social) benefit with the marginal social cost or, if there is a budget constraint, the net marginal benefits may be required to exceed a threshold when there are multiple projects. Few organizations are without budget constraints, which can lead to a decision metric requiring that a critical ratio be exceeded; in this case, the ratio of benefits to costs must exceed a threshold (Bellinger, 2007; Weinstein & Zeckhauser, 1973).
- Risked benefit-cost analysis (RBCA) introduces numerous alternatives for the performance metric based on equation (11.2). The standard approach is that a social decision maker who can pool risk across many projects will choose to maximize the expected net present value (Arrow & Lind, 1970) as given by equation (11.2), either omitting the error term or assuming that its expectation is zero. This metric may be used whether

or not risk-based valuation has been used for the component measures, such as consumer or producer surplus. The increasingly frequent practice of reporting the statistical distribution of net present value and letting a decision maker choose an appropriate option or level of stringency implies that a decision maker will use his own risk preferences, however obtained, to make a selection that may not be based on expected value. Concern for dynamic uncertainty when there are irreversible commitments, as with unknown prices but fixed long-term investment, leads to decision criteria in which there is a value of information that may be informed by waiting. The standard model in such settings is stochastic, expected value optimization in which net benefits may need to exceed a positive threshold, the option value of waiting, in order to take action (Dixit & Pindyck, 1994).

11.9.2 *Preferences and Behavioral Response Metrics*

Preferences and resulting behavioral responses are the core of benefit-cost analysis. The monetized meta-choice implied in benefit-cost analysis means that values, sometimes revealed through choices or surveys, are measured in monetary terms. The literature on the monetization of values implicit in choices is vast, some of it under topics such as welfare economics (e.g. Just et al., 2008) or benefit-cost analysis (Zerbe, 2008). Preferences and behavioral responses may be modeled as rational, often using an expected utility approach (Boardman et al., 2011). Much current research is focused on behavioral economics involving an explicit or implicit non-expected utility approach where responses may be consistent but not be rational (DellaVigna, 2009; Machina, 1987). Risk preferences can provide an important distinction between DBCA and RBCA. In RBCA and its associated equation (11.2), consumer preferences are frequently modeled as risk averse whereby some adjustment to expected cost or benefit is added to the valuation for a consumer (or producer). While some functional forms are commonly used, there is little agreement on the appropriate model and empirical parameters of risk for a consumer (Eeckhoudt et al., 2005). At the same time, for some problems, risk-based preference adjustment may or may not be empirically significant (Farrow, 2015). Although the estimation and monetization of risk preferences is a research frontier, advances in valuation have led to relatively standard values for a statistical life in benefit-cost analyses, as well as for other risk-based outcomes (Boardman et al., 2011).

Preferences over time is a long-standing dimension of concern. A declining exponential weight on values in the future (discounted values) has long been standard, but behavioral economics and concern for uncertainty are altering that consensus. Behavioral economic studies have found that people may overweight (relative to exponential discounting) the near term and also the distant future (DellaVigna, 2009). Further, debate remains about the impact of incorporating uncertainty into discount factors, although a typical result with uncertainty is that the future is discounted less than with exponential discounting (Arrow et al., 2013; Gollier, 2012).

11.10 Variability and Randomness

Variability in impacts and valuations are one source of risk in the outcome metric. For instance, an unconditional mean value for an impact may ignore predictable conditionality of the mean on socioeconomic and other conditioning factors. This is discussed in the literature in the context of a benefit transfer function that includes adjustments for varying conditions instead of using an unconditional value (Boyle et al., 2010). Similarly, if impacts are estimated separately from values, then the impacts may depend on various conditioning factors, such as distance or income and add further variability risk.

Fundamental to an RBCA is the estimation of probability, or more accurately, a set of probabilities and outcomes represented by π_i and the associated outcomes in equation (11.2). In some DHS settings, the probability distributions may be reasonably estimated based on historical data that are sufficient to inform policy. Probabilities of floods of a given size may be such a case. In other DHS security settings, the estimation of probability distributions is more problematic, although not necessary impossible. Complications may involve the availability of classified information, the fact that some probabilities may be endogenous (as through game theory) where the probability of attack may depend on the defender's chosen strategies, and the infrequent nature of observed outcomes (Farrow, 2007). Some analysts have chosen not to estimate probabilities, but instead estimate a DBCA (or variations thereof) and find a break-even risk where the DBCA becomes positive as is frequently used in DHS regulatory analyses (Farrow & Shapiro, 2009; Mueller & Stewart, 2011). In some settings, DHS uses estimates of probabilities, but they are not available to outside researchers. Probability estimation remains a fundamental challenge for RBCA analysis.

The impact of stochastic randomness or pure error in an RBCA appears in several ways. Models for components of equation (11.2) are often built from parameterized models and sometimes from regression coefficients. Perhaps clearest in the case of regression coefficients, those coefficients have a statistical distribution based on the random error in the underlying estimation equation. Such randomness can be included in Monte Carlo type analyses and adds risk to the outcome measure even if the distribution is symmetric about its mean.

Seldom considered in an RBCA is the degree of model fit associated with a stochastic error term, which may be added as noted in equation (11.2). For instance, homeland security models may have a larger random error component than a flood model. Such added variability is a component of the risk of the dependent metric. When Monte Carlo simulation methods are used, analysts may be able to include a measure of the pure error term in the RBCA if they are willing to place an estimate on the accuracy of the model (Farrow, 2012).

11.11 Conclusion

This chapter emphasizes the ubiquity of risk in decision modeling in general and in benefit-cost modeling in particular. The meta-choice of benefit-cost analysis implies a set of framing assumptions. It is useful to understand the distinctions between the several major types of descriptive and prescriptive analyses used to inform homeland security decisions. The model choice can introduce specification error. If a benefit-cost analysis is chosen as the appropriate analysis, then risk, associated with variability in outcomes, affects numerous components of the analysis. The analyst must choose whether or not to: a) use a partial or general equilibrium setting; b) deviate from standard aggregation and constraint assumptions; c) include conditioning factors that may shift impacts or values; d) include risk preferences in valuation; e) include uncertainty in discounting; f) use explicit probability distributions; and g) incorporate the accuracy of the model itself in the variability of the outcome. Taking all these elements into consideration may lead to the paralysis of analysis. The challenge to analysts is to incorporate those risk considerations that are central to the problem without foundering in a morass of analytical uncertainty.

Appendix Table 11.1. *Models and assumptions*

Assumptions	Impact Analysis	Standard Decision Analysis	Advanced Decision Analysis	Std. Benefit-Cost Analysis	Advanced Benefit-Cost Analysis	Econ. Consequence Analysis	Cost-Effect. Analysis	Other models, Notes
Who has standing	Citizens or as otherwise defined	Citizens or as otherwise defined	Citizens or as otherwise defined	Citizens or as otherwise defined	Citizens or as otherwise defined	Citizens or as otherwise defined	Citizens or as otherwise defined	
Whose preferences	Not specified	Decision maker (s) DM	Decision maker (s)	Society	Society	Not specified	Society	
Aggregate preferences	Adding up of natural units; impact may be distributional info.	None	None or Some group approach	Equal weight SWF	Unequal weight SWF	Adding up of dollar impacts with other dimensions as mental model	Equal weight SWF (including quantity outcome)	
Weights	None	DM or crowd dollar weighted	DM weighted	"crowd dollar weighted," surplus	"crowd dollar weighted," surplus; subjective distributional	Dollar weighted by price	Mixture dollar weighted and DM or expert weighted	
Independent or interdependent personal preferences	Independent	Independent	Independent	Independent	Independent	Independent	Independent	Taking account of system risk or interdependent preferences
Extent of economic interaction across markets	None	Limited	Usually Limited but could be large	Usually limited but can be large	Usually limited but can be large	Often large, General Eq.	Usually limited but can be large	
Time	Unlikely to discount	Constant discounting or no discounting	Constant or no discount	Discounting (constant)	Constant or Variable discounting	Some static, some dynamics/ resiliency		

(continued)

Appendix Table 11.1 (*cont.*)

Assumptions	Impact Analysis	Standard Decision Analysis	Advanced Decision Analysis	Std. Benefit-Cost Analysis	Advanced Benefit-Cost Analysis	Econ. Consequence Analysis	Cost-Effect. Analysis	Other models, Notes
Risk Analysis metric (s)	Impacts in natural units	Expected Value	Weighted cardinal utility, non-expected utility	Monetized social expected net present value ESNPV	Risk preference adjusted ESNPV	Impacts in monetary units valued at price	ESNPV cost per unit (issue with multiple outcomes)	Movement toward estimation of statistical distribution of outcomes
Risk preference	Not specified	Risk neutral or risk values of DM	Risk neutral or risk values of DM	Risk Neutral	Risk pref. of consumers or firm	Risk neutral	Usually risk neutral	Can apply to all of consumers, producers, DM
Objective (implicit or explicit)	Not specified	Max EV or EU	Max EU (or non EU approaches)	Max expected Total surplus	Max risk adjusted total surplus	Max expected dollar value, such as GDP	Min Cost	
Risk Decision rules	Decision maker or stakeholders apply mental model to aggregate	Choose highest EV, EU	Highest expected utility	Choose largest sum of "surplus" as EV.	Choose largest sum of "surplus" as EV; possible to let decision maker choose	GDP and possibly other dimensions	Choose least cost (benefit assumed)	
Social welfare function	Not specified	None	None	Additive, unweighted	Additive, weighted	Additive, unweighted	Least cost of reaching given objective (or most output for given budget)	

: all models tend but need not be consistent (i.e., all could apply standing to the same group of people)
: models vary relatively consistently across model types
: models tend to but need not make different assumptions
Source: Author

Appendix Table 11.2. *Assumptions for deterministic (standard) benefit-cost analysis*

Based on the first and second theorems of welfare economics being correct: 1) a PC society will be Pareto Optimal; 2) PO point is not unique

Topic	Assumptions
Consumer preferences and optimum	"Rational" Complete ordering Transitive Continuous Non-satiation Individual (no interpersonal utility) Max utility subject to budget constraint
Firm optimum	Given production function Input prices Output price Max profit
Government activity	Responds to market failures including public goods; access to either lump sum or optimal taxation
Nature of goods or services	Homogenous goods, no externalities (see second best), no taxation that distorts markets.
Market equilibrium	Perfectly competitive equilibrium in all input and output markets
Time	Discount at a constant rate
Risk	Standard: none Advanced: numerous concerns for risk
Scope of analysis (partial or general equilibrium)	Include those actions where dQi/DPj not equal to zero (could lead to partial or GE analysis)
Social welfare function	Additive in individual utility dW/dUi = a (constant and uniform) dUi/dYi= b (marginal utility of income a constant and uniform) dUi/dUj = 0, no interpersonal utility
Typical decision context: choosing among several alternatives with no budget constraint	Undertake the alternative with the largest PVNB. Alternative decision rules exist as for constraints or scaling of project.

NOTES

1. Text in italics is a restatement of 1995 principles, while plain type is the 2007 revision, which highlights additional guidance for regulatory purposes.
2. DHS clearly does not consider this list exhaustive as it states in the same source: "Additional threats and hazards, such as droughts, heat waves, winter storms, rain storms, and different types of technological/accidental or human-caused hazards, can also pose a risk to jurisdictions across the country and should be considered, as appropriate, in preparedness planning. Non-influenza diseases with pandemic potential and other animal diseases should also be considered. In addition, assessment participants identified a number of events for possible inclusion in future iterations of the SNRA, including electric grid failure, plant disease outbreak, and transportation system failure" (DHS, 2011).
3. Standard assumptions for a DBCA are in Appendix 2; see also Boardman et al. (2011).

REFERENCES

Adler, M. (2012). *Well being and fair distribution*. Oxford: Oxford University Press.

Arrow, K., & Lind, R. (1970). Uncertainty and the evaluation of public investment decisions. *American Economic Review*, *60*, 364–378.

Arrow, K., et al. (2013, July 26). Determining benefits and costs for future generations. *Science*, *341*(6144): 349–350.

Bellinger, W. (2007). *The economic analysis of public policy*. New York: Routledge.

Boardman, A., Greenberg, D., Vining, A., & Weimer, D. (2011). *Cost-benefit analysis*, 4th ed. Boston: Prentice Hall.

Boadway, R., & Bruce, N. (1984). *Welfare economics*. Oxford: Basil Blackwell.

Boyle, K., Kuminoff, N., Parmeter, C., & Pope, J. (2010). The benefit-transfer challenge. *Annual Review of Resource Economics*, *2*, 161–182.

Chetty, R. (2009). Sufficient statistics for welfare analysis: A bridge between structural and reduced form methods. In K. Arrow & T. Bresnahan (Eds.), *Annual review of economics*, Volume 1. Palo Alto, CA: Annuals Reviews.

Clemen, R. T., & Reilley, T. (2001). *Making hard decisions*. Belmont, CA: Duxbury Press, 2001.

Council on Environmental Quality (CEQ). (2005). Regulations for Implementing the Procedural Elements of the National Environmental Policy Act., 40 CFR Parts 1500–1508.

DellaVigna, S. (2009). Psychology and economics: Evidence from the field. *Journal of Economic Literature*, *47*(2), 315–372.

Dixit, A., & Pindyck, R. (1994). *Investment under uncertainty*. Princeton, NJ: Princeton University Press.

Dreze, J., & Stern, N. (1987). The theory of cost-benefit analysis. In A. J. Auerbach & M. Feldstein (Eds.) *Handbook of public economics* (chapter 14). Amsterdam: North-Holland and Elsevier Science Publishers.

Eeckhoudt, L., Gollier, C., & Schlesinger, H. (2005). *Economic and financial decisions under risk*. Princeton, NJ: Princeton University Press.

Farrow, S. (2007, January). The economics of homeland security expenditures: Foundational expected cost-effectiveness approaches. *Contemporary Economic Policy*, pp. 14–26.

Farrow, S. (2012). A missing error term in benefit-cost analysis. *Environmental Science and Technology*, *46*(5), 2523–2528.

Farrow, S. (2015). A comparison of key benefit estimation issues for natural hazards and terrorism: Ex ante / ex post valuation and endogenous risk. In C. Mansfield & V. K. Smith (Eds.) *The benefit-cost analysis of security focused regulations.* Cheltenham, UK: Edward Elgar.

Farrow, S., & Shapiro, S. (2009). The benefit-cost analysis of homeland security expenditures. *Journal of Homeland Security and Emergency Management*, *6*(1), 1–20.

Farrow, S., & Viscusi, W. K. (2011). Principles and standards for the benefit-cost analysis of safety. *Journal of Benefit-Cost Analysis*, *2*(3), #5.

Farrow, S., & Zerbe, R. (2012). *Principles and standards for benefit-cost analysis.* Cheltenham, UK: Edward Elgar.

Florio, M. (2014). *Applied welfare economics: Cost-benefit analysis of projects and policies*. New York: Routledge.

Gollier, C. (2012). *Pricing the planet's future: The economics of discounting in an uncertain world.* Princeton, NJ: Princeton University Press.

Goulder, L., & Williams, R. (2003). The substantial bias from ignoring general equilibrium effects in estimating excess burden, and a practical solution. *Journal of Political Economy*, *111*(4), 898–927.

Haimes, Y. (2005). *Risk modeling, assessment and management.* Hoboken, NJ: Wiley.

Just, R., Hueth, D., & Schmitz, A. (Eds.). (2008). *Applied welfare economics.* Cheltenham, UK: Edward Elgar.

Keeney, R. L., & Raiffa, H. (1976). *Decision making with multiple objectives preferences and value tradeoffs*. New York: Wiley.

Loomis, J. (2012). Incorporating distributional issues into benefit-cost analysis. In S. Farrow & R. Zerbe (Eds.), *Principles and standards for benefit-cost analysis.* Cheltenham, UK: Edward Elgar.

Machina, M. (1987). Choice under uncertainty: Problems solved and unsolved. *Economic Perspectives*, *1*(1), 121–154.

Mueller, J., & Stewart, M. (2011). *Terror, security and money: Balancing the risks, benefits and costs of homeland security.* Oxford: Oxford University Press.

Nyborg, K. (2012). *The ethics and politics of environmental cost-benefit analysis.* London: Routledge.

Office of Management and Budget (OMB). (2007). *Principles for risk assessment.* Washington, DC: Author.

Office of Management and Budget (OMB). (2014). Circulars A-94, A-4, A-11. Washington, DC: Author.

Rose, A. (2014). Macroeconomic consequence of terrorist attacks: Estimation for the analysis of policies and rules. In C. Mansfield & V. K. Smith (Eds.) *The benefit-cost analysis of security focused regulations.* Cheltenham, UK: Edward Elgar.

Santos, J. (2006). Inoperability input-output modeling of disruptions to interdependent economic systems. *Systems Engineering, 9*(1), 20–34.

U.S. Department of Homeland Security (DHS). (2011). *Risk management fundamentals.* Washington, DC: Author. Available at www.dhs.gov/xlibrary/assets/rma-risk-management-fundamentals.pdf.

Vining, A., & Boardman, A. (2006). Metachoice in policy analysis. *Journal of Comparative Policy Analysis*, *8*(1), 77–87.
Weinstein, M., & Zeckhauser, R. (1973). Critical ratios and efficient allocation. *Journal of Public Economics*, *2*(2), 147–157.
Zerbe, R., Jr. (Ed.). (2008). *Benefit-cost analysis*. Cheltenham, UK: Edward Elgar.

12

Enhancing Post-disaster Economic Resilience: Public-Private Partnership for Insuring Terrorism

Howard Kunreuther and Erwann Michel-Kerjan

12.1 Introduction

Terrorism risk poses dynamic challenges to our national security as the threat is continuously evolving. Significant efforts have been undertaken to prevent another large-scale attack on U.S. soil since September 11, 2001 (9/11). But if a terrorist organization were to succeed in perpetrating such an attack in the near future, a critical question is who will pay for the economic consequences. Insurance can play a central role here, as it has for many years (Willis & Al-Shahery, 2014). But the private sector cannot provide insurance protection on its own. The nature of the terrorist threat requires a public-private partnership.

This chapter is structured as follows. Sections 12.2 and 12.3 discuss the insurability of the terrorism risk and explain why insurers and reinsurers who were unconcerned with providing coverage prior to 9/11 then pivoted after experiencing severe losses from the 9/11 attacks to feel that terrorism was an uninsurable risk. Section 12.4 describes the Terrorism Risk Insurance Act (TRIA) of 2002 (renewed several times, most recently in January 2015), a financial arrangement between the federal government and the private insurance market that has made commercial terrorism coverage available. We perform a series of empirical analyses under the 2015 design of the TRIA program to demonstrate who will pay what portion of the loss for several plausible attack scenarios across the United States. As we show, insurers will pay all of the insured losses today unless they are truly catastrophic, in which case taxpayers will contribute to the cost.

12.2 Challenges in Insuring Terrorism

Insurance markets function best when the losses associated with a particular risk are independent of each other and the insurer has accurate information on the annual likelihood of the relevant events and the resulting damage should a disaster occur. By selling a large number of policies for a given a risk and applying the law of large numbers on independent risks, an insurer should be able to estimate with an acceptable degree of precision the claim payments it expects to make during a given period of time.

To illustrate, consider an insurer who offers a fire insurance policy to a set of identical homes each valued at $100,000. Based on past data, the insurer estimates that the likelihood that a home will be destroyed by fire next year is 1/1,000. In this case, the expected annual loss for each home would be $100 (i.e., 1/1000 × $100,000). Assume that the risk of a fire to any home is uncorrelated with the risk of fire to any other home (that is, fire is an independent risk). If the insurer issued only one policy to cover the full loss from a fire, then there would be a variance of approximately $100 associated with its expected annual loss.[1] As the number of policies (n) issued increases, the variance of the expected annual loss, or the mean loss per policy, decreases in proportion to n. Thus, if $n = 10$, the variance of the mean loss will be approximately $10. When $n = 100$ the variance decreases to $1, and with $n = 1{,}000$ the variance is $0.10. It is thus not necessary to issue a very large number of policies to significantly reduce the variability of expected annual losses per policy if the risks are independent.

This model of insurance works well for risks such as fire (excluding wildfire) and automobile accidents where the losses are normally independent and there are sufficient data to estimate the risk accurately. Terrorism risk does not satisfy these conditions, so it is more problematic to insure. Foremost among these characteristics are the high level of uncertainty about the probabilities and potential catastrophic nature of losses (including business interruption), high correlation of risk across entities exposed to loss, and interdependencies, such as would arise if a building that had been damaged by a bomb collapsed on another structure that was not a direct target of the terrorists.

12.2.1 *Dynamic Uncertainty and Time Scale*

The likelihood of terrorist attack scenarios on a particular city during a pre-specified time period is extremely difficult, if not impossible, to estimate. Since terrorists are likely to design their strategy as a function of their own resources and their knowledge of the vulnerability of the entity they want to attack, the nature of the risk is continuously evolving. This *dynamic uncertainty* makes the likelihood of future terrorist events extremely difficult to estimate (Michel-Kerjan, 2003).

Consider a dynamic game where the actions of the terrorist groups are dependent on the actions taken by those threatened by the terrorists. From the terrorists' point of view, they must determine what targets to attack and the commitment of resources to specific activities. Their decisions will be influenced by the types of protective measures undertaken by those at risk (Richardson et al., 2005).

More specifically, terrorists may respond to security measures by shifting their attention to more vulnerable targets. Keohane and Zeckhauser (2003) analyze the relationships between the actions of potential victims to reduce losses from a terrorist attack and the behavior of terrorists. Establishing observable protective measures against a given mode of attack on a specific building (for example) should lead the terrorists to direct their attention elsewhere and hence reduce the probability of an attack against that building. However, shielding that piece of property increases the likelihood of an attack on an unprotected structure.[2]

Rather than investing in additional security measures, firms may prefer to move their operations from large cities to less-populated areas to reduce the likelihood of an attack. Of course, terrorists may choose these less-protected regions as targets if there is heightened security in the urban areas. Terrorists also may change the nature of their attacks if protective measures in place make the likelihood of success of the original option much lower than another course of action (e.g., switching from hijacking to bombing a plane). This substitution effect has to be considered when evaluating the effectiveness of specific policies aimed at curbing terrorism (Bier et al., 2007) and makes it difficult to estimate the risk associated with future attacks and hence the ability to price terrorism insurance.

Large historical databases on losses from natural hazards are in the public domain. Modeling firms have utilized these data in conjunction with estimates by scientists and engineers on the likelihood and consequences of future disasters in specific locations. However, data on terrorist groups' activities and current threats are normally undisclosed for national security reasons. Moreover, while some time-series data on terrorist acts over the past years are in the public domain, they may not reflect the changing planned activities of terrorist groups today. The nature of terrorist organizations, their motives, targets, and modus operandi have radically changed in the past 20 years and may do so again over the next 20 years.[3]

The challenges in estimating the likelihood of an attack have led to insurers' use of deterministic approaches – "what-if" scenarios – rather than probability-based approaches. A typical scenario, and one that we will examine, is a 10-ton truck bomb exploding in a major metropolitan area. Rating agencies use this scenario to stress test the capacity of insurers to pay claims.

12.2.2 *Potential for Catastrophic Losses*

Another limitation of terrorism risk insurability is the potential for devastating losses to different lines of insurance. The insured losses from 9/11 illustrate the high degree of risk correlation between different lines of insurance coverage. The 9/11 attacks affected commercial property, caused business interruption, and led to significant claims from other lines of coverage: workers' compensation, life, health, disability, and general liability insurance. More than 3,000 people were killed and 2,250 others were injured. The attacks inflicted damage estimated at nearly $80 billion, about $32.4 billion of which was covered by insurance (Hartwig, 2004). Figure 12.1 depicts the composition of the $32.4 billion total insured loss estimates due to these terrorist attacks.

Reinsurers – many of whom are more geographically diversified globally than are primary insurers – paid for about two-thirds of the insured claims of 9/11. As a result, they withdrew their coverage against losses from terrorist attacks, leading most insurers to conclude that the terrorism risk was uninsurable.

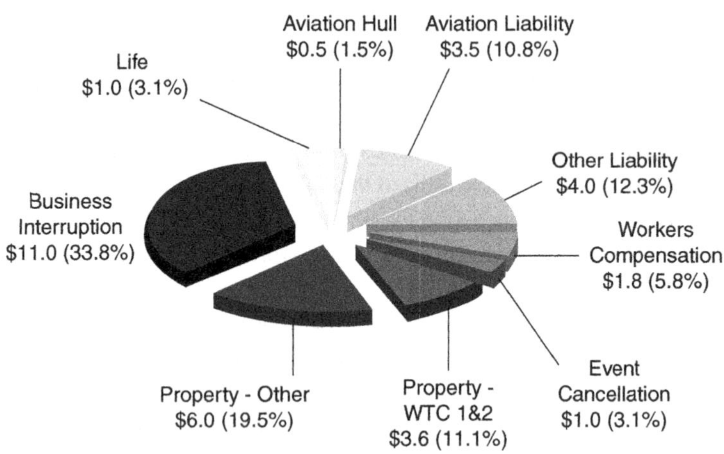

Figure 12.1. Composition of 9/11 insured loss estimates (total: $32.4 billion in 2001 prices), by line ($ billion).

Sources: Wharton Risk Center (2005) using data from the Insurance Information Institute

Attacks using chemical, biological, radiological, and nuclear (CBRN) weapons have the potential to inflict even larger insured losses, especially on workers' compensation and business interruption lines. For example, the bombing of a chlorine tank in Washington, DC, could kill and injure hundreds of thousands of people. Nuclear attacks could have a much more severe impact. Indeed, there is evidence that terrorist groups explored the possibility of obtaining a nuclear device to build "suitcase nuclear bombs" (Kunreuther et al., 2014).

12.2.3 *Interdependencies*

The vulnerability of one organization, critical economic sector, and/or country depends to some extent not only on its own choice of security investments, but also on the actions of other agents. This concept of *interdependent security* implies that failures of a weak link in a connected system could have devastating impacts on all parts of it, and that as a result there may be suboptimal investment in the individual components (Heal & Kunreuther, 2005; Kunreuther & Heal, 2003). The existence of such interdependencies provides another challenge to insurers in determining how much terrorism coverage to offer and what premium to charge.

Interdependencies do not require proximity. In the case of the 9/11 attacks, security failures at Boston's Logan Airport led to the tragedies at the World Trade Center (WTC), at the Pentagon, and in rural Pennsylvania. Firms located in the WTC could have done nothing on their own to prevent these aircraft from crashing into the towers.

Except for very specific policies (e.g., contingent business income coverage), terrorism insurance normally does not cover losses unless the insured is the direct target of an attack (Godard et al., 2002). For example, the Federal Aviation Administration's (FAA) subsequent ban of takeoffs of all civilian aircraft regardless of destination caused business interruption losses to the city of Chicago. In March 2004, the city was denied insurance compensation for losses that resulted from the FAA's decision. The specific clause of the insurance contract for business interruption specified that it would cover only losses that were the "direct result of a peril not excluded," thus imposing a limitation that excludes interdependent effects due to the response to an attack (U.S. District Court, 2004).

12.3 Decision Processes of Insurers Regarding Terrorism Insurance

A large body of cognitive psychology and behavioral decision research over the past 30 years has revealed that individuals and organizations often make decisions by combining *intuitive* thinking with *deliberative* thinking. In his thought-provoking book *Thinking, Fast and Slow*, Nobel Laureate Daniel Kahneman has characterized the differences between these two modes of thinking. *Intuitive* thinking operates automatically and quickly with little or no effort and no voluntary control. It is often guided by emotional reactions and simple rules of thumb that have been acquired by personal experience. *Deliberative* thinking allocates attention to effortful and intentional mental activities where individuals undertake trade-offs and recognize relevant interdependencies and the need for coordination (Kahneman, 2011).

Choices are normally made by combining these two modes of thinking and generally result in good decisions when individuals have considerable past experience as a basis for their actions. However, with respect to low-probability, high-consequence events such as terrorism, there is a tendency to either ignore a potential disaster or overreact to a recent one. More specifically, individuals often exhibit systematic biases such as the *availability heuristic*, where the judged likelihood of an event depends on its salience and memorability (Tversky & Kahneman, 1973). There is thus a tendency to ignore rare risks until after a catastrophic event occurs.

12.3.1 *Impact of Availability Bias on Insurer Behavior*

Prior to 9/11, terrorism exclusions in commercial property and casualty policies in the U.S. insurance market were extremely rare because losses from terrorism had historically been small and, to a large degree, uncorrelated. Attacks of domestic origin were isolated, carried out by groups or individuals with disparate agendas. Thus the country did not face a concerted domestic terrorism threat, as did countries such as France, Israel, Spain, and the UK (Kunreuther & Michel-Kerjan, 2004).

In fact, insurance losses from terrorism were viewed as so improbable that the risk was not explicitly mentioned or priced in any standard policy, and it was never excluded from so-called all-risk policies with the exception of some marine cargo, aviation, and political risk policies. Even the first attack on the WTC in 1993 that killed six people and caused $725 million in insured damages, and the Oklahoma City bombing of 1995 that killed 168 people but wrought no insured property losses since the building was self-insured, were not seen as threatening to insurers. Neither disaster led them to revise their view of terrorism as a peril worth considering when pricing a commercial insurance policy. The insurers viewed the likelihood of a major terrorism loss as below their threshold level of concern, so they did not feel the need to consider the risk separately when pricing coverage for commercial insurance prior to 9/11 (Kunreuther et al., 2013).

As mentioned previously, reinsurers (most of them European) were responsible for a large portion of the claims from 9/11.[4] Coming on top of a series of catastrophic natural disasters over the previous decade and portfolio losses due to stock market declines, their capital base was severely diminished. Furthermore, reinsurers' appetite for new capital to provide coverage against terrorism risk was sharply curtailed. The few who marketed policies charged extremely high rates for very limited protection. This directly affected insurance supply. Most insurers stopped covering terrorism in areas they perceived as high risk unless they were forced to include it in their policies, as with workers' compensation. When coverage was offered, the prices were likely to increase significantly over what they were prior to 9/11 and coverage limits were reduced.

Take the case of insuring Chicago's O'Hare Airport. Prior to 9/11, the airport had $750 million of terrorism insurance coverage at an annual premium of $125,000. After the terrorist attacks, insurers offered the airport only $150 million of coverage at an annual premium of $6.9 million. The airport was forced to purchase this coverage in order to operate (Jaffee & Russell, 2003). Golden Gate Park in San Francisco, CA, was unable to obtain terrorism coverage and its non-terrorism coverage was reduced from $125 million to $25 million. Yet the premiums for this reduced amount of protection increased from $500,000 in 2001 to $1.1 million in 2002 (Smetters, 2004).

12.3.2 *Role of Ambiguity on Insurer Behavior*

Only after 9/11 did the insurance industry seriously look at its exposure to future losses. Most insurers determined that they could not offer terrorism insurance because the uncertainties surrounding the likelihood and consequences of another terrorist attack were so significant. Those who offered coverage charged extremely high premiums; their behavior is consistent with empirical studies on how underwriters and actuaries would price coverage when they faced risks with ambiguous probabilities and uncertain losses relative to a well-specified risk.

A recent Web-based experiment provided actuaries and underwriters in insurance companies with scenarios in which they seek advice and request probability forecasts from different groups of experts and then must determine what price to charge for coverage for flood and wind damage from hurricanes. The average premium that insurers would charge was approximately 30% higher for coverage against either of these risks if the probability of damage was ambiguous rather than well-specified and if the experts were conflicted over their estimates. The data reveal that they would likely charge more in the case of conflict ambiguity (that is, experts disagree on point estimates) than imprecise ambiguity (that is, experts agree on a range of probabilities, recognizing that they cannot estimate the probability of the event precisely) (Cabantous et al., 2011).

12.4 The Terrorism Risk Insurance Act (TRIA)

Given the retrenchment of the insurance and reinsurance markets for terrorism coverage and the demand by commercial enterprises for terrorism coverage to meet lending requirements or to protect against the potential for future losses, the Terrorism Risk Insurance Act (TRIA) was passed in 2002. TRIA established a public-private risk-sharing arrangement between the federal government, the insurance industry, and commercial policyholders.

First, TRIA instituted a *mandatory offer requirement* whereby all U.S. primary insurance companies had to offer coverage against terrorism risk for specified commercial lines of insurance on the same terms and conditions as other perils provided by their commercial insurance policies. Firms are not required to purchase this coverage unless mandated by state law normally the case for workers' compensation insurance coverage.

TRIA also established a risk-sharing mechanism between the insurance industry, the federal government, and all commercial insurance policyholders in the United States for covering insured losses from future terrorist attacks. The Act was extended in 2005 for two years, in 2007 for seven years, and in January 2015 for another six years until December 31, 2020, with an increasing portion of the losses from a terrorist attack covered by the insurance industry and commercial policyholders each time the legislation was renewed.

12.4.1 *Structure of the New TRIA Partnership*[5]

Under TRIA's 2015 renewed design, losses from events certified by the federal government as terrorism and that result in more than $200 million in insured losses in TRIA-eligible lines of business would be shared as follows.

- Commercial policyholders are responsible for paying any losses within their standard insurance policy deductibles.
- Commercial insurance companies then provide coverage for all losses in excess of these policy deductibles, provided that total insurance industry losses did not exceed $100 billion (often referred to as the TRIA "program cap").
- The federal government reinsures the commercial insurer's terrorism losses in excess of a TRIA deductible (D^*), with the deductible reset

annually so that it equals 20% of that insurer's prior year direct earned premium (DEP) for the lines covered under the program. D^* has increased from 1% in 2002 to 20% since 2007 and remained at that level when the program was renewed in 2015.

- Losses in excess of each insurer's deductible (D) would be shared between the insurance company and the federal government with the insurer covering 20% of the excess loss, and the federal government taking on the remining 80%.
- Should total insurance industry losses exceed $100 billion, primary insurers are responsible for reimbursing policyholders only for their proportionate share of losses up to $100 billion and Congress shall determine the procedure and source of any payments for the uninsured losses.
- After an insured event occurs, the federal government with recoup part or all of its payments under TRIA by levying surcharges on all commercially insured policyholders in subsequent years. More specifically, under a *mandatory recoupment mechanism*, the federal government is required to recoup 140% of its payments below the *insurance industry marketplace aggregate retention* ("retention") – set at $27.5 billion in 2015, gradually increasing to $37.5 billion by 2020 – and above the aggregate insurers' uncompensated outlays (i.e., insurer losses within the deductible and coinsurance) during the calendar year.[6] In the event that uncompensated insurer outlays across the insurance industry exceed that industry retention level, the U.S. Treasury has the option to recoup any and all federal payments through a discretionary recoupment mechanism.

Figure 12.2.a depicts the public-private loss sharing for a representative insurer covering terrorism for its commercial policyholders (either through workers' compensation or property insurance) under the current TRIA arrangement when total insured losses are less than $100 billion. If a terrorism loss incurred by an insurance company (i) is less than its TRIA deductible amount (D_i), as determined as a percentage of its prior year Direct Earned Premium (DEP) in TRIA-eligible lines, the insurer does not receive any reimbursement from the federal government. This situation is illustrated by an insured loss of L_1 where the insurer's payment is represented by the dark area on the left side of Figure 12.2.a.

When the insured loss from a certified terrorist attack in 2018 exceeds the program trigger and is above the insurer's deductible, as depicted by L_2 in Figure 12.2.a, the insurer pays the entire claim and the federal government reimburses the insurer for 82% of the losses above its deductible. Hence, the insurer ends up paying 18%. The lighter area on the right side of the figure represents the federal government's payment. D_i plays an important role in determining loss sharing between insurers and the federal government and can amount to very large sums for many

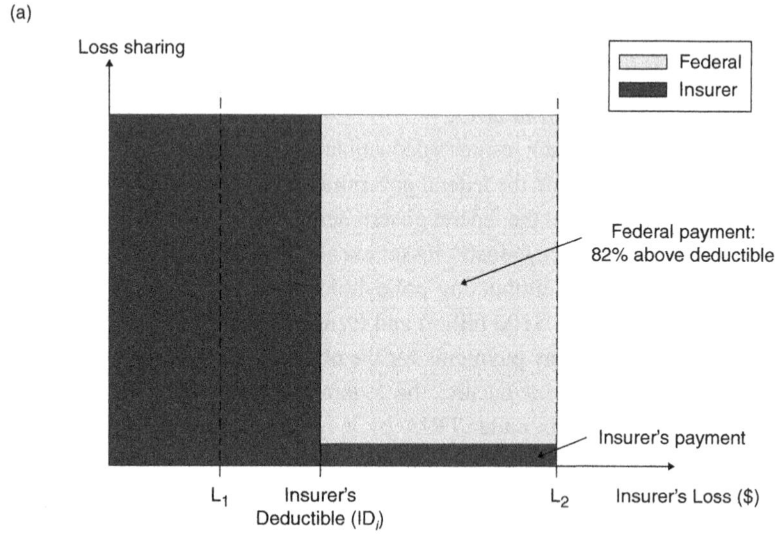

Figure 12.2a. Loss sharing between an insurer and the federal government in 2018.

insurers. Should a terrorist attack occur in 2018, insurers will be responsible for losses equal to 18% of their DEP in 2017.

If the entire insurance industry suffers terrorism losses on its U.S. portfolio that requires the government to cover a portion of its claims, then these outlays shall be fully or partially recouped *ex post*, as described earlier. Insurers levy this surcharge against all commercial property and casualty policyholders, whether or not they had purchased terrorism insurance, and transfer the collected funds to the Treasury.

Figure 12.2.b depicts loss sharing arrangement at the end of the renewal period, 2020, under the current TRIA legislation. All the insurers whose policyholders suffer an insured terrorist attack are depicted in the bottom portion of the figure, all commercial policyholders are in the middle area. The taxpayers are in the upper portion of the figure after the federal government has reimbursed all insurers for 80% of their claims payments above their individual TRIA deductible level.

In the example considered here, since the total insured loss L for the entire insurance industry is greater than \$37.5 billion but total losses retained by insurers within their deductibles and coinsurance requirements are below the market aggregate retention, a portion of the federal outlays is subject to the mandatory recoupment. Should the federal government elect to exercise its authority to levy a discretionary recoupment surcharge against commercial

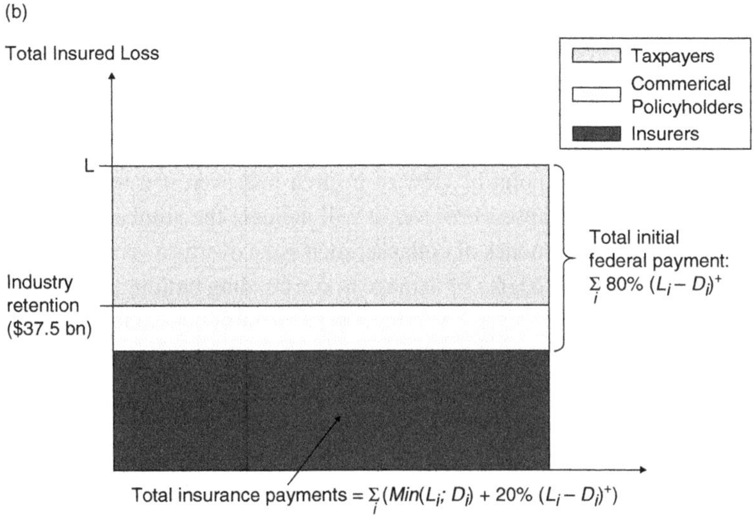

Figure 12.2b. Overall loss sharing under TRIA 2015–2020.

policyholders to fund federal outlays not covered by the mandatory recoupment mechanism, there would be a one-for-one reallocation of loss from all US taxpayers to commercial policyholders.

12.4.2 *Estimating Losses from Terrorist Attack Scenarios*

Insurers have to determine what premiums to charge their clients for terrorism coverage, which requires them to estimate the likelihood of different losses occurring. Insurers can address these problems by constructing deterministic scenarios that reveal their impact on potential losses if they offer coverage to all of their commercial policyholders and determine what their own losses are likely to be from terrorist attacks of different magnitudes in any part of the country. In this way, insurers can determine their ability to cover these losses with their current surplus as a function of their risk portfolio. They can then determine how much coverage they want to provide to commercial firms in a particular region so as not to be over-exposed.

How the exposure is diversified geographically and across industries also plays a key role. For example, a $1 billion exposure in a given city will be treated by TRIA differently than an exposure of $100 million in each of 10 different cities.

In the case of an explosion of a bomb in an urban area, damage is a function of the attack size and building type. The explosion of a bomb can be expressed as a TNT-equivalent. The size of this charge can be thought of as the intensity of the event. Damage to the target building results from the resulting shock wave, the subsequent pressure wave, and fire. The target building may sustain total damage from the point of view of insured loss even if it remains standing. If the building collapses, however, it will increase the number of fatalities. Furthermore, different modes of collapse, such as an overturn versus a pancake collapse, will affect the degree of damage to surrounding buildings and thus the total area affected by the event. The buildings surrounding the target building are also likely to be damaged by the resulting shock and pressure waves and/ or by falling or flying debris.

In our report, *TRIA after 2014* (Kunreuther et al., 2014), we analyze loss sharing using three attack scenarios in four major cities using the terrorism model of the modeling firm Risk Management Solutions (RMS). The three scenarios reflect different modes of attacks: a) 10-ton truck bomb; b) 1-ton sarin chemical agent release; and c) 1-kiloton nuclear weapon. For each of the three attack modes, key high-profile targets were identified in the central business districts of Chicago, Houston, Los Angeles, and New York. The RMS models estimated property losses from commercial insurance, and casualty losses arising from workers' compensation coverage (Kunreuther et al., 2014).

Since terrorism cannot be excluded from workers' compensation insurance and must be purchased by commercial enterprises, a 100% take-up rate for terrorism-related workers' compensation losses is assumed. There are no available data on terrorism coverage for all commercial firms operating in the United States, but insurance brokers Marsh & McLennan (2013) and AON Benfield (2013) have both published terrorism insurance take-up rates for their clients. This level has remained stable over the past five years ranging from a low of 40% in the chemical industry to a high of 80% in the media industry (Marsh, 2013). The analysis that follows assumes a 50% take-up rate for terrorism insurance for the property lines, recognizing that the actual percentage may vary from one city to another as well by the type of firm.

Building on the range of possible terrorist attack scenarios, Table 12.1 details how changing the size of the loss from $5 billion to $100 billion affects the distribution of payments in one specific metropolitan area (New York, NY) under the new TRIA loss-sharing arrangements. The other damage scenarios are extrapolated from the loss ranges using the RMS models.

Table 12.1. *Amounts of Losses Paid by Various Stakeholders under New TRIA Loss-Sharing Arrangements (Illustration with New York City) (in billion dollars)*

Loss	5.0	15.0	25.0	32.0	50.0	100.0
Property	3.0	9.0	15.0	19.2	30.0	60.0
Workers Comp	2.0	6.0	10.0	12.8	20.0	40.0
Non-Insured	1.5	4.5	7.5	9.6	15.0	30.0
Total Insured	3.5	10.5	17.5	22.4	35.0	70.0
Insurers	3.1	8.5	13.6	16.7	23.0	33.2
Commercial Policyholders	0.5	2.6	5.2	7.5	15.9	5.8
Federal Taxpayers	0	0	0	0	0	31.1

Consider a $5 billion loss, with $3 billion for property damage and business interruption and $2 billion workers' compensation with 50% of the property damage being uninsured (i.e., $1.5 billion uninsured). As shown in Table 12.1, insurers would cover $3.14 billion – or 90% of the total insured losses. The rest will be recouped by the federal government against all commercial policyholders, whether they are insured against terrorism or not.[7] In this scenario and others shown in Table 12.1, the taxpayers will not pay anything for the losses. For a severe $50 billion loss, insurers would pay $23.04 billion – or 66% of the total insured losses. Commercial policyholders will pay the remaining 34% of the insured losses. An important question is whether the uninsured firms will be able to cover their losses. In a recent study, we show that larger firms are more likely to purchase terrorism insurance. The uninsured smaller firms may have more difficulty recovering from a terrorist attack than the larger ones given their financial constraints (Michel-Kerjan et al., 2015).

The amounts paid by the relevant stakeholders as a function of losses to New York City from terrorist attacks are depicted in Figure 12.3. The two vertical lines in these figures represent the projected $25 billion loss from a sarin attack and $32 billion from a 10-ton truck bomb attack. Several findings are worth noting:

- The federal government will not be responsible for any payments until the total insured losses from a terrorist attack exceed $59 billion.
- When total insured and non-insured losses reach $100 billion, insurers will ultimately be responsible for approximately $33 billion in

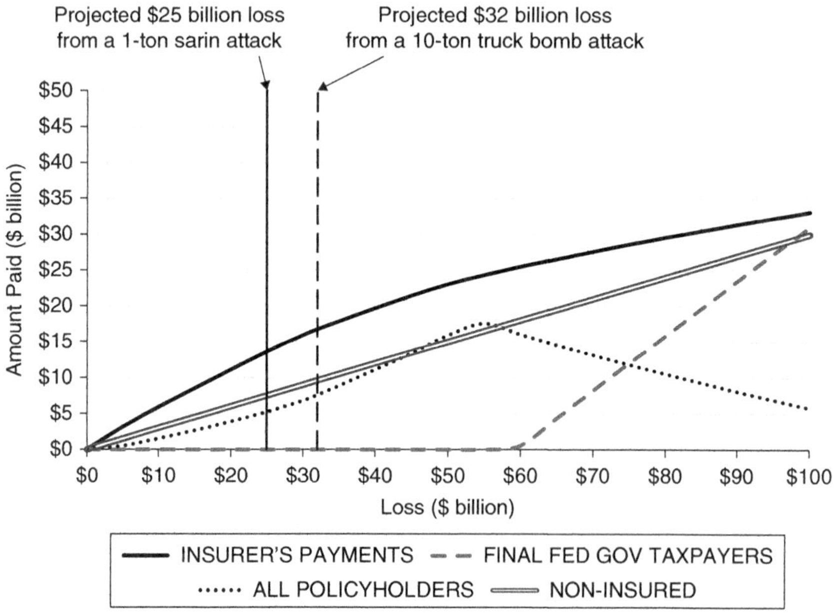

Figure 12.3. Amount paid by stakeholders for different losses from terrorist attacks to New York City under the new TRIA legislation (with a $37.5 billion industry retention).

payments, in federal taxpayers almost $31 billion, the commercial policyholders more than $5.8 billion. The remaining $30 billion would be uninsured.

- Commercial policyholders will always be assessed more than $10 billion when total losses from terrorist attacks in New York City are in the $38 billion to $82 billion range. The maximum they would pay – $17.9 billion – is reached when losses are $54 billion.

In analyzing the terrorism attack scenarios, we have focused solely on the insurable losses under the scenario and not the broader economic losses that would have to be addressed.

12.5 Conclusion

TRIA has made terrorism an insurable risk by protecting insurers against a catastrophic event through a loss-sharing arrangement where the federal government provides financial assistance if insurance industry claims exceed a certain level, and recoups some or all of their payments over time rather than immediately after the event to protect insurers' financial position. Because they pay these claims without charging the insurers a premium for this reinsurance coverage, there is no need for the government to estimate the likelihood of these catastrophic events occurring and legitimation is provided for recouping 140% of their federal payments via the mandatory recoupment provision in TRIA.

The private-public partnership between the insurance industry and the federal government through the passage of TRIA recognizes the need for addressing the two major concerns that insurers have with respect to providing protection against extreme events: the uncertainty in estimating the probability of catastrophic events occurring, and the magnitude of the resulting losses that can significantly reduce their firms' surplus.

As TRIA was set to expire in 2014, there was considerable discussion in Congress as to whether the program, launched as a temporary partnership, should indeed be renewed. Some argued that free reinsurance was a gift to insurance firms: any recoupment from the federal government would not be against insurers, but against all firms that had a commercial insurance policy, whether or not they had terrorism coverage. Another point raised in favor of not renewing TRIA was that since the private market had been exposed to a large risk under the program, this implied that they could handle this risk without a federal backstop.

Those favoring the renewal of TRIA contended that the program had worked relatively well since the taxpayers had not been asked to contribute anything and firms had access to affordable terrorism insurance. It was not clear, either, that not renewing TRIA would actually result in less exposure to the American taxpayers. If TRIA were to expire, and unless reinsurers reentered the U.S. market with much more capacity and at a price considered reasonable by insurers, most primary carriers would exclude this risk from their portfolio everywhere they could. Those that provide it would charge much higher premiums to take into account expensive capital they need to set aside to meet regulatory and/or rating agencies' requirements. Under extreme pressure from the media and interest groups, the federal government would be asked to step in, especially if the attack occurs during an election year (Michel-Kerjan, 2012).

The tension was clear when Congress failed to renew TRIA at the end of 2014 (Kunreuther & Michel-Kerjan, 2014). For a few weeks, America was uncovered again. TRIA was renewed later in January 2015.

The way the federal government and the insurers (and their reinsurers) will react to a terrorist attack on U.S. soil will depend on its size. Should an attack inflict minor economic losses, as was the case in the 2013 Boston Marathon bombing, the insurance market is not likely to be affected. However, another large-scale attack such as 9/11, or more frequent small-scale attacks, may cause a further retrenchment by insurers with restrictions in capacity and coverage. Some reasons that terrorism insurance prices have decreased in recent years is the absence of any significant attack, competitive pressure on general insurance rates, and concerted efforts by insurers over the past decade to bring their exposures into line with existing TRIA parameters (GAO, 2014; Michel-Kerjan, 2013).

NOTES

1. The variance for a single loss L with probability p is $Lp\ (1-p)$. If $L = \$100{,}000$ and $p = 1/1{,}000$, then $Lp(1-p) = \$100{,}000\ (1/1{,}000)(999/1{,}000)$, or \$99.90.
2. One exception would be if terrorist groups attack trophy buildings to prove that they can inflict damage to well-protected structures.
3. The IRA in the UK in the 1990s was very different than Al Qaeda in the 2000s, which itself is very different than ISIS today. For instance, ISIS has much more financial power than any terrorist organization has ever had (Michel-Kerjan, 2015).
4. Testimony of Jacques Dubois, chairman and CEO of Swiss Re America on behalf of Swiss Re before the U.S. Senate Committee on Banking, Housing, and Urban Affairs, May 18, 2004.
5. This section draws on Kunreuther et al., 2014.
6. The recoupment rate was driven by the budget scoring of the 2007 extension of TRIA to make the legislation budget neutral.
7. Note that under TRIA, the federal government recoups 133% of its payments from the commercial policyholders.

REFERENCES

AON Benfield (2013). Response to the U.S. Treasury and the Presidential Working Group: Terrorism (Re)insurance.

Bier, V., Oliveros, S., & Samuelson, L. (2007). Choosing what to protect. *Risk Analysis*, *27*(3).

Cabantous, L., Hilton, D., Kunreuther H., & Michel-Kerjan, E. (2011). Is imprecise knowledge better than conflicting expertise? Evidence from insurers' decisions in the United States. *Journal of Risk and Uncertainty*, *42*(93), 211–232.

Godard, O., Henry, C., Lagadec, P., & Michel-Kerjan, E. (2002). *Treatise on new risks. Precaution, crisis management and insurance (in French) (p. 620)*. Paris: Editions Gallimard, Folio-Actuel.

Government Accountability Office (GAO). (2014, May). *Terrorism insurance: Treasury needs to collect and analyze data to better understand fiscal exposure and clarify guidance*. GAO-14–445, Washington DC.

Hartwig, R. (2004). *The fate of TRIA: Is terrorism an insurance risk?* New York: Insurance Information Institute.

Heal, G., & Kunreuther, H. (2005). IDS models for airline security. *Journal of Conflict Resolution*, *49*(2), 201–217.

Jaffee, D., & Russell, T. (2003). Market under stress: The case of extreme event insurance. In R. Arnott, B. Greenwald, R. Kanbur, & B. Nalebuff (Eds.), *Economics for an imperfect world: Essays in honor of Joseph E. Stiglitz* (pp. 35–52). Cambridge, MA: MIT Press.

Kahneman, D. (2011). *Thinking, fast and slow.* New York: Farrar, Straus and Giroux.

Keohane, N., & Zeckhauser, R. (2003). The ecology of terror defense. *Journal of Risk and Uncertainty*, *26*(2/3), 201–229.

Kunreuther, H., & Heal, G. (2003). Interdependent security. *Journal of Risk and Uncertainty*, *26*(2/3), 231–249.

Kunreuther, H., & Michel-Kerjan, E. (2004). Policy Watch: Challenges for terrorism insurance in the United States. *Journal of Economic Perspectives*, *18*(4), 201–214.

Kunreuther, H., & Michel-Kerjan, E. (2014, December 9). The Unexpected Threat to Super Bowl XLIX. *Bloomberg Businessweek.*

Kunreuther, H., Michel-Kerjan, E., Lewis, C., Muir-Wood, R., & Woo, G. (2014). *TRIA after 2014.* Philadelphia: Wharton Risk Management Center, University of Pennsylvania.

Kunreuther, H., Pauly, M. V., & McMorrow, S. (2013). *Insurance and behavioral economics: Improving decisions in the most misunderstood industry.* New York: Cambridge University Press.

Marsh & McLennan Companies. (2013, September 18). Terrorism risk insurance. Report to the Federal Advisory Committee on Insurance, Department of Treasury.

Michel-Kerjan, E. (2003). Large-scale terrorism: Risk sharing and public policy. *Revue d'Economie Politique*, *113*(5), 625–648.

Michel-Kerjan, E. (2012, September 11). Testimony before the U.S. House of Representatives. Committee on Financial Services. *TRIA at Ten Years: The Future of the Terrorism Risk Insurance Program.*

Michel-Kerjan, E. (2013, September 25). Testimony before the U.S. Senate. Committee on Banking, Housing, and Urban Affairs. *Reauthorizing TRIA: The state of the terrorism risk insurance market.* Washington, DC.

Michel-Kerjan, E. (2015). Effective risk response needs a prepared mindset. *Nature*, *517*, 413.

Michel-Kerjan, E., Raschky, P., & Kunreuther, H. (2015). Corporate demand for insurance: New evidence from the U.S. property and terrorism insurance markets. *Journal of Risk and Insurance*, *82*(3), 505–530.

Richardson, H. W., Gordon, P., & Moore, II, J. E., eds. (2005). *The economic impacts of terrorist attacks.* Cheltenham, UK: Edward Elgar.

Smetters, K. (2004). Insuring against terrorism: The policy challenge. In R. Litan & R. Herring (Eds.), *Brookings-Wharton papers on financial services* (pp. 139–182).

Tversky, A., & Kahneman, D. (1973). Availability: A heuristic for judging frequency and probability. *Cognitive Psychology*, *5*(2), 207–232.

U.S. District Court for the Northern District of Illinois, Eastern Division. (2004, March 11). *City of Chicago v. Factory Mutual Insurance Company*. 02C7023.

Wharton Risk Center. (2005). *TRIA and beyond: Terrorism risk financing in the U.S.* Philadelphia: The Wharton School, University of Pennsylvania.

Willis, H., & Al-Shahery, O. (2014). *The National Security Perspectives on Terrorism Risk Insurance in the United States*. RAND Corporation.

13

Economic Impacts of Changes in Wait Times at U.S. Ports of Entry

*Bryan Roberts, Adam Rose, Nathaniel Heatwole, Dan Wei, Misak Avetisyan, Fynnwin Prager, Charles Baschnagel, and Isaac Maya**

13.1 Introduction

Inspections of people and vehicles at U.S. border crossings are vital to homeland security. The benefits of these activities are the avoided consequences associated with the movement of contraband, unauthorized border crossers, and potential terrorists into the United States. At the same time, inspections incur various types of costs. Construction and operational costs associated with inspection facilities are a significant federal expenditure. Moreover, inspections generate various spillover effects related to the delays in the flows of passengers and cargo across U.S. borders. On the passenger side, they decrease the amount of tourism and business travel into and out of the country, and thus impact the location of related spending. Those people who do make the trip incur delays that cost them time. On the freight side, delays translate into increases in various explicit transportation costs, such as additional fuel, as well as implicit costs, such as the value of lost time. Reducing wait times at ports of entry (POEs), through the addition of U.S. Customs and Border Protection (CBP) officers would lessen these negative spillover effects, though it would also incur additional demands on the federal budget.

* This research was supported by the U.S. Department of Homeland Security (DHS) through the National Center for Risk and Economic Analysis of Terrorism Events (CREATE) under Cooperative Agreement No. 2010-ST-061-RE0001. We wish to thank the U.S. Customs and Border Protection for access to primary data on travel and wait times at major border crossings. We are grateful to the following people for responding to questions during the course of the research and commenting on earlier drafts of this chapter: Kevin McAleenan, Jennifer Pennese, Adam Wolf, Troy Riley, and William Kallin. We thank Oswin Chan, Katherine Lu, Steve McGonegal, Timothy Beggs, and Omeed Baghelai for their helpful assistance. We also thank Georgia Harrigan of the DHS Office of University Programs for helpful suggestions along the way. However, any opinions, findings, and conclusions or recommendations in this document are those of the authors and do not necessarily reflect views of the DHS or the University of Southern California.

This study estimates the macroeconomic impacts on the U.S. economy of changes in primary inspection wait times for inbound traffic at major U.S. border crossings. In addition to the changes in the direct spillover costs previously noted, it also examines various types of economic ripple effects of changed wait times. These include multiplier effects from tourist and business travel expenditures, as well as freight cost general equilibrium effects, which refer to complex interactions in relation to competitive changes in imports and exports of intermediate (unfinished) and final (finished) goods. We evaluate impacts on passenger and commercial vehicles, but not for pedestrians crossing the U.S. border.

The analysis is based on extensive primary data provided by CBP, federal government publications, commercial vendors, and the professional literature on the topic. At the microeconomic level, the study applies economic analysis and operations research methods. At the macroeconomic level, it applies computable general equilibrium (CGE) analysis, where the data are extensive and the issues are complex, and the more practical tool of input-output (I-O) analysis, where the subject of inquiry is less demanding. The results indicate that changes in CBP staffing would have significantly positive net impacts on U.S. gross domestic product (GDP), trade balances, and employment. Moreover, the ensuing reductions in the opportunity cost of waiting by passengers and truck drivers are also significant. These results should prove useful to policy makers who must decide on the overall level of resources dedicated to border inspections, managers who must decide how these resources are allocated across particular inspection sites, and the general public and its representative organizations who incur the costs but also the benefits associated with inspections.

13.2 Methodological Overview

CBP inspects vehicles, passengers, and freight shipments as they enter the United States at POEs. If the customs officer at a primary inspection booth has concerns about a vehicle or passenger, it will be referred to secondary inspection, where a more intensive examination will be made. We evaluate only the economic impacts associated with primary inspections as secondary inspections affect only a minority of traffic. Figure 13.1 provides a schematic description of the three basic components of the study for which quantified estimates are made: passenger vehicle traffic at land border crossings, commercial vehicle traffic at land border crossings, and passenger flows at international airports. For each component, specific methodological approaches and assumptions were made with respect to both microeconomic and macroeconomic analysis.

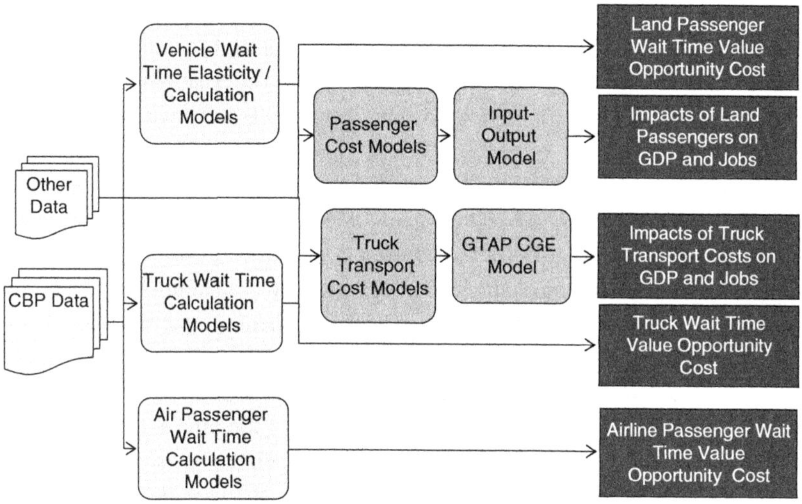

Figure 13.1. Schematic of the methodology.

13.2.1 *Microeconomic Analysis*

The analysis at the level of individual POEs was carried out using CBP data and operations research and economic analysis methods. For passenger vehicle traffic at land border crossings, two key parameters are needed: how wait time changes with the number of primary inspection booths, and how the number of cross-border trips changes with wait time. We quantify the response of wait time to booths opening by developing a theoretical model of a saturated queuing system that can be applied to any border crossing and use the model and CBP data to quantify how wait time changes with the number of booths open. We also evaluate the results of a staffing experiment at the San Ysidro crossing in July 2012 in which staffing was significantly increased above its normal level. We find that the actual outcomes of the July experiment are consistent with what the theoretical model predicts, which provides some validation for the model.

We quantify the response of cross-border trips to wait time using outcomes of the San Ysidro experiment, which yielded an elasticity value that is consistent with values of the short-run elasticity of travel volume to travel time in the literature (see Litman, 2012; Wardman, 2012).

For commercial vehicle flows at land border crossings, we adapt the theoretical model developed for passenger vehicles to quantify how wait time changes with more booths open. No empirical evidence is available on how the number of commercial vehicle trips changes with wait time, so we make the conservative assumption that this elasticity equals zero. For passenger flows at airports, we monetize the value of time spent waiting in passport inspection queues and use the price elasticity of demand for international air travel to translate a change in the monetized value of time into a change in the number of trips.

The core economic impacts that we quantify are change in the monetized value of time waited, and net change in income (GDP) and employment associated with change in cross-border trip flows. For all three flows, we measure the opportunity cost value of change in time waited by translating the changes in time waited into a dollar value of time spent in inspection queues. To quantify the income and employment impacts of net change in passenger vehicle flows, change in wait time is converted into changes in passenger vehicle traffic using the elasticity of vehicle trips with respect to wait time, and then in turn into changes in tourist and business travel expenditures on the basis of average per-person visit spending. We measure the impacts on the U.S. economy of both increased levels of Canadian and Mexican visitors to the United States resulting from a lower wait time, which increases spending in the U.S. economy, and increased levels of U.S. visitors to Canada and Mexico, which reduces spending in the U.S. economy. The latter flow increases because American residents who travel to Canada or Mexico experience a reduction in wait time upon their return. Our income and employment impact estimates thus permit quantification of the net impact on the U.S. economy. A similar approach is followed for international air passenger flows.

For commercial vehicle flows, a logistical analysis of the commercial vehicle inspection process is used to estimate the effect of explicit transportation costs (e.g., customs broker fees, fuel), and implicit spillover costs (e.g., the value of lost time). On the U.S.–Mexico border, there is also a cost associated with the drayage operation (which affects the vast majority of shipments), or the unloading of cargo from the truck that transported it within Mexico, to another truck that transports it across the border, to yet another truck that transports it within the United States The drayage operation is a unique feature of the southern border that is not applicable to shipments across the northern border.

These various costs are estimated using data from the literature, combined with information from interviews with personnel at CBP, customs broker firms, and trucking carriers. Estimates of the rate at which carriers value their time are available in the literature (e.g., NCHRP, 1999), along with information related to the logistics of how freight moves through the border. Interviews with subject matter experts (e.g., customs brokers) were then used to fill in any data gaps. Because of the many significant uncertainties in the specification of transportation-related costs, sensitivity analysis is used to explore the implications of changes to significant input variables (see Roberts et al., 2013 for the sensitivity analysis results).

13.2.2 *Macroeconomic Analysis*

A number of non-price factors affects the movement of goods across international borders, some of which have been analyzed. Trade facilitation involves activities that improve trade capacities of various countries. Wilson and colleagues (2003) refer to it as the improvement of logistics at both customs offices and ports of entry, as well as the design of more efficient administration and procedures. Arvis and colleagues (2012) introduce the Logistics Performance Index (LPI), which considers the effect of cargo clearance effectiveness, transport and information technology infrastructure quality for logistics, ease and reasonable pricing of shipping services, timely shipping to destination, etc. We focus on the reduction of wait times at ports of entry through increased staffing at checkpoints, although reduced paperwork and simplification of trade procedures can also lower transport costs and have a significant impact on timely delivery of intermediate and final consumption goods.

For commercial vehicle flow, the impacts of changes in wait times on freight transportation are estimated with the use of a CGE model. This is a multi-market model of behavioral responses of individual producers and consumers to price signals within the limits of available labor, capital, and natural resources (Rose, 1995). CGE is a state-of-the-art approach to economic consequence analysis. It overcomes the major limitations of I-O by allowing for nonlinearities such as input substitution, including behavioral content, and providing an explicit role for prices and markets (Dixon & Rimmer, 2002).

The more sophisticated CGE approach is necessitated by the complexities of international trade. Changes in wait times translate into changes in transportation costs, which, in turn, translate into changes in the relative competitiveness of U.S. imports and exports. Reduced wait times of goods entering the United States make them cheaper and spur U.S. imports. However, the vast majority of the imports are unfinished (intermediate) goods, i.e., goods used in the production of finished (final or consumer) goods. Hence, changes in wait time have the effect of lowering the cost of production in the United States and making U.S. exports to all countries more competitive. This stimulates U.S. exports worldwide and increases U.S. GDP, personal income, and employment. Evaluating the extent to which the negative effect of increased import competitiveness for U.S. major trading partners is offset by the effect of increased U.S. export competitiveness requires the use of a sophisticated model.

We utilized a version of the Global Trade Analysis Project (GTAP, 2012) CGE model that allows substitution among modes of transport (Avetisyan & Hertel, 2015). The GTAP system was developed in conjunction with the U.S. International Trade Commission (ITC) and the World Trade Organization (WTO), and is the most widely used international CGE model today. The model consists of 129 countries, each comprised of 57 industry commodity groupings, and incorporates the import/export trade linkages between them.

To evaluate the regional and national impacts of changes in tourist and business spending associated with the changes in wait times for passenger vehicle traffic, we used an I-O analysis approach. The I-O approach models the economy as a set of integrated, linear supply chains and is the most frequently used tool of economic impact analysis at the regional level (Miller & Blair, 2009). It is a useful model for examining simple demand-side stimulants, such as increased travel expenditures.

Our approach evaluates impacts at the regional level by first translating changes in passenger vehicle traffic into estimates of changes in the numbers of Canadian and Mexican travelers. The change in number of travelers is then translated into change in their direct spending on individual product and service categories, ranging from restaurants to hotels. The I-O model then translates this direct stimulus into total economic activity by computing all of the rounds of supply-chain effects. The sum total is a multiple of the initial direct stimulus (hence the term "multiplier" effect). We limit our analysis to the demand-side, or upstream linkage effects of suppliers to hotels, restaurants, etc. Supply-side, or downstream linkages are not applicable to the final goods and services associated with consumer expenditures because they are "final," i.e. require no further processing.

A similar approach is applied to analyzing the national economic impacts of travel spending by foreign visitors to the United States, but, in this case, the stimulus is to the entire U.S. economy. However, we also factor in the diversion of some spending to other countries stemming from the increased attractiveness of foreign travel to U.S. residents resulting from decreased wait times they would confront upon their return.

The empirical version of the I-O models we used was obtained from the Impact Analysis for Planning (IMPLAN) System (MIG, 2012). This system consists of an extensive economic database, algorithms for generating I-O models for any country or county group, and algorithms for performing impact analysis. IMPLAN is the most widely used source of I-O tables and related data.

The national impacts of changes in expenditures of air travelers due to wait time reduction at airport passport inspection sites are estimated with the use of the USCGE Model, which has the characteristics of the CGE modeling approach discussed earlier. However, unlike GTAP, this is a single-country model, sufficient in this case because trade competitiveness issues are not pertinent for passenger travel.

13.3 Data

Data were obtained from CBP, other government sources, businesses, and consultants on key variables such as wait times, freight and passenger volume, tourist and business travel spending, and transportation costs. These data consist of statistics related to flows of passenger vehicles, commercial vehicles, and air travelers through land border crossings and international airports in the United States. The passenger vehicle data set consists of hourly observations for fiscal year 2010 through 2012 on the number of vehicles and passengers processed, average wait time, average number of primary inspection booths open, and average inspection time. There are three types of lanes at land border crossings; we limit our analysis to regular lanes because analysis showed that it would always be optimal for CBP to add an officer to these lanes. Data for commercial vehicles are similar to that for passenger vehicles.[1] Finally, airport data consist of observations by flight, with data for the number of passengers on the flight, total time waited by the flight's passengers, and maximum time waited by a passenger on the flight (with each of these variables provided for three separate groups of passengers: U.S. citizens, U.S. legal residents, and non-U.S. non-immigrants), and the average number of booths open during the time interval when the flight's passengers were processed.

For freight shipments on commercial vehicles, data on the freight travel distances and freight costs were specified using data from the literature, government documents, industry group publications, and interviews with subject matter experts. Many of these data required additional refinement for use in our model, such as converting all per-mile costs to hourly costs so that they can be readily linked to the changes in wait time at the border.

Nearly all of the macroeconomic data used were contained in the IMPLAN I-O models, the GTAP CGE model, and the USCGE model. The major exception was data on tourist expenditures, obtained from the U.S. Bureau of Economic Analysis and travel industry sources.

13.4 Overview of Results

We analyzed the economic impacts of changes in wait times at 13 crossings at six major land ports along the southern border of the United States and seven crossings at three major land ports along the northern border, as well as at 14 inspection sites at four major airports (see Table 13.1). A majority of the land crossings process both passenger vehicles and commercial vehicles.[2] Simulations were run for the addition of one CBP officer to each crossing for one year. Each additional officer was assumed to work eight hours per day for 153 days of the year, reflecting actual CBP staffing practices. The FY 2012 levels of vehicle and passenger traffic were taken as the base from which changes were calculated.

The results are presented separately for each of the components because they involve different estimating methods, data, assumptions, and stakeholder groups. This enables us to sum the appropriate results for the impacts on the U.S. economy and its residents separately from the impacts on other countries. The results are presented in Table 13.1 for impacts on the United States, as well as for Canada and Mexico (see italicized entries).

The net impacts on the U.S. economy as a whole for the +1 staffing case are:

- $72.0 million in GDP[3]
- 1,183 annual jobs
- $22.8 million in value of time gained for U.S. residents

The value of wait time saved in the +1 staffing case is approximately $550,000 per staff member on average for the land ports and airports evaluated in this study. Given that the simulation represents the addition of 43 staff members (one at each of 17 land passenger ports, 12 land freight ports, and 14 airport inspection sites), each new staff member on average generates 27 additional jobs throughout the U.S. economy.

The ratio of economy-wide employment gains to one additional CBP officer might appear high at first, but is reasonable if placed in perspective. The ratio presented for CBP staffing changes is less like a standard industry multiplier for ordinary business activity and more like an action at a pressure point in the economy, in this case to alleviate a bottleneck. Hence, we would expect it to have a higher than average "multiplier." The situation is more akin to investment analysis in critical facilities. For example, a recent congressional study evaluating FEMA hazard mitigation grants found a benefit-cost ratio of more than 100 for a few million dollars of investment in burying electric power lines underground to avoid hundreds of millions of dollars of business interruption losses from a major electricity outage due to severe storms (MMC, 2005).

Most of the economic gains and losses from CBP staffing changes will accrue to the regions surrounding the POEs. For example, adding one CBP officer at each of the 17 passenger land crossings is projected to lead to an increase in GDP of $61.8 million and employment gains of 1,053 jobs in the

United States as a whole. However, 80% of the GDP gains and 94% of the employment gains are captured by the POE regions alone. The regional gains are a relatively high proportion despite the fact that the multiplier values are higher for national impacts. The main reason for this outcome is the diversion of some spending to other countries stemming from the increased attractiveness of foreign travel to U.S. residents from decreased wait times. This subtraction of spending diverted to foreign countries to yield net impacts is nationwide and not limited to the POE regions.

13.5 Microeconomic Analysis of Changes in Wait Times

For all three types of border crossings (passenger vehicles, commercial vehicles, and passport inspection of air travelers), we estimate the impact on wait time of adding one CBP officer to primary inspection, and use this estimate to quantify the decrease in wait time that would have been experienced by the volume of those being inspected in FY 2012. We then convert this into a monetary value of time saved using a standard opportunity cost methodology. For passenger vehicles and international air travelers, we also quantify the new number of cross-border trips that a decreased wait time induces. For commercial vehicles, changes in wait times are used to infer changes in freight travel costs (see Section 13.6) and in international competitiveness (see Section 13.7). The increase in passenger vehicle travel and international air travel is then used to estimate regional and national macroeconomic impacts in Section 13.8.

Analysis was carried out for passenger vehicle traffic at 17 land border crossings (see Table 13.2), international air arrivals at 14 inspection sites at four international airports (see Table 13.3), and commercial vehicle traffic at 12 land border crossings (see Table 13.4).[4] Only crossings that have significant volumes of traffic and nonzero wait time outcomes were selected. The land crossings account for a large majority of cross-border traffic and wait time experienced.[5]

13.5.1 *Passenger Vehicle Lanes at Land Border Crossings: Impact on Wait Time of Adding an Officer*

Quantifying how wait time changes when staffing is added to or subtracted from a border crossing is challenging, because there is considerable adaptive behavior – traffic arrival mounts up and falls off at a crossing, and staff is added and then subtracted in reaction to the traffic flow. This endogenous behavioral reaction prevents the use of simple regression techniques to estimate the relationship between wait time and staffing.[6] We rely instead on two approaches

Table 13.1. *Economic impacts of decreases in wait times at selected U.S. land and air ports of entry (POEs) (+1 customs officer at each POE)*

		GDP (million 2011$)	Employment (jobs)	Value of time saved (million 2011$)
Ground Passenger Travel	Value of lowered wait time for U.S. residents	n.a.	n.a.	$17.0
	Value of lowered wait time for Canadian and Mexican residents	n.a.	n.a.	$10.2
	Net impact on port region and U.S. GDP and employment	$61.8	1,053	n.a.
Air Passenger Travel	Value of lowered wait time for U.S. residents	n.a.	n.a.	$5.8*
	Value of lowered wait time for foreign travelers	n.a.	n.a.	$9.2*
	Net impact on U.S. GDP and employment*	$6.0*	87*	n.a.
Truck Freight Transportation	Value of lowered wait time for U.S. truckers	n.a.	n.a.	n.c.
	Value of lowered wait time for Canadian and Mexican truckers	n.a.	n.a.	$29.1
	Net impact on U.S. GDP and employment	$4.2	43	n.a.
	Net impacts on the Canadian and Mexican economies	$6.8	n.c.	n.a.
Total Canada and Mexico		n.c.	n.c.	$43.9
Total United States		$72.0	1,183	$22.8

n.a. – not applicable; n.c. – not calculated

Note: values for Canada and Mexico are italicized.

* Average of the lower-bound and upper-bound results presented in Table 13.3 and Table 13.9.

to quantifying the elasticity of wait time to the addition of an officer to primary inspection. First, we evaluated the results of a staffing experiment at the San Ysidro POE that was run on the weekend of July 21–22, 2012. This experiment increased inspection booths open in regular lanes by roughly 35% above the levels normally experienced on weekend hours, and wait time fell by 62%, implying an elasticity value of roughly -1.8. Second, we develop a simple model of a saturated queuing system with multiple booths and derive a formula for how wait time changes in response to adding one booth.[7] This formula contains variables for which empirical data are collected by CBP, and it can be applied to historical hourly data for any border crossing to predict how wait time would change if one booth was added in a given hour. The July experiment at San Ysidro provides some validation of this theoretical model.[8]

Adding one officer to a border crossing's staff creates eight additional hours of primary inspection capability on the days that the officer works on enforcement activities, which is 153 days in a year. For each border crossing, we identified the eight most congested contiguous hours of each day in 2012, assumed that an additional officer would be deployed to these hours, and used the theoretical formula to determine how wait time would have changed in each hour of these eight-hour stretches if an additional officer had been available. We then calculated an average change in wait time for each eight-hour stretch, and averaged across all eight-hour stretches in 2012. Column A in Table 13.2 gives the average wait time value for each crossing in 2012, and column C gives the average percentage change in wait time resulting from the addition of one officer.

13.5.2 *Passenger Vehicle Lanes at Land Border Crossings: Value of Time Saved for Existing Traffic*

We calculated the value of time saved annually for existing passenger vehicle traffic by applying the estimated percentage change in wait time in column C of Table 13.2 to the total amount of time waited by vehicle passengers at a crossing in the identified eight-hour stretches in 2012 multiplied by the ratio (153/365) to capture the fact that the additional officer is only available for 153 days of the year. This yields a time savings value in hours that we then monetized using a methodology established by the U.S. Department of Transportation (2011), which sets the value of time for U.S.-resident business travelers at the median pretax wage plus benefits, and the value of time for personal/leisure travelers at the median pretax wage without benefits multiplied by 0.6. These values are also multiplied by 1.47 to reflect the disutility of waiting in queues.[9] Values for business and personal travel time were then

Table 13.2. *Impacts of +1 customs staffing for passenger vehicles at U.S. land border crossings*

Results for Regular Lanes		Average wait time in minutes in 2012	(as % of San Ysidro value)	Average change in wait time from adding one officer to eight most congested hours	Change in the monetary value of time waited by passengers in 2012 (million $U.S.)			Long-run cross-border trip-wait time elasticity	Number of new trips (increment to 2012 level)	
Port	Crossing				U.S. residents	Canadian/ Mexican residents	Total		U.S. resi-dents	Canadian/ Mexican residents
		(A)	(B)	(C)	(D)	(E)	(F)	(G)	(H)	(I)
Calexico	Calexico/ East	41	55%	−35%	$1.6	$0.3	$1.9	−0.55	52,370	40,536
	Calexico/ West	48	66%	−16%	$1.3	$0.2	$1.4	−0.66	43,946	28,960
El Paso	Ysleta	26	35%	−37%	$0.9	$0.2	$1.1	−0.35	29,953	32,456
	Paso Del Norte	28	38%	−37%	$1.0	$0.2	$1.2	−0.38	32,079	34,761
	Bridge of the Americas	33	44%	−31%	$1.4	$0.3	$1.8	−0.44	49,369	53,495
Laredo	Lincoln-Juarez	23	31%	−28%	$1.0	$0.2	$1.2	−0.31	31,941	28,841
	Convent St.	21	28%	−48%	$0.7	$0.1	$0.8	−0.28	20,480	18,493

Nogales	Mariposa	20	27%	–55%	$0.8	$0.2	$1.0	–0.27	25,034	33,185
	Deconcini	23	31%	–46%	$0.7	$0.2	$0.9	–0.31	24,300	32,212
San Ysidro	San Ysidro	74	100%	–8%	$2.4	$0.2	$2.5	–1.00	77,155	24,631
Buffalo-Niagara Falls	Rainbow Bridge	7	9%	–33%	$0.4	$0.8	$1.3	–0.09	11,909	21,318
	Lewiston Bridge	14	19%	–38%	$0.8	$1.5	$2.2	–0.19	21,859	39,129
	Peace Bridge	7	9%	–39%	$0.5	$0.9	$1.4	–0.09	12,784	22,884
Blaine	Peace Arch	18	24%	–49%	$1.4	$2.3	$3.7	–0.24	43,417	65,125
	Pacific Highway	17	23%	–62%	$1.2	$1.9	$3.1	–0.23	36,930	55,394
Detroit	Windsor Tunnel	5	6%	–73%	$0.5	$0.3	$0.8	–0.06	16,745	10,807
	Ambassador Bridge	4	5%	–63%	$0.5	$0.4	$0.9	–0.05	16,473	10,631

aggregated into a composite value of time based on the breakdown of cross-border travelers by purpose of trip.[10] Results presented here use nationwide U.S. wage values (results using the median wage of the region in which the border crossing is situated generally did not differ significantly).

For nonimmigrant Canadian nationals, the value of time for both business and personal travelers is based on the median nationwide Canadian wage rate converted into U.S. dollars using the 2011 commercial exchange rate. For non-immigrant Mexican nationals, the value of time for both business and personal travelers is based on multiplying the U.S. nationwide median wage by the ratio of Mexican to U.S. per capita income in 2011, with Mexican per capita income valued in nominal U.S. dollars using the commercial exchange rate.

Columns D and E in Table 13.2 present the value of saved wait time for U.S. residents and Canadian/Mexican residents. Savings to U.S. residents from the addition of one officer equal on average roughly $1.2 million at southern border crossings and $0.8 million at northern border crossings. Savings to Mexican residents at southern border crossings equal on average about $0.2 million, and to Canadian residents at northern border crossings about $1.2 million.

Passenger Vehicle Lanes at Land Border Crossings: New Cross-Border Trips Induced by Lower Wait Time

Lower wait time reduces the monetary and nonmonetary costs associated with making a cross-border trip and therefore presumably increases the number of trips. During the July experiment at San Ysidro, the number of trips rose by roughly 30% above baseline values for weekends in 2012 in response to a fall in wait time of 62% below baseline values, suggesting an ultra-short-run elasticity of trip demand to wait time of –0.5. This elasticity is consistent with typical values of the short-run elasticity of travel demand with respect to travel time estimated in other studies, which ranges between 0 and –1, and averages roughly –0.5 (see Litman, 2012; Wardman, 2012).

Studies also suggest that the long-run elasticity is roughly twice the short-run elasticity. Given that the elasticity value resulting from the San Ysidro experiment equals the typical value suggested by the literature of –0.5, we use a value of –0.5 for the short-run elasticity for the San Ysidro crossing, and a long-run elasticity value of –1.

For other crossings, change in wait time presumably has less of an impact on generating new trips, because the level of wait time is significantly less than at San Ysidro (see Table 13.2, columns A and B). We do not have any controlled experiments available for other border crossings, and we do not have a formal methodology that permits estimating these elasticities. In the absence of experiments and a formal methodology, we assume that the trip wait time elasticity at another border crossing is directly proportional to the ratio of the average wait time at that crossing to the average wait time at San Ysidro (see column G in Table 13.3).

Vehicles processed by CBP at a border crossing include flows of Mexican or Canadian residents who are making trips into the United States, and U.S. residents who are returning from trips to Mexico or Canada. Our economic impact analysis requires estimates of the changes in these flows broken down separately. We use CBP data on the nationality of vehicle passengers processed at a port of entry to approximate the proportion of the flow that is U.S. residents with the proportion of the flow that is U.S. citizens and U.S. legal permanent residents, and the proportion of the flow that is Mexican or Canadian residents with the proportion of the flow that is Mexican or Canadian citizens.[11] Change in the number of new trips by U.S.-resident and Canada/Mexico-resident passengers is given in columns H and I of Table 13.2, respectively.[12]

Commercial Vehicles at Land Border Crossings: Change in Wait Time and Cross-Border Trips

For commercial vehicles, we analyzed the impact of an additional officer on wait time using the same methodology as for passenger vehicles.[13] However, unlike the case of passenger vehicles, we assume that the number of commercial vehicle trips across the southern and northern borders does not increase if wait time is reduced. It is not obvious how the impact of wait time on the number of commercial vehicle trips should be modeled. Shipping companies might plan a given number of trips regardless of the wait time, so that the primary impact of an increase or fall in wait time is to the shipping cost. At some point, an increasing wait time level will necessarily impact the number of commercial vehicle trips, but it is not clear that current levels are so high that this will be the case.[14] Given the lack of any empirical evidence, we conservatively assume that these elasticities equal zero.

The methodology used to value the time saved by commercial vehicles is quite different from that used to value time saved by passenger vehicles, and is described in detail in Section 13.6.

International Air Arrivals at Airports: Change in Wait Time and Cross-Border Trips

CBP must carry out inspection of passengers who arrive on international flights at U.S. airports, who arrive at a U.S. airport, disembark from the plane, walk to a passport control inspection area, and enter either the queue for U.S. citizens (USCs) and legal permanent residents (LPRs), or the queue for nonimmigrant foreign nationals (NIMs). For each international flight arriving at a U.S. airport,

Table 13.3. *Impacts of +1 customs officer staffing for air passenger arrivals at U.S. international airports*

Airport/Terminal		+1 Customs Officer Deployed to Eight Most Congested Hours of the Day									
		Change in Average Wait Time	Change in Value of Total Wait Time (million U.S. dollars)				Change in Number of International Air Trips by Type of Passenger in Response to Change in Wait Time				
			Results Without Taking into Account Baggage Delivery		Results Taking into Account Baggage Delivery		Lower-Bound Estimates: Consideration of Baggage Wait Time				
			U.S. Residents	Foreign Residents	U.S. Residents	Foreign Residents	U.S. Resident Passengers	Foreign Resident Passengers	U.S. Resident Passengers	Foreign Resident Passengers
Chicago O'Hare (ORD)		–4.7%	$1.7	$1.6	$0.5	$0.7	203	15	211	42
New York (JFK)	American	–5.7%	$0.8	$1.1	$0.2	$0.6	101	5	245	16
	British	–7.6%	$0.4	$0.9	$0.1	$0.5	48	3	137	10
	Delta	–6.4%	$0.9	$0.7	$0.3	$0.4	155	8	159	11
	Terminal 1	–4.9%	$0.6	$1.6	$0.2	$0.9	95	5	362	24
	Terminal 4	–3.5%	$1.2	$1.1	$0.4	$0.6	230	11	276	19
Los Angeles (LAX)	Satellite 2	–6.6%	$0.5	$0.9	$0.1	$0.4	46	4	108	11
	Satellite 5	–9.4%	$0.3	$0.5	$0.1	$0.3	33	3	88	9
	Satellite 7	–9.2%	$0.5	$0.4	$0.1	$0.2	37	3	53	5

	Terminal 4	−13.7%	$0.2	$0.1	$0.0	$0.0	18	2	19	2
	Tom Bradley	−3.1%	$0.5	$0.8	$0.1	$0.3	62	5	99	10
Miami (MIA)	Central Terminal	−2.7%	$0.9	$1.1	$0.2	$0.5	176	12	140	8
	South Terminal	−5.7%	$0.4	$1.2	$0.1	$0.6	58	4	298	18
	North Terminal	−2.4%	$0.1	$0.2	$0.0	$0.1	16	1	39	2
TOTAL			***$9.0***	***$12.2***	***$2.4***	***$6.1***	**1,278**	**79**	**2,234**	**187**

Source: Roberts et al. (2014).

CBP collects data for USC, LPR, and NIM passengers separately on how many passengers were processed at passport control, the total amount of time passengers waited, the maximum time a passenger waited, and how many inspection booths were open on average to process the flight's passengers.

Passenger arrivals from a particular flight arriving at a passport control site tend to arrive as a group. Over the course of a day, flights tend to arrive in clusters at a terminal, and their passengers arrive at an inspection site in discrete bunches that overlap, so that passengers from different flights are intermingled in the queues. We have developed a simple algebraic methodology that relates the average wait time a flight's passengers experience to the number of booths open at a site, and use the methodology to quantify change in wait time resulting from adding an officer to the site. This methodology can also be applied to flights that are not in clusters, as well as to the first flight in a cluster. As in the case of land border crossings, we assume that an additional officer works for 153 days a year, and on each day worked, the officer is deployed to the eight busiest hours of the day, which are determined by the number of passengers on arriving flights. Table 13.3 shows how the average wait time changes at each of the 14 inspection sites evaluated in our study.

International Air Arrivals at Airports: Value of Time Saved for Existing Traffic

A key issue with respect to evaluating the impact of an extra officer on time waited in passport inspection queues is whether a lowered passport inspection wait time translates into a lower time required to exit the airport, or simply redistributes wait time between passport inspection and the baggage wait area. After clearing passport inspection, those passengers with checked baggage must collect their baggage, and a wait might be involved at this stage. We did not have data on the time that a flight's baggage arrived at the baggage wait area, and we were thus not able to quantify the degree to which a lowered passport inspection time simply redistributed wait time between the two processes. In order to generate lower- and upper-bound estimates, we evaluated two scenarios: (a) upper-bound estimates that assume that a lower passport inspection wait time translates one-to-one into a lower airport exit time for all passengers, and (b) lower-bound estimates that assume that only passengers with passport inspection wait times higher than 30 minutes have a reduced airport exit time due to lower passport inspection wait times.

Table 13.3 gives results on how the value of time changes for the two passenger groups under the lower- and upper-bound scenarios.[15] We calculate the value of time saved annually for existing airline passenger arrivals by setting a base volume of wait time equal to the total amount of time waited at the four passport control sites in the identified eight-hour stretches in 2012. We then apply the estimated percentage change in wait time resulting from one additional officer to this value, and multiply by the ratio (153/365) to capture the fact that the additional officer is only available for 153 days of the year.

We follow the methodology established by the U.S. DOT (2011) for valuing the time of air travelers, which sets significantly higher values of time for air travel than for travelers in ground vehicles. We also follow the recommendations of Robinson (2007) and adjust this value to reflect the disutility of waiting in border-crossing queues. For U.S. citizens and LPRs, we use an appropriately adjusted value of the nationwide median wage. For NIMs, we first construct a weighted average of the ratios of per capita income in various countries valued in U.S. dollars at the commercial exchange rate to U.S. per capita income, with weights being the fraction of overseas visitors arriving by air to the United States from a particular country in 2011.[16] This ratio is then multiplied by the value for a U.S. citizen/LPR air traveler's time.

International Air Arrivals at Airports: New International Air Trips Resulting from Change in Wait Time

Change in passport inspection wait time may cause the number of international air trips to change, because, as in the case of land border crossings, the costs associated with having to go through passport inspection at airports may deter some international travelers from coming to the United States, as well as deter some U.S. residents from traveling abroad (as they would undergo inspection upon their return). We quantify the change in air passenger trips resulting from a decrease in passport control wait time at airports by monetizing the value of time waited, determining how this monetized value changes with the addition of one officer, and using the price elasticity of demand for international air travel to convert change in monetized value into change in the number of international air trips.[17] The numbers of new trips given in Table 13.3 are very small in comparison with the number of passengers processed at the 14 inspection sites in 2012, which are in the hundreds of thousands or millions (see also Prager et al., 2015).

13.6 Impacts of Reduced Wait Times on Truck Transportation Costs

This section estimates the changes in transportation costs for trucks using a logistical analysis of the border inspection process.[18]

13.6.1 *Volumes of Truck Traffic at the Border*

The number of trucks that traveled through each of the POEs during FY 2012 is given in Table 13.4, both in total and for only the eight most congested hours

Table 13.4. *Truck volumes and estimated wait time changes for adding one customs officer at each U.S. land port of entry*

	Port of Entry	Crossing	No. of Trucks Processed[a]				Wait Time (Average), Default Case (minutes)		Change in Wait Time During Eight Most Congested Hours,+1 Officer Case[b,c]	
			Annual Truck Trips (FY 2012)	Eight Most Congested Hours of Each Day Only	Per Block of Eight Most Congested Hours (average)	Per Block of Eight Most Congested Hours x 153 Days (average)[d]	Overall	Eight Most Congested Hours of Each Day Only	Minutes	Percent
Southern Border	**Calexico**	**Calexico/ East**	320,482	211,515	680	104,040	25.3	28.8	–20.0	–69.6%
	El Paso	**Ysleta**	360,470	203,124	653	99,909	10.1	13.1	–9.5	–72.9%
		Bridge of the Americas	290,220	199,316	645	98,685	14.6	17.5	–10.2	–58.4%
	Laredo	**Columbia Solidarity**	215,701	145,353	709	108,477	7.4	8.6	–8.2	–95.5%
		World Trade Bridge	1,356,418	841,894	2,319	354,807	23.8	29.0	–20.0	–69.0%
	Nogales	**Mariposa**	298,730	230,501	688	105,264	30.4	37.5	–20.5	–54.6%
	Otay Mesa	**Otay Mesa**	644,925	443,118	1,221	186,813	29.4	34.2	–13.0	–38.1%
Subtotal (Southern Ports)			**3,486,946**	**2,274,821**	**n.a.**	**1,057,995**	**n.a.**	**n.a.**	**n.a.**	**n.a.**

Northern Border	**Blaine**	**Pacific Highway**	343,396	170,679	466	71,298	11.0	16.4	–13.6	–83.0%
	Buffalo–Niagara Falls	**Lewiston Bridge**	309,365	82,187	225	34,425	2.3	8.1	–5.8	–71.0%
		Peace Bridge	625,651	256,453	701	107,253	5.3	11.2	–5.3	–47.1%
	Detroit	**Windsor Tunnel**	39,186	24,719	67.5	10,328	3.5	4.6	–4.6	–100.0%
		Ambassador Bridge	1,425,757	574,819	1,571	240,363	4.6	7.0	–4.3	–61.2%
Subtotal (Northern ports)			**2,743,355**	**1,108,857**	**n.a.**	**463,667**	**n.a.**	**n.a.**	**n.a.**	**n.a.**
Total (all ports)			**6,230,301**	**3,383,678**	**n.a.**	**1,521,662**	**n.a.**	**n.a.**	**n.a.**	**n.a.**

[a] Includes only those days for which data are available.

[b] Relative to the default/status quo values.

[c] Optimally allocated to the eight most congested hours of each day.

[d] Note that these are the only trucks that encounter wait time reductions.

of each day. Table 13.4 also lists the total number of trucks processed per block of eight most congested hours (excluding those days when the truck volume is zero – perhaps because the port was not open on that particular day). The additional CBP officer is assumed to be added to the eight most congested hours of each day, and to work 153 days (shifts) per year. Accordingly, the number of trucks that will encounter a reduced wait time at the border is the total number of trucks processed per block of eight most congested hours multiplied by 153 days.

We assume that only those trucks moving through the border during the hours and days when the additional customs officer is on duty will encounter a lower wait time. While the wait time reductions induced by the additional officer's presence can extend beyond the eight hours when the officer is actually on duty (because they help thin out the queue), we lack a basis by which to assess these "spillover" wait time reduction effects, so we conservatively assume that they are zero.

13.6.2 *Wait Times for Trucks at the Border*

Our model does not include the time spent in or waiting for a secondary inspection, nor do we include the processing time for the primary inspection itself. Accordingly, the total time needed for the border crossing is simply the wait time for the primary inspection, and the change in the total border crossing time is then the change in the wait time. The wait times for trucks in the default case and the changes in the wait times induced by the additional CBP officer are summarized by POE in Table 13.4. Note that although our model uses hours as the calculational unit for time, the wait times in Table 13.4 are given in minutes, as the magnitudes are such that this makes for a more intuitive unit for display.

13.6.3 *Truck Travel Distances*

We use the simplifying assumption that all trucks crossing the border are registered in the country from which they depart (Canada or Mexico), for lack of other data. The "overall" truck transport distances (from origin to destination) are estimated by the proxy of the straight-line distances between the population centroids of the United States and the country of origin of the goods (Canada or Mexico).[19] This is, however, only an approximation to the true average distance traveled by freight shipments crossing U.S. borders.[20] Using the distance calculator from the U.S. National Hurricane Center (NHC, 2010), the total transport distances are 674 miles for Canada–U.S. shipments, and 1,240 miles for Mexico–U.S. shipments.

We assume the distance traveled within each country is proportional to the (average) trip distance for domestic shipments (not imports) within that country, which we estimate as the ratio of domestic freight movement in ton-kilometers to domestic freight movement in tons. Using data obtained from the North American Transportation Statistics Database (NATSD, 2012), in 2010 (the most recent year for which data are available for all three countries), these estimated trip distances for domestic shipments are 279 km for Canada, 469 km for Mexico, and 251 km for the United States.

The ratio of the domestic distance for Mexico to that for the United States is (469 km/251 km) = 1.87, so we assign 87% more of the total Mexico–U.S. transport distance (1,240 miles – see earlier) to Mexico. For Mexico–U.S. shipments, we estimate that 808 miles occurs in Mexico and 432 miles occurs in the United States. Applying this procedure to Canada–U.S. shipments yields 355 miles traveled in Canada and 319 miles traveled in the United States. Additionally, for Canada, in 2009, trucks weighing 4.5 metric tons or more drove some 1,969 million vehicle-miles for "trips across Canada and United States border" (Statistics Canada, 2010). With a total of 5,020,633 truck trips into the United States from Canada in 2009 (BTS, 2012), and assuming that all of these trucks are Canadian-registered (see earlier), this equates to 392 miles traveled per trip in Canada. The average estimated trip distance in Canada, then, is 373.5 miles.

13.6.4 *Truck Operating Costs*

Fuel consumed is the ratio of the distance traveled to the truck fuel efficiency. Similarly, the distance traveled by the truck is the product of the truck's speed and the total travel time. We use average truck speeds of 40 miles/hour when away from the border (ATRI, 2011), and 5 miles/hour when in the queue and during the border crossing. For U.S. truck fuel efficiency, we use the regression equation presented by Schrank and colleagues (2011), based on average truck speed.

While not addressing 18-wheel trucks specifically, the American Transportation Statistics Database (NATSD, 2012) does include (average) fuel efficiency data for "new light-duty trucks." These data for 2002–2007 (the only years for which these data are available for all three countries) indicate that Canadian trucks are around 1% more fuel efficient than U.S. trucks, and that Mexican trucks are around 9% more fuel efficient than U.S. trucks, so we make the appropriate adjustments. For the cost of diesel fuel, we use \$3.42/gallon for Mexico, \$4.03/gallon for the United States, and \$4.94/gallon for Canada (all deflated from 2013\$ to 2011\$), as obtained from MyTravelCost.com (2013).

So that they can be readily linked to the reduced wait times at the border, in cases where truck operating cost data from the literature are specified on a per mile (rather than per hour) basis, we convert them to hourly costs using an overall average truck speed of 40 miles/hour (ATRI, 2011). The average hourly wage for "heavy and tractor-trailer truck drivers" in the United States is $19.15/hour, according to the U.S. Bureau of Labor Statistics (BLS, 2012). The American Transportation Research Institute gives figures of $18.74/hour excluding benefits, and $24.68/hour with benefits (ATRI, 2011). Workers in Canada's "transportation and warehousing" industry earn, on average, $22.99/hour (Statistics Canada, 2012). Transport Canada (2005) also provides numerous truck driver wage estimates, working out to around $23.10/hour, on average. Averaging all of these values together gives $20.86/hour for U.S. truck drivers and $23.05/hour for truck drivers in Canada. Driver wage data for Mexico could not be found in the literature, so this was estimated by adjusting the wage rates for the United States and Canada, using the relative GDPs per capita of the two countries (GDP data from the World Bank, 2013), working out to $4.48/hour (average value).

The Minnesota Department of Transportation suggests a (average) cost of $12.80/hour for truck repairs, maintenance, tires, and depreciation (MDOT, 2003). The American Trucking Research Institute proposes $23.95/hour (ATRI, 2011). Boyer (1997) posits $25.28/hour for vehicle depreciation, licensing, interest, tires, and maintenance, which we have inflated to 2011 dollars. After excluding driver wages (i.e., with the average wage rate subtracted out), Schrank and colleagues (2011) determine a value of $76.85/hour for vehicle depreciation, interest, insurance, general maintenance, tires, and repairs (Ellis, 2009). The average operating cost for U.S. trucks is therefore about $34.72/hour. Data from Transport Canada (2005) indicate an average cost of about $77.60/hour for truck repairs, cleaning, tires, depreciation, licenses, profit margin, interest, insurance, and administration. For truck operating costs in Mexico, we adapt the values for the United States and Canada using relative per capita GDP, yielding $11.37/hour, on average.

Customs broker fees (per truck trip) are generally $50–$150 on the northern border, and $20–$30 on the southern border, where the process is more repetitive and organized (Gould, 2012). We use the range midpoints, which are $25 for the southern border and $100 for the northern border (we also assume that the magnitude of these fees does not depend on the wait times at the border). Based on interviews conducted with various trucking carriers, Haralambides and Londono-Kent (2004) suggest a (average) loading/unloading cost of about $154/trip associated with the drayage operation at the U.S.–Mexico border, which we adopt.

13.6.5 *Changes in Truck Transportation Costs*

The total truck transportation cost (from origin to destination) is equal to the sum of the border-related transport costs and those transport costs that are incurred away from the border. The change in the total truck transport cost is then the change in the border-related transport costs, as this is the only component of the total transport cost (origin to destination) that changes with wait time.

Table 13.5 summarizes the freight transportation cost results for adding one additional customs officer at each POE during the eight most congested hours of each day (for 153 days out of the year). The changes in truck transportation costs range from –0.252% (for Blaine) to –0.053% (for El Paso-Ysleta), with the change in the total transportation costs (sum over all trucks) ranging from –$2.74 M (for Laredo-World Trade Bridge) to –$0.09 M (for Detroit-Windsor Tunnel). On the southern border, the overall change in truck transport costs (weighted average across all southern ports by number of trucks) is –0.096% (–$6.42 M in aggregate). On the northern border, the results are quite comparable, with an overall change in truck transport costs of about –0.092% (–$5.25 M in aggregate). The total change in truck transportation costs (sum over all trucks and both borders) is about –$11.67 M annually.

13.6.6 *Truck Opportunity Cost of Time Savings*

In addition to the various "out-of-pocket" transportation costs, border delays impose an opportunity cost of time. A National Cooperative Highway Research Program study (NCHRP, 1999) estimated that the value of time for commercial vehicles ranged from $144–$193/hour (experiments conducted in 1995), so we use the midpoint, or about $272/hour in 2011$. Adjusting this value based on the relative GDPs per capita yields $285/hour for Canada and $56.80/hour for Mexico.

The changes in the opportunity cost of time for trucks caused by adding one additional customs officer to each POE during the eight most congested hours of each day are summarized in Table 13.5. Note that all trucks entering the United States are assumed to be foreign-registered (see this section, subsection 3), and hence why Table 13.5 does not include opportunity cost values for U.S. trucks. The total value of the change in the opportunity cost of time for Canadian trucks (sum over all northern ports in Table 13.5) is about –$13.35 M. For Mexican trucks, the total change in the opportunity cost of time (sum over all southern ports in Table 13.5) is comparable, at about –$15.74 M, a significant portion of which comes from only a single crossing (the World Trade Bridge at Laredo). The total annual opportunity cost of time savings, considering all inbound truck traffic (all POEs), is roughly $29.09 M.

Table 13.5. *Changes in truck transport costs and opportunity costs of time for adding one customs officer to each U.S. land port of entry*

	Port of Entry	Crossing	Changes in Total Truck Transport Costs[a,b]		Changes in Opportunity Cost of Time[a,b]	
			Total (all trucks)	Percent	Canada (all trucks)	Mexico (all trucks)
Southern Border	**Calexico**	Calexico/East	–$0.81 M	–0.131%	n.a.	–$1.97 M
	El Paso	Ysleta	–$0.37 M	–0.053%		–$0.90 M
		Bridge of the Americas	–$0.39 M	–0.070%		–$0.95 M
	Laredo	Columbia Solidarity	–$0.34 M	–0.083%		–$0.84 M
		World Trade Bridge	–$2.74 M	–0.105%		–$6.72 M
	Nogales	Mariposa	–$0.83 M	–0.145%		–$2.04 M
	Otay Mesa	Otay Mesa	–$0.94 M	–0.076%		–$2.30 M
Subtotal (Southern ports)			**–$6.42 M**	**n.a.**	**n.a.**	**–$15.74 M**
Northern Border	**Blaine**	Pacific Highway	–$1.81 M	–0.252%	–$4.61 M	n.a.
	Buffalo-Niagara Falls	Lewiston Bridge	–$0.37 M	–0.057%	–$0.94 M	
		Peace Bridge	–$1.06 M	–0.081%	–$2.69 M	
	Detroit	Windsor Tunnel	–$0.09 M	–0.109%	–$0.23 M	
		Ambassador Bridge	–$1.92 M	–0.065%	–$4.89 M	
Subtotal (Northern ports)			**–$5.25 M**	**n.a.**	**–$13.35 M**	**n.a.**
Total (all ports)			**–$11.67 M**	**n.a.**	**–$13.35 M**	**–$15.74 M**

[a] Relative to the default/status quo values.
[b] Optimally allocated to the eight most congested hours of each day.
Note that opportunity cost data are not presented for U.S. truckers, as all entering trucks are assumed to be foreign-registered (see Section 13.6.3).

13.7 National Competitiveness and Macroeconomic Impacts of Changes in Freight Transportation Costs

The majority of U.S. trade with its largest trading partners, Canada and Mexico, is conducted via land transport through its northern and southern borders, respectively. In 1994, the United States, Canada, and Mexico signed the North American Free Trade Agreement (NAFTA) to eliminate various trade duties and restrictions, and created the largest free trade zone in the world. According to the Office of the U.S. Trade Representative (2012), NAFTA has better connected 450 million people who produce goods and services worth $17 trillion. In 2009, the total value of the United States' trade with its NAFTA partners was $1.6 trillion, with exports and imports worth $397 billion and $438 billion, respectively.

Although the ratification of the free trade agreement eliminated trade restrictions and tariffs, it only moderately improved border-crossing procedures (e.g., inspections/processing).[21] Time is one of the most important factors affecting trade between various countries. It is therefore intuitive to assume that the longer a product travels from the origin to the destination country, the higher its shipping or transport cost. Shipping time may also vary due to other factors, such as wait times at border crossings and other checkpoints, quality of logistics infrastructure, etc. The relationship between time and trade has been addressed in a number of studies.

We use the GTAP CGE model to evaluate the impact on the U.S. economy of the change in transportation cost quantified in Section 13.6. As discussed in Section 13.2, the more sophisticated CGE approach is necessitated by the complexities of international trade.

13.7.1 *Trade and Transport in GTAP*

International trade and transport in the Global Trade Analysis Project (GTAP) model are represented by "margin" services (shipping services, or transport costs) and merchandise goods. These data are contained in a "trade matrix," which includes bilateral flows of only non-margin components, while the margins preserve the balance between global imports (CIF values) and exports (FOB values).[22] The difference between imports and exports of global merchandise represents global exports of transport services (transport margins). In GTAP, global imports are defined by the sum of merchandise trade valued at CIF prices; similarly, the sum of merchandise trade valued at FOB prices represents global exports.

In the GTAP model, the origins and destinations of traded goods are specified. However, this is not the case for transport services, which are grouped into a single global transport services industry, and then allocated to the various importing countries using the share of exports of traded merchandise (non-margins) for each country in the global exports of traded goods. More specifically, when the source country, *r*, exports a commodity to the country of destination, *s*, the export commodity (at FOB price) is combined with the composite international transport good (a mix of air, water, and other transport modes), thereby generating the CIF price of the commodity in the destination country. While the transport input is a fixed-composite good, the model allows for substitution among modes within the composite transport good. In the GTAP model, the modal substitution is determined by a constant elasticity of substitution (CES) production function. The CES, with a typical value of 0.9–2.8, governs modal choice changes in response to variations in the relative costs.[23]

13.7.2 *Results and Comparisons*

We utilize the GTAP model to analyze the economic impacts of a reduction in transport costs, due to changes in CBP staffing, for all goods shipped through the five northern and seven southern ports of entry. For each of these POEs, we use the reduction in total truck freight costs presented in Section 13.6 as the major input to the GTAP model.

An example of our macroeconomic results is presented for the Blaine, Washington, Michigan, POE, which is projected to have a 0.25% reduction of truck transport costs. Overall, the results indicate a very small positive increase in U.S. GDP of $0.79 million in 2011 dollars. As expected, the reduction in transport costs for shipping goods from Canada to the United States results in the increase of total imports to the United States. Our findings demonstrate a link between reduced transport costs and growth in trade, which is consistent with Djankov and colleagues (2010) and Hummels and colleagues (2013). In the United States, the change in the trade balance is positive, because exports grow more than imports. In Canada, the change in the trade balance is negative, because imports grow more than exports. Also, the GDP of Canada increases by $2.47 million due to increased exports. Note that, while the United States implements and pays for the changes that result in the reductions in border-crossing times, it also gains from reductions in transport costs because intermediate goods shipments reduce the cost of production in the United States, thereby making U.S. exports more competitive worldwide. This is consistent with the results of Minor and colleagues (2008) illustrating that countries that reduce trade time gain significant benefits. There are also changes in the U.S. exports to all regions of the world due to the increased imports of cheaper intermediate products from Canada, which lower production costs in U.S. manufacturing.

Since the additional customs officers reduce the wait time for imports to the United States, our primary focus is on this type of trade flow, as opposed to exports. In our analysis, we consider both imported intermediate goods and imported commodities for final consumption. We apply the share of imported goods used for final consumption in total demand to distinguish between these two types of imports. Specifically, if the share of final consumption imports in total demand is less than 35%, we classify these imports as intermediate commodity imports. We then adjust intermediate imports based on the characteristics of specific commodities in the GTAP model. Some of the U.S. truck imports from Canada with a significantly large share of intermediate consumption are 99%, 94%, 86%, and 84% for ferrous metals, petroleum and coal products, metal products, and transport equipment, respectively.

We summarize GDP and employment impacts of +1 changes in staffing at all POEs in Table 13.6 for all major southern and northern POE crossings. Overall, Canada is predicted to gain most from the +1 change in CBP staffing, with a GDP increase of $5.3 million, more than 3.4 times that of Mexico and nearly 1.3 times that of the United States. The results stem from the much greater reductions in wait times at Canadian truck POEs at Blaine and Windsor Tunnel than at any of the southern border POEs.

Overall, we find that adding one CBP officer at each land border crossing, on average per crossing, would generate an increase in U.S. GDP of $350,000 and 3.58 additional jobs (over and above the additional border agents). Both Canada and Mexico would reap net macroeconomic gains as well, with those of Canada exceeding U.S. gains by several fold. Note also that the implicit multiplier of 3.58 jobs generated for every new customs agent is not an ordinary economic impact multiplier calculated under normal (average) conditions, but rather one related to alleviating a bottleneck (at the margin), so is not surprisingly high (see also Avetisyan et al., 2015).

13.8 Regional and National Macroeconomic Impacts of Changes in Tourism and Business Travel

In this section, we estimate the macroeconomic impacts of changes in tourist and business travel resulting from CBP inspection staffing changes at the passenger land POEs and at 14 terminals of four international airports. The analysis for the land POEs is based on estimates of change in Canadian and Mexican land passenger entries into the United States presented in Section 13.5. The change in the number of these travelers was then translated into changes in

Table 13.6. *Changes in U.S. GDP and employment for adding one customs officer to each U.S. land freight port of entry*

	Port of Entry	Crossing	Change in transport cost (percent)	Change in U.S. Employment (no. jobs)	Change in U.S. GDP (million 2011 U.S.$)	Change in Canada GDP (million 2011 U.S.$)	Change in Mexico GDP (million 2011 U.S.$)
Southern Border	**Calexico**	Calexico/East	−0.13%	4	0.414	−0.050	0.346
	El Paso	Ysleta	−0.05%	2	0.166	−0.020	0.138
		Bridge of the Americas	−0.07%	2	0.218	−0.026	0.183
	Laredo	Columbia Solidarity	−0.08%	3	0.263	−0.032	0.220
		World Trade Bridge	−0.11%	3	0.331	−0.040	0.277
	Nogales	Mariposa	−0.15%	5	0.459	−0.055	0.384
	Otay Mesa	Otay Mesa	−0.08%	2	0.241	−0.029	0.201
Subtotal			**n.a.**	**21**	**2.093**	**−0.253**	**1.750**
Northern Border	**Blaine**	Pacific Highway	−0.25%	9	0.899	2.473	−0.087
	Buffalo-Niagara Falls	Lewiston Bridge	−0.06%	2	0.150	0.618	−0.025
		Peace Bridge	−0.08%	3	0.300	0.773	−0.025
	Detroit	Windsor Tunnel	−0.11%	5	0.450	1.082	−0.037
		Ambassador Bridge	−0.07%	3	0.300	0.618	−0.025
Subtotal			**n.a.**	**22**	**2.099**	**5.564**	**−0.199**
Totals			**n.a.**	**43**	**4.192**	**5.311**	**1.551**

their direct spending on goods and services. This section analyzes the total, or "multiplier" effect of the changes in direct tourism and business travel spending. The impact analysis is performed for the regional economy of each land POE, the rest of the United States, and the United States as whole. We used an I-O analysis approach to evaluating impacts of land passenger travel, rather than a CGE model, for several reasons. The major one is that for passenger travel, we need to use a regional model, because most of the spending takes place in the border region; however, CGE models for our border regions were unavailable. At the same time, I-O models are adequate to compute multiplier impacts of small changes in travel expenditures, which are not likely to result in any significant price changes. Many studies have successfully utilized I-O models to estimate tourism impacts on local and national economies (Madsen & Zhang, 2010; Pavlakovich-Kochi, & Charney, 2008; SANDAG, 2006).

The economic impact analysis of the CBP inspection staffing changes at the four international airports is based on estimates of change in the number of trips by U.S. and foreign residents traveling for leisure and business purposes by air as presented in Table 13.3 in Section 13.6. The changes in numbers of in-bound and out-bound air travelers were translated into changes in their spending on goods and services. A CGE model of the United States is applied to estimate the indirect and total impacts.

13.8.1 *Input-Output Analysis*

An I-O table consists of a set of accounts of all transactions between sectors in a given year. An I-O model can be constructed from these accounts by assuming a proportional relationship between inputs and outputs. It is especially adept at estimating ripple, or multiplier effects.

For this study, we employ the most widely used source of regional and national I-O tables in the United States, the Impact Analysis for Planning (IMPLAN) System (MIG, 2012). The IMPLAN sectoring scheme is based on the North American Industrial Classification System (NAICS), which disaggregates the economy into 440 sectors. In the I-O models constructed for this study, we aggregate the 440 IMPLAN sectors into 86 three-digit NAICS sectors.

We use the standard demand-side version of the I-O model, which evaluates how a change in final demand in a sector (or many sectors) would generate changes in "upstream" demands for successive rounds of production inputs. Tourism is a final service, so there are essentially no forward linkage effects. Overall, I-O is adequate to our task because there are not likely to be any significant price interaction effects in the case of land passenger travel.

A total of eight land passenger POEs was selected to be analyzed. To designate the POE regions, we adopted the following decision rules: 1) for a POE that is located in/near big cities, the regional impact analysis covers the entire metropolitan statistical areas (MSAs); 2) for those POEs that are not near any MSAs, the POE region only includes the county within which the POE is located.

13.8.2 *Further CGE Analysis*

We use the USCGE model to identify the macroeconomic impacts of changes in business and tourism air travel. This model has been used in more than a dozen published studies simulating the economic consequences of terrorist attacks (Rose et al., 2007, 2009), epidemics (Oladosu et al., 2013), climate policy (see, e.g., Oladosu & Rose, 2007; Prager, 2013), and trade-offs between urban security and commerce (Rose et al., 2014) among others. Data on 57 producing sectors are combined with data on nine household income groups, two categories each of government (State/Local and Federal), and external accounts (Rest of the United States and Rest of the World). The USCGE model is programmed and solved with the General Algebraic Modeling Software (GAMS).

Four simulations are run in the USCGE model, capturing the economy-wide effects of the direct impact values presented in Table 13.7 (for which lower-bound estimates are provided).

A. Increased spending by foreign visitors (for both leisure and business travel) to the United States;
B. Decreased spending in the United States due to increased U.S. residents' leisure travel abroad;
C. Decreased spending in the United States due to increased U.S. residents' business travel abroad;
D. Increased spending by U.S. leisure and business travelers on U.S. airlines to travel abroad.

Simulation A is modeled as an increase in exports for the domestic sectors in which foreign visitors spend, including: Petroleum, Other Non-durable Manufacturing, Air Transport, Rail Transport, Other Transport, Private Transit, Hotels and Restaurants, Personal Services, Other Business Services, Entertainment, and Medical Services. Foreign visitors who spend money in the United States are effectively exporting goods to their country of residence. Exports are increased in the USCGE model, and are expected to stimulate domestic production. In turn, prices are reduced and hence domestic demand, output, and employment are all stimulated.[24]

Simulation B is modeled as a reduction in household spending on tourism and travel-related sectors. The CGE model consumption function distributes these spending changes across sectors with respect to relevant variables, such as price changes. *Simulation C* is modeled as a reduction in business demand for sectors that business travelers spend in: Air Transport, Rail Transport, Other Transport, Private Transit, Retail Trade, Hotels and Restaurants, Personal Services, Other Business Services, Entertainment, and Medical Services. We constrained sector productivity as a device to reflect reductions in business demand for these sectors. Direct impacts on each sector were identified separately and run as a simultaneous simulation to calculate the combined aggregate impacts.

Simulation D is modeled as an increase in sales to U.S. households and businesses from Air Transport sector. We use a "phantom subsidy" approach, which is called such because it is not actually spent by government, and does not change the basic price of the good to the consumer (see Dixon & Rimmer, 2002). We stimulate demand for Air Transport by reducing the indirect tax on that sector.

13.8.3 *Methodology and Data Sources for Estimating Changes in Travel Expenditures*

Expenditure Changes of Land Travelers

When we calculate the economic impact of changed tourist spending resulting from wait time reductions for foreign travelers, we also consider the offsetting impacts resulting from changes in U.S. resident travels. Reduced wait time for returning Americans to reenter the United States from their travels to Canada and Mexico can also result in more trips to these two countries, and thus reduced domestic travel plans or other expenditure domestically by the Americans to offset their increased spending abroad.

To estimate the economic impacts of CBP staffing changes to both the POE region and to the rest of the United States, we first calculate the proportion of direct spending taking place in the POE regions vs. that taking place in other parts of the United States. Due to the lack of data, we use the average of the data from the SANDAG study (5.6%) and the Arizona Office of Tourism study (17.39%), which is 11.5%, for all the other land POEs, to represent the proportion of the direct tourist spending that takes place outside of the POE region by the foreign travelers. In the SANDAG study, the implicit assumption used for the offsetting impact is that roughly 68% of the additional spending from the foregone trips to Mexico of the Americans will take place in San Diego County. We use the same percentage number in our study for POEs that are located in large MSAs, and 50% for other POEs.

The average spending per trip by Canadian and Mexican visitors traveling to the United States is affected by various factors, such as the transportation mode, destination, and length of stay. Since our analysis focuses on eight POEs that spread over three U.S.–Canada border states (NY, WA, MI) and three U.S.–Mexico border states (CA, AZ, TX), we focus the calculation of average spending per visitor on these six states.

The major data source we use to compute average spending by Canadian travelers who visit NY, WA, and MI is the Statistics Canada report *International Travel 2010*. To compute the average per-person visit spending by Mexican visitors to California, we use the California Tourism Industry 2011 report on International Visitor Profiles (Mexico). The calculations for Arizona are largely based on University of Arizona (2008). For Texas, we refined data in report by University of Texas – Pan American (2005).

The major data source we use to compute average spending by American travelers to Canada is the Statistics Canada report *International Travel 2010*. As for the average spending by Americans who travel to Mexico, we use the SANDAG (2006) study. For other southern land POEs, we use the ratio of per capita income in another location to that of San Diego.

In order to distribute the total tourist expenditures among various consumption categories, we first obtained Year 2010 data on tourist expenditures on goods and services from the BEA Travel and Tourism Satellite Account (BEA, 2011) by visitor type. We then mapped the BEA data to the relevant industries in our 86-sector scheme.

Expenditure Changes of Air Travelers

We obtained per-visitor expenditure data by spending category for both oversea leisure and business visitors from the Department of Commerce (U.S. DOC, 2013a). The average foreign leisure visitor spent $3,061 per U.S. trip in 2012, of which about 45% was spent on international airfare. The second largest portion of expenditure is related to shopping, gifts, and other purchases. The average foreign business traveler spent $4,108 per U.S. trip in 2012, of which nearly 60% was spent on international airfare. The second largest expenditure for business travelers was lodging.

For international airfare expenditures, we account for the percentage of service provided by the U.S. air carriers. We compute the market shares of U.S. and foreign carriers for both U.S. citizens/residents and foreign visitors using international

Table 13.7. *Changes in expenditure by U.S. residents resulting from the addition of one customs officer per U.S. air terminal port of entry (in millions of 2011$)*

Spending Change	Lower Bound		Upper Bound	
	Leisure	Business	Leisure	Business
Increased Spending by Foreign Visitors to the United States	$4.7	$0.5	$12.9	$1.2
Total Reduced Domestic Spending (due to Increased U.S. Resident Travel Abroad)	–$3.2	–$0.4	–$14.5	–$1.2
Increased Spending on U.S. Airlines to Travel Abroad	$0.8	$0.1	$3.2	$0.3

airport arrival data provided by CBP for calendar year 2012. Based on the total counts for the four airports, in 2012, 50.6% of the U.S. international air travelers and 34.3% of the foreign visitors chose U.S. airline services. These percentages are applied to the total international airfare expenditures by visitor type to estimate the portions of expenditures that go to the U.S. airlines.

The first row of Table 13.7 presents the estimates on changes in foreign visitor expenditures. When one additional officer is added in each of the 14 terminals, the direct foreign visitor expenditures will increase in the lower-bound scenario by about $4.7 million and $0.5 million for leisure visitors and business travelers, respectively. For the upper-bound case, the direct expenditure change estimates are about 2.7 times the corresponding estimates in the lower-bound case.

Reduction in CBP passport inspection wait times also increases international leisure and business travel by U.S. residents. We assume that, given budget constraints, increased travel expenditures abroad by U.S. residents results in reduced domestic spending of the same amount. Based on the U.S. Department of Commerce report (U.S. DOC, 2013b), the average American leisure traveler spent $2,612 per foreign trip in 2012, while the average business traveler spent $4,538 per foreign trip. The total change in U.S. resident expenditures on foreign travel is calculated by multiplying the expenditures per visitor by the total increases in returning U.S. residents due to wait time reduction.

The second row of Table 13.7 presents the estimate for the total reduction in domestic spending by the U.S. resident travelers who increase their international travel. The third row of Table 13.7 presents the estimate for increased revenue to the U.S. airline industry from the increased international travel by U.S. residents.

We assume that U.S. resident leisure travelers will reduce their domestic spending proportionally across all spending categories to offset their increased international travel expenditures, while U.S. residents traveling abroad for business purposes will reduce only their domestic business travel expenditures as an offset.

13.8.4 *Simulation Results*

Impact Results for Wait Time Changes at Passenger Land POEs

The direct inputs into our analysis consisted first of data on increased Canadian and Mexican travelers and their spending in the United States. The total number of additional passengers projected corresponding to the +1 additional CBP officer scenario was 552,857, ranging from the low of slightly more than 10,000 passengers at each of the Detroit crossings to a high of 65,000 at the Blaine Peace Arch Crossing. Average per-person trip spending was \$176, ranging from a low of \$131 at the two Detroit crossings to a high of \$217 at several southern border crossings. Increased total spending was projected to be \$97.3 million, with a low of slightly less than \$1.5 million at the two Detroit crossings and a high of \$11.6 million at the Bridge of the Americas Crossing near El Paso. We also estimated the increased number of American passengers traveling to Canada and Mexico and their corresponding reduced spending in the United States.

Total GDP impacts in the passenger port regions totaled \$49.7 million, but varied significantly across the crossings. GDP impacts were negative at the San Ysidro, Windsor Tunnel, and Ambassador Bridge crossings, because more American travelers ventured to Canada and Mexico than foreign travelers to the United States, and the spending by American visitors, especially at the northern border crossings, is much greater. The San Ysidro, Windsor Tunnel, and Ambassador Bridge port regions are projected to incur losses of around \$1 million each, while all other port regions would gain, with the Bridge of the Americas crossing gaining the most at \$7.6 million. The largest negative impacts on the rest of the United States are associated with the increased flow of Americans crossing at San Ysidro, and the greatest positive stimulus is associated with a greater influx of travelers from Mexico at the Bridge of the Americas crossing. The net impact on the rest of the United States is \$12.1 million, for a net total of \$61.8 million for all crossings.[25]

Table 13.8 presents the summary results on gross output, GDP, and employment impacts for all the passenger land crossings aggregated together. Adding

Table 13.8. *Summary of economic impacts resulting from adding one customs officer at each of 17 U.S. passenger land ports of entry (POEs)*

Region	Impact Type	Output Impacts (M 2011$)	GDP Impacts (M 2011$)	Employment Impacts (jobs)
All Eight POE Regions	Impacts of Increased Spending by Canadians/Mexicans	145.5	79.4	1,488
	Impacts of Reduced Spending by Americans	–51.3	–29.7	–499
	Net Impacts	94.1	49.7	990
Rest of the United States	Impacts of Increased Spending by Canadians/Mexicans	136.4	72.6	881
	Impacts of Reduced Spending by Americans	–108.9	–60.5	–818
	Net Impacts	27.5	12.1	64
U.S. Total	Impacts of Increased Spending by Canadians/Mexicans	281.9	152.0	2,370
	Impacts of Reduced Spending by Americans	–160.2	–90.2	–1,316
	Net Impacts	121.6	61.8	1,053

one CBP officer at each of the 17 selected passenger land POEs can lead to an increase in gross output of $121.6 million and an increase in GDP of $61.8 million. The employment gains are estimated to be 1,053 jobs. The vast majority of gains are captured by the POE regions themselves. The regional gains account for a high proportion, despite the fact that the multiplier values are higher for national impacts. The main reason for this outcome is the diversion of some spending to other countries stemming from the increased attractiveness of foreign travel to U.S. residents from decreased wait times. This subtraction to yield net impacts is nationwide and not limited to the POE regions. Not surprisingly, the sectors receiving the greatest stimulus are those that provide goods and services to tourists, such as Food Services and Drinking Places, Accommodations, Amusement and Recreation Services, etc.

Impact Results for Wait Time Changes at Airline Passenger POEs
Table 13.9 presents the GDP and employment impacts of each of the four Simulations described in Section 13.8.2, and the aggregate impacts for the four Simulations combined. There is an overall stimulus to the U.S. economy (GDP and jobs) from reducing wait times. Each Simulation reflects a different component of the overall economic impact. When these impacts are combined, there is a net positive impact on the U.S. economy in terms of GDP and employment. For the +1 CBP officer at each POE lower-bound case, the result is a net $3.7 million increase in GDP and 68 additional jobs. The upper-bound case renders increases of $8.3 million in GDP and 106 jobs. The results we present in Table 13.1 represent the average of the lower-bound and upper-bound estimates presented in Table 13.9.

13.9 Study Limitations

Note that our analysis is subject to several caveats and limitations, which can be addressed in future research, regarding the range of impacts that are included and how our results can be used, including:

- Our results pertain to adding one officer at each of the 43 primary inspection sites included in the study. However, the relationship between the change in officers and wait time is not linear and is also not symmetric for additions and subtractions. Scaling the average impact values based on +1 officer scenarios for more than one officer will overestimate the true impact values.[26]
- Our results do not take into account constraints on the ability to deploy extra primary inspection officers to border crossings. In particular, we do not take into account the availability of infrastructure (traffic lanes and/or

Table 13.9. *Summary of economic impacts on the U.S. resulting from adding one customs officer at 14 passport inspection sites at four U.S. international airports*

Simulation	Direct Impact	GDP Impacts (M 2011$)		Employment Impacts (jobs)	
		Lower bound	Upper bound	Lower bound	Upper bound
A	Increased spending by foreign visitors to the United States	$4.0	$10.8	71	192
B	Decreased spending in the United States due to increased U.S. leisure travel abroad	–$0.3	–$2.6	–2	–44
C	Decreased spending in the United States due to increased U.S. business travel abroad	–$1.5	–$6.5	–20	–122
D	Increased spend by U.S. leisure and business travelers on U.S. airlines to travel abroad	$1.5	$6.5	18	79
Total Impact		**$3.7**	**$8.3**	**68**	**106**

inspection booths) that are required to permit an extra officer to conduct inspection. Infrastructure constraints may be unlikely to inhibit the deployment of one extra officer at a border crossing, but they will become more of an issue as greater numbers of officers are added.

- Our results are based on FY 2012 volumes of passenger and commercial vehicles and international flight arrivals entering the United States, and we assume they do not rise or fall due to factors such as continued economic recovery, changes in gasoline prices, and other determinants of travel volumes. We also do not quantify how the number of commercial vehicle trips changes in response to wait time. A longer or shorter wait time may deter or encourage these trips, respectively.
- Our analysis of the value of wait time savings assumes that the value of a minute waited in a queue is the same regardless of how many minutes have been waited. However, the disutility of waiting in a queue may be nonlinear,

and the disutility of an extra minute waited may rise with the total amount of time waited. If disutility is nonlinear, then our methodology underestimates the value of time waited in queues, and the underestimation becomes more severe as the wait time increases.

- Our analysis only applies to short-run impacts.
- We do not quantify how increased passport inspection capacity at an airport might increase the flight-processing capacity of the airport, and thus the scheduling of new flights.
- We do not quantify the formation of new businesses or closing of existing businesses in the border region if wait time falls or rises, respectively. Change in wait time for commercial vehicles may encourage or discourage business formation.
- We do not quantify the impact of lower wait time on cross-border supply chain productivity by, for example, reducing the need to hold inventories, or by capturing or losing economies of scale. Improved supply chain performance would reduce total costs of production.
- Our estimate of distance traveled by freight shipments, which is based on the average location of people in the three countries as represented by their population centroids, may not be a good proxy for the actual average distance traveled.
- Our analysis of international air travelers makes two strong assumptions. First, it requires travelers to be aware of what passport inspection wait times typically are. Second, it requires travelers to take the wait time into account when forming an estimate of the opportunity cost of a trip, which means that wait time is salient and plays an active role in their decision about where and when to travel. However, it might be the case that international air travelers do not take passport inspection waits into account when making travel decisions.
- Our lower-bound estimates for international air travelers are based on the assumption of a standard baggage delivery time of 30 minutes. To the extent that average baggage delivery time is less than this, and/or synchronized with the movement of passengers through passport inspection lines, our lower-bound estimates are too conservative. No data on actual baggage delivery time are available, as these data are kept by airlines and have not been made public.

13.10 Conclusions and Possibilities for Future Research

Our research has quantified the economic benefits of adding primary inspection officers to U.S. land border crossings and airports that process large traffic flows, and our results suggest that primary inspection wait times at U.S. border crossings have significant economic impacts, and that the addition of a primary inspection officer at the typical large-volume border crossing brings about significant economic benefits.

Decisions on the overall level of CBP staffing and allocation of staffing across inspection sites should ultimately be informed by benefit-cost analyses, and this requires comparing the benefits of extra staff to the cost of extra staff. To support optimal decisions, analysis will need to describe the cost and benefit schedules associated with deploying a wider range of extra officers. Development of these schedules will require explicitly taking into consideration infrastructure constraints. Our research is thus a first step toward a more complete cost-benefit framework that would prove of greatest utility to decision makers.

We also recommend that future studies address the outcomes of border inspections that are related to CBP's law enforcement missions, which are the justification for imposing inspections of arrivals to the United States. Using data on enforcement outcomes, it would be possible to analyze the enforcement returns to primary inspections. Enforcement outcomes include interdiction of illegal drugs, counterfeit goods and other contraband, and apprehension of unauthorized border crossers and those wanted on criminal charges. The return to the U.S. public of these outcomes can potentially be monetized by developing valuations of the consequences of violation of relevant laws.[27]

If both of these research directions are pursued, then it might be possible to quantitatively characterize all of the key tradeoffs that characterize a border crossing and the relevant marginal cost and benefit schedules associated with them. This could provide important support to making optimal decisions on border-crossing resource allocation and deployment policies.

NOTES

1. Available commercial vehicle data do not distinguish the total number of commercial vehicle trips through regular lanes and FAST lanes. We analyzed regular lane wait times, and used the total number of trips and lanes as proxies for the number of regular lane trips and regular lanes open, respectively.
2. CBP staffs inspection processes at all U.S. border crossings but generally does not own the infrastructure (buildings, bridges, etc.) comprising a crossing. Infrastructure can be owned by the federal government, state or local governments, or private owners.
3. GDP measures the value of all final goods and services produced in the economy that are transacted through markets. It is a net measure of economic production because it does not include intermediate goods (goods to produce other goods). Thus it differs from gross output, which corresponds to total sales revenue, and which does double-count intermediate goods. We also emphasized that GDP

and employment are measured in two different units and therefore should not be added together to arrive at a single indicator of macroeconomic impacts.

4. Three of the airports have multiple sites at which passport control is carried out. We conducted analysis for all sites.
5. The selected land crossings account for 68% of the total passenger vehicle traffic crossing U.S. land borders, and 79% of the total wait time experienced by passenger vehicles.
6. Regression analysis would require identifying suitable instruments to control for simultaneity, and no instruments have yet been found.
7. We cannot use results from queuing theory to derive a relationship, because queuing theory is only useful for systems that are not "saturated," so that queues are not building up in a deterministic fashion over time. The dominant patterns of queue build-ups and build-downs at border crossings are rush hour patterns, which are deterministic queuing outcomes.
8. We used the theoretical formula to predict how wait time should have changed at San Ysidro if a 35% increase in staffing took place on the 2012 survey dates and found that the predictions were very close to how wait time actually changed.
9. This value has been estimated in the transportation economics literature and is recommended by Robinson (2007) to adjust for disutility of waiting in border-crossing queues.
10. For passengers crossing the northern border, data from Statistics Canada (2011) on Canadian and American nationals crossing the border by automobile are used. For travel across the southern border, the only available evidence available on trip distributions is from a survey of several thousand travelers crossing in the San Diego region in 2004 and 2005 that supported the SANDAG (2006) study.
11. Some U.S. citizens and legal permanent residents actually reside in Mexico or Canada, and the increase in the number of trips they make into the United States would stimulate economic activity in the United States rather than their country of residence. In border regions where such populations are quite significant, the estimates developed here will significantly understate the impact on the U.S. economy of lowered wait time.
12. We note that when wait time falls, the resulting increase in trips across the border will produce an increase in wait time that offsets the initial reduction in wait time; this will in turn reduce the initial increase in traffic, and so forth. We do not attempt to control for this "rebound" effect in our analysis, but do note that the results for the San Ysidro experiment reflect this effect.
13. The queuing systems are identical. For commercial vehicles, it would always be optimal for CBP to add an extra officer to regular lanes rather than to FAST lanes, which are the trusted-traveler equivalent for commercial vehicles, so we present results only for regular lanes.
14. Very high levels of wait time could also have longer-run impacts on decisions about where to locate production plants and supply chain activity.
15. As one would expect a priori, average passport inspection wait time levels are systematically higher for nonimmigrant foreign nationals than for U.S. citizens and legal permanent residents.
16. These weights are constructed from data obtained from Office of Travel and Tourism Industries (2012).
17. We do not quantify any increase in airport capacity to process flights that might result from increased primary inspection capacity, and any related increase in the number of flights and passengers arriving at the airport.

18. Unless otherwise noted, all values cited in this section are given in 2011 dollars and are per truck trip. Inflation adjustments are made using the Producer Price Index (PPI-all commodities), and conversions from Canadian to U.S. dollars are made using the Bank of Canada's (2013) currency converter (www.bankofcanada.ca/rates/exchange/).
19. Data on the locations and populations of various subnational administrative units (in the year 2000) were obtained from NASA's Socioeconomic Data and Applications Center (SEDAC, 2005). A large portion of Canada's population is clustered far to the south, and combined with Canada's notable "dip" to the south in the province of Ontario, this results in the Canadian centroid being located in the state of Michigan.
20. The average location of people is an imperfect proxy for the movement of goods. Moreover, freight movements travel by road and likely deviate from the straight-line path.
21. See Taylor and colleagues (2003) for an analysis of the U.S.–Canada border.
22. Free on board (FOB) value does not include the cost of shipping and insurance. The seller is required to deliver merchandise on board of a vessel specified by the buyer.
23. The GTAP model has three transport industries: Other Transport, Water Transport, and Air Transport. In order to model a scenario that impacts only the truck transport industry, it is necessary to disaggregate the Other Transport industry into a truck-transport component and a non-truck-transport component (which contains rail transport, pipelines, auxiliary transport activities, and travel agencies). We draw on external data sources such as GTAP (2012) and also use the sectoring bridge between the U.S. I-O table and GTAP to break the Other Transport industry into these two components.
24. Note that when a previously endogenous variable like Exports of Air Transport is made exogenous, so that it can be "shocked," the number of equations and unknowns must be rebalanced by making a previously exogenous variable endogenous. In this case, the small country assumption is relaxed for the Air Transport sector to enable the model to solve feasibly. This approach is used for Simulation A only.
25. The 2006 SANDAG study analyzed the economic impacts due to the total delay times at the San Diego–Baja California border, on both the U.S. economy and the Mexican economy, based on a 2004 survey of roughly 3,500 cross-border travelers in the San Diego area. According to the survey, only 30% of the passengers who entered the San Ysidro POE were American residents. In our study, if we use the SANDAG assumption that 30% of passengers are American residents who went to Mexico and are coming back, rather than the 76% that is based on the CBP FY 2012 data, the GDP impacts for San Ysidro would change from -$5.4 M to $18.0 M, and the employment impacts would change from 71 job losses to 240 job gains. Total GDP impacts for 17 land POEs would increase from $61.8 M to $85.2 M, or an increase of 38%. Total employment impacts for 17 land POEs would increase from 1,053 job gains to 1,365 job gains, or an increase of 30%.
26. The marginal impact of extra officers on wait time declines with the number of officers deployed to a crossing. The first additional officer will reduce wait time the most, because the officer will be assigned to the most congested hours of the day. The second additional officer will reduce wait time somewhat less, because the most congested hours of the day that this officer will be assigned to will have somewhat lower wait time due to the addition of the first officer, and so on. At

some point, if enough officers are deployed to a border crossing, and if enough infrastructure is available to permit their deployment, wait time will equal zero for all hours of the year.

27. Our preliminary analyses demonstrate that at the San Ysidro border crossing, the average amount of time in which a primary inspection of passenger vehicle is carried out falls as wait time rises, and that as the average inspection time falls, the percentage of passenger vehicles referred to secondary inspection also falls. These results suggest that as the processing volume rises at a border crossing, pressure builds on primary inspection officers to speed up inspection to keep queues at acceptable lengths, and that this has an impact on enforcement outcomes. We recommend additional statistical analysis to convincingly identify relationships between processing volume, wait time, and enforcement outcomes.

REFERENCES

American Transportation Research Institute (ATRI) (2011). An analysis of the operational costs of trucking: 2011 Update. www.atri-online.org/research/results/Op_Costs_2011_Update_one_page_summary.pdf.

Arvis, J.-F., Mustra, M., Ojala, L., Shepherd, B., and Saslavsky, D. (2012). Connecting to Compete 2012 – Trade Logistics in the Global Economy – The Logistics Performance Index and Its Indicators. International Trade Department, The World Bank, Washington, DC.

Avetisyan, M., & Hertel, T. (2015). Impacts of logistics improvement on international trade and modal choice. Working paper, Texas Tech University.

Avetisyan, M., Heatwole, N., Rose, A., & Roberts, B. (2015). Competitiveness and macroeconomic impacts of reduced wait times at U.S. land freight border crossings. *Transportation Research A 78*, 84–101.

Boyer, K. D. (1997). American trucking, NAFTA, and the cost of distance. *Annals of the American Academy of Political and Social Science 553*, 55–65.

California Tourism Industry (2011). *International Visitor Profiles (Mexico)*. http://industry.visitcalifornia.com/Industry/Research/InternationalTravelandTourismStatistics/.

Dixon, P., & Rimmer, M. T. (2002). *Dynamic general equilibrium modeling for forecasting and policy: A practical guide and documentation of MONASH*. Amsterdam, The Netherlands: North-Holland.

Djankov, S., Freund, C., & Pham, C. S. (2010). Trading on time. *The Review of Economics and Statistics 92*(1), 166–173.

Ellis, D. (2009). Cost per hour and value of time calculations for passenger vehicles and commercial trucks for use in the urban mobility report. College Station: Texas Transportation Institute.

Ghaddar, S., & Brown, C. J. (2005). *The economic Impact of Mexican visitors along the U.S.–Mexico border: A research synthesis.*

Global Trade Analysis Project (GTAP) Model (2012). www.gtap.agecon.purdue.edu/models/current.asp.

Gould, T. (2012). *President, Tom Gould Customs*, Los Angeles, CA. Personal communication.

Haralambides, H. E., & Londono-Kent, M. P. (2004). Supply chain bottlenecks: Border crossing inefficiencies between Mexico and the United States. *International Journal of Transport Economics, 31*(2), 171–183.

Hummels, D. L., & Schaur, G. (2013). Time as a trade barrier. *The American Economic Review, 103*(7), 2935–2959.

Litman, T. (2012). Understanding transport demands and elasticities: How prices and other factors affect travel behavior. Victoria Transport Policy Institute.

Madsen, B. & Zhang, J. (2010). Towards a new framework for accounting and modeling the regional and local impacts of tourism. *Economic Systems Research, 22*(4), 313–340.

Miller, R., & Blair, P. (2009). *Input-output analysis: foundations and extensions, 2nd Ed.* Englewood Cliffs, NJ: Prentice-Hall.

Minnesota Department of Transportation (MDOT) (2003). *The per-mile costs of operating automobiles and trucks.* Report no. MN/RC 2003–19. G Barnes and P Langworthy, authors.

Minor, P., & Tsigas, M. (2008). Impacts of better trade facilitation in developing countries: analysis with a new GTAP database for the value of time in trade. Nathan Associates Inc. www.gtap.agecon.purdue.edu/resources/download/4036.pdf.

Multihazard Mitigation Council (MMC). (2005). *Natural hazard mitigation saves: Independent study to assess the future benefits of hazard mitigation activities.* Report to the Federal Emergency Management Agency by the Applied Technology Council under contract to the Multihazard Mitigation Council.

MyTravelCost.com. (2013). Petrol prices around the world. www.mytravelcost.com/petrol-prices/, accessed on January 26, 2013.

National Cooperative Highway Research Program (NCHRP). (1999). *Valuation of travel-time savings and predictability in congested conditions for highway user-cost estimation.* Report 431. Washington, DC: National Academy Press.

North American Transportation Statistics Database (NATSD). (2012). Joint venture of various U.S., Mexico, and Canada government agencies. http://nats.sct.gob.mx/index_en.html.

OECD. (2003). Quantitative assessments of the benefits of trade facilitation. Paris.

Office of Travel and Tourism Industries. (2012). Profile of overseas travelers to the United States: 2011 inbound." U.S. Department of Commerce, International Trade Administration.

Office of the U.S. Trade Representative. (2012). *North American Free Trade Agreement (NAFTA).* www.ustr.gov/trade-agreements/free-trade-agreements/north-american-free-trade-agreement-nafta.

Oladosu, G., & Rose, A. (2007). Income distribution impacts of climate change mitigation policy in the Susquehanna River Basin. *Energy Economics, 29*(4), 520–544.

Oladosu, G., Rose, A., & Lee, B. (2013). Economic impacts of potential foot and mouth disease agro-terrorism in the United States: A computable general equilibrium analysis. *Journal of Biothreats and Biodefense*, S12: 001.

Pavlakovich-Kochi, V., & Charney, A. H. (2008). *Mexican visitors to Arizona: Visitor characteristics and economic impacts 2007–08.* Report to the Arizona Office of Tourism.

Prager, F. (2013). The economic and political impacts of U.S. federal carbon emissions trading policy across households, sectors and states. Unpublished PhD dissertation, Price School of Public Policy, University of Southern California, Los Angeles, CA.

Prager, F., Rose, A., Wei, D., Roberts, B., & Baschnagel, C. (2015). Economy-wide impacts of reduced wait times at U.S. international airports. *Research in Transportation Business and Management 16*, 112–120.

Roberts, B., Heatwole, N., Wei, D., Avetisyan, M., Chan, O., Rose, A., & Maya, I. (2013). *The impact on the U.S. economy of changes in wait times at ports of entry.* Report to U.S. Customs and Border Protection, Center for Risk and Economic Analysis of Terrorism Events (CREATE), University of Southern California.

Roberts, B., McGonegal, S., Prager, F., Wei, D., Rose, A., Baschnagel, C., Beggs, T., & Baghelai, O. (2014). *Analysis of primary inspection wait time at U.S. ports of entry*. Report to U.S. Customs and Border Protection, Center for Risk and Economic Analysis of Terrorism Events (CREATE), University of Southern California.

Robinson, L. (2007). Valuing travelers' time for border crossing and related Activities. Final report prepared for Customs and Border Protection, Department of Homeland Security.

Rose, A. (1995). Input-output economics and computable general equilibrium models. *Structural Change and Economic Dynamics*, *6*, 295–304.

Rose, A., Avetisyan, M., & Chatterjee, S. (2014). A framework for analyzing the economic tradeoffs between urban commerce and security. *Risk Analysis*, *34*(5), 1554–1579.

Rose, A., Oladosu, G., Lee, B., & Beeler Asay, G. (2009). The economic impacts of the 2001 terrorist attacks on the World Trade Center: A computable general equilibrium analysis. *Peace Economics, Peace Science, and Public Policy*, *15*, Article 6.

Rose, A., Oladosu, G., & Liao, S. (2007). Business interruption impacts of a terrorist attack on the electric power system of Los Angeles: customer resilience to a total blackout. *Risk Analysis*, *27*, 13–31.

San Diego Association of Governments (SANDAG). (2006). *Economic impacts of wait times at the San Diego–Baja California border*.

Schrank, D., Lomax, T., & Eisele, B. (2011). *2011 urban mobility report*. College Station: Texas Transportation Institute.

Socioeconomic Data and Applications Center (SEDAC) (2005). Gridded population of the world (GPW) – centroids. Version 3. U.S. National Aeronautics and Space Administration. http://sedac.ciesin.columbia.edu/data/set/gpw-v3-centroids/.

Statistics Canada. (2010). *Canadian vehicle survey: Annual – 2009*. Ottawa: Statistics Canada.

Statistics Canada. (2011). *International travel 2010*. www5.statcan.gc.ca/access_acces/alternative_alternatif.action?l=eng&loc=http://www.statcan.gc.ca/pub/66-201-x/66-201-x2010000-eng.pdf&t=International%20Travel.

Statistics Canada. (2012). Earnings, average hourly for hourly paid employees, by industry. www.statcan.gc.ca/start-debut-eng.html.

Taylor J., Robideaux, D., & Jackson, G. (2003). The U.S.–Canada border: Cost impacts, causes, and short to long term management options. Washington, DC: U.S. Department of Transportation.

Transport Canada. (2005). *Operating costs of trucks in Canada – 2005*. Ottawa: Transport Canada.

U.S. Bureau of Economic Analysis (BEA). (2012). *Travel and tourism satellite account*. http://tinet.ita.doc.gov/research/programs/satellite/index.html.

U.S. Bureau of Labor Statistics (BLS). (2012). May 2011 national occupational employment and wage estimates. www.bls.gov/oes/current/oes_nat.htm.

U.S. Bureau of Transportation Statistics (BTS). (2012). Border crossing/entry data: Query detailed statistics. http://transborder.bts.gov/programs/international/transborder/TBDR_BC/TBDR_BCQ.html.

U.S. Department of Transportation (U.S. DOT). (2011). Revised departmental guidance on valuation of travel time in economic analysis. memorandum.

U.S. National Hurricane Center (NHC). (2010). Latitude/longitude distance calculator. www.nhc.noaa.gov/gccalc.shtml.

Wardman, M. (2012). Review and meta-analysis of U.K. time elasticities of travel demand. *Transportation, 39*, 465–490.

World Bank. (2013). *World databank*. http://databank.worldbank.org/data/home.aspx.

14

Organizational Decision Processes

Gregory S. Parnell and Barry Ezell

"Ultimately, a company's value is just the sum of the decisions it makes and executes." (Blenko et al., 2010)

14.1 Introduction

As the opening quote illustrates, decision making and execution are two important organizational capabilities. This is equally true for corporations and government organizations. Senior decision makers establish the mission and vision of the organization. Decision makers have the authority and responsibility to determine the strategies to perform the mission and strive toward the vision of the organization. In addition, these decision makers determine the processes that the organization uses to make decisions. This chapter focuses on the organization's decision process and identifies the attributes of a quality decision and quality organization decision processes.

Why does an organization need a decision process? First, strategic decisions are challenging. Many of the most important strategic decisions are challenging because they involve conflicting objectives of decision makers and stakeholders, complex alternatives, and many uncertainties about the future. Second, many stakeholders are involved in the decision or are impacted by the execution of the decision. The interaction with decision makers and stakeholders is critical to making informed decisions. Third, decision execution may be more successful if the implementers are involved in the decision process.

In this chapter, we view the decision process using concepts from the field of decision analysis. Decision analysis is "a philosophy and a social-technical process to *create value for decision makers and stakeholders* facing difficult decisions involving multiple stakeholders, multiple (possibly conflicting) objectives, complex alternatives, important uncertainties, and significant consequences. Decision analysis is founded on an axiomatic decision theory and uses insights from the study of decision making" (Parnell et al., 2013). The purpose of decision analysis is to provide insights to decision makers. Decision processes are the enterprise/organizational processes designed to provide the essential information needed for fact-based decision making.

This chapter is organized as follows. First, we distinguish between decision quality and organizational decision quality. Second, we describe the attributes of a decision-driven organization that makes quality decisions. Third, we describe the organizational decision quality criteria used for the Society for Decision Professional's Raiffa-Howard Award. Fourth, we present a state homeland security decision process and discuss how the decision process was structured, the interaction with decision makers and stakeholders, and how the results were used. Finally, we summarize this chapter.

14.2 Decision Quality and Organizational Decision Quality

Before we begin the discussion of decision processes, we need to introduce some basic concepts: decisions, outcomes, decision quality, and organizational decision quality. Decision quality is an assessment of each decision and organizational decision quality is an assessment of the decision processes of the organization or enterprise. Decision analysis defines decision quality at the time of the decision and *not* by the decision outcome.

14.2.1 *Decisions versus Outcomes*

A decision is an irrevocable allocation of resources (Howard, 1988). At the time of the decision, there are usually implementation uncertainties. An outcome is the result of a decision after the uncertainties have been resolved. Based on our preferences, the outcome may be good or bad. Figure 14.1 shows the four basic possibilities. We can illustrate using a driving example. If we are well rested and avoid distractions (good decision), we expect to arrive at our destination safely (good outcome). However, we could still have an accident (bad outcome) due to another driver's actions or bad weather. Likewise, if we drive under the influence of alcohol and/or use our cell phone excessively, we might expect to have an accident (bad outcome). However, we may be lucky and make it home safely (good outcome).

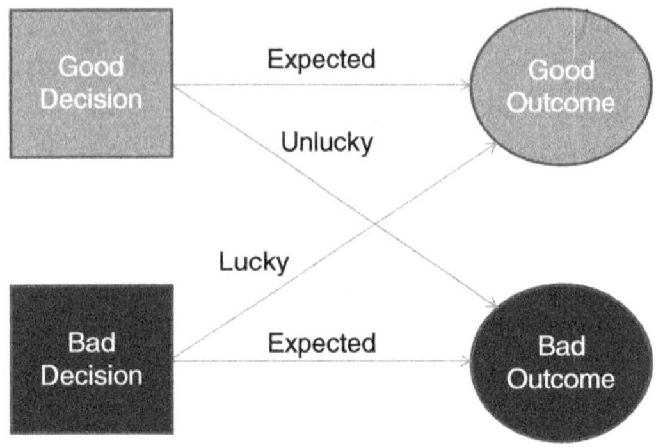

Figure 14.1. Decision versus outcome.

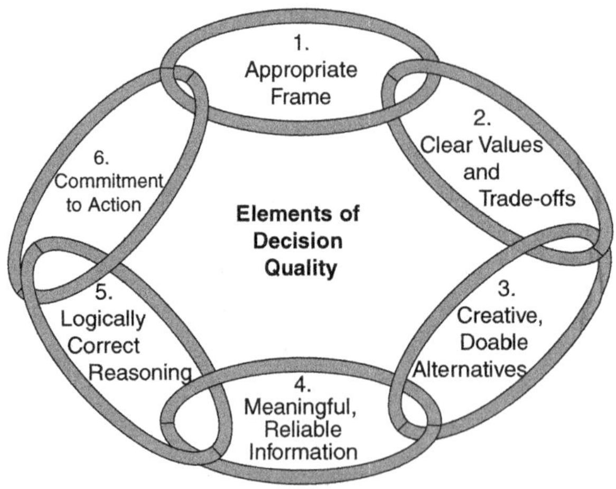

Figure 14.2. Six elements of decision quality (modified from Matheson & Matheson, 1998).

14.2.2 *Decision Quality*

A good decision can be described with a Decision Quality Chain (Matheson & Matheson, 1998). The chain is shown in Figure 14.2. (The order has been changed from the original order to place identification of values before the development of alternatives. This reflects an important concept from Value-Focused Thinking (Keeney, 1992) that we will develop a better set of alternatives if we begin by thinking about our values and objectives and then generate alternatives.) A good decision requires high quality in each of six essential elements: an appropriate frame; clear values and trade-offs; creative, doable alternatives; meaningful, reliable information; logically correct reasoning; and commitment to action. The chain symbolism implies that decision quality is as good as the weakest link. Next, we describe each link in the chain.

1. **Frame** – The frame is how we view the decision opportunity or problem. A good decision requires an appropriate frame that clearly specifies the purpose of the decision, the scope of the decision, and the perspectives of all the decision maker(s) and stakeholders. The best way to construct a good frame is to communicate with the decision makers and stakeholders using interviews, focus groups, and/or surveys.
2. **Values** – A good decision requires value metrics for comparing possible outcomes of the decision to be clearly specified. The value metrics should represent the decision makers' and stakeholders' preferences and the trade-offs between competing values should be explicitly stated. For corporations, the value metrics may focus primarily on economic value. However, social and environmental objectives are becoming increasingly important. For public programs, a wider range of value metrics is usually required to meet the needs of stakeholders.
3. **Alternatives** – A good decision requires a set of diverse alternatives that offer the potential to create value for shareholders and stakeholders, each of which is feasible and compelling and which together span the full range of possible actions. Limiting the quality and number of alternatives limits the potential value of the decision.
4. **Information** – A good decision requires meaningful and reliable information, which usually includes identifying the major uncertainties and, in decision analysis and risk analysis, properly identifying future scenarios and/or assessing probability distributions. The ideal here is not to collect *all* of the information that might be relevant, but, rather, to collect information only up to the point where the cost of obtaining additional information exceeds the value of that information for making the decision.
5. **Logical reasoning** – A good decision uses logically correct reasoning based on sound mathematics. For decision analysis, the axioms of decision theory provide a sound mathematical foundation. The use of decision techniques that are not based on sound mathematics is a fundamental flaw.
6. **Commitment to action** – A good decision is one that the decision makers and stakeholders are prepared to execute in a timely manner. Without commitment to action, we may have clarity of intention, but we have very little chance of achieving the potential value of the decision. Decision analysts have found that commitment to action is enhanced by early and active participation of the key stakeholders (including decision implementers) in the decision process.

14.2.3 *The Decision-Driven Organization*

Bain & Company introduce the concept of the Decision-Driven Organization and emphasize the importance of decision making. They rate Decision Effectiveness using a questionnaire with "responses to assess decision quality (whether decisions proved to be right more often than not), speed (whether decisions were made faster or slower than competitors), yield (how well decisions were translated into action), and effort (the time, trouble, and expense required for each key decision)" (Blenko et al., 2010). Decision Effectiveness is descriptive and focuses primarily on decision outcomes. Only one of the elements (effort) can be assessed at the time of the decision.

A more useful idea to develop quality decisions is the Decision-Driven Organization. The Decision-Driven Organization recommends six steps that focus on decision making and not reorganization as the key to improving organizational performance (Blenko et al., 2010):

1. **Identify your organization's key decisions.** One of the first steps is the identification of key decisions. The main idea is making decision identification a deliberate step and not an ad hoc process.
2. **Determine where in the organization those decisions should happen.** The organizational strategy, the type of decision (product development, operations, etc.), and the planning horizon (long range, midrange, or short range) will be factors in determining where and at what level in the organization the decision should be made.
3. **Organize the macrostructure around sources of value.** The organization should begin by understanding the sources of value and the decisions that create value. Some customer segments may be providing more value than other segments. It may be possible to reduce the low-value segments and increase the high-value segments.
4. **Figure out what level of authority decision makers need.** Reorganizations are not the only way to implement strategy. It is more important that decision makers are given the responsibility and the authority to make the decisions and implement these decisions.
5. **Align other elements of the organizational system, such as incentives, information flow, and processes, with those related to decision making.** Once the decisions have been identified and the decision makers empowered, the next step is to design incentives for managers and workers, information flow (from people and information systems), and processes to support the decision makers.

6. **Help managers develop the skills and behaviors necessary to make and execute decisions quickly and well.** Education and training are essential to develop the skills and behaviors of managers. Successful companies have extensive training programs for all levels of management.

14.2.4 *Organizational Decision Quality*

Organizational decision quality is an assessment of the organization's decision-making culture and practice. According to Spetzler (2007), an enterprise has Organizational Decision Quality (ODQ) if it meets the following criteria:

- "It routinely makes high quality decisions. It correctly recognizes and declares decisions, issues, and challenges; it understands the behavioral traps inherent in natural decision-making and knows how to compensate for them; it frames decisions appropriately; and, it addresses decision making with the balance of content, analytic rigor, and facilitation necessary to pursue the actions most likely to deliver optimal value.
- "It has a broad array of tools, techniques, and processes in place and decision makers and staff are skilled in using them appropriately.
- "Its decision makers understand their roles and have thoroughly developed the knowledge and skills their roles demand.
- "The organization is well aligned internally. It has a common language for and a solid understanding of what constitutes decision quality; it has a strong common desire to achieve it on a regular basis, and it 'walks the talk.'
- "The organization continually endeavors to learn and improve its decision competency."

14.3 Enabling Organizational Decision Quality

ODQ has four objectives. First, an organization should use appropriate decision processes: high-quality decisions with appropriate decision processes focused on value. Second, the organization should invest in decision education and training. The organization should have a broad array of tools, techniques, and processes with skilled DMs and staff. Third, the decision processes should be aligned with other organizational processes. Finally, the organization should have processes that support learning and continuous improvement. We will discuss each of these objectives in the following sections.

14.3.1 *Appropriate Decision Processes*

ODQ results in high-quality decisions. We have already defined quality decisions. What kind of decision processes contribute to ODQ? Parnell and colleagues (2013) identify the importance of selecting the appropriate decision-analysis process based on content, analytic, and organizational complexities. Figure 14.3 presents four decision processes (Parnell et al., 2013) that are commonly used. Parnell and colleagues recommend two best-practice decision processes (Dialogue Decision Process and Decision Conferencing) and identify two flawed processes (Strictly Analytical Process and the Advocacy Process). The Dialogue Decision Process and Decision Conferencing are recommended because they appropriately engage the decision makers and the stakeholders in the decision process for the content, analytic, and organizational complexities.

The Dialogue Decision Process (Spetzler, 2007) has been successfully applied within hundreds of organizations to achieve good decision making in the face of content, analytic, and organizational complexities. This process centers on a structured dialogue between two teams: 1) a team of decision makers, the Decision Board; and 2) a team created to assist the decision makers, the Decision Team. The structured dialogue occurs in the following planned interactions.

1. **Framing**. At this first meeting, the decision problem or opportunity frame is presented based on research, interviews, focus groups, and surveys of decision makers and stakeholders. The decision board approves the frame.
2. **Values**. After the frame is approved, the decision team meets with the stakeholders to identify the value metrics to be used to evaluate the alternatives. At the second meeting, the value metrics are approved.
3. **Alternatives**. After the values are approved, the decision team works with stakeholders and subject matter experts to develop a set of creative, doable alternatives that span the set of possible decisions. The alternative set is presented and approved at the third meeting.
4. **Evaluation**. To prepare for the next meeting, the alternatives are evaluated using the value metrics. This step may require the uses of models and simulations. After the initial alternative evaluation, the decision team revises the alternatives to create hybrid alternatives (alternatives that combine the features of two or more alternatives) that create more potential value than the initial set of alternatives. The evaluation of the original alternatives and the hybrid alternatives is presented to the decision makers at the fourth meeting.
5. **Decision**. If the decision makers are satisfied with the frame, values, alternatives, and evaluation, a decision is made and implementation begins. If

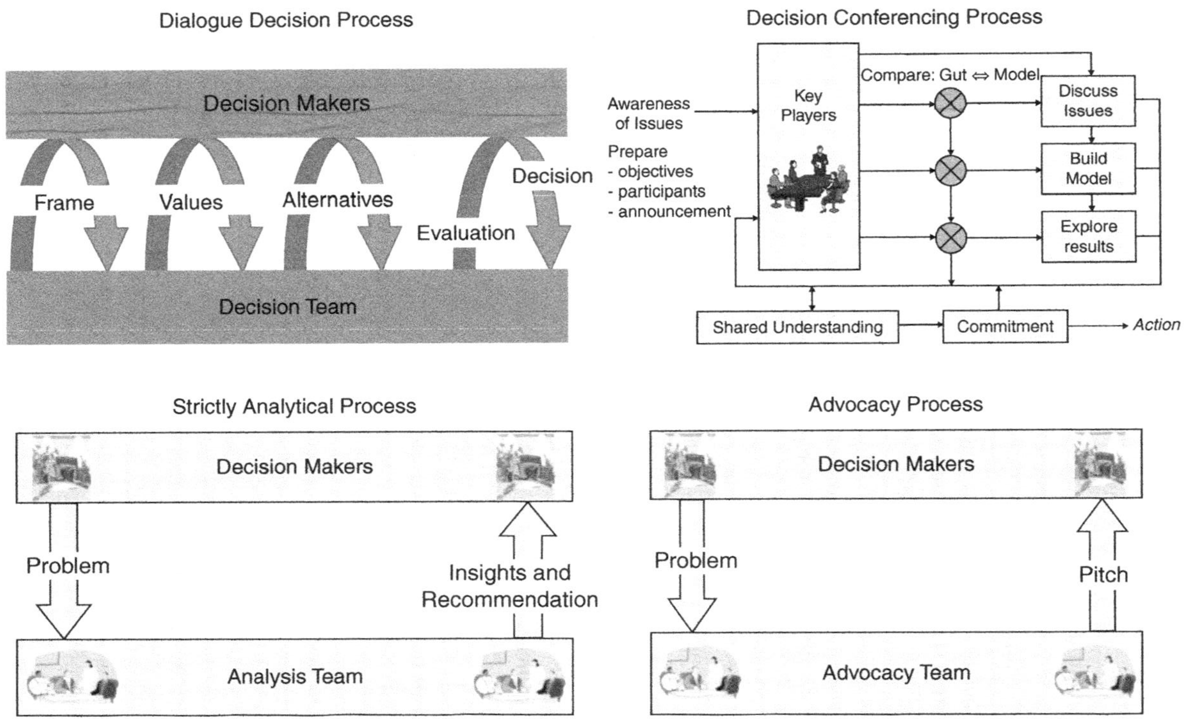

Figure 14.3. Four common decision processes.

not, one or more of the steps are repeated. If appropriate, an additional meeting may be held to review the implementation plan.

The Dialogue Decision process is designed to meet the requirements of a quality decision and uses some of the same terms. This process is appropriate for decisions with significant content, analytic, and organizational complexity. However, the process requires training of the decision team members and the decision board. In addition, it requires time and management discipline to enforce the process on major decisions.

Another best-practice decision process is Decision Conferencing (Parnell et al., 2013). Decision conferencing is most useful for decision situations in which there is a clear sense of urgency so that all key participants are willing to devote the time and effort required and in which the chief challenge is in gaining the agreement and commitment of the decision makers rather than in overcoming difficult content and analytic challenges.

In a Decision Conference, decision makers and stakeholders work together in an intensive two-to-three-day session under the guidance of skilled facilitators to identify key issues, build models, evaluate alternatives, reach decisions, and plan implementation. Time permitting, preparation for the meeting can include identifying key issues, defining objectives, and developing a value model to help evaluate and compare alternatives. The remaining required information is elicited directly from the participants rather than derived from external data sources.

Phillips (2007) identifies three reasons why Decision Conferencing is valuable: 1) it helps decision makers and stakeholders create a shared understanding of the issues; 2) it develops a sense of common purpose among the participants; and 3) it fosters a commitment among the participants on the way forward.

The Decision Conference process requires a trained facilitator(s) in order to keep the key decision makers and stakeholders focused on the problem, develop a credible model, reach a decision, and obtain an implementable solution.

Next, we describe two common types of decision processes that have major flaws. The focus of a strictly analytical process is on the analytic complexity of the decisions and not the content and organizational complexity. Typically, this approach uses complex analytic models to compare the existing alternatives. Generation of additional alternatives may or may not receive appropriate attention. Interaction with decision makers and key stakeholders typically occurs only at the start and end of the process. The final presentation provides analytic insights about the alternatives. But since the decision makers and stakeholders have not participated in the process (organizational complexity), they may raise issues or alternatives that were not considered in the analysis (content complexity). Furthermore, they may not be willing to accept the results of an analysis that they consider a "black box" and do not understand. This process has some unique flaws. First, the analysts tend to focus on the model (analytic complexity) and not the problem (content complexity) or the organizational issues (organizational complexity). Second, features of the problem that are not captured in the model tend to be ignored.

The second flawed type of decision process is the advocacy process. In this process, an advocate (who may be just one person or a group of people) prepares a presentation to the decision makers with the goal of convincing them to undertake a particular course of action. The advocate is a "champion" of the recommended course of action, either because he or she believes that it is best for the organization or he or she has been told to recommend this alternative by a senior leader. The advocacy process has the benefit that this advocate is highly motivated to achieve successful implementation.

The advocacy process shares with the strictly analytical process the flaw that decision makers are presented with a recommendation that they were not involved in exploring and evaluating. Their role in the advocacy process is to try to identify the weaknesses of the recommended course of action and to either accept or reject the "pitch." The advocacy process has three additional unique flaws. First, the features of the problem that the advocated alternative does not address are not included in the presentation (content complexity). Second, possible alternatives to the recommended course of action are generally not presented for consideration to avoid unwanted distraction from the advocated course of action. Third, the analysis supporting the advocated actions tends to focus on its upside and minimize its downside, with information that is unfavorable to the recommendation being downplayed or omitted.

14.3.2 *Education and Training*

Education and training are critical enablers in the decision-driven organization to achieve organizational decision quality. Every organization requires education in the technical, management, and financial professions that are critical to the organization's mission. Without competent people, mission accomplishment is at risk. However, professional training in decision making, decision techniques, and decision processes is essential for an organization to achieve the ODQ objectives. The training should include courses for decision makers, technical experts, and analysts.

The senior decision makers must understand the decision processes that are required for quality decisions, the resources required to perform these processes, and the types of analytical products from these processes. Technical experts must also understand the types of processes, their level of participation, and the data requirements that they are responsible for in each process. The decision team members must understand and be able to perform the techniques required to provide the analysis to support a quality decision.

14.3.3 *Alignment with Organizational Processes*

Decision processes should be aligned with the business processes of the organization. Three major business processes should be considered: strategic planning, project management, and budgeting.

Strategic planning is the process senior leaders use to make resource allocation decisions to achieve the mission and strive to move toward the vision of the organization. Increasingly, organizations are considering the Triple Bottom Line (profit, people, and planet) (Elkington, 1997) in their strategic planning. The development of strategy and the strategic decisions to implement strategy require sound decision processes. Many times decision makers begin their strategy process with a SWOT (Strengths, Weaknesses, Opportunities, and Threats) analysis. In addition, organizations should consider a decision audit (Blenko et al., 2010). The identification of the strategic decisions to be made is an important first step in the strategy process.

One of the most important strategic organizational processes is project management. Project management is a key process used to manage a one-of-a-kind activity. Projects are used to meet strategic objectives (Meredith et al., 2014). Most strategic changes are implemented with projects. Project management is critical to organization success. Let's consider a major oil-and-gas example. "ChevronTexaco has spent much time and resources developing … a world-class project system. The roots of the Chevron Texaco Project Development and Execution Process (CPDEP) go back to the 1980s, when the company had what it thought were good processes and good projects. It started looking at ways to determine if these projects were in fact leading projects. Chevron's CEO at the time, after seeing the results of benchmarking, said it was a very dark day whenever the project we all felt was stellar turned out to be one that was very poor compared with industry averages" (Gregory, 2002).

"In 1997 Chevron formed a group called Project Resources Company. … Within Project Resources, we have project professionals assigned to specific projects within strategic business units. Project Resources also has a shared services group made up of professionals who consult with project teams and

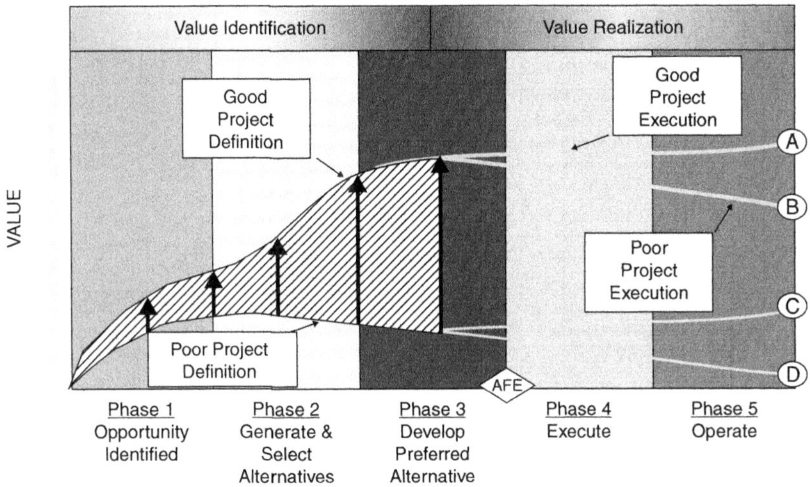

Figure 14.4. Value creation in CPDEP.

business units to assist in meeting their objectives. Project Resources is the owner of the CPDEP process, entrusted to create and deploy tools and systems across the company" (Gregory, 2002).

The steps in the CPDEP process are shown in Figure 14.4 (Lavingia). The process is divided into two parts: value identification and value realization. Decision analysis plays an important role in the three phases of value identification: identify opportunity; generate and select alternatives; and develop the preferred opportunity. The chart makes two important points. First, good project definition can result in a preferred alternative with high potential value. Second, good project execution is required to realize the value.

Chevron uses project management to manage all five phases of the CPDEP (Decision Making in an Uncertain World: A Chevron Case Study). The CPDEP is a stage-and-gate process. Each phase has primary operations activities that produce information; clearly defined deliverables; a decision to proceed, exit, or recycle. The study shows how "Chevron has progressed from a slow moving system of communications, based on functional lines in sequential stages, to an improved way of working, based on decision-making in empowered interdisciplinary teams. This approach allows quick, incisive thinking and adoption of appropriate technical solutions to oil extraction." The team used decision analysis to identify the alternatives, uncertainties, and consequences and to determine the Net Present Value of the alternatives.

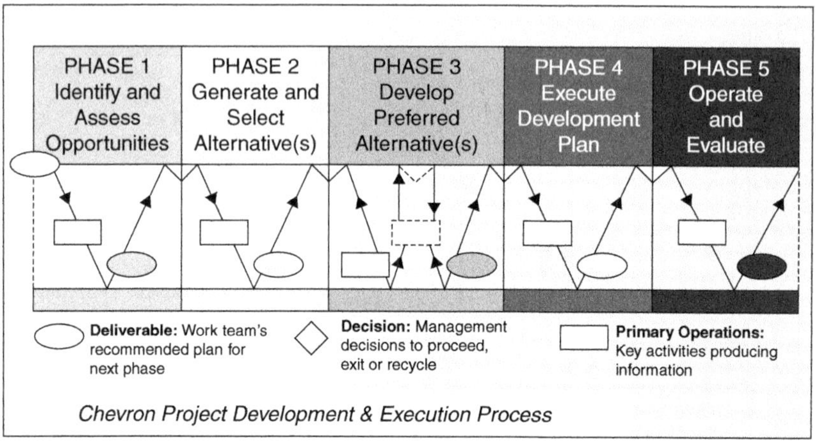

The third critical business process is budgeting. Budgeting is closely related to the strategy process and the project management process. To achieve their strategy, senior managers must ensure that their strategies are funded to enable implementation. Likewise, the financial implications of the go-no go project gate decisions must be included in the budget. However, the budget process can also impact strategy and project implementation. A x% increase or decrease can accelerate or slow down strategy implementation and project progress. Again, a quality decision process can impact these important decisions.

14.3.4 *Learning and Continuous Improvement*

The last element of the ODQ is learning and continuous improvement. The culture of continuous improvement must be driven and reinforced by the senior leaders of the organization. The organization should embrace change that has the potential to create more value. New analysis techniques should be developed and implemented. The organization needs to invest in learning from each decision and each implementation. To have maximum impact, this learning needs to be incorporated into the training programs.

In the next two sections, we will present private company example of ODQ and a homeland security example of a sound decision process.

14.4 Raiffa-Howard Award for Organizational Decision Quality

The Society for Decision Professionals' (SDP) mission is to bring decision quality to important and complex decisions faced by organizations, individuals, and society (Society for Decision Professionals). The SDP awarded the inaugural Raiffa-Howard Award to Chevron Corporation for achieving excellence in applying the principles of Decision Quality (DQ) throughout the organization (Raiffa-Howard Award). The award is given based on a careful evaluation by an independent panel of experienced decision professionals. The examining panel determined that "Chevron deserves the award based on a review of Chevron's adoption in six categories.

- The Culture makes DQ an organizational competency.
  - The Chevron culture is infused with DQ and reinforces DQ. DQ is pervasive at Chevron – widely applied across businesses, functions and geographies. It is deeply ingrained and embedded in the expected Capital Stewardship behaviors (with over 25,000 employees trained), and is integrated into corporate processes and standards. It is part of the language and Chevron executives say, 'DQ is in our DNA.' Chevron is committed to a set of values and behaviors termed The Chevron Way. DQ fits well with The Chevron Way.
- Decision makers have the knowledge, clear roles, and behaviors to use DQ constructively.
  - Decision makers are trained in DQ, understand DQ and expect DQ evidence in their decision recommendations. Many thousands of Chevron executives, managers and staff members have been trained in DQ. Executives now expect DQ and they value what it does for them. They have defined clear decision-making roles and frequently use decision boards for major decisions.
- Professional Decision Support staff members have the appropriate capabilities and the trust and respect of the decision makers.
  - Over 250 decision professionals have Decision Analyst in their job title. About 600 professionals are active members in the internal Chevron Decision Analysis community. Chevron is decentralized; therefore, most of the decision professionals are embedded in major projects, operating companies, and business units. They have a few core groups with highly experienced decision professionals who lead training, provide mentoring and contribute special expertise when needed. They are committed to learning and to sustaining their professional excellence.
- Effective fit-for-purpose Decision Processes engage the right people and have the rigor to achieve DQ.
  - The major processes – The Chevron Way (The Chevron Way), Capital Stewardship, and Project Management System – sustain the culture and further embed DQ in the fabric of the organization. Chevron's assurance

activities include review of DQ. In addition, Chevron's systematic look-back process identifies learning opportunities.

- Tools fit the decision needs, are theoretically sound, and are adequate to meet many decision needs.
  - Chevron uses a broad array of tools and is exploring new ones.

14.5 Commonwealth of Virginia Decision Process for Homeland Security Grant Allocation, 2009–Present

The Department of Homeland Security (DHS) through the Federal Emergency Management Agency (FEMA) provides grant allocations to selected areas to fund counterterrorism and emergency preparedness projects. For the Commonwealth of Virginia, these grant programs are administered through the Virginia Department of Emergency Management (VDEM). As part of the decision process, VDEM must submit an investment justification to FEMA for all funded projects. Investment justifications require funded projects to be consistent with federal, state, and local planning documents. This can be challenging because goals, objectives, and priorities shift as new guidance is issued and as projects are implemented and project managers work toward closing risk and capability gaps. Stakeholder perspectives, decision context, and access to information make grant allocation a complex sociotechnical challenge.

In order to address the challenges, VDEM required a more systematic and objective decision process. While subjectivity can never be totally eliminated from this process, it can be mitigated by establishing a scoring methodology that can be used to rank each proposal within a portfolio relative to others based on senior leadership priorities.

The Commonwealth of Virginia Homeland Security Decision Model was developed by Old Dominion University's (ODU) Virginia Modeling, Analysis and Simulation Center (VMASC) under the direction and guidance of the Virginia Department of Emergency Management (VDEM) and the Office of Veterans Affairs and Homeland Security, from 2009 to present.[1] The decision process was developed to aid Commonwealth stakeholders and senior executive decision makers in funding allocations for the Virginia Homeland Security Grant Program (HSGP).

Senior executive and statewide stakeholder involvement was critical to develop the decision criteria and buy-in from the community. This required fully engaging senior leadership at every step of a Dialogue Decision Process. Figure 14.5 displays the four-step process. Additionally, the model had to be kept as simple and straightforward as possible to allow for enough flexibility that modifications could be easily made as the decision-making environment

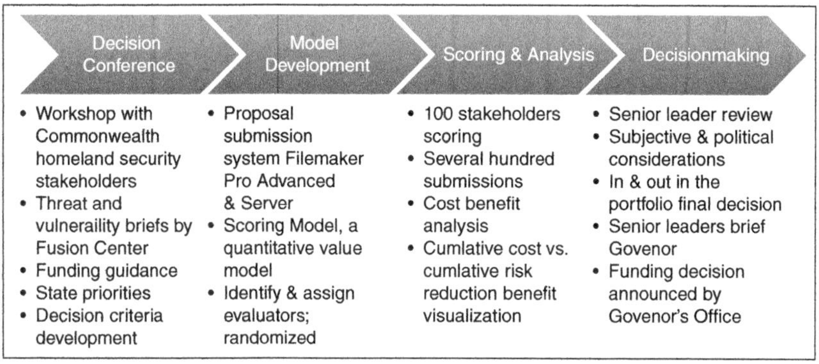

Figure 14.5. The Commonwealth of Virginia decision process to inform homeland security grant allocation decision. The methodology was originally developed in 2009 and has evolved each year at the local, region, and state levels.

changed, as well as to support model usefulness beyond a single round of decision making.

Over the past six years, decision criteria have changed as a result of many factors such as gap closure, reduction in grant dollars, transition from purchasing new capability to sustaining capability, and elections. The decision process has been useful in that it reflects the values and preferences of the Commonwealth stakeholder community. The table that follows shows decision criteria in 2009 and how they have evolved into the 2015 decision criteria. The trend shows the interaction with stakeholders has made a significant improvement in the quality of the measures.

2009 Urban Areas Security Initiative (UASI) Region	2013 state only	2015 state only
How well does this investment justification (IJ) support VA homeland security strategy goals/ objectives? How well does this IJ support Hampton Roads (HR) homeland security strategy goals/ objectives?	Evaluation of risk Necessity of the project Mitigation efficacy Results evaluation Project management control measures Viability of long-term sustainment plan Support for declared state priorities Viability of project management plan	Risk: Threat, vulnerability and consequence Sustainment Project management • Accountability • Efficacy Necessity • Capability linkages • Project validation • Vetted

(*continued*)

How much near-term impact will this IJ have if it meets its FY10 objectives? How much extended-term impact with this IJ have if it meets its multiyear objectives? How strongly does this IJ correlate to “declared” target capability(s)? How viable is the FY10 project management plan for this IJ? How viable is the long-term sustainment approach for this IJ? How well does this IJ accommodate “scale back”? If not selected/funded, how likely is this IJ to be picked up/ funded by other sources? If not selected/funded, how significant are the regional security implications?		Regionalism • Scope • Resource sharing • Multi-disciplined • Whole of government Collaboration
7-point scale regional vote No further definition	Senior executive at state level (no other stakeholders) Definitions added Value functionsPortfolio Decision Analysis Web-based data submission systems	Senior executive at state-level 70 stakeholders Value hierarchy and value functions Scored by dozens of experts Portfolio decision analysis Approved by senior committee and governor

The decision process has been effective since it included all six elements of decision quality.

1. **Frame** – The first step in the process, the Decision Conference, defined the decision frame for each year’s decisions by establishing the priorities and guidance for UASI regions as well as at the state and federal levels.

2. **Values –** The value hierarchy and the value functions defined the values of the decision makers and stakeholders. In 2009–2012, a handful of senior experts established the decision criteria. By 2015, statewide stakeholders were also included in the process.
3. **Alternatives** – A diverse set of creative alternatives was developed for each year. In 2009, alternatives continued to address known capability gaps. By 2015, alternatives dealt with how and what capabilities to sustain.
4. **Information** – The best available information was used to score the alternatives. Threat and vulnerability information was provided by federal partners as well as information from State Fusion Center. In addition, UASI regions relied on analytic products to provide information to help score alternatives.
5. **Logical reasoning** – Multiple objective decision analysis provided the logically sound mathematics to evaluate the alternatives and the portfolio. The difficult trade space for regional UASIs and the state was cost, sustaining capability, and closing remaining gaps.
6. **Commitment to action** – Commitment to action was enhanced by early and active participation of the decision implementers in the decision process. Senior leaders provided the guidance and ensured regional UASI and statewide participation. In addition, the VDEM staff continues to monitor projects under development and assist FEMA for periodic audits.

14.6 Conclusions

Today's organizations face significant challenges including the Triple Bottom Line of profits (or cost-effectiveness for a public organization), people, and planet. The potential benefits of Organizational Decision Quality are significant for private and public organizations.

According to the SDP, "demonstrated benefits are apparent from Organizational Decision Quality. Chevron has outperformed its peer group in total shareholder return for both the last five- and 10-year periods. Chevron executives state that DQ is an important contributor to this performance. The upstream business of Chevron (about 90% of all capital expenditures) requires big investments (often many billions) that need to be productive for decades. These decisions require facing huge uncertainties. Chevron meets this challenge with DQ" (Raiffia-Howard Award).

For public organizations, including homeland security organizations, the same types of decision processes can be used to engage experts, stakeholders, and decision makers to create and implement cost-effective programs. This chapter described the successful decision process implemented for the Urban Areas Security Initiative for Hampton Roads (since 2009) and the Commonwealth of Virginia (since 2012).

NOTE

1. The methodology began in 2009 supporting the Urban Areas Security Initiative for Hampton Roads. The Commonwealth adopted the support in 2012, and it is still in use today.

BIBLIOGRAPHY

Blenko, M. W., Mankins, M. C., & Rogers, P. (2010). The decision-driven organization. *Harvard Business Review*.

The Chevron Way. (n.d.). Retrieved December 28, 2014, from Chevron Corporation: www.chevron.com/about/chevronway/.

Decision-making in an uncertain world: A Chevron case study. (n.d.). Retrieved December 28, 2014, from Business Case Studies: http://businesscasestudies.co.uk/chevron/decision-making-in-an-uncertain-world/#axzz3NDPqSBNF.

Elkington, J. (1997). *Cannibals with forks: The triple-bottom-line of the 21st Century.* Gabriola Island: New Society Publishers.

Gregory, J. (2002). Chevron Texaco project development and execution process. In N. R. Committee for Oversight and Assessment of U.S. Department of Energy Project Management. *Proceedings of government/industry forum: The owner's role in project management and preproject planning.* Washington, DC: National Academy Press.

Howard, R. (1988). Decision analysis practice and promise. *Management Science*, *34*, 679–695.

Keeney, R. (1992). *Value-focused thinking: A creative path to decision making.* Cambridge, MA: Harvard University Press.

Lavingia, N. J. (n.d.). *Business success through excellence in project management.* Retrieved December 28, 2014, from Critical Facilities Roundtable: www.cfroundtable.org/ldc/040706/excellence.pdf.

Matheson, D., & Matheson, J. (1998). *The smart organization.* Boston, MA: Harvard Business School Press.

Meredith, J. R., Mantel, S. J., Jr., & Shafer, S. M. (2014). *Project management: A managerial approach* (9th ed.). Wiley & Sons.

Parnell, G., Bresnick, T., Tani, S., & Johnson, E. (2013). *Handbook of decision analysis.* Hoboken, NJ: Wiley & Sons.

Phillips, L. (2007). Decision Conferencing. In W. Edwards, R. Miles, & D. von Winterfeldt, *Advances in Decision Analysis* (pp. 375–399). Cambridge: Cambridge University Press.

Raiffia-Howard Award. (2014, October). Retrieved December 28, 2014, from Society for Decision Professionals: http://decisionprofessionals.com/news/Chevron-Receives-the-First-Raiffa-Howard-Award-for-Organizational-Decision-Quality.

Society for Decision Professionals. (2015). Retrieved April 14, 2015, from Society for Decision Professionals: http://decisionprofessionals.com/.

Spetzler, C. S. (2007). In W. Edwards, R. Miles, & D. von Winterfeldt (Eds.) *Advances in decision analysis* (pp. 451–468). Cambridge: Cambridge University Press.

15

A Value Model for Evaluating Homeland Security Decisions

Ralph L. Keeney and Detlof von Winterfeldt

15.1 Introduction

In 2007, Michael Chertoff, then secretary of the U.S. Department of Homeland Security (DHS), characterized the task of the Department as follows:

> We have to identify and prioritize the risks – understanding the threat, the vulnerability and the consequences. And then we have to apply our resources in a cost-effective manner. (Chertoff, 2007)

His successors have confirmed this risk management philosophy of the DHS, and many activities are under way to develop appropriate risk assessment and management approaches, models, and tools at the Department. As part of the Department's risk management strategy, the Office of Risk Management and Analysis was created, whose purpose is to "ensure that risk information and analysis are provided to inform a full range of homeland security decisions" (DHS, 2010).

This risk assessment and management framework includes estimating the risks we face from terrorism and making appropriate risk reduction investments to counter terrorism. As several authors have pointed out (Ezell et al., 2010; Willis et al., 2005), this, in turn, requires the estimation of:

1. threat (probability of various types and targets of attempts of terrorist attacks) and threat reduction due to preventive and deterring countermeasures;

This research was partially supported by the U.S. Department of Homeland Security through the National Center for Risk and Economic Analysis of Terrorist Events (CREATE) under grant No. 2007-ST-061-RE0001. Any opinions, findings, and conclusions or recommendations in this document are those of the authors and do not necessarily reflect the views of the U.S. Department of Homeland Security.

2. vulnerability (probability of a successful attack, given an attempt) and vulnerability reduction due to protective countermeasures;
3. consequences (including lives lost, direct and indirect economic impacts) given a successful attack and the reduction of consequences due to response and recovery improvements; and
4. costs and other consequences of counterterrorism policies and actions.

Items 1 and 2 can, in principle, be addressed with tools like probabilistic risk analysis, though the practical difficulties are significant (see, e.g., National Research Council, 2008). Items 3 and 4 involve describing the possible consequences of terrorist attacks and the costs and side effects of countermeasures in a way that reflects the nation's values and concerns. This requires identifying the nation's objectives pertaining to terrorism. It is also important to specify metrics for each objective to describe possible consequences of terrorist actions and of counterterrorism strategies.

A complete set of homeland security objectives is the common foundation about which almost all individuals can agree. Essentially, everyone agrees with objectives such as minimizing the number of fatalities due to terrorism, limiting the damage to infrastructure, minimizing inconvenience to the U.S. public, protecting civil liberties, and minimizing costs of terrorism protection. There will naturally be differences about the relative importance of achieving different objectives, but usually not about the objectives themselves. Thus, the objectives tend to unify the nation and the Department by establishing a common value framework.

A model that is constructed from all the relevant objectives, metrics for each of the fundamental objectives, and judgments representing value tradeoffs between objectives, attitudes toward risk, and any synergies in achieving different objectives is called a value model. Once constructed, one component of the value model is a utility function over the strategic objectives that can be used to evaluate alternative actions. As a value model provides a clear understanding of what we want to achieve regarding homeland security, it helps to focus and coordinate all efforts to achieve it. With a value model, the DHS should be able to better invest its $60 billion annual budget. Fortunately, the amount of effort to construct a high-quality value model is very small compared to the effort to evaluate alternatives to best achieve the purposes of homeland security. Indeed, using the objectives in the value model, this evaluation should be more consistent, less time-consuming, and easier to explain to others.

This chapter describes how to construct a value model for homeland security. As a practical matter, such a value model would need to be constructed by working with the DHS staff and endorsed by DHS policy makers, so we will refer to it as a DHS value model. Section 15.2 introduces the idea of a value model in the terrorism context. Section 15.3 outlines our methodology and describes a preliminary and illustrative, but conceptually complete DHS value model developed using this methodology. Specifically, we describe how we identified DHS objectives, organized them, developed metrics for them, and provided an initial prioritization using value trade-offs. Section 15.4 discusses uses of such a value model to evaluate both priorities and potential actions of DHS. Section 15.5 outlines how DHS could build on the illustrative value model developed here to incorporate a complete set of national interests and to assess the necessary value trade-offs.

15.2 Construction of a Value Model

To evaluate terrorism risks and to assess the effectiveness of actions to counter terrorism, it is now fairly standard within DHS (see Bier & von Winterfeldt, 2007; Willis et al., 2005) to define risk using the following model:

$$R(Terrorist\ Attack) = p(Attempted\ Attack) * q(Success \mid Attempt) * [-u(Consequences)], \quad (15.1)$$

where R is the risk of a terrorist attack, p is the probability of an attempted attack in a given time period (or frequency, if there is a possibility of multiple attacks in the time period), q is the probability that the attempt is successful, and u is a utility function for possible consequences as described later in this section. The utility function u is constructed such that $u(x) > u(y)$ if the consequences x are preferred to the consequences y. In the terrorism context, we almost always deal with negative consequences and risks that involve disutilities. To assure that the risk is calculated based on disutilities, we assign a negative sign to u.

The assessment of p's and q's can be done using risk analysis methods (see Ezell et al., 2010; von Winterfeldt & O'Sullivan, 2006; Willis et al., 2005). In practice, this is a very difficult task and the appropriate tools are still debated in the risk analysis community (e.g., National Research Council, 2008).

The assessment and evaluation of consequences is less controversial, and this is where the development of a value model is important and useful. It allows us to use equation (15.1) to appraise proposed counterterrorism actions. Any such countermeasure, call it C, should have an effect on one or more of p, q, and u (due to its effect on the consequences). For a desirable counterterrorism action, p, q, and consequences will likely change, so $p_C < p$ (due to preventive actions), $q_C < q$ (due to protective action), and $u_C > u$ (due to protective and response and recovery actions). The difference between R_C and R is the benefit of the countermeasures.

15.2.1 *The Concept of a Value Model*

A value model is similar to any other model in that it is constructed with logic, judgments, and information. The main difference between a value model and other models is its focus on value judgments instead of on facts and probabilities. Most other models try to describe or predict consequences answering questions like the following: If a specific alternative is chosen or if a specific scenario occurs, what are the possible consequences and their relative likelihoods? A value model tries to answer the question of what is the relative desirability of the various possible consequences that may occur.

A value model is constructed in four steps. First, a complete list of the relevant objectives must be identified and then organized. Second, a metric must be identified for each fundamental objective that can be used to measure the degree to which that objective is achieved by various alternatives. Third, an aggregation rule must be selected to combine achievements on each of the objectives. Fourth, value trade-offs must be assessed that indicate the amount of achievement on one objective that is equivalent in value to a specified amount of achievement on a second objective. We briefly summarize the steps later. More details are found in Keeney (2007) and Keeney and von Winterfeldt (2007).

15.2.2 *Identifying Objectives*

To compile an appropriate list of objectives requires the initial creative step by an individual or individuals to write down anything that they care about or value in a specific decision context. Our decision context of concern is all alternatives appropriate to consider for better protecting the citizens of our nation from terrorism. Each item on this list is termed a value. As values are stated in numerous ways, we want to convert them to a common form. Specifically, we translate each value into a corresponding objective that is specified using a verb and an object. For instance, for the value *safety*, the corresponding objective would be *ensure safety*. For a value stated as *long security lines at airports*, the corresponding objective might be *minimize waiting time at airport security*.

There is a substantial amount of experience indicating that creating a complete list of objectives for decisions is very difficult. Benjamin Franklin (1772), in a letter to his colleague Joseph Priestley, said this about difficult decisions: "they are difficult, chiefly because while we have them under consideration, all the reasons pro and con are not present to the mind at the same time."

Recently, experimental work has shown that for several important individual decisions (e.g., selecting a work internship), essentially none of the individuals could identify a complete set of their own objectives and that the average number of objectives identified was less than 50% of those later identified as personally relevant from a complete list provided by the experimenters (Bond et al., 2008). Furthermore, the decision makers subsequently indicated that the objectives that they failed to identify were roughly as important as those that they had identified. To facilitate individuals developing a more complete list of objectives, numerous devices have been shown to stimulate thinking (Bond et al., 2010; Keeney, 1992). As a high-level summary, identifying an appropriate complete list of objectives cannot be done simply by making a list; rather, it requires hard, introspective thought about the decision situation in several sessions over a period of time, facilitated by a person familiar with developing objectives.

Once what is considered to be a complete list of objectives is identified, they are organized into four types: strategic objectives that are the ultimate objectives of an organization that guide all of its decisions; fundamental objectives that characterize the ultimate intent of any specific decision; means objectives that influence the degree of achievement of other objectives (specifically strategic or fundamental objectives); and process objectives that concern how the decision is made rather than what decision is made. In this chapter, our concern is with the strategic objectives of homeland security, so we focus on these and the process and means objectives contributing to the achievement of the strategic objectives.

15.2.3 *Specifying Metrics to Measure Objectives*

Identifying a metric is required to measure consequences that indicate performance with respect to each strategic objective. Sometimes metrics are easy to identify. If the objective is to minimize economic cost, the metric can simply be millions of dollars. However, if the objective is to minimize waiting time at airport security, selecting an appropriate metric requires more thought. Metrics to consider include the average time it takes to get through security, the number of people taking more than 20 minutes to pass through security, the percentage of passengers waiting in line more than 20 minutes before they get to the security check, and the extra time passengers allocate to get through security. Different security line arrangements may be better using some of these metrics and worse using others. Hence, value judgments are made in selecting a metric, so careful judgment must be used to choose metrics that measure the real concerns as closely as possible.

15.2.4 *Combining Achievement on Different Objectives*

Combining the achievement on different objectives requires an equation that, in essence, allows one to appropriately add up the positive impacts (pros) and subtract the negative impacts (cons) on different objectives to create an overall indicator of the desirability of the combined consequences. Suppose there are two objectives O_1 and O_2 with corresponding metrics X_1 and X_2, and let x_1 and x_2 be the specific impacts in terms of X_1 and X_2. An indicator of the desirability or undesirability of any consequence (x_1,x_2) is wanted. Such an indicator is simply a number called the utility of (x_1,x_2) and written $u(x_1,x_2)$. We construct this indicator such that $u(x_1,x_2)$ is greater than $u(x_1',x_2')$ if and only if (x_1,x_2) is preferred to (x_1',x_2'). Furthermore, in situations where there is uncertainty about which consequences (x_1,x_2) might occur, as is almost always the case, it simplifies any analysis as the expectation (i.e., average) of the utilities of the possible consequences logically indicates relative desirability. Fortunately, von Neumann and Morgenstern (1947), Savage (1954), and Pratt and colleagues (1964) have provided a logically sound foundation for utility theory that has such a property. This concept has been thoroughly extended to consequences with multiple objectives (Fishburn, 1965; Keeney & Raiffa, 1976). One possible utility function for the case of two objectives is

$$u(x_1,x_2) = k_1u_1(x_1) + k_2u_2(x_2), \tag{15.2}$$

where u_1 and u_2 are single-objective utility functions and the k's are parameters that are calculated from assessed value trade-offs.

15.2.5 *Value Trade-offs*

The aspect of a value model where there is the greatest potential for disagreements concerns the value trade-offs. Almost everyone's set of strategic objectives for homeland security would include the following two objectives: minimize loss of life and minimize costs, which might naturally be measured by the number of fatalities and costs in millions of dollars. As some potential alternatives would cost more and potentially result in less life lost and other alternatives would cost less and potentially result in relatively more life lost, a value trade-off between these two objectives must be addressed either explicitly or implicitly. In other words, a value judgment must be made about the relative importance of reducing the risks of fatalities due to terrorism and the cost of reducing those risks. For example, is it worth $50 million to reduce the risk of an anthrax attack that would cause 20 fatalities by one chance in 1,000? From such a value judgment, one can calculate how much additional cost is appropriate if it could reduce the number of expected fatalities by one in a terrorist attack. Providing a reasonable judgment to address this issue is one of the most difficult value trade-offs that we face.

15.3 Construction of a DHS Value Model

The four steps described in Section 15.2 are used here to construct a homeland security value model. Sequentially, we cover identifying objectives, choosing metrics for each, selecting an aggregation rule to combine achievement on each of the objectives, and specifying the value trade-offs among the objectives. It is important to point out that our value model is preliminary and illustrative. Section 15.5 outlines the effort appropriate to construct a value model with legitimacy for DHS.

15.3.1 *Identifying and Organizing DHS Objectives*

Several sources were used to develop the initial list of potential homeland security objectives. These included a selective literature review, interviews with individuals knowledgeable about homeland security, a formal questionnaire to individuals working on homeland security problems, and our own experience with homeland security issues and value modeling.

The selective literature review involved searching for indications of objectives in the professional articles and popular press reports that we read about terrorism and counterterrorism activities. When a news article stressed delays for passengers at an airport or a concern about recording conversations on telephones, we listed values concerning inconvenience and civil liberties on the objectives list. If a professional article analyzed the risks of suicide bombers in malls, we added objectives concerning public fear and the negative business impacts to the list of values. The same procedures were used for our own work. For example, in analyzing a proposed requirement that all commercial airlines should include an anti-surface-to-air-missile system on their airplanes (von Winterfeldt and O'Sullivan 2006), objectives included *minimize life-cycle cost of surface-to-air missile countermeasures* and *minimize national economic consequences due to a missile attack*, which suggested strategic homeland security objectives of *minimize life-cycle costs of anti-terrorism measures* and *minimize national economic impact of terrorism.*

In discussions with several individuals in DHS and in professions concerned with homeland security, we asked individuals to tell us any objectives that they thought were important to homeland security. We specifically challenged them to think of objectives that might be missed by many others who had not thought as deeply about the issue.

Our most structured activity for identifying objectives was a questionnaire developed and sent to managers and researchers on projects at CREATE, the first DHS University Center of Excellence at the University of Southern California. The questionnaire, shown in Table 15.1, provided additional lines for responses as required. Collectively, the 12 responses provided 209 objectives, mostly written in the form of a verb and a noun (i.e., object) as requested.

To structure these objectives and others added from our literature search and discussions, we began by placing similar objectives into categories defined by a general higher-level objective. Examples included: disrupt terrorist capabilities, improve decision making within DHS, enhance communication, ensure health and safety, minimize social impacts, and protect civil liberties. Within these categories, we could easily combine objectives that were essentially the same but stated differently. Then, following procedures discussed in Keeney (1992, 2007), we organized the higher-level objectives, using cause–effect relationships and hierarchical relationships, into the strategic objectives of homeland security and the means objectives and process objectives that help to achieve the strategic objectives. Figure 15.1 indicates the high-level process, means, and strategic objectives and their relationships.

It is useful to elaborate on the highest-level strategic objectives in Figure 15.1. Viewed from the current time, we want to reduce the current threat by deterring potential terrorist attempts prior to the initiation of an attack and to foil any attack that has been initiated. If an attempt is successful, there are many homeland security objectives under the high-level objective to reduce consequences of attacks. Obviously, any and all efforts of our government to pursue counterterrorism activities, in addition to influencing the objectives, have other, significant consequences. Regarding these counterterrorism activities, we would naturally like to minimize their costs and the many direct consequences of these actions in terms of objectives that mirror the objectives of terrorist attacks. This is not odd as both are the results of our national principles of life, liberty, and the pursuit of happiness. One of these potential consequences leads us to specify another high-level objective as to protect civil liberties. Over the longer term, we want to minimize future terrorism threats.

Table 15.1. *Questionnaire for identifying DHS objectives*

Your Name:

Please provide a list of all of the objectives and consequence metrics that should be important to DHS and the nation.

Please separate these objectives and consequence metrics between those that are relevant to the impacts of terrorist events (e.g., fatalities) and those that are relevant to the impacts of countermeasures (e.g., privacy or trade impacts).

List as many objectives and consequence metrics as you can. Experience shows that after people have listed all the objectives that they initially think of, they still can identify several more objectives with additional hard thinking.

A useful way to state objectives is with a verb and an noun (e.g., the object of preference).

The following are example objectives (with potential consequence measures in brackets).

a. minimize deaths from terrorist acts (number of people killed in an attack)

b. limit business disruptions (aggregate loss of revenue to businesses due to disruption)

c. avoid inconveniences to U.S. citizens (person-days of unproductive waiting time due to security measures)

Note that there can be several consequence metrics for each objective.

Your objectives and consequence metrics for evaluating the impacts of terrorist attacks

1

2

3

4

5

6

7

8

9

10

Your objectives and consequence metrics for evaluating the impacts of counterterrorism decisions

1

2

3

4

5

6

7

8

9

10

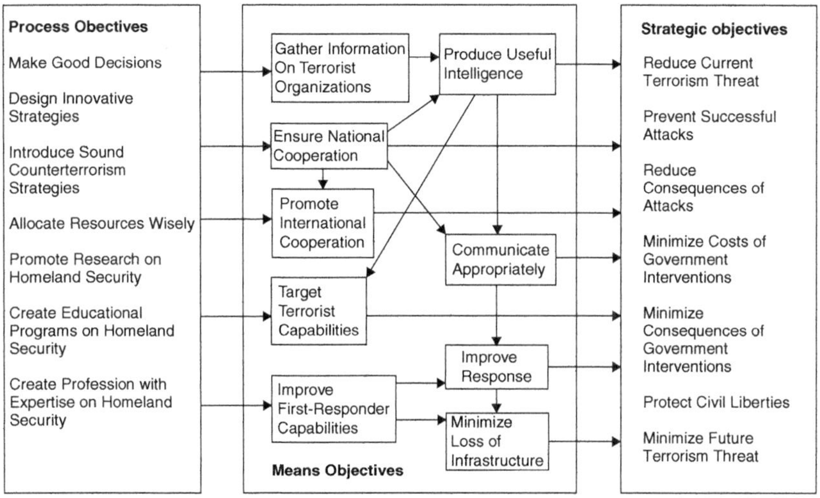

Figure 15.1. Homeland security means-ends objectives network.

15.3.2 *Selecting Metrics to Measure DHS Objectives*

A metric is required to indicate the degree to which each of the associated strategic objectives is met. Selecting any metric is a decision that uses judgments to balance how well it indicates performance on the objective, the availability of data and information to evaluate alternatives in terms of the metric, and the ease with which policy makers understand the meaning of different levels of achievement with the metric. In this study, we selected the metrics ourselves using our experience with similar value-modeling projects.

Table 15.2 presents the strategic objectives, along with metrics for all of the lower-level objectives in the table. A set of metrics for the lower-level objectives is a metric for the related higher-level objective. The value trade-offs for all of the lower-level objectives in Table 15.2 are discussed later in this section.

The metrics for consequences of terrorist attacks and for consequences of government counterterrorism are mostly obvious choices. For concerns of illness and injuries, a threshold is needed to define what is meant by the term *serious*. In these cases, we chose *serious* to mean that the consequences would prohibit normal life function for at least one year. For metrics concerning fear and disruption of lifestyle, thresholds were also used. Feeling fear was defined to mean at least once weekly for a period of a year and a disrupted lifestyle required altering basic routines such as working, shopping, entertainment, and travel. The hours of inconvenience are spread over numerous individuals. For environmental impact, we consider the impact on the habitat in a destroyed area in the future.

Table 15.2. *Homeland security strategic objectives*

Objectives	Metrics	Unit Value Trade-off($ millions)
Reduce Current Terrorism Threat	**p See section 15.4.1**	
Prevent Successful Attacks	**q See section 15.4.1**	
Reduce Consequences of Attacks		
Minimize Fatalities	X_1 Number of fatalities	6–12
Minimize Morbidity		
Illnesses	X_2 number of serious illnesses	0.1–0.5
injuries	X_3 number of serious injuries	0.1–0.5
Minimize Economic Costs		
direct costs		
to individuals	X_4 millions of dollars	1
to businesses	X_5 millions of dollars	1
to society	X_6 millions of dollars	1
indirect costs	X_7 millions of dollars of gdp	1
Minimize Social Impacts		
jobs lost	X_8 number of jobs lost	0–0.2
fear	X_9 number of citizens feeling fear	0.001–0.010
personal disruption of lifestyle	X_{10} number of citizens with disrupted lifestyles	0.0005–.0.005
inconvenience	X_{11} waste of 100,000 hours	1–5
Minimize Environmental Impact	X_{12} square miles of habitat destroyed	1–10
Minimize Violent Crime	X_{13} number of criminal participants	0.01–0.05
Minimize Direct Costs of Government Interventions	X_{14} millions of dollars	1
Minimize Consequences of Government Interventions		
Minimize Fatalities	X_{15} number of fatalities	0–10
Minimize Morbidity		
illnesses	X_{16} number of serious illnesses	0.1–0.5
injuries	X_{17} number of serious injuries	0.1–0.5

Table 15.2 (*cont.*)

Objectives	Metrics	Unit Value Trade-off($ millions)
Minimize Economic Costs		
direct costs		
to individuals	X_{18} millions of dollars	1
to businesses	X_{19} millions of dollars	1
to society	X_{20} millions of dollars	1
indirect costs	X_{21} millions of dollars of gdp	1
Minimize Social Impacts		
jobs lost	X_{22} number of job lost	0.05–0.2
fear	X_{23} number of citizens feeling fear	0.001–0.01
personal disruption of lifestyle	X_{24} number of citizens with disrupted lifestyles	0-0005.–0.005
inconvenience	X_{25} waste of 100,000 hours	1–5
Minimize Environmental Impact	X_{26} square miles of habitat destroyed	1–10
Minimize Violent Crime	X_{27} number of criminal participants	0.01–0.05
Protect Civil Liberties		
Minimize Violations of Privacy	X_{28} number of incidents	0.001–0.005
Minimize Violations of Civil Rights	number of false arrests or false public accusations	0.005–0.020
Minimize Retaliation against Innocent People	X_{30} number of hate crimes against innocent people	0.1–0.2
Minimize Future Terrorism Threat		
Minimize Recruitment of Terrorists	X_{31}number of new recruits	0.02–0.2
Minimize Economic Support for Terrorism	X_{32} millions of dollars provided to terrorist organizations	1–10
Minimize Public Support for Terrorism		
in the Moslem world	X_{33} percent of Moslem world public sympathetic to terrorism	1–10
in the United States	X_{34} percent of U.S. public sympathetic to terrorism	1–10

15.3.3 *Combining Impacts of Different DHS Objectives*

Combining impacts on different objectives requires using a utility function as discussed in Section 15.2. The important message of this section is that an appropriate DHS utility function should be additive over the several objectives and linear in impacts on each objective. As explained, this additive linear function is often reasonable.

The 34 lowest-level strategic objectives in Table 15.2 define in detail the high-level strategic objectives. The measurement of possible consequences of DHS alternatives occurs at this detailed level. Hence, the DHS utility function is defined over the set of lowest-level metrics. Notationally for the illustrative DHS utility function, there are 34 strategic objectives $O_1,\ldots,O_{34}$ measured respectively by the metrics $X_1,\ldots, X_{34}$, where x_i is an impact on X_i, $i = 1,\ldots, 34$, so $(x_1,\ldots,x_{34})$ is a consequence. The appropriate DHS utility function is

$$u(x_1,\ldots,x_{34}) = -(k_1x_1+k_2x_2+\cdots+k_{34}x_{34}), \tag{15.3}$$

where the parameters k_i, $i = 1,\ldots, 34$ are calculated from value judgments about the value trade-offs between objectives discussed in Section 15.3.4. For all of the metrics from X_1 to X_{34}, the natural metric used results in a larger consequence being worse than a smaller consequence, which is why a negative sign precedes the consequences corresponding for those metrics in the utility function in (15.3). This convention is necessary to have a greater utility associated with the preferred consequences. An appropriate measure of disutility, used in calculating risk R in (1), is –u.

The Appropriateness of Additivity. There are mathematical conditions, discussed in detail in Fishburn (1965) and Keeney and Raiffa (1976), that present the formal logic for a utility function to be of the additive form of (15.3). Here, we offer two logical arguments to support the case that the DHS utility function over strategic objectives should be additive. The first provides the intuitive logic and the second the structural logic.

In non-mathematical terms, the appropriateness of adding the values of achievements on separate objectives follows if one is not additionally concerned about the achievements on any combinations of objectives. For example, when evaluating investment costs and operational costs of counterterrorism alternatives, this can usually be done by creating a life-cycle cost measure (which is additive in the two terms), without considering how well the alternatives reduce terrorism consequences. These concerns would be addressed by other objectives.

If the value of the performance on different objectives is non-additive, the reason is often obvious. For example, if one objective is to "increase the availability of a radiological detection device," measured in percent of downtime, and another one is to "increase the effectiveness of the device," measured in the percent of correct identification of a radiological source, it is clear that the two objectives are multiplicative, instead of additive. For example, a device that is available only 10% of the time would essentially be useless no matter how well it performed when available. In cases like this, a new single objective can be created that combines the two non-additive ones. For example, we could use "expected performance" as a measure for the device, which is the product of availability and effectiveness.

Another grounds for non-additivity occurs when an emergent value is not directly captured by the existing individual objectives. For example, we may define separate objectives for children and adult fatalities of terrorist attacks. Because each death matters, these should be additive losses, though there are good reasons to put more weight on children. However, if a large number of individuals die as a result of an attack, there is an induced "horror and fear" concern that may lead to a much greater loss evaluation than indicated by using an additive model that valued deaths of individual adults and children only. The Beslan preschool attack by Chechen rebels is an example. In such cases, one should create a new objective concerning "induced fear and horror" of a terrorist attack that captures that emergent value, so together with the original objectives, the additive utility function is appropriate.

The structural logic for an additive utility function can be illustrated by using a simple example. Suppose that there are only two strategic objectives pertaining to homeland security: O_1 = minimize fatalities and O_2 = minimize economic damage in the country. Further suppose that these are measured respectively by the metrics X_1 = number of fatalities and X_2 = total cost in millions of dollars caused by terrorism. Now suppose that instead of the additive utility function, the non-additive utility function

$$u(x_1,x_2) = k_1u_1(x_1) + k_2u_2(x_2) + k_3u_1(x_1)u_2(x_2), \qquad (15.4)$$

is suggested as a possible utility function. Clearly the impacts due to fatalities x_1 are incorporated by $k_1u_1(x_1)$ and the impacts due to the economic cost x_2 are incorporated by $k_2u_2(x_2)$. If the third factor $k_3u_1(x_1)u_2(x_2)$ in (15.4) is included, it should be measuring something different and in addition to fatalities and economic costs attributable to terrorism. However, the first two factors in (15.4) include all impacts of terrorism that ultimately matter given that the set of the two strategic objectives is complete. Thus, the third term is not necessary and hence inappropriate. If an understanding for the inclusion of this factor turned out to have merit, this would indicate that the set of the original two strategic objectives was not complete. It would then be appropriate to identify what additional strategic objective O_3 was originally missing and captured by this third term. Next, an appropriate metric X_3 for that third objective should be selected with x_3 indicating impacts in terms of that metric. Then, an appropriate utility function would be

$$u(x_1,x_2,x_3)=k_1u_1(x_1)+k_2u_2(x_2)+k_3u_3(x_3), \tag{15.5}$$

where u_3 is a utility function that evaluates impacts in terms of objective O_3.

The Appropriateness of Linearity. Three general situations lead to the conclusion that linearity is reasonable for the single-objective utility functions. One situation is when the metric for a strategic objective measures something that is of value in itself, as opposed to being of value for its uses. This case is relevant for evaluating metrics such as the number of fatalities. Clearly each life saved is valuable in and of itself. It was of course valuable to the person whose life was lost and to the family and friends of that individual. It may be reasonable to count each life saved or lost as equal to any other for evaluating homeland security decisions. Alternatively, it might be argued that a child should count more than an adult. In that case, we could have separate objectives for children and adult fatalities and each would be linear.

A second situation that leads to linearity is when the possible consequences are described as estimates of expected impact with respect to those consequences. When expected impacts are used as consequences, then linearity is implicitly assumed because, for instance, a consequence of 25 fatalities for sure would be equivalent to a 50-50 chance of either 10 or 40 fatalities because each circumstance has the same expectation, namely 25 fatalities. For such situations, it would not make sense to build the value model with components that were not linear in consequences.

The third situation where linearity is reasonable is when the same objective with its corresponding metric is relevant to numerous domains. Regarding homeland security activities, the objective of minimizing economic costs applies to numerous federal government activities. Hence, even though the economic costs of terrorism or counterterrorism activities could be large, they are still small compared to the costs of federal government. If an appropriate utility function for federal government costs were created, it may not be linear and may be concave to exhibit risk aversion. However, that utility function should be smooth and cover a range of trillions of dollars. Over any relatively small section of that range, it follows from a mathematical result known as Taylor's formula that a linear function would be an excellent approximation. Thus, a linear utility function is reasonable over a range of at least a few hundred billion dollars, which is several times the annual budget of DHS.

An example may help clarify why linearity is reasonable for impacts on strategic homeland security objectives. When considering the deaths that may be caused by a suicide bomber detonating himself in a crowded shopping mall, one may think that it is worth $2 billion to avoid the first death, another $500 million to avoid the second death, and maybe $100 million to avoid the third. The dollar value to avoid additional deaths keeps dropping to the level where each additional death counts the same. Maybe in this case this occurs at around 25 deaths where each additional fatality seems like it should be worth $10 million to avoid. Preferences such as these are certainly not linear in the number of deaths prior to 25. However, such preferences can be reasonable. The reason for these nonlinear preferences has to do with the indirect impacts attributable to the deaths caused by terrorists. In this example, assume that these indirect effects are the number of people feeling fear, the number of people who have their freedom to move freely and safely curtailed, and the indirect economic impacts of fewer people visiting shopping malls. For the values just described, the biggest indirect impact would occur with the first death, less with the second, and so forth. It is the enhanced concern that terrorist attacks such as this might occur that results in changes to people's behavior and feelings. The difference between 25 and 35 deaths from a suicide bomber in a shopping mall might not induce any additional indirect impacts. However, each of those deaths from 25 to 35 would be critically important to the families and friends of those who died and of course to the individuals themselves.

If we have a complete set of strategic objectives included in a utility function, one objective would concern deaths, another would concern the indirect costs, a third the number of individuals feeling fear, and a fourth the number of individuals whose freedom to move freely and safely was curtailed. Each of the indirect impacts should be captured explicitly by the second through fourth objectives, and the first objective concerning deaths should account only for an individual's death and its devastating impact on family and friends. With this understanding, each of the statistical fatalities prior to the twenty-fifth fatality should be just as valuable as, but no more valuable than, those fatalities that followed the twenty-fifth. As a result, the appropriate utility function for evaluating the number of statistical fatalities should be linear in fatalities.

When the overall consequences of a terrorist attack are captured with a full range of strategic objectives, the calculations could still show that it is worth $2 billion to avoid the first death due to a suicide bomber in a shopping mall. The reasoning would be something like the following: the indirect costs caused would be $1.5 billion, the loss of freedom of numerous people to move around as they pleased would be assessed to be equivalent in value to $400 million, the fear caused to another large number of individuals would be equivalent in value to $90 million, and the first statistical fatality is worth $10 million to avoid.

15.3.4 *Assessing Value Trade-offs*

Once a homeland security utility function that has the additive linear form is constructed, the final step is to assess value trade-offs from which the k_i parameters in (15.3) are calculated. Here, we clarify what is meant by value trade-offs, discuss the procedures to assess them, and use those procedures to specify first-cut estimates of the parameters.

Understanding value trade-offs. A value trade-off quantifies the relative value of achievement on two different objectives. It specifies how much achievement on the first objective one is willing to give up in order to achieve a specified amount on the second objective. In lay terms, it is an even swap of achievement on the first objective for achievement on the second objective. An illustration will be useful.

Consider two of the objectives of homeland security: O_1, minimize fatalities, and O_2, minimize serious injuries. Let us measure these using the metrics X for the number of fatalities and Y for the number of serious injuries, which will be defined as an injury prohibiting normal life function for one year. Figure 15.2 illustrates consequences concerning these two objectives by (x, y), where (10, 400) means 10 fatalities and 400 serious injuries.

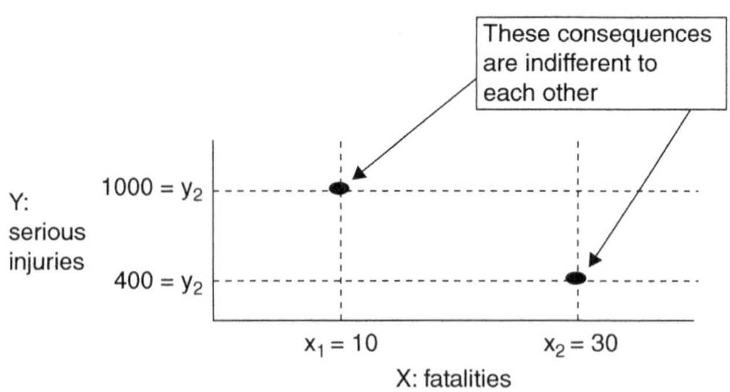

Figure 15.2. Aid to interpreting value trade-offs.

In considering the value trade-off between fatalities and serious injuries, suppose we start with consequence (30, 400). If it was judged that a reduction of fatalities from 30 to 10 would be an overall improvement if serious injuries did not increase to as many as 1,000 and would be equally bad if they increased to exactly 1,000. This would mean that (30, 400) is indifferent to (10, 1,000), so the value trade-off would be 600 serious injuries to avoid 20 fatalities.

Finding pairs of consequences that are indifferent is how value trade-offs are assessed. And of course, value judgments (i.e., the indifferences) are the basis to specify the value trade-offs. An assessed indifference pair of consequences has four different but consistent interpretations:

- From (x_1, y_2), an increase in X to x_2 is compensated for in terms of value by a decrease in Y to y_1.
- From (x_2, y_1), an increase in Y to y_2 is compensated for in terms of value by a decrease in X to x_1.
- From (x_1, y_1), an increase in X to x_2 and an increase in Y to y_2 are equally valued (i.e., equally undesirable).
- From (x_2, y_2), a decrease in X to x_1 and a decrease in Y to y_1 are equally valued (i.e., equally desirable).

As the DHS utility function in (15.3) is additive and linear, another consistent interpretation of a value trade-off is to compare the relative value of a unit change on each of the objectives. Since 20 fatalities is deemed equivalent in value to 600 serious injuries, we can say that avoiding one fatality is 30 times as valuable as avoiding one serious injury or that one fatality is equally as valuable as avoiding 30 serious injuries. Alternatively, we can say that avoiding one fatality would compensate for an increase of 30 serious injuries. When a value trade-off uses a unit consequence (e.g., one fatality in this case) on one of the metrics, it is referred to as a unit value trade-off.

Specifying value trade-offs. Value trade-offs are based on value judgments and should be grounded on sound information and reasoning. In the example given earlier, the meaning of a fatality is clear and the meaning of a serious injury should be unambiguously defined. As a serious injury is defined to have a duration of a year, we imagine that almost everyone would conclude that one fatality is more important (i.e., worse) than one serious injury and perhaps most would say that it is worse than at least 20 serious injuries, although there is no correct answer as this is a value judgment. Also, many people might say that 200 serious injuries are worse than one fatality. If these value judgments seem appropriate, they imply that the value trade-off between fatalities and serious injuries is that one fatality is indifferent to y serious injuries, where y is between 20 and 200.

Information can help us select a reasonable value trade-off. It would be useful to understand what the quality of life of an individual would be in a year with this serious injury relative to that individual's quality of life without that injury. Useful work on this would include gathering information from individuals who have been seriously injured for a year and professionals who treat and routinely work with such individuals. Suppose it was decided that on average the quality of life of the seriously injured individual is 60% compared to the quality of life with no injury. Other information could provide a good estimate of the years of life lost by those who were fatalities. Suppose the data suggested 35 years of life lost on average, but when adjusted for possible degraded quality of life in later years, that about 32 years of quality life were lost with each fatality. Thus, the serious injury is equivalent to a 40% loss of one quality-life year and a fatality is equivalent to a 100% loss for 32 quality-life years. With the assumption that all quality-life years are equal, which is probably reasonable given the precision of value judgments and the differences among individuals, this indicates that a serious injury on average results in the loss of 0.4 quality-life years and a fatality results in the loss of 32 quality-life years, which suggests a value trade-off that 80 serious injuries is equivalently as undesirable as one fatality.

As discussed in Section 15.2.5, any DHS utility function must incorporate the difficult-to-make value trade-off between the two objectives: to minimize loss of life and to minimize costs. To make a reasoned and reasonable value judgment about this value trade-off requires understanding three basic facts. First, the actual value trade-off being addressed is between very small risks of life loss to many individuals and the costs of reducing these risks. These small risks to many individuals have direct implications for the expected number of lives lost, and since our concern for lives lost is the fundamental reason that we care about the original risks, this trade-off is better understood by considering the value trade-off between loss of lives and costs. Second, this value trade-off cannot be avoided, so it is appropriate to address it explicitly. Third, as there is no way to avoid it, addressing it is not immoral.

To illustrate that the value trade-off concerns small risks to many individuals and costs, this simple story paraphrased from Keeney (1995) is useful. Suppose an accident in a coal mine traps one miner, Kirk Eastman. He has enough water and air to survive for a week. A quick appraisal indicates that it would cost $25 million to drill a tunnel and rescue Kirk, an effort that is sure to succeed. The decision is made to proceed. However, just before beginning work on the tunnel, an individual knowledgeable about mine safety raises the following dilemma: "Coal mining is a risky occupation, and accidents all too frequently occur in mines because of tunnel collapses. If we spend the $25 million to strengthen support systems, we would expect fewer accidents over the next five years and statistics suggest that the lives of four miners would therefore be saved. Should we spend $25 million to save the life of one miner when we could spend the same amount to save four miners?"

Perhaps $25 million should be spent for each purpose, but if only one of them can be funded, many people would choose to rescue Kirk. There is, of course, no right or wrong answer to this. Rescuing Kirk Eastman would avoid what is referred to as an identifiable fatality, one where it is known exactly whose life is saved or lost. Saving four miners who would have been in accidents that did not occur because of safety investments would avoid four "statistical fatalities." In the former case, everyone would know who, that is, Kirk, was saved, whereas in the latter case, this could never be known. Because of this distinction, it may be appropriate to assign a different economic value trade-off (and most people suggest a smaller one) to saving a statistical life than to saving an identifiable life.

Considering terrorist attacks and counterterrorism decisions, the lives that are saved or lost are statistical fatalities. A program to detect and stop potential terrorist attacks before they occur would lower the risks of life loss by a small amount for large numbers of people. Using data, information, and logical reasoning, it may be possible to estimate the expected number of statistical lives saved, but it would never be known whose lives were saved. Thus, when we consider value trade-offs between lives lost and economic costs, we are in essence valuing the reduction of small risks to numerous individuals. If a risk to each of one million individuals is reduced from three chances in a million of dying in a given year to two chances in a million of dying in that year, the expected number of lives saved is one, which is defined as one statistical life saved.

Regarding most counterterrorism activities, spending more money could reduce specific risks of loss of life to some individuals. More safety features on aircraft, more security checks, hardening of potential targets of terrorists, and funding more activities to identify and intercept potential terrorists all cost money and, chosen well, would reduce life-threatening risks. Hence, an inherent component of almost all counterterrorism decisions is to make a value trade-off between the economic costs and loss of life, meaning the loss of statistical lives. Even though many people are uncomfortable with the concept of trading off economic costs against potential fatalities, there is no way to avoid them. We could act like it isn't necessary or avoid thinking about it, but any actions will imply that a certain value trade-off was made.

Some people consider it immoral to trade off loss of life versus economic costs. However, most moral theories hold that an action is not immoral when there are no alternatives, so making such a value trade-off is not immoral. Furthermore, many moral theories argue that refraining from actions that would save lives is immoral. To the extent that analysis can lead to better decisions that will save more lives, it is perhaps immoral not to explicitly address the crucial value trade-offs between costs and statistical fatalities. Thinking carefully about that value trade-off and making it explicitly may indeed improve the analysis and save lives.

The federal government has numerous programs that are intended to avoid fatalities. With many of these programs, such as highway safety programs and various health programs, fatalities can be avoided for an investment of much less than $10 million per statistical life saved. Another relevant fact follows from considering where the money comes from that the government uses to avoid public fatalities. It naturally comes from members of the public in the form of various taxes. Analyses have shown that on average individuals and families spend their money in ways that make their lives safer. In other words, the likelihood of dying in any one year is less if individuals have more disposable income. The analyses of this phenomenon indicate that an additional statistical life is lost for each $8.3 million to $15.0 million taken from individuals depending on whether it is taken in equal amounts from all families or proportional to a family's income (Keeney, 1997). Collectively, these facts suggest that a reasonable value trade-off is in the range of $6 million to $12 million per statistical fatality, because spending more money to avoid statistical fatalities from terrorism would lead to more statistical fatalities from other causes, which are borne by the same people.

Preliminary set of value trade-offs. The parameters in the homeland security utility function (15.3) indicate the relative importance of a unit of impact on each of the metrics. Being relative, we can arbitrarily choose the parameter for one metric and specify the other parameters with respect to it. As economic costs are essentially a continuous scale, whereas many of the other metrics involve numbers of things, it is convenient to report the value trade-offs relative to economic costs. Specifically, we will use the unit impact in terms of the direct economic costs to society as our basis for comparison, so we will assign a parameter of 1 to a $1 million direct economic cost to society.

Table 15.2 shows ranges of unit value trade-offs for the objectives and metrics that we identified for DHS decision making. These unit value trade-offs specify the economic cost of terrorism countermeasures that is equally valued (i.e., equally less desirable) as a unit consequence of each metric. For example, a value trade-off of $200,000 for a serious injury means that the equivalent cost of a statistical injury to society is $200,000 or that avoiding one such statistical injury would be worth an investment of $200,000. In Table 15.2, we assigned a range of $100,000 and $500,000 to the unit value trade-off for a serious injury, defined as an injury serious enough to prohibit normal life function for one year. This range, which is primarily for illustrative purposes, is based on literature surveys of trade-offs for injuries and illnesses of various severities and the logic relating injuries to statistical fatalities as discussed later.

Let us suppose that $8 million is chosen as the value trade-off for a statistical life. It is easy to link sets of the value trade-offs. For example, assume that we have valued 80 serious injuries equivalent to one statistical fatality and one statistical fatality equivalent to an economic cost of $8 million. Then, 80 serious injuries must be equivalent in value to a cost of $8 million, which leads to the unit value trade-off that $100,000 is equivalent to one serious injury.

This indicates that consistency checks are useful and easy to employ. Rather than calculate the value trade-off between serious injuries and economic costs, information and logic could have been used to specify it. Suppose that were done and it suggested a cost of $120,000 was equivalent to a serious injury for a year. This is not so different from the calculated $100,000 and it suggests that either assessment is reasonable. Furthermore, it provides a basis, along with the two interrelated value trade-offs, to develop a completely consistent set of the three by adjusting some or all of them slightly. For example, with further reflection on the relevant information, and perhaps with the gathering of additional information, it may be decided that $9 million is equivalent to one statistical fatality, one statistical fatality is equivalent to 75 serious injuries, and one serious injury is equivalent to $120,000, which are one consistent set of these value trade-offs.

Some value trade-offs are more difficult to assess because there is very little literature addressing them directly. In these cases, we used basic reasoning to provide ranges. Experiencing extended periods of fear is an example. The metric is the number of people experiencing significant episodes of fear, by which we mean a strong emotional reaction at least once a week over a period of a year. This would be a state in which one certainly would have a degraded lifestyle and perhaps have to obtain psychological or psychiatric help. The unit is one person who experiences such significant fear. For illustration we use a range of $1,000 to $10,000 of equivalent cost. The $10,000 level would be equivalent to a person who would need substantial psychological treatment; the $1,000 level would be equivalent to a person who routinely worries about terrorism events, but this does not rise to a condition requiring treatment.

We illustrate the specification of value trade-offs with two additional examples: disruption of lifestyles and inconveniences, both consequences of terrorist acts. Since terrorist acts are usually conducted in public places to instill fear, some people may choose to reduce their activities in large public places like shopping malls or sports events. This clearly disrupts their lifestyle. How much is it worth to avoid such disruption to an individual for one year? We give a range of $500 to $5,000. This is partially based on calibrating the value trade-off against the "fear" trade-off, which suggests that, while there is no substitution for fear, there are substitution mechanisms for the disrupted lifestyles, for example, going to smaller venues, shopping by Internet, etc. This leads us to conclude that the value trade-off for fear is higher than that for disruption. The upper bound of the value trade-off for disruption is based on consideration of what it would take to compensate one individual for a year of disruption of the kind we are talking about. It seems to us that $5,000 would be a generous compensation.

The value trade-off for inconveniences was developed quite differently. In this case, the trade-off is for 100,000 hours of wasted time, spread over many people, due, for example, to standing in airport security lines, added time to obtain visas, etc. Our proposed value trade-off range is $1 million to $5 million per 100,000 hours, which translates into $10 to $50 per hour. The lower bound comes from studies of productivity losses during electricity outages (between $3 and $10 per outage hour per person; see LaCommare & Eto, 2004; von Winterfeldt et al., 2004). However, during electricity outages, one can still pursue some useful activities, especially during daylight hours, suggesting the lower bound of $10/hour/person. The higher number is an estimate based on the assumption that many of the targeted waste hours are by professionals and that they are truly lost productive hours. For example, there is not much else that one can do while waiting in an airport security line.

Many of the metrics for consequences of terrorist attacks, consequences of counterterrorism efforts, and civil liberties indicate the number of individuals who have been impacted in specific ways. For each of these, a threshold level of significant impact, analogous to that discussed for illnesses, must be chosen in order to estimate the number of individuals experiencing that impact. To assess the relative value of deterring a potential terrorist attack or foiling an attack in progress, information about the possible consequences of such a successful attack and its likelihood is needed, for which the value can be calculated from the consequences of attacks as illustrated in Section 15.4.1. For the metrics for future terrorism threats, information is needed about how each of these influences the likelihood of attacks.

15.4 Uses of a Homeland Security Value Model

The DHS value model can be used to improve homeland security decision making in numerous ways. All follow from the same principle: if you better understand what you are trying to achieve and why, it is more likely that you will achieve it. The DHS value model unambiguously outlines what DHS wants to achieve.

15.4.1 *Evaluating Terrorism Risks*

The DHS value model (15.3) can be used to evaluate of terrorist risks as defined in Section 15.2. For this purpose, we need to assess p, q, and the consequences of a successful terrorist attack, specified under the strategic objective *reduce consequences of attacks*. Note that in this assessment we do not need to consider the consequences under the remaining strategic objectives. However, we have to consider that all assessments of probabilities and consequences are relative to the status quo, i.e., with current countermeasures and interventions in place. To illustrate, recognize that in Table 15.2, O_i, X_i, and x_i, $i = 1,\ldots, 13$, represent the set of 13 lower-level strategic objectives under the major strategic objective *reduce consequences of attacks*. Hence, if the consequences of a terrorist attack are $x = (x_1,\ldots,x_{13})$ and the utility function is $u(x) = u(x_1,\ldots,x_{13})$, the risk of that terrorist attack as defined in (15.1) can be represented as

$$\text{R(Terrorist Attack)} = \text{p(Attempted Attack)} * \text{q(Success}|\text{Attempt)} * [-u(x_1,\cdots,x_{13})], \tag{15.6}$$

where $u(x_1, \ldots,x_{13})$, is the DHS utility function (15.3).

Using (15.6), it is possible to evaluate the relative risks of different types of terrorist attacks, call them TA_j, $j = 1,\ldots,J$. The consequences of each attack, if successful, can be referred to as x_j, $j = 1,\ldots,J$, where $x_j = (x_{1j}, x_{2j},\ldots,x_{13j})$. Because of uncertainty, we might want to assess a probability function $p_j(x)$ describing the possible consequences of TA_j, from which an expected utility u_j can be calculated. From (15.3), the utility of no terrorist attack is 0, so all utilities of terrorist attacks will be less than zero. The absolute magnitude of u_j then indicates the (dis)utility of the terrorist attack TA_j if it successfully occurs and the magnitude of $R(TA_j)$, which we can write as R_j, indicates the disutility of the attempted terrorist attack TA_j. The relative magnitudes of the u_js indicate the relative significance of successful TA_js, and the relative magnitudes of the R_js indicate the relative importance of additionally protection against terrorist attacks TA_j.

15.4.2 *Evaluating the Benefits of Countermeasures*

We define the risk of terrorism given that countermeasure C has been implemented as follows:

$$\begin{aligned}&R_C(\text{Terrorist Attack with Countermeasure C})\\&=p_C(\text{Attempted Attack}*q_C(\text{Success}|\text{Attempt})*[-u(x_C)],\end{aligned} \tag{15.7}$$

where p_C is the probability of the attempted attack given the countermeasure C, q_C is the probability that the attack is successful given the countermeasure C, x_C is the consequence of such an attack, as defined in Section 15.4.1, if successful given the countermeasure C, and R_C(Terrorist Attack) is the disutility of that terrorist attack given that countermeasure C has been implemented. In general we would expect $p_C \leq p$, $q_C \leq q$, and $u(x_C) > u(x)$.

The benefit B of a countermeasure to terrorist attack TA can be measured by

$$B(\text{Countermeasure}) = R(TA) - R_C(TA) \tag{15.8}$$

The decisions necessary to manage terrorism are very complex. Hence, it is not surprising that it is a difficult task to gather all of the information necessary to systematically appraise terrorism management alternatives as outlined here. However, it has been done with some success in several DHS contexts. Examples include assessing the risks of bioterrorism (DHS, 2006), the risk of surface-to-air-missile attacks against commercial aircraft (von Winterfeldt & O'Sullivan, 2006), and evaluating the benefits of the Western Hemisphere Travel Initiative (LaTourette & Willis, 2007). Most previous studies used relatively simple value models, which employed a small set of objectives and consequence measures (usually X_1 to X_6), and value trade-offs justified by the literature.

Fortunately, the amount of effort to construct a high-quality value model is very small compared to the effort to estimate the possible consequences x and x_C necessary to evaluate counterterrorism alternatives of homeland security. Since the value model is applicable across terrorism risks and across countermeasures, it would be a very effective allocation of effort to develop a DHS value model and use it, sometimes with appropriate adaptation for specific risk and benefit assessments. With such use, the evaluation of terrorism management alternatives should be internally more consistent, require less time, and be easier to explain to others.

15.4.3 *Developing a Severity Index for Terrorist Consequences*

The same objectives, metrics, and value trade-offs in Table 15.2 provide the basis for creating a severity index for both possible and any actual terrorist attacks. Specifically, an index for the severity SI of a successful terrorist attack resulting in consequence $(x_1,\ldots,x_{13})$ is

$$SI(x_1,\cdots,x_{13})=k_1x_1+k_2x_2+\cdots+k_{13}x_{13}, \tag{15.9}$$

where a larger SI indicates a greater severity. Also, the magnitude of severity indicated using (15.9) is linear, meaning that a severity calculated as 200 is 10 times as great as one calculated to be 20.

To illustrate how the severity index could be used, suppose a terrorist attack occurred that resulted in 30 fatalities, 200 seriously injured individuals, and $20 million in direct costs collectively to individuals and businesses, that caused fear to 50,000 citizens, and that led to $75 million in indirect economic costs. Furthermore, suppose that the other consequences of terrorist attacks listed in Table 15.2 were essentially negligible. Thus, $x_1 = 30$, $x_3 = 200$, $x_4 + x_5 = 20$, $x_7 = 75$, and $x_9 = 50{,}000$, with the other $x_i = 0$. For this illustration, we will also assume that the relevant unit value trade-offs are the midpoints of the illustrative ranges in Table 15.2, so $k_1 = 9$, $k_3 = 0.3$. $k_4 = k_5 = 1$, $k_7 = 1$, and $k_9 = 0.0055$. Substituting into (15.9), the severity index SI for this attack, call it A, is

$$\begin{aligned} SI_A &= 9(30)+0.3(200)+1(20)+1(75)+0.0055(50{,}000) \\ &= 270+60+20+75+275 \\ &= 700. \end{aligned} \tag{15.10}$$

From the calculation leading to (15.10), it is easy to see the relative contributions of the different consequences to the overall severity. In this case, fatalities and fear each account for just under 40% of the total severity.

The same concepts and formulation can be used to evaluate the seriousness of possible different types of terrorist attacks in the future, such as a suicide bomber, destruction of a government building, a dirty bomb in a harbor, or a small nuclear explosion near a large concentration of people. For these calculations, estimates are needed of what the possible consequences may be. There would naturally be uncertainties about the possible consequences. Since the severity index (15.9) is linear in each of the consequences, the expected severity of a future attack can be calculated by using the expected consequences of each of the metrics that describe that attack in (15.9).

As an illustration, suppose that the only consequences of the four types of attacks mentioned earlier are fatalities x_1 and direct economic costs to society x_6. Table 15.3 labels the four possible attacks as B, G, D, and N, describes them in terms of their consequences, and calculates their severity index with (15.9) using midpoint unit value trade-offs $k_1 = 9$, $k_6 = 1$. As these severities are based

Table 15.3. *Illustrative calculations for severity indices*

Terrorist attack	X_1 = fatalities ($k_1 = 9$)	X_6 = direct costs to society($mil) ($k_6 = 1$)	SI (severity index)	MSI (magnitude severity index)
B: Suicide bomb in public transportation system	15	40	175	2.2
G: Destroy government building with truck bomb	600	1,500	6,900	3.8
D: Dirty bomb in the Los Angeles/Long Beach harbor area	20	50,000	50,180	4.7
N: Small nuclear explosion near population center	40,000	5,000	365,000	5.6

on the linear scale, it is correct to interpret that the dirty bomb in a harbor has a severity magnitude of about 7 times the destruction of a government building.

Severity scales for natural catastrophes, such as earthquakes and hurricanes, use a log scale (i.e., magnitude scale) rather than a linear scale as in (15.9). For example, an earthquake of Richter magnitude 7 has 8 times the destructive potential of an earthquake of Richter magnitude 6. A log scale allows one to include magnitudes of two events, where one event is a billion times larger than the other, on a 1 to 10 scale. To be on a par with these natural indices, it is easy to take the log of the severity calculated with the SI index to create a magnitude severity index. We could refer to this index as MSI, which is simply calculated as

$$\text{MSI}_A = \log(\text{SI}_A) \tag{15.11}$$

for any terrorist attack A. Applying (15.11) to the attacks in Table 15.3 yields the MSI results shown there. Aside from providing numbers for severity that are in the same ranges as those for natural catastrophes, these numbers are easier to refer to and capture the magnitudes of potential terrorist attacks without being overly precise about differences that are very difficult to estimate.

15.4.4 *Improving the Quality of DHS Decision Processes*

As the short-term and long-term effects of using a DHS value model are somewhat different, we discussed the two cases separately.

Short-Term Effects. The DHS value model provides a solid foundation to construct an appropriate value model for any specific decision, such as protecting our borders, installing new security systems at airports, responding to terrorist attacks, or managing a terrorist watch list. For any specific homeland security decision context, one should consider how each objective in the DHS value model might be influenced by potential choices for the specific decision. From this, you essentially follow the process outlined in Section 15.3. You need to identify all relevant objectives, structure them into both a fundamental objectives hierarchy and a means-ends objectives network, choose metrics for the objectives, and then weight the fundamental objectives. Appropriate fundamental objectives for a specific decision can be logically developed from the strategic objectives of DHS. Appropriate weights for the specific value model can be logically derived from the weights used in the DHS strategic value model.

Once you have identified all the objectives for a specific decision, they can be used for creating alternatives, developing a decision model, indicating what data and information are needed, evaluating alternatives, and communicating about the decision.

All objectives, but particularly all of the means objectives, can be used to stimulate the creation of alternatives. For each objective, one asks what could be done that might better achieve this objective. Consider a possible terrorist attack that was intended to release a serious pollutant that would harm people if inhaled. Alternatives could be developed that directly influence means objectives such as minimizing pollutant emissions, minimizing pollutant concentrations, or minimizing human exposure to pollutants that could contribute to better achieving the fundamental objective of minimizing illness.

For complex decisions, a decision model is useful to relate each of the alternatives to the consequences that will possibly occur. To guide the development of the model, the means-ends objectives network is helpful. This network breaks down these relationships in a manner consistent with the data and information that must be gathered. Using our example, information can be collected on where and how much pollutant might be released. Then either an atmospheric model, for air pollutants, or a water flow model, for water pollutants, can usefully relate the emissions to concentrations. Next, a population model can relate concentrations to exposures. And finally a medical model can relate exposures to illnesses. At this stage, the alternatives can be analyzed to indicate how each performs using the utility function.

Obviously, sensitivity analysis can indicate what changes occur in the relative evaluation of alternatives and how significant they are when input information is varied. By viewing different objectives or sets of objectives in the utility function, one can identify the relative pros and cons of each of the alternatives in terms of how well the alternatives meet the various objectives. As it is achieving the objectives that is of concern to people, it is useful to describe the pros and cons of the alternatives in communicating to the public.

For some DHS decisions, the formality suggested earlier in this chapter is probably not practical or useful. In these cases, it is still critically important to understand the objectives of the decision context and to create innovative alternatives to better achieve them. Thus, even if no model is explicitly used for evaluating alternatives, the quality of the decision-making process can be enhanced by listing of objectives explicitly and using them to create alternatives and to understand their pros and cons.

Long-Term Effects. Three key means objectives in the DHS means-ends objectives network in Figure 15.1 contribute to long-term quality decision processes. These are to promote research on homeland security, create educational programs on homeland security, and create a profession with expertise on homeland security. The DHS value model can be used as an outline of the substance of concern for each of these. We want quality research that will help us better understand how well proposed alternatives will measure up in terms of specified objectives and that will suggest other alternatives that might be even better. We want educational programs to educate a substantial group of individuals who understand the body of knowledge concerning all of the relationships among homeland security objectives and the alternatives that influence the achievement of these, as well as their significance. Some of these individuals will hopefully join a growing group of homeland security professionals who contribute to enlarging our collective knowledge about terrorism and guide coherent effective counterterrorism programs to reduce terrorism as a threat and bring about an enhanced quality of life.

15.5 Conclusions

This chapter makes and supports the following two points:

1. Constructing and using a legitimate value model for homeland security could make a substantial contribution to improving homeland security. Hence, constructing such a value model should be a high-priority task. Its existence would guide the thinking, decisions, and actions of homeland security and promote different groups and individuals to work toward accomplishing the same national goals.
2. The knowledge, techniques, and experience necessary to construct a high-quality value model for homeland security currently exist. This chapter summarizes the knowledge and techniques required to construct a preliminary homeland security value model and illustrates its use. This demonstrates that a value model for homeland security that meets a high standard of relevance and legitimacy can be constructed and how to do it.

A homeland security value model should be constructed by a project with a diverse team of informed members, including DHS policy makers, managers, stakeholders, and a few experts on value methodology. Their task would be to carefully conduct the procedures described in this chapter. Specifically, it should identify a complete set of strategic objectives for counterterrorism, specify good metrics for each of these objectives, verify that the additive-linear form of utility function is appropriate or adapt it as necessary, establish a logical set of value trade-offs, and document the entire effort. Throughout the project, the team should frequently gather and exchange information with personnel in any other governmental departments, local or national, concerned with homeland security, and frequently consult the public to ensure that its concerns are addressed.

REFERENCES

Bier, V., & von Winterfeldt, D. (Eds.) (2007). *Terrorism risk analysis. Special Issue of the Journal Risk Analysis, 27*, 503–634.

Bond, S. D., Carlson, K. A., & Keeney, R. L. (2008). Generating objectives: Can decision makers articulate what they want? *Management Science, 54*(1), 56–70.

Bond, S. D., Carlson, K. A., & Keeney R. L. (2010). Improving the generation of decision objectives, *Decision Analysis, 7*(3), 238–255.

Chertoff, M. (2007, August). Speech at the Center for Risk and Economic Analysis of Terrorism Events, University of Southern California.

Ezell, B., Bennett, S. P., von Winterfeldt, D., Sokolowski J., & Collins, A. J. (2010). Probabilistic risk analysis and terrorism risks. *Risk Analysis, 30*, 575–589.

Fishburn, P. C. (1965). Independence in utility theory with whole product sets. *Operations Research, 13*, 8–45.

Franklin, B. (1772). Letter to Joseph Priestley. Reprinted in W. Isaacson (Ed.) (2005). *A Benjamin Franklin reader*. New York: Simon and Schuster.

Keeney, R. L. (1992). *Value-focused thinking: A path to creative decisionmaking.* Cambridge, MA: Harvard University Press.

Keeney, R. L. (1995). Understanding life-threatening risks. *Risk Analysis, 15*, 627–637.

Keeney, R. L. (1997). Estimating fatalities induced by the economic costs of regulations. *Journal of Risk and Uncertainty, 14*, 5–23.

Keeney, R. L. (2007). Developing objectives and attributes. In W. Edwards, R. F. Miles, & D. von Winterfeldt (Eds.), *Advances in decision analysis* (pp. 104–128). Cambridge: Cambridge University Press.

Keeney, R. L., & Raiffa, H. (1976). *Decisions with multiple objectives.* New York: Wiley. Reprinted in 1993, Cambridge University Press.

Keeney, R. L., & von Winterfeldt, D. (2007). Practical value models. In W. Edwards, R. F. Miles, & D. von Winterfeldt (Eds.), *Advances in decision analysis* (pp. 232–252). Cambridge: Cambridge University Press.

LaCommare, K. H., & Eto, J. H. (2004). Understanding the cost of power interruptions to U.S. electricity customers. Technical Report LBNL-55718, Lawrence Berkeley Laboratory, Berkeley, CA.

LaTourette, T., & Willis, H. (2007). Using probabilistic terrorism risk modeling for regulatory cost-benefit analysis. WR-487-487-IEC. The RAND Corporation, Pittsburg, PA.

National Research Council (2008). *Bioterrorism risk analysis: A call for change.* Washington, DC: National Academy Press.

Pratt, J. W., Raiffa, H., & Schlaifer, R. (1964). The foundations of decision under uncertainty: An elementary exposition. *American Statistical Association Journal*, pp. 353–375.

Savage, L. J. (1954). *The foundations of statistics*. New York: Wiley.

U.S. Department of Homeland Security (DHS). (2006). *Bioterrorism risk analysis*. Biological Threat Characterization Center, National Biodefense and Countermeasures Center, Fort Dedrick, MD.

U.S. Department of Homeland Security (DHS). (2010). Website of the Office of Risk Management and Analysis. http://dhs.gov/xabout/structure/gc_1185203978952.shtm, viewed November 23, 2010.

von Neumann, J., & Morgenstern, O. (1947). *Theory of games and economic behavior*, 2nd edition. Princeton, NJ: Princeton University Press.

von Winterfeldt, D., Eppel, T., Adams, J., Neutra, R., & DelPizzo, V. (2004). Managing potential health risks from electric powerlines: A decision analysis caught in controversy. *Risk Analysis*, *24*, 1487–1502.

von Winterfeldt, D., & O'Sullivan, T. (2006). Should we protect commercial airplanes against surface-to-air missile attacks by terrorists? *Decision Analysis*, *3*, 63–75.

Willis, H., Morral, A. R., Kelly, T. K., & Medby, J. J. (2005). *Estimating terrorism risks*. Technical Report, RAND Corporation, Pittsburgh, PA.

16

Identifying, Structuring, and Comparing the Objectives of Al Qaeda and ISIL

Johannes Siebert and Greg Keeney

16.1 It Is Important to Understand Terrorist Objectives

Islamic terrorist groups have dominated the world stage for more than a decade as one of the most severe threats to Western society. They kill, maim, and instill fear into large numbers of individuals, destroy infrastructure and culture heritage, and cause long-term economic and political damage to the nations in which they operate in order to pursue their political, military, and religious objectives. Waugh (1983) argued that "the importance of assessing the goals and value priorities, as well as the capabilities and strengths of commitment, of challenging terrorist organizations, is clearly necessary for the determination of government bargaining strategies."

In this context, McCormick (2003) (based on the previous work of Tullock, 1974; Hacker, 1976; Price, 1977; Corsi, 1981; Waterman, 1981; Sandler et al., 1983; Waugh, 1983; Muller & Opp, 1986; Sandler & Lapan, 1988; Lapan & Sandler, 1993; Enders & Sandler, 1995) defined terrorism "as an instrumental activity designed to achieve or help achieve a specified set of long-run and short-run objectives."

Keeney (1992) proposed several methods to identify objectives through discussion with the decision maker. However, it is very difficult and potentially life threatening to obtain access to terrorist leaders. If access were even possible, terrorists would almost certainly not assist their adversaries and would refuse to cooperate in such a procedure. Therefore, the methods Keeney (1992) suggested cannot be applied to such an organization. In this chapter, we provide an overview about suitable methods to identify objectives of organizations who are or who are not willing to cooperate.

The fight against Al Qaeda led by the United States over the past decade has accomplished clear victories. Al Qaeda's leader, Osama Bin Laden, was located and killed, and its leadership was systematically shattered. Al Qaeda is still active in many countries; however, the significance and influence of the organization has been reduced, which has contributed to a safer world. Simultaneously, the power vacuum created by a poorly run Iraqi government and military has given a strong foothold to the Islamic State of Iraq and the Levant (ISIL). ISIL has expanded explosively across Iraq and Syria, and threatens its neighbors and the West. Waugh (1983) pointed out that not "all terrorist organizations do … have the same objectives." Therefore, it is questionable whether methods that proved effective in the fight against Al Qaeda will also be effective in combating ISIL. To answer this question, it is not only necessary to identify the objectives. It is also necessary to compare the objectives of these terrorist groups and identify differences. Developing a strategy to fight ISIL that will achieve long-term success should evaluate and consider the differences between ISIL and Al Qaeda. In this chapter, we summarize and illustrate the results of the detailed comparison of Al Qaeda's and ISIL's objectives carried out by Siebert and colleagues (*submitted*).

In Section 16.2, we provide an overview about techniques to identify the objectives of both organizations that are deemed friendly and those that are deemed foes. We summarize how objectives can be structured and how two objectives hierarchies can be compared systematically. In Section 16.3, we illustrate in a case study and compare the objectives hierarchies of Al Qaeda and ISIL. In Section 16.4, we draw conclusions regarding the methodology, Al Qaeda's and ISIL's objectives, and counterterrorism strategies.

16.2 Identifying, Structuring, and Comparing Objectives

16.2.1 *Identifying Objectives*

In his book *Value-Focused Thinking: A Path to Creative Decisionmaking*, Keeney recommends 10 techniques to identify objectives, illustrated in Table 16.1: (1992, 57).

These techniques are based on a discussion between a facilitator (e.g., a decision analyst) and the decision makers and stakeholders in a decision situation. Keeney (1992) suggests to begin such a discussion by asking the decision maker, "What would you like to achieve in this situation?" Actively formulating the desired outcome(s) provides insight into what the decision maker wants, values, or should want. In the following steps, the thinking process can be inspired by considering alternatives, problems and shortcomings, consequences, goals, constraints, and guidelines.

These techniques require that decision analyst and decision maker cooperate. In extreme cases, the "opponent organization" has no interest in cooperation; it

Table 16.1. *Devices to use in identifying objectives*

1. A wish list
2. Alternatives
3. Problems and shortcomings
4. Consequences
5. Goals, constrains, and guidelines
6. Different perspectives
7. Strategic objectives
8. Generic objectives
9. Structuring objectives
10. Quantifying objectives

is even hostile and is not an official entity, e.g., a criminal or terrorist organization. In such cases, the only way to identify values and objectives is to analyze information provided by these organizations consciously or unconsciously. G. Keeney and von Winterfeldt (2010) developed a method for identifying objectives of a terrorist organization by using the available primary written sources along with secondary sources.

To elicit the values and objectives of adversaries, a researcher can rely on both primary and secondary sources of information. For avoiding doubt, primary sources are defined as "original works of research or raw data without interpretation or pronouncements that represents an official opinion or position" (Cooper & Schindler, 2008, 104). In the context of obtaining the values and objectives of an adversary, there are two classes of primary sources of information: direct statements and observed actions of the adversary organization and their associates. The religious and spiritual leaders of terrorist organizations often give speeches that are intended to address both their followers and their enemies. Since these leaders are interested in proliferating their message across the globe, they frequently turn to the Internet to maximize their reach. Social media tools are particularly popular and powerful for achieving this goal. Separately, yet related, many followers of terrorist organizations use social media tools to share their own views and to recruit new followers. Values and objectives from terrorist adversaries can also be extracted during interrogation sessions.

Actions, in addition to statements, are another form of primary source. By observing the actions of an adversary's organization and its associates, one is able to elicit and infer values and objectives. For example, if an Islamic terrorist group executes prisoners by beheading, one could infer objectives such as

Kill and Frighten Individuals or *Create a Brutal and Radical Brand*. In addition to recent actions, it is also possible to analyze the history of an organization and pivotal moments and actions in its past.

"Secondary sources are interpretations of primary data" (Cooper & Schindler, 2008, 104). In the context of obtaining values and objectives of adversaries, we rely on two classes of secondary sources of information. One class is information obtained from experts: expert interviews conducted pro rata and expert interviews that are publicly available. The other class consists of news reports and articles based on primary and secondary sources of information. Note that experts may carry personal bias or have their own objectives vis-à-vis their interview, and media groups often lean toward supporting a specific political agenda, while also depending on advertising and sales. Evaluating a combination of several sources may help identify inconsistencies or bias. When conducting such an analysis, it is also important to keep possible differences between short-term objectives, and long-term objectives and values in mind. It is more likely that long-term decisions reflect the values of an organization than short-term decisions.

In academic literature, there are two projects documented in which researchers identified objectives for terrorist groups. G. Keeney and von Winterfeldt (2010) examined Al Qaeda. They analyzed primary documents from books and the Internet (e.g., written speeches and statements). When using Internet sources, they pursued a structured approach using a bookmark tagging system to systematically categorize various news articles and video clips about Al Qaeda. They identified 82 tags and ranked them by frequency. They identified five strategic, nine fundamental, and 19 means objectives of Al Qaeda. As a final step, they arranged the fundamental and strategic objectives in a detailed means-ends network. More detail is provided in Section 16.3.1.

Siebert and colleagues (2016) extended the method of G. Keeney and von Winterfeldt (2010) by including structured interviews with experts to elicit the strategic, fundamental, and means objectives of the leadership and the followers of the terrorist group ISIL. The authors derived a means-ends network that reveals the objectives and their relationships. The evaluation of written materials and the structured interviews were carried out independently by different researchers. Their results were then merged. More details on the procedure can be found in Section 16.3.1. They found that a more complete and less biased objective determination was achieved by conducting two separate analyses of the findings by two researches. After eliciting the objectives, a means-ends network was constructed that revealed the objectives and the interrelationships of the leaders and followers of ISIL.

Table 16.2. *Four-step procedure to guide the top-down approach for creating an objectives hierarchy*

Step 1	Create a list of strategic, fundamental, and means objectives
Step 2	Ensure that the strategic objectives cover all values supported by underlying fundamental objectives
Step 3	Ensure all fundamental objectives are supported by underlying means objectives
Step 4	Organize means objectives to have direct relationship with one or more fundamental

Source: Siebert et al. *submitted.*

16.2.2 *Structuring Objectives*

Based on a comprehensive list of objectives, a decision analyst can obtain an overview of the values of an organization that is generally fairly complete. At first, this list is merely an unrelated set of data. Structuring these objectives by means of a detailed analysis, however, reveals the relationships between objectives and creates order, enabling the derivation of metrics for multiattribute utility models. In this process, it is essential to distinguish between strategic, fundamental, and means objectives.

G. Keeney and von Winterfeldt (2010) stated that "strategic objectives provide guidance for all decisions. They serve as the mechanism by which leaders can guide decisions made by different individuals and groups within an organization." In contrast, "fundamental objectives concern the ends that decision makers value in a specific decision context" (Keeney, 1994), i.e., "Fundamental objectives provide guidance for specific decisions that are usually carried out over the medium to long term" (G. Keeney & von Winterfeldt, 2010) and concern an essential reason for making a decision. Means objectives are the specific actions and activities that serve fundamental and strategic objectives. Thus, strategic, fundamental, and means objectives form a hierarchy. The usually long-term strategic objectives can be achieved by pursuing fundamental objectives. Means objectives concern actual actions and operations that are pursued in order to accomplish fundamental and in the end strategic objectives.

Top-down or bottom-up approaches can be used to structure objectives. The top-down approach can be formulated as a procedure of four consecutive steps as illustrated in Table 16.2.

The bottom-up approach is typically more labor-intensive. By revealing important patterns of relationships between objectives, one can successively identify different layers of objectives. When determining whether an objective is linked to a higher-level objective, an analyst must evaluate whether the objective directly supports achieving the higher-level objective. For example, *Increasing the Number of Fighters and Followers* directly supports being able to *Eliminate the Current Rulers in Iraq and the Levant*, which in turn directly supports being able to achieve the objective of *Control and Govern the Islamic State*. Such means-ends chains provide the basis to derive the aforementioned objectives hierarchy. In the first step, means objectives are clustered and linked to fundamental objectives. In the second step, fundamental objectives are linked to strategic objectives. The two approaches are not mutually exclusive, and it is useful to combine features of both approaches in order to conduct a thorough analysis.

The analyst's interpretation of objectives might influence the result in both approaches, and therefore a "derived objectives hierarchy should never be considered as the only possible configuration" (Siebert et al., 2016). It is quite possible that there are even differences in objectives hierarchies created by different decision analysts using the identical list of objectives. However, Siebert and colleagues demonstrated that two objectives hierarchies developed from different source material and using different analysts can be very similar.

The final goal of these structuring procedures is the formulation of a means-ends network that unveils all important relations between objectives and visualizes the complex interactions between objectives. A means-ends network can include a one-to-many set of relationships such as illustrated in Figure 16.1.

If mostly single relationships are considered, we refer to objectives hierarchies such as the hierarchy used for structuring the objectives of ISIL (Siebert et al. 2016), shown in Figure 16.3 in Section 16.3.1. The purpose of such an illustration is to define clear relationships in clusters, and to distinguish between strategic, fundamental, and means objectives. Objectives in a lower level of the objectives hierarchy contribute often to more than one objective of the higher level. Therefore, a lower-level objective is assigned to the higher-level objective it contributes to most.

16.2.3 *Comparing Objectives Hierarchies*

An objectives hierarchy should provide a comprehensive and detailed structure that guides the majority of decision situations for an organization.

However, objective hierarchies are not set in stone as organizational objectives can change over time. Leadership change could cause a shift in strategy, resulting in all decision maker members of an organization being required to use the new objectives hierarchy. Understanding the differences between past and present strategies is important to successfully clarify and enact a change in behavior. With the example of a sudden appearance of a new competitive

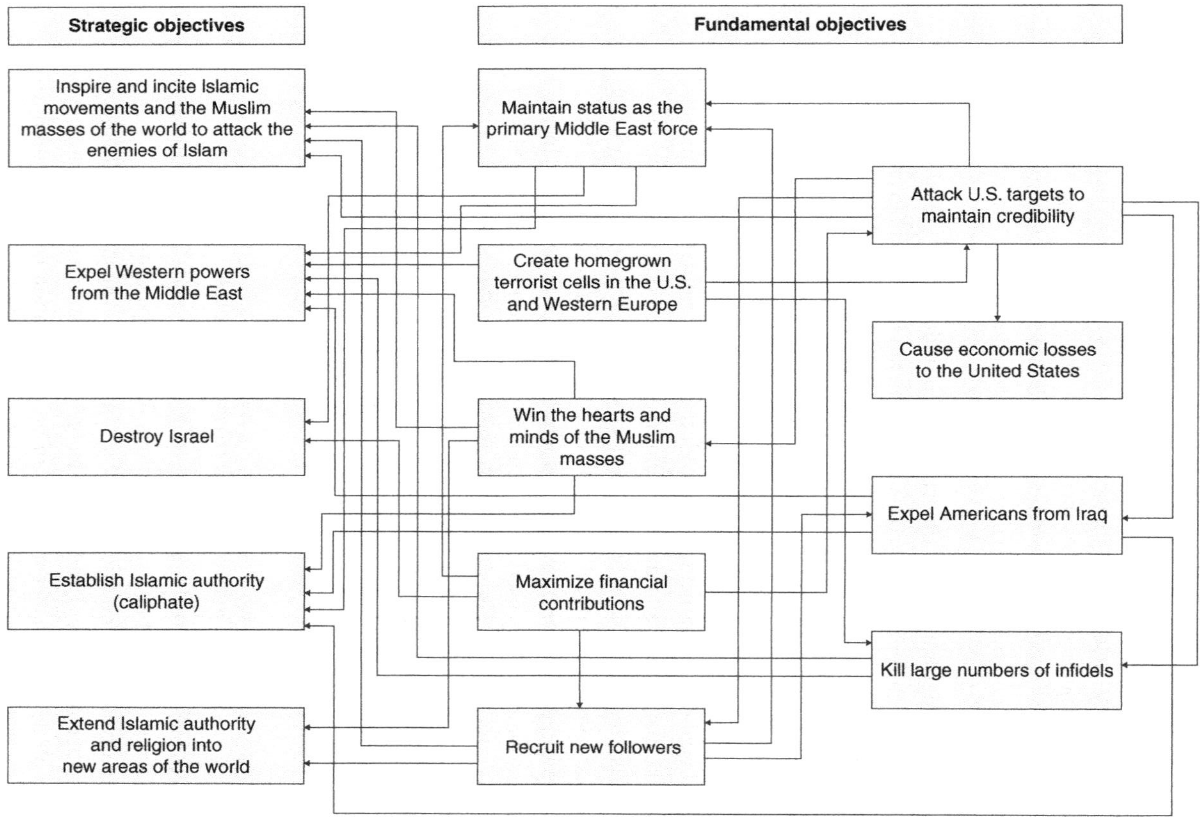

Figure 16.1. Relationships between fundamental and strategic objectives of Al Qaeda (an arrow from A to B means "A is significant for achieving B").
Source: G. Keeney and von Winterfeldt (2010).

Table 16.3. *Four-step procedure to guide the systematical comparison of the objectives of two decision makers*

Step 1	Decision analyst(s) involved in the separate objectives hierarchies should exchange information to develop a common understanding of these hierarchies
Step 2	Decision analyst(s) should ensure that the terms *strategic*, *fundamental*, and *means objectives* are used consistently for all hierarchies and the same number of layers are used
Step 3	Decision analyst(s) should compare the objectives layer by layer starting with the strategic objectives because these represent the long-term goals of the decision maker
Step 4	Decision analyst(s) should step back and compare the objectives hierarchies on a macro level. He or she should ensure that their analysis captures all relevant similarities and differences

Source: Siebert et al. *submitted.*

or hostile organization, applying a generic strategy to deal with a changed competitive landscape might not be effective, e.g., the strategies to combat Al Qaeda should not be transferred to ISIL without sufficient analysis. Therefore, it is necessary in such cases to systematically identify differences in two or more objectives hierarchies.

For this purpose, Siebert and colleagues (*submitted*) proposed a four-step procedure, which is illustrated in Table 16.3.

16.3 Case Study: Comparing Objectives of Two Terrorist Groups

In this section, we illustrate the systematic comparison of two objectives hierarchies by summarizing the results of the comparison of the two Sunni terrorist groups Al Qaeda and ISIL. In the following, *italics* are used for citing objectives identified in the papers by G. Keeney, John, Siebert, and von Winterfeldt.

16.3.1 *Determining the Objectives of Al Qaeda and ISIL*

After the terrorist attacks of September 11, 2001, Al Qaeda emerged as the predominant Islamic terrorist group, enjoying an unprecedented effect on Western society, receiving global attention, and spreading fear throughout many nations. Al Qaeda is rumored to be founded at a meeting in 1988, toward the end of the Soviet occupation of Afghanistan, by Osama Bin Laden, Aiman az-Zawahiri, and Sayyed Imam Al-Sharif (Rollins, 2011; Wander, 2008). The terrorist group originated from a network called Maktab al Khidamat (also known as Al Khifah), which was founded by Osama Bin Laden and aimed at recruiting fighters for the Afghan war against the Soviet invasion (Rollins, 2011). Al Qaeda consisted of a base of 10,000 to 20,000 participants in the 1980s. Osama Bin Laden's goal was to topple secular, pro-West Arab leaders (Rollins, 2011) and to support fighters for the liberation of unrightfully occupied Muslim territory (Zelin, 2014), goals that were carried out through many different terrorist attacks in the decade of the 1990s. The aims of the group advanced toward an attempt to establish an Islamic state throughout the mid-2000s. Killing Osama Bin Laden in 2011, along with many other senior leaders of the group (Zelin, 2014), resulted in the general notion that Al Qaeda was shattered and defeated. Despite its reduced size and influence, Al Qaeda is still working to achieve its goals, and can be considered a competitor of ISIL (Pop, 2015).

An insurgency in Syria and Iraq in 2004 was the starting point for ISIL. Under its first leader, Abu Musab al Zarqawi, ISIL was affiliated with Al Qaeda. However, a schism occurred when ISIL began to develop extremely brutal forms of terrorism that included beheadings and attacks on Shia Muslims. After al Zarqawi was killed by a U.S. bomb strike, ISIL's new leader, Aby Bakr al Bagdadi, developed a new strategy, which was focused on establishing a regional caliphate as soon as possible. This differs significantly from the Al Qaeda strategy of worldwide attacks on Westerners. Additionally, there is a large difference in the groups' outlooks on brutal and gruesome killings; Al Qaeda leaders believe these should be avoided, the ISIL leadership uses such acts as a key component of its strategy. Sunni extremists and former Baathist military officers dominate ISIL, and followers are predominantly recruited from disenfranchised Sunni youths in Iraq and Syria. However, a large number of foreign fighters from Arab and Western countries, including religious converts, criminals, and adventurers, have also joined ISIL.

The withdrawal of U.S. troops from Iraq caused a power vacuum. ISIL took advantage of the opportunity and filled the void with powerful military force. At the end of 2014, ISIL had about 30,000 troops under its command (Cockburn, 2014). This included about 5,000–10,000 foreign fighters (Ackermann, 2014; Altman, 2014; Lund, 2013). ISIL has also shown prowess as a revenue-generating machine, drawing income from taxation, oil sales, smuggling operations, and the theft and sale of culturally important antiques, at times exceeding $1 million of revenue per day (Ackermann, 2014). This income finances military operations and has allowed ISIL to create a semblance of limited governmental administration activities.

At the end of August 2014, President Obama stated, "we don't have a strategy yet for Islamic State" (*Washington Times*, August 28, 2014). A starting point to systematically create a strategy to combat a hostile organization is to

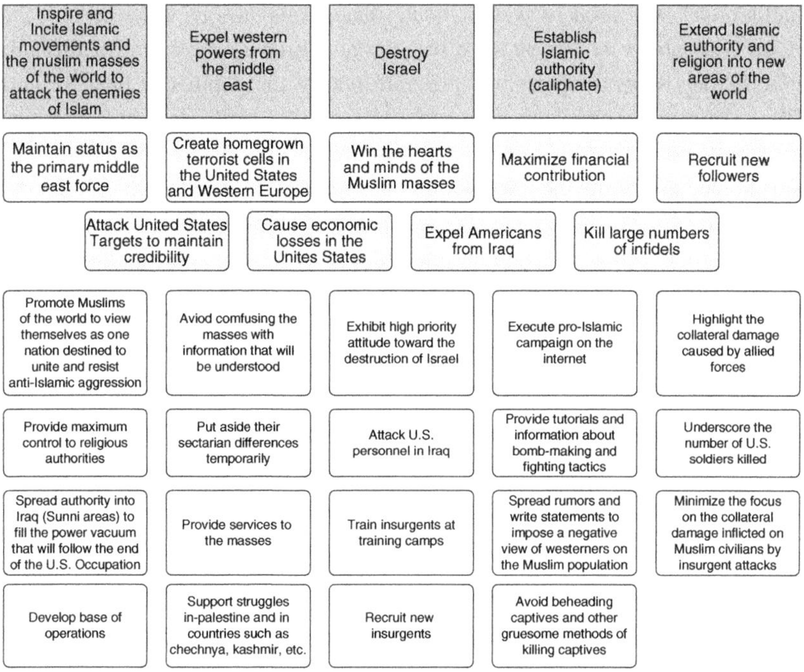

Figure 16.2. Al Qaeda's strategic, fundamental, and means objectives.
Source: G. Keeney and von Winterfeldt (2010).

determine its objectives. Any strategy that is implemented will not be successful unless it prevents the hostile organization from achieving its objectives. For a deeper understanding of ISIL, it is important to not only identify its objectives, but also to analyze its differences compared to Al Qaeda.

G. Keeney and von Winterfeldt (2010) analyzed primary documents from books and the Internet (e.g., written speeches and statements by Al Qaeda leaders), and thus identified five strategic, nine fundamental, and 19 means objectives of Al Qaeda; they arranged the fundamental and strategic objectives in a detailed means-ends network (see Figure 16.1 in Section 16.2.2). The means objectives were illustrated in an additional table. Figure 16.2 summarizes Al Qaeda's objectives in an objective hierarchy, and Figure 16.3 does the same for ISIL. The strategic objectives are listed on the upper layer (dark grey), the fundamental objectives are listed in the middle layer (light grey), and the means objectives are listed in the lower layer (white).

With the case study of Al Qaeda, G. Keeney and von Winterfeldt relied on books and the Internet for their sources. A bookmark tagging system called "del.ici.ous" was used to systematically categorize various news articles and video clips about Al Qaeda from the digital sources. As a result of this process, 82 tags were used, and the program ranked the popularity of these tags. The most popular tags concerned top Al Qaeda leaders, field of operations, speeches, and tactics. Sources were almost always identified with multiple tags. This helped providing the information base we used to view in which direction the most recent Al Qaeda messages seemed to be leaning. This method enabled the analysis to remain current and organized, with regards to the most recent information available, providing insights into the organization's thinking (G. Keeney & von Winterfeldt, 2010).

Three of Al Qaeda's five strategic objectives reveal a strong religious motivation. These are *Extend Islamic Authority and Religion into New Areas of the World*, *Establish Islamic Authority (Caliphate)*, and *Inspire and Incite Islamic Movements and the Muslim Masses of the World to Attack the Enemies of Islam*. The other two strategic objectives are special military goals, namely, *Expel Western Powers from the Middle East* and *Destroy Israel*; both serve to eventually control the Middle East. Four of the nine fundamental objectives concern Al Qaeda's enemies. These are *Create Homegrown Terrorist Cells in the United States and Western Europe*, *Cause Economic Loss in the United States*, *Expel Americans from Iraq*, and *Kill Large Numbers of Infidels*. The three objectives *Maintain Status as the Primary Middle East Force*, *Attack U.S. Targets to Maintain Credibility*, and *Win the Hearts and Minds of the Muslim Masses* aim at gaining ever more power and consolidating Al Qaeda's position as a dominant force in supporting Islamic interests. The remaining two objectives are supporting activities for continued operations, i.e., *Maximize Financial Contributions* and *Recruit New Followers*.

Siebert and colleagues (2016) used the method of G. Keeney and von Winterfeldt (2010) and added expert interviews as a new source of information. Both efforts for gaining information were conducted independently and separately by two analysts who used the same decision analysis concepts and techniques, but did not share or exchange information in the first phase of the project. Both analysts spent roughly two months (September and October 2014) collecting and sorting information with one of the two approaches.

One approach required analyzing transcripts of interviews with 59 subject matter experts (SMEs). Typical questions were, for example, *What are ISIL's strategic objectives?* and *Why is ISIL so attractive to its followers?* The staff of the National Security Institute (NSI) conducted and transcribed the interviews and produced a 554-page document. The interview discussions covered a broad range of historical and political issues related to ISIL. U.S. Special Operations Command Central (SOCCENT) staff in consultation with a broader research team identified and selected the SMEs, who were experts on terrorism, Islam, the Middle East, international politics, and related fields. Many SMEs had been raised in the Middle East, and others had worked in the region and/or based their statements on interviews with ISIL followers and contacts. Most SME interviews were with single subject matter experts, and some interviews were conducted in small groups.

The other approach used publicly available primary and secondary sources. The primary sources used were transcripts of 12 speeches by ISIL leaders (Al Baghdadi and Al Adnani). The secondary sources used were expert interviews and articles from U.S. and German media, with a focus on reputable news organizations. We began our review of our source material in early September 2014, to identify the status of information on ISIL at this time. The majority of our source material was published between June and October 2014; however, newer developments were taken into account during the writing of this chapter. Primary and secondary sources were not simply combined but critically compared and thus used for validation.

In a second phase, again without interaction, means-ends networks that visually summarize and structure the objectives and their relationships were derived from the information collected in the first phase. In a third phase, the two analysts compared their results and as in other instances of combining objectives hierarchies from different stakeholders (e.g., Keeney et al., 1995), we combined unique objectives from both hierarchies into a single objectives hierarchy.

The 35 objectives (four strategic, five fundamental, two key means, and 24 means objectives) that ISIL pursues are illustrated in Figure 16.3. Two of ISIL's strategic objectives are political objectives, and the other two are religious objectives. The group wants to *Establish a Caliphate in Iraq and the Levant*, *Control, Govern the Islamic State*, *Recreate the Power and Glory of (Sunni) Islam*, and *Expand Islam and Sharia Law Worldwide*. *Establish[ing] a Caliphate in Iraq and the Levant* is supported by the primary fundamental objective of *Function as a State and Provide Services*. Successfully *Control[ing] and Govern[ing] the Islamic State* first requires ISIL to *Eliminate the Current Rulers in Iraq and the Levant*. ISIL needs to pursue the fundamental objectives of *Implement a Pure and Strict Version of Islam and Give Meaning to the Lives of Sunnis* in order *to Recreate the Power and Glory of (Sunni) Islam*, and ISIL must *Purge the World of Anti-Islamic Forces* in order to *Expand Islam and Sharia Law Worldwide.*

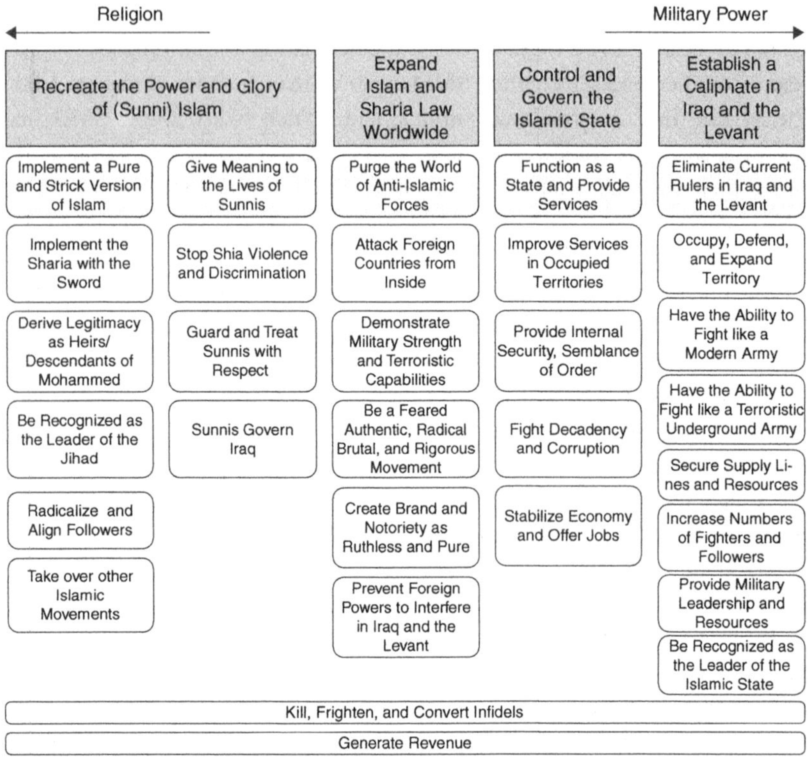

Figure 16.3. ISIL strategic, fundamental, and means objectives.
Source: Siebert and colleagues (2016).

16.3.2 *Comparing the Objectives of Al Qaeda and ISIL*

High-level similarities and differences between Al Qaeda's and ISIL's objectives are presented in this section. A detailed review, including expert statements that support our positions, can be found in Siebert and colleagues (*submitted*). We compare each groups' fundamental objectives related to religion, gaining power, establishing a caliphate, and killing enemies.

Objectives Related to Religion

Al Qaeda and ISIL both pursue a strategic religious objective. Al Qaeda wants to *Extend Islamic Authority and Religion into New Areas of the World* and ISIL wants to *Expand Islam and Sharia Worldwide*. Both extremist organizations use terrorism as the means to accomplish their objectives; however, there are key differences in their attitude regarding sharia law and literal interpretation of the Koran. Al Qaeda wants to represent all Muslims, whereas ISIL considers only Sunnis true followers of Islam.

Al Qaeda wants to *Win the Hearts and Minds of the Muslim Masses.* Therefore, it tries to develop a base that is suitable for followers from all Muslim sects. For that reason, Al Qaeda claims that it is willing to *Put Aside Sectarian Differences Temporarily*. In contrast, ISIL accepts only Sunnis as true followers of Islam and fuels sectarian conflicts as can be inferred from its strategic objective: *Recreate the Power and Glory of (Sunni) Islam.* ISIL uses sectarian differences as a mechanism for organizational growth, and clearly proclaims Shi'a followers as enemies; *Give Meaning to the Lives of Sunnis* and *Stop Shi'a Violence and Discrimination widens the gap between the two sects.*

Objectives Related to Territory and Power

Expel[ing] Western Powers from the Middle East is an important strategic objective for Al Qaeda and a related fundamental objective is to *Expel Americans from Iraq*. These objectives of Al Qaeda were already identified by G. Keeney and von Winterfeldt in 2010. Al Qaeda accomplished these objectives, to a certain degree, as the vast majority of U.S. troops have been withdrawn from Iraq and Afghanistan. As a result of the shifted landscape and military posture of the United States, ISIL does not consider the objective *Expel Western Powers from the Middle East* "strategic" and pursues instead the fundamental objective to *Eliminate Current Rulers in Iraq and the Levant.* This objective formerly served the strategic objective to *Establish a Caliphate*, but is now related to the strategic objective to *Control the Islamic State*.

Al Qaeda has called for and successfully executed attacks against the West in order to achieve the three fundamental objectives: *Attack U.S. Targets to Maintain Credibility*, *Create Homegrown Terrorist Cells in the United States and Western Europe*, and *Cause Economic Losses in the United States.* These fundamental objectives are related to Al Qaeda's strategic objective *Inspire and Incite Islamic Movements and the Muslim Masses of the World to Attack the Enemies of Islam.*

ISIL's approach is primarily focused on local command and control, and as a result, instead pursues the more general fundamental objective *Purge the World of Anti-Islamic Forces*, which does not focus as specifically on the United States. The beheadings of Japanese journalists in January 2015 and the immolation of a Jordanian pilot in February 2015 underscore this change of fundamental objectives. Despite a somewhat less ambitious strategy for attacks on U.S. soil, the increased regional instability as a result of ISIL's localized focus may still have negative long-term economic consequences for the United States and other Western nations.

Al Qaeda's specific strategic objective to *Destroy Israel* is not an objective of ISIL. However, *Attack[ing] Westerners and Jews* is important for some ISIL followers (Siebert et al., 2016). The focus on enemies that are in close proximity to territory controlled by ISIL has enabled it to concentrate its resources to maximize effectiveness.

Our research suggests that ISIL does generally not cooperate with or rely on other groups in the area it controls (However, recent events as of June 2015 indicate that ISIL may now be interested in limited cooperation with groups operating outside of Syria and Iraq, e.g., Boko Haram). Al Qaeda, on the other hand, is not opposed to partnering with other terrorist organizations to expand its reach. Different goals pertaining to command and control of the Islamic State are a key differentiator between the two terrorist organizations.

Objectives Related to Establishing a Caliphate

Both Al Qaeda and ISIL have strategic objectives that seek to *Establish a Caliphate.* Our research suggests that Al Qaeda uses this objective in a primarily ideologically fashion, in order to raise attention and positively impact recruitment efforts. ISIL's timeline and focus on accomplishing this objective is immediate. Al Qaeda's fundamental objective to *Maintain Status as the Primary Middle East Force* is far less ambitious than ISIL's objectives of *Functioning as a State and Providing Services* and *Being Recognized as the Leader of the Islamic State.*

Instead of talking about a caliphate 200 years from now, ISIL created a caliphate. Abrahms (2008) argued that "the 'opportunistic' nature of Al Qaeda's political platform has led scholars to question the movement's dedication to achieving it." A possible explanation for this difference may be that historically caliphates were dominated by Shias.

The different methods and degrees of dedication to *Establish a Caliphate* are reflected in the different fundamental and means objectives too. Related to the fundamental objective of ISIL, *Function as a State and Provide Services,* are four means objectives: *Improve Service in Occupied Territories, Provide Internal Security and Semblance of Order, Fight Decadency and Corruption,* and *Stabilize Economy and Offer Jobs.* Seen in isolation, these means objectives appear as an integral part of building and maintaining any functioning government. In the case of ISIL, they actively claim to be "saviors to the people." In contrast, Al Qaeda does not intend to *Establish a Caliphate* in the short term and thus employs a more general means objective to *Provide Service to the Masses.*

The different approaches toward *Establish[ing] a Caliphate* seem to be correlated with the ability to recruit followers. Osama Bin Laden was never in the position to declare a caliphate, and was not able to raise an army; ISIL has declared a caliphate, and this has helped increase its ability to recruit large numbers of fighters.

Objectives Related to Killing Enemies

Al Qaeda and ISIL are indeed both brutal and violent terrorist groups, and both intend to *Kill Infidels*. Key differences separate the organizations. Primarily, ISIL regularly executes Muslims, including Sunni Muslims, and Al Qaeda does not. Al Qaeda wants to address all Muslims and refrains from killing any follower of Islam. In addition, Al Qaeda wants to *Avoid Beheading Captives and Other Gruesome Methods of Killing Captives* and to *Minimize the Focus on the Collateral Damage Inflicted on Muslim Civilians by Insurgent Attacks.* Both are important means to *Win the Hearts and Minds of the Muslim Masses*. This is perfectly in line with Al Qaeda's long-term strategy for establishing a caliphate that is truly supported by its citizens.

In contrast, ISIL creates fear with gruesome beheadings of Westerners and believes that this may stop the Western intervention in Iraq and the Levant. It uses social media to spread these messages of fear and maximize its notoriety and perceived prowess. This also in turn helps recruit new followers. Killing other Muslims and beheadings are believed to be one of the primary reasons for why Al Qaeda and ISIL split. ISIL does not want to *Win the Hearts and Minds of the Muslim Masses*. Instead, it wants to *Be a feared, Authentic, Radical, Brutal, and Rigorous Movement* and *Create Brand and Notoriety as Ruthless and Pure*.

16.4 Conclusion

16.4.1 *Methodology*

We have provided an overview about suitable methods to identify objectives of organizations and distinguish differences between organizations that are willing to cooperate (one's own organization, organization with partial congruence of interests) and that are not willing to cooperate (competitive organization with contradictory interests and adversary organization with hostile interests). If an organization is willing to cooperate, we recommend using interview techniques carried out with decision maker and stakeholders proposed by Keeney (1992). If an organization is not willing to cooperate, we recommend using primary and secondary sources of information; direct statements of an adversary organization and its associates' interviews with experts, and written secondary sources such as news reports and articles.

The identification and structuring of objectives can be carried out by multiple decision analysts. In order to reduce the influence of one decision analyst on another, analysts can work independently and then combine information. The resulting objectives hierarchies can be validated externally with experts answering questionnaires. Such a questionnaire should investigate the completeness of the set of objectives as well as the classification of objectives as strategic, fundamental, and means. The simplest initiative would likely include a single decision analyst exploring direct statements and actions available on the Internet. A more robust initiative could include two separate approaches pursued simultaneously with combined results, followed externally validating the results with experts as proposed by Siebert and colleagues (2016). We recommend spending more time analyzing and evaluating when the stakes are high.

16.4.2 *Main Differences and Similarities between Al Qaeda's and ISIL's Objectives*

Both ISIL and Al Qaeda want to spread Islam around the world and enforce the introduction of sharia, and both aim at reviving a caliphate. However, there are important differences between them with respect to their religious attitudes, and the time schedule and means used for achieving their goals.

ISIL interprets the Koran literally and adheres uncompromisingly to sharia law and considers only Sunnis true Muslims, whereas Al Qaeda addresses all Muslims and is willing to put aside sectarian differences. Al Qaeda wants to expel Western powers from the Middle East, to sever the ties between the Muslim world and the West, and pays lip service to establishing a caliphate. ISIL wants to extinguish any non-Islamic culture and to establish a caliphate immediately. ISIL has captured and controlled large areas of land and has declared that its caliphate now exists. As a result, its objectives relate to governance. ISIL's preferred method of killing for fanfare is much more brutal than Al Qaeda. ISIL's sectarian jihad, its cruel public beheadings, and successful attacks deep within enemy territory are used specifically as a catalyst for recruiting more followers.

Al Qaeda's strategy focuses on attacking enemies that are far away, with special emphasis on the United States and Israel. In contrast, ISIL focuses its resources on its immediate surroundings in Iraq and the Levant. The immediate threat to American lives by ISIL may be less than that by Al Qaeda, but ISIL's local jihad and rapid expansion in scale and scope of the sectarian civil war may create far long-term economic damages, and will challenge the United States militarily and politically.

In summary, both Al Qaeda and ISIL are clearly violent and extremely terroristic groups. However, as described earlier, there are many key differences in their strategic, fundamental, and means objectives and the mean accepted for achieving these objectives. These differences contributed to the schism between the organizations.

16.4.3 *Different Counterterrorism Strategies*

Al Qaeda and ISIL pursue different strategies, and as a result, counterterrorism strategies that proved effective in the fight against Al Qaeda will not necessarily prove successful in the fight against ISIL.

As mentioned previously, an important difference between the strategies of Al Qaeda and ISIL surrounds their positions on the time-horizon for establishing a caliphate and territorial expansion. Al Qaeda views building the caliphate as a long-term goal, and it did not and does not have control over substantial territories in the Middle East. ISIL wants to establish a caliphate immediately, and indeed already controls a pseudo-government spanning significant territory in Syria and Iraq.

Al Qaeda's primary operations are surreptitious, e.g., attacking checkpoints, police stations, and Western troops with suicide bombers or small command units. Surviving attackers escape and evade capture often with the help of local civilian support. Additionally, Al Qaeda has pursued a strategy of planning and executing attacks from afar, such as the September 11 attacks in New York and Washington, DC. As a result of Al-Qaeda's objectives and strategies, our countermeasures of destroying its command structure proved successful; it significantly erodes its ability to organize, recruit, finance, and carry out attacks against Westerners.

ISIL considers its territorial expansion as the nucleus of a caliphate. It instills tremendous fear in the population by "governing with the sword," and large numbers of civilians fearfully cooperate with ISIL simply as a means of survival. ISIL's occupation has engulfed oil production facilities, which it keeps in operation through the use of force, and the organization provides some services to the civilians living underneath it as would a legitimate government. Therefore, Western powers should strive to win the hearts and minds of the chastised populace in order to encourage defection and revolt. Separately, because ISIL operates more like a military organization, as compared to Al-Qaeda, precision airstrikes aided by forward air controllers on the ground could effectively target and eliminate large clusters of militant targets, supplies, and equipment.

With any post-insurgency strategy, military and political planners must have a strategy for the aftermath of the conflict and the power void that would occur when ISIL is defeated. The ISIL expansion is a result of both military and political (American and Iraqi) choices made during the drawdown of the conflict. Additionally, as ISIL's size, sphere of influence, and strategy changes over time, our strategies must too be reassessed based on thorough analysis.

Authors' Note

Our comparison between ISIL and Al Qaeda is based on the objectives of ISIL at the beginning of 2015. Later developments were not considered. Given the nature of the continuously changing landscape, we believe that our methodology would be most valuable if used at frequent intervals, in order to help ascertain timely and accurate information for decision makers and policy leaders.

REFERENCES

Abrahms, M. (2008). What terrorists really want. Terrorist motives and counterterrorism strategy. *International Security, 32*(4), 78–105.

Ackermann, S. (2014). Foreign jihadists flocking to Iraq and Syria on "unprecedented scale" – UN. October 30, 2014, accessed through www.theguardian.com/world/2014/oct/30/foreign-jihadist-iraq-syria-unprecedented-un-ISIL.

Altman, A. (2014). How many foreign fighters have Joined ISIL? September 16, 2014, accessed through http://op-talk.blogs.nytimes.com/2014/09/16/how-many-foreign-fighters-have-joined-ISIL/.

Cockburn, P. (2014). War with ISIL: Islamic militants have army of 200,000, claims senior Kurdish leader. November 16, 2014, http://www.independent.co.uk/news/world/middle-east/war-with-ISIL-islamic-militants-have-army-of-200000-claims-kurdish-leader-9863418.html.

Cooper, D. R., & Schindler, P. S. (2008). *Business research methods* (10th edn.). Boston: McGraw-Hill.

Corsi J. R. (1981). Terrorism as a desperate game. Fear, bargaining, and communication in the terrorist event. *Journal of Conflict Resolution, 25*, 47–85.

Enders, W., & Sandler, T. (1995). Terrorism: Theory and applications. In K. Hartley & T. Sandler (Eds.), *Handbook of defense economics* (pp. 129–148). Amsterdam: Elsevier.

Hacker, F. J. (1976). *Crusaders, criminals, crazies: Terror and terrorism in our time*. New York: Norton.

Keeney, G. L., & von Winterfeldt, D. (2010). Identifying and structuring the objectives of terrorists. *Risk Analysis, 30*(12), 1803–1816.

Keeney, R. L. (1992). *Value-focused thinking: A path to creative decisionmaking*. Cambridge, MA: Harvard University Press.

Keeney, R. L. (1994). Using values in operations research. *Operations Research, 42*(5), 793–813.

Lapan H. E., & Sandler, T. (1993). Terrorism and signalling. *European Journal of Political Economy, 9*, 383–397.

Lund, A. (2013). Who are the foreign fighters in Syria? An interview with Aaron Y. Zelin, December 2014, accessed through http://carnegieendowment.org/syriaincrISIL/?fa=53811.

McCormick, G. H. (2003). Terrorist decision making. *Annual Review of Political Science, 6*, 473–507.

Muller, E. N., & Opp, K. D. (1986). Rational choice and rebellious collective action. *American Political Science Review, 80*, 471–488.

Pop, V. (2015). Al Qaeda in fierce competition with Islamic State, EU official says. *Wall Street Journal*, online, released: March 21, 2015, accessed: June 10, 2015, through: http://www.wsj.com/articles/al-qaeda-in-fierce-competition-with-islamic-state-eu-official-says-1426931563.

Price E. H., Jr. (1977). The strategy and tactics of revolutionary terrorism. *Comparative Studies in Society and History, 19*, 52–66.

Rollins, J. (2011). Al Qaeda and affiliates: Historical perspective, global presence and implications for U.S. policy, Congressional Research Service, accessed through http://assets.opencrs.com/rpts/R41070_20100205.pdf.

Sandler, T., & Lapan, H. E. (1988). The calculus of dissent: An analysis of terrorists' choice of targets. *Synthese, 76*, 245–261.

Sandler T., Tschirhart, J. T., & Cauley, J. (1983). A theoretical analysis of transnational terrorism. *American Political Science Review, 77*, 36–54.

Siebert, J., Keeney, G., Rosoff, H., & von Winterfeldt, D. (*submitted*). Comparative analysis of terrorists' objectives hierarchies.

Siebert J., von Winterfeldt, D., & John, R. (2016). Identifying and structuring the objectives of the "Islamic State of Iraq and the Levant" and its followers. *Decision Analysis* (INFORMS), *13*(1), 26–50, http://dx.doi.org/10.1287/deca.2015.0324.

Tullock, G. (1974). *The social dilemma: The economics of war and revolution.* Blacksburg: University of Virginia.

Wander, A. (2008). A history of terror: Al-Qaeda 1988–2008, *The Guardian*, online, accessed: July 13, 2008, through www.theguardian.com/world/2008/jul/13/history.alqaida.

Washington Times. (2014). Obama confesses: "We don't have a strategy yet" for Islamic State, accessed: August 28, 2014, through www.washingtontimes.com/news/2014/aug/28/obama-admits-isil-dilemma-we-dont-have-strategy-ye/?page=all.

Waterman, H. (1981). Reasons and reason: Collective political activity in comparative and historical perspective. *World Politics, 33*, 554–589.

Waugh, W. L. (1983). The value in violence: Organization and political objectives of terrorist groups. *Conflict Quarterly, 3*, 5–19.

Zelin, A. (2014, June). The war between ISIS and al-Qaeda for supremacy of the global jihadist movement, Research Notes, The Washington Institute for Near East Policy, Number 20.

17

Multi-objective Decision Making: Expected Utility vs. Some Widely Used (and Flawed) Methods

Ali E. Abbas

17.1 Introduction

To begin, it is important to recall the six elements of decision quality (see, e.g., Howard & Abbas, 2015). The first element is the decision maker (and the stakeholders that are affected by the decision). There is no decision without a decision maker or stakeholders. Failing to consider all people affected by the decision can result in several disputes after the decision is made. The second element is the frame of the decision, which involves identifying the problem; identifying the different perspectives involved, and determining the boundaries of the decision (what is taken as given, what really needs to be decided now, and what will be decided later). Failing to identify the right frame is common and can lead to poor decisions. The third element of decision quality is the proper identification of feasible alternatives: without a set of alternatives, there is no decision to be made. If you have only one alternative, then you should just do it. And if you have not selected an appropriate set of alternatives to analyze, then the alternative that results from the analysis might not be the best one at hand. The fourth element of decision quality involves identifying the preferences of the decision maker. This involves identifying appropriate values and trade-offs among the different objectives. The fifth element is one that most enterprises spend a lot of time and money on: this is the information element that captures the information and uncertainties about the decision in a meaningful way. Uncertainty is represented by probability, and updating of information is conducted using Bayes' rule. Decision analysis provides a method to quantify information and to determine the economic feasibility of gathering more information. Many organizations attempt to continue gathering information to minimize all uncertainty, but, indeed, gathering too much information (as companies often do) might not be the most prudent thing to do. Finally, the sixth element of decision quality is one that is quite often underestimated in

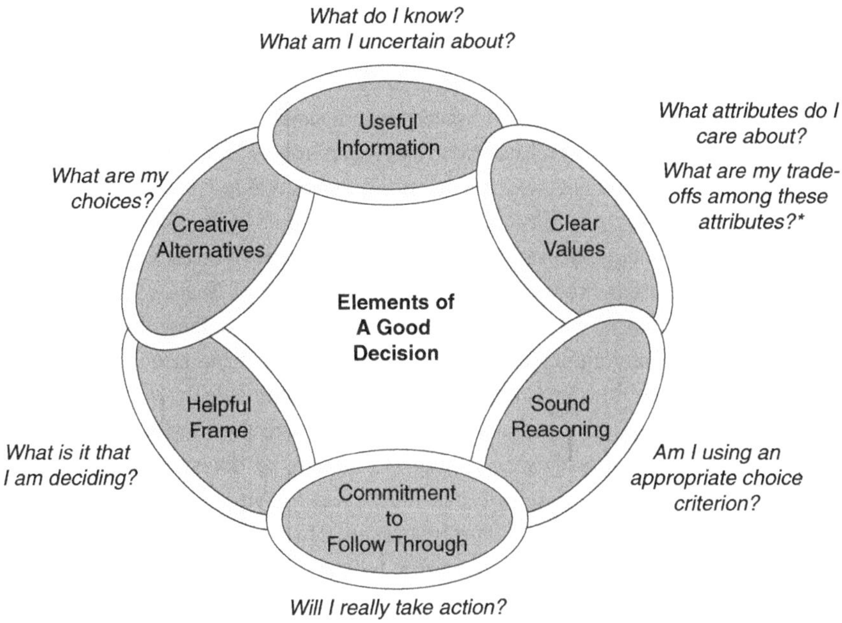

Figure 17.1. Elements of decision quality (Abbas & Howard, 2015).

organizations and results in poor decision making despite the efforts conducted with each of the previous elements. This sixth element is the logic or the choice criterion by which the decision is made. The norms of rational decision making imply that the choice criterion is to choose the alternative with the highest expected utility (von Neumann & Morgenstern, 1947). Figure 17.1 illustrates the six elements of decision quality.

Many enterprises think about the elements of decision quality and spend a lot of time and effort gathering data and information, but then use arbitrary choice criteria for making a decision. The crux of decision making lies in the idea that the quality of a decision does not rest on the outcome obtained, but on the quality of the analysis and thought.

This chapter illustrates how getting the logic wrong results in poor decision making and arbitrary decisions. Most arbitrary methods of decision making are motivated by three factors:

(i) Simplicity;
(ii) Aversion to using probability to describe uncertainty; and
(iii) Reluctance to assign explicit preferences and trade-offs in decision problems.

As we shall see, however, these reasons, while they might seem appealing at first, lend themselves to incorrect methods of decision making and also lead to unintentional implied beliefs and preferences: *Simplicity is not an excuse for using a wrong decision criterion.* Failing to incorporate uncertainty can result in making the wrong decision, and the reluctance to assert preferences and then using an arbitrary choice criterion results not only in poor decision making, but also in absurd preferences and trade-offs. One of the most widely used methods that has been motivated by these criteria and that has had a large amount of literature criticism is the Analytic Hierarchy Process (AHP) (see, e.g., Dyer, 1990; Saaty, 1990).

This chapter starts with three methods of decision making that are still widely used in many enterprises, and highlights some of the issues with their use. The first is the min-max regret criterion for making a decision, the second involves using a majority vote in systems design, and the third is the widely used (and widely misused) weight-and-rate method of decision making. For each of these methods, an example based on a real-life situation is presented to illustrate the normative issues that might result. This chapter concludes with a discussion of the expected utility criterion for making a decision and methods for constructing utility functions in decisions with multiple attributes.

17.2 The Majority-Vote Choice Criterion over the Individual Attributes

The majority-vote criterion is a choice criterion that is popular because of its simplicity. But when applied independently to attributes of a multiattribute problem, it can lead to errors in decision making. This section illustrates an example of what might occur when this criterion is applied to decide on a particular alternative in a multiattribute decision. This particular example applies to any type of choice between alternatives where multiple criteria are sought. It is disguised within the context of an automobile manufacturer.

17.2.1 *Numerical Example: The Majority-Vote Method over Attributes*

Consider the following example, where an automobile manufacturer is choosing among design alternatives. The different alternatives include engine size (number of cylinders, four or six), vehicle size (family sedan or compact), and transmission type (manual or automatic).

The manufacturer would like to conduct a market survey before making the design selection decision, and so it decides to ask customers about their preferences for each of the design attributes. For example, it interviews three segments:

1. The family sedan folk,
2. The big city folk, and
3. The sports vehicle folk.

The company asks each segment to circle its preference using the following questionnaire:

Q1: Vehicle Size: Which do you prefer: family sedan or compact?
Q2: Engine Size: Which do you prefer: four cylinders or six cylinders?
Q3: Transmission Type: Which do you prefer: automatic or manual transmission?

17.2.2 *Results of the Market Survey*

Group 1: The Family Sedan Folk

The family sedan folk express their preferences as follows:

- They prefer a family sedan to a compact vehicle (they will not purchase a compact car).
- They prefer six cylinders to four (but might purchase a four-cylinder engine to reduce fuel consumption).
- They prefer automatic transmission to manual transmission (but they might purchase a vehicle with manual transmission).

Group 2: The Big City Folk

The big city folk express their preferences as follows:

- They prefer a compact vehicle to a family sedan (but might buy a sedan).
- They prefer a four-cylinder engine to a six-cylinder one (they will not purchase a six-cylinder engine because of fuel consumption).
- They prefer automatic transmission to manual transmission (but might purchase manual transmission).

Group 3: The Sports Folk

The sports vehicle folk express their preferences as follows:

- They prefer a compact vehicle to a family-size sedan.
- They prefer a six-cylinder engine to a four-cylinder engine.
- They prefer manual transmission to automatic transmission (they will not purchase automatic transmission because they prefer the feeling of manual control).

Given the survey results, the manufacturer then constructs a matrix of preferences as shown in Table 17.1.

Table 17.1. *Results of the market survey*

Vehicle Size	Engine Size	Transmission Type
1 Group Prefers Family Sedan	1 Group Prefers 4 Cylinder	2 Groups Prefer Automatic
2 Groups Prefer a Compact Car	2 Groups prefer 6 Cylinder	1 Group Prefers Manual

The manufacturer tallies the responses and observes that:

- Two groups prefer a compact car and only one group prefers a family sedan, so it chooses a compact vehicle design.
- Two groups prefer a six-cylinder engine and only one group prefers a four-cylinder engine, so it chooses a six-cylinder engine for the design.
- Two groups prefer automatic transmission and only one group prefers manual transmission, so it chooses automatic transmission.

Using this choice criterion, the manufacturer designs:

> *A compact vehicle with a six-cylinder engine and automatic transmission*, a vehicle that none of its surveyed segments will purchase: the family sedan folk will not buy it because it is compact, the big city folk will not buy it because it has six cylinders, and the sports folk will not buy it because it has automatic transmission.

This example is based on a real story of how an automobile manufacturer incorporated survey data and used a choice criterion for making the final design. The product was deemed a failure for the manufacturer and named the worst vehicle of the year.

The previous example illustrates the need for a sound decision-making process and a normative method for making decisions even when incorporating survey data. Many enterprises spend a lot of time and effort gathering information, as did this manufacturer, but do not spend nearly as much time or effort on the methodology by which they should make the decision. As we shall see, examples of such poor decision-making methods abide.

17.3 The Min-Max Regret Criterion

The min-max criterion was discussed in Savage (1951, 1972). The first reference motivates this choice criterion as follows: (i) it does not require the explicit assignment of probability to represent uncertainty; and (ii) no explicit statements about trade-offs, risk aversion (Arrow, 1965; Pratt, 1964), or preferences are required. Savage (1951) mentions:

> *In fact, it [the min-max rule] is the only rule of comparable generality proposed since Bayes' was published in 1763.*

The min-max criterion suggests using the alternative that provides the least amount of maximum regret experienced. Common forms of widely used methods of incorporating regret in the valuation of prospects would include subtracting a certain amount from the valuation of some prospects due to the experienced regret, i.e., a prospect would be valued less (even before the uncertainty is revealed) because you could have obtained a better outcome had you chosen a different alternative.

This type of modeling may provide descriptive representations that mimic how people actually behave in the world, and may be used as a predictor of how people behave. But, as we shall see, incorporating regret into the valuation of outcomes and then using what is known as the "min-max" criterion can lead to errors in decision making. While numerous papers have previously criticized the min-max regret criterion based on its arbitrariness, this section illustrates a new criticism of min-max regret based on rank reversal. The following (disguised) example uses a modified version of the famous Party Problem (Howard & Abbas, 2015) to illustrate how using the "min-max regret" criterion may lead to what is known as rank reversal. The example extends directly to any other types of decisions, but the purpose is to present the main concept in a simple form.

17.3.1 *Numeric Example: Rank Reversal with Min-Max Regret*

Joan is interested in having a party. She has three location alternatives from which to choose: Indoors, Outdoors, and the Porch. She is uncertain about the weather, which could be Sun or Rain. Joan faces six prospects and assigns a dollar value for each prospect as follows:

- Outdoors-Sun has a value of $100.
- Outdoors-Rain has a value of $0.
- Porch-Sun has a value of $70.
- Porch-Rain has a value of $10.
- Indoors-Sun has a value of $40.
- Indoors-Rain has a value of $50.

Figure 17.2 shows the decision tree for these prospects. Note that the tree does not include any probabilities or utility values for the monetary amounts.

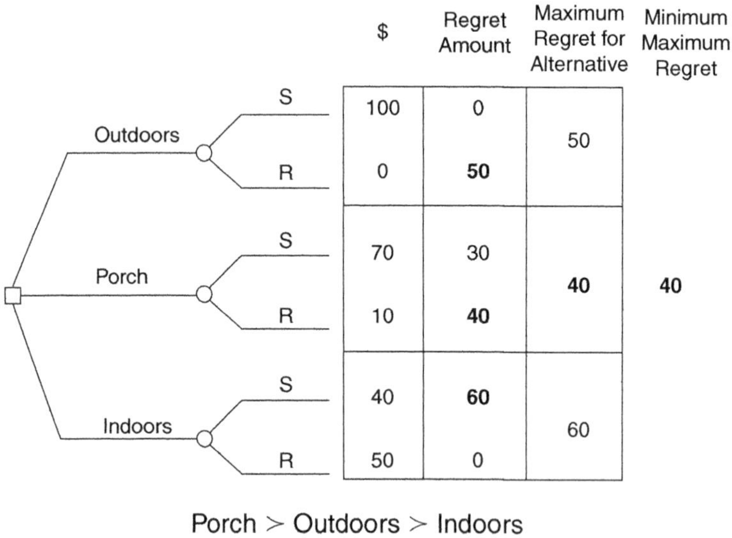

Figure 17.2. Using the min-max regret criterion with three alternatives.

The regret principle for each prospect would then consider:

a) The value of any prospect (for a chosen alternative and a given outcome for the uncertainty) less
b) The best prospect that could have been materialized (with the same outcome of the uncertainty but with any other alternative).

For example, the Outdoor-Rain party has an outcome of the weather uncertainty, Rain, and a chosen alternative, Outdoors. Its value is \$0. This prospect would then be compared to the best prospect that can be materialized when the weather is Rain regardless of the chosen alternative. This would correspond to the prospect Indoors-Rain, which has a value of \$50, because this is the best prospect that can be materialized when the weather is Rain.

The amount of regret for each prospect is then the difference between the highest value that can be materialized with the outcome of the uncertainty less the value of that prospect. The regret amount for the prospect Outdoor-Rain would then be \$50-\$0 = \$50. Similarly, Indoors-Sun (whose value is \$40) would be compared to Outdoor-Sun (whose value is \$100) because going Outdoors provides the highest value with the outcome Sun. The difference in value between the two prospects (the regret amount) is \$100-\$60=\$40. Likewise, Porch-Sun would be compared to Outdoor-Sun to give a regret amount of \$30, and Porch-Rain would be compared to Indoors-Rain to give a regret amount of

$40. The regret associated with Indoors-Rain is zero because going Indoors is the best you could have done if the weather is Rain, and Outdoors-Sun also has zero regret because of the same argument – going Outdoors is the best you can do when the weather is Sun.

The next step determines the maximum regret that can be obtained for any alternative:

- The Outdoors alternative has a maximum regret of $50.
- The Porch alternative has a maximum regret of $40.
- The Indoors alternative has a maximum regret of $60.

The final step is to choose the alternative with the lowest maximum regret. In this example, the alternative with lowest maximum regret is the alternative Porch, which has a maximum regret amount of $40. The second best alternative according to this criterion would be Outdoors, with a maximum regret amount of $50. The third best alternative would be Indoors, which has the highest maximum regret, $60.

The preference order for the alternatives is

$$\text{Porch} \succ \text{Outdoors} \succ \text{Indoors}$$

So far this sounds simple, and, indeed, many people, including Savage, advocate the use of the min-max principle. Let us now illustrate why this min-max regret principle causes a problem.

Suppose that Joan lost the Indoors alternative, and as a result has the decision tree shown in Figure 17.3. One would expect that her preferences would then be to have the party on the Porch and then the Outdoors alternative, because there is no reason why Porch is not the best alternative anymore if she loses the Indoors alternative (a less-preferred alternative).

But let us carry out the analysis one more time given we have lost the Indoors alternative. Figure 17.3 illustrates the tree, the regret amounts, and the maximum regret for each alternative.

As we can see now, the alternative with the lowest maximum regret is Outdoors, which has a maximum regret of $10. The Porch alternative has a maximum regret of $30. Therefore, the min-max choice criterion would recommend the alternative Outdoors, and so the ranking of the alternatives would be

$$\text{Outdoors} \succ \text{Porch}$$

Preferences have reversed by removing an alternative that was not even at the top of the list of available alternatives. This phenomenon is known as rank

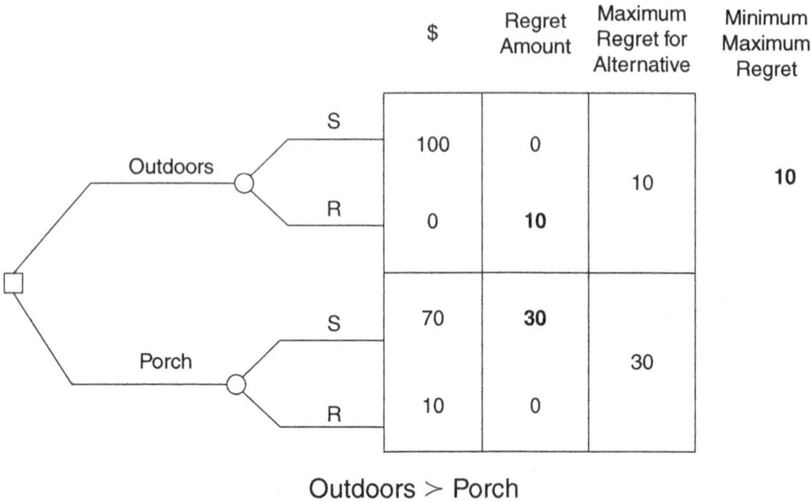

Figure 17.3. Decision tree without the Indoors alternative.

reversal. Rank reversal is a salient feature whenever comparisons are made (here we compared a prospect to the best that could have occurred with the materialization of the uncertainty). The problem with incorporating regret is that you can no longer rank (or value) a prospect independently without considering other prospects. This is why removing an alternative may lead to a change in rank.

Would you want to use a choice criterion that is prone to this type of rank-reversal behavior? Would you not worry about whether including additional alternatives would have changed the rank? How many alternatives do you include? What is the limit? The min-max regret criterion has gained appeal because of its simplicity and because (descriptively) people may experience regret in their lives. While this is a descriptive phenomenon, it is by no means a recommendation you would make to your beloved ones: "to experience regret." Rather, you would ask them to learn from their actions and move on to make better decisions in the future. In fact, as we mentioned earlier, the quality of the outcome should not determine the quality of the decision that is made. Therefore, incorporating some form of regret once you have made a good decision is meaningless. To best summarize this learning, let us propose the phrase "do not live your life based on what was or what could have been."

Of course the example presented here can be scaled to any type of decision with uncertainties. For example, a project manager may be considering investing in design alternatives: low upgrades, moderate upgrades, and major upgrades, each with different costs, and an uncertainty may be whether the product will be obsolete in a few years. The same type of rank-reversal phenomenon may occur.

17.4 The "Weight and Rate" Choice Criterion

"Weight and rate" is probably one of the most widely used (and widely misused) methods of decision making. Weight and rate is a simple method of ranking prospects of a multiattribute decision by assigning a weight to each attribute and then assigning a rate to the desirability of each attribute within a given prospect. The weight-and-rate scoring function has the form

$$S(x_1,\ldots,x_n)=\sum_{i=1}^{n} w_i S_i(x_i),$$

where x_i is the level of the attribute, S_i is a score assigned to the level of attribute x_i, and w_i is its corresponding weight. The scores are usually normalized to have the same range (e.g., 0 to 100 and the weights are non-negative and sum to one (i.e., $w_i \geq 0, \sum_{i=1}^{n} w_i = 1$). To illustrate how weight and rate is often used, consider the following example.

17.4.1 *Example: A Typical (Erroneous) Application of Weight and Rate in Systems Engineering*

A firm is deciding on the type of vehicle it would like to launch. The firm considers some important higher-level factors such as complexity of the design, the probabilities of both marketing and technical success of the product, and the geographic location of the manufacturing platform. The firm further describes that the complexity factor would be determined by whether it used state-of-the-art proven technology or a technology that has been previously incorporated. The platform on which the vehicle would be rated is based on geographical location.

Motivated by simplicity, the firm uses the weight-and-rate approach as follows:

1) Three major categories: complexity, probability (marketing success and technical success), geographic location.

2) Weights: The firm is reluctant to assign weights, and so it assigns equal weights of 1/3 to each category. It further divides the probability category into two subcategories (one for marketing success and one for technical success) with equal weights.
3) For both the complexity and geographical location categories, the firm decides on a scale 0 to 100 so it would be compatible with the probability scales in percentages.

The firm uses a weight-and-rate scoring function as follows

$$\begin{aligned}\text{Score} = \ & \frac{1}{3} \times \text{Complexity} \\ & + \frac{1}{3}\left(\frac{1}{2}\text{Prob(Techincal Success)} + \frac{1}{2}\text{ProbMarketing Success)}\right) \\ & + \frac{1}{3} \times \text{Geographic Location}\end{aligned}$$

For example, the firm might rate using a proven technology as 30 out of 100, technical success for a given design as 50%, marketing success as 70%, and a particular geographical location as 85%.

This formulation might seem reasonable, so what is the problem with using it?

First, weight and rate implies additive preferences over the individual attributes. This means that your preference for the levels of any of the attributes does not depend on the remaining attributes (this condition is known as preferential independence as discussed in Debreu (1960)). Preferential independence has a more general form of

$$P(x_1,...,x_n) = \sum_{i=1}^{n} g_i(x_i),$$

for some functions g_i that need not be normalized to the same range or even monotone. In this particular example, the choice of geographic location might have an effect on the technology used. One location might be less sophisticated than the other, and so both location and technology need to be considered simultaneously. Care must therefore be taken when choosing an additive form or preferences.

Besides the restriction on the range of the scoring functions, an additive preference function requires direct value attributes into its argument (see

Howard, 1992, and Howard & Abbas, 2015, for the distinction between direct and indirect values). Direct-value attributes are those that contribute directly to the deterministic ordering of the (deterministic) prospects of the decision. Examples of direct-value attributes for an oil company would include money and safety. Market share, however, is in most cases not a direct-value attribute, but an indirect-value attribute. We care about market share because it leads (probabilistically) to more profit, but if we were guaranteed the same amount of profit, then we would not necessarily care about market share. Many applications of weight-and-rate analysis do not take this distinction between direct and indirect values into account and rather use arbitrary attributes. Probability of success is an indirect value attribute; you care directly about success, and indirectly about its probability. Probability of success is incorporated into the expectation operator.

The third issue with weight-and-rate analysis above is that (even if applicable and the attributes are set correctly), it applies to deterministic decision problems. If there is uncertainty, then utility values should be assigned to the prospects using a utility function over value (see, e.g., Matheson & Abbas, 2005). Instead, numerous papers use probability (such as probability of success or reliability of a product) as an attribute within the weight-and-rate function. When uncertainty is present, the best decision is the one that maximizes the expected utility, and not the utility of an expectation or the utility of a probability.

The fourth issue with how weight and rate is quite often used is that there may be situations where a clear value function can rank the prospects and yet this physical or economic relationship is ignored. Think, for example, about energy generation from a turbine pump where the energy is proportional to mass and angular velocity. Two prospects with different mass and different velocities should be ranked based on their energy contribution and the physical relationship and not on an arbitrary weight-and-rate function.

The fifth issue with many cases of weight-and-rate applications is that arbitrary scales are used to score the attributes. Such measures use constructed scales (such as complexity of the technology as an attribute) and ask decision makers to assign an arbitrary scale (e.g., complexity on a scale 0 to 100) where 0 is low complexity and 100 is high complexity. Many problems arise with this formulation. First, we need to make sure that complexity is indeed a direct-value attribute. Second, these constructed scales do not have a clear interpretation (What does two or three really mean? And how do you determine it?). It is not recommended to assign an arbitrary score (e.g., 0 to 100) for the performance of each objective. It is much better to preserve the original units of the indicators to help assign weights in a meaningful way and to combine them into an aggregate meaningful metric. The units of the attributes are important and should not be ignored because they will help provide meaningful interpretations for the metric and for the weights. It is best to stick with the actual units of the attributes at hand. Otherwise, you might also end up with constructed scales that do not provide enough resolution. For example, what does 2.2 or 2.5 really mean?

Besides the arbitrary choice of attributes and arbitrary constructed scales that are quite often used with weight and rate, there is often arbitrariness in assigning weights. In many enterprises that use the weight-and-rate methodology, people tend to avoid the process of assigning weights and assign them equally. This is a problem because assigning equal weights arbitrarily is an assertion of equal trade-offs among the attributes. To illustrate, consider the more restrictive form of weight and rate" that is widely used:

$$P(x_1,\ldots,x_n) = \sum_{i=1}^{n} w_i x_i,$$

which implies constant substitution between the levels between the attributes because across a contour of constant preference, we have $dP = 0$ and so

$$\frac{dx_i}{dx_i} = -\frac{w_j}{w_i}.$$

This constant trade-off might be the case for some problems, but it certainly need not apply to all preference structures. Think of the attributes of air you breathe in year 1 vs. year 2 or trade-offs between remaining lifetime and wealth. In both of these cases, as the levels of one attribute drop below certain values, the amount of substitution for another attribute would vary significantly. In fact, weights need not appear at all in the utility formulation (copulas), and even if the firm used an additive utility, the weights do have a clear interpretation in terms of lottery assessments.

To think about this situation from a decision analytic perspective, the firm would need to identify the frame of the decision (e.g., whether it should continue this line of vehicles or even consider a new line), the available design alternatives, the uncertainties it has (such as probabilities of success or better formulate them in terms of contributions to demand for the product to provide clarity on what success really means), the direct values and the utility values over deterministic prospects characterized by direct-value attributes, and then use a choice criterion (the expected utility criterion) for choosing among the design alternatives. In so doing, the firm should use a corporate utility function for the choice of the alternatives (Howard & Abbas, 2015) and consider the effects of any incentive structures on the managerial choices (Abbas et al., 2009; Abbas & Matheson, 2009).

17.5 Multiattribute Value and Multiattribute Utility Functions

Having discussed some flawed methods of decision making, we now explain the expected utility criterion and the rationale for its use.

17.5.1 *The Rationale for Expected Utility Maximization*

The axioms of expected utility theory were introduced by von Neumann and Morgenstern (1947). They were later rephrased and modified in various settings. Here we use the axioms of expected utility proposed by Howard (1992), and, for more information, see Howard and Abbas (2015).

These axioms (rules) are:

1) **The Probability Rule:** This rule states that (i) I can characterize each decision alternative I face in terms of the set of possible prospects that may occur; and (ii) I can assign a probability of receiving each prospect for every alternative I choose.
   Put very generally, this rule states that we can draw the decision tree with all of its nodes and numbers.
2) **The Order Rule:** This rule states that I can rank order the deterministic prospects that I have characterized in a list from the most preferred to the least preferred. Ties are allowed, but preferences must be transitive, i.e., if I prefer prospect A to prospect$B$, and if I prefer prospect B to prospect C, then I must prefer prospect A to prospect C.
3) **The Equivalence Rule** (von Neumann Morgenstern utility rule): This rule states that given three ordered prospects with strict preference, $A \succ B \succ C$, I can assign a preference probability, p_B, that would make me indifferent between receiving prospect B for certain, or a binary deal having a probability p_B of receiving A and $1 - p_B$ of receiving C.
4) **The Substitution Rule:** This rule states that whenever I face a prospect B, for which I have stated a preference probability p_B of receiving A and $1 - p_B$ of receiving C in the Equivalence Rule, I would be indifferent between receiving this binary deal and receiving prospect B.

This rule allows us to make various substitutions of a binary deal (with prospects A and C) and the deterministic prospect, B, whenever either occurs in the decision tree.

5) **The Choice Rule:** This rule states that if I face two binary decision alternatives, "L1" and "L2," both yielding the same prospects A and B (prospect A is preferred to B), and if "L1" has a higher probability of getting A, then I should choose "L1" over "L2."

Figure 17.4(a) shows a set of prospects, $R_1,\ldots, R_n$, that have been characterized by the Probability Rule and their corresponding probabilities, $p_1,\ldots, p_n$, for a given decision alternative. The figure also shows the Order Rule ranking (assume R_1 is the best prospect and R_n is the worst). Figure 17.4(b) shows an example of a von Neumann and Morgenstern utility assessment, where a preference probability, U_j, is assigned to prospect R_j in terms of the best and worst prospects, R_1 and R_n. Figure 17.4(c) shows the substitutions made for the prospects in terms of their utilities (preference probabilities), and Figure 17.4(d) shows the equivalent lottery after multiplying the probabilities in Figure 17.4(a) by the von Neumann and Morgenstern utilities. Note that the probability of the best prospect in Figure 17.4(d) is equal to the expected utility of the original alternative, $\sum_{j=1}^{n} p_j U_j$.

A decision maker who follows the five rules mentioned earlier will be indifferent between receiving the lotteries of Figures 17.4(a) and 17.4(d). Thus *any decision alternative can be reduced into an equivalent binary alternative using von Neumann and Morgenstern utility (preference probability) assessments.* Furthermore, *the expected utility of the original alternative is equal to the probability of getting the best prospect in the equivalent binary alternative.* Since any other alternative can also be reduced into a binary alternative with the same prospects, R_1 and R_n (but with different probabilities of achieving them), the Choice Rule determines the best decision alternative: we choose the decision alternative (lottery), with the highest expected utility.

We now observe the following:

1. **Expected utility maximization is a direct result of the rules**: Not any arbitrary numeric measure representing some form of preferences can be used in this formulation. The measure whose expected value is to be maximized over the set of alternatives must be the von Neumann Morgenstern utility (that reduces the lottery into an equivalent binary lottery). If we choose to optimize any other arbitrary measure, such as minimizing the "value at risk" or minimizing the "maximum regret," or any arbitrary "risk

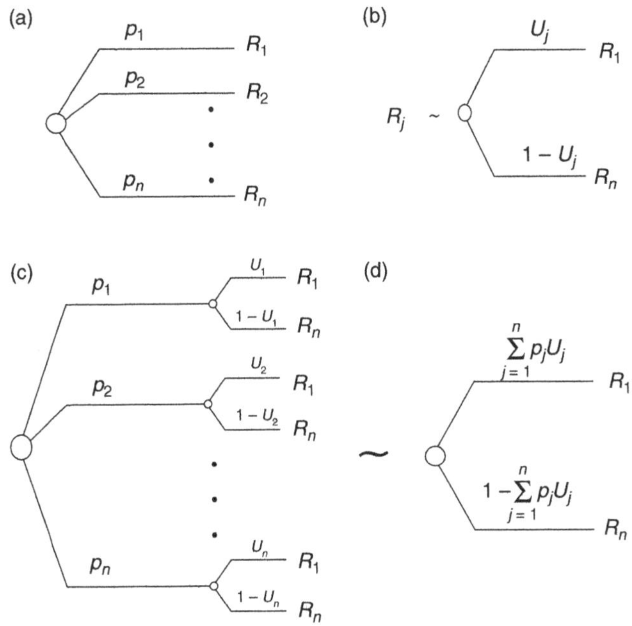

Figure 17.4. Interpretation of expected utility of a lottery as the probability of the best outcome of an equivalent lottery containing only two outcomes – the best and the worst.

measure" or "score," then we would be violating one of these rules. Which rule would you like to violate in your decision making?

2. **Utility functions are not always needed:** In some cases (such as *deterministic deals*), we have no uncertainty in the decision problem. In other cases, we may have uncertainty, but also have *deterministic dominance*, where the worst possible prospect that can be obtained by one alternative is better than the best possible prospect obtained by another. We may also have *first-order stochastic dominance conditions*, where we always have a higher probability of exceeding any monetary amount with one alternative than with another. *In all these cases, we do not need utility values* since the preference between these lotteries can be determined by the Order Rule ranking alone (such as preferring more of an attribute to less). Our focus will be the more general case where such dominance conditions do not necessarily exist and where a utility function is indeed required.
3. **The same rules apply to multiple attributes:** In many cases, prospects can be characterized by multiple attributes, such as health state and wealth.

The formulation we discussed previously is not restricted to prospects characterized by a single attribute. It simply assigns the utility value to a prospect using the equivalence rule interpretation regardless of how it is characterized.

4. **No major restrictions on the functional form of the utility function:** This formulation does not pose restrictions on the shape or the functional form of the utility (or multiattribute utility) function, except as imposed by the Equivalence Rule and that the higher the utility value, the more preferred is the prospect.

The requirements for the expected utility criterion are clear: for each alternative, we determine the probability of each prospect as well as its utility value. If there is data or results of a market survey, we use the data to update our belief (probability distribution) for the different prospects. To determine the best design alternative for the automobile manufacturer, as an example, the manufacturer should have determined the expected utility of each design alternative. If the manufacturer is interested in profit, then it would calculate

$$\text{Profit} = \text{Revenue} - \text{Cost}$$

It would then assign a probability distribution over profit. This distribution would be determined by demand for the product as a function of its price, as well as the uncertainty about the costs (manufacturing, marketing, etc.). The best design decision is the one with the highest expected utility of profit. The firm may use the market survey to update its demand distribution for each design alternative using Bayesian analysis (Abbas et al., 2012; Sun et al., 2013; Traverso & Abbas, 2009). The actual value of the market survey in this decision can also be calculated using value of information analysis using Bayesian updating (see, e.g., Howard, 1966).

17.5.2 *Expected Utility Formulations for Decisions with Multiple Attributes*

It is often the case that prospects of a decision can be characterized by more than a single attribute. A medical decision, for example, may involve health state and wealth, and the purchase of a new home may involve attributes such as comfort level, view, and price. The same rules we discussed earlier still apply: we still need to characterize the prospects in terms of the multiple attributes and assign a probability to them; we need to order the prospects; we need to assign a utility value to the prospects; we need to follow the Substitution Rule; and we need to follow the Choice Rule.

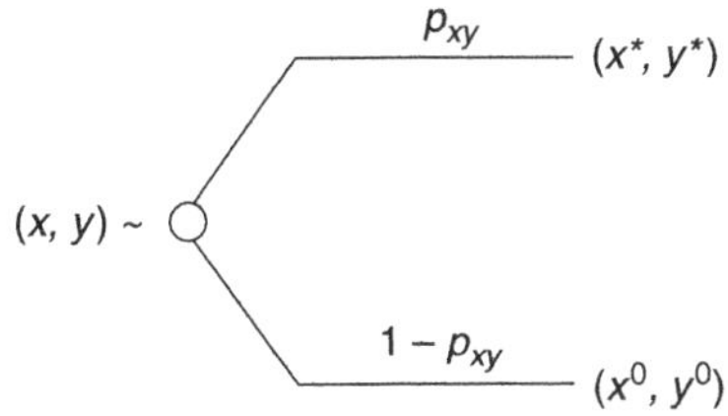

Figure 17.5. The Equivalence Rule is the same with multiple attributes.

Now suppose that the prospects are characterized in terms of some level (x, y) of two attributes, X,Y such as salary (100K$) and vacation time (days). In principle, we can still order the prospects in accordance with the Order Rule without much difficulty. For example, if the decision maker prefers more vacation and more salary to less, then the prospect (4,6) is preferred to either (2,6) or (4,3). But if we face two prospects, one having a higher salary but less vacation, such as (4,6) and (6,5), then the Order Rule ranking requires a bit more thought. In these situations, it is helpful to use a value function to rank order the prospects.

The value function captures the preference ordering of the prospects and maps each prospect (x, y) to some numeric value $V(x, y)$. The higher the magnitude of this value function, the more preferred is the prospect. Thus the value function helps with the Order Rule ranking when multiple attributes are present. When there is no uncertainty, this value function determines the best decision alternative. The next section discusses value functions in more detail.

The Equivalence Rule is also straightforward when multiple attributes are present. For two attributes, X and Y, let (x^*, y^*) be the most preferred prospect we face and (x°, y°) be the least preferred. For each prospect (x, y), the Equivalence Rule implies that we can state a preference probability, p_{xy}, of receiving the best prospect (x^*, y^*) and a probability $1 - p_{xy}$ of receiving the worst prospect (x°, y°), that makes us indifferent to receiving (x, y) for certain (Figure 17.5).

If there is a relatively small number of prospects, then we can state a preference probability, p_{xy}, for each prospect individually by direct assessment and the problem is solved. For salary and vacation time, as an example, we would need to assign a preference probability of getting (2,3) or a binary deal giving (4,6) or (0,0). Once these utility values are assigned, the best decision alternative is the one with the highest expected utility. However, if the alternatives we face have numerous prospects (such as a continuum of prospects on the domain of the multiple attributes), then this utility elicitation task may be tedious. We therefore need some tools to help us assign utility values in these situations. The assessment of such multiattribute utility functions will be the focus of the next few sections.

17.5.3 *Deterministic Multiattribute Decision Problems (Value Functions)*

When there is no uncertainty, each decision alternative leads to a single prospect. The problem is reduced to ranking the prospects in accordance with the Order Rule. As we discussed, we do not need utility values to determine the best decision alternative in this case since this task can be achieved using a more general value function. This is often referred to as the "*deterministic phase of a decision analysis*" (Matheson & Howard, 1968). In principle, the actual measure that we use in the value function output is not important if we are merely interested in ranking the prospects (provided that the higher the magnitude of the value function, the more preferred is the prospect). Any monotonic transformation of the value function would also suffice.

When the attributes are not all represented monetarily (such as the attribute health state), we need further clarity on the constructed scale that is used. In particular, health state on a numeric scale, for example, is often determined by time–trade-off questions that convert health prospects into equivalent healthy life time.

To simplify the construction of the value function, and to eliminate double counting, we need to think carefully about the attributes we use in the characterization of the prospects. It is helpful to make a distinction between "*direct and indirect value attributes*" (Howard, 1992). *Indirect value attributes will not contribute to a change in the Order Rule ranking of the prospects.*

To further clarify the distinction between direct and indirect values, consider the following example. In buying a house, we may place a direct value on the view, because it affects desirability, and on the total cost associated with it. However, we would only place an indirect value on the cost of maintenance because it is only a part of the total cost, an attribute that we have placed direct value on. In principle, we do not care about the breakdown of the total cost in terms of maintenance, insurance, or utilities, provided the total cost remains the same. If we faced two houses identical in every aspect but having different maintenance costs yet the same total costs, then we would be indifferent between them. Note, however, that if we were very proud to have a house with low maintenance cost and it would give us great pleasure to tell our friends about it, then we would place a direct value on maintenance cost, and it would be another attribute of direct value. Once the direct-value attributes are clarified, a value function is then used to rank the prospects. The best (deterministic) decision alternative is the one whose prospect ranks highest using this value measure.

An appropriate value function is one that captures the qualitative and quantitative features of the problem. The following example from (Abbas & Howard, 2005) presents a value function of a "*peanut butter and jelly sandwich.*" The attributes involved are thickness of slices of bread, b, thickness of peanut butter, p, thickness of jelly, j, and fraction of thickness of peanut butter to jelly, f,

$$V(p,j,b,f) = V_{\max} \frac{pjbf}{(p^* j^* b^* f^*)^2}(2p^* - p)(2j^* - j)(2b^* - b)(2f^* - f), \quad (17.1)$$

where p^*, j^*, b^*, f^* are the optimal values of p, b, j, f (respectively) and $V_{\max}$ is the dollar amount that the decision maker is willing to pay for the optimal sandwich. Note that the value function is not necessarily increasing with any of the attributes on the entire domain, and that there is dependence in preferences among the attributes. For example, the preference for the thickness of peanut butter will depend on the thickness of jelly.

Contours of prospects providing the same value within the value function are known as "*isopreference contours*." We should be indifferent between any two prospects that lie on the same isopreference contour (these prospects rank equally according to the Order Rule ranking).

The value function also determines the trade-offs between the levels of attributes. For example, we can tell the amount of increase in peanut butter that would make us just indifferent to a decrease in thickness of bread by observing two points on the same isopreference contour. In principle, the increase (or percentage increase) in one attribute that is needed to compensate for a decrease (or percentage decrease) in the other, to achieve the same value, varies across the domain of the attributes. Of course the same concept applies to any other attributes. The purpose is to present it in a simple and commonly understood form.

One type of value functions that is popular because of its simplicity is the additive value function. This value function implies that the increase in one attribute required to compensate for a decrease in the other is constant across the entire domain of the attributes. For two attributes, this value function has the form

$$V(x,y) = w_x x + w_y y, \quad (17.2)$$

where w_x, w_y are weight parameters. This value function, however, comes with the strong implication that preferences for levels of attribute X do not depend on the level of attribute Y. I.e. for two points $(x_1, y_1), (x_2, y_2)$ on the same isopreference contour,

$$\frac{y_2 - y_1}{x_2 - x_1} = -\frac{w_x}{w_y},$$

which means that the increase in attribute X needed to compensate for a decrease in attribute Y does not depend on the levels of either X or Y. This value function is similar to the weight-and-rate function except that the attributes and the weights assigned to them are meaningful.

Another value function that is monotonically increasing with each of the attributes and may be of practical interest is one where the fractional increase in one attribute required to compensate for a fractional decrease in another is constant across the domain. This value function has the form

$$V(x, y) = xy^{\eta}, \tag{17.3}$$

where η is a trade-off parameter. It can be shown by direct derivatives across an isopreference contour that

$$\eta = -\frac{dx / x}{dy / y}.$$

This also means that for small percentage changes, the percentage increase in attribute X needed to compensate for a percentage decrease in attribute Y is constant.

17.5.4 *Constructing Multiattribute Utility Functions Using Value Functions*

While the value function is sufficient to determine the best decision alternative in deterministic decision problems, we need to assign utility values to the prospects when uncertainty is present to maximize the expected utility. This section illustrates how to assign a multiattribute utility function to the prospects once the value function is specified.

Let $V_{xy} = V(x, y)$ be the magnitude of the value function for any prospect (x, y), and $V^* = V(x^*, y^*)$, $V^0 = V(x^0, y^0)$, be the magnitude of the value

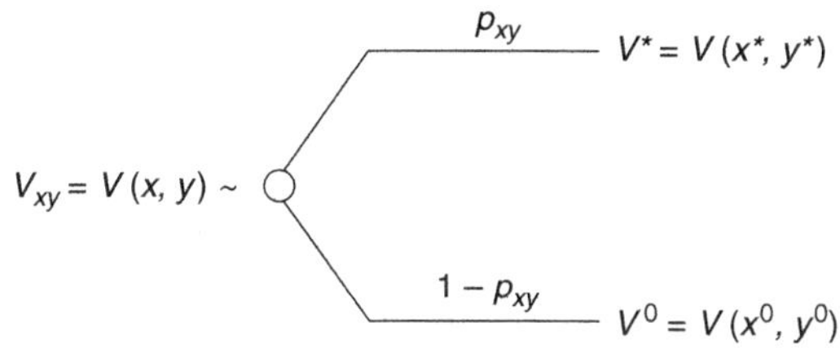

Figure 17.6. Equivalence Rule using a value function for multiple measures.

function for the most and least preferred prospects (respectively). The preference probability for any prospect (x, y) can now be determined by simply asking the decision maker for his indifference between the two deals in Figure 17.6. Note that the decision maker need not even be concerned about the actual values of X and Y. This assessment is simply a utility function (preference probability) assessment over a single measure, V in terms of the best value, V^*, and the worst, V^0. Note that this measure need not be money, but, as is well known, money is simpler to assign a utility function to in comparison to other constructed scales. It also enables the calculation of the certain equivalent and the value of information on the uncertainties of interest.

Let U_V be the utility function we assign over the value measure. Given a prospect (x, y) whose value measure is $V(x, y)$, the utility of (x, y), namely $U(x, y)$, is equal to

$$U(x, y) = U_V(V(x, y)). \tag{17.4}$$

The problem is now reduced to a one-dimensional utility assessment over the value measure (Matheson & Abbas, 2005).

Having a value function limits the individual utility functions that we can assign over the individual attributes. For example, if we have assigned a value function, $V(x, y)$, and a utility function over value, $U_V(V)$, then the conditional utility function of attribute X at the maximum value of attribute Y, must be of the form

$$U(x, y^*) = U_V(V(x, y^*)).$$

Similarly, the conditional utility function at any other instantiation of Y can be obtained by direct substitution. We do not have another degree of freedom to assign individual utility functions over the attributes once the value function and the utility function over value have been specified.

Example: On Fates Comparable to Death

The following example is adapted from Howard (1980) on making trade-offs concerning situations where a decision maker is exposed to fates comparable to death (such as outcomes of medical surgery). The decision maker provides a value function over two attributes: consumption, X, and health state, Y. The health state is a disability level normalized from 0 (instant, painless death) to 1 (current health with no disability). The decision maker assigns a value function over consumption and health states as

$$V(x,y)=xy^{\eta}, \tag{17.5}$$

where x is expressed in dollars, y is the health state, and η determines the trade-off between x and y. The value function units are total well-being. One method to determine the multiattribute utility function is to assess a utility function, $U_V(V)$, over the value function. For example, if $U_V(V)$were exponential, then

$$U_V(V)=U_V(V(x,y))=1-e^{-\gamma V(x,y)}=1-e^{-\gamma xy^{\eta}}, \tag{17.6}$$

where $\gamma=\gamma_V^U$ is the risk aversion coefficient toward value.

17.5.5 *Basic Expansion Theorems for Multiattribute Utility Functions*

Some people may prefer to construct the utility function without the use of a value function, by making direct conditional utility assessments over each attribute at certain instantiations of the remaining (complement) attributes. The remainder of this chapter is concerned with this assessment method. This section derives a general expansion theorem for multiattribute utility functions. We follow Abbas (2011) to derive this basic expansion theorem and then illustrate its use in constructing multiattribute utility functions with conditional utility assessments.

Basic Definitions

We consider multiattribute utility functions with n attributes, $X_1,X_2,\dots,X_n$. We use the lower case, x_i to denote an instantiation of attribute X_i and use the vector $(x_1,\dots,x_n)$, to represent a prospect of the decision. We use $\overline{X}_i$ to denote the set of complement (remaining) attributes to X_i, and use $\overline{x}_i$ to denote a certain instantiation of this complement. For simplicity, we also refer to a consequence, $(x_1,\dots,x_n)$, as $(x_i,\overline{x}_i)$, or $(x_i,x_j,\overline{x}_{ij})$, where $\overline{x}_{ij}$ is an instantiation of the complement attributes to X_i and X_j. We assume that the multiattribute utility function, $U(x_1,\dots,x_n)$, is (i) continuous; (ii) bounded; and (iii) for each attribute,

X_i, there exist two reference values, x_i^0 and x_i^*, for which $U(x_i^*, \bar{x}_i) > U(x_i^0, \bar{x}_i)$ at all reference values of the complement attributes. We refer to this class of utility functions as class S. Since the utility function is bounded, we assume hereon, with no further loss of generality, that it is normalized to range from zero to one.

Define a normalized conditional utility function for attribute X as

$$U(x \mid \bar{x}) = \frac{U(x, \bar{x}) - U(x^0, \bar{x})}{U(x^*, \bar{x}) - U(x^0, \bar{x})}, \tag{17.7}$$

and note that $U(x^0 \mid \bar{x}) = 0$ and $U(x^* \mid \bar{x}) = 1$.

Further define a normalized "conditional disutility function," $\bar{U}(x \mid \bar{x})$, as

$$\bar{U}(x \mid \bar{x}) = 1 - U(x \mid \bar{x}). \tag{17.8}$$

Expansion of a Utility Function around a Single Attribute

To express a multiattribute utility function in terms of the normalized conditional assessment, $U(x \mid \bar{x})$, of attribute X, we rearrange (17.7) and substitute into (17.8) to get

$$U(x, \bar{x}) = U(x^*, \bar{x})U(x \mid \bar{x}) + U(x^0, \bar{x})\bar{U}(x \mid \bar{x}). \tag{17.9}$$

We refer to this expression as "*expanding the utility function around X*." The pattern associated with (17.9) is that when X appears as x^* in the term $U(x^*, \bar{x})$, it is followed by the conditional assessment of X on its complement. When it appears as x^0 in $U(x^0, \bar{x})$, it is followed by the conditional disutility of X on its complement.

Expansion of a Utility Function around Multiple Attributes

Three basic terms result from the expansion (17.9):

$$U(x^*, \bar{x}), U(x^0, \bar{x}), \quad \text{and} \quad U(x \mid \bar{x}).$$

Suppose we further expand the two terms $U(x^*, \bar{x})$ and $U(x^0, \bar{x})$ with respect to a second attribute, Y (note that we do not yet expand the third term $U(x \mid \bar{x})$). This formulation would imply

$$U(x^*, \bar{x}) = U(x^*, y^*, \overline{xy})U(y \mid x^*, \overline{xy}) + U(x^*, y^0, \overline{xy})\bar{U}(y \mid x^*, \overline{xy}) \quad \text{and} \tag{17.10}$$

$$U(x^0,\overline{x}) = U(x^0,y^*,\overline{xy})U(y \mid x^*,\overline{xy}) + U(x^0,y^0,\overline{xy})\bar{U}(y \mid x^0,\overline{xy}). \tag{17.11}$$

Substituting from (17.10) and (17.11) into (17.9) gives

$$\begin{aligned} U(x,y,\overline{xy}) = {} & U(x^*,y^*,\overline{xy})U(x \mid \overline{x})U(y \mid x^*,\overline{xy}) \\ & +U(x^*,y^0,\overline{xy})U(x \mid \overline{x})\bar{U}(y \mid x^*,\overline{xy}) \\ & +U(x^0,y^*,\overline{xy})\bar{U}(x \mid \overline{x})U(y \mid x^0,\overline{xy}) \\ & +U(x^0,y^0,\overline{xy})\bar{U}(x \mid \overline{x})\bar{U}(y \mid x^0,\overline{xy}). \end{aligned} \tag{17.12}$$

We have chosen to write (17.12) without any simplification to illustrate the pattern that is evolving, and to highlight some important points about this expansion:

1) The terms $U(x^*,y^*,\overline{xy})$, $U(x^*,y^0,\overline{xy})$, $U(x^0,y^*,\overline{xy})$, $U(x^0,y^0,\overline{xy})$ in the right-hand side require a joint utility assessment of the attributes that have not been expanded, namely $\overline{xy}$, at all possible combinations of the minimum and maximum values of the attributes that have. We refer to these terms as the *unexpanded terms* and write them as $U(x_K^{*0},\overline{x}_K)$.
2) When the instantiation of X is x^* in the unexpanded terms, it is followed by a conditional utility assessment $U(x \mid \overline{x})$, and when its instantiation is x^0, it is followed by a conditional disutility assessment, $\bar{U}(x \mid \overline{x})$. The same applies to attribute Y and any other attribute we wish to expand.
3) Equation (17.12) implies an expansion order: we first expand around X and then Y.
4) The assessments needed for the basic expansion theorem require conditioning a utility assessment on the level of another attribute (or attributes). This might make the assessment complicated. The next section explains a condition to simplify this task.

Simplifying the Assessments with Utility Independence Assertions

One way to simplify the conditional utility assessments and the functional form of the basic expansion theorem is to condition the assessments on a smaller number of attributes. To do that, we can use what is known as a utility independence assumption (Keeney & Raiffa, 1976).

A set of attributes X_K *is utility independent of another set* X_I given X_D, written $(X_K \ UI \ X_I \mid X_D)$, with $X_I \cup X_D = \overline{X}_K$, *if preferences for lotteries over* X_K do not depend on the instantiations of X_I.

If attribute X is utility independent of its complement attributes, then this is equivalent to saying that the conditional utility assessment of any attribute satisfies

$$U(x \mid \bar{x}) = U(x \mid \bar{x}^0), \tag{17.13}$$

which implies that the normalized utility function of X does not depend on the level of the remaining attributes. Therefore, a single-attribute assessment of X at a fixed instantiation of the complement is all that is needed in terms of utility assessments over X. Therefore, the basic expansion theorem reduces to a multilinear combination of single-attribute utility assessments. Keeney and Raiffa (1976) show that this case where every attribute is utility independent of its complement leads to the multilinear form,

$$\begin{aligned} U(x_1,\dots,x_n) = {} & \sum_{i=1}^{n} k_i U_i(x_i) \\ & + \sum_{i=1}^{n} \sum_{j>i} k_{ij} U_i(x_i) U_j(x_j) + \cdots + k_{123\dots n} U_1(x_1) U_2(x_2) \dots U_n(x_n). \end{aligned} \tag{17.14}$$

where the scaling constants satisfy

$$\begin{aligned} k_i &= U(x_i^*, \bar{x}_i^o), && i = 1,..,n \\ k_{ij} &= U(x_i^*, x_j^*, \bar{x}_{ij}^o) - k_{x_i} - k_{x_j}, && i = 1,..,n,\ j > i \\ k_{ijk} &= U(x_i^*, x_j^*, x_l^*, \bar{x}_{ijl}^o) - k_{x_i x_j} - k_{x_i x_l} - k_{x_j x_l} - k_{x_i} - k_{x_j} - k_{x_l}, && i = 1,..,n,\ l > j > i \\ k_{ij\dots n} &= U(x_i^*, x_j^*, \dots, x_n^*) - \sum_i k_{x_1 \dots x_{i-1} x_{i+1} \dots n} - \cdots - \sum_i \sum_{j>i} k_{x_i x_j} - \sum_i k_{x_i}. \end{aligned}$$

A stronger independence condition also holds if every subset of the attributes is utility independent of its complement, a condition known as mutual utility independence (Keeney and Raiffa 1976) and so the utility function has either the multiplicative form

$$1 + kU(x_1,\dots,x_n) = \prod_{i=1}^{n} [1 + k k_i U_i(x_i)], \tag{17.15}$$

or the additive form

$$U(x_1,\dots,x_n) = \sum_{i=1}^{n} k_i U_i(x_i). \tag{17.16}$$

The simplicity of constructing multiattribute utility functions using single-attribute utility assessments has popularized the use of the functional forms (17.14), (17.15), and (17.16) in many decision analysis applications, but they do come with strong implications. Note that the assumption of mutual utility independence for two attributes, health and wealth, as an example, would imply that our preferences for decision alternatives leading to different health states will not change with our wealth state, and that our preference for decision alternatives leading to lotteries over money will not change with our health state.

When Utility Independence Conditions Do Not Hold

If utility independence conditions do not hold, then we are faced with the need to incorporate utility dependence among the attributes into our analysis as implied by the basic expansion theorem. Several authors have discussed this issue and have presented methods to incorporate utility dependence (see, e.g., Abbas, 2009; Abbas & Bell, 2011, 2012; Abbas & Howard, 2005; Bell, 1979; Farquhar, 1975; Fishburn, 1974, 1977; Keeney & Raiffa, 1976; Kirkwood, 1976).

In particular, Farquhar (1975) proposes a decomposition theorem for a class of multiattribute utility functions with certain utility independence structures known as fractional hypercubes. He shows that special cases of this formulation reduce to the multilinear form. Kirkwood (1976) incorporates utility dependence by defining "parametrically dependent" conditional utility functions. Bell (1979) incorporates utility dependence by assessing conditional utility functions at the boundaries and deriving the remaining utility values by interpolation. Keeney and Raiffa (1976) take an alternate approach and reformulates the attributes of a decision problem to derive a new set of attributes with a lesser degree of utility dependence.

Abbas (2004) defines utility copulas for attribute dominance utility functions, and Abbas (2009) extends this idea to more general utility functions. Utility copula functions express any continuous multiattribute utility function, $U(x_1,...,x_n)$, that is strictly increasing with each argument for at least one reference value of the remaining (complement) attributes in terms of individual von Neumann-Morgenstern normalized utility assessments of each objective.

The multiattribute utility function can be represented by the utility copula formulation using the relation

$$U(x_1,...,x_n) = C(U(x_1 \mid \bar{x}_1^{\lambda_1}),...,U(x_n \mid \bar{x}_n^{\lambda_n})), \tag{17.17}$$

where $U(x_i \mid \bar{x}_i^{\lambda_i})$ represents a utility function of an individual objective at a specific reference value $\bar{x}_i^{\lambda_i}$ of the set of complement objectives to objective X_i. In our work, we use the superscript * to denote the maximum values of the attributes and the superscript $\circ$ to denote the minimum values.

Because utility functions are invariant to a linear transformation, a necessary condition that needs to be satisfied for consistency of the resulting utility surface with the individual utility assessments is that the multiattribute utility function be a linear transform of the individual utility functions at the reference values of the complement attributes at which they are assessed. This condition is known as the marginal property of utility copulas.

We now highlight two special classes of utility copula functions:

Class 1 utility copulas satisfy

$$C_1(1,\ldots,1,v_i,1,\ldots,1) = a_i v_i + b_i, \quad i = 1,\ldots,n. \tag{17.18}$$

A special case of utility copula functions that will be useful in our formulation is the Archimedean utility copula (Abbas, 2009) of the form

$$C(v_1,\ldots,v_n) = a\psi^{-1}\Big[\prod_{i=1}^{n} \psi(l_i + (1-l_i)v_i)\Big] + b, \tag{17.19}$$

where, $0 \le l_i < 1$, $a = 1/(1-\psi^{-1}[\prod_{i=1}^{n} \psi(l_i)])$, $b = 1-a$, and the function ψ satisfies the following conditions: (i) $\psi(v)$ is continuous on the domain $v \in [0,1]$; (ii) $\psi(v)$ is strictly increasing on the domain $v \in [0,1]$; (iii) and $\psi(1) = 1$.

In related work, Abbas and Bell (2011, 2012) extend the notion of utility independence, where preferences over lotteries do not change, to situations where preferences over lotteries can change, but only once as the level of another attribute varies. They define a family of utility functions that incorporates utility dependence and allows for a richer class than the functional forms of utility independence.

17.6 Conclusion

The primary purpose of this chapter was to emphasize the need for a sound decision-making system. Numerous methods of decision making are widely used in enterprises, but care must be taken when choosing a particular methodology. The aim of this chapter was to draw attention to these widely used methods and to highlight some of the issues with their use. The examples used in this chapter have been disguised, but the methodology used and the rationale presented for making a decision are all based on real situations in systems engineering in large enterprises, one for the design of a vehicle, another for spacecraft design, and a third for border security situations. Neither party knew that there was an error with how it was thinking about the decision, and they both liked their respective approaches because of simplicity. Numerous other flawed methods also prevail and have not been discussed in this chapter, such as the analytical hierarchy process (see Dyer, 1990, for a discussion of its issues) and the use of fuzzy logic to represent probabilities (see Lindley, 1982, for a discussion of the normative issues of representing uncertainty without probability).

Another purpose of this chapter is to provide rationale for the use of the expected utility criterion and methods for constructing multiattribute utility functions in decisions with multiple attributes. We discussed how using a value function can reduce the multiattribute utility analysis into a one-dimensional utility function over value, and we also discussed other methods for constructing utility functions using the basic expansion theorem and its simplification using independence conditions. Finally, we reviewed some methods that have been proposed when utility independence conditions do not hold.

The choice criterion for making rational decisions is clear: expected utility maximization. The job of the analyst is not to propose an arbitrary method (because of its simplicity), but to help the enterprise provide the appropriate inputs to the expected utility decision model.

REFERENCES

Abbas, A. E. (2004). *Entropy methods in decision analysis.* PhD Thesis. Stanford University.

Abbas, A. E. (2009). Multiattribute utility copulas. *Operations Research, 57*(6), 1367–1383.

Abbas, A. E. (2011). General decompositions of multiattribute utility functions. *Journal of Multicriteria Decision Analysis, 17*(1, 2), 37–59.

Abbas, A. E. (2015, June). Perspectives on some widely used methods of multi-objective decision making. Proceedings of the IIE Annual Conference, Nashville, TN.

Abbas, A. E. (2017). *Foundations of multiattribute utility.* Cambridge: Cambridge University Press.

Abbas, A. E., & Aczel, J. (2010). The role of some functional equations in decision analysis. *Decision Analysis, 7*(2), 215–228.

Abbas, A. E., & Bell, D. E. (2011). One-switch independence for multiattribute utility functions. *Operations Research, 59*(3), 764–771.

Abbas, A. E., & Bell, D. E. (2012). One-switch conditions for multiattribute utility functions. *Operations Research, 60*(5), 1199–1212.

Abbas, A., Hazelrigg, G., & Al-Kindi , M. (2012). Bayesian inference for the demand of engineering products. ASME DTM International Design Engineering Technical Conferences and Computers and Information in Engineering Conference.

Abbas, A. E., & Howard, R. A. (2005). Attribute dominance utility. *Decision Analysis, 2*(4), 185–206.

Abbas, A. E., & Matheson, J. E. (2009). Normative decision making with multiattribute performance targets. *Journal of Multi-criteria Decision Analysis, 16*(3–4), 67–78.

Abbas, A. E., Matheson, J. E., & Bordley, R. F. (2009). Effective utility functions induced by organizational target-based incentives. *Managerial and Decision Economics, 30*(4), 235–251.

Arrow, K. J. (1965). The theory of risk aversion. Lecture 2 in Aspects of the Theory of Risk-Bearing. *Yrjo Jahnssonin Saatio*, Helsinki.

Bell, D. E. (1979). Multiattribute utility functions: Decompositions using interpolation. *Management Science, 25*, 744–753.

Debreu, G. (1960). Topological methods in cardinal utility theory. In K. Arrow, S. Karlin, & P. Suppes (Eds.) *Mathematical methods in the social sciences* (pp. 16–26). Stanford, CA: Stanford University Press.

Dyer, J. (1990). Remarks on the analytic hierarchy process. *Management Science, 36*(3), 249–258.

Farquhar, P. H. (1975). A fractional hypercube decomposition theorem for multiattribute utility functions. *Operations Research, 23*, 941–967.

Fishburn, P. C. (1974). Von Neumann-Morgenstern utility functions on two attributes. *Operations Research, 2*, 35–45.

Fishburn, P. C. (1977). Approximations of two-attribute utility functions. *Mathematics of Operations Research, 2*, 30–44.

Howard, R. A. (1966). Information value theory. *IEEE Transactions on Systems Science and Cybernetics, 2*(1), 22–26.

Howard, R. A. (1980). *On making life and death decisions.* In R. A. Howard & J. E. Matheson (Eds.) *The principles and applications of decision analysis*, vol. II. Menlo Park, CA: Strategic Decisions Group.

Howard, R. A. (1992). In praise of the old time religion. In W. Edwards (Ed.) *Utility theories: Measurements and applications.* New York: Springer.

Howard, R. A., & Abbas, A. E. (2015). *Foundations of decision analysis*, 1st Edition. New York: Pearson.

Keeney, R. L., & Raiffa, H. (1976). *Decisions with multiple objectives.* New York: Wiley.

Lindley, D. (1982). Scoring rules and the inevitability of probability. *International Statistical Review, 50*(1), 1–11.

Matheson, J. E., & Abbas, A. E. (2005). Utility transversality: A value-based approach. *Journal of Multi-criteria Decision Analysis, 13*(5–6), 229–238.

Matheson, J. E., & Howard, R. A. (1968). An introduction to decision analysis. In R. A. Howard and J. E. Matheson (Eds.), *The principles and applications of decision analysis.* Menlo Park, CA: Strategic Decisions Group.

Pratt, J. (1964). Risk aversion in the small and in the large. *Econometrica, 32*, 122–136.

Saaty, T. (1990). How to make a decision: The analytic hierarchy process. *European Journal of Operational Research, 48*, 9–26.

Savage, L. (1951). The theory of statistical decision, *Journal of the American Statistical Association*, *46*, 55–67.

Savage, L. (1972). *The foundations of statistics*. New York: Dover Publications.

Sun, Z., & Abbas, A. E. (2013). Bayesian updating on price elasticity of uncertain demand. *IEEE International Systems Conference (SysCon)*, 222–228.

Traverso, M. G., & Abbas, A. E. (2009). Demand curve prediction via Bayesian probability assignment over a functional space. *Winter Simulation Conference*, 2971–2976.

von Neumann, J., & Morgenstern, J. (1947). *Theory of games and economic behavior.* Princeton, NJ: Princeton University Press.

18

Achieving Multiple Objectives with Limited Resources: Using Utility Theory and Control Theory

Ali E. Abbas and Dušan M. Stipanović

18.1 Introduction

The problem of maximizing the utility of multiple objectives given limited resources is a common problem in homeland security applications. Examples abide. Consider, for example, the assignment of (a limited resource of) security officers at an airport whose objectives might be twofold: to survey the greatest number of passengers, as well as to minimize the waiting time of passengers. Another example might be the allocation of unmanned aerial systems (UAS) in search and rescue missions in the event of a disaster. The number of UAS might be limited and their objectives might be: (i) to maximize the surveillance of a given area through their scanning detectors; (ii) to maintain close proximity to each other to maintain communication links and so that more than one might be called on; and (iii) to avoid collision with each other. Other examples of allocation of resources in security applications include the routing of police cars in a large city to maximize the surveillance of an area, and yet to stay in reasonable proximity of each other in case they are called on by an agent.

The purpose of this chapter is to explain a general methodology for addressing these types of problems where multiple objectives are sought and where a limited number of resources is available. The main idea of the proposed approach will be to first determine the multiple objectives and assess the trade-offs among them in a meaningful way using advances in utility theory (Abbas, 2009b, 2013), and then to use multi-objective control theoretic procedures to derive the control strategies that maximize the multiple objectives (Stipanović, Melikyan, & Hovakimyan, 2010). As we shall see, this approach will enable a closed-form expression for the decision strategies, a method to control multiple agents achieving the objectives, and a simple implementation of the control.

The feasibility of this approach is made possible by capturing the different objectives (such as surveillance and collision avoidance) using new constructs, called Archimedean multiattribute utility copulas (Abbas, 2009b; Abbas & Sun, n.d.), that have the flexibility to model a large variety of trade-offs among multiple objectives. The control strategies are then obtained through the optimization procedure, which provides explicit solutions and thus requires minimal computational time when implemented. The only computation needed is related to updating feedback information, which may be the position of a target being followed.

A fundamental difference between the proposed approach and traditional control procedures that use Lyapunov-type functions is that here we assess the multiattribute utility function itself and then use it in the control procedure instead of trying to find the Lyapunov function itself and then solving for the control.

After explaining the basic procedure and the properties of the proposed approach, this chapter will discuss an application to the design of control strategies (paths) to guide the trajectories of (a limited number of) unmanned aerial systems (UAS) to help them meet specific objectives. The objectives considered include maximizing the surveillance of a given area, tracking (to remain in close proximity), and collision avoidance. The new strategies will enable vehicles (or defenders in case of attacker-defender models) to make autonomous, fast, and accurate normative decisions based on large amounts of data that may be incomplete (or corrupted by noise or human inefficiencies). While accomplishing multiple objectives such as surveillance, target recognition, and tracking, the vehicles must avoid collisions and keep the communication links reliable at all times. Experimental results for the performance of the UAS using the proposed approach are also provided.

18.2 Multiattribute Utility Copulas

Before we proceed with the basic formulation of the problem, it is important to explain the formulation by which multiple objectives are incorporated into the utility function. The approach will capture the value (and utility) of achieving each objective individually and then assigning trade-offs among the objectives using an appropriate multiattribute utility function. The multiattribute utility functions we incorporate use the concept of a utility copula function (Abbas, 2009b) that we illustrate later in this chapter.

Utility copula functions express any continuous multiattribute utility function, $U(x_1,\ldots,x_m)$, that is, strictly increasing with each argument for at least one reference value of the remaining (complement) attributes in terms of individual von Neumann-Morgenstern normalized utility assessments of each objective using the relation (Abbas, 2009b).

$$U(x_1,\ldots,x_m) = C(U_1(x_1),\ldots,U_m(x_m)),$$

where C is a utility copula function and the functions $U_i(x_i)$ are von Neumann-Morgenstern utility functions for each attribute at a specific reference value of the complement attributes.

A special case of utility copula functions that will be useful in our formulation is the Archimedean utility copula (Abbas, 2009b) form

$$C(v_1,\ldots,v_m) = a\psi^{-1}\Big[\prod_{i=1}^{m}\psi(l_i + (1-l_i)v_i)\Big] + b, \tag{18.1}$$

where, $0 \le l_i < 1$, $a = 1/(1-\psi^{-1}[\prod_{i=1}^{m}\psi(l_i)])$, $b = 1-a$, and the function ψ satisfies the following conditions: (i) $\psi(v)$ is continuous on the domain $v \in [0,1]$; (ii) $\psi(v)$ is strictly increasing on the domain $v \in [0,1]$; (iii) and $\psi(1) = 1$.

The Archimedean utility copula will be used to construct the objective function for the control. Let us now highlight a few of its properties. It was shown in Stipanović, Tomlin, and Leitmann (2012) that the Archimedean form (1) can be used to construct a differentiable over-approximation of the maximum in the following way. We have

$$\max(v_1,\ldots,v_m) \le -\ln\left(\frac{C(e^{-v_1},\ldots,e^{-v_m})-1}{a(1-L_M)}+1\right), \tag{18.2}$$

where $L_M = \max(l_1,\ldots,l_m)$.

Furthermore, by defining the following function

$$\rho(x) = -\ln(C(e^{-v_1(x)},\ldots,e^{-v_m(x)})) \tag{18.3}$$

it can be shown that $\rho \le \epsilon$ implies that each individual value $v_j \le \epsilon_j$ where

$$\epsilon = \underset{j\in\{1,\ldots,m\}}{\arg\min}\{-\ln(1-a(1-l_j)(1-e^{-\epsilon_j}))\} \tag{18.4}$$

So if each threshold value ϵ_j is chosen such that $v_j \le \epsilon_j$ means that the j-th objective is accomplished, then the condition $\rho \le \epsilon$ as a single constraint guarantees that all m objectives are accomplished. This is an important result that will help with the control strategies as it guarantees that when the overall Archimedean utility function falls below a certain amount, ϵ, then each of the individual objectives also falls below a certain threshold ϵ_j, and so the objectives are all met.

The main essence of the optimization approach is to construct differentiable approximations for the minimum and maximum values of the utility function and then use a control design procedure that is similar to the control Lyapunov function design (Bacciotti & Rosier, 2005; Freeman & Kokotović, 1996) The similarity is that these differentiable approximations of the minimum and maximum are being used in the optimization instead of Lyapunov functions as agents' goal functions.

We will now show how Archimedean utility copulas (Abbas, 2009a, 2009b) can be used to construct approximations of the minimum and maximum. To achieve this, we define the following functional forms:

$$\begin{aligned} U_1(l,v) &\equiv U_1(l_1,\ldots,l_m,v_1,\ldots,v_m) = \psi^{-1}\left(\prod_{i=1}^{m} \psi(l_i+(1-l_i)v_i)\right), \\ U_2(l,v) &\equiv U_2(l_1,\ldots,l_m,v_1,\ldots,v_m) = \psi^{-1}\left(\prod_{i=1}^{m} \psi^{1/m}(l_i+(1-l_i)v_i)\right), \end{aligned} \tag{18.5}$$

where $\psi(\cdot):[0,1]\to[0,1]$ is a strictly increasing function of its scalar argument on the interval $[0,1]$. Scalar parameters $l_i \in [0,1)$, $i \in \{1,\ldots,m\}$ are nonnegative real numbers strictly less than one. Scalar variables (objectives) $v_i \in [0,1]$, $i \in \{1,\ldots,m\}$ are assumed to take values between zero and one.

It was shown in Stipanović, Valicka, and Abbas (2014) that we can construct an approximation from below of the maximum, a_M, of the Archimedean form using multiattribute functions as

$$a_M \geq -\ln\left(\frac{1-U_2(l_1,\ldots,l_m,1-e^{-a_1},\ldots,1-e^{-a_m})}{1-L_M}\right) \equiv \underline{\beta}(a). \tag{18.6}$$

As for an approximation from above of the maximum, we can also use a multiattribute related function as

$$a_M \leq -\ln\left(\frac{U_1(l_1,\ldots,l_m,e^{-a_1},\ldots,e^{-a_m})-L_M}{1-L_M}\right) \equiv \bar{\beta}(a). \tag{18.7}$$

Therefore, the maximum value of the Archimedean form a_M can be bound as

$$\underline{\beta}(a) \le a_M \le \overline{\beta}(a)$$

Now let us show how multiattribute copulas can be used to satisfy multiple objectives where each objective is accomplished when $a_i \le \varepsilon_i$ for some positive numbers ε_i, $i \in \{1,\ldots,m\}$ which are determined by the particular applications of objectives.

Recall that $\overline{\beta}(a)$ is an upper bound on the maximum. Therefore, by minimizing $\overline{\beta}(a)$, we are guaranteeing that the maximum of the utility function will fall below a minimum and each objective will be met. Therefore, the goal would be to solve the following minimization problem:

$$\min\left\{\overline{\beta}(a)\right\} = \min\left\{-\ln\left(\frac{U_1(l_1,\ldots,l_m,e^{-a_1},\ldots,e^{-a_m}) - L_M}{1-L_M}\right)\right\}$$

subject to: 1) system dynamics
2) actuaction (or control) constraints
3) other constaints such as limited resources and cost (18.8)

The individual thresholds $\varepsilon_1,\ldots,\varepsilon_m$ are set depending on the objectives. The ultimate goal of optimization (8) is to obtain a $\overline{\beta}(a) \le \min\{\varepsilon_1,\ldots,\varepsilon_m\}$ by calculating the control law of the dynamic system, which would result in accomplishing all objectives because each objective would fall below its threshold minimum accomplishment value.

Now it is important to notice that the optimization problem (8) is not static, but dynamic due to system dynamics which are governed by a control law that is the ultimate outcome of the optimization process. Thus we actually have $a(t)$ where t denotes the time the control is applied, and instead of static formulations, we consider a dynamic optimization of the following form:

$$\min\left\{\frac{d\overline{\beta}(a(t))}{dt}\right\}$$

subject to: 1) system dynamics
2) actuaction (or control) constraints
3) other constaints such as limited resources and cost (18.9)

with the same goal, that is, $\overline{\beta}(a) \le \min\{\varepsilon_1,\ldots,\varepsilon_m\}$.

At any time t, the optimization problem can be applied. This dynamic optimization process resembles the well-known dynamic programming approach of Bellman (2003), yet there is an essential difference between the two because we do not have to solve for the *value function*, which is known to be extremely numerically complex. Instead, the value function is constructed using the objectives and the assessed trade-offs among them. Furthermore, as it will be also shown later, the optimization problem (9) results in explicit closed forms for control laws in a number of complicated and challenging scenarios. Thus these control laws can be easily and efficiently implemented on actual dynamic systems, which may be, for example, autonomous vehicles.

18.3 Case Study: Maximizing the Utility of Multiple Objectives with UAS

We now apply our problem formulation to the problem of efficiently computing paths (control strategies) for a limited number of UAS that seek to maximize surveillance of a particular area. In so doing, they seek to avoid collision, and also seek to have continuous tracking. This is a problem of particular importance particularly to the southern border. The problem would be extremely difficult to solve given the traditional dynamic programming approach. Furthermore, it would require several constraints on the minimum distance between the UAS (to avoid collision) and also a constraint on the maximum distance to maintain tracking.

Instead, we incorporate surveillance, collision avoidance, and tracking as objectives and assign trade-offs among the objectives using an Archimedean utility copula. The design of path strategies for each UAS in our formulation will be based on optimizing the UAS' individual goal functions constructed using multiattribute utility copulas and trade-offs among the objectives. If an Archimedean utility copula form is adopted, the approach will guarantee that the multiple objectives will be accomplished when the aggregate utility falls below a certain value. The guarantees are based on the fact that the particular types of Archimedean multiattribute utility functions have the property that achieving a threshold level of the copula function implies a minimum threshold level on each of the objectives has been achieved. Furthermore, as we discussed, Archimedean utility copulas can be constructed as appropriate approximations of the minimum and maximum functions.

18.3.1 *Choosing the Attributes and Individual Goal Functions for the UAS Control Problem*

Let us consider a surveillance problem with a limited number of vehicles. This particular scenario considers vehicles with several objectives. This section describes how to incorporate some of the objectives mathematically into the corresponding objective functions.

Objective Function for Surveillance

The first objective is the surveillance. By this we mean that the vehicles' sensors have footprints whose union is much smaller than the area to be covered. Thus they have to move around to survey the most amount of space in the given domain. For our approach, the domains need not be connected or even convex. The overall surveillance is the area covered by the sensor footprints. It is an integral of the total (new) area covered by the sensors.

Objective Function for Collision Avoidance

The second objective we consider is safety, that is, vehicles have to avoid collisions between themselves as well as with any static obstacles in their trajectory. This will be achieved by a goal function for collision avoidance related to the Euclidean distance between each vehicle and a neighboring vehicle (or object).

Avoiding collisions among agents is captured by a collision avoidance objective function related to an objective of avoiding collisions between two agents i and j with their position coordinates represented by state vectors $x_i \in \mathbb{R}^{n_i}$ and $x_j \in \mathbb{R}^{n_j}$, which is given by

$$v_{ij}(x) = \left(\min \left\{ 0, \frac{x_i - x_j \;{}^2 - R^2}{x_i - x_j \;{}^2 - r^2} \right\} \right)^2, \tag{18.10}$$

where x is a concatenation of all state variables and $R > r > 0$. The sensing or detection regions of the agents are assumed to be of radius R and the avoidance or unsafe regions (in this case determined by the smallest safe distance between the agents) are given by radius r as depicted in Figure 18.1. For example, for aircraft, the radius r is standardly taken as five nautical miles. These collision avoidance functions were inspired by the idea of avoidance control (Leitmann & Skowronski, 1977), which was originally introduced as a novel Lyapunov-based approach to differential games.

Similarly, a collision avoidance function to represent an objective for agent i to avoid an obstacle denoted by a positive integer l may be formulated as

$$v_{il}(x) = \left(\min \left\{ 0, \frac{(x_i^p - x_l^o)^T P_{il} (x_i^p - x_l^o) - R_{il}^2}{(x_i^p - x_l^o)^T P_{il} (x_i^p - x_l^o) - r_{il}^2} \right\} \right)^2, \tag{18.11}$$

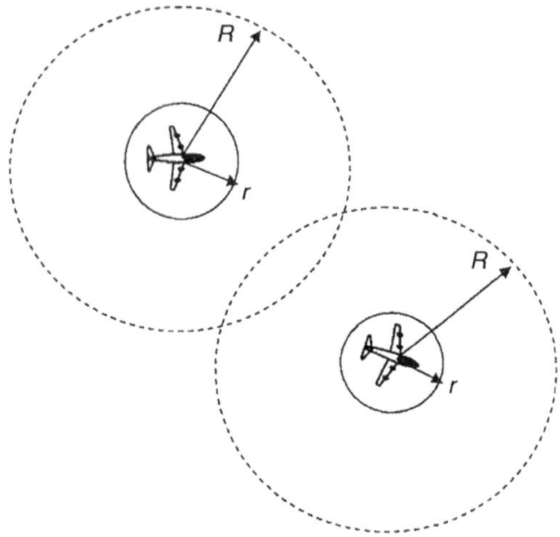

Figure 18.1. Aircraft with collision and detection regions.

where $R_{il} > r_{il} > 0$ are positive scalars and P_{il} is a positive definite matrix chosen so that the set

$\{z : (z - x_j^o)^T P_{ij}(z - x_j^o) \leq r_{ij}^2\}$ includes obstacle l, and a value for R_{il} is determined based on the sensing capabilities of an agent i.

Objective Function for Proximity

The third objective is keeping the information links between the vehicles reliable. Assuming the communication links are wireless, that means that, during the operation, the vehicles have to be close enough since the proximity guarantees wireless links to be reliable.

During the operation, we assume that the agents exchange information via wireless communication links. In order to keep these links reliable, the agents need to stay close. This requirement is formulated in terms of distances between the agents being smaller than particular distances that guarantee reliable communication links. This means that any two agents i and j are accomplishing other objectives (such as coverage) while trying to keep mutual distance to be less or equal to a distance $\widehat{R}_{ij}$ that allows them to communicate without interruptions. To formulate a proximity objective for agents i and j to be distance wise closer than $\widehat{R}_{ij}$, we use the following function (Stipanović, Valicka, Tomlin, & Bewley, 2013):

$$v_{ij}^p(x_i^p, x_j^p) = \max\{0,\ x_i^p - x_j^p\ {}^2 - \widehat{R}_{ij}^2\}^2. \tag{18.12}$$

One can immediately notice that the safety and the proximity objectives are, technically speaking, opposing objectives: one objective requires a minimum distance to avoid collision, while the other requires a maximum distance to keep within proximity for the information links. This implies that there must be a compromise so the vehicles never get too close or too far. This is the purpose of assigning trade-offs among the objectives using the Archimedean copulas. As we shall see in the last section, where we present experimental results, the ability to represent these objectives with general forms of trade-offs makes this problem feasible in terms of practical implementation of control laws on unmanned vehicles, especially when one of the objectives is surveillance.

18.4 Control Strategies Based on the Archimedean Utility Copula

Our design is similar to what is known as the control Lyapunov functions approach (Bacciotti & Rosier, 2005; Freeman & Kokotović, 1996), where, instead of standard Lyapunov functions, we use functional forms based on multiattribute copulas and objective functions that provide some ideas on constructing control strategies based on approximation functions. This way we bypass a bottleneck of the Lyapunov function design, which avoids numerous difficulties of computing optimal control laws (Bardi & Capuzzo-Dolcetta, 1997; Falcone & Ferretti, 1994; Mitchell, Bayen, & Tomlin, 2005).

Since our agents are in motion, they are modeled as dynamic systems, that is, each agent is represented by a set of differential equations

$$\dot{x}(t) = f(t, x(t), u(t)), \quad x(t_0) = x_0 \in \mathbb{R}^n, \; t_0 \in [0, +\infty) \tag{18.13}$$

where the state variable vector-valued function $x(\cdot)$ with $x(t) \in \mathbb{R}^n$ is assumed to be a solution if it is an absolutely continuous function (Kolmogorov & Fomin, 1975) satisfying equation (18.13) almost everywhere and the initial condition $x(t_0) = x_0$ at given initial time t_0.

The input control $u(\cdot) : [0, +\infty) \rightarrow \mathbf{U}$ is restricted to a set of *admissible* functions $\mathcal{U}$ with the image being a compact set denoted as $\mathbf{U}$, $\mathbf{U} \subset \mathbb{R}^m$. We refer to the set of admissible controls in a standard way as a set of measurable functions (Bressan & Piccoli, 2007; Filippov, 1988). Some standard conditions which function $f(\cdot,\cdot,\cdot) : [0, +\infty) \times \mathbb{R}^n \times \mathbf{U} \rightarrow \mathbb{R}^n$ and admissible controls should satisfy in order to assure existence and uniqueness of solutions of (18.13) can be found in Bressan and Piccoli (2007), Coddington and Levinson (1955), and Filippov (1988). It is important to note that we do not need the solutions of (18.13) to be unique; however, uniqueness simplifies the presentation.

We assume that a goal is mathematically formulated as some value function $v(t,x) \leq \epsilon$ where $v(\cdot)$ is a differentiable goal function, yet since we are dealing with a dynamically changing scenario, we have to include time in this condition. Thus, we say that a goal defined by the corresponding goal function $v(\cdot)$ is said to be accomplished at time T if a trajectory $x(\cdot)$ of system (18.13) satisfies $v(T,x(T)) \leq \epsilon$ (Stipanović et al, 2012). For the case of multiple objectives, the goal function $v(t,x)$ is designed as an approximation of the maximum of the utility copula function, namely, $\overline{\beta}(a(t,x))$. The same type of procedure applies to the case of achieving a single goal or multiple goals using the Archimedean copula.

Then one way to satisfy this goal is to decrease values of function $v(\cdot)$ along the trajectories of the dynamic system (18.13) in time until it satisfies the goal condition at some time $t = T$ when $v(T,x(T)) \leq \epsilon$.

Since the goal function is differentiable, a design of a control strategy is based on the following minimization procedure for each vehicle that is similar to the control Lyapunov design (Bacciotti & Rosier, 2005; Freeman & Kokotović, 1996):

$$\begin{aligned} \hat{u}(t,x) &= \arg\min_{u_c \in \mathbf{U}} \left\{ \frac{\partial v(t,x)}{\partial t} + \frac{\partial v(t,x)}{\partial x} \dot{x} \right\} \\ &= \arg\min_{u_c \in \mathbf{U}} \left\{ \frac{\partial v(t,x)}{\partial t} + \frac{\partial v(t,x)}{\partial x} f(t,x,u_c) \right\} \\ &= \arg\min_{u_c \in \mathbf{U}} \left\{ \frac{\partial v(t,x)}{\partial x} f(t,x,u_c) \right\} \end{aligned} \tag{18.14}$$

where $\partial v(t,x)/\partial t + \dfrac{\partial v(t,x)}{\partial x} f(t,x,u_c)$ is the time derivative of $v(t,x(t))$ computed along the trajectories $x(t)$ of (18.13) and control value $u_c = u(t)$. Variable u_c is an auxiliary variable that takes values in the compact set $\mathbf{U}$.

We assume that a control exists that satisfies (18.13) and is *admissible*, that is, $u(\cdot) \equiv \hat{u}(\cdot, x(\cdot)) \in \mathcal{U}$. In the case of time-invariant systems where $\dot{x}(t) = f(x(t), u(t))$ with goal function $v(\cdot)$, (18.14) simplifies to

$$\hat{u}(x) = \arg\min_{u_c \in \mathbf{U}} \left\{ \frac{\partial v(x)}{\partial x} f(x,u_c) \right\} \tag{18.15}$$

so that $u(t) = \hat{u}(x(t))$.

In order to obtain guarantees that a control strategy accomplishes the goal in dynamic scenarios, we use differential inequalities (Stipanović et al.,

2010). Assuming that $v(t_0, x(_0)) > \epsilon$, that is, at initial time, the goal is not achieved, we design a comparison function $G(\cdot,\cdot,\cdot):[0,+\infty)\times\mathbb{R}^n\times\mathbb{R}\to\mathbb{R}$ that satisfies the following inequality:

$$\frac{\partial v(t,x)}{\partial t}+\frac{\partial v(t,x)}{\partial x}f(t,x,\hat{u}(t,x))\leq G(t,x,v(t,x)) \tag{18.16}$$

for any $(t,x)\in[t_0,+\infty)\times\mathbb{R}^n$. The choice of $G(\cdot,\cdot,\cdot)$ is very much problem dependent and often done by choosing equality sign in (18.16) as shown in Stipanović and colleagues (2012) and (Stipanović et al., 2014). Then, if the maximal solution $\overline{z}(t)$ (which requires much less restrictive conditions on existence than regular solutions (Filippov, 1998; Lakshmikantham & Leela, 1969) of the corresponding comparison system

$$\dot{z}(t)=G(t,x(t),z(t)),\;\; z(t_0)=z_0 \tag{18.17}$$

with initial condition $z_0 > v(t_0, x_0)$, satisfies $\overline{z}(T)\leq\epsilon$ at some T, $T > t_0$, then the goal is accomplished (Stipanović et al., 2012).

Our approach is particularly well suited for multiple-agent, dynamic systems where the dynamics of the agents are affine in control, which is the case of dynamic models of many different vehicles as shown in Mastellone, Stipanović, Graunke, Intlekofer, and Spong (2008), Rodríguez-Seda, Tang, Spong, and Stipanović (2014), Siegwart and Nourbakhsh (2004), and Spong, Hutchinson, and Vidyasagar (2005). In mathematical terms, that means that the agents are modeled as

$$\frac{dx_i(t)}{dt}=f_i(x_i(t),u_i(t))=h_i(x_i(t))+g_i(x_i(t))u_i(t), x_i(t_0)=x_{i0}, \forall i\in\{1,\ldots,M\} \tag{18.18}$$

where M is the number of agents, $x_i(t)$ denotes the i-th agent's state, and $u_i(t)$ denotes its *admissible* control at time t, respectively. One of the well-known models that is used for differential drive vehicles is the nonlinear unicycle model defined by:

$$g_i(x_i)=\begin{bmatrix}\cos(x_{i1}^p) & 0\\ \cos(x_{i2}^p) & 0\\ 0 & 1\end{bmatrix}, \quad h(x_i)\equiv 0 \tag{18.19}$$

where (x_{i1}^p, x_{i2}^p) denote the Cartesian coordinates of the i-th vehicle and also the first two state-variables. The third state-variable is the heading angle and the inputs are speed and angular velocity, respectively. The inputs are assumed to be measurable functions that satisfy the following bounds:

$$\| u_i(t) \| \leq \mu_i,\ \forall t \in [t_0, +\infty),\ \ \mu_i > 0,\ \ \forall i \in \{1, \ldots, M\}, \tag{18.20}$$

where μ_i's are given positive real numbers chosen to impose important implementation restrictions on the control laws. Each agent is assumed to have its own goal function $v_i(\cdot)$ which arguments are objective functions. Then due to the fact that the dynamic systems (18.18) are affine in control, the control laws for the agents may be computed as $u_i(t) = \hat{u}_i(x(t)), i \in \{1, \ldots, m\}$, where

$$\begin{aligned} \hat{u}_i(x) &= \underset{\|u_i\| \leq \mu_i}{\arg\min} \left\{ \sum_{j=1}^{m} \frac{\partial v_i(x)}{\partial x_j} h_j(x_j) + \sum_{j=1}^{m} \frac{\partial v_i(x)}{\partial x_j} g_j(x_j) u_j \right\} \\ &= \underset{\|u_i\| \leq \mu_i}{\arg\min} \left\{ \frac{\partial v_i(x)}{\partial x_i} g_i(x_i) u_i \right\} = -\mu_i w_i(x),\ \ i \in \{1, ..., m\}, \end{aligned} \tag{18.21}$$

with $w_i(x) = g_i^T(x_i) \dfrac{\partial v_i(x)}{\partial x_i}^T \left\| \dfrac{\partial v_i(x)}{\partial x_i} g_i(x_i) \right\|^{-1}$ if $\left\| \dfrac{\partial v_i(x)}{\partial x_i} g_i(x_i) \right\| \neq 0$. When $\left\| \dfrac{\partial v_i(x)}{\partial x_i} g_i(x_i) \right\| = 0$ we can set $w_i(x) = k_i$ where k_i is any constant vector of appropriate dimension that satisfies $k_i \leq 1$ (Stipanović et al., 2012).

Finally, it is important to note that our control design results in the *closed-form* solutions and thus requires no additional computation when being implemented as shown in Valicka, Rekoske, Stipanović, and Abbas (2014). This is due to the fact that the goal functions in equation (18.21) are known for each agent and thus their gradients can be precomputed and explicitly substituted before the implementation.

18.5 Experiments

In order to validate our control design based on multiattribute copulas, a number of experiments were conducted with ground unmanned vehicles in the Mechatronics Laboratory at the University of Illinois at Urbana-Champaign

(a) (b)

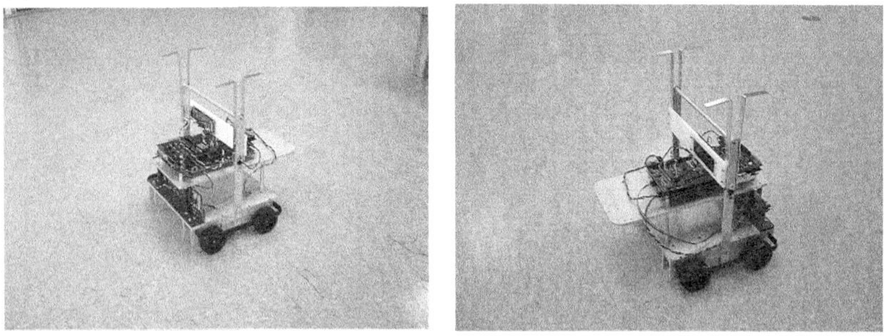

Front view. Back view.

Figure 18.2. One of the differential drive robots used in experiments.

as reported in Valicka and colleagues (2014). These vehicles are so-called differential drives, which are known to be non-holonomic (Siegwart & Nourbakhsh, 2004) and modeled using nonlinear dynamics (Spong, Hutchinson, & Vidyasagar, 2005) and thus very specific and challenging to control. One of the vehicles used in the experiments is depicted in Figure 18.2. The control incorporated the three objectives using an Archimedean utility copula.

It is important to stress that the control law design as in equation (18.21) does not suffer from "the curse of dimension" issues, that is, its complexity does not depend on the number of vehicles and number of objectives since the control laws are computed explicitly and thus are *scalable*. To illustrate this, we will provide results from two experiments with three and four robots, respectively, achieving three objectives. These three objectives were collision avoidance (both inter-vehicle and obstacle), surveillance of a square domain (an approximate length being 4.5 meters.), and that the vehicles move as a swarm, that is, always stay close enough so that the wireless communication links stay reliable.

We refer to these three objectives as collision avoidance, surveillance or coverage control, and swarming, respectively, and we refer to the experiment with three vehicles as Experiment 1 and a four-vehicle experiment as Experiment 2. The vehicles are denoted as objects 1, 2, and 3 in Experiment 1 and as 1, 2, 3, and 4, in Experiment 2. Two static obstacles are denoted as objects 4 and 5 in Experiment 1 and 5 and 6 in Experiment 2. Differential drive vehicles were modeled using nonlinear and non-holonomic unicycle models

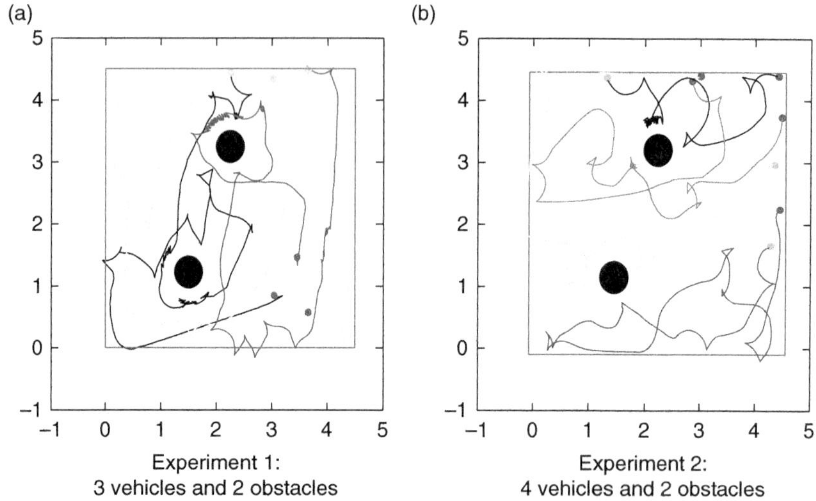

Figure 18.3. Vehicles' trajectories.

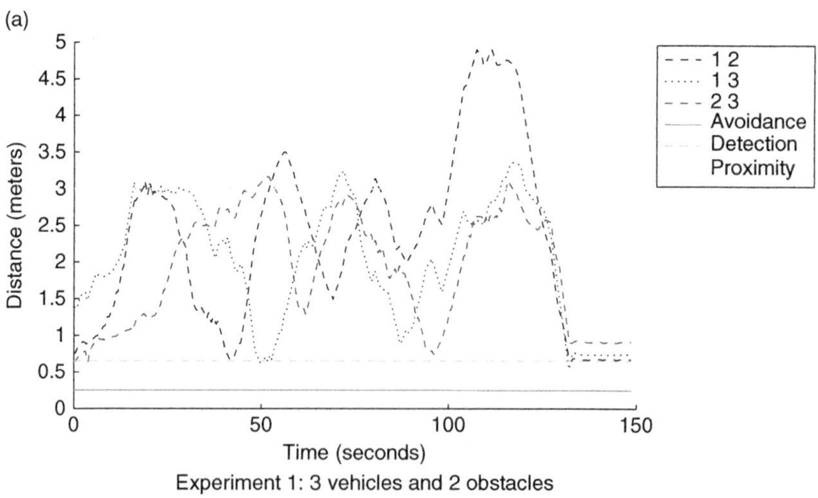

Figure 18.4. Pairwise distances between the vehicles.

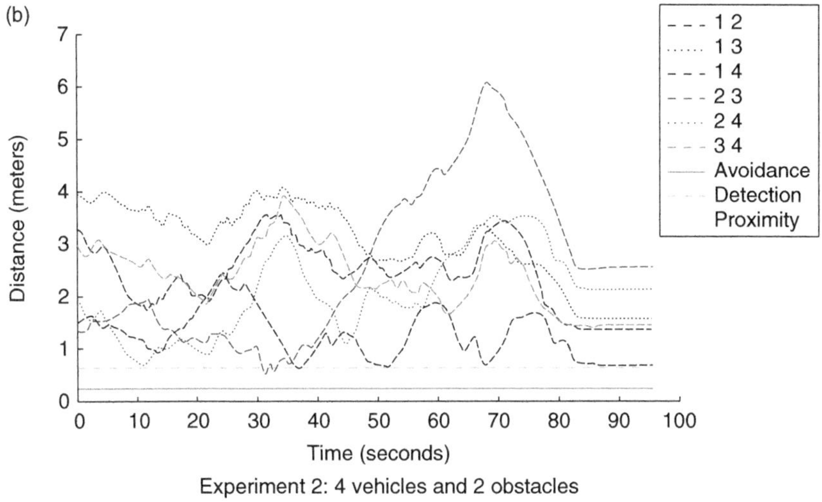

Figure 18.4 (*cont.*)

(Stipanović et al., 2014), and multiattribute utility copulas were used to design their control strategies as shown in Section 18.4. In both experiments, two static obstacles were placed in the rectangular domain that was to be surveilled by the vehicles. Vehicles' trajectories while accomplishing the objectives are depicted in Figure 18.3.

Vehicles' initial positions were depicted in green and final in red. In terms of collision avoidance and proximity, pairwise distances between the vehicles are depicted in Figure 18.4. There one may observe that the collision avoidance was never violated as being a strict, that is, safety objective and proximity may be violated for relatively short periods of time since it is only a "soft" constraint.

Then in Figure 18.5, pairwise distances between the vehicles and the obstacles are depicted that show that the minimum separation is maintained not only between the vehicles, but also with respect to the obstacles.

The robots satisfactorily covered, that is, surveilled the entire square domain after 149 seconds in Experiment 1 and after 97 seconds in Experiment 2, which is another proof of the control design being scalable since it provided faster coverage with more vehicles with absolutely no modifications in the cases of three and four vehicles.

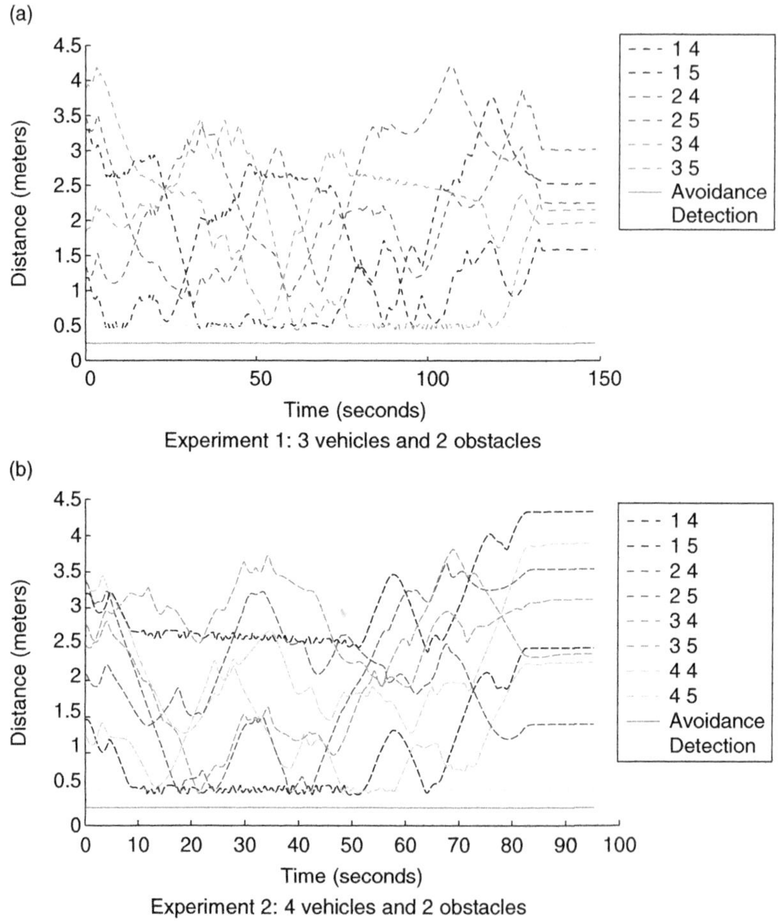

Figure 18.5. Distances between the vehicles and the obstacles.

18.6 Conclusion

This chapter proposed a method to allocate limited resources for the purpose of achieving multiple objectives. The main idea expresses the utility of achieving the objectives using an Archimedean utility copula. The approximation of the maximum of the Archimedean utility copula can be used as a value function in the dynamic system formulation. The approach results in a closed-form expression for the control law that is scalable with the dimensionality of the problem. The proposed approach can be applied to numerous applications where the purpose is to achieve multiple objectives with limited resources.

BIBLIOGRAPHY

Abbas, A. E. (2009a). Invariant multiattribute utility functions. *Theory and Decision, 68*, 69–99.

Abbas, A. E. (2009b). Multiattribute utility copulas. *Operations Research, 57*, 1367–1383.

Abbas, A. E. (2013). Utility copula functions matching all boundary assessments. *Operations Research, 61*(2), 359–371.

Abbas, A. E., & Sun, Z. (n.d.) Multiattribute utility functions satisfying mutual preferential independence. *Operations Research, 63*(2), 378–393.

Bacciotti, A., & Rosier, L. (2005). *Liapunov functions and stability in control theory.* Berlin, Germany: Springer-Verlag.

Bardi, M., & Capuzzo-Dolcetta, I. (1997). *Optimal control and viscosity solutions of Hamilton-Jacobi-Bellman equations.* Boston, MA: Birkhäuser.

Bellman, R. (2003). *Dynamic programming. Mineola*, NY: Dover Publications; Reprint edition.

Bressan, A., & Piccoli, B. (2007). *Introduction to the mathematical theory of control.* Springfield, MO: American Institute of Mathematical Sciences.

Coddington, E. A., & Levinson, N. (1955). *Theory of ordinary differential equations.* New York: McGraw Hill.

Falcone, M., & Ferretti, R. (1994). Discrete time high-order schemes for viscosity solutions of Hamilton-Jacobi-Bellman equations. *Numerische Mathematik, 67*, 315–344.

Filippov, A. F. (1988). *Differential equations with discontinuous righthand sides.* Dordrecht, The Netherlands: Kluwer Academic Publishers.

Filippov, V. V. (1998). *Basic topological structures of ordinary differential equations. Dordrecht*, The Netherlands: Kluwer Academic Publishers.

Freeman, R. A., & Kokotović, P. V. (1996). *Robust nonlinear control design: State space and Lyapunov technique.* Boston, MA: Birkhäuser.

Kolmogorov, A. N., & Fomin, S. V. (1975). *Introductory real analysis.* New York: Dover Publications.

Lakshmikantham, V., & Leela, S. (1969). *Differential and integral inequalities: Theory and applications.* New York: Academic Press.

Leitmann, G., & Skowronski, J. (1977). Avoidance control. *Journal of Optimization Theory and Applications, 23*, 581–591.

Mastellone, S., Stipanović, D. M., Graunke, C., Intlekofer, K., & Spong, M. W. (2008). Formation control and collision avoidance for multi-agent nonholonomic systems: Theory and experiments. *International Journal of Robotics Research, 13*, 107–126.

Mitchell, I., Bayen, A. M., & Tomlin, C. J. J. (2005). A time-dependent Hamilton-Jacobi formulation of reachable sets for continuous dynamic games. *IEEE Transactions on Automatic Control, 50*, 947–957.

Rodríguez-Seda, E. J., Tang, C., Spong, M. W., & Stipanović, D. M. (2014). Trajectory tracking with collision avoidance for nonholonomic vehicles with acceleration constraints and limited sensing. *International Journal of Robotics Research, 33*, 1569–1592.

Siegwart, R., & Nourbakhsh, I. R. (2004). *Introduction to autonomous mobile robots.* Boston, MA: MIT Press.

Spong, M. W., Hutchinson, S., & Vidyasagar, M. (2005). *Robot modeling and control.* Hoboken, NJ: John Wiley & Sons.

Stipanović, D. M., Melikyan, A., & Hovakimyan, N. (2010). Guaranteed strategies for nonlinear multi-player pursuit-evasion games. *International Game Theory Review, 12*, 1–17.

Stipanović, D. M., Tomlin, C. J., & Leitmann, G. (2012). Monotone approximations of minimum and maximum functions and multi-objective problems. *Applied Mathematics & Optimization, 66*, 455–473.

Stipanović, D. M., Valicka, C., & Abbas, A. E. (2014). Control strategies for players in pursuit-evasion games based on their preferences. *International Game Theory Review, 16*, 1–20.

Stipanović, D., M., Valicka, C., Tomlin, C. J., & Bewley, T. R. (2013). Safe and reliable coverage control. *Numerical Algebra, Control and Optimization, 3*, 31–48.

Valicka, C., Rekoske, R. A., Stipanović, D. M., & Abbas, A. E. (2014). Multiattribute utility copulas for multi-objective coverage control. *Paladyn, Journal of Behavioral Robotics, 5*, 12–34.

19

Defender-Attacker Decision Tree Analysis to Combat Terrorism

Ryan J. B. Garcia and Detlof von Winterfeldt

19.1 Introduction

Risk and decision analysis models and tools have been widely used to estimate and manage the risks from natural disasters and technological accidents (Bedford & Cooke, 2001; Bier & Cox, 2007; Pate-Cornell, 2007). Thus, after the terrorist attacks on September 11, 2001, it was natural to apply risk and decision analysis to terrorism events as well (Ezell, Bennett, von Winterfeldt, Sokolowski, & Collins, 2010; Pate-Cornell & Guikema, 2002; Willis, Morral, Kelly, & Medby, 2005). Former Secretary Michael Chertoff advocated risk-based approaches to decision making during his tenure at the U.S. Department of Homeland Security (DHS). In a speech given at the University of Southern California, Secretary Chertoff stated:

> We have to identify and prioritize risks – understanding the threat, the vulnerability and the consequence. And then we have to apply our resources in a cost-effective manner.[1]

Translating the traditional risk and decision-analysis paradigm to combating terrorism presents unique challenges. First, the notion of a terrorist threat is quite different from the notion of a rare, but random occurrence of a natural disaster or a technological accident. Unlike nature or accidental failures of technological systems, an intelligent adversary seeks to exploit vulnerabilities and tries to carry out an attack at a time and place that is likely to produce the maximum amount of damage. Furthermore, intelligent adversaries are likely to adapt to defenses and in some cases even use them to their advantage. Therefore, some analysts have concluded that traditional risk and decision-analysis models are not applicable to terrorism, looking instead for new models and tools in game theory and related disciplines (Merrick & Parnell, 2011; National Research Council, 2008; Parnell, Smith, & Moxley, 2009).

Extensive-form games provide a useful way to represent terrorism events as a sequence of defender-attacker actions. These games are usually characterized by two opposing players, where player 1 (the defender) chooses actions that reduce risks by altering the decisions of player 2 (the attacker), intermingled with and followed by random events. Merrick and Parnell (2011) provide an overview of nine versions of defender-attacker decision models. These models include versions where both the defender and the attacker are maximizing their expected utilities (the standard extended form of the game, also known as the Stackelberg security game (Tambe, 2011)) and versions where the defender's decision, the attacker's decision, or both are considered uncertain.

In this chapter, we examine a version of the defender-attacker decision model that begins with a defender's decision to select from among several possible preventive or protective options and is followed by the attacker's choice of mode and target of attack. Because the defender has only incomplete knowledge about the attacker's motivations and capabilities, the model does not assume that the attacker's decisions are predetermined by a known expected utility calculation. Instead, we operationalize the uncertainty related to the attacker's decision as a probability distribution that is conditional on the defender's action.

There has been some debate about the value of game theoretic vs. decision theoretic models in the terrorism risk analysis literature. The advantage of fully game theoretic models, such as the Stackelberg security game, is the symmetry between defenders and attackers and the elegance in modeling both sides' states of knowledge and utility functions. A disadvantage is that the attacker's state of knowledge and utility functions are extremely hard to quantify. Some solutions are also unnecessarily conservative, especially when assuming full rationality of the attacker. Conversely, the advantage of the decision-theoretic formulation is that it creates a standard decision analysis model for the defender and simplifies the attacker's model as a probability distribution over attack modes and targets. The resulting solution algorithms are extremely simple and efficient. The disadvantage is that it creates an additional burden on the researcher to assess the attacker's conditional probabilities of attack, modes, and targets.

An early implementation of the decision-theoretic framework was an analysis of the costs and benefits of the proposed installation of Man-Portable Air Defense Systems (MANPADS or surface-to-air missiles) countermeasures to protect commercial airplanes in the United States (von Winterfeldt & O'Sullivan, 2006). This application was carried out in close collaboration with the decision makers at the Department of Homeland Security, who had initially been skeptical of the approach, but later commented favorably on it (von Winterfeldt & O'Sullivan, 2006). The analysis suggested that MANPADS countermeasures are not currently cost-effective, a conclusion that ultimately led to the cancellation of this program. Another implementation involved the choice of alternative radiological detection devices (Bakir & von Winterfeldt, 2011). In both cases, each player only had one move (choice of a defense, followed by a choice of an attack mode), with several random events following these initial decisions. Recognizing this limitation, von Winterfeldt and O'Sullivan (2006) noted the following:

> By far the most important research to be done is on the effects of countermeasures on terrorist motivations and intent to attack commercial airplanes in the United States. One would assume that the installation of countermeasures would significantly reduce the probability of an attack using infrared guided MANPADS, but it is also reasonable to assume that the probability of attacks using non-IR-guided MANPADS, RPGs, high-caliber rifles, mortars or other current or future weapons would increase. ... Research on these dynamic effects, using sequential decision trees with changing probabilities of terrorist attack modes, may shed light on this issue.

This chapter takes up the challenge these recommendations present for future research by extending the defender-attacker decision tree approach in four major ways: first, it includes another defensive decision after an attack occurs; second, it considers both multiple attack modes and multiple defenses; third, it examines the effect of weapons substitution once attackers know the defensive choice; and fourth, it ends the game by the most appropriate defender choice depending on the attacker's actions.

19.2 Introduction to Defender-Attacker Decision Trees

Decision trees are one of the oldest tools of decision and risk analysis and are covered in nearly all textbooks of decision analysis (Clemen & Reilly, 2013; Howard & Abbas, 2015; Raiffa, 1968; von Winterfeldt & Edwards, 1986). Decision trees start with a decision node, followed by one or more chance or decision nodes, and end with a chance node. Each path through a decision tree results in an outcome, which is conceptualized as a sure thing (usually a monetary equivalent or a utility).

Decision trees are analyzed or "solved" from the end node to the root node, by taking expectations at chance nodes and maximizing expected utilities or minimizing expected costs at decision nodes (Raiffa, 1968). One then calculates the expected utilities or expected costs at the first decision node and recommends the decision with the highest expected utility or the lowest expected cost. If the consequences at the end of a decision tree have multiple attributes, one first calculates the utilities of the consequences, using an appropriate multiattribute model. Keeney and von Winterfeldt (2007) discuss two types of practical models for this purpose – a multiattribute value model and an equivalent cost model. In this chapter, we will use an equivalent cost model in which we price out non-monetary consequences.

In adversarial situations, we can no longer assume that a decision maker faces a series of random events. Instead, the decision maker faces potential attacks by a hostile agent, who by definition desires to exploit the defender's vulnerabilities, counters the decision maker's defensive actions, and, in general, pursues specific goals to negatively affect the decision maker. Game theorists have explored this paradigm – with substantial success – in the context of terrorism risk assessment and management (Tambe, 2011).

As an alternative to the traditional game-theoretic formulation, one can adapt the decision tree framework to terrorism problems by implementing defender-attacker interactions as seen from the defender's point of view. Within this paradigm, the defender maximizes his or her expected utility, but the attackers are not conceptualized as rational decision makers. Instead, the attackers' decisions are assumed to be uncertain because the defender does not know the motivations or capabilities of the attackers with any degree of precision.

Figure 19.1 shows a schematic representation of such a decision tree. The decision tree ends with outcomes that are assumed to be known or can be estimated with some confidence. They usually have multiple consequences (e.g., the outcomes of a successful MANPADS attack involve deaths, injuries, lost airplanes, and losses to the economy), and we assume in this chapter that they can be converted into an equivalent cost by pricing out non-monetary attributes.

The model starts with the defender's decision node, including the decision not to invest in any defensive action. The defender's decision alternatives are denoted as "Defender Decision D_i." This decision is followed by the attacker's choice of an attack (mode and target of attack or not to attack at all). Instead of characterizing this decision by a decision node, we characterize it as a chance node with branches "Attacker Decision A_{ij}" to denote that they are conditional on the defender's choice of alternative D_i. The main mechanisms for changing the attacker's decision probabilities are through deterrence, rendering certain attack modes ineffective, or by reducing the consequences of an attack. We thus capture the adaptation of the attacker to the defender's actions without assuming that the attacker is fully informed or rationally maximizes his or her expected utility.

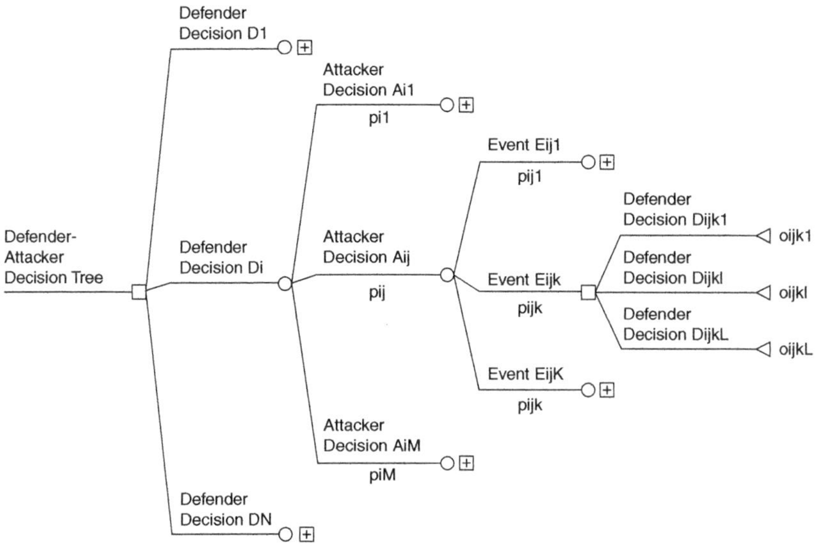

Figure 19.1. Schematic defender-attacker decision tree.
(Squares denote decision nodes, circles denote chance nodes, and triangles denote end nodes; o_{ijkl} denote the outcomes at the end of the tree; a "+" sign indicates that the tree is continued, though not necessarily symmetrically.)

The combination of defense and attack decisions is usually followed by a sequence of random events (Is the attack successful? What are its consequences?). In the simplest form of a defender-attacker decision tree, this is the end of the tree and consequences are assessed. In more complex forms, the tree continues and subsequent defender (and sometimes attacker) actions are considered. In Figure 19.1, we added a final defensive decision node to the tree (with branches D_{ijkl}), but no subsequent events or attacker responses. The idea is that after an attack occurs, the defender will choose a defense that is most appropriate for the specific attack mode.

The schematic decision tree in Figure 19.1 does not depict all possible decision-event combinations. With the exception of the i-j-k-l pathway, we truncated the tree at an earlier stage to simplify its representation. Note, however, that the truncated branches of the tree continue and that this extension does not necessarily have to be a symmetric representation of the i-j-k-l pathway, but can take different turns and terminate at different nodes. This model is solved in the same way as any multistage decision tree, by determining the decision that maximizes the defender's expected utility or minimizes the respective expected cost at the root node of the tree. Considering expected cost (EC) minimization, we seek to determine the initial defender decision D_i such that

$$\min_i\{EC(D_i)\} = \min_i\left\{\sum_{j,k} p_{ij}\, p_{ijk} \min_l[EC(D_{ijkl})]\right\},$$

where the D_{ijkl} are the possible second stage defender decisions in response to the initial defender decision D_i, an attack A_{ij}, and event E_{ijk}.

19.3 The MANPADS Threat

Man-portable air defense systems (MANPADS) that use laser-guided or heat-seeking surface-to-air missiles pose a significant threat to commercial aircraft (Bolkcom & Elias, 2006; GAO, 2004; Hunter, 2002; Phelps, 2003; Shanker, 2002). This threat has led to significant research and development efforts into measures designed to effectively counter MANPADS and other weapons systems that pose a danger to commercial air traffic. Between 2004 and 2005, the U.S. Department of Homeland Security created a multimillion-dollar program to develop direct infrared countermeasures (DIRCMs) designed to deflect a heat-seeking MANPADS missile by jamming its heat-seeking ability. In a previous study, von Winterfeldt and O'Sullivan (2006) employed a decision tree analysis that indicated DIRCMS are currently too expensive to provide a cost-effective protection against MANPADS. The model was developed in collaboration with and delivered to the MANPADS Countermeasures Office of the DHS so the DHS could populate it with classified probability and cost estimates. Partially as a result of this analysis, the DHS cancelled the MANPADS countermeasures program in 2008.

One of the concerns von Winterfeldt and O'Sullivan expressed was that an implementation of MANPADS countermeasures would likely lead terrorists who want to attack airplanes to shift their targets and use of weapons. While infrared MANPADS pose the highest-profile threat, terrorists can utilize other weapons to bring down commercial airliners. For example, terrorists can use rocket-propelled grenades (RPGs) and large-caliber rifles to attack aircraft or aircraft facilities (Bennett, 2003; Chow et al., 2005; Grau, 2008; O' Sullivan, 2004; Violence Policy Center, 2002).

The study by von Winterfeldt and O'Sullivan recognized the important effects that the availability of these alternative weapons might have on the cost-effectiveness of DIRCMs and suggested extending this line of study to include multiple threats as well as multiple possibilities to counter these threats.

In this chapter, we focus on a terrorist attack using one of three types of weapons: Infrared (IR) MANPADS, Command Line of Sight and Laser-Beam Rider (CLOS/LBR) MANPADS, and rocket-propelled grenades (RPGs). These weapons have been chosen based on the different types of threats they pose, their availability to terrorist organizations, and the utility of their use in a potential attack against an aircraft. IR MANPADS are shoulder-fired surface-to-air missiles that use an infrared sensor to lock on to the heat signature of an aircraft's engine. CLOS MANPADS are remotely guided to the target via radio frequency and LBR MANPADS use a laser built into the shoulder-mounted firing tube to guide the missile to the aircraft. CLOS/LBR MANPADS are the most advanced types of surface-to-air missiles and are used by today's modern military. One of these weapon systems, a Russian-build BUK missile, was suspected to have downed a Malaysian passenger jet over the Ukraine in July 2014.[2] Finally, an RPG is an explosive round similar to a grenade that utilizes thrust for delivery to the target, but does not have an automatic guidance system.

Countermeasures used in the model were chosen based on the most effective and practicable means to reduce the threat of these weapons. In-flight deployable flares are generally considered the most cost-effective means to counter IR-based guidance systems. Yet, flares have been ruled out as a potential IR MANPADS countermeasure for civilian aircraft due to the fire hazard they pose in urban areas. Given this limitation, DIRCMs have the potential to be the most effective way to protect against infrared guided missiles. CLOS/LBR MANPADS and RPGs, however, do not rely on onboard homing devices and are not affected by DIRCMs. At this time, the only effective means to counter CLOS/LBR MANPADS is to harden an airplane to increase its survivability if attacked.

While MANPADS are capable of striking aircraft at up to 15,000 feet, an RPG is a close-range, explosive projectile that is more readily used for attacking an aircraft upon landing, immediately after takeoff, or while it is on the ground. Therefore, we have proposed an increase in perimeter control – a significant increase in security patrols and surveillance equipment in areas that may be used to launch an attack within the immediate perimeter of an airport – as the most effective immediate countermeasure against short- to medium-range weapons. CLOS and LBR MANPADS are grouped together because they require the same type of countermeasure. Given that non-IR MANPADS are not likely to be thwarted before they are fired (they have an effective range of about 50 miles from the airport) or while in flight, we have suggested that hardening of an aircraft's fuselage may be the only effective way to increase the survivability of an attack with these types of weapons.

Starting with the general framework of von Winterfeldt and O'Sullivan's previous model, this chapter makes several important changes and additions to account for the increase in potential weapons available and the countermeasures that would be necessary to thwart such attacks. Most notably, the extended model includes variables that take into consideration the fact that implementing one set of countermeasures might make weapon substitution attractive to terrorists and that countermeasures designed for one weapon might affect the parameters associated with another weapon. Also, based on statements made by government officials and predictions offered by analysts, this model takes into account the fact that a decision maker might be retroactively forced to implement countermeasures if a successful attack occurs when there were no previous countermeasures in place to defend against the particular weapon that was used in the attack[3].

19.4 Decision Tree

Figure 19.2 shows the portion of the decision tree detailing the defender's decision not to install any countermeasures, followed by the attacker's decision to attack with IR MANPADS. Additionally, it depicts the decision to install DIRCMs, followed by the attacker's decision to use RPGs. These two parts of the tree are further detailed by the events related to interdiction, hit, and crash. The final defender decision is in response to an attack. If the defender is attacked, a reasonable response is to install the countermeasure that defends against a repeat attack of the same kind. All other truncated branches of the tree are similarly expanded as indicated by a "+" in Figure 19.2.

Note that if the defender decides not to install countermeasures and an IR MANPADS attack occurs, the defender responds by installing DIRCMs. If the attackers had used CLOS/LBR MANPADS, the defender response would be to harden the airplanes; if the attack were with RPGs, the defender would choose perimeter control. Probabilities in the model are conditional on all preceding decisions or events as shown under the respective branches. These include the probability of an attack with a given weapon over 10 years, the effectiveness of deterrence, the probability of interdiction if there is an attack, and the effectiveness in preventing a hit or fatal crash. At the end of the tree, the consequences are estimated using a parametric approach indicated by an "x."

Table 19.1 shows the base case parameters and their ranges as used in this analysis. The ranges are wide and include all parameter values that are

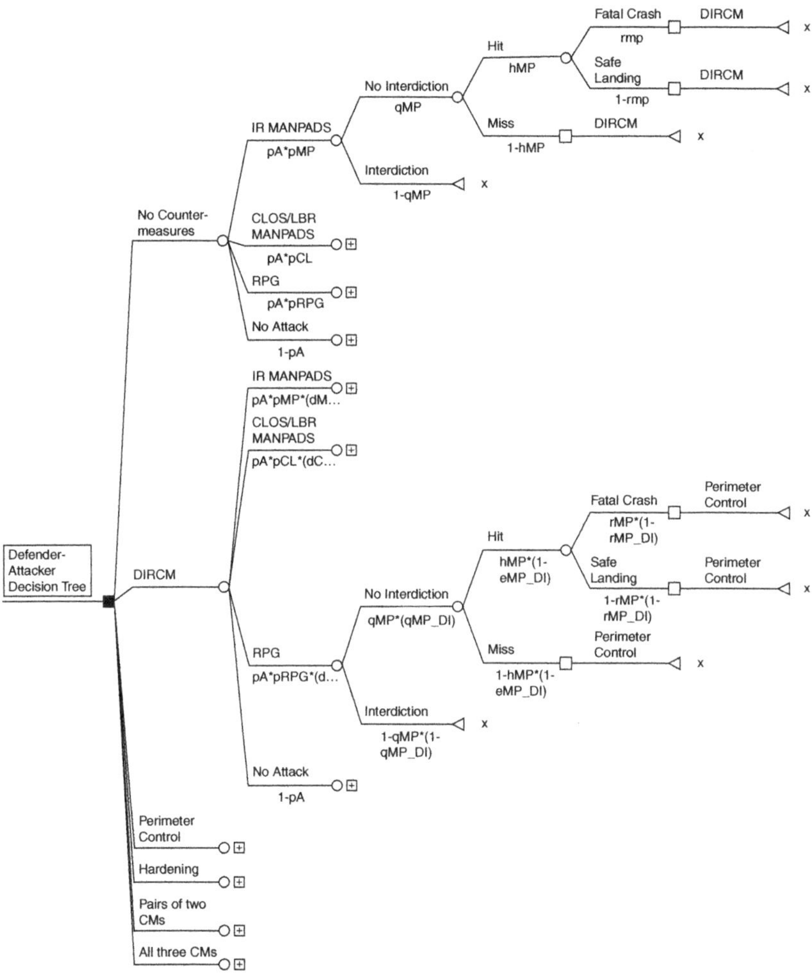

Figure 19.2. Decision tree.
(Circles represent chance nodes, squares represent decision nodes, and triangles represent terminal nodes; the probabilities and consequences are shown as model parameters; if a node is not a terminal node, the tree is continued with a tree like the one above, indicated by a "+".)

considered reasonable by experts in this area. The base case values are our best guesses from information in the open source literature. Better information about some of these parameters exists, but this information is sensitive or classified.

Table 19.1. *Base case parameters and ranges*

	Var. Name	Description	Base Case	Min	Max
Probabilities	pA	Attempted Attack	0.5	0	1
	pCL	Attack with CLOS/ LBR|Attempt	0.1	0	1
	pMP	Attack with MANPADS|Attempt	0.6	0	1
	pRPG	Attack with an RPG|Attempt	0.3	0	1
	qCL	Interdiction|Attempt with CLOS/LBR	0	0	1
	qMP	Interdiction|Attempt with MANPADS	0	0	1
	qRPG	Interdiction|Attempt with RPG	0.1	0	1
	hCL	Hit|Attack CLOS/LBR	0.8	0	1
	hMP	Hit|Attack with MANPADS	0.8	0	1
	hRPG	Hit|Attack with RPG	0.5	0	1
	rCL	Crash|Hit by CLOS/LBR	0.8	0.5	1
	rMP	Crash|Hit by MANPAD	0.8	0.5	1
	rRPG	Crash|Hit RPG	0.5	0	1
Deterrence Effect					
	dCL_H	Deterrence of Hardening on CLOS/LBR	0	0	0.5
	dMP_H	Deterrence of Hardening on MANPADS	0	0	0.5
	dRPG_H	Deterrence of Hardening on RPG	0	0	0.5
	dMP_DI	Deterrence of DIRCM on MANPADS	0.5	0	1
	dCL_DI	Deterrence of DIRCM on CLOS/LBR	0	0	0.5
	dRPG_DI	Deterrence of DIRCM on RPG	0	0	0.5
	dRPG_PC	Deterrence of Perimeter Control on RPG	0.5	0	1
	dCL_PC	Deterrence of Perimeter Control on CLOS/LBR	0	0	0.5
	dMP_PC	Deterrence of Perimeter Control on IR	0	0	0.5

Table 19.1 (*cont.*)

	Var. Name	Description	Base Case	Min	Max
Substitution Effect					
	sRPG_DI	Substitution Effect of DIRCM on RPG usage	2	1	2
	sCL_DI	Substitution Effect of DIRCM on CLOS/LBR usage	1	1	2
	sMP_PC	Substitution Effect of Perimeter Control on MP usage	1	1	2
	sCL_PC	Substitution Effect of Perimeter Control on CLOS/LBR usage	1	1	2
Effectiveness of Countermeasures					
	qRPG_PC	Interdiction Effectiveness of Perimeter Control on RPG	0.5	0	1
	qCL_PC	Interdiction Effectiveness of Perimeter Control on CLOS/LBR	0	0	0.5
	qMP_PC	Interdiction Effectiveness of Perimeter Control on MP	0	0	0.5
	eMP_DI	Diversion/Destruction Effectiveness of DIRCM on MP	0.8	0.5	1
	rCL_H	Crash Reduction Effectiveness of Hardening on CLOS/LBR	0.25	0	0.5
	rMP_H	Crash Reduction Effectiveness of Hardening on MP	0.25	0	0.5
	rRPG_H	Crash Reduction Effectiveness of Hardening on RPG	0.25	0	0.5
Consequences					
	LL	Lives Lost|Crash	200	0	500
	CP	Cost of Plane (millions)	200	0	500
	ELFC	Economic Loss|Fatal Crash (billions)	100	0	200

(continued)

Table 19.1 (*cont.*)

	Var. Name	Description	Base Case	Min	Max
	pcEL_SL	Percent of Economic Loss|Hit and Safe Landing	0.25	0	0.5
	pcEL_M	Percent of Economic Loss|Miss	0.1	0	0.25
	CH	Cost of Hardening (billions)	15	0	30
	CDI	Cost of DIRCM (billions)	30	0	60
	CPC	Cost of Perimeter Control	5	0	20
Trade-offs					
		Value of Life (millions)	5	0	10

Throughout this chapter, lowercase letters denote the type of variable and uppercase letters denote which type of weapon/countermeasure the variable refers to. An "Attempt with RPG" occurs with probability pRPG, an "Attempt with an IR MANPADS" occurs with probability pMP, and an "Attempt with CLOS/LBR MANPADS" occurs with probability pCL. Conditional propositions are denoted by the variable, followed by an underscore, followed by the conditioning variable. For example, the effectiveness of DIRCMs to avert a hit by an IR MANPADS is denoted as eMP_DI. Consequences are denoted in uppercase letters, e.g., "Loss of Life" is denoted as LL.

Many of the base case estimates and ranges were adapted from von Winterfeldt and O'Sullivan. In the following, we discuss only the specific parameters that are new and/or different from their model.

Probabilities. The probability of an attack with some kind of weapon is assigned a base case of pA = 0.5 and a range from 0 to 1. IR MANPADS are easily available to terrorists and can be smuggled into the United States through a variety of avenues. Therefore, an IR MANPADS is the most likely form of the three attack modes. We quantify the base case probability of an IR MANPADS attack as pMP = 0.6, which is conditional on one of the three attack modes occurring (von Winterfeldt and O'Sullivan used an unconditional base case probability of an attack with a MANPADS at 0.25, slightly lower than the 0.3 probability in the present study). RPGs can also be obtained easily, but need to be deployed closer to the target and have less chance of hitting it. We assign a conditional probability of using an RPG, given an attack, as half of pMP (RPG = 0.3). Modern CLOS/LBR weapons are not easy to obtain and are thus much less likely to be used than either an RPG or an IR MANPADS. In the base case, we assign it a conditional probability that is one-sixth the probability of using an IR MANPAD (pCL = 0.1). These ratios of 6:3:1 are consistently applied to pMP, pRPG, and pCL, except when deterrence and substitution effects occur.

Once an attempt with one of the three weapons types is made, there is a chance that it may be interdicted. Since RPGs have to be delivered at close range, we assign the highest interdiction probability to this mode of attack. Yet, even in this case, we only expect a fraction of possible attempts to be interdicted and assign a base case of qRPG = 0.1. The base case interdiction probabilities for IR and CLOS/LBR missiles are qIR = qCL = 0. Since these weapons are all subsets of MANPADS, they are capable of striking targets at altitudes of up to 15,000 feet. Given the general flight trajectory of commercial aircraft, these weapons are effective at a radial range of approximately 50 miles from most airports. This equates into a firing area of 7,800 square miles or more, rendering perimeter controls essentially ineffective (DelBoca, 2003).

After the surface-to-air weapon is discharged, it will either hit the plane or miss it. IR and CLOS/LBR MANPADS have a relatively high chance of hitting or damaging the plane, with a base case of hMP = hCL = 0.8. RPGs have a lower probability of hitting the plane because they don't have an automatic targeting system. We used a base case probability of hRPG = 0.5. Once the plane is hit, the damage can lead either to a fatal crash or to a safe landing. The 2003 IR MANPADS attack on a DHL plane in Baghdad demonstrated that there is a some chance of survival when an IR or CLOS/LBR missile strikes a plane. We assign the base case probabilities of a fatal crash given a hit as rMP = rCL = 0.8.

Deterrence Effect. The event nodes following the choice to install a given countermeasure or package of countermeasures are exactly the same as the event nodes in the sub-tree resulting from the choice "No Countermeasures," but choosing to install countermeasures will change the likelihood of the attacker's decisions through a deterrence effect. Deterrence can be achieved through the threat of retribution or through denial of success of an attempted attack. In this context, we are exclusively referring to deterrence by denial, or at least by a demonstrated reduction of the likelihood of success of an attack. For example, if DIRCMs are deployed, the likelihood of a successful IR MANPADS attack is reduced substantially due to the ability of the DIRCM to destroy or divert the IR MANPADS. This reduced likelihood of success should deter terrorists from attempting an IR MANPADS attack. In the sub-tree that starts with implementing DIRCMs, deterrence is parameterized with a factor dMP_DI, the effectiveness of reducing the probability of an IR MANPADS attack by using DIRCMs. By multiplying the non-deterred probability pMP by (1-dMP_DI), we obtain an effective new probability of the use of IR MANPADS conditional on the implementation of DIRCMs (see also Figure 19.2).

Similarly, the use of RPGs can be deterred by perimeter control (dRPG_PC) and CLOS/LBR MANPADS can be deterred by hardening the plane d(CL_H). In the base case model, we assume that only two deterrence effects are operating: DIRCMs deter the use of IR MANPADS (dMP_DI = 0.5) and perimeter control deters RPGs (dRPG_PC = 0.5). The other deterrence effects are assumed to be 0. For example, perimeter control will not deter terrorists from using IR or CLOS/LBR MANPADS because they can be effectively fired a substantial distance from the controlled area around an airport.

Substitution Effect. While the use of countermeasures provides some degree of deterrence, they may also increase the likelihood of other types of attacks through a substitution effect. For example, DIRCMs may reduce the use of IR MANPADS, but lead to an increase in the probability of using RPGs or CLOS/LBR MANPADS. In the current model, we assume that the main substitution effect is from IR MANPADS to RPGs when employing DIRCMs. The assigned substitution multiplier is sRPG_DI = 2, which doubles the probability of using RPGs when DIRCMs are implemented.

As a result of deterrence and substitution, the probabilities of the attacks change with the countermeasures that are implemented. These changes are captured in the model to distribute the probabilities of the three attack modes among pA with the remaining probability (1-pA) calculated as the residual.

Effectiveness of Countermeasures. Moving past the "Attempt" event node, countermeasures will also affect the probability of interdiction, hitting the plane, and crashing when hit. For example, DIRCMs will primarily reduce the probability of the plane getting hit by an IR MANPADS. Perimeter control will increase the probability of interdicting attempts for weapons that need to be fired at close range (RPGs). Hardening the plane will increase the probability of surviving a hit with any of the aforementioned weapons.

Considering interdiction first, we assume in the base case that only perimeter control affects interdiction and only for interdicting RPGs (p = 0.10 of interdiction without perimeter control). Even with perimeter control, interdiction is unlikely due to the vast terrain that would need to be controlled. We set the effectiveness of interdiction for RPGs at qRPG_PC = 0.50 (increasing the interdiction probability from 0.10 to 0.15). For the other two weapons, perimeter control will not increase the interdiction probability by much (if at all) because they can be deployed a substantial distance from the airport perimeter.

The variable eMP_DI stands for the effectiveness of the deflection/destruction effectiveness of defending against an IR MANPADS using a DIRCM.

Like von Winterfeldt and O'Sullivan, we use a base case effectiveness of 80%. DIRCMs are not effective against CLOS/LBR missiles or RPGs.

The last class of parameterized effectiveness variables is the crash reduction effectiveness. These variables represent the reduction in crash probability given a hit and are only nonzero for the "Hardening" countermeasure. Hardening the fuselage will reduce the probability of a crash given a hit with any weapon by some degree. We are using a base case of rMP_H = rCL_H = rRPG_H = 0.25. Different missile types have different effects on the plane's fuselage, but we cannot make any meaningful distinctions of the effects of hardening on the probability of a fatal crash. IR MANPADS explode near one of the engines (prior to hitting it, when the heat signature exceeds a given limit) to provide maximum damage. Due to the automatic guiding system, CLOS/LBR MANPADS are most likely to hit the center of the plane. Conversely, due to the lack of a guidance system, RPGs will hit the plane wherever they were aimed.

Consequences. The consequences of following a particular path through the decision tree are assessed at the terminal nodes of each branch (indicated by a triangle in Figures 19.1 and 19.2). The consequences that we considered in this analysis are:

1. Loss of lives due to a crash (LL)
2. Cost of the plane (CP)
3. Economic losses to the airline industry and the overall economy (EL)
4. Costs of the countermeasures (CC, depending on the countermeasure)

As with the probabilities, the model utilizes a highly parameterized approach and covers a very wide range of losses and costs. The fifth panel of Table 19.1 summarizes the base case values and ranges. The parameterized model utilizes a range of between zero and 500 deaths (Lives Lost base case = 200), a cost of the loss of the airplane between zero and $500 million (base case = $200 million), and estimated losses to the airline industry and the overall U.S. economy at between zero and $200 billion (base case = $100 billion). The high end of the economic loss (EL) was motivated by recent analyses that estimated the costs of a 9/11-like attack on the airline industry. For example, Gordon and colleagues estimate that the economic costs for the two years following the 9/11 attack could be as high as $250 billion (Gordon, Kim, Moore, Park, & Richardson, 2006). Their estimate includes the cost of an initial shutdown of the airline industry and a subsequent drop in passenger volume of about 20% in the first year and 10% in the second year. The low end was motivated by Santos and Haimes, who estimated that a drop of 10% in airline passenger volume cost the economy up to $80 billion/year (Santos & Haimes, 2004).

Unsuccessful MANPADS attacks also have economic costs because they may create fear of future attacks and lead to a reduction in passenger traffic. For hits with safe landings, the model utilizes a loss reduction parameter (EL_SL) that has a base case of 25% with a range from 0% to 50% of the economic costs of a hit and fatal crash. For misses, the model uses a loss reduction parameter EL_M with a base case of 10% and a range from 0% to 25%.

Estimates of the aggregate costs of DIRCM countermeasures are controversial. The RAND Corporation estimates a capital cost of $10 billion and $2.5 billion in annual operations and maintenance costs, assuming that all large commercial airliners (5,000 passenger planes) are equipped with countermeasures. This would lead to an undiscounted $35 billion life-cycle cost over 10 years. Airline industry officials quote much higher costs of up to $100 billion and promoters of MANPADS countermeasures quote much lower costs of $10 billion or less. To accommodate this range of opinions, the 10-year cost of the system is parameterized between $0 billion and $60 billion (base case = $30 billion).

The cost of hardening was estimated using the capital costs and the additional fuel costs that result from the increase in weight of the plane. The capital costs are quite uncertain, so we consider a base case of investing $1 million in each of 5,000 large planes to harden its fuselage, which amounts to $5 billion. At a weight increase of 1,000 kg, the increase in fuel consumption per 1,000 plane miles is about 50 kg or 17 gallons of kerosene (Weselby, 2012). Given that each plane flies about 1.75 million miles per year (*USA Today*, 2012) and that we would need to harden 5,000 planes, we infer that the increase in fuel cost would be about $6.6 billion over 10 years for an increase in weight of 1,000 kg. The actual weight increase could be between 1,000 kg and 2,000 kg, so the fuel cost would be between $6.6 billion and $13.3 billion. These estimates generate a base case of total 10-year life-cycle cost of $15 billion (fuel costs of $10 billion plus the capital costs of $5 billion).

Perimeter expenses are somewhat easier to estimate. For the base case, we add 50 officers to the airport security detail to monitor and control the perimeter of each of the 86 major airports with more than 1 million passengers per year. Assuming $125,000/year of personnel costs including overhead, we arrive at an estimate of $5.4 billion for 10 years. For the base case, we used $5 billion, to avoid the impression of over-precision of this rough calculation.

With this information, one can calculate an overall equivalent cost (EC) at each end node of the decision tree as the weighted sum of the component costs:

$$EC_j = \sum_{i=1}^{n} c_i x_{ij}$$

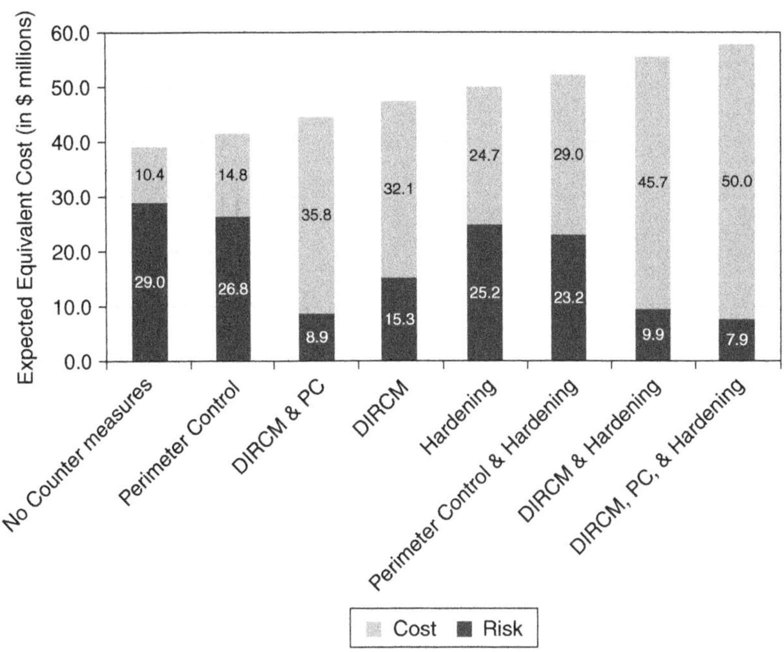

Figure 19.3. Direct cost vs. equivalent costs of the risks for eight alternatives to defend against attacks on commercial airplanes.

where EC_j is the equivalent cost for the j-th end node in the tree, c_i is the equivalent cost of one unit of consequence i, and x_{ij} is the ith consequence for the j-th end node in the decision tree.

We considered discounting the equivalent costs, but this would involve separate evaluations of an attack for each of the possible time periods (from 1 to 10 years), as well as separate probability assessments for each of the 10 time periods. While this does not pose any technical difficulties, it appears to be a technical detail that would scale the calculations for each alternative in a similar way and would therefore add little insight. Thus, the results are presented in undiscounted 10-year equivalent costs.

19.5 Analysis of the Decision Tree

Figure 19.3 shows the expected equivalent cost of the eight alternatives for the base case analysis of the decision tree, ordered from left to right by their total expected equivalent cost. For each alternative, it shows the residual risk (in terms of expected equivalent costs) in dark shading and the cost of the countermeasures

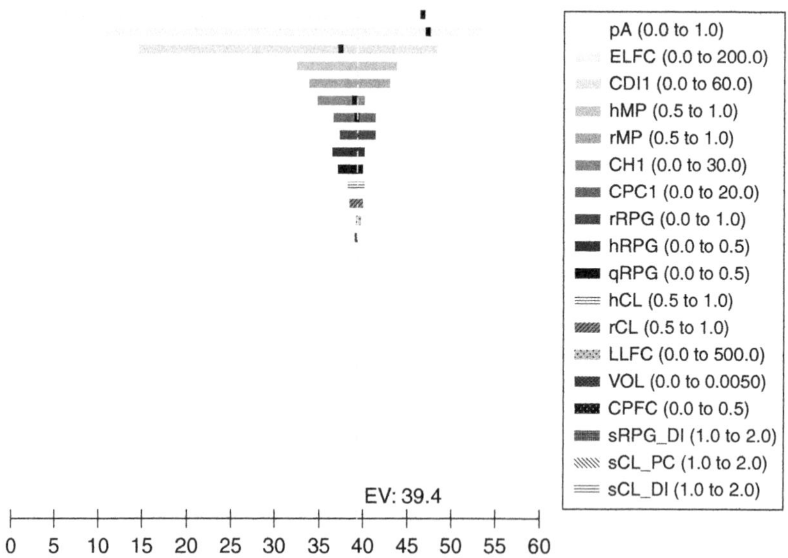

Figure 19.4. Tornado diagram of selected input variables for the decision tree analysis.

(including the cost of countermeasure installed after an attack) in light shading. The two alternatives with the lowest expected equivalent total cost are No Countermeasures and Perimeter Control. While perimeter control adds $5 billion, this cost is partially offset by the risk reduction this countermeasure produces. The third lowest cost is combination of DIRCMs and perimeter control, which also has one of the lowest residual risks. It is noteworthy that, should the defender decide to use DIRCMs, it is also advisable to implement perimeter control because the substitution effect will lead to a higher likelihood of using RPGs when DIRCMs are used as the only defense.

Figure 19.4 is a tornado diagram that displays the change in expected equivalent cost of the optimal decision as a function of varying 25 of the 42 input variables through the range of values shown in Table 19.1 (von Winterfeldt & O'Sullivan, 2006). These 25 variables were chosen because of their relevance for decision making and their identification as important factors by previous analyses. The larger the range covered by the horizontal bar, the larger the influence the variable has on the overall expected cost. The tornado diagram shows that the probability of an attack (pA), the economic loss in a fatal crash (ELFC), and the cost of DIRCMs are the most influential variables in determining the overall expected equivalent cost of the decision. This result is consistent with the finding of von Winterfeldt and O'Sullivan.

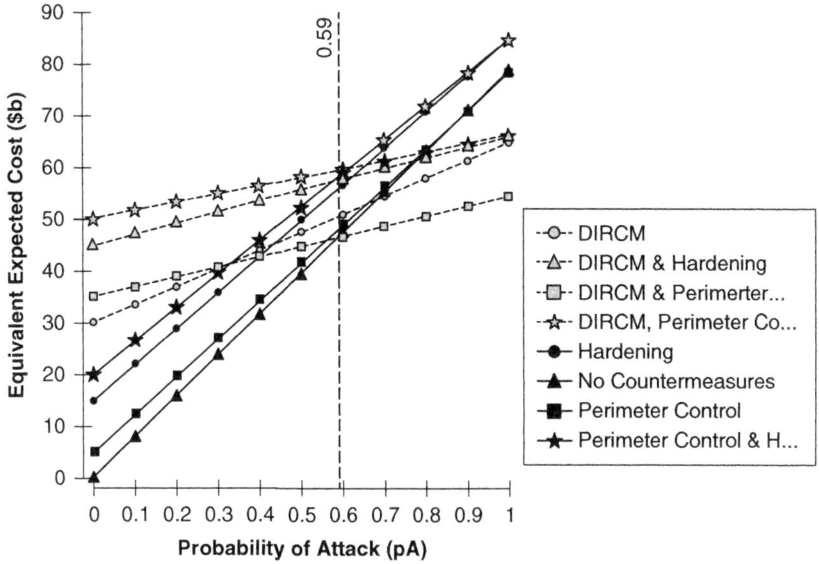

Figure 19.5. One-way sensitivity analysis on the probability of an attack.

Figure 19.5 shows the one-way sensitivity analyses of the probability of an attack, varied through the whole range from 0 to 1. Similar results are obtained for one-way sensitivity analyses of the loss to the economy and the cost of DIRCMS: No Countermeasures and Perimeter Control have the lowest expected equivalent costs when the probability of attack is low; DIRCMs combined with perimeter control have the lowest expected equivalent cost when the probability of attack is high. This is in contrast to the findings of von Winterfeldt and O'Sullivan, who suggested that DIRCMs alone would be cost-effective if the probability of an attack was high.

In order to analyze the joint effects of the three most influential types of variables, several sensitivity analyses were conducted varying two parameters at the same time. Figure 19.6 demonstrates how the optimal decision changes as a function of an attempted attack in 10 years (pA) and the loss to the economy of a fatal crash (ELFC), if we assume that the costs of DIRCMs are set to its base case at $30 billion. For low economic losses and probabilities of attack, No Countermeasures is the optimal choice. For higher economic losses and probabilities of attack, DIRCMs and Perimeter Control is the optimal choice. None of the other countermeasures or their combinations is optimal for any combination of these two variables. Interestingly, Perimeter Control never becomes an optimal choice, even though its expected equivalent costs are very close to No Countermeasures. Figure 19.7 shows the same two-way analysis assuming

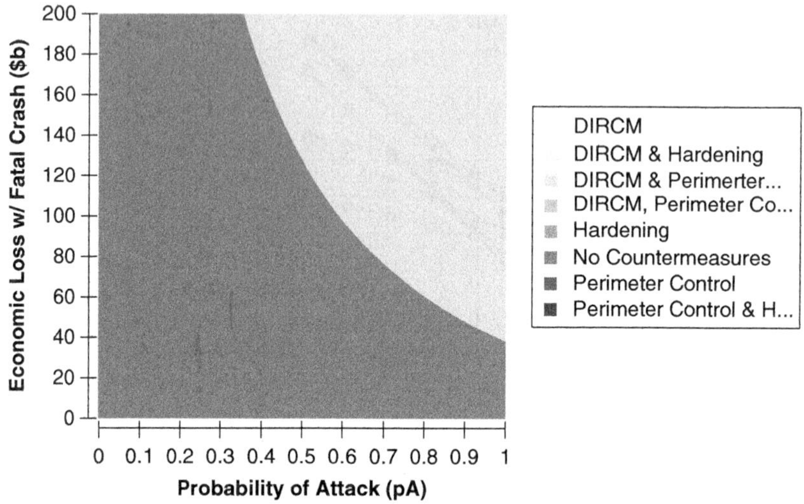

Figure 19.6. Two-way sensitivity analysis – probability of an attack and the loss to the economy (cost of DIRCMs set at $30 billion).

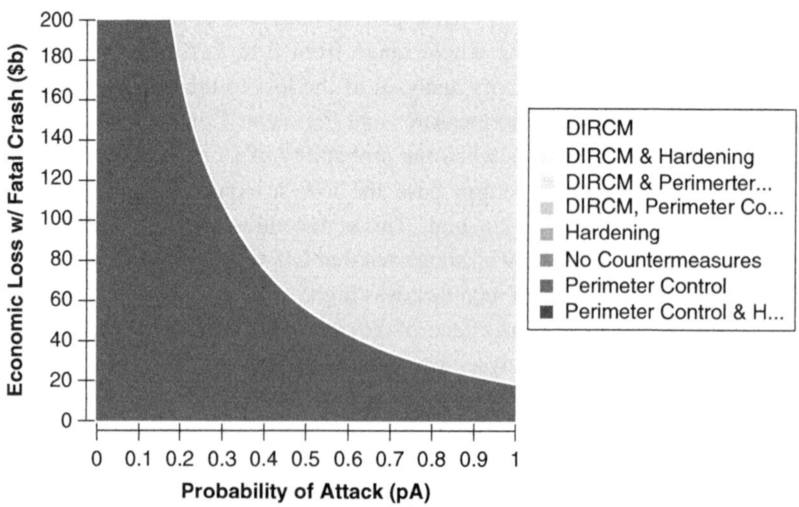

Figure 19.7. Two-way sensitivity analysis – probability of an attack and the loss to the economy (cost of DIRCMs set at $10 billion).

DIRCM costs of $10 billion. In this case, the area favoring countermeasures increases, but for most reasonable values of pA and ELFC the best decision is still to choose No Countermeasures. However, if countermeasures are chosen, it should be the combined choice of DIRCMs and Perimeter Control.

19.6 Summary and Conclusion

This study describes an approach to terrorism risk analysis called defender-attacker decision tree analysis, and applies it to the problem of defending commercial airplanes against surface-to-air attacks from terrorists. Unlike game-theoretic solutions to this problem, we do not assume that the attackers make rational choices that maximize their expected utilities. Instead, the model operationalizes the defenders' limited knowledge about the attackers' capabilities and motivations by assigning conditional probability distributions to the attackers' attack modes. Defenders, on the other hand, are assumed to act rationally and maximize their expected utility or minimize their expected costs. Once an attack occurs, however, defenders are assumed to choose the action that defends best against that type of attack. This response may or not be rational in an economic sense, but it certainly reflects the political reality of responses to disasters.

In a previous study, von Winterfeldt and O'Sullivan (2006) applied a very simple two-stage version of defender-attacker decision tree analysis to evaluate the cost-effectiveness of DIRCMs to defend against IR MANPADS attacks by terrorists. This chapter extends this analysis by examining multiple modes of attack, multiple defenses, multiple stages of defense, and substitution effects. The extended analysis indicates that of the 42 variables identified to affect the choice of countermeasures, three types of variables mattered most: the probability of an attempted attack, the loss to the economy from a fatal crash, and the cost of DIRCMs. For low to moderate probabilities of attack and economic losses, no countermeasures is the best option with perimeter control as a close second. Once the probability of an attack and the economic consequences reach a certain threshold, DIRCMs become an attractive choice. Yet, if one chooses DIRCMs, one should also choose perimeter control due to the substitution effect, which makes the use of RPGs more likely. This is an important caveat to the main recommendations in the von Winterfeldt and O'Sullivan paper and it demonstrates the value of evaluating multiple countermeasures simultaneously.

The main innovation in this analysis is that it examined the combined effect of deterrence, substitution, and countermeasure effectiveness (on reducing interdiction, hit, and crash probabilities) on the total expected equivalent costs of the decision under analysis. Another innovation was that it examined not only single countermeasures, but also packages of two and three countermeasures. The results indicate that substitution effects are important and that combinations of two countermeasures may be preferred to single countermeasures, even though the combined costs are likely to be larger.

Acknowledgments

Detlof von Winterfeldt acknowledges support from the U.S. Department of Homeland Security (DHS) through the National Center for Risk and Economic Analysis of Terrorism Events (CREATE) at the University of Southern California (USC) under award number 2010-ST- 061-RE0001. Additional support for developing the decision tree approach was provided by the National Science Foundation under grant SES-1314644. However, any opinions, findings, and conclusions or recommendations in this document are those of the authors and do not necessarily reflect views of the U.S. Department of Homeland Security, the National Science Foundation, the University of Southern California, or CREATE.

NOTES

1. From a speech given at the University of Southern California. August 17, 2006.
2. *USA Today*. August 11, 2015. Quoting a statement by the Dutch Safety Board, which investigated this accident, on the same day.
3. Rep. John Mica, R- FL, chairman of the House Aviation Subcommittee, was quoted in *USA Today* as stating, “In a time of tight budgets both for the government and the airlines, it’s very difficult. It’s going to take an aircraft being brought down, and then it will be required” (Hall M. Project to Protect Airliners in Doubt. *USA Today*. July 18, 2006).

REFERENCES

Bakir, N. O., & von Winterfeldt, D. (2011). Is better nuclear weapons detection capability justified? *Journal of Homeland Security and Emergency Management*, *8*(1), 1–18.

Bedford, T., & Cooke, R. (2001). *Probabilistic risk analysis: Foundations and methods*. Cambridge, UK: Cambridge University Press.

Bennett, R. K. (2003, August 14). Think the unthinkable. *Tech Central Station*. www.techcentralstation.com/081403D.html.

Bier, V., & Cox, L., Jr. (2007). Probabilistic risk analysis for engineered systems. In W. Edwards, R. F. Miles, & D. von Winterfeldt (Eds.), *Advances in decision analysis* (pp. 279–310). New York: Cambridge University Press.

Bolkcom, C., & Elias, B. (2006, February 16). Homeland security: Protecting airliners from terrorist missiles. National Defense. Foreign Affairs, Defense, and Trade Division: Congressional Research Service (CRS).

Chow, J., Chiesa, J., Dreyer, P., Eisman, P., Karasik, T. W., Kvitky, J., Linger, S., Ochmanek, D., & Shirley. C. (2005). *Protecting commercial aviation against the shoulder-fired missile threat.* Santa Monica, CA: The RAND Corporation.

Clemen, R., & Reilly, T. (2013). *Making hard decisions.* Mason, OH: South-Western Cengage Learning.

DelBoca, R. (2003). Washington, DC: Prepared testimony before the House Transportation and Infrastructure Committee, Aviation Subcommittee.

Ezell, C. E., Bennett, S. P., von Winterfeldt, D., Sokolowski, J., & Collins, A. J. (2010). Probabilistic risk analysis and terrorism risk. *Risk Analysis, 30*(4), 575–589.

Gordon. P., Kim, S., Moore, J. E., Park, H., & Richardson H. (2006). The economic impacts of a terrorist attack on the U.S. aviation system. *Risk Analysis, 27*(3), 505–512.

Grau, L. R. (2008, May–August). The RPG-7 on the battlefields of today and tomorrow. *Infantry.*

Howard, R. A., & Abbas, A. E. (2015). *Foundations of decision analysis.* New York: Pearson Education.

Hunter, T. B. (2002, November 28). The proliferation of MANPADS. *Jane's Intelligence Review.*

Keeney, R. L., & von Winterfeldt, D. (2007). Practical value models. In E. Edwards, R. F. Miles, & D. von Winterfeldt (Eds.), *Advances in decision analysis* (pp. 232–252). New York: Cambridge University Press.

Merrick, J., & Parnell, G. (2011). A comparative analysis of PRA and intelligent adversary methods for counterterrorism risk management. *Risk Analysis, 31*(9), 1488–1510.

National Research Council. (2008). *Department of Homeland Security bioterrorism risk assessment: A call for change.* Washington, DC: National Academy Press.

O'Sullivan, T M. (2004, November). External terrorist threats to civilian airliners: A summary risk analysis of MANPADS, other ballistic weapons risks, future threats, and possible countermeasures policies. University of Southern California: Center for Risk and Economic Analysis of Terrorism Events (CREATE). Revised February 2005.

Parnell, G. S., Smith, C. M., & Moxley, F. I. (2009). Intelligent adversary risk analysis: A bioterrorism risk management model. *Risk Analysis, 30*, 32–48.

Pate-Cornell, M. E. (2007). The engineering risk analysis method and some applications. In W. Edwards, R. F. Miles, & D. von Winterfeldt (Eds.), *Advances in decision analysis* (pp. 302–324). New York: Cambridge University Press.

Pate-Cornell, E., & Guikema, S. (2002). Probabilistic modeling of terrorist threats: A systematic analysis approach to setting priorities among countermeasures. *Military Operations Research, 7*(4), 5–20.

Phelps, M. (2003, January). Do SAMs pose a real threat to civil aviation? *Aviation International News.*

Raiffa, H. (1968). *Decision analysis.* Reading, MA: Addison-Wesley.

Santos, J. R., & Haimes, Y. Y. (2004). Modeling the demand reduction input-output (I-O) inoperability due to terrorism on interconnected structures. *Risk Analysis*, *24* 1437–1451.

Shanker. (2002, November 29). Ideal terror weapons: Portable, plentiful missiles. *New York Times*.

Tambe, M. (2011). *Security and game theory: Algorithms, deployed systems, lessons learned*. Cambridge, UK: Cambridge University Press.

U.S. Government Accountability Office (GAO). (2004). The Department of Homeland Security needs to fully adopt a knowledge-based approach to its Counter-MANPADS Development Program. Washington, DC: Author, GAO-04-341R.

USA Today. (2012). www.usatoday.com/story/travel/columnist/cox/2012/11/19/ask-the-captain-how-far-does-a-jet-fly-during-its-lifetime/1712269/.

Violence Policy Center. (2002). Sitting ducks: The threat to chemical and refinery industry from 50 caliber sniper rifles. www.vpc.org/50caliber.html.

von Winterfeldt, D., & Edwards, W. (1986). *Decision analysis and behavioral research*. New York: Cambridge University Press.

von Winterfeldt, D., & O'Sullivan, T. (2006). Should we protect commercial airplanes against surface-to-air missile attacks by terrorists? *Decision Analysis*, *3*, 63–75.

Weselby, H. (2012). *Fuel burn vs. maintenance cost*. Washington, DC: IIATA Maintenance Cost Conference.

Willis, H. H., Morral, A. R., Kelly, T. K., & Medby, J. J. (2005). *Estimating terrorism risk*. MG-388-RC, RAND Corporation. Santa Monica, CA.

20

Decision Making for Bioterror Preparedness: Examples from Smallpox Vaccination Policy

Edward H. Kaplan and Lawrence M. Wein

20.1 Introduction

While the possible terrorist use of biological agents against civilian targets was a subject of concern before September 11, 2001 (Alibek, 1999; Miller, Engelberg, & Broad, 2001), the deliberate mailings of anthrax-laden letters and resultant infections, deaths, and panic that ensued in October of that same year provided a bioterror "proof of concept." Policy makers in the United States and abroad suddenly found themselves faced with numerous decisions regarding bioterror preparedness, and what to do about smallpox quickly topped the list of concerns. Smallpox vaccination policy provides a very interesting case study for those interested in the use of OR/MS methods in confronting bioterror and health issues more generally, especially since OR modeling influentially informed the policy-making process in this instance.

This chapter reports some of the simple, yet powerful arguments employed in the smallpox vaccination policy debate. First described are the issues and positions held during this debate, emphasizing the clash between the "bio" and the "terror" in "bioterror." Following this, three simple choice models that were used repeatedly in the arguments are considered. A key common idea across these examples is the importance of focusing on the consequences of decisions made rather than the details of what is believed to be the most likely smallpox scenario, as will become clear shortly. Concluding comments follow.

Reprinted by permission, Kaplan, E.H. and Wein, L.M. (2004), Decision making for bioterror preparedness: examples from smallpox vaccination policy. In F. Sainfort, M.L. Brandeau, & W.P. Pierskalla (Eds.), *Handbook of operations research and health care: Methods and applications* (pp. 519–536). Copyright 2004, Kluwer Academic Publishers.

20.2 Vaccination Policy in the Large and in the Small(pox)

The eradication of smallpox surely stands as one of the greatest public health achievements of the past century (Fenner, Henderson, Arita, Jezek, & Ladnyi, 1988). However, fear that smallpox could be employed as a weapon of bioterror prompted public health and security officials in the United States and elsewhere to revisit the issue (Alibek, 1999; Miller et al., 2001; Tucker, 2001). As of November 2001, the Centers for Disease Control and Prevention (CDC), acting on the advice of the Advisory Committee on Immunization Practices (ACIP), had developed a smallpox emergency response policy (Centers for Disease Control and Prevention, 2001). Absent any analysis other than analogy to the procedures employed by the World Health Organization (WHO) in the global smallpox eradication campaign, the CDC plan called for a ring vaccination response: confirmed smallpox cases would be isolated, their close contacts would be traced and vaccinated, contacts found febrile would be quarantined, while asymptomatic contacts would monitor and report their temperatures to local public health officials. Only if this policy failed to contain a smallpox outbreak would an (unspecified) "broader" vaccination response be initiated.

It did not take long for this policy to come under intense public scrutiny. In an influential article published in the *New England Journal of Medicine*, Dr. William Bicknell, the former commissioner of health for the State of Massachusetts, argued the case for voluntary pre-attack vaccination against smallpox (Bicknell, 2002). Noting that widespread vaccination would greatly reduce the value (to terrorists) of a smallpox attack, Bicknell further noted the lack of population immunity against smallpox in the United States today (in contrast with the much higher levels of immunity due to prior vaccination campaigns and survival from disease in those countries where the WHO finally eradicated smallpox). Were ring vaccination employed, Bicknell worried, "An epidemic is highly likely to outrun the vaccinators."

In this same issue of the *New England Journal of Medicine*, Dr. Anthony Fauci, the highly respected director of the National Institute of Allergy and Infectious Diseases (NIAID), voiced similar concerns (Fauci, 2002). He wrote that "there is considerable skepticism about the feasibility of this strategy because of the possibility of simultaneous attacks in multiple cities." Fauci also noted the difference in the degree of population immunity against smallpox between the countries of the WHO campaign and the United States today, the logistical difficulties posed by ring vaccination, and the strategic advantage pre-attack vaccination would offer by removing the incentive for a smallpox attack.

However, both Drs. Bicknell and Fauci recognized an important rationale for limiting vaccination, namely, the potential for severe complications, including death, that could result from widespread use of the *vaccinia* anti-orthopox vaccine (Neff, Lane, Fulginiti, & Henderson, 2002). While Dr. Bicknell (2002) argued that one could proceed with a careful pre-attack vaccination plan that would screen out those at greatest risk of complications, Dr. Fauci (2002) called for "an open and public dialogue on the advantages and disadvantages of universal voluntary vaccination, as well as on the smallpox response plan of the CDC."

Such a public dialogue indeed unfurled, with much of the attention focused on a series of ACIP meetings that took place in June 2002. Meanwhile, at the invitation of Dr. Ellis McKenzie from the Fogarty International Center of the National Institutes of Health, the authors (jointly with David Craft) had developed a mathematical model that embedded the logistics of alternative emergency response policies into a smallpox disease transmission model (Kaplan, Craft, & Wein, 2002). The model and its attendant results had yet to be published in June, though they had been presented at a small number of public and private scientific meetings.

On June 15, ACIP held a public forum on smallpox vaccination policy at the Institute of Medicine in Washington, DC (Institute of Medicine, 2002). Following introductory remarks by Dr. Fauci and others, a review of the clinical features of smallpox, and the historical rationale for the CDC's proposed response policy, the basic results from our model were presented. As detailed in Kaplan and colleagues (2002), the bottom line was clear: unless faced with a very small attack with an agent of only mild infectiousness, ring vaccination would result in many more deaths – perhaps by a factor of 200 – than would immediate mass vaccination in the area of the attack. Any degree of pre-attack vaccination would make any post-attack policy work better, but absent such pre-attack vaccination, the ring policy would almost surely fail (Institute of Medicine, 2002, 18–20; Kaplan et al., 2002). On the basis of this presentation at the Institute of Medicine forum, the first author was invited to brief government officials at the White House Conference Center on July 9.

Also presented at the public forum was Dr. Alan Zelicoff's detailed account of a smallpox outbreak in Aralsk, Kazakhstan in 1971 (Institute of Medicine, 2002, 20–21, 11]. This outbreak is important to understand, for it occurred among passengers on a research vessel that passed by the former Soviet bioweapons testing site at Vozrozhdeniye Island in the Aral Sea. As Dr. Zelicoff explained, "It is probably the case that smallpox was aerosolized, which answers the age-old question of whether or not smallpox is in fact aerosolizable and infectious in that state."

In spite of these findings, on June 20, ACIP recommended vaccinating only 15,000 first responders nationwide (Altman, 2002). However, this recommendation was quickly dismissed. On July 7, the *New York Times* ran a front-page story reporting that the federal Department of Health and Human Services was leaning toward recommending vaccinations for about 500,000 health care and emergency workers (Broad, 2002). The story also summarized our main findings and Dr. Zelicoff's report, and provided the sense that a shift in thinking regarding the threat of smallpox along with control options was beginning to take hold.

On the afternoon of July 7, the first author received a phone call with an invitation to appear the next morning on the *Today Show*. In a lengthy interview with Katie Couric, the lessons from our analysis regarding the prospects for control of a smallpox attack were discussed. The next day, our results were presented in a three-hour, closed-door meeting with representatives from the Office of the Vice President, National Security Council, Office of Homeland Security, Council of Economic Advisors, Health and Human Services, and other government agencies.

This meeting revealed clearly two competing schools of thought – the "bio" and the "terror" in bioterror – that have surfaced repeatedly in the smallpox debate. Adherents to the "bio" view consider smallpox preparation and response as just another public health problem, and, not surprisingly, the bulk of the "bios" can be found in the medical and public health community, as well as in agencies such as the Department of Health and Human Services and the Centers for Disease Control. "Bios" place great weight on historical smallpox outbreaks and control methods, and worry deeply about possible vaccine complications. This latter concern is consistent with the principle of *primum non nocere* (first do no harm), which typifies the beliefs of many health and medical practitioners.

However, those steeped in military strategy and homeland security, economics, public administration, and policy making more generally, including some with medical training, expressed the more utilitarian view of getting the greatest good for the greatest number. For example, widespread vaccination against smallpox now could be viewed purely as a matter of strategy by depriving terrorists of an effective weapon. Or, minimizing the time required to bring an outbreak under control could be viewed as an important objective. In considering what to do about a smallpox *bioterror attack*, the historical record is less important than the terrorist imagination.

That these points of view conflict is obvious: to get the greatest good for the greatest number, it might be required to first do *some* harm. To vaccinate widely either before or after an attack, one must face the risk of vaccine complications, including death.

Our analysis of emergency response to a smallpox attack appeared in the online *Proceedings of the National Academy of Sciences* on July 12 (Kaplan et al., 2002). On July 14, the first author traveled to Israel, and over the course of two weeks met with leading Israeli officials to discuss smallpox response policy in a series of public and private meetings and presentations (Siegel-Itzkovitch, 2002). Israel is an interesting contrast to the United States, for essentially the same debate transpired there, but over a much shorter time period.

Our results were presented to CDC staff in Atlanta on August 9. This was a helpful visit, for many had misinterpreted (Kaplan et al., 2002) as a call for immediate pre-attack vaccination of the entire United States. While CDC officials continued to disagree with some assumptions and findings of that paper, these disagreements were now much more informed. Indeed, by the end of September, the CDC for the first time released a smallpox mass vaccination clinic guide (Centers for Disease Control and Prevention, 2002), with detailed instructions for clinic operations as well as the estimated number of persons necessary for operating such clinics, with an eye toward rapidly making such clinics operational in the event of a smallpox attack.

Following months of deliberation, on Friday, December 13, 2002, President Bush announced his smallpox vaccination policy (Stevenson & Stolberg, 2002). During phase 1, 500,000 members of the military were ordered to receive the vaccine; as commander in chief, the president was also vaccinated. In addition, 500,000 first responders – those most likely to come into contact with smallpox cases in the event of an attack and those responsible for vaccinating the general public in the event of an emergency – were slated for vaccination. The general public would only receive the vaccine in the event of an actual attack during this phase; indeed, the president stated specifically that neither his family nor his staff would be vaccinated. The second phase of this plan called for the vaccination of roughly 10 million doctors, nurses, police, fire, and other emergency personnel during 2003. Finally, starting in 2004, phase 3 of the policy allowed for the voluntary vaccination of the general public. As with the first phase of this plan, the entire public would be eligible for vaccine in the event of an attack.

Thus, from June through December 2002, smallpox vaccination policy in the United States evolved from ACIP's recommendation to vaccinate only 15,000 first responders to a much broader plan, and from commitment to a limited ring vaccination response in the event of an attack to preparation for rapid mass vaccination of the affected population in an emergency. Mathematical models played a key role in the arguments for this change in policy. Next, we review some of the simple, yet effective modeling arguments that were employed to good effect in the debate over smallpox vaccination policy.

20.3 Natural Outbreak or Bioterror Attack? Ring versus Mass Vaccination

As mentioned previously, ring vaccination was deployed in the World Health Organization's global smallpox eradication program. This policy worked well in areas of *moderate to high immunity*, for controlling *natural outbreaks*, and in populations that had *low mobility* (though even under these circumstances, the value added by ring vaccination has been questioned (Kaplan & Wein, 2002). It is important to consider each of these points. First, before ring vaccination was deployed to "mop up" the last vestiges of smallpox worldwide, large-scale vaccination campaigns had already increased the degree of immunity in many countries to the point where between 25% and 75% of the population was already immune from smallpox (Fenner et al., 1988). Moreover, while smallpox killed roughly 30% of its victims, the surviving 70% were effectively immune to reinfection. Such immunity effectively reduces the infectiousness of the virus, since there are fewer persons who can be infected, making it much more difficult for the infection to spread – and much easier for ring vaccination to contain the infection.

Second, the outbreaks in question were all *natural*. One can think of a natural outbreak as infections as yet undetected by prior control efforts. Typically, such "leftover" infections numbered in the single digits. By contrast, a bioterror *attack* would involve the deliberate infection of as many persons as possible. The potential for an aerosol attack makes an initial outbreak of hundreds or even thousands of infections feasible (e.g., by deployment in a congested area such as a train station, airport, sporting event, concert etc.), if not highly likely. Thus, in preparing for response to a smallpox *bioterror attack*, it is important to decide the *outer level of risk* against which one wishes to defend. It makes little sense to focus only on policies that work for limited outbreaks of the form ring vaccination was designed to contain.

Third, smallpox was largely endemic in the rural areas of countries where sanitation and other health conditions were generally poor. Person-to-person contact in such immobile populations is thus somewhat limited. For example, in villages that are spaced 30 miles apart, the imposition of a 20-mile travel quarantine seems sensible – it could prevent disease from leaving one village and entering another. Incidentally, the November 2001 CDC emergency response plan stated that asymptomatic contacts of smallpox cases would also be restricted to traveling no further than 20 miles from their place of residence (Centers for Disease Control & Prevention, 2001, Guide C, p. C-7); one can only marvel at the presumed effectiveness of such a plan when applied to New York or any large American (or other) city!

In Kaplan and colleagues (2002), the view taken was deliberately rather different than that promulgated by ACIP and the CDC. Rather than consider only small outbreaks, a large attack that infects 1,000 persons in a city of 10 million persons was considered as a base case, though several parameters were varied over wide ranges in sensitivity analyses. For example, the number of secondary infections transmitted by each infected person early in the outbreak (the so-called reproductive number denoted by R_0) was assigned a value of 3 in this base case (at the lower end of the range estimated recently in Gani and Leach (2001) and thus favorable to ring vaccination), though this too was varied from 1 through 20 in sensitivity analyses.

Others have criticized such assumptions as overly pessimistic, preferring instead to focus on more likely scenarios. For example, Halloran and colleagues state that "we explored what we consider to be the most likely method of attack, namely, a few infected individuals moving through the community" (Halloran, Longini, Nizam, & Yang, 2002). In evaluating the choice between ring versus mass vaccination, then, a question of modeling strategy emerges – should one focus on the "most likely" scenario, a "worst case" scenario, or some combination of these or other possibilities? This choice is important, for without conducting any analysis, it seems clear that ring vaccination is optimal for small outbreaks with low levels of infectiousness while mass vaccination is optimal for large outbreaks with an infectious agent.

Consider the following simple argument: It is not sensible to focus on the "most likely method of attack" and decide on a response policy based solely on such a scenario, even if one believes that, should an attack occur at all, the odds are overwhelming that only a small attack would transpire. Rather, one should consider the *consequences* of different response policies when the attack scenario itself is uncertain. The idea is not to guess what will happen and then pretend that is what will transpire. The idea is to choose a robust response policy that will deliver good results *no matter what*.

To fix ideas, consider two possible attack scenarios, which we refer to as Big (e.g., aerosol attack with highly infectious virus) and Small (e.g., a single infected terrorist trying to spread smallpox by mingling in the population). Also, consider two possible responses – ring (or *traced*) vaccination (TV) and mass vaccination (MV). Using the models reported in Kaplan and colleagues (2002), one can estimate the expected number of deaths that would occur for

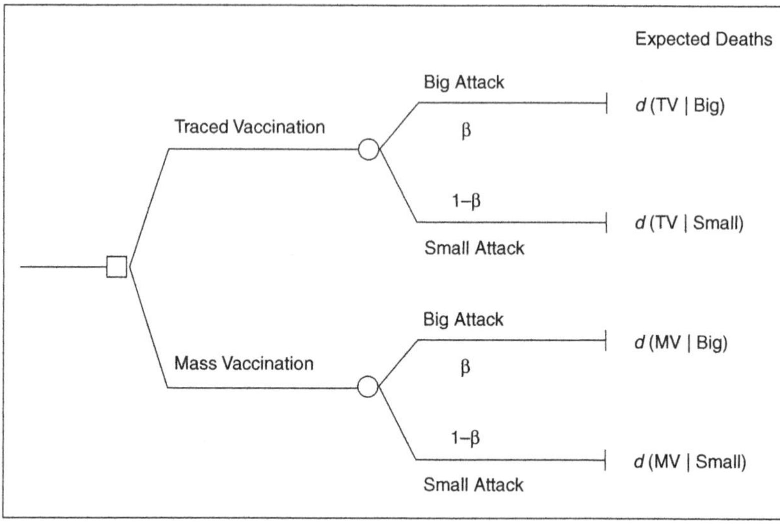

Figure 20.1. Decision tree: choosing a response strategy.

each of the four response strategy/attack scenario combinations. For a given strategy/scenario pair, let *d*(Stategy|Scenario) denote the expected number of deaths that would result (where Strategy ∈ {TV,MV} and Scenario ∈ {Big, Small}).

The decision problem is to choose a response strategy – TV or MV – once an attack has been detected. However, at the time one would face this decision, it would not be known whether the attack was Big or Small. Letting β denote the conditional probability that that attack is Big (given that an attack has occurred), the choice problem is represented by the decision tree shown in Figure 20.1.

Presuming that the objective is to minimize the expected number of deaths, the optimal decision is to choose MV providing

$$\beta\, d(\text{MV}|\text{Big}) + (1-\beta)\, d(\text{MV}|\text{Small}) < \beta\, d(\text{TV}|\text{Big}) + (1-\beta)\, d(\text{TV}|\text{Small})$$

which is equivalent to selecting MV if

$$\beta > \frac{d(\text{MV}|\text{Small}) - d(\text{TV}|\text{Small})}{\big(d(\text{MV}|\text{Small}) - d(\text{TV}|\text{Small})\big) + \big(d(\text{TV}|\text{Big}) - d(\text{MV}|\text{Big})\big)}$$

As a numerical example, let a Big attack be specified by 1,000 initial infections with $R_0 = 3$, and a small attack correspond to a single initial infection with $R_0 = 1.2$. Then using the model in Kaplan and colleagues (2002) (with all other parameters set at the base case values reported there), one obtains the following "payoff matrix" for the number of deaths that would result from considering all Strategy/Scenario pairs:

Expected Deaths

	Big Attack	Small Attack
Choose TV	110,000	2.3
Choose MV	560	10.4

From these figures, it is apparent that the death-minimizing choice is to respond with MV if the probability of a Big attack exceeds

$$\frac{10.4-2.3}{(10.4-2.3)+(110{,}000-560)} = 7.4\times10^{-5}$$

This means that even if one is 99.99% certain that the attack is Small, it is still optimal (to a death minimizer) to respond with mass vaccination. Worded differently, even though one would almost surely employ TV with perfect information regarding the size of the attack, absent such information one should respond with MV.

Though simple, this example makes a very powerful point. It is a mistake to first envision the most likely mode of attack (as argued in Halloran et al., 2002), and then design a response strategy that is optimal for such an attack scenario. As argued in Kaplan and colleagues (2002), the consequences of choosing the wrong response policy are highly asymmetric (as is clear from the payoff matrix shown earlier). In the face of a Small attack, choosing MV over TV leads to only a few expected incremental deaths (8.1 in the example given previously). In the event of a Big attack, however, responding with TV would result in such a disaster that it is sensible to avoid this possibility altogether by responding with MV in the event of an attack.

20.4 Build the Button Now: Vaccinating the Vaccinators

The previous section suggested that mass vaccination is the expected death-minimizing response in the event of a smallpox attack. However, in order to implement mass vaccination after an attack, those charged with vaccinating the

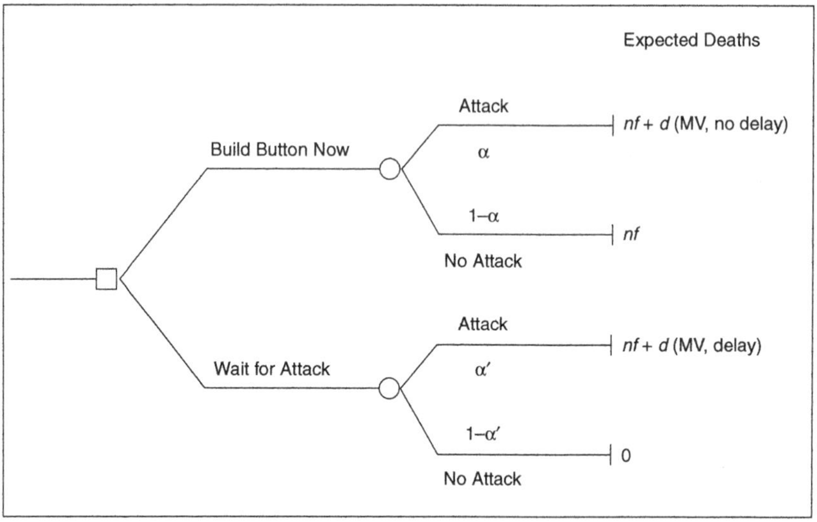

Figure 20.2. Decision tree: build button now or wait for attack.

public (as well as those responsible for receiving and treating initial smallpox cases) must themselves be immunized. Should one wait until an attack occurs before "vaccinating the vaccinators," or instead should one "build the button now?" That is, should one vaccinate a sufficient number of vaccinators now so that, in the event of an attack, localized mass vaccination of the population in the area of the attack could begin immediately?

In Kaplan and colleagues (2002), it is argued that response delay is equivalent to increasing the size of an attack, since infections would continue to spread before actions to control the epidemic could be launched, increasing the total number of persons infected at the time response operations begin. However, as discussed earlier, vaccination against smallpox is not risk free. Thus, unlike the decision regarding which response strategy to employ upon detecting an attack, choosing how many vaccinators (and other first responders) to vaccinate now must depend on the probability of an attack.

A simple version of this decision problem is illustrated in Figure 20.2. If one opts to "build the button now" and vaccinate n first responders, a fraction f of whom is expected to die from the vaccine itself, then one should expect nf deaths among first responders in the absence of an attack.

However, if an attack does occur (which it will with probability α), then in addition one would expect d(MV, no delay) deaths to occur in the population. If instead one opts to wait for an attack before "vaccinating the vaccinators," then there will be no deaths in the absence of an attack, but $nf + d(\text{MV, delay})$ if an attack does occur (which will occur with probability $\alpha' > \alpha$, assuming that an attack would be *less* likely given evidence of serious preparation).

Again choosing death minimization as the objective, it is sensible to vaccinate n vaccinators now if

$$\alpha > \frac{nf}{nf + \left(\frac{\alpha'}{\alpha} d(\text{MV, delay}) - d(\text{MV, no delay})\right)}$$

Since $\alpha' > \alpha$, a sufficient condition to build the button now is given by equating α' to α in our expression, which means preparing now if

$$\alpha > \frac{nf}{nf + \left(d(\text{MV, delay}) - d(\text{MV, no delay})\right)}.$$

As an illustration, in Kaplan and colleagues (2002), there were 5,000 full-time vaccinators, which would translate to 15,000 vaccinators each working eight-hour shifts. Presuming a Big attack, that mass vaccination would be employed in such an event, and that failing to prepare in advance would contribute an incremental one day of response delay to the base case studied in Kaplan and colleagues (2002), $d(\text{MV, no delay}) = 560$ deaths while $d(\text{MV, delay}) = 653$ deaths. Given that vaccination itself carries a 1 in 1 million risk of death, with these figures, it makes sense to build the button now if the probability of an attack when such preparations are undertaken exceeds

$$\frac{15{,}000 \times 10^{-6}}{15{,}000 \times 10^{-6} + (653 - 560)} = 1.6 \times 10^{-4}.$$

To ascertain whether this is a large or small threshold for the unknown probability of attack, it is important to note that the correct time scale to consider is the duration of time over which vaccination against smallpox remains effective. This is on the order of 5 to 10 years, so if you believe the likelihood of an attack over a 5-to-10-year period from now exceeds 1.6 in 10,000, it is prudent to prepare now.

Israel's decision making can be contrasted with U.S. policy in light of the results presented earlier in this chapter. While U.S. officials continue debating what to do about smallpox, Israel quickly (i.e., over a period of several weeks) decided to “build its button now.” Approximately 15,000 Israeli first responders have already been vaccinated, while a plan for locating clinics and rapidly vaccinating the Israeli public has already been determined (Sa’ar, 2002). Israel’s policy of “vaccinating the vaccinators” stands in contrast to preparations in the United States (National Public Radio, 2002).

In the United States, the CDC has released a vaccination clinic guide (Centers for Disease Control and Prevention, 2002). Extrapolating from this guide suggests that to staff a sufficient number of vaccination clinics to vaccinate the entire U.S. population within 10 days of an outbreak requires the vaccination now of approximately 1.25 million doctors, ambulance drivers, police officers, etc. To reach the population within five days would double this requirement to roughly 2.5 million. Given these simple back-of-the-envelope calculations, it remains difficult to understand why the government is currently calling only for the vaccination of 510,000 hospital workers.

20.5 Why Wait? Pre- versus Post-attack Vaccination

In view of the benefits that accrue from preparing in advance for a smallpox attack, many have argued that perhaps we should “go all the way” and vaccinate the entire population now. Whether it is sensible to do so again depends on the vaccine fatality rate, the risk of attack, and the consequences of an attack should one occur. As discussed earlier, the consequences of an attack further depend on the particular response policy employed, thus whether to vaccinate the entire population prior to an attack depends in a crucial way on the policy one would employ to control an epidemic post-attack.

This decision problem is shown in Figure 20.3. If the entire population of N persons were vaccinated, one would expect Nf deaths as a result. If instead one opts to only vaccinate post-attack, assuming that a mass vaccination response is employed, one would expect $d(\text{MV})$ deaths. An attack will occur with probability α if no vaccination occurs pre-attack. We ignore vaccine fatalities among first responders as these would be negligible relative to vaccine fatalities in the general population, let alone deaths in the event of a smallpox attack. We also assume a perfect vaccine and 100% vaccination coverage; departures from both of these assumptions could easily be incorporated without changing the basic result.

To a death minimizer, it makes sense to vaccinate the entire population pre-attack if

$$Nf < \alpha\, d(\text{MV})$$

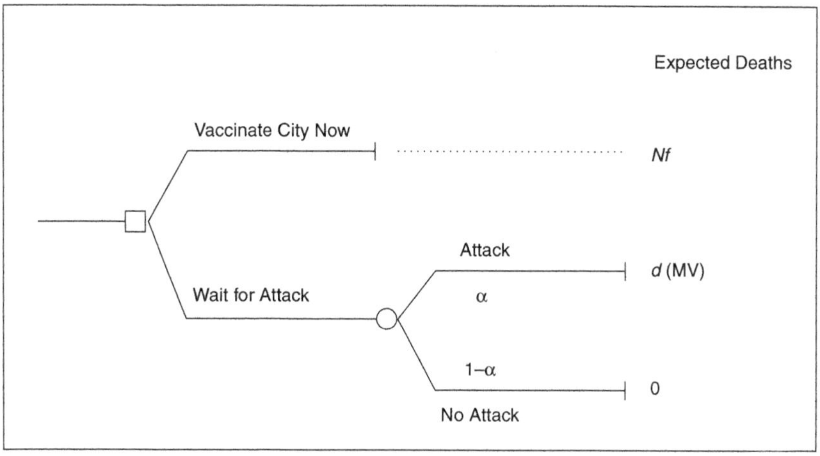

Figure 20.3. Decision tree: pre- versus post-attack vaccination.

which again yields a threshold on the attack probability – it is sensible to act now if

$$\alpha > \frac{Nf}{d(\mathrm{MV})}.$$

Consider again the base case of a Big attack from Kaplan and colleagues (2002), where $N = 10$ million, $f = 1$ per million, and $d(\mathrm{MV}) = 560$. With these parameters, it would only make sense to vaccinate the entire population now if the probability of attack exceeds 10/560 or about 1.8%.

While some would consider this a small probability upon recalling that it represents the likelihood of an attack over a 5-to-10-year time horizon, note that it is two orders of magnitude higher than the attack threshold justifying the immediate vaccination of first responders. Also, note that if traced vaccination were employed in response to an outbreak, then the threshold would drop to 1/11,000 due to the expected 110,000 deaths that would result were TV invoked post-attack.

This simple model suggests an explanation for much of the controversy in the smallpox debate. Whatever post-attack strategy is employed, if one is highly confident that the resulting outbreak can be quickly controlled, then there is little pressure to vaccinate the population now. Alternatively, if one harbors doubts regarding the ability to contain a smallpox attack, pre-attack vaccination becomes a more attractive option. Clearly, citizens differ in their assessments of government's ability to provide protection from smallpox post-attack. Citizens also differ in their views regarding the likelihood that such an attack could occur. Given such heterogeneity in views regarding risks and consequences, it is hardly surprising that many seek access to smallpox vaccine now, while others prefer to wait.

While these arguments suggest that it is not sensible to vaccinate the entire country if the risk of an attack remains below about 1.8%, this model assumes a common valuation of risks and benefits. However, in a democracy such as the United States, people assume different risks and benefits all the time. While there is roughly a 1 per 1 million chance of death from a smallpox vaccination, those of us who drive face a 145 per 1 million risk of death each year from road accidents. Some people smoke, some drink and use illicit drugs, and some jump out of airplanes for sport. As Dr. Bicknell would argue, why shouldn't individuals be afforded a choice in this instance as well (Bicknell, 2002), especially since as previously noted, any amount of pre-attack vaccination makes any post-attack policy work better? This position presupposes that individuals are truly knowledgeable about the risks of being vaccinated or not (or can be made aware of these risks via communication of the facts), are capable of rational decision making, and thus the only differences among individuals are their preferences/beliefs regarding risks and benefits.

20.6 Conclusions

Smallpox was eliminated from the planet more than 20 years ago, but it could return as a weapon of bioterror. Barring an accident at one of the two known smallpox repositories (in Atlanta and Novosibirsk, Siberia), the appearance of smallpox anywhere in the world must be considered the result of a deliberate release. While the WHO's eradication campaign provided a wealth of data pertaining to the spread and control of natural outbreaks, there are simply no data at all that describe how smallpox would spread as the result of a bioterror attack. Some have assumed that deliberate outbreaks would mimic their natural counterparts, insinuating that methods believed successful in the past would surely work again in the future. By contrast, the arguments of this chapter do not suggest that the way to prepare for a disaster is to assume that it cannot occur (nor for that matter do they suggest that the way to evaluate a policy is to assume that it works, as ACIP did in recommending ring vaccination in response to a smallpox attack). Rather, simply allowing for the possibility that a large bioterror attack could occur greatly changes the optimal preparedness and response strategy, even if the probability assigned to such an attack remains very small. The consequences of decisions matter more than the details of the "most likely scenario" in considering what to do about smallpox and other bioterror agents. Simple models have served well to make this argument in the ongoing smallpox debate.

In spite of such arguments and the president's policy decision of December 2002, the United States is still unprepared for a smallpox bioterror attack. Nationwide, fewer than 40,000 first responders had volunteered to receive the smallpox vaccine as of December 2003, far short of the 10.5 million that were to have been immunized by the end of phase 2. As detailed elsewhere (Bicknell & Bloem, 2003), there are numerous administrative, political, and public health reasons for this state of affairs. Whether or not the pre-event smallpox vaccination program will recover remains to be seen.

Acknowledgments

E. H. K. was supported in part by the Center for Interdisciplinary Research on AIDS via Grant MH/DA56826 from the U.S. National Institutes of Mental Health and Drug Abuse.

REFERENCES

Alibek, K. (1999). *Biohazard.* New York: Random House.

Altman, L. K. (2002, June 22). Smallpox proposal raises ethical issues. *New York Times.*

Bicknell, W. J. (2002). The case for voluntary smallpox vaccination. *New England Journal of Medicine, 346,* 1323–1325.

Bicknell, W. J., & Bloem, K. D. (2003, September 5). *Smallpox and bioterrorism: Why the plan to protect the nation is stalled and what to do.* Cato Institute: Briefing Paper #85.

Broad, W. J. (2002, July 7). U.S. to vaccinate 500,000 workers against smallpox. *New York Times.*

Centers for Disease Control and Prevention. (2001). *CDC interim smallpox response plan and guidelines, draft 2.0.* Atlanta, GA: Author.

Centers for Disease Control and Prevention. (2002). *Smallpox vaccination clinic guide.* Atlanta, GA: Author.

Fauci, A. S. (2002). Smallpox vaccination policy – the need for dialogue. *New England Journal of Medicine, 346,* 1319–1320.

Fenner, F., Henderson, D. A., Arita, I., Jezek, Z., & Ladnyi, I. D. (1988). *Smallpox and its eradication.* Geneva: World Health Organization.

Gani, R., & Leach, S. (2001). Transmission potential of smallpox in contemporary populations. *Nature, 414,* 748–751.

Halloran, M. E., Longini Jr., I. M., Nizam, A., & Yang, Y (2002). Containing bioterrorist smallpox. *Science, 298,* 1428–1432.

Institute of Medicine. (2002). *Scientific and policy considerations in developing smallpox vaccination options.* Washington, DC: National Academies Press.

Kaplan, E. H., Craft, D. L., & Wein, L. M. (2002). Emergency response to a smallpox attack: the case for mass vaccination. *Proceedings of the National Academy of Sciences of the USA*, *99*, 10395–10440.

Kaplan, E. H, & Wein, L. M. (2002). Smallpox eradication in west and central Africa: Surveillance-containment or herd immunity? *Epidemiology*, *14*, 1–4.

Miller, J., Engelberg, S., & Broad, W. (2001). *Germs*. New York: Simon and Schuster.

National Public Radio. (2002, November 1). Profile: Smallpox preparedness of Israel and the U.S. *All Things Considered*, www.npr.org/programs/atc/transcripts/2002/nov/021101.northam.html, Accessed December 7, 2002.

Neff, J. M., Lane, J. M., Fulginiti, V. A., & Henderson, D. A. (2002). Contact vaccinia – transmission of vaccinia from smallpox vaccination. *Journal of the American Medical Association*, *288*, 1901–1905.

Sa'ar, R. (2002, October 18). Ministry presents smallpox crisis plan. *Ha'aretz*.

Siegel-Itzkovitch, J. (2002, July 28). Should only infected people be vaccinated? *Jerusalem Post*.

Stevenson, R. W., & Stolberg, S. G. (2002, December 14). Bush lays out plan on smallpox shots; military is first. *New York Times*.

Tucker, J. B. (2001). *Scourge: The once and future threat of smallpox*. New York: Atlantic Monthly Press.

Zelicoff, A. P. (2002). An epidemiological analysis of the 1971 smallpox outbreak in Aralsk, Kazakhstan. In J. B. Tucker & R. A. Zilinskas (Eds.), *The 1971 smallpox epidemic in Aralsk, Kazakhstan, and the Soviet biological warfare program* (pp. 12–21). Occasional Paper No. 9. Washington, DC: Monterey Institute of International Studies, Center for Nonproliferation Studies.

21

Stackelberg Security Games (SSG) Basics and Application Overview

Bo An and Milind Tambe

21.1 Introduction

Security is a critical concern around the world that arises in protecting our ports, airports, transportation or other critical national infrastructure, curtailing the illegal flow of drugs, weapons, and money, and suppressing urban crime, as well as in protecting wildlife, fish, and forests from poachers and smugglers; and it arises in problems ranging from physical to cyber-physical systems. In all of these challenges, we have limited security resources that prevent full security coverage at all times. Instead, limited security resources must be deployed intelligently, taking into account differences in priorities of targets requiring security coverage, the responses of adversaries to the security posture, and potential uncertainty over the types, capabilities, knowledge, and priorities of adversaries faced.

Game theory is well suited to adversarial reasoning for security resource allocation and scheduling problems. Casting the problem as a Bayesian Stackelberg game in consideration of uncertainties, we have developed new algorithms for efficiently solving such games to provide randomized patrolling or inspection strategies. These algorithms have led to some initial successes in this arena, leading to advances over previous approaches in security resource scheduling and allocation, e.g., by addressing key weaknesses of predictability of human schedulers. These algorithms are now deployed in multiple applications: ARMOR has been deployed at the Los Angeles International Airport (LAX) since 2007 to randomize checkpoints on the roadways entering the airport and canine patrol routes within the airport terminals (Pita et al., 2008); IRIS is a game-theoretic scheduler for randomized deployment of the U.S. Federal Air Marshal Service (FAMS) requiring significant scale-up in underlying algorithms that has been in use since 2009 (Tsai, Rathi, Kiekintveld, Ordóñez, & Tambe, 2009); and PROTECT, which requires further scale-up, is deployed for generating randomized patrol schedules for the U.S. Coast Guard in Boston, New York, Los Angeles, and other ports around the United States (An et al., 2013; An, Pita,

Shieh, Tambe, Kiekintveld, & Marecki, 2011; Shieh et al., 2012). Furthermore, TRUSTS is being evaluated for deterring fare evasion, suppressing urban crime, and counterterrorism within the Los Angeles Metro System (Jiang, Nguyen, Tambe, & Procaccia, 2013; Zin & Tambe, 2012), and GUARDS was earlier tested by the U.S. Transportation Security Administration (TSA) for security inside the airport (Pita, Tambe, Kiekintveld, Cullen, & Steigerwald, 2011). Moreover, recent work on "green security games" has led to testing our decision aids for protection of fisheries with the U.S. Coast Guard and protection of wildlife at sites in multiple countries, and "opportunistic crime security games" have focused on suppressing urban crime. These initial successes point the way to major future applications in a wide range of security arenas, with major research challenges in scaling up our game-theoretic algorithms, to addressing human adversaries' bounded rationality and uncertainties in action execution and observation, as well as in preference elicitation and multi-agent learning.

This chapter will discuss applications of security games, and outline research challenges in security games, including algorithms for scaling up security games, as well as for handling significant adversarial uncertainty and learning models of human adversary behaviors. The rest of this chapter is organized as follows. We start with introducing Stackelberg security games (SSG) basics in Section 21.2. Section 21.3 categorizes different types of security game models. Section 21.4 discusses both deployed and emerging security applications. Section 21.5 overviews some applications of security games beyond the security domains. Section 21.6 concludes with open research issues and future research directions.

21.2 Stackelberg Security Games (SSG) Basics

A generic Stackelberg game has two players, a *leader* and a *follower*. A leader commits to a strategy first, and then a follower optimizes her reward, considering the action chosen by the leader (von Stengel & Zamir, 2004). The two players in a Stackelberg game need not represent individuals, but could also be groups that cooperate to execute a joint strategy, such as a police force or a terrorist organization. Each player has a set of possible *pure strategies*, or the actions that she can execute. A *mixed strategy* allows a player to play a probability distribution over pure strategies. Payoffs for each player are defined over all possible pure-strategy outcomes for both the players. The payoff functions are extended to mixed strategies by taking the expectation over pure-strategy outcomes. The follower can observe the leader's strategy, and then act in a way to optimize his own payoffs.

To see the advantage of being the leader in a Stackelberg game, consider the game with the payoffs as shown in Table 21.1. The leader is the row player

Table 21.1. *Payoff table for example Stackelberg game*

	c	d
a	3,1	5,0
b	2,0	4,2

and the follower is the column player. The only pure-strategy Nash equilibrium for this game is when the leader plays a and the follower plays c, which gives the leader a payoff of 3; in fact, for the leader, playing b is strictly dominated.

However, in this game, if the leader can commit to playing b before the follower chooses his strategy, then the leader will obtain a payoff of 4, since the follower would then play d to ensure a higher payoff for himself. If the leader commits to a mixed strategy of playing a and b with equal (0.5) probability, then the follower will play d, leading to a higher expected payoff for the leader of 4.5. As we can see from this example, the equilibrium strategy in the Stackelberg game can be in fact different from the Nash equilibria.

Stackelberg games are used to model the attacker-defender strategic interaction in security domains, and this class of Stackelberg games (with certain restrictions on payoffs) (Yin, Korzhyk, Kiekintveld, Conitzer, & Tambe, 2010) is called *Stackelberg security games*. In the Stackelberg security game framework, the security force (defender) is modeled as the leader and the terrorist adversary (attacker) is in the role of the follower. The defender commits to a mixed (randomized) strategy, whereas the attacker conducts surveillance of these mixed strategies and responds with a pure strategy of an attack on a target. Thus, the Stackelberg game framework is a natural approximation of real-world security scenarios. In contrast, the surveillance activity of the attacker cannot be modeled in the simultaneous move games with the Nash equilibrium solution concept. The objective is to find the optimal mixed strategy for the defender.

21.3 Categorizing Security Games

With progress in security games research and the expanding set of applications, it is valuable to consider categorizing this work into three separate areas. These categories are driven by applications, but they also impact the types of games (e.g., single-shot vs. repeated games) considered and the research issues that arise. Specifically, the three categories are:

(i) infrastructure security games; (ii) green security games; (iii) opportunistic crime security games. We discuss each category next.

21.3.1 *Infrastructure Security Games*

This type of games and their applications are where the original research on security games was initiated. Key characteristics of these games include the following:

- *Application characteristics*: These games are focused on applications of protecting infrastructure, such as ports, airports, trains, flights, and so on; the goal is often assisting agencies engaged in counterterrorism. Notice that the infrastructure being protected tends to be static, and little changes in a few months, e.g., an airport being protected may have new construction once in two or three years. The activities in the infrastructure are regulated by well-established schedules of movement of people or goods. Furthermore, the targets being protected often have a discrete structure, e.g., terminals at an airport, individual flights, individual trains, etc.
- *Overall characteristics of the defender and adversary play*: These games are single-shot games. The defender does play her strategy repeatedly, i.e., the defender commits to a mixed strategy in this security game. This mixed strategy may get played for months at a time. However, a single attack by an adversary ends the game. The game could potentially restart after such an attack, but it is not set up as a repeated game as in the game categories described later in this chapter.
- *Adversary characteristics*: The games assume that the adversaries are highly strategic and that they may attack after careful planning and surveillance. These carefully planned attacks have high consequences. Furthermore, since these attacks are a result of careful planning with the anticipation of high consequences, attackers commit to these plans of attacks and are not considered to opportunistically move from target to target.
- *Defender characteristics*: The defender does not repeatedly update her strategies. In these domains, there may be just a few attacks that may occur, but these tend to be rare; there are not a very large number of attacks that occur repeatedly. As a result, traditionally, no machine learning is used for the defender to update her strategies over time.

21.3.2 *Green Security Games*

This type of games and their applications are focused on trying to protect the environment; we adopt the term from "green criminology."[1]

- *Application characteristics*: These games are focused on applications of protecting the environment, including forests, fish, and wildlife. The goal is thus often to assist security agencies against poachers, illegal fishermen, or those illegally cutting trees in national parks in countries around the world. Unlike infrastructure security games, animals or fish being protected may move around in geographical space, introducing new dimensions of complexity. Finally, the targets being protected are spread out over vast, open geographical spaces, e.g., protecting trees from illegal cutting in large forest regions.

- *Overall characteristics of the defender and adversary play*: These games are not single-shot games. Unfortunately, the adversary often conducts multiple repeated "attacks," e.g., poaching animals repeatedly. Thus, a single illegal activity does not end the game. Instead, after obtaining reports, e.g., over a month of illegal activities, the defender often re-plans her security activities. In other words, these are repeated security games where the defender plays a mixed strategy, while the attacker attacks multiple times, and then the defender re-plans and plays a new mixed strategy and the cycle repeats. Notice also that the illegal activities of concern here may be conducted by multiple individuals, and thus multiple adversaries are active at any one point.
- *Adversary characteristics*: As mentioned earlier, the adversaries are engaged in repeated illegal activities; the consequences of failure or success are not as severe as in the case of counterterrorism. As a result, every single attack (illegal action) cannot be carried out with the most detailed surveillance and planning. The adversaries will hence exhibit more bounded rationality and bounded surveillance in these domains. Nonetheless, these domains are not ones where illegal activities can be conducted opportunistically (as in the opportunistic crime security games discussed later). This is because in these green security games, the adversaries often have to act in extremely dangerous places (e.g., deep in forests, protecting themselves from wild animals), and thus given the risks involved, they cannot take an entirely opportunistic approach.
- *Defender characteristics*: Since this is a repeated game setting, the defender repeatedly updates her strategies. Machine learning can now be used in this work for the defender to update her strategies over time, given that attack data are available over time. The presence of large amounts of such attack data is very unfortunate in that very large numbers of crimes against the environment are recorded in real life, but the silver lining is that the defender can improve her strategy exploiting these data.

21.3.3 *Opportunistic Crime Security Games*

This type of games and their applications are focused on trying to combat opportunistic crime. Such opportunistic crime may include criminals engaged in thefts such as snatching cell phones in metros or stealing student laptops from libraries.

- *Application characteristics*: These games focus on applications involving protecting the public against opportunistic crime. The goal is thus often to assist security agencies in protecting public's property such as cell phones, laptops, or other valuables. Here, human crowds may move around based on scheduled activities, e.g., office hours in downtown settings, or class timings on a university campus, and thus the focus of what needs to be protected may shift on a regular schedule. At least in urban settings, these games focus on specific, limited geographical areas as opposed to vast, open spaces as involved in "green security games."
- *Overall characteristics of the defender and adversary play*: While these games are not explicitly formulated as repeated games, the adversary may conduct or attempt to conduct multiple "attacks" (thefts) in any one round of the game. Thus, the defender commits to a mixed strategy, but a single attack by a single attacker does not end the game. Instead, multiple attackers may be active at a time, conducting multiple thefts while the defender attempts to stop these thefts from taking place.
- *Adversary characteristics*: Once again, the adversaries are engaged in repeated illegal activities, and the consequences of failure or success are not as severe as in the case of counterterrorism. As a result, given that every single attack (illegal action) cannot be carried out with the most detailed surveillance and planning, the adversaries may act even less strategically, and exhibit more bounded rationality and bounded surveillance in these domains. Furthermore, the adversaries are not as committed to detailed plans and are flexible in their execution of their plans, as targets of opportunity present themselves.
- *Defender characteristics*: Available crime data can be used to aid the defender in planning and adapting the defense strategy in response to the criminals' actions. Machine-learning techniques are particularly applicable here.

Even though we have categorized the research and applications of security games in these three categories, not everything is very clearly divided in this fashion. Further research may reveal other categories or generate subcategories of these three categories.

21.4 Deployed and Emerging Security Applications

As we discussed at the beginning of this chapter, the past several years have witnessed the successful application of multi-agent systems in allocating

Table 21.2. *Deployed key applications and associated research problems*

Major Applications	Research challenges and major references
ARMOR for Los Angeles International Airport	Algorithms for scheduling check points and canine units (Paruchuri, Pearce, Marecki, Tambe, Ordóñez, & Kraus, 2008; Pita et al., 2008)
IRIS for U.S. Federal Air Marshals Service	Algorithms for solving large-scale games with complex scheduling constraints (Jain, Kardes, Kiekintveld, Ordóñez, & Tambe, 2010; Tsai et al., 2009)
PROTECT for U.S. Coast Guard	Scheduling boat patrol, modeling human behavior (An et al., 2013; Fang, Jiang, & Tambe, 2013; Shieh et al., 2012)
GUARDS for U.S. Transportation Security Agency	Dealing with heterogeneous defender activities (Pita et al, 2011)
TRUSTS for Urban Security in Transit Systems	Deal with spatial and temporal travel constraints and interruptions (Jiang et al., 2013; Zin & Tambe, 2012)
PAWS for protecting wildlife	Learning behavior model from poaching data (Fang, Stone, & Tambe, 2015; Yang, Ford, Tambe, & Lemieux, 2014)
Protecting public events	Continuous and infinite strategy space (Yin, An, & Jain, 2014; Yin, Xu, Gan, An, & Jiang, 2015)

limited resources to protect critical infrastructures (An, Brown, Vorobeychik, & Tambe, 2013; An, Jain, Tambe, & Kiekintveld, 2011; An et al., 2013; An et al., 2011; Basilico, Gatti, & Amigoni, 2009; Jain, Tsai, Pita, Kiekintveld, Rathi, Tambe, & Ordóñez, 2010; Jiang et al., 2013; Korzhyk, Conitzer, & Parr, 2010; Pita et al., 2011; Tambe & An, 2012; Zin & Tambe, 2012). In the rest of this section, we describe in detail the application of the Stackelberg game framework in multiple significant security domains. Table 21.2 outlines different applications and associated research problems, as well as major references.

21.4.1 *Infrastructure Security Game Applications*

ARMOR for Los Angeles International Airport

Los Angeles International Airport (LAX) is the largest destination airport in the United States and serves 60–70 million passengers per year. The LAX police use diverse measures to protect the airport, which include vehicular checkpoints and police units patrolling with canines. The eight different terminals at LAX have very different characteristics, like physical size, passenger loads, foot traffic, or international versus domestic flights. Furthermore, the numbers of available vehicle checkpoints and canine units are limited by resource constraints. Thus it is challenging to optimally allocate these resources to improve their effectiveness while avoiding patterns in the scheduled deployments.

The Assistant for Randomized Monitoring over Routes (ARMOR) system focuses on two of the security measures at LAX (checkpoints and canine patrols) and optimizes security resource allocation using Bayesian Stackelberg games. Take the vehicle checkpoints model as an example. Assume that there are n roads; the police's strategy is placing m < n checkpoints on these roads where m is the maximum number of checkpoints. The adversary may choose to attack through one of these roads. ARMOR models different types of attackers with different payoff functions, representing different capabilities and preferences for the attacker. ARMOR uses Decomposed Optimal Bayesian Stackelberg Solver (DOBSS) to compute the defender's optimal strategy (Paruchuri et al., 2008). ARMOR has been successfully deployed since August 2007 at LAX to randomize checkpoints on the roadways entering the airport and canine patrol routes within the airport terminals (Pita et al., 2008).

IRIS for U.S. Federal Air Marshals Service

The U.S. Federal Air Marshals Service (FAMS) allocates air marshals to flights originating in and departing from the United States to dissuade potential aggressors and prevent an attack. Flights are of different importance based on a variety of factors such as the numbers of passengers, the population of source/destination, international flights from different countries, and special events that can change the risks for particular flights at certain times. Security resource allocation in this domain is significantly more challenging than for ARMOR: a limited number of FAMS need to be scheduled to cover thousands of commercial flights each day. Furthermore, these FAMS must be scheduled on tours of flights that obey various constraints (e.g., the time required to board, fly, and disembark). Therefore, we face significant computational challenge while generating the optimal scheduling policy that meets these scheduling constraints.

Against this background, the Intelligent Randomization in Scheduling (IRIS) system has been developed and has been deployed by FAMS since October 2009 to randomize schedules of air marshals on international flights. In IRIS, the targets are the set of n flights and the attacker could choose to attack one of these flights. FAMS can assign m < n air marshals to protect these flights. Since the number of possible schedules exponentially increases with the number of flights and resources, DOBSS is no longer applicable to the FAMS domain. Instead, IRIS uses the much faster ASPEN algorithm (Jain et al., 2010) to generate the schedule for thousands of commercial flights per day.

Figure 21.1. USCG boats patrolling the ports of Boston and NY. (a) PROTECT is being used in Boston. (b) Extend PROTECT to NY.

IRIS also uses an attribute-based preference elicitation system to determine reward values for the Stackelberg game model.

PROTECT for U.S. Coast Guard

The U.S. Coast Guard's (USCG) mission includes maritime security of the U.S. coasts, ports, and inland waterways – a security domain that faces increased risks due to threats such as terrorism and drug trafficking. Given a particular port and the variety of critical infrastructure that an adversary may attack within that port, the USCG conducts patrols to protect this infrastructure; however, while the adversary has the opportunity to observe patrol patterns, limited security resources imply that USCG patrols cannot be at every location 24/7. To assist the USCG in allocating its patrolling resources, the Port Resilience Operational/Tactical Enforcement to Combat Terrorism (PROTECT) model is being designed to enhance maritime security and has been in use at the port of Boston since April 2011 (Figure 21.1). Similar to previous applications ARMOR and IRIS, PROTECT uses an attacker-defender Stackelberg game framework, with the USCG as the defender against terrorist adversaries that conduct surveillance before potentially launching an attack.

PROTECT is currently deployed in the ports of Boston, New York, Los Angeles/Long Beach, and several others (An et al., 2013). Indeed, the goal now is to deploy PROTECT at ports nationwide. Furthermore, beyond just port protection, PROTECT has been extended to protect ferry systems (Figure 21.2) such as the Staten Island ferry in New York (Fang et al., 2013).

While PROTECT builds on previous work, it offers some key innovations. First, to improve PROTECT's efficiency, a compact representation of the defender's strategy space is used by exploiting equivalence and dominance.

Figure 21.2. Protecting ferries with patrol boats.

Second, the evaluation of PROTECT for the first time provides real-world data: (i) comparison of human-generated vs PROTECT security schedules; and (ii) results from an adversarial perspective team's (human mock attackers) analysis.

GUARDS for U.S. Transportation Security Agency

The U.S. Transportation Security Administration (TSA) is tasked with protecting the nation's more than 400 airports. To aid the TSA in scheduling resources to protect airports, a new application called Game-theoretic Unpredictable and Randomly Deployed Security (GUARDS) has been developed. While GUARDS also utilizes Stackelberg games as ARMOR and IRIS, GUARDS faces three key challenges Pita et al., 2011: 1) reasoning about hundreds of heterogeneous security activities; 2) reasoning over diverse potential threats; and 3) developing a system designed for hundreds of end-users. To address those challenges, GUARDS created a new game-theoretic framework that allows for heterogeneous defender activities and compact modeling of a large number of threats and developed an efficient solution technique based on general-purpose Stackelberg game solvers. GUARDS was originally tested at an undisclosed airport and further results are awaited (Pita et al., 2011).

TRUSTS for Urban Security in Transit Systems

TRUSTS focuses on three major security challenges: deterring fare evasion, suppressing crime, and counterterrorism. Significant focus in TRUSTS has been on

(a) (b)

Los Angeles Metro — Barrier-free entrance

Figure 21.3. TRUSTS for transit systems.

deterring fare evasion. Specifically, in some urban transit systems, including the Los Angeles Metro Rail system, passengers are legally required to purchase tickets before entering, but are not physically forced to do so (Figure 21.3). Instead, patrol units move about through the transit system, inspecting tickets of passengers, who face fines for fare evasion. This setting yields the problem of computing optimal patrol strategies to deter fare evasion and hence maximize revenue. The Tactical Randomization for Urban Security in Transit Systems (TRUSTS) system models the patrolling problem as a leader-follower Stackelberg game (Jiang et al., 2013; Zin & Tambe, 2012). Urban transit systems, however, present unique computational challenges since there are exponentially many possible patrol strategies, each subject to both the spatial and temporal constraints of travel within the transit network under consideration. To overcome this challenge, TRUSTS uses a compact representation that captures the spatial as well as temporal structure of the domain. The system has been evaluated using real-world ridership data from the Los Angeles Metro Rail system.

One key finding from initial tests was that the schedules generated by officers were often interrupted. Interruptions occurred because in frequent interactions with the public, sometimes officers would get stopped by lost travelers, or sometimes they would need to arrest someone. Such interruptions mean that the schedules now need to be highly dynamic. To that end, a new generation of Stackelberg game–based scheduling algorithms – using Markov Decision Problems – was designed. This led to schedules now being loaded onto smartphones and given to officers. The schedules are then automatically updated on the smartphones if interruptions occur (Luber, Yin, Fave, Jiang, Tambe, & Sullivan, 2013).

Protecting Public Events

Security games can also be applied to resource allocation for protecting public events. Public events in major cities are prime terrorism targets since they usually provide easy access to a large number of targets for the adversary. There have been some successful terrorist attacks on large, public events in the United States and Europe in the past few years, e.g., the recent Boston Marathon bombings on April 15, 2013, and the July 7, 2005, London bombings. Intelligent deployment of limited security resources to protect such events is therefore extremely important and challenging since the importance of targets changes over time. For example, the value of targets along a marathon track changes over time with the changing number of participants and spectators at any specific area over the course of the race. In addition, since the attacker may attack at any time and the defender can relocate resources among targets at any time, the strategy space of each agent is continuous and infinite. Furthermore, due to the relative infrequency of such events, the attacker does not get the opportunity to conduct surveillance and respond to a distribution of defender strategies. In this case, a pure defender strategy sampled from the optimal mixed strategy does not necessarily outperform the one-shot optimal pure strategy in terms of ex-post payoff. Algorithms have been proposed to compute the optimal pure defender strategy despite the infinite strategy space and time-varying target values (Yin et al., 2014; Yin et al., 2015).

21.4.2 *Green Security Game Applications*

A number of newer applications are focused on suppressing environmental crime. One of those is protecting forests (Johnson, Fang, Yang, Tambe, & Albers, 2012), where we must protect a continuous forest area from extractors. Since the attacker's behavior (e.g., extracting important resources from the forest) could be effected by spatial considerations, it is critical for the defender to incorporate spatial considerations into her enforcement decisions (Albers, 2010).

Another area of interest is protecting endangered species. Endangered species poaching is reaching critical levels as the populations of these species plummet to unsustainable numbers. The global tiger population, for example, has dropped more than 95% from the start of the 1900s and has resulted in three out of nine species' extinctions. Depending on the area and animals poached, motivations for poaching range from profit to sustenance, with the former being more common when profitable species such as tigers, elephants, and rhinos are the targets. To counter poaching efforts and to rebuild the species' populations, countries have set up protected wildlife reserves and conservation agencies

Figure 21.4. Computational game theory can play a role in suppressing environmental crime.

tasked with defending these large reserves. Because of the size of the reserves and the common lack of law enforcement resources, conservation agencies are at a significant disadvantage when it comes to deterring and capturing poachers. Agencies use patrolling as a primary method of securing the parks. Due to their limited resources, however, patrol managers must carefully create patrols that account for many different variables (e.g., limited patrol units to send out, multiple locations that poachers can attack at varying distances to the outpost). Protection Assistant for Wildlife Security (PAWS) aims to assist conservation agencies in their critical role of patrol creation by predicting where poachers will attack and optimizing patrol routes to cover those areas (Figure 21.4; Yang et al., 2014).

Another emerging application domain is that of ensuring the sustainability of fish resources. Marine fisheries are acknowledged to be some of the most important food resources for countries around the world. As reported by the World Wildlife Fund for Nature (WWF), cod are currently at risk from overfishing in the UK, Canada, and most other Atlantic countries. Global cod catch has suffered a 70% drop over the past 30 years, and, if this trend continues, the world's cod stocks will disappear in 15 years. Illegal, unreported, and unregulated (IUU) fishing is one of the major threats to the sustainability of ocean fish resources. As estimated by the National Oceanic and Atmospheric Administration (NOAA), IUU fishing produces between 11 million and 26 million tons of seafood annually, representing as much as 40% of the total catch in some fisheries. The driver behind IUU fishing is high economic profit and low chance of seizure. It is impossible

Figure 21.5. U.S. Coast Guard personnel on a mission to protect fisheries.

to maintain a 24/7 presence to prevent IUU fishing everywhere due to the limited patrolling resources. Hence the allocation of the patrolling resources becomes a key challenge for security agencies like the U.S. Coast Guard (Figure 21.5).

21.4.3 *Opportunistic Crime Security Game Applications*

A notable characteristic of urban crime, distinct from organized terrorist attacks, is that most urban crimes are opportunistic in nature, i.e., criminals do not plan their attacks in detail; rather, they seek opportunities for committing crime and are agile in their execution of the crime (Short, D'Orsogna, Pasour, Tita, Brantingham, Bertozzi, & Chayes, 2008; Zhang, Jiang, Short, Brantingham, & Tambe, 2014). Thus, the Opportunistic Crime Security Game (OCSG) model is a good fit for addressing crime in urban settings.

A particular category of urban crime is crime on transit systems such as phone snatching.

OCSG has been validated in trials on the LA Metro systems with good results (Fave et al., 2014). The trials ran for two days with each test consisting of a two-hour patrol involving two teams of two security officers. Each team had to patrol seven stations of a particular LA Metro train line using schedules generated using the OCSG framework. Figure 21.6 shows such probabilities and correlates them to the crime statistics for each of the 14 stations to patrol. In the figure, the x-axis enumerates the 14 stations to patrol. The bar graphs (y-axis on the right) show, for each station, the total number of crimes that happened during 2012 and 2013. Finally, the line graph shows the different coverage probabilities calculated for each station (y-axis on the left). In the figure, the stations with a larger coverage probability (stations 5 to 10) are either the stations with a large number of crimes (stations 5 and 8) or the adjacent stations (stations 6, 7, 9, and 10). The latter stations are given a large coverage probability because the OCSG model anticipates the possibility that criminals will choose stations 6, 7, 9, and 10 anticipating that stations 5 and 8 will be frequently patrolled by security officers. Hence, these coverage probabilities show how the OCSG

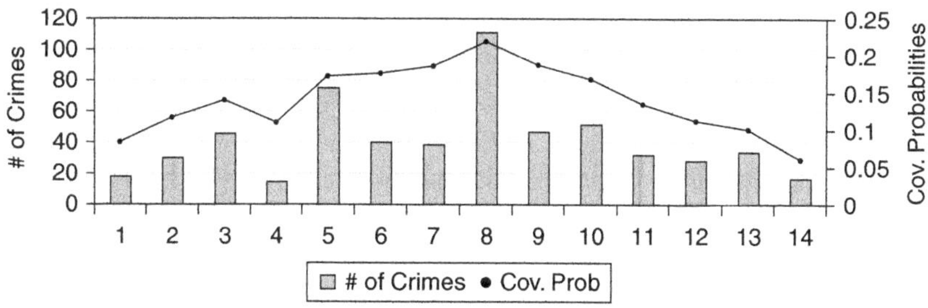

Figure 21.6. Crime statistics and coverage probabilities.

model allows schedulers to build effective, real-world patrol schedules. During the tests, the officers were able to write five citations and make two arrests. In general, they were able to understand and follow the schedule easily.

There are many other possible applications of OCSG, including scheduling police patrols and providing security at crowded places such as fairs.

21.5 Applications of Security Games beyond Security

The security games model is rich enough to capture many real-world scenarios of defender and adversary interaction beyond the physical security settings discussed here. One interesting work, called audit games (Blocki, Christin, Datta, Procaccia, & Sinha, 2013, 2015) enhances the security games model with choice of punishments in order to capture scenarios of security and privacy policy enforcement in large organizations. Large organizations (such as Google, Facebook, and hospitals) hold enormous amounts of privacy-sensitive data. These organizations mandate their employees to adhere to certain privacy policies when accessing data. Auditing of access logs is used by organizations to check for policy-violating accesses and then the violators are punished. Auditing often requires human help to investigate suspicious cases, and thereby arises the problem of allocating few resources to the huge number of cases to investigate. Another relevant question in this domain is how much should the organization punish in case of a violation?

The audit game models the adversary as an agent that performs certain tasks (e.g., accesses to private data), and a subset of these tasks is policy violations. The auditor inspects a subset of the tasks, and detects violations from the inspected set. As punishments do affect the behavior of the adversary, it is critical for the auditor to choose the right level of punishment. As a consequence, the choice of a punishment level is added to the action space of the auditor. However, punishment is not free for the auditor, the intuition being that a high punishment level creates a hostile work environment, leading to lack in productivity of employees that results in loss for the organization (auditor). As a consequence, the auditor cannot impose infinite punishment and deter any adversary. The auditor's cost for a punishment level is modeled as a loss proportional to the choice of the punishment level. The auditor moves first by committing to an inspection and punishment strategy, followed by the best response of the adversary. The resultant Stackelberg equilibrium optimization turns out to be non-convex due to the punishment variable. The authors present efficient algorithms for various types of scheduling constraints.

There are various other interesting applications such as using a security game model to randomize regression testing in the domain of software testing (Kukreja, Halfond, & Tambe, 2013). In software testing, due to time constraints, often not all the tests for a software product can be run. Thus, this leads to the problem of using few resources (test) to protect against attacks (software bugs); the security game model is a natural fit for such a scenario. Another interesting application (Li & Conitzer, 2013) is in the domain of testing students. Often the set of all exam questions for large-scale exams (e.g., driver's license test) is known a priori. However, if this set is large, the test takers cannot memorize all the answers. Then, the examiner must choose questions randomly to best test the students on the questions-answers they did not memorize. This can also be posed as a Stackelberg game. Other applications include the scenario of scheduling randomized patrols for improving security of transit networks in Singapore (Varakantham, Lau, & Yuan, 2013).

21.6 Major Research Issues

In this section, we highlight some key research challenges, including scalability, robustness, and human adversary modeling. Readers can refer to other chapters (e.g., basic solution concepts and algorithms, linear programming methods, addressing the unpredictable human element, and learning) of this book for detailed discussions of the research issues outlined next.

Scalability: The first research challenge is improving the scalability of our algorithms for solving Stackelberg (security) games. The strategy space of both the defender and the attacker in these games may exponentially increase with the number of security activities, attacks, and resources. As we scale up to larger domains, it is critical to develop newer algorithms that scale up significantly beyond the limits of the current state of the art of Bayesian Stackelberg solvers. Driven by the growing complexity of applications, a sequence of algorithms for solving security games has been developed, including DOBSS (Paruchuri et al., 2008), ERASER (Jain et al., 2010), and ASPEN (Jain et al., 2010). However, existing algorithms still cannot scale up to very large-scale domains such as scheduling randomized checkpoints in cities. In such graph-based security games, the strategy space of the defender grows exponentially with the number of available resources and the strategy space of the attacker grows exponentially with the size of the road network considered. Three key ideas have been used to design efficient algorithms, including (i) Marginals (compact representation); (ii) Incremental strategy generation; and (iii) Incremental constraint generation. For patrolling domains with spatiotemporal constraints, defender mixed strategies can be compactly represented as fractional flows. This approach has recently been applied to efficiently compute fare-enforcement patrols in urban transit systems (Jiang, Yin, Kraus, Zhang, & Tambe, n.d.; Zin & Tambe, 2012) and boat patrols for protecting ferries (Fang et al., 2013). To schedule checkpoints, an incremental strategy generation approach called "double oracle" is applied that does not require the enumeration of the entire strategy space for either of the players (Jain, Korzhyk, Vanek, Pechoucek, Conitzer, & Tambe, 2011; Jain, Tambe, & Conitzer, n.d.). When solving large-scale security games, we may need to deal with large (even infinite) set of constraints and the constraint generation approach was recently applied (Nguyen, Yadav, An, Tambe, & Boutilier, 2014). The key idea is to sample a subset of constraints and gradually expand this set by adding violated constraints to the relaxed problem until convergence to the optimal solution.

Robustness: The second challenge is improving solutions' robustness. Classical game theory solution concepts often make assumptions on the knowledge, rationality, and capability of players. Unfortunately, those assumptions could be wrong in real-world scenarios. Therefore, while computing the defender's optimal strategy, algorithms should take into account various uncertainties faced in the domain, including payoff noise (Kiekintveld, Marecki, & Tambe, 2011), execution/observation error (Yin, Jain, Tambe, & Ordóñez, 2011), and uncertain capability (An, Tambe, Ordóñez, Shieh, & Kiekintveld, 2011). For observation uncertainty, it is typically assumed that the attacker has perfect knowledge of the defender's randomized strategy or can learn the defender's strategy after conducting a fixed period of surveillance. In consideration of surveillance cost, these assumptions are clearly simplistic since attackers may act with partial knowledge of the defender's strategies and may dynamically decide whether to attack or conduct more surveillance. Security game models with limited observation (An et al., 2013; An, Kempe, Kiekintveld, Shieh, Singh, Tambe, & Vorobeychik, 2012) have been proposed in which the attacker either makes limited number of observations or dynamically determines a place to stop surveillance. Since the belief state space exponentially increases with observation length, it is still computationally challenging to solve large games in consideration of limited observation.

Bounded Rationality: One required research direction with respect to robustness is addressing bounded rationality of human adversaries, which is a fundamental problem that can affect the performance of our game-theoretic solutions. Recently, there has been some research on applying ideas (e.g., prospect theory (Kahneman & Tvesky, 1979) and quantal response (McKelvey & Palfrey, 1995)) from social science or behavioral game theory within security game algorithms (Nguyen, Yang, Azaria, Kraus, & Tambe, 2013; Pita, Jain, Tambe, Ordóñez, & Kraus, 2010; Yang, Jiang, Tambe, & Ordóñez, 2013; Yang, Kiekintveld, Ordóñez, Tambe, & John, 2011). Previous work usually applies existing frameworks and sets the parameters of these frameworks by experimental tuning or learning. However, in real-world security domains, we may have very limited data, or may have only limited information on the biases displayed by adversaries. Recently, monotonic maximin (Jiang et al., 2013) was proposed as a robust solution concept to Stackelberg security games with boundedly rational adversaries. It tries to optimize defender utility against the worst-case monotonic adversary behavior, where monotonicity is the property that actions with higher expected utility are played with higher probability. An open research challenge is to combine such robust-optimization approaches with available behavior data. Furthermore, since real-world human adversaries are sometimes distributed across coalitions of socially, culturally, and cognitively biased agents acting behind a veil of uncertainty, we may need significant interdisciplinary research to build in social, cultural, and cognitive biases into our adversary models.

Planning and Learning: The challenges of planning and learning pertain to new security domains such as patrolling metro systems and forests. In these domains, schedules might be interrupted due to some unexpected events (e.g., writing a citation in a train line or boarding an illegal fisherman's boat). As a consequence, incorporating planning within security games is necessary to model stochastic decision making. Furthermore, in the real world, a schedule's spatial and temporal constraints are typically continuous dimensions. Hence, more expressive models need to be derived to represent such schedules. So far, data available in most domains have been principally used to define the game's matrices. However, in the newer domains, the planned schedules are frequently interrupted. Hence, the information about the locations and times of interruptions, as well as the information about the actual interactions between attackers and defenders, can also be used. The former can be used to represent the uncertainty of the environment, whereas the latter can be used to learn the attacker's behavior. An online planning algorithm for the protector in resource conservation games is proposed to use the information gained by observing the extractor's actions (Qian, Haskell, Jiang, & Tambe, 2014). The resource conservation game is modeled as a repeated game and then is casted as a POMDP.

Integrating Dynamic Information: The Transportation Security Administration (TSA) is launching a new concept to address its current challenges by integrating and quantifying risk information to comprehensively analyze risk on a per-flight basis. Rather than using a one-size-fits-all approach, the idea is to dynamically tailor the security by integrating information across five threat vectors: passengers, checked baggage, cargo, aircraft operator, and airport/perimeter. This approach will provide a more comprehensive picture of total flight risk, enabling the TSA to reduce risk by allowing for more dynamic allocation of resources. This is a massive project that would radically impact domestic aviation security in the United States. The security games framework is being used as part of this project to quantitatively test and validate the concepts used in this new aviation security concept of the future. This may have a significant impact on how this new concept of dynamic security gets operationalized.

NOTE

1. We use the term *green security games* also to avoid any confusion that may come about given that terms related to the environment and security have been adopted for other uses. For example, the term *environmental security*, broadly speaking, refers to threats posed to humans due to environmental issues, e.g., climate change or shortage of food. The term *environmental criminology*, on the other hand, refers to analysis and understanding of how different environments affect crime.

REFERENCES

Albers, H. (2010). Spatial modeling of extraction and enforcement in developing country protected areas. *Resource and Energy Economics*, *32*(2), 165–179.

An, B., Brown, M., Vorobeychik, Y., & Tambe, M. (2013). Security games with surveillance cost and optimal timing of attack execution. In *Proceedings of the 12th International Conference on Autonomous Agents and Multiagent Systems (AAMAS)* (pp. 223–230).

An, B., Jain, M., Tambe, M., & Kiekintveld, C. (2011). Mixed-initiative optimization in security games: A preliminary report. In *Proceedings of the AAAI Spring Symposium on Help Me Help You: Bridging the Gaps in Human–Agent Collaboration* (pp. 8–11).

An, B., Kempe, D., Kiekintveld, C., Shieh, E., Singh, S., Tambe, M., & Vorobeychik, Y. (2012). Security games with limited surveillance. In *Proceedings of the 26th Conference on Artificial Intelligence (AAAI)* (pp. 1241–1248).

An, B., Ordóñez, F., Tambe, M., Shieh, E., Yang, R., Baldwin, C., DiRenzo, J., Moretti, K., Maule, B., & Meyer, G. (2013). A deployed quantal response-based patrol planning system for the U.S. Coast Guard. *Interfaces*, *43*(5), 400–420.

An, B., Pita, J., Shieh, E., Tambe, M., Kiekintveld, C., & Marecki, J. (2011, March). GUARDS and PROTECT: Next generation applications of security games. *SIGECOM*, *10*, 31–34.

An, B., Tambe, M., Ordóñez, F., Shieh, E., & Kiekintveld, C. (2011) Refinement of strong Stackelberg equilibria in security games. In *Proceedings of the 25th Conference on Artificial Intelligence* (pp. 587–593).

Basilico, N., Gatti, N., & Amigoni, F. (2009). Leader-follower strategies for robotic patrolling in environments with arbitrary topologies. In *Proceedings of the 8th International Conference on Autonomous Agents and Multiagent Systems (AAMAS)* (pp. 500–503).

Blocki, J., Christin, N., Datta, A., Procaccia, A. D., & Sinha, A. (2013). Audit games. In *Proceedings of the 23rd International Joint Conference on Artificial Intelligence* (pp. 41–47).

Blocki, J., Christin, N., Datta, A., Procaccia, A. D., & Sinha, A. (2015). Audit games with multiple defender resources. In *AAAI Conference on Artificial Intelligence (AAAI)* (pp. 791–797).

Fang, F., Jiang, A. X., & Tambe, M. (2013). Optimal patrol strategy for protecting moving targets with multiple mobile resources. In *Proceedings of the 12th International Conference on Autonomous Agents and Multiagent Systems (AAMAS)* (pp. 957–964).

Fang, F., Nguyen, T. H., Pickles, R., Lam, W. Y., Clements, G. R., An, B., Singh, A., Tambe, M., & Lemieux, A. (2016). Deploying PAWS: Field optimization of the protection assistant for wildlife security. In *IAAI* (pp. 3966–3973).

Fang, F., Stone, P., & Tambe, M. (2015). When security games go green: Designing defender strategies to prevent poaching and illegal fishing. In *Proceedings of the 24th International Joint Conference on Artificial Intelligence (IJCAI)* (pp. 2589–2595).

Fave, F. M., Brown, M., Zhang, C., Shieh, E., Jiang, A., Rosoff, H., Tambe, M., & Sullivan, J. (2014). Security games in the field: Deployments on a transit system. In F. Dalpiaz, J. Dix, & M. van Riemsdijk (Eds.), *Engineering multi-agent systems*, volume 8758 of *Lecture Notes in Computer Science*, pages 103–126.

Jain, M., Kardes, E., Kiekintveld, C., Ordóñez, F., & Tambe, M. (2010). Security games with arbitrary schedules: A branch and price approach. In *Proceedings of the 24th AAAI Conference on Artificial Intelligence* (pp. 792–797).

Jain, M., Korzhyk, D., Vanek, O., Pechoucek, M., Conitzer, V., & Tambe, M. (2011). A double oracle algorithm for zero-sum security games on graphs. In *Proceedings of the 10th International Conference on Autonomous Agents and Multiagent Systems (AAMAS)* (pp. 327–334).

Jain, M., Tambe, M., & Conitzer, V. (n.d.). Security scheduling for real-world networks. In *Proceedings of the 12th International Conference on Autonomous Agents and Multiagent Systems (AAMAS)* (pp. 215–222).

Jain, M., Tsai, J., Pita, J., Kiekintveld, C., Rathi, S., Tambe, M., & Ordóñez, F. (2010). Software assistants for randomized patrol planning for the LAX airport police and the federal air marshal service. *Interfaces*, *40*(4), 267–290.

Jiang, A. X., Nguyen, T. H., Tambe, M., & Procaccia, A. D. (2013). Monotonic maximin: A robust Stackelberg solution against boundedly rational followers. In *Proceedings of the Conference on Decision and Game Theory for Security (GameSec)* (pp. 119–139).

Jiang, A., Yin, Z., Kraus, S., Zhang, C., & Tambe, M. (n.d.) Game-theoretic randomization for security patrolling with dynamic execution uncertainty. In *Proceedings of the 12th International Conference on Autonomous Agents and Multiagent Systems (AAMAS)* (pp. 207–214).

Johnson, M., Fang, F., Yang, R., Tambe, M., & Albers, H. (2012). Patrolling to maximize pristine forest area. In *Proceedings of the AAAI Spring Symposium on Game Theory for Security, Sustainability and Health.*

Kahneman, D., & Tvesky, A. (1979). Prospect theory: An analysis of decision under risk. *Econometrica, 47*(2), 263–291.

Kiekintveld, C., Marecki, J., & Tambe, M. (2011). Approximation methods for infinite Bayesian Stackelberg games: Modeling distributional uncertainty. In *Proceedings of the 10th International Conference on Autonomous Agents and Multiagent Systems (AAMAS)* (pp. 1005–1012).

Korzhyk, D., Conitzer, V., & Parr, R. (2010). Complexity of computing optimal Stackelberg strategies in security resource allocation games. In *Proceedings of the 24th AAAI Conference on Artificial Intelligence* (pp. 805–810).

Kukreja, N., Halfond, W. G., & Tambe, M. (2013). Randomizing regression tests using game theory. In *Automated Software Engineering (ASE), 2013 IEEE/ACM 28th International Conference on Artificial Intelligence* (pp. 616–621).

Li, Y., & Conitzer, V. (2013). Game-theoretic question selection for tests. In *Proceedings of the 23rd International Joint Conference on Artificial Intelligence* (pp. 254–262).

Luber, S., Yin, Z., Fave, F. D., Jiang, A. X., Tambe, M., & Sullivan, J. P. (2013). Game-theoretic patrol strategies for transit systems: The trusts system and its mobile app (demonstration). In *International Conference on Autonomous Agents and Multiagent Systems (AAMAS)[Demonstrations Track]* (pp. 1377–1378).

McKelvey, R. D., & Palfrey, T. R. (1995). Quantal response equilibria for normal form games. *Games and Economic Behavior, 10*(1), 6–38.

Nguyen, T., Yadav, A., An, B., Tambe, M., & Boutilier, C. (2014). Regret-based optimization and preference elicitation for Stackelberg security games with uncertainty. In *Proceedings of the 28th International Joint Conference on Artificial Intelligence* (pp. 756–762).

Nguyen, T. H., Yang, R., Azaria, A., Kraus, S., & Tambe, M. (2013). Analyzing the effectiveness of adversary modeling in security games. In *Conference on Artificial Intelligence (AAAI)* (pp. 718–724).

Paruchuri, P., Pearce, J. P., Marecki, J., Tambe, M., Ordóñez, F., & Kraus, S. (2008). Playing games with security: An efficient exact algorithm for Bayesian Stackelberg games. In *Proceedings of the 7th International Conference on Autonomous Agents and Multiagent Systems (AAMAS)* (pp. 895–902).

Pita, J., Jain, M., Tambe, M., Ordóñez, F., & Kraus, S. (2010). Robust solutions to Stackelberg games: Addressing bounded rationality and limited observations in human cognition. *Artificial Intelligence, 174*(15) 1142–1171.

Pita, J., Jain, M., Western, C., Portway, C., Tambe, M., Ordóñez, F., Kraus, S., & Parachuri, P. (2008). Deployed ARMOR protection: The application of a game-theoretic model for security at the Los Angeles International Airport. In *Proceedings of the 7th International Conference on Autonomous Agents and Multiagent Systems (AAMAS)* (pp. 125–132).

Pita, J., Tambe, M., Kiekintveld, C., Cullen, S., & Steigerwald, E. (2011). GUARDS – game theoretic security allocation on a national scale. In *Proceedings of the 10th International Conference on Autonomous Agents and Multiagent Systems (AAMAS)* (pp. 37–44).

Qian, Y., Haskell, W. B., Jiang, A. X., & Tambe, M. (2014). Online planning for optimal protector strategies in resource conservation games. In *Proceedings of the 2014 International Conference on Autonomous Agents and Multiagent Systems* (pp. 733–740).

Shieh, E., An, B., Yang, R., Tambe, M., Baldwin, C., DiRenzo, J., Maule, B., & Meyer, G. (2012). PROTECT: An application of computational game theory for the security of the ports of the United States. In *Proceedings of the 26th AAAI Conference on Artificial Intelligence (AAAI)* (pp. 2173–2179).

Short, M. B., D'Orsogna, M. R., Pasour, V. B., Tita, G. E., Brantingham, P. J., Bertozzi, A. L., & Chayes, L. B. (2008). A statistical model of criminal behavior. *Mathematical Models and Methods in Applied Sciences*, *18*, 1249–1267.

Tambe, M., & An, B. (2012). Game theory for security: A real-world challenge problem for multiagent systems and beyond. In *Proceedings of the AAAI Spring Symposium on Game Theory for Security, Sustainability and Health* (pp. 69–74).

Tsai, J., Rathi, S., Kiekintveld, C., Ordóñez, F., & Tambe, M. (2009). IRIS: A tool for strategic security allocation in transportation networks. In *Proceedings of the 8th International Conference on Autonomous Agents and Multiagent Systems (AAMAS)* (pp. 37–44).

Varakantham, P., Lau, H. C., & Yuan, Z. (2013). Scalable randomized patrolling for securing rapid transit networks. In *Conference on Innovative Applications of Artificial Intelligence (IAAI)* (pp. 1563–1568).

von Stengel, B., & Zamir, S. (2004). Leadership with commitment to mixed strategies. Technical Report LSE-CDAM-2004-01, Centre for Discrete and Applicable Mathematics, London School of Economics and Political Science, London, UK.

Yang, R., Ford, B. J., Tambe, M., & Lemieux, A. (2014). Adaptive resource allocation for wildlife protection against illegal poachers. In *Proceedings of the 11th International Conference on Autonomous Agents and Multiagent Systems (AAMAS)* (pp. 453–460).

Yang, R., Jiang, A. X., Tambe, M., & Ordóñez, F. (2013). Scaling-up security games with boundedly rational adversaries: A cutting-plane approach. In *Proceedings of the 22nd International Joint Conference on Artificial Intelligence (IJCAI)* (pp. 404–410).

Yang, R., Kiekintveld, C., Ordóñez, F., Tambe, M., & John, R. (2011). Improving resource allocation strategy against human adversaries in security games. In *Proceedings of the 22nd International Joint Conference on Artificial Intelligence (IJCAI)* (pp. 458–464).

Yin, Y., An, B., & Jain, M. (2014). Game-theoretic resource allocation for protecting large public events. In *Proceedings of the 28th Conference on Artificial Intelligence (AAAI)* (pp. 826–834).

Yin, Y., Jain, M., Tambe, M., & Ordóñez, F. (2011). Risk-averse strategies for security games with execution and observational uncertainty. In *Proceedings of the 25th AAAI Conference on Artificial Intelligence (AAAI)* (pp. 758–763).

Yin, Y., Xu, H., Gan, J., An, B., & Jiang, A. (2015). Computing optimal mixed strategies for security games with dynamic payoffs. In *Proceedings of the 24th International Joint Conference on Artificial Intelligence (IJCAI)* (pp. 681–688).

Yin, Z., Korzhyk, D., Kiekintveld, C., Conitzer, V., & Tambe, M (2010). Stackelberg vs. Nash in security games: interchangeability, equivalence, and uniqueness. In *Proceedings of the 9th International Conference on Autonomous Agents and Multiagent Systems (AAMAS)* (pp. 1139–1146).

Yin, Z., & Tambe, M. (2012). A unified method for handling discrete and continuous uncertainty in Bayesian Stackelberg games. In *Proceedings of the 11th International Conference on Autonomous Agents and Multiagent Systems (AAMAS)* (pp. 855–862).

Zhang, C., Jiang, A. X., Short, M. B., Brantingham, P. J., & Tambe, M. (2014). Defending against opportunistic criminals: new game-theoretic frameworks and algorithms. In *Decision and game theory for security* (pp. 3–22). Springer.

22

Basic Solution Concepts and Algorithms for Stackelberg Security Games

Christopher Kiekintveld and Manish Jain

22.1 Introduction

As described in other chapters of this book, security games have become a widely used paradigm for optimizing security decisions for many different types of domains (Tambe, 2011). All of these applications are based on analysis performed using formal game-theoretic models of the underlying interactions between attackers and defenders. The scale and complexity of these models demands that we use computational methods to analyze them, and a great deal of research has focused on developing more efficient and scalable algorithms for different types of security games.

In this chapter, we will introduce some of the most basic models of security games, and overview some of the solution algorithms that have been developed to solve them. We start with the most fundamental model of the task of allocating limited security resources against an intelligent attacker, and present a basic mathematical programming formulation[1] for calculating the optimal defender strategy and attacker response. This model captures the inherent trade-off between allocating resources to protect high-value targets, while at the same time using randomization so that the policy is not predictable and easily exploited by an attacker.

We also present three types of extensions to the basic model, and several algorithms for these cases. The first extension introduces a compact representation that is suitable for many different types of security games more efficiently than standard models. This approach represents the payoff functions based on the values for a set of targets, and using marginal probabilities for assigning resources to each target to represent the defender's strategy space (marginals are the total probability of covering each target using any resource). The simplest versions of security games using compact representations can be solved using very fast (polynomial-time) algorithms.

Next, we introduce two extensions that model different types of uncertainty that the defender may have about the attacker. The first uses a Bayesian framework that represents uncertainty about different types of possible attackers using probability distributions. The second uses intervals to represent uncertainty about the possible range of attacker payoff values, rather than specifying exact values. Finally, we introduce extensions to the framework that model different types of defender resources, and scheduling constraints on those resources.

Later chapters in this book will discuss more complex models of security games, and more sophisticated solution algorithms that account for many different assumptions, including complex strategy spaces, different forms of uncertainty, human opponents, limitations on observation, repeated interactions, and so forth. However, most of these more complex models build on the foundational models and solution concepts introduced in this chapter.

22.2 Basic Security Games

Game-theoretic models are formal mathematical descriptions of interactions between multiple decision makers; these interactions are called games (Fudenberg & Tirole, 1991; Osbourne & Rubinstein, 1994). The most basic type of game is a *normal-form game* (NFG) (also called *strategic form*). An NFG defines the set of players, a set of actions for each player, and the payoffs (or utilities) for each player for every possible outcome, where an outcome is a vector specifying a single action chosen by each player.

One common way to represent small games visually is using a *payoff matrix*. Figure 22.1 shows an example of a payoff matrix for a very simple security game. The game has two players, the *defender* and the *attacker.* The defender has one resource available (e.g., a police officer) that can be deployed to protect either terminal 1 or terminal 2 at an airport. The attacker has the choice of attacking either terminal 1 or terminal 2, and does not know which terminal is protected and which one is not. The numbers in each cell represent the payoff value for each player, with the defender's payoff listed first and the attacker's listed second. The idea of the game is that the attacker is trying to avoid the defender, while the defender is trying to go to the same location as the attacker to prevent the attack. The magnitudes of the values for terminal 1 are generally higher than those for terminal 2, indicating that this terminal is more important.

If we fix the strategy played by one player in this game, we can define the *best response* for the opponent to be the strategy that results in the maximum payoff against the fixed strategy. For example, if the police choose to protect terminal 1, the best response for the adversary is to attack terminal 2, resulting

A Simple Security Game

Attacker (columns) / **Defender** (rows)

	Terminal #1	Terminal #2
Terminal #1	5, –3	–1, 1
Terminal #2	–3, 5	2, –1

Figure 22.1. An example of a very simple security game.

in payoffs of −1 and 1 for the police and the attacker, respectively. One of the fundamental solution concepts in game theory is *Nash equilibrium.* A strategy profile (i.e., a strategy for each player) is a Nash equilibrium if every player is playing a best response to the strategies of the other players.

Game theory uses the notion of a *mixed strategy* to include the possibility that players can randomize their strategy choices, which is important in adversarial games. A mixed strategy assigns a probability for playing each of a player's individual "pure" strategies. For example, here the police could go to each terminal 50% of the time. The payoffs are extended to mixed strategies using the expected value, and the concepts of best response and Nash equilibrium can also be generalized to mixed strategies in the straightforward way.

The standard assumption for normal-form games is that players move simultaneously, without knowledge of the other player's strategy before they choose their own strategy. Simultaneous move games (and the corresponding concept of Nash equilibrium) is based on this assumption. However, in security domains, attackers often have the capability to observe and learn about the defensive strategy before they plan an attack. We can model this by using a variation of the Nash equilibrium concept called *Stackelberg equilibrium.* In a Stackelberg game, one player (in our case the defender) moves first and commits to playing a specific mixed strategy. The other player (the attacker) observes this strategy and can choose to play a best response to the known strategy.

22.2.1 *Stackelberg Security Games: Formal Definition*

We now formally define Stackelberg security games (SSG) (Kiekintveld, Jain, Tsai, Pita, Tambe, & Ordóñez, 2009; Paruchuri, Pearce, Marecki, Tambe, Ordóñez, & Kraus, 2008). An SSG has two players, a defender, Θ, and an attacker, Ψ. These players need not represent individuals, but could also be groups that cooperate to execute a joint strategy, such as a police force or terrorist organization. Each player has a set of possible *pure strategies*, denoted $\sigma_\Theta \in \Sigma_\Theta$ and $\sigma_\Psi \in \Sigma_\Psi$. A *mixed strategy* allows a player to play a probability distribution over pure strategies, denoted $\delta_\Theta \in \Delta_\Theta$ and $\delta_\Psi \in \Delta_\Psi$. Payoffs for each player are defined over all possible joint pure-strategy outcomes: $\Omega_\Theta: \Sigma_\Psi \times \Sigma_\Theta \rightarrow R$ for the defender and similarly for each attacker. The payoff functions are extended to mixed strategies in the standard way by taking the expectation over pure-strategy outcomes. The attacker can observe the defender's strategy and choose an optimal strategy in response, so the attacker's best response is a function that maps from defender strategies into attacker strategies: $F_\Psi: \Delta_\Theta \rightarrow \Delta_\Psi$.

Stackelberg equilibrium is a variation of Nash equilibrium for Stackelberg games. There are two types of Stackelberg equilibria that identify unique outcomes, first proposed by Leitmann (1978), and typically called "strong" and "weak" after Breton and colleagues (1988). The strong version assumes that the attacker will break ties in favor of the defender, while the weak form assumes that the attacker will break ties to minimize the payoff for the defender. Strong Stackelberg equilibria (SSE) are known to exist in all Stackelberg games (Basar & Olsder, 1995), while weak Stackelberg equilibria may not exist. There is also a formal argument for why strong equilibria are more useful. The defender can induce the attacker to play the SSE by playing a strategy arbitrarily close to the SSE such that the attacker strictly prefers this solution to any alternative (von Stengel & Zamir, 2004). We adopt strong Stackelberg equilibrium as our solution concept in part for these reasons, and also because it is the most commonly used in related literature (Osbourne & Rubinstein, 1994; Paruchuri et al., 2008).

Definition 1 *A set of strategies* $(\delta_\Theta, F\Psi)$ *forms a* strong Stackelberg equilibrium *(SSE) if it satisfies the following:*

1. *The leader plays a best response:*
   $\Omega_\Theta(\delta_\Theta, F_\Psi(\delta_\Theta)) \geq \Omega_\Theta(\delta'_\Theta, F_\Psi(\delta'_\Theta)) \forall \delta'_\Theta \in \Delta_\Theta$.
2. *The follower plays a best response:*
   $\Omega_\Psi(\delta_\Theta, F_\Psi(\delta_\Theta)) \geq \Omega_\Psi(\delta_\Theta, \delta_\Psi) \forall \delta_\Theta \in \Delta_\Theta, \delta_\Psi \in \Delta_\Psi$.
3. *The follower breaks ties optimally for the leader:*
   $\Omega_\Theta(\delta_\Theta, F_\Theta(\delta_\Theta)) \geq \Omega_\Theta(\delta_\Theta, \delta_\Psi) \forall \delta_\Theta \in \Delta_\Theta, \delta_\Psi \in \Delta^*_\Psi(\delta_\Theta)$, *where*
   $\Delta^*_\Psi(\delta_\Theta)$ *is the set of follower best responses, as earlier.*

Whether the Stackelberg leader benefits from the ability to commit depends on whether commitment to mixed strategies is allowed. Committing to a pure strategy can be either good or bad for the leader; for example, in the "Rock,

Paper, Scissors" game, forcing commitment to a pure strategy would guarantee a loss. However, the ability to commit to a mixed strategy always weakly increases the leader's payoffs in equilibrium profiles (von Stengel & Zamir, 2004). In the context of a Stackelberg security game, a deterministic policy is a liability for the defender (the leader), but a credible randomized security policy is an advantage. Our model allows commitment to mixed strategies by the defender.

22.2.2 *Solving Stackelberg Security Games*

One of the simplest approaches for solving basic Stackelberg security games is to apply the *multiple-LPs* method (which was first introduced by Contizer and colleagues (2006) for solving general Stackelberg games). This approach uses linear programming (LP) to specify the problem of finding the optimal strategy for the defender to play as a series of optimization problems. These optimization problems can be solved using general-purpose optimization software such as CPLEX and Gurobi.

The multiple-LPs approach uses one linear program for each attacker pure strategy. The LP finds the optimal mixed strategy for the defender, assuming that the attacker plays according to the specified strategy. The optimal strategy for the defender is found by selecting the best strategy returned by any of the individual LPs. Each linear program can be solved in polynomial time in the size of the game matrix, and the number of LPs to solve is equal to the number of attacker strategies. Therefore, the overall algorithm runs in polynomial time. The LP formulation is as follows, where $p_\Theta(\sigma_\Theta)$ denotes the probability of playing action σ_Θ in the defender's mixed strategy.

$$\max_{\sigma_\Psi \in \Sigma_\Psi} \max_{\delta_\Theta} \sum_{\sigma_\Theta \in \Sigma_\Theta} p_\Theta(\sigma_\Theta) \cdot \Omega_\Theta(\sigma_\Theta, \sigma_\Psi) \tag{22.1}$$

$$s.t. \sum_{\sigma_\Theta \in \Sigma_\Theta} p_\Theta(\sigma_\Theta) \cdot \Omega_\Psi(\sigma_\Theta, \sigma_\Psi) \geq \sum_{\sigma_\Theta \in \Sigma_\Theta} p_\Theta(\sigma_\Theta) \cdot \Omega_\Psi(\sigma_\Theta, \sigma'_\Psi) \forall \sigma'_\Psi \in \Sigma_\Psi \tag{22.2}$$

$$\sum_{\sigma_\Theta \in \Sigma_\Theta} p_\Theta(\sigma_\Theta) = 1 \tag{22.3}$$

$$p_\Theta(\sigma_\Theta) \geq 0 \forall \sigma_\Theta \in \Sigma_\Theta \tag{22.4}$$

22.2.3 *Compact Stackelberg Security Games*

The base SSG model specifies that each player has a set of actions from which to select. These actions can represent any type of attack/defense action,

Table 22.1. *Example payoffs for an attack on a target*

	Covered	Uncovered
Defender	5	−20
Attacker	−10	30

including arbitrarily complex plans. However, for some types of scenarios, this can lead to extremely large (exponential) numbers of pure strategies. For example, if the defender's actions represent schedules for a large number of defensive resources or sequential patrolling plans with sequences of actions, the number of pure strategies is exponentially large.

In many interesting cases, we can use a compact representation of security games that captures the important characteristics of the game while avoiding the exponential size of the basic model. We now introduce a compact representation for security games for scheduling multiple defender resources. Let $T = \{t_1, \ldots, t_n\}$ be a set of *targets* that may be attacked, which also corresponds to the set of pure strategies for the attacker. The defender has a set of resources available to *cover* these targets, labeled $R = \{r_1, \ldots, r_m\}$. For now we assume that these resources are identical, and every resource can be assigned to protect any target. We also assume that these resources are *fully protective*, so that assigning just one resource to a target provides the maximum possible protecting and there is no benefit to assigning additional resources.

We define four payoffs for each target to specify the possible outcomes for an attack on the target, as shown in Table 22.1. There are two cases, depending on whether the target is covered by the defender. The defender's payoff for an uncovered attack denoted by $U_\Theta^u(t)$, and for a covered attack $U_\Theta^c(t)$. Similarly, $U_\Psi^u(t)$ and $U_\Psi^c(t)$ are the payoffs of the attacker. In case of security games, $U_\Theta^c(t) > U_\Theta^u(t)$ (it is always favorable to the defender to cover a target). Similarly, $U_\Psi^u(t) > U_\Psi^c(t)$ (it is always better for the attacker if the attacked target is not protected).

A crucial feature of the model is that payoffs depend only on the target attacked, and whether it is covered by the defender. The payoffs do *not* depend on the remaining aspects of the schedule, such as whether any un-attacked target is covered or which specific defense resource provides coverage. For example, if an adversary succeeds in attacking terminal 1, the penalty for the defender is the same whether the defender was guarding terminal 2 or 3. Therefore, from a payoff perspective, many resource allocations by the defender are identical. We exploit this by summarizing the payoff-relevant aspects of the defender's strategy in a *coverage vector*, C, that gives the probability that each target is covered, c_t. We can now represent the defender's strategy using the coverage vector, which has linear size with respect to the number of targets. The number of payoffs needed to specify the game is also linear in the number of targets. The attacker's strategies can be restricted to pure strategies specifying which target is attacked without loss of generality because a strong Stackelberg equilibrium (SSE) solution still exists under this restriction (Paruchuri et al., 2008). We can calculate the expected payoffs for both players using only this compact representation, so it can also be used in solution algorithms.

22.2.4 *Solving Compact SSG*

Several polynomial-time algorithms have been developed for special cases of compact SSG. In particular, these algorithms assume that all defender resources are identical (i.e., it does not matter which resource is used to protect a specific target, unconstrained (i.e., any resource can be used to protect any target), and fully protective (i.e., assigning more than one resource to a target does not improve the protection). The first polynomial algorithm for this case was the ORIGAMI method introduced by Kiekintveld and colleagues (2009), which can be implemented directly or formulated as a mathematical program. ORIGAMI has $O(n^2)$ worst-case complexity. An improved algorithm presented in Lerma and colleagues (2011) can solve the same class of problems as ORIGAMI in $O(n \cdot \log(n))$ time. A simplified version of this algorithm runs in linear time, $O(n)$ for games with the additional properties that the attacker's payoff for attacking a protected target is 0 (i.e., $U_{\Psi}^{c}(t) = 0$), and that no target is covered with probability 1. This algorithm is known as LTRA for Linear-Time Resource Allocation.

Problem Analysis

We present here the analysis and algorithm for LTRA since it is the fastest and simplest known method, and it provides most of the intuition for the generalized algorithm (which is described in detail in Lerma, Kreinovich, & Kiekintveld, 2011). The original analysis for ORIGAMI showed that for any Stackelberg security game with identical, unconstrained resources that cover a single target and a payoff structure such that $U_{\Theta}^{u}(t) < U_{\Theta}^{c}(t)$ and $U_{\Psi}^{u}(t) > U_{\Psi}^{c}(t)$ for all targets t, we can find an SSE solution without considering the defender's payoffs. To do so, we need to find a coverage vector that (1) minimizes the maximum expected payoff for the attacker and (2) maximizes the number of targets that have the same maximum payoff for the attacker. If we define the set of targets that are best responses for the attacker given some coverage vector C as the *attack set*, then condition 2 states that we are looking for the largest possible attack set, given that we have already minimized the attacker payoff for targets within this set. Recall that in an SSE, that attacker will break ties in favor of the defender, so the defender will receive the best possible payoff for any target in the attack set. Therefore, we do not need to account for the defender's payoffs directly; they are maximized implicitly by minimizing the attacker's payoffs and maximizing the size of the attack set. We note that these conditions are not sufficient for SSE in cases where the specified payoff structure does not hold, if defender resources cover multiple targets or have more complex scheduling constraints.

We now further analyze the problem, leading to a linear-time algorithm for finding SSE. The attacker seeks to maximize the expected value of a successful attack:

$$\arg\max_i (1 - c_i) \cdot U_\Psi^u(t_i), \tag{22.5}$$

while the defender chooses a coverage vector to minimize the attacker's expected payoff. Let t_{i_o} denote the optimal target to attack, so we have for every target t_i:

$$(1 - c_{i_o}) \cdot U_\Psi^u(t_{i_o}) \geq (1 - c_i) \cdot U_\Psi^u(t_i). \tag{22.6}$$

Now, assume that for some i this inequality is strict and that $c_i > 0$. In this case, we could decrease c_i and increase the probability c_j for all j such that

$$(1 - c_j) \cdot U_\Psi^u(t_j) = (1 - c_{i_o}) \cdot U_\Psi^u(t_{i_o}), \tag{22.7}$$

thus decreasing the expected payoff of the attacker.

Therefore, for the minimizing coverage vector, all targets can be divided into two groups:

- either the expected value for attacking the target is equal to the optimal value,
- or the expected value is less than the optimal value and the coverage probability assigned to the target is 0.

In other words, the optimal solution will have the property that the attacker's expected value for all targets with positive coverage probability is equal to some constant q:

$$(1 - c_i) \cdot U_\Psi^u(t_i) = q. \tag{22.8}$$

For any target t_i with $c_i > 0$ we can calculate the necessary value of c_i as:

$$c_i = 1 - \frac{q}{U_\Psi^u(t_i)}. \tag{22.9}$$

For all other targets $U_\Psi^u(t_i) < q$, and therefore

$$1 - \frac{q}{U_\Psi^u(t_i)} < 0. \tag{22.10}$$

Summarizing: once we know q, we can find all the probabilities c_i by using the formula

$$c_i = \max\left(1 - \frac{q}{U_\Psi^u(t_i)}, 0\right). \tag{22.11}$$

For each target t_i, this formula requires a constant number of computational steps. After q is computed, we can compute all the probabilities c_i in time $O(n)$.

We have shown that to find an optimal coverage vector, it is sufficient to find the constant q, which can be used to directly compute the coverage probabilities for each target. This constant can be found from the condition that $\sum_{i=1}^{n} c_i = m$, i.e., that

$$\sum_{i=1}^{n} \max\left(1 - \frac{q}{U_\Psi^u(t_i)}, 0\right) = m. \tag{22.12}$$

The left-hand side of this equality decreases as q increases. So, if for some q, the resulting sum is smaller than m, this means that the optimal value q_o is smaller than q: $q_o < q$; similarly, if for some q, the resulting sum is larger than m, this means that the optimal value q_o is larger than q: $q_o > q$.

The structure of the optimal covering vector can be clarified if we sort the targets in order of descending attacker payoffs for successful attacks, so that:

$$U_\Psi^u\left(t_{(1)}\right) \geq \cdots \geq U_\Psi^u\left(t_{(n-1)}\right) \geq U_\Psi^u\left(t_{(n)}\right). \tag{22.13}$$

We can also add

$$U_\Psi^u\left(t_{(0)}\right) \stackrel{\text{def}}{=} +\infty \text{ and } U_\Psi^u\left(t_{(n+1)}\right) \stackrel{\text{def}}{=} 0, \text{ then}$$

$$U_\Psi^u\left(t_{(0)}\right) \geq \cdots \geq U_\Psi^u\left(t_{(n)}\right) \geq U_\Psi^u\left(t_{(n+1)}\right). \tag{22.14}$$

The values $U_\Psi^u\left(t_{(i)}\right)$ divide the real line into intervals $\left[U_\Psi^u\left(t_{(i+1)}\right), U_\Psi^u\left(t_{(i)}\right)\right]$, so the threshold constant q must be in one of these intervals, i.e., between $U_\Psi^u\left(t_{(k+1)}\right)$ and $U_\Psi^u\left(t_{(k)}\right)$ for some k. In this case, according to the above formula for c_i, all targets with a value greater than q (i.e., the targets $t_{(1)}, t_{(2)}, \ldots, t_{(k)}$ in the above ordering) will be protected with positive probability, and all targets with value smaller than q (i.e., targets $t_{(k+1)}, t_{(k+2)}, \ldots$) are left unprotected. Let k denote the index of last target that has positive probability. Given the constraint that the coverage probabilities add to m, we can write:

$$\sum_{i=1}^{k}\left(1-\frac{q}{U_\Psi^u\left(t_{(i)}\right)}\right)=m, \tag{22.15}$$

$$k-m=q\cdot\sum_{i=1}^{k}\frac{1}{U_\Psi^u\left(t_{(i)}\right)}, \tag{22.16}$$

$$q=\frac{k-m}{\sum_{i=1}^{k}\frac{1}{U_\Psi^u\left(t_{(i)}\right)}}. \tag{22.17}$$

So, instead of selecting q, we can simply select a threshold value k. Once we have found this k, we can then compute the threshold value q by using the above formula and then use this q to find the optimal coverage probabilities.

For the optimal value $k = k_o$, the corresponding value q is located in the interval $\left[U_\Psi^u\left(t_{(k+1)}\right), U_\Psi^u\left(t_{(k)}\right)\right]$. If for some k, the value q computed by the above formula is smaller than $U_\Psi^u\left(t_{(k+1)}\right)$, this means that we are trying to cover too few targets; with the same q, we can cover more, so the optimal value k_o should be larger: $k > k_o$.

Similarly, if for some k, the value q computed by the above formula is larger than $U_\Psi^u\left(t_{(k)}\right)$, this means that we are trying to cover too many targets, so the optimal value k_o should be smaller: $k_o < k$.

LTRA Algorithm

We now present a linear-time algorithm called Linear-Time Resource Allocation (LTRA) (Lerma et al., 2011) for finding optimal SSE solutions based on the analysis in the previous section. The algorithm uses bisection to find the value of the constant q, which is used to calculate the coverage for each target. The algorithm maintains three lists of targets:

- the list T^c of the targets t_i that we are sure will be covered with a positive probability $c_i > 0$;
- the list T^u of the targets t_i that we are sure will not be covered ($c_i = 0$);
- the list $T^?$ of the targets t_i that we do not yet know whether they will be covered.

In the beginning, we set $T^c = T^u = \emptyset$ and

$$T^? = \{t_1, t_2, \ldots, t_n\}. \tag{22.18}$$

At each stage, we update the value

$$S^c = \sum_{t_i \in T^c} \frac{1}{U_\Psi^u(t_i)}. \tag{22.19}$$

In the beginning, since $T^c = \emptyset$, we take $S^c = 0$. At each iteration, we do the following:

- First, we compute the median d of the values $U_\Psi^u(t_i)$ corresponding to all "undecided" targets $t_i \in T^?$.
- Considering the elements of the undecided set $T^?$ we divide them into two subsets

$$T^+ = \{t_i : U_\Psi^u(t_i) \geq d\}, \quad T^- = \{t_i : U_\Psi^u(t_i) < d\}. \tag{22.20}$$

  In the set T^+, we find the target t_k+ with the smallest value of $U_\Psi^u(t_i)$; in the set T^-, we find the target t_k- with the largest value of $U_\Psi^u(t_i)$.

- We then compute

$$S^+ = \sum_{t_i \in T^+} \frac{1}{U_\Psi^u(t_i)},$$
$$s^+ = S^c + S^+, \text{ and } q = \frac{k-d}{s^+}. \tag{22.21}$$

- If $q < U_\Psi^u(t_{k^-})$, then, as we have argued in our analysis, this means that we are trying to cover too few targets, so definitely all the elements from the set T^+ should be covered. Thus, we replace T^c with $T^c \cup T^+$, $T^?$ with T^-, and S^c with s^+.
- If $q > U_\Psi^u(t_{k^+})$, this means that we are trying to cover too many targets, so definitely all the elements from the set T^- should not be covered. Thus, we replace T^u with $T^u \cup T^-$ and $T^?$ with T^+ (and keep S^c unchanged).
- Finally, if $U_\Psi^u(t_{k^-}) \leq q \leq U_\Psi^u(t_{k^+})$, this means that this q is optimal.

The algorithm iterates until it finds the optimal value q. Once we get the optimal value q, we can find the optimal coverage probabilities as $c_i = \max\left(1 - \frac{q}{U_\Psi^u(t_i)}, 0\right)$.

In each iteration we can compute the median in linear time (Cormen, Leiserson, Rivest, & Stein, 2009), and all other operations with the set $T^?$ also take time $\mathcal{T}$ linear in the number of elements $|T^?|$ of this set $T^?$: $\mathrm{T} \leq C \cdot |T^?|$ for some C. We start with the set $T^?$ of size n. On the next iteration, we have a set of size $n/2$, then $n/4$, etc. Thus, the overall computation time is $\leq C \cdot (n + n/2 + n/4 + \cdots) \leq C \cdot 2n$, i.e., linear in n.

22.3 Security Games with Uncertainty

We now discuss extensions of the basic SSG model that allow us to model uncertainty about the attacker in two different ways. The first model is a Bayesian game, where the defender has uncertainty about the type of the attacker. *Type of an attacker* refers to a situation where there indeed may be attackers with different preferences (i.e., payoff matrices), or we choose to model an attacker with multiple payoff matrices. In the second model, the defender has uncertainty about the attacker's payoffs that is expressed using intervals.

22.3.1 *Bayesian Stackelberg Security Games*

The first method for modeling uncertainty about the attacker in security games is based on Bayesian games. We assume that there can be a (finite) set of possible attacker types, and for each type there is a different matrix of payoff values for both the attacker and defender. In addition, there is a known probability distribution over the possible attacker types. This form of Bayesian Stackelberg security game is particularly useful for representing uncertainty about different categories of possible attackers, such as a sophisticated terrorist organization or a local criminal gang.

Formally, the set of attacker types is denoted by Γ. Each type γ is represented by a different payoff matrix. The leader does not know the follower's type. The goal is to *find the optimal mixed strategy* for the leader to commit to, given that each follower type will know the mixed strategy of the leader when choosing its own strategy. Payoffs for each type are defined as before over all possible joint pure-strategy outcomes: $\Omega_\Theta : \Sigma_\Psi^\Gamma \times \Sigma_\Theta \to \mathcal{R}$ for the defender and similarly for each attacker type. The leader's best response is now a weighted best response to the followers' responses, where the weights are based on the probability of occurrence of each type. The strategy of each attacker type γ becomes: $F_{\Psi}^{\gamma} : \Delta_{\Theta} \rightarrow \Delta_{\Psi}^{\gamma}$, which still satisfies constraints 2 and 3 in Definition 1.

Bayesian SSG can be solved using the multiple-LPs method, but this requires performing a Harsanyi transformation (Harsanyi & Selten, 1972) that can result in a game that is exponentially larger than the original Bayesian game. DOBSS is a mixed-integer programming (MIP) formulation of the defender's optimization problem for a Bayesian SSG that works directly on the Bayesian game, avoiding the Harsanyi transformation (Paruchuri et al., 2008). Therefore, it is much faster in practice than using multiple-LPs for this problem. The formulation of the MIP is presented next. In the formulation, $q_{\sigma\Psi}^{\gamma}$ is a binary variable representing whether an attacker of type γ chooses action $\sigma\Psi$. In addition, p^{γ} is the probability that the attacker is of type γ, and M is an arbitrarily large constant.

$$\max_{q_{\sigma\Psi}^{\gamma}} \max_{\delta_{\Theta}} \sum_{\sigma_{\Theta}\in\Sigma_{\Theta}} \sum_{\gamma\in\Gamma} \sum_{\sigma_{\Psi}\in\Sigma_{\Psi}} p_{\Theta}(\sigma_{\Theta}) \cdot \Omega_{\Theta}\left(\sigma_{\Theta}^{\gamma}, \sigma_{\Psi}^{\gamma}\right) \cdot q_{\sigma_{\Psi}}^{\gamma} \cdot p^{\gamma} \tag{22.22}$$

$$0 \leq a^{\gamma} - \sum_{\sigma_{\Theta}\in\Sigma_{\Theta}} p_{\Theta}(\sigma_{\Theta}) \cdot \Omega_{\Psi}\left(\sigma_{\Theta}^{\gamma}, \sigma_{\Psi}^{\gamma}\right) \leq \left(1 - q_{\sigma_{\Psi}}^{\gamma}\right) \cdot M \qquad \forall \gamma \in \Gamma, \sigma_{\Psi} \in \Sigma_{\Psi} \tag{22.23}$$

$$\sum_{\sigma_{\Theta}\in\Sigma_{\Theta}} p_{\Theta}(\sigma_{\Theta}) = 1 \tag{22.24}$$

$$\sum_{\sigma_{\Psi}\in\Sigma_{\Psi}} q_{\sigma_{\Psi}}^{\gamma} = 1 \quad \forall \gamma \in \Gamma \tag{22.25}$$

$$p_{\Theta}(\sigma_{\Theta}) \geq 0 \quad \forall \sigma_{\Theta} \in \Sigma_{\Theta} \tag{22.26}$$

$$q_{\sigma\Psi}^{\gamma} \in \{0,1\} \quad \forall \gamma \in \Gamma, \forall \sigma_{\Psi} \in \Sigma_{\Psi} \tag{22.27}$$

Another faster algorithm for solving such general Bayesian Stackelberg games called HBGS has also been proposed by Jain and colleagues (2011); however, it builds on the key insights provided by DOBSS. A faster algorithm called ERASER for solving compact representations for Bayesian Stackelberg security games based on marginals has also been proposed (Kiekintveld et al., 2009); we shall discuss an extension of ERASER in the next section when talking about scheduling constraints. Similarly, Yin and colleagues provided a unified approach to handle uncertainty not only over discrete follower types, but also other key continuously distributed real-world uncertainty, due to leader's execution error, the follower's observation error, and continuous payoff uncertainty (Yin & Tambe, 2012). We now discuss models where uncertainty is specified as an interval.

22.3.2 *Interval Security Games*

In an interval security game (ISG) (Kiekintveld, Islam, & Kreinovich, 2013), we model uncertainty about the attacker using intervals instead of probability distributions. The solution concept takes a worst-case approach to the uncertainty, rather than maximizing expected value. Specifically, the defender knows only that the attacker's payoffs are in some interval of possible values, and tries to maximize the worst-case outcome for any realization of payoffs consistent with these intervals. Modeling uncertainty using intervals is common in robust optimization (Ben-Tal & Nemirovski, 2002), and this idea has also been used to develop a notion of equilibrium in game theory based on robust optimization (Aghassi & Bertsimas, 2006). The interval-based approach has some advantages over Bayesian methods for modeling uncertainty. It is simpler for domain experts to understand and specify a model based on interval uncertainty because the model does not require eliciting detailed information about probability distributions. In many cases, an interval model is a more natural and effective way to represent the game. Interval models also have considerable computational advantages over Bayesian models (Kiekintveld et al., 2013). While Bayesian models are NP-hard, we introduce a polynomial-time approximation algorithm for interval representations that provides tight bounds on solution quality. For large security games, our algorithm may be the only computationally feasible approach for handling uncertainty.

ISG Model

We now formalize the *interval security games* (ISG) model, which extends the standard security game model so that the defender has uncertainty about the attacker's payoffs that is represented using intervals. We still assume that both the attacker and defender know their *own* payoffs with certainty. In addition, we do not need to model the attacker as having uncertainty about the defender's payoffs because the attacker is able to directly observe the strategy of the defender, and therefore does not require knowledge of the payoffs to predict the defender's strategy.

In the model, rather than having a single value representing the attacker's payoffs for the two cases of a successful and unsuccessful attack $\left(U_{\Psi}^{u}(t)\right.$ and $\left.U_{\Psi}^{c}(t)\right)$, we have pairs of values that represent the maximum and minimum possible payoffs for a successful or unsuccessful attack on target t_i. We denote these values using the notation $U_{\Psi}^{u,max}(t)$ and $U_{\Psi}^{u,min}(t)$ for the uncovered case, and $U_{\Psi}^{c,max}(t)$ and $U_{\Psi}^{c,min}(t)$ for the covered case. The idea is that the defender knows only that the attacker's payoffs lie within a range of possible values, and not the precise value. The defender does not have information about the distribution of payoffs within these intervals, and therefore cannot compute an expected payoff. Therefore, we cannot apply the concept of strong Stackelberg equilibrium. Instead, we follow the literature on robust optimization and take a worst-case approach. The defender's goal in our framework is to select a coverage vector, C, that maximizes the defenders worst-case payoff over all of the possible ways that the attacker payoffs could be chosen from the defined intervals.

Analysis of ISG

We define the *potential attack set* for a coverage strategy C to be the set of all targets that *could* give the attacker the maximum expected value, for any realization of attacker payoffs consistent with the payoff intervals. For every target, the attacker has a range of expected payoffs:

$$v^{max}(t_i) = c_i \cdot U_{\Psi}^{c,max}(t_i) + (1 - c_i) \cdot U_{\Psi}^{u,max}(t_i). \tag{22.28}$$

$$v^{min}(t_i) = c_i \cdot U_{\Psi}^{c,min}(t_i) + (1 - c_i) \cdot U_{\Psi}^{u,min}(t_i). \tag{22.29}$$

Observe that for a given coverage vector C, the attacker can guarantee a payoff of at least the maximum of the minimum values over all targets; let us denote this value by $R = \max_{t_i} v^{min}(t_i)$. Given the value of R, we can identify the targets that could be attacked for some realization of the payoff values. Any target t_i with a maximum expected value $v^{max}(t_i) \geq R$ could be the best target for the attacker to attack. To see this, suppose that the attacker's payoff for t_i is the maximum value in the interval, and the payoffs for all other targets are the minimal values, so that the best possible value for attacking any target other than t_i is R. Therefore, the potential attack set, $\Lambda(C)$, is defined as:

$$\Lambda(C) = \{t_i : v^{max}(t_i) \geq R\}. \tag{22.30}$$

The defender's expected payoff for each target is:

$$d_i = c_i \cdot U_{\Theta}^{c}(t_i) + (1 - c_i) \cdot U_{\Theta}^{u}(t_i). \tag{22.31}$$

The defender's objective is to select a strategy C to maximize the worst-case payoff over the targets in the potential attack set:

$$\max_{C}\left(\min_{t_i \in \Lambda(C)} d_i\right). \tag{22.32}$$

Note that this objective function takes a worst-case approach to uncertainty about the attacker's payoffs, but still takes the expected payoffs with respect to the realizations of the coverage probabilities. This problem cannot be solved using linear programming because the set of targets $t_i \in \Lambda(C)$ depends on C.

The main idea of our approach is to transform the optimization problem specified in the equations above into a series of feasibility problems. Our first observation is that the defender's maximum possible expected payoff increases monotonically as the number of available resources m is increased. This follows as a consequence of the fact that the defender's set of possible coverage strategies is strictly larger for larger m. Using this observation, we can frame the problem as a binary search in the space of defender expected payoffs. In each iteration, we test whether the defender payoff at the midpoint is feasible given the number of resources available. If it is not, the maximum payoff must be smaller than the midpoint. If it is feasible, the maximum payoff is greater than or equal to the midpoint. Using this strategy, we can approximate the maximum payoff to within a very small tolerance.

To implement this approach, we need to analyze the problem to determine whether a given defender payoff (denoted by D^*) is feasible given the resources available, m. Since we are interested in worst-case outcomes, this means that we need to guarantee that the defender will achieve *at least* D^* for any attacker payoffs in the known intervals. For D^* to be guaranteed by a coverage strategy C, one of the following two conditions must hold for every target:

1. The target is in the potential attack set $\Lambda(C)$, but the defender's expected payoff if the target is attacked is at least D^*.
2. The target is not in the potential attack set $\Lambda(C)$.

We now derive constraints that satisfy these conditions for each target using the minimum amount of resources (i.e., coverage probability). We can calculate the coverage required to satisfy condition 1 for each target (if it is in Λ) from the equation for the defender's payoff. The minimal coverage for target t_i is given by:

$$c_i^1 = \max\left(0, 1 - \frac{D^* - U_\Theta^u(t_i)}{U_\Theta^c(t_i) - U_\Theta^u(t_i)}\right). \tag{22.33}$$

The problem now reduces to finding the potential attack set that minimizes the overall coverage probability required to meet conditions 1 and 2 for all targets. A naïve approach would be to enumerate all of the possible attack sets and calculate the minimum coverage for each such set. For any given set, we can calculate the value of R directly, and then calculate the minimal coverage required for each target in Λ from Equation 22.33. For a given value of R, we can also calculate the minimum coverage that would be required on each target t_i so that the target is *not* in Λ, which requires that the maximum possible expected payoff for the attacking t_i is less than R. We calculate the minimum coverage as follows, using the maximum attacker payoffs from the possible intervals:

$$c_i^2 = \max\left(0, 1 - \frac{R - U_{\Psi}^{u,max}(t_i)}{U_{\Psi}^{c,max}(t_i) - U_{\Psi}^{u,max}(t_i)}\right). \tag{22.34}$$

By summing the values of c_i^1 for targets in Λ and c_i^2 for the remaining targets, we get the minimum coverage required to guarantee D^* for this potential attack set. Unfortunately, the number of such sets is exponential in the number of targets, so enumerating them is inefficient. To avoid this problem, we make another observation that allows us to efficiently explore the candidate solutions. For every set Λ there is a target, which we label $\hat{t}$, that has the maximum minimum expected payoff, R. This is the target that defines the value of R. Since there are only n targets, we can test each target as a candidate for $\hat{t}$ and construct a coverage vector that meets the necessary constraints using minimal resources. If the solution is feasible for any one of the n targets that are candidates for $\hat{t}$, then the value of D^* is feasible. In the following section, we describe an algorithm that uses this solution strategy to efficiently approximate the optimal coverage vector C for the defender.

ISG Algorithm

We now describe in detail our algorithm for solving an ISG (Kiekintveld et al., 2013). The pseudocode is given in Algorithms 1 and 2. Algorithm 1 implements binary search in the space of possible defender payoffs. The feasibility check is presented in Algorithm 2. The goal is to construct a solution that will guarantee the defender D^* while using the minimum resources; if we can construct a solution that uses less than the available resources m, we have found a feasible solution. The strategy is to divide the search into n possible cases, each of which corresponds to a different assumption about which target will have the maximum guaranteed payoff for the attacker,

R. The algorithm iterates through each choice of t_i as a candidate for this special target $\hat{t}$. For each case, the algorithm constructs a coverage vector using minimal coverage probability that guarantees the defender D^* based on conditions 1 and 2.

First, for $\hat{t}$ the target is part of the potential attack set in this solution based on our assumption that it will have the maximum minimum expected payoff. Therefore, if this target is attacked, it must give the defender an expected payoff of at least D^*, as calculated in equation (22.33). We take the value $c_{\hat{t}}^1$ necessary to ensure D^* and use this to calculate the value of R. This value of R is as high as possible because we use the minimum coverage. We do not need to consider adding additional coverage to $\hat{t}$ to decrease the value of R because this could only increase the coverage needed for any other target. To see this, note that the values of c_i^1 are independent of R and the values of c_i^2 increase monotonically as R decreases.

Now that we have the value of R, we calculate the value of c_i^2 for every other target. We calculate the minimum coverage required to satisfy one of these two conditions by taking (c_i^1, c_i^2) for each target. One final condition must be met for each target for our initial assumption to hold: the target we are assuming is $\hat{t}$ must have the maximum minimum expected payoff. We enforce this by adding an additional constraint on the coverage probability assigned to each target so that the maximum attacker payoff for the target is less than the calculated value of R for $\hat{t}$:

$$c_i^3 = \max\left(0, 1 - \frac{R - U_\Psi^{u,min}(t_i)}{U_\Psi^{c,min}(t_i) - U_\Psi^{u,min}(t_i)}\right). \tag{22.35}$$

These values increase monotonically as R decreases. The overall minimum coverage for each target is given by $\max\left(c_i^3, \min\left(c_i^1, c_i^2\right)\right)$. We sum these coverages over all targets and compare this with the available resources m to determine whether this selection of $\hat{t}$ yields a feasible solution. If this case does not yield a feasible solution, the algorithm continues testing the other targets as candidates for $\hat{t}$. As soon as a feasible solution is found, the subroutine terminates and the binary search continues.

The worst-case complexity of the algorithm is $O(n^2 \cdot \log(1/\varepsilon))$ where ε is the error tolerance parameter for the binary search. Each feasibility check requires one iteration to test each target as $\hat{t}$, and each iteration does several constant-time operations on each target to determine the minimal coverage. Therefore, the feasibility check has complexity $O(n^2)$. Binary search requires $O(\log(1/\varepsilon))$ iterations to converge within ε, giving the overall complexity of $O(n^2 \cdot \log(1/\varepsilon))$.

Algorithm 1 ISG Solver

for all $t_i \in T$ **do**
  $c_i \leftarrow 0$
end for
maxPayoff $\leftarrow 0$
maxPayoff $\leftarrow \min_{t_i \in T} U_\Theta^u(t_i)$
while *maxPayoff* – *minPayoff* > ε **do**
  midPoint ← (*maxPayoff* – *minPayoff*)/2
  if feasibilityCheck(*midPoint, m, C*) **then**
    minPayoff ← *midpoint*
  Else
    maxPayoff ← *MidPoint*
  end if
end while
return *C*

22.4 Security Games with Scheduling Constraints

We now discuss extensions of the basic SSG model that allow us to model scheduling constraints on the defender's resources. It is imperative to model such scheduling constraints in many real-world domains. For example, U.S. carriers fly more than 27,000 domestic and 2,000 international flights daily, presenting a massive scheduling challenge for the Federal Air Marshals Service (FAMS). The air marshal resources are constrained by the network of departure and arrival flights, and additional policy constraints imposed by the FAMS (e.g., they need to return to the city where they started and so on).

Algorithm 2 feasibilityCheck

for all $t_i \in T$ **do**

$$c_i^1 \leftarrow \max(0, 1 - \frac{midPoint - U_\Theta^u(t_i)}{U_\Theta^c(t_i) - U_\Theta^u(t_i)})$$

end for
for all $t_i \in T$ **do**
  $totalCov \leftarrow c_i^1$
  $c_i \leftarrow c_i^1$
  if $c_i > 1$ **then**
    GOTO next t_i
  end if
  $R \leftarrow \left(c_i^1 \cdot U_\Psi^{c,min}(t_i)\right) + \left(\left(1 - c_i^1\right) \cdot U_\Psi^{u,min}(t_i)\right) - \epsilon'$

for all $t_j \in \{T \setminus t_i\}$ **do**

$$c_j^2 \leftarrow \max\left(0, 1 - \frac{R - U_\Psi^{u,max}(t_j)}{U_\Psi^{c,max}(t_j) - U_\Psi^{u,max}(t_j)}\right)$$

$$c_j^3 \leftarrow \max\left(0, 1 - \frac{R - U_\Psi^{u,min}(t_j)}{U_\Psi^{c,min}(t_j) - U_\Psi^{u,min}(t_j)}\right)$$

$minCov \leftarrow max(c_i^3, min(c_i^1, c_i^2))$
if $minCov < 0 \parallel minCov > 1$ **then**
GOTO next t_i
end if
$totalCov \leftarrow totalCov + minCov$
$c_j \leftarrow minCov$
end for
if $totalCov \leq m$ **then**
return TRUE, C
end if
end for
return FALSE

We now present algorithms for *Security Problems with ARbitrary Schedules* (SPARS), drawing on techniques used to solve very large mixed-integer programs (Jain, Kardes, Kiekintveld, Ordóñez, & Tambe, 2010). The main algorithm here is Accelerated SPars ENgine (ASPEN), which is based on the branch and price framework (Jain et al., 2010). The first novel feature of ASPEN is that it uses column generation to avoid representing the full (exponential) strategy space for the defender. To this end, it provides a novel decomposition of SPARS into a master problem and a network flow sub-problem. Second, ASPEN uses a novel branch-and-bound method for searching the space of attacker strategies, achieving significant performance improvements by integrating branching criteria and bounds generated with ORIGAMI algorithm (Kiekintveld et al., 2009).

22.4.1 *Generalized Model with Scheduling Constraints: SPARS*

Here also, the attacker's pure strategy space $\mathcal{A}$ is the set of targets T that could be attacked, $T = \{t_1, t_2, \ldots, t_n\}$.[2] The corresponding mixed strategy a = $\langle a_i \rangle$ is a vector where a_i represents the probability of attacking t_i. The defender allocates resources of different types $\lambda \in \Lambda$ to protect targets, with the number of available resources given by $R = \{r_1, r_2, \ldots, r_{|\Lambda|}\}$. Each resource can be assigned to a *schedule* covering multiple targets, $s \subseteq T$, so the set of all legal schedules $S \subseteq \wp(T)$. There is a set of legal schedules for each λ, $S_\lambda \subseteq S$.

The defender's pure strategies are the set of *joint schedules* that assign each resource to at most one schedule. Additionally, we assume that a target may be covered by at most 1 resource in a joint schedule (though this can be generalized). A joint schedule **j** can be represented by the vector $\mathbf{P_j} = \langle P_{jt} \rangle \in \{0, 1\}^n$ where P_{jt} represents whether target t is covered in joint schedule **j.** The set of all feasible joint schedules is denoted by **J.** We define a mapping M from **j** to $\mathbf{P_j}$ as: $M(\mathbf{j}) = \langle P_{jt} \rangle$, where $P_{jt} = 1$ if $t \in \cup_{s \in \mathbf{j}} s$; 0 otherwise. The defender's mixed strategy **x** specifies the probabilities of playing each $\mathbf{j} \in \mathbf{J}$, where each individual probability is denoted by x_j. Let $\mathbf{c} = \langle c_t \rangle$ be the vector of coverage probabilities corresponding to **x**, where $c_t = \sum_{\mathbf{j} \in \mathbf{J}} P_{jt} x_j$ is the marginal probability of covering t. The payoff definitions and the rest of the notation used here is same as before.

Example: Consider a FAMS game with five targets (flights), $T = \{t_1, \ldots, t_5\}$, and three marshals of the same type, $r_1 = 3$. Let the set of feasible schedules be $S_1 = \{\{t_1, t_2\}, \{t_2, t_3\}, \{t_3, t_4\}, \{t_4, t_5\}, \{t_1, t_5\}\}$. The set of feasible joint schedules is shown next, where column $\mathbf{J}_1$ represents the joint schedule $\{\{t_1, t_2\}, \{t_3, t_4\}\}$.

$$P = \begin{array}{c} \\ t_1: \\ t_2: \\ t_3: \\ t_4: \\ t_5: \end{array} \begin{array}{c} \begin{array}{ccccc} J_1 & J_2 & J_3 & J_4 & J_5 \end{array} \\ \begin{bmatrix} 1 & 1 & 1 & 1 & 0 \\ 1 & 1 & 1 & 0 & 1 \\ 1 & 1 & 0 & 1 & 1 \\ 1 & 0 & 1 & 1 & 1 \\ 0 & 1 & 1 & 1 & 1 \end{bmatrix} \end{array}$$

Each joint schedule in **J** assigns only two air marshals in this example, since no more than one FAM is allowed on any flight. Thus, the third air marshal will remain unused. Suppose all of the targets have identical payoffs $U_d^c(t) = 1$, $U_d^u(t) = -5$, $U_a^c(t) = -1$ and $U_a^c(t) = 5$. In this case, the optimal strategy for the defender randomizes uniformly across the joint schedules, $\mathbf{x} = \langle .2, .2, .2, .2, .2 \rangle$, resulting in coverage $\mathbf{c} = \langle .8, .8, .8, .8, .8 \rangle$. All pure strategies have equal payoffs for the attacker, given this coverage vector.

SPARS problems can be formulated as mixed-integer programs in which adversary strategies are represented by integer variables a with $a_t = 1$ if target t is attacked and 0 otherwise. Two key computational challenges arise in this formulation. First, the space of possible strategies (joint schedules) for the defender suffers from combinatorial explosion: a FAMS problem with 100 flights, schedules with three flights, and 10 air marshals has up to 100,000 schedules and $\binom{100000}{10}$ joint schedules. Second, integer variables are a well-known challenge for optimization. Branch and Price (Barnhart, Johnson, Nemhauser, Savelsbergh, & Vance, 1994) is a framework for solving very large optimization problems that combines branch and bound search with column generation to mitigate both of these problems. This method operates on joint schedules (and not marginal probabilities, like ERASER-C), so it is able to handle arbitrary scheduling constraints directly.

An example of branch and price for our problem is shown in Figure 22.2, with the root representing the original problem. Branches to the left (gray nodes) set exactly one variable t_i in a to 1 and the rest to zero, resulting in a linear program that gives a lower bound on the overall solution quality. Branches to the right fix variable t_i to zero, leaving the remaining variables unconstrained. An upper bound on solution quality computed for each white node can be used to terminate execution without exploring all of the possible integer assignments. Solving the linear programs in each gray node normally requires enumerating all joint schedules for the defender. Column generation (i.e., pricing) is a technique that avoids this by iteratively solving a restricted *master problem*, which includes only a small subset of the variables, and a *slave problem*, which identifies new variables to include in the master problem to improve the solution.

Unfortunately, branch and price is not an "out of the box approach" and it has only recently begun to be applied in game-theoretic settings (Halvorson, Conitzer, & Parr, 2009). We introduce a novel master-slave decomposition to facilitate column generation for SPARS, including a network flow formulation of the slave problem. We also show experimentally that conventional linear programming relaxations used for branch and bound perform poorly in this domain, and we replace them with novel techniques based on fast algorithms for security games without scheduling constraints.

ASPEN Column Generation

The linear programs at each leaf in Figure 22.2 are decomposed into *master* and *slave* problems for column generation (see Algorithm 3). The master solves for the defender strategy **x**, given a restricted set of columns (i.e., joint schedules) **P**. The objective function for the slave is updated based on the solution of the master, and the slave is solved to identify the best new column to add to the master problem, using *reduced costs* (explained later). If no column can improve the solution the algorithm terminates.

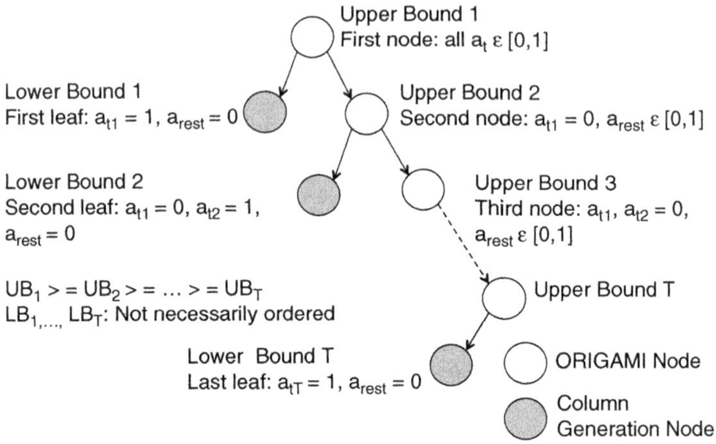

Figure 22.2. Working of branch and price.

Algorithm 3 Column generation

1. Initialize P
2. Solve master problem
3. Calculate reduced cost coefficients from solution
4. Update objective of slave problem with coefficients
5. Solve slave problem

if optimal solution obtained, **then**

   6. Return (**x, P**)

Else

   7. Extract new column and add to **P**

   8. Repeat from Step 2

end if

Master Problem: The master problem (equations (22.36) to (22.41)) solves for the probability vector **x** that maximizes the defender reward (Table 22.2 describes the notation). This master problem operates directly on columns of **P**, and the coverage vector **c** is computed from these columns as **Px**. Constraints (22.37)–(22.39) enforce the SSE conditions that the players choose mutual best responses, mirroring similar constraints in ERASER-C. The defender expected payoff for target t is given by the tth component of the column vector $\mathrm{DPx} + \mathrm{U}^u_d$ and denoted $\left(\mathrm{DPx} + \mathrm{U}^u_d\right)_t$. Similarly, the attacker payoff for target t is given by $\left(\mathrm{APx} + \mathrm{U}^u_a\right)_t$. Constraints (22.37) and (22.38) are active only for the single target t^* attacked ($a_{t^*} = 1$). This target must be a best-response, due to constraint (22.39).

Table 22.2. *Notation*

Variable	Definition	Dimension
P	Mapping between T and J	$\|T\| \times \|J\|$
x	Probability distribution over **J**	$\|J\| \times 1$
a	Attack vector	$\|T\| \times 1$
d	Defender reward	–
k	Adversary reward	–
d	Column vector of d	$\|T\| \times 1$
k	Column vector of k	$\|T\| \times 1$
D	Diag. matrix of $U_d^c(t) - U_d^u(t)$	$\|T\| \times \|T\|$
A	Diag. matrix of $U_a^c(t) - U_a^u(t)$	$\|T\| \times \|T\|$
U_d^u	Vector of values $U_d^u(t)$	$\|T\| \times 1$
U_a^u	Vector of values $U_a^u(t)$	$\|T \times 1$
M	Huge positive constant	–

$$\max \quad d \tag{22.36}$$

$$\text{s.t. } \mathrm{d} - \mathrm{DPx} + \mathrm{U}_d^u \le (1-a)M \tag{22.37}$$

$$\mathrm{k} - \mathrm{APx} + \mathrm{U}_a^u \le (1-a)M \tag{22.38}$$

$$\mathrm{APx} + \mathrm{U}_a^u \le \mathrm{k} \tag{22.39}$$

$$\sum_{j \in J} x_j = 1 \tag{22.40}$$

$$\mathrm{x}, \mathrm{a} \le 0 \tag{22.41}$$

Slave Problem: The slave problem finds the best column to add to the current columns in **P**. This is done using *reduced cost*, which captures the total change in the defender payoff if a candidate column is added to **P**. The candidate column with minimum reduced cost improves the objective value the most (Bertsimas & Tsitsiklis, 1994). The reduced cost $\overline{c}_j$ of variable x_j, associated with column $\mathbf{P}_j$, is given in equation (22.42), where **w**, **y**, **z**, and h are dual variables of master constraints (22.37), (22.38), (22.39), and (22.40), respectively. The dual variable measures the influence of the associated constraint on the objective, and can be calculated using standard techniques.

$$\overline{c}_j = \mathrm{w}^T\left(\mathrm{DP_j}\right) + \mathrm{y}^T\left(\mathrm{AP_j}\right) - \mathrm{z}^T\left(\mathrm{AP_j}\right) - h \tag{22.42}$$

An inefficient approach would be to iterate through all of the columns and calculate each reduced cost to identify the best column to add. Instead, we formulate a minimum cost network flow (MCNF) problem (Bertsimas & Tsitsiklis, 1994) that efficiently finds the optimal column. Feasible flows in the network map to feasible joint schedules in the SPARS problem, so the scheduling constraints are captured by this formulation. For a SPARS instance we construct the MCNF graph G as follows.

A source node source$_\lambda$ with supply r_λ is created for each defender type $\lambda \in \Lambda$. A single sink node has demand $\sum_{\lambda \in \Lambda} r_\lambda$. Targets in schedule s for resource λ are represented by pairs of nodes $(a_{s_\lambda,t}, b_{s_\lambda,t})$ with a connecting link (so each target corresponds to many nodes). For every schedule $s_\lambda \in S_\lambda$ we add a path from the source to the sink: $\langle \text{source}_\lambda, a_s, t_{i_1}, b_s, t_{i_1}, a_s, t_{i_2}, \ldots, b_s, t_{i_L}, \text{sink} \rangle$. The capacities on all links are set to 1, and the default costs to 0. A dummy flow with infinite capacity is added to represent the possibility that some resources are unassigned. The number of resources assigned to t in a column $\mathbf{P}_\mathbf{j}$ is computed as: assigned $(t) = \Sigma_{s \in S}$ flow[link($a_{s,t}$, $b_{s,t}$)]. Constraints are added to G so assigned $(t) \leq 1$ for all targets t.

A partial graph G for our earlier example is shown in Figure 22.3, showing paths for three of the five schedules. The paths correspond to schedules $\{t_1, t_2\}$, $\{t_2, t_3\}$, and $\{t_1, t_5\}$. The supply and demand are both 3, corresponding to the number of available FAMS. Double-bordered boxes mark the flows used to compute assigned(t_1) and assigned(t_2). Every joint schedule corresponds to a feasible flow in G. For example, the joint schedule $\{\{t_2, t_3\}, \{t_1, t_5\}\}$ has a flow of one unit each through the paths corresponding to schedules $\{t_2, t_3\}$ and $\{t_1, t_5\}$, and a flow of 1 through the dummy. Similarly, any feasible flow through the graph G corresponds to a feasible joint schedule, since all resource constraints are satisfied.

It remains to define link costs such that the cost of a flow is the reduced cost for the joint schedule. We decompose $\bar{c}_j$ into a sum of cost coefficients per target, $\hat{c}_t$, so that $\hat{c}_t$ can be placed on links $(a_{s,t}, b_{s,t})$ for all targets t. $\hat{c}_t$ is defined as $w_t.D_t + y_t.A_t - z_t.A_t$ where w_t, y_t and z_t are t^{th} components of $\mathbf{w}$, $\mathbf{y}$, and $\mathbf{z}$. D_t is equal to $U_d^c(t) - U_d^u(t)$ and $A_t = U_a^c(t) - U_a^u(t)$. The overall objective given here for the MCNF problem sums the contributions of the reduced cost from each individual flow and subtracts the dual variable h. If this is non-negative, no column can improve the master solution, otherwise the optimal column (identified by the flow) is added to the master and the process iterates.

Improving Branch and Bounds

ASPEN uses branch and bound to search over the space of possible attacker strategies. A standard technique in branch and price is to use LP relaxation,

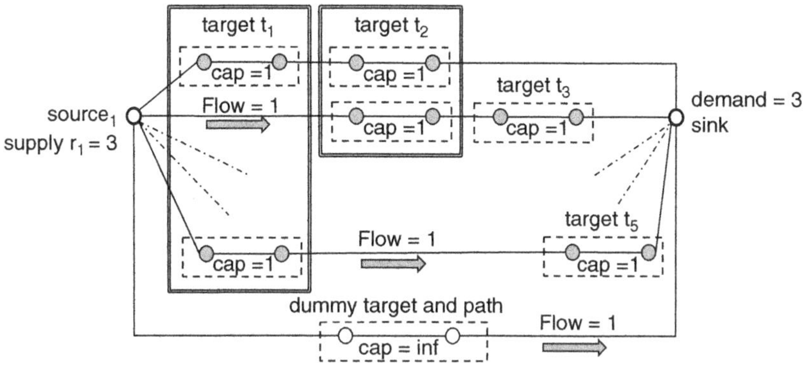

Figure 22.3. Example network graph.

i.e., allow the integer variables to take on arbitrary values, to give an optimistic bound on the objective value of the original MIP for each internal node. Hence, ASPEN uses a variant of ORIGAMI, called ORIGAMIS (Jain et al., 2010), a novel branch and bound heuristic for SPARS which is an efficient solution method for security games without scheduling constraints and heterogeneous resources.

While we do not describe ORIGAMI-S here, it builds on ORIGAMI and minimizes the attacker's maximum payoff. It implements constraints similar to network flow constraints to capture the scheduling constraints in SPARS. ORIGAMI-S is solved once at the beginning of ASPEN, and targets in the attack set are sorted by expected defender reward. The maximum value is an initial upper bound on the defender reward. The first leaf node that ASPEN evaluates corresponds to this maximum valued target (i.e., setting its attack value to 1), and a solution is found using column generation. This solution is a lower bound of the optimal solution, and the algorithm stops if this lower bound meets the ORIGAMI-S upper bound. Otherwise, a new upper bound from the ORIGAMI-S solution is obtained by choosing the second-highest defender payoff from targets in the attack set, and ASPEN evaluates the corresponding leaf node. This process continues until the upper bound is met, or the available nodes in the search tree are exhausted. Further details and correctness proofs can be found in Jain and colleagues (2010).

22.4.2 *Compact Representations with Scheduling Constraints*

We now describe ERASER-C ("Constrained"), an extension of ERASER (Kiekintveld et al., 2009) that adds the capability to represent certain kinds of resource and scheduling constraints, while operating on the compact representation. Here again, these constraints are motivated by the FAMS domain. It demonstrates that the basic idea of using a compact representation of the defender's strategy space and the payoff functions for both players is still useful in this setting when resources are heterogeneous.

The first extension allows resources to be assigned to *schedules* covering multiple targets. The set of legal schedules $S = \{s_1 \ldots s_l\}$ is a subset of the power set of the targets, with restrictions on this set representing scheduling constraints. We define the relationship between targets and schedules with the function $M: S \times T \rightarrow \{0, 1\}$, which evaluates to 1 if and only if t is covered in s. The defender's strategy is now an assignment of resources to schedules, rather than targets. A second extension introduces *resource types*, $\Omega = \{\omega_1, \ldots, \omega_v\}$, each with the capability to cover a different subset of S. The number of available resources of each type is given by the function $\mathcal{R}(\omega)$. Coverage capabilities for each type are given by the function $Ca: S \times \Omega \rightarrow \{0, 1\}$, which is 1 if the type is able to cover the given schedule and 0 otherwise.[3]

The combination of schedules and resource types captures key elements of the FAMS domain. Air marshals are resources, and flights are potential targets, with payoff values defined by risk analysis of the flight. However, a marshal cannot be on all possible flights due to location and timing constraints. Legal schedules can be used to define the set of possible flights that an air marshal could feasibly fly, given these constraints. Resource types are used to define the initial state (notably, location) of a marshal, which defines a subset of legal schedules that any given marshal could fly.

The effect of adding scheduling and resource coverage constraints is to constrain the space of feasible coverage vectors. Consider an example with a single resource defending three targets. There are two legal schedules, covering targets {1, 2} and {2, 3}. Given only these schedules, it is not possible to implement a coverage vector that places 50% probability on both targets 1 and 3, with no coverage of target 2.

An MILP implementing ERASER-C is presented in equations (22.43–22.54). The MILP is very similar to the original ERASER, but enforces additional constraints on the legal schedules and coverage for each resource type. The q variables represent the total probability that is assigned to each schedule by all resource types, and the h variables are the probability assigned to a schedule by a specific type of resource. In equations (22.53) and (22.54), Z is a large constant, relative to the maximum payoff.

Constraint (22.44) restricts the attack vector to binary variables, corresponding to pure strategies for the attacker. Constraints (22.45) through (22.47) restrict the defender's strategy so that no target is assigned probability greater than 1. The coverage of each schedule must sum to the contributions of the individual resource types, specified in (22.49). The mapping between the coverage of schedules and coverage of targets is enforced in constraint (22.50). Constraint (22.51) restricts the schedule so that only the available number of resources of each type is used. No probability may be assigned to disallowed schedules for each resource type, which is explicitly enforced by constraint (22.52). The final three constraints specify the maximization performed by both the attacker and defender, exactly as in the ERASER (see Kiekintveld et al., 2009).

$$\max \quad d \tag{22.43}$$

$$a_t \in \{0,1\} \quad \forall t \in T \tag{22.44}$$

$$c_t \in [0,1] \quad \forall t \in T \tag{22.45}$$

$$q_s \in [0,1] \quad \forall s \in S \tag{22.46}$$

$$h_{s,\omega} \in [0,1] \quad \forall s \in S \tag{22.47}$$

$$\sum_{t \in T} a_t = 1 \tag{22.48}$$

$$\sum_{\omega \in \Omega} h_{s,\omega} = q_s \quad \forall s \in S \tag{22.49}$$

$$\sum_{s \in S} q_s M(s,t) = c_t \quad \forall t \in T \tag{22.50}$$

$$\sum_{s \in S} h_{s,\omega} Ca(s,\omega) \leq \mathcal{R}(\omega) \quad \forall \omega \in \Omega \tag{22.51}$$

$$h_{s,\omega} \leq Ca(s,\omega) \quad \forall s,\omega \in S \times \Omega \tag{22.52}$$

$$d - U_{\Theta}(t,C) \leq (1-a_t) \cdot Z \forall t \in T \tag{22.53}$$

$$0 \leq k - U_{\Psi}(t,C) \leq (1-a_t) \cdot Z \forall t \in T \tag{22.54}$$

The theorem detailing the proof of correctness of the ERASER-C is described in Kiekintveld and colleagues (2009).

22.5 Conclusion

In this chapter, we have described formulations and algorithms for computing strong Stackelberg equilibrium for Stackelberg games. We introduced basic methods as well as advances made for solving larger models: models with larger strategy spaces as well as models with differing forms of uncertainty. We talked about methods that relied on compact representations, exploited target/attacker type independence as well as made use of large-scale optimization techniques. We also described algorithms that dealt with complex scheduling constraints present in real-world domains. These algorithms have not only been applied in deployed real-world applications, but have also laid the foundation for the development of many advanced algorithms. Additionally, while this chapter dealt with rational adversaries, subsequent chapters will describe research related to incorporating behavioral aspects of human decision making in game-theoretic models.

NOTES

1. Mathematical programming refers to a set of methods used in many fields to specify optimization problems using mathematical equations. Readers unfamiliar with optimization methods (e.g., linear programming) and tools for solving optimization problems (e.g., CPLEX) may benefit from going through some basic tutorial materials on the subject. There are many tutorials and textbooks available on these topics, so we do not cover this background material here.
2. In this section, we follow the notation from the original paper on SPARS, which has some slight differences from the notation presented earlier.
3. Our current implementation uses complete matrices to represent M and Ca, but sparse representations could offer additional performance improvements.

REFERENCES

Aghassi, M. & Bertsimas, D. (2006). Robust game theory. *Mathematical Programming, 107*, 231–273.

Barnhart, C., Johnson, E., Nemhauser, E., Savelsbergh, M., & Vance, P. (1994). Branch and price: Column generation for solving huge integer programs. *Operations Research*, *46*, 316–329.

Basar, T., & Olsder, G. H. (1995). *Dynamic noncooperative game theory.* 2nd edition. San Diego, CA: Academic Press.

Ben-Tal, A., & Nemirovski, A. (2002). Robust optimization – methodology and applications. *Mathematical Programming*, *92*, 453–480.

Bertsimas, D., & Tsitsiklis, J. N. (1994). *Introduction to linear optimization.* Athena Scientific.

Breton, M., Alg, A., & Haurie, A. (1988). Sequential Stackelberg equilibria in two-person games. *Optimization Theory and Applications*, *59*(1), 71–97.

Conitzer, V., & Sandholm, T. (2006). Computing the optimal strategy to commit to. *ACM EC-06*: 82–90.

Cormen, T. H., Leiserson, C. E., Rivest, R. L., & Stein, C. (2009). *Introduction to algorithms. 3rd edition.* Cambridge, MA: MIT Press.
Fudenberg, D., & Tirole, J. (1991). *Game theory.* Cambridge, MA: MIT Press.
Halvorson, E., Conitzer, V., & Parr, R. (2009). Multi-step multi-sensor hider-seeker games. *IJCAI*, 336–341.
Harsanyi, J. C., & Selten, R. (1972). A generalized Nash solution for two-person bargaining games with incomplete information. *Management Science*, *18*(5), 80–106.
Jain, M., Kardes, E., Kiekintveld, C., Ordóñez, F., & Tambe, M. (2010). Security games with arbitrary schedules: A branch and price approach. *AAAI.*
Jain, M., Kiekintveld, C., & Tambe, M. (2011). Quality-bounded solutions for finite Bayesian Stackelberg games: Scaling up. In *Proceedings of the International Conference on Autonomous Agents and Multiagent Systems (AAMAS).*
Kiekintveld, C., Islam, T., & Kreinovich, V. (2013). Security games with interval uncertainty. In *Twelfth International Conference on Autonomous Agents and Multiagent Systems (AAMAS).*
Kiekintveld, C., Jain, M., Tsai, J., Pita, J., Tambe, M., & Ordóñez, F. (2009). Computing optimal randomized resource allocations for massive security games. In *AAMAS-09.*
Leitmann, G. (1978). On generalized Stackelberg strategies. *Optimization Theory and Applications, 26*(4), 637–643.
Lerma, O., Kreinovich, V., & Kiekintveld, C. (2011). Linear-time resource allocation in security games with identical fully protective resources. In *AAAI Workshop on Applied Adversarial Reasoning and Risk Modeling.*
Osbourne, M. J., & Rubinstein, A. (1994). *A course in game theory.* Cambridge, MA: MIT Press.
Paruchuri, P., Pearce, J. P., Marecki, J., Tambe, M., Ordóñez, F., & Kraus, S. (2008). Playing games with security: An efficient exact algorithm for Bayesian Stackelberg games. In *AAMAS-08*, 895–902.
Tambe, M. (2011). *Security and game theory: Algorithms, deployed systems, lessons learned.* Cambridge: Cambridge University Press.
von Stengel, B., & Zamir, S. (2004). Leadership with commitment to mixed strategies. Technical Report LSE-CDAM-2004-01, CDAM Research Report.
Yin, Z, & Tambe, M. (2012). A unified method for handling discrete and continuous uncertainty in Bayesian Stackelberg games. In *International Conference on Autonomous Agents and Multiagent Systems (AAMAS).*

23

Mixed-Integer Optimization Methods for Solving Stackelberg Security Games

William Haskell and Fernando Ordóñez

23.1 Introduction

The key to being able to use Stackelberg game models in practical applications is the ability to solve problem sizes that accurately represent real problems. Much work on Stackelberg games in security applications is based on the observation that these games can be cast as mixed-integer optimization problems. Although mixed-integer optimization problems are theoretically difficult, large instances can still be solved efficiently with modern optimization solvers. In this chapter, we aim to review and discuss different mixed-integer formulations of Stackelberg games and the use of exact and heuristic optimization methods to solve these Stackelberg games.

We refer to the Stackelberg games in this chapter as Stackelberg security games (SSGs) because they represent security problems. The defender corresponds to the leader and the attacker to the follower of this game. These games are played over a set of targets that we denote $T = \{1, \ldots, |T|\}$. In this setting, the defender acts first and chooses a mixed strategy over the set of patrol schedules for all its assets on T. We denote the set of possible patrol schedules as J, and x_j is the probability that patrol schedule $j \in J$ is selected. Next, the attacker observes the defender's mixed strategy and forms its attack strategy. We use a_t to denote the probability with which target $t \in T$ is attacked. We use the vectors $\mathbf{x} = (x_j)_{j \in J}$ and $\mathbf{a} = (a_t)_{t \in T}$ to denote the complete mixed strategies of the leader and the follower, respectively.

The payoffs to the defender and the attacker depend on whether the defender covered a target that was attacked; these payoffs do not depend on the type of the defender resource that was covering the target. Let $U_d^c(t)$ denote the payoff to the defender if target t is attacked while it is covered. If target t is attacked when it is not covered, then the defender payoff is $U_d^u(t)$. The attacker payoff is $U_d^c(t)$ when it attacks a covered target, and $U_d^u(t)$ when it attacks an uncovered target. Naturally $U_d^c(t) > U_d^u(t)$ and $U_a^c(t) < U_a^u(t)$, covering a target is better for the defender and worse for the attacker. The rewards for the defender (attacker) of the combined actions $j \in J$ and $t \in T$ are denoted by

$$R_{jt} = \begin{cases} U_d^c(t) & t \in j \\ U_d^u(t) & t \notin j, \end{cases}$$

and

$$C_{jt} = \begin{cases} U_a^c(t) & t \in j \\ U_a^u(t) & t \notin j. \end{cases}$$

For the upcoming optimization problem, let e be a vector of all 1's, then the bi-level program that determines the optimal strategy $\mathbf{x}$ for the leader is

$$\begin{aligned} \max_{\mathrm{x,a}} & \sum_{j \in J, t \in T} R_{jt} x_j a_t \\ \text{s.t.} \;\; & e'\mathrm{x} = 1, \; \mathrm{x} \geq 0 \\ a_t = & \arg\max_{e'\mathrm{a}=1, \mathrm{a} \geq 0} \sum_{j \in J, k \in T} C_{jk} x_j a_k. \end{aligned}$$

This problem finds the leader's mixed strategy that maximizes its expected reward assuming that the adversary breaks ties in favor of the leader (the strong Stackelberg equilibrium solution). We note that although the follower has more information when making its decision, the leader in a Stackelberg game can only improve from a Nash equilibrium solution. The leader can always play a Nash equilibrium solution, forcing the follower to also play Nash. Hence, the leader will only deviate if it is to its advantage.

For example, suppose the defender has to choose among n possible entry points to conduct enhanced surveillance. Subsequently, the attacker, knowing the frequency with which enhanced surveillance is conducted in each entry point, then decides which of the n entry points to attack. In this game, both players' actions are drawn from $J = T = \{1 \ldots n\}$. If the defender can select up to k of the n entry points in which to conduct enhanced surveillance while the attacker selects one of the entry points to attack, then the set of actions of the attacker remains $T = \{1 \ldots n\}$, while the actions for the defender include all subsets of up to k elements of $T = \{1 \ldots n\}$, and there are $\sum_{i=0}^{k} \binom{n}{i}$ possible actions for the defender.

For every fixed leader mixed strategy x, the inner optimization solved by the attacker is a linear programming problem over the simplex. We know that there is always an optimal extreme point for such a problem, and all extreme points of the simplex correspond to pure strategies. Thus, there is no loss of generality in assuming that the adversary selects only pure strategies. By using this observation, we can transform the leader's bi-level optimization problem into a mixed-integer programming problem, as long as the adversary has a linear objective. This equivalent formulation requires a sufficiently large constant M, an auxiliary variable d to represent the reward obtained by the defender, and an auxiliary variable q to represent the optimal reward for the attacker. The problem is expressed as follows:

$$\begin{aligned} \max_{x,a,d,q}\ & d \\ \text{s.t.}\ & \sum_{j\in J} R_{jt}x_j \geq d - M(1-a_t) \quad t\in T \\ & \sum_{j\in J} C_{jt}x_j \leq q \quad t\in T \\ & \sum_{j\in J} C_{jt}x_j \geq q - M(1-a_t) \quad t\in T \\ & e'x = 1, x \geq 0 \\ & e'a = 1, a \in \{0,1\}^{|T|}. \end{aligned} \tag{23.1}$$

The first set of constraints in this formulation is a linearization of the defender's reward given the attacker's pure strategy. The second set of constraints ensures that q is an upper bound on the attacker's reward for any pure strategy, and the third makes q the optimal reward for the attacker for some pure strategy. The foregoing representation of SSGs is the basis for the ERASER model and algorithm developed in Kiekintveld and colleagues (2009). That work considered multiple adversaries and constraints over the set of joint defender strategies.

A number of variations of our formulation have been considered. For instance, a Bayesian formulation that allows for multiple adversaries with different payoff matrices and sets of actions as well as multiple security resources was introduced in Paruchuri and colleagues (2008). A Bayesian formulation in Yin and colleagues (2011) allows for uncertainty over adversary types, rationality, and observation capabilities. Efficient algorithms have been developed to solve Bayesian Stackelberg games. These algorithms exploit the special structure present in the security domain, such as: (1) the rewards for every player take only two values, depending on whether an attack is successful; (2) a network structure exists between targets (Jain, Kardes, Kiekintveld, Ordónez, & Tambe, 2010; Jain, Tsai, Pita, Kiekintveld, Rathi, Ordóñez, & Tambe, 2010; Kiekintveld et al. 2009).

In the rest of this chapter, we focus on the optimization models used to represent SSGs where players are rational and have perfect information. We emphasize the efficiency of different formulations and the mixed-integer optimization methods used to solve these formulations as they scale up to large size. We review a few of the large-scale optimization methods that have been applied to solve SSGs. We begin by presenting a review of different MIP formulations of SSGs and then discuss their relative merits in the next section. In Section 23.3, we discuss large-scale linear optimization methods and their use in SSGs. Then we present heuristic approaches used in solving SSGs in Section 23.4. We finish this chapter summarizing some key points in Section 23.5.

23.2 Mixed-Integer Programming Formulations of SSG

We begin with the case where the defender is facing a set of adversary types L, where attacker $l \in L$ is faced with probability p_l and corresponds to payoff matrices $\left[R^l_{jt}\right]_{j\in J, t\in T}$ and $\left[C^l_{jt}\right]_{j\in J, t\in T}$. By keeping track of the possible actions of each adversary separately with attack vectors a^l, we can express this problem as a Bayesian Stackelberg game:

$$\begin{aligned}
\max_{x,a,d,q} \ & \sum_{l\in L} p_l d^l \\
\text{s.t.} \quad & \sum_{j\in J} R^l_{jt} x_j \ge d^l - M\left(1-a^l_t\right) && t\in T, l\in L \\
& \sum_{j\in J} C^l_{jt} x_j \le q^l && t\in T, l\in L \\
& \sum_{j\in J} C^l_{jt} x_j \ge q^l - M\left(1-a^l_t\right) && t\in T, l\in L \\
& e'x = 1, x \ge 0 \\
& e'a^l = 1, a^l \in \{0,1\}^{|T|} && l\in L.
\end{aligned} \tag{23.2}$$

In this formulation, the defender mixed strategy $\boldsymbol{x}$ is chosen to maximize the expected reward taking the average over the $|L|$ adversaries. As mentioned earlier, this formulation is the basis of the ERASER formulation introduced in Kiekintveld and colleagues (2009). The ERASER model also included linear constraints on $\boldsymbol{x}$ to define feasible patrolling schedules.

Two equivalent formulations for this problem have been proposed previously. The DOBBS problem formulation given next was introduced in Paruchuri and colleagues (2008):

$$
\begin{aligned}
\max_{z,a,q} & \sum_{l\in L}\sum_{j\in J}\sum_{t\in T} p_l R^l_{jt} z_{jtl} & & \\
\text{s.t.} \quad & \sum_{j\in J}\sum_{t\in T} z_{jtl} = 1 & & l \in L \\
& \sum_{j\in J} z_{jtl} = a^l_t & & t\in T, l\in L \\
& \sum_{t\in T} z_{jtl} = \sum_{t\in T} z_{jtl} & & j\in J, l\in L \\
& 0 \le q^l - \sum_{j\in J}\sum_{k\in T} C^l_{jt} z_{jkl} \le M\left(1-a^l_t\right) & & t\in T, l\in L \\
& e' b^l = 1, \mathrm{b}^l \in \{0,1\}^{|T|} & & l\in L \\
& z_{jtl} \ge 0 & & t\in T, j\in J, l\in L.
\end{aligned}
\tag{23.3}
$$

This problem is obtained by introducing variables z_{jtl} to represent the product $z_{jtl} = x_j a^l_t$, which is used to linearize the bilinear defender objective. The first three constraints in this formulation ensure that variable $\boldsymbol{z}$ corresponds to the product of $\boldsymbol{x}$ and $\boldsymbol{a}$, with the remaining constraints being equivalent to the constraints in (23.2). The last formulation we present does not require the use of big M parameters and was introduced in Casorrán-Amilburu and colleagues (2015):

$$
\begin{aligned}
\max_{z,a} & \sum_{l\in L}\sum_{j\in J}\sum_{t\in T} p_l R^l_{jt} z_{jtl} & & \\
\text{s.t.} \quad & \sum_{j\in J}\sum_{t\in T} z_{jtl} = 1 & & l \in L \\
& \sum_{j\in J} z_{jtl} = a^l_t \quad t\in T, & & l\in L \\
& \sum_{t\in T} z_{jtl} = \sum_{t\in T} z_{jt1} & & j\in J,\; l\in L \\
& \sum_{j\in J}\left(C^l_{jt} - C^l_{jk}\right) z_{jtl} \ge 0 & & t,k\in T, l\in L \\
& e' a^l = 1, a^l \in \{0,1\}^{|T|} & & l\in L \\
& z_{jtl} \ge 0 & & j\in J, t\in T, l\in L.
\end{aligned}
\tag{23.4}
$$

Although these three formulations solve the same problem, it is not clear a priori if one of these formulations provides some benefit in solution time. The theoretical work in Casorrán-Amilburu and colleagues (2015) has shown that the LP relaxation of (23.4) is contained in the LP relaxation of (23.3), which suggests that it should provide a smaller initial gap and lead to faster solution times. In fact, Casorrán-Amilburu and colleagues (2015) have also shown that the LP relaxation of (23.4) has all integer extreme points when there is only one adversary. However, problem (23.4) has more variables and constraints than problem (23.2), which makes solving each LP relaxation within a branch and bound solution procedure more burdensome. On the other hand, (23.2) has fewer variables and constraints, but the use of big M constants can lead to large gaps and slow convergence in solving the integer problem. It is therefore unclear which formulation will be more effective in practice.

23.2.1 *Exploiting the Payoff Structure of Security Applications*

If the defender can place k security resources on a set T of potential targets, then the total combined set of actions J the defender can execute becomes prohibitively large for current computing capabilities. We note that in a security application, it is only relevant whether a security resource meets the attacker at a target. If the follower attacks a protected target, then the defender receives a reward while the attacker receives a penalty. When the follower attacks an unprotected target, then the defender receives a penalty and the attacker a reward. If we assume the rewards and penalties only depend on the target being attacked/protected and not on what the other defender resources are doing, then we assume that the payoff matrices have the following structure:

$$R^l_{jt} = \begin{cases} U^c_d(t) & t \in j \\ U^u_d(t) & t \notin j \end{cases} \qquad C^l_{jt} = \begin{cases} U^c_a(t) & t \in j \\ U^u_a(t) & t \notin j. \end{cases}$$

Here $t \in j$ denotes the fact that defender strategy $j \in J$ protects (places a security resource) on target t, and vice versa. We assume that at most one security resource can be assigned to one target.

23.2.2 *Models*

We note that $\sum_{i \in X} R^l_{ij} x_i = \sum_{i \in X: j \in i} R^l_{dj} x_i + \sum_{i \in X: j \notin i} P^l_{dj} x_i$. In addition, we define $y_j = \Sigma_{i \in X: j \in i} x_i$ to be the total frequency of coverage at target j. We can use this notation to express the objective as follows:

$$\sum_{i \in X} R^l_{ij} x_i = R^l_{dj} \sum_{i \in X: j \in i} x_i + P^l_{dj} + P^l_{dj} \sum_{i \in X: ii} x_i = R^l_{dj} y_j + P^l_{dj} \left(1 - y_j\right).$$

We can use the frequency of coverage variable y_i to represent model (SGM) more concisely for SSGs as follows:

$$
\begin{aligned}
\max_{x,a,d,q} \quad & \sum_{l \in L} p_l d^l \\
\text{s.t.} \quad & d^l \le R^l_{dj} y_j + P^l_{dj}\left(1-y_j\right) + M\left(1-q^l_j\right) && j \in Q, l \in L \\
& 0 \le a^l - P^l_{aj} y_j - R^l_{aj}\left(1-y_j\right) \le M\left(1-q^l_j\right) && j \in Q, l \in L \\
& \sum_{j \in Q} q^l_j = 1, q^l \in \{0,1\}^{|Q|} && l \in L \\
& \sum_{j \in Q} y_j = m, 0 \le y \le 1 \\
& \sum_{i \in X} x_i = 1, x \ge 0 \\
& y_j = \sum_{i \in X : j \in i} x_i && j \in Q
\end{aligned}
$$

The constraints on coverage y (in the fourth constraint given earlier) follow from the fact that the defender must cover m targets out of a total of n and from the constraints on $\boldsymbol{x}$. If we can omit the last two constraints, then we have an optimization problem that does not consider the large defender action set X and can be thus efficiently solved:

$$
(\text{SGM}-\text{SSG}) \qquad
\begin{aligned}
\max_{x,a,d,q} \quad & \sum_{l \in L} p_l d^l \\
\text{s.t.} \quad & d^l \le R^l_{dj} y_j + P^l_{dj}\left(1-y_j\right) + M\left(1-q^l_j\right) && j \in Q, l \in L \\
& 0 \le a^l - P^l_{aj} y_j - R^l_{aj}\left(1-y_j\right) \le M\left(1-q^l_j\right) && j \in Q, l \in L \\
& \sum_{j \in Q} q^l_j = 1, \; q^l \in \{0,1\}^{|Q|} && l \in L \\
& \sum_{j \in Q} y_j = m, \; 0 \le y \le 1
\end{aligned}
$$

The difficulty arises in being able to recover the mixed strategy $\boldsymbol{x}$ from a feasible vector of frequency of coverage at each target. This is always possible if there are no additional constraints and if all possible combinations of m targets are feasible defender strategies. This problem is the first formulation we will compare for SSGs.

We now derive new models that attempt to exploit the payoff matrix structure of SSGs for MIPSG and DOBBS. We need to see how the objective

$\sum_{l\in L}\sum_{i\in X}\sum_{j\in Q} p_l R^l_{ij} z_{ijl}$ can be expressed under the specific structure of SSGs, and we write

$$\sum_{j\in Q}\sum_{i\in x} R^l_{ij} z_{ijl} = \sum_{j\in Q}\left(R^l_{dj} \sum_{i\in X: j\in i} z_{ijl} + P^l_{dj} \sum_{i\in X: j\notin i} z_{ijl} \right).$$

To find a concise representation that does not use the set X, we should write $\sum_{i\in X: j\in i} z_{ijl}$ in a concise manner. Let us define $w_{jl} = \sum_{i\in X: j\in i} z_{ijl}$ as the sum of all z_{ijl} for all strategies i that cover target j. We then define $\tilde{w}_{jkl} = \sum_{i\in X: j\in i} z_{ikl}$ to be the sum of z_{ijl} over strategies i that cover target k. Note that this means that $w_{jl} = \tilde{w}_{jjl}$ and $\sum_{i\in X: j\notin i} z_{ijl} + \tilde{w}_{jkl} = q^l_k$ for any $j, k \in Q$ and $l \in L$. Using this notation we can then write

$$\begin{aligned}
\sum_{l\in L}\sum_{j\in Q}\sum_{i\in X} p_l R^l_{ij} z_{ijl} &= \sum_{l\in L}\sum_{j\in Q} p_l \left(R^l_{dj} w_{jl} + P^l_{dj}\left(q^l_j - w_{jl}\right)\right) \\
&= \sum_{l\in L}\sum_{j\in Q} p_l \left(R^l_{dj} + P^l_{dj}\right) w_{jl} + p_l P^l_{dj} q^l_j .
\end{aligned}$$

This objective along with appropriate constraints on the new variables w_{jl} should create a problem from which the variables z_{ijl} can be recovered. In fact, the following problem can be shown to be equivalent to DOBBS:

$$(\text{DOBBS}-\text{SSG}) \qquad \begin{aligned}
\max_{z,q}\ & \sum_{l\in L}\sum_{j\in Q} p_l \left(R^l_{dj} + P^l_{dj}\right) \tilde{w}_{jjl} + p_l P^l_{dj} q^l_j & \\
\text{s.t.}\ & \sum_{j\in Q} q^l_j = 1 & l \in L \\
& q^l_j \in \{0,1\} & j \in Q, l \in L \\
& 0 \le \tilde{w}_{jkl} \le q^l_k & j,k \in Q, l \in L \\
& y_j - 1 + q^l_k \le \tilde{w}_{jkl} \le y_j + 1 - q^l_k & j,k \in Q, l \in L \\
& \sum_{j\in Q} y_j = m, 0 \le y \le 1 & \\
& 0 \le a^l - P^l_{aj} y_j - R^l_{aj}\left(1 - y_j\right) \le M\left(1 - q^l_j\right) & j \in Q, l \in L
\end{aligned}$$

The last constraint comes from the derivation of the standard DOBBS follower optimal response constraint $0 \le a^l - \sum_{i\in X}\sum_{k\in Q} C^l_{ij} z_{ikl} \le M\left(1 - q^l_j\right)$ $j \in Q, l \in L$:

$$\sum_{i\in X}\sum_{k\in Q} C^l_{ij} z_{ikl} = \sum_{k\in Q}\sum_{i\in X:j\in i} P^l_{aj} z_{ikl} + \sum_{i\in X:j\notin i} R^l_{aj} z_{ikl}$$
$$= \sum_{k\in Q}\left(P^l_{aj}\tilde{w}_{jkl} + R^l_{aj}\left(q^l_k - \tilde{w}_{jkl}\right)\right)$$
$$= P^l_{aj} y_j + R^l_{aj}\left(1 - y_j\right)$$

A similar derivation for the MIPSG optimal response constraint $\sum_{i\in X}\left(C^l_{ij} - C^l_{ik}\right) z_{ijl} \geq 0 \;\; j,k \in Q, l \in L$ reveals that we need to consider the $\tilde{w}_{jkl}$ variables:

$$\sum_{i\in X}\left(C^l_{ij} - C^l_{ik}\right) z_{ijl} = \left(P^l_{aj} w_{jl} + R^l_{aj}\left(q^l_j - w_{jl}\right)\right) - \left(P^l_{ak}\tilde{w}_{kjl} + R^l_{ak}\left(q^l_j - \tilde{w}_{kjl}\right)\right).$$

Therefore, in this situation the equivalent MIPSG problem is

$$(\text{MIPSG} - \text{SSG}) \quad \begin{array}{lll} \max_{z,q} & \sum_{l\in L}\sum_{j\in Q} p_l\left(R^l_{dj} + P^l_{dj}\right)\tilde{w}_{jjl} + p_l P^l_{dj} q^l_j & \\ \text{s.t.} & \sum_{j\in Q} q^l_j = 1 & l \in L \\ & q^l_j \in \{0,1\} & j \in Q, l \in L \\ & 0 \leq \tilde{w}_{jkl} \leq q^l_k & j,k \in Q, l \in L \\ & y_j - 1 + q^l_k \leq \tilde{w}_{jkl} \leq y_j + 1 - q^l_k & j,k \in Q, l \in L \\ & \sum_{j\in Q} y_j = m, 0 \leq y \leq 1 & \\ & \left(P^l_{aj} - R^l_{aj}\right)\tilde{w}_{jjl} + R^l_{aj} q^l_j \geq \left(P^l_{ak} - R^l_{ak}\right)\tilde{w}_{kjl} + R^l_{ak} q^l_j & j,k \in Q, l \in L \end{array}$$

23.2.3 *Computational Results*

Preliminary computational results to empirically compare the difficulty in solving these formulations have appeared in Lagos and colleagues (2015) and Casorrán-Amilburu and colleagues (2015). These results, reported next, show that the D2 formulation can be solved faster and for larger problems due to having fewer variables and constraints. However, this benefit is reduced for larger instances and also when the security game payoff matrices have larger variability.

23.3 Methods for Large-Scale Linear Programming

We begin with a brief review of large-scale linear programming methods. We consider three fundamental approaches: column generation, cut generation,

Table 23.1. *Preliminary computation results from Lagos and colleagues (2015)*

Models	No. Resources	No. Instances	Avg. Sol Time	StdDev. Sol Time
D2	1	10	0.6	0.3
	2	10	1.7	0.3
	3	10	5.8	1.6
	4	367	307.9	962.2
	5	10	145.1	52.8
	6	9	335.7	98.9
	7	1	852.6	-
	Avg.	417	283.9	906.9
MIPSG	1	10	0.4	0.2
	2	10	4.2	0.9
	3	10	59.4	22.8
	4	282	389.8	612.9
	5	2	567.3	154.1
	Avg.	314	355.7	591.2

and duality decomposition. Column generation is designed for linear programming problems with a large number of variables; cut generation handles a large number of constraints; and duality decomposition exploits special structure to decompose a single large linear programming problem into several smaller ones. See Bertsimas and Tsitsiklis (1997, chapter 6) for a more extensive discussion of column generation and constraint generation for linear programming, and Ruszczynski (2006, section 4.4) for a review of duality decomposition. We complete this section discussing the use of column generation and cut generation to solve SSGs.

We first discuss column generation. Consider the standard form linear programming problem

$$\min_{x \in \mathbb{R}^n} \{c'x : Ax = b, x \geq 0\}, \tag{23.5}$$

with problem data $c \in \mathbb{R}^n$, $A \in \mathbb{R}^{m \times n}$, and $b \in \mathbb{R}^m$. In problem (23.5), there are n variables and m constraints. Suppose that the number of constraints m is manageable but that the number of variables n poses computational difficulties. We will see in the security games that follow that n is often exponential in the problem size and thus we cannot enumerate all of the variables. For example, the number of decision variables may be equal to the number of possible defender strategies.

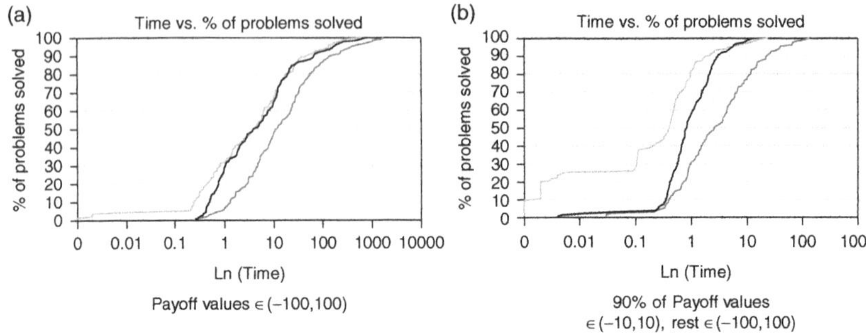

Figure 23.1. Solution time comparison of MIPSG and DOBBS, from Lagos and colleagues (2015).

In general, problem (23.5) is solved using the Simplex algorithm. The Simplex algorithm is an iterative procedure that moves from an extreme point of the feasible region $P = \{x \in \mathbb{R}n \mid Ax = b, x \geq 0\}$ to an adjacent extreme point with lower cost, until no further improvements are possible. See Bertsimas and Tsitsiklis (1997, chapter 3) for a detailed derivation of the Simplex algorithm. Each extreme point in P corresponds to a "basis," which is a subset of size m of the decision variables denoted $\{B(1), \ldots, B(m)\} \subset \{1, \ldots, n\}$. Corresponding to a basis $\{B(1), \ldots, B(m)\}$, we also define $B = [A_{B(1)}, \ldots, A_{B(m)}] \in \mathbb{R}^{m \times m}$ to be the matrix of columns corresponding to the basic indices and $c_B = (c_{B(1)}, \ldots, c_{B(m)}) \in \mathbb{R}^m$ to be the vector of costs corresponding to basic indices. A basis $\{B(1), \ldots, B(m)\}$ determines a "basic feasible solution" of P defined by

$$Ax = b, x_i = 0, \forall i \notin \{B(1), \ldots, B(m)\}, \quad x \geq 0,$$

or equivalently $Bx_B = b$ where $x_B = (x_{B(1)}, \ldots, x_{B(m)})$. Notice that this system consists of n equations in n unknowns. We know that basic feasible solutions correspond to the extreme points of P, and the Simplex algorithm returns an optimal basic feasible solution whenever one exists.

When n is large, we can use the column generation algorithm to solve problem (23.5). We start with a small subset $I \subset \{1, \ldots, n\}$ (generally $|I|$ is much smaller than n) and then consider the restricted problem:

$$\min_{x \in \mathbb{R}^{|I|}} \left\{ \sum_{i \in I} c_i x_i : \sum_{i \in I} A_i x_i = b, x \geq 0 \right\}. \tag{23.6}$$

Problem (23.6) is obtained from problem (23.5) by setting $x_i = 0$ for all $i \notin I$. It is clear that the optimal value of problem (23.6) is greater than the optimal value of problem (23.5), since any element in the feasible region of problem (23.6) corresponds to an element in the feasible region of problem (23.5) with the additional constraints $x_i = 0$ for all $i \notin I$. The idea is that I is small enough so that problem (23.6) is computationally tractable.

Once an instance of problem (23.6) is solved, we have an optimal basis $\{B(1), \ldots, B(m)\}$, a corresponding matrix $B = [A_{B(1)}, \ldots, A_{B(m)}]$, and cost vector $c_B = (c_{B(1)}, \ldots, c_{B(m)})$ corresponding to the basic indices. The next step is to determine if any further variables in $\{1, \ldots, n\} \backslash I$ should be added to I to obtain a new instance of problem (23.6). For this analysis, we compute the reduced costs

$$c_i - c_B' B^{-1} A_i$$

for every $i \in \{1, \ldots, n\} \backslash I$. The reduced costs have the following interpretation. For $i \in \{1, \ldots, n\} \backslash I$, $c_i - c'_B B^{-1} A_i$ is the rate of change in the cost if variable x_i is increased by one, the variables corresponding to $\{B(1), \ldots, B(m)\}$ are adjusted to maintain feasibility via

$$Bx_B + A_i = 0,$$

or $x_B = -B^{-1} A_i$, and all other variables are held constant. In the Simplex algorithm, we search for the variable with the most negative reduced cost by solving

$$\min_{i \in \{1,\ldots,n\}} \left\{ c_i - c_B' B^{-1} A_i \right\}. \tag{23.7}$$

If the quantity in equation (23.7) is nonnegative, then there is no need to add any more variables since we already have an optimal solution to problem (23.5). However, if equation (23.7) is negative, then we have found an entering variable that can reduce the cost. In this case, we update the set I to include the new variable and then we solve a new instance of problem (23.6).

Column generation solves a sequence of instances of problem (23.6), until it is determined that there is no longer any need to augment the set I by adding more variables. This determination is made by evaluating the expression in (23.7). We point out that, in general, there are no convergence rate guarantees for column generation, i.e., we do not know a priori how long we will need to run column generation before finding an optimal solution of the original problem (23.5). Further, for a general problem (23.5), the minimization in equation (23.7) requires enumeration of $c_i - c'_B B^{-1} A_i$ over all $i = 1, \ldots, n$. However, it is often the case that the minimization in equation (23.7) can be made more efficient by exploiting special structure of problem (23.5).

In Bertsimas and Tsitsiklis (1997, section 6.2), column generation is applied to the cutting stock problem – which is a classic example of the column generation technique. Here, the problem of computing the most negative reduced cost in equation (23.7) is itself shown to be an integer programming problem. In the cutting stock problem, we imagine a paper company that must produce rolls of paper of different widths to meet customer demand. The company has a single supply of large rolls of paper of width W, while customer demand requires b_i rolls of smaller widths $w_i \leq W$ for $i = 1, \ldots, m$. The large roll of width W can be cut into smaller rolls in several different patterns. Patterns are represented by the columns $A_j \in \mathbb{R}^m$ where a_{ij} is the number of rolls of width w_i produced by pattern j. Feasible patterns must satisfy $\sum_{i=1}^{m} a_{ij} e_i \leq W$ so that they can be cut from the single large roll. We let n be the number of possible patterns, which in general is very large.

The paper company would like to minimize the number of large rolls necessary to meet customer demand, and it solves

$$\min \sum_{j=1}^{n} x_j \tag{23.8}$$

$$\text{s.t.} \sum_{i=1}^{n} a_{ij} x_j = b_i, \quad \forall_i = 1, \ldots, m, \tag{23.9}$$

$$x_j \geq 0, \qquad \forall_i = 1, \ldots, n. \tag{23.10}$$

Since n is quite large, it is difficult to solve problems (23.8)–(23.10) directly because it has too many variables. If we apply the column generation technique, then given a basis matrix B we want to solve

$$\min_{j=1,\ldots,n} \{c_j - c_B' B^{-1} A_j\} = \min_{j=1,\ldots,n} \{1 - p' A_j\} = 1 - \min_{j=1,\ldots,n} \{p' A_j\}$$

to identify a new pattern, where we use $c_j = 1$ for all $j = 1, \ldots, n$ and $p = c'_B B^{-1}$. Identifying an entering variable thus amounts to solving the maximization problem $\max_{j=1,\ldots,n} \{p' A_j\}$, which for this problem has the following special structure:

$$\max \sum_{i=1}^{m} p_i a_i \tag{23.11}$$

$$\text{s.t.} \sum_{i=1}^{m} a_i w_i \leq W, \tag{23.12}$$

$$a_i \leq 0, \quad \forall_i = 1, \ldots, m. \tag{23.13}$$

Typically, integrality constraints ($a_i \in \mathbb{Z}_+$ for all $i = 1, \ldots, m$) are included in problem (23.11)–(13) as well.

To make this example specific, suppose large rolls have width $W = 6$ and the smaller rolls have widths $w_1 = 1$, $w_2 = 2$, and $w_3 = 3$. Also let the customer demand for the smaller rolls be $b_1 = 5$, $b_2 = 5$, and $b_3 = 5$. We initialize problems (23.8)–(23.10) with the restricted problem

$$\min \sum_{j=1}^{3} x_j \tag{23.14}$$

$$\text{s.t.} \begin{bmatrix} 1 \\ 0 \\ 0 \end{bmatrix} x_1 + \begin{bmatrix} 0 \\ 1 \\ 0 \end{bmatrix} x_2 + \begin{bmatrix} 0 \\ 0 \\ 1 \end{bmatrix} x_3 = \begin{bmatrix} 5 \\ 5 \\ 5 \end{bmatrix} \tag{23.15}$$

$$x_1, x_2, x_3 \geq 0, \tag{23.16}$$

where we use the initial patterns A_1, A_2, and A_3, which correspond to cutting exactly one smaller roll of each width from the larger roll and discarding the rest of the material. Clearly, these patterns are suboptimal, but they readily provide a convenient initial basis. Solving problems (23.14)–(23.16) gives $(x_1, x_2, x_3) = (5, 5, 5)$, and

$$p = (1,1,1) \begin{bmatrix} 1 & 0 & 0 \\ 0 & 1 & 0 \\ 0 & 0 & 1 \end{bmatrix} = (1,1,1).$$

The resulting instance of problems (23.11)–(23.13) is then

$$\max a_1 + a_2 + a_3 \tag{23.17}$$

$$\text{s.t. } a_1 + 2a_2 + 3a_3 \leq 6, \tag{23.18}$$

$$a_1, a_2, a_3 \geq 0, \tag{23.19}$$

which gives the new pattern $(a_1, a_2, a_3) = (6, 0, 0)$. In this pattern, we cut the large roll of width W into six smaller rolls of width $w_1 = 1$. In the next iteration, we solve

$$\min \sum_{j=1}^{3} x_j \tag{23.20}$$

$$\text{s.t.} \begin{bmatrix} 1 \\ 0 \\ 0 \end{bmatrix} x_1 + \begin{bmatrix} 0 \\ 1 \\ 0 \end{bmatrix} x_2 + \begin{bmatrix} 0 \\ 0 \\ 1 \end{bmatrix} x_3 + \begin{bmatrix} 6 \\ 0 \\ 0 \end{bmatrix} x_4 = \begin{bmatrix} 5 \\ 5 \\ 5 \end{bmatrix} \tag{23.21}$$

$$x_1, x_2, x_3, x_4 \geq 0, \tag{23.22}$$

starting from the previous basis B. Problems (23.20)–(23.22) now include the new pattern identified in problems (23.17)–(23.19).

Next, we describe cut generation on the linear programming dual to problem (23.5):

$$\min_{p \in \mathbb{R}^m} \left\{ p'b : p'A \leq c \right\}, \tag{23.23}$$

see Bertsimas and Tsitsiklis (1997, chapter 4) for a thorough discussion of linear programming duality. Problem (23.23) is a linear programming problem with inequality constraints only. Note that problem (23.23) has m variables and n inequality constraints. In general, it is more natural to apply cut generation to inequality-constrained linear programming problems rather than equality-constrained linear programming problems. In fact, the cut generation procedure we are about to describe is more general than linear programming and can in principle be applied to any optimization problem with a large number of constraints. For example, Kortanek and No (2013) develop a cutting-plane algorithm for semi-infinite convex optimization problems (convex optimization problems with finitely many variables and infinitely many constraints).

Suppose that the number of variables m is manageable but that the number of constraints n in problem (23.23) is very large and presents computational difficulties. For example, in a later security game we will see a polyhedral feasible region specified by a very large number of constraints. Cut generation works by choosing a subset $J \subset \{1, \ldots, m\}$ of the constraints and then solving the relaxed problem

$$\min_{p \in \mathbb{R}^m} \left\{ p'b : p'A_j \leq c, \forall_j \in J \right\}. \tag{23.24}$$

The idea is that the number of constraints in problem (23.24) is manageable. We observe that the optimal value of problem (23.24) is less than the optimal value of problem (23.23), since the feasible region of problem (23.23) is contained in the feasible region of problem (23.24).

Given an optimal solution p^* to problem (23.24), the next step is to identify any violated constraints of problem (23.23). This step is called "cut generation" since we are looking for constraints in problem (23.23) that are necessary to add to J. For problem (23.23), we solve the cut generation problem

$$\min_{j\in\{1,\dots,m\}}\left\{c_j-\left(p^*\right)' A_j\right\}. \tag{23.25}$$

If equation (23.25) is nonnegative, then p^* is actually feasible for the original problem (23.23). However, if equation (23.25) is strictly negative, then we have found a violated constraint. We then add this constraint to J and resolve problem (23.24) using the larger constraint index set J. Just like column generation, cut generation solves a sequence of linear programming problems. In cut generation, we are building a tighter and tighter approximation of the feasible region to the original problem (23.23).

Column generation and cut generation are "dual" to each other. This observation is confirmed by noting that applying column generation to problem (23.5) is the same as applying cut generation to its dual problem (23.23). Notice that equations (23.7) and (23.25) are identical to each other if we set $p = c'_B B^{-1}$.

After seeing column generation and cut generation, it is natural to ask if there is a procedure that can handle both a large number of variables and a large number of constraints simultaneously. In general, there is no universal procedure for addressing the case where there is both a large number of variables and a large number of constraints. However, some problem specific methods are available. For instance, de Farias and Van Roy (2004) combine basis function approximation and constraint sampling to give approximate solutions for the linear programming approach to Markov decision processes. Duality decomposition is another approach for handling a large number of variables and constraints for linear programming problems with special structure.

To illustrate, consider

$$\min_{x\in\mathbb{R}^n,\, y\in\mathbb{R}^q} \quad c'x+d'y \tag{23.26}$$

$$\text{s.t. } Ax+By=b, \tag{23.27}$$

$$Dx=g, \tag{23.28}$$

$$Fy=h, \tag{23.29}$$

$$x,y\geq 0. \tag{23.30}$$

Problems (23.26)–(23.30) have a special structure that can be exploited by duality decomposition. In particular, we see that there are two groups of variables in problems (23.26)–(23.30), x and y. These variables only appear together in constraint (23.27), if this constraint were omitted, then we would actually have two separate linear programming problems: one in the variables x alone and one in the variables y alone.

Based on this observation, we will "decompose" around the constraint (23.27). Introduce the Lagrange multiplier $p \in \mathbb{R}^m$ for constraint (23.27), and define the new problem

$$\begin{aligned} g(p) \triangleq \min_{x\in\mathbb{R}^n, y\in\mathbb{R}^q} \; & c'x + d'y + p'(b - Ax - By) \\ \text{s.t. } & Dx = g, \\ & Fy = h, \\ & x, y \geq 0, \end{aligned}$$

as a function of p. We interpret the term $p'(b - Ax - By)$ that appears in the objective of $g(p)$ as a "pricing" term that replaces the hard constraint $Ax + By = b$. By inspection, we see that $g(p)$ decomposes into two smaller linear programming problems

$$\begin{aligned} g_x(p) = \min_{x\in\mathbb{R}^n} \; & c'x - p'Ax \\ \text{s.t. } & Dx = g, \\ & x \geq 0, \end{aligned}$$

corresponding to x and

$$\begin{aligned} g_y(p) = \min_{y\in\mathbb{R}^q} \; & d'y - p'By \\ \text{s.t. } & Fy = h, \\ & y \geq 0, \end{aligned}$$

corresponding to y. The dual to problems (23.26)–(23.30) is then:

$$\max_{p\in\mathbb{R}^m} \left\{ p'b + g_x(p) + g_y(p) \right\}. \tag{23.31}$$

We can solve problem (23.31) via the subgradient method, for example. Just like column generation and cut generation, we solve problems (23.26)–(23.30) by solving a sequence of linear programming problems. In the case of duality decomposition, we update the pricing term p by solving the dual problem

(23.31). See Bertsimas and Tsitsiklis (1997, section 4.10) for a discussion of general linear programming duality and Ruszczynski (2006, section 7.1.3) for the theory of the dual subgradient method.

23.3.1 *Column generation*

In this section, we illustrate the application of the column generation technique to security games. We focus on the work in Jain and colleagues (2010), which studies massive Stackelberg security games for defending infrastructure. In this setting, defender resources must obey scheduling constraints that limit their ability to move from one target to the next. The equilibrium condition for this class of games corresponds to an extremely large-scale optimization problem. In particular, there is a decision variable in this optimization corresponding to every possible defender strategy. Since the number of possible schedules for defender resources is exponential in the number of targets, we see that there is an exponential number of such decision variables. Standard algorithms are not able to scale up to solve this problem. In response, Jain and colleagues (2010) combine column generation with branch and bound to compute a Stackelberg equilibrium of this game. This procedure is known as branch and price and was first developed for large-scale, mixed-integer programming problems in Barnhart and colleagues (1998).

This Stackelberg game is played over a set of targets that we denote as $T = \{1, \ldots, |T|\}$. The defender acts first and chooses a mixed strategy over the set of patrol schedules on T for its assets. Then, the attacker observes the defender's mixed strategy and chooses a target in T to attack.

The defender has a set of resources indexed by Λ, where a specific resource type is denoted $\lambda \in \Lambda$, and the number of available resources of each type is $R = \{r_1, r_2, \ldots, r_{|\Lambda|}\}$. Each resource type can be assigned to a schedule $s \subseteq T$, which covers multiple targets in T. The set of all available schedules is $S \subseteq \wp(T)$, where $\wp(T)$ is the power set of the set of targets T (the set of all subsets of targets), and $S_\lambda \subseteq S$ is the set of available schedules for each resource type $\lambda \in \Lambda$. The defender's pure strategies consist of joint schedules where each of its assets is assigned to a schedule. For this model, we impose the assumption that each target may be covered by at most one resource in a joint schedule. A joint schedule j can be represented compactly by a vector $\boldsymbol{P}_j = \langle P_{jt} \rangle \in \{0,1\}^n$ where $P_{jt} = 1$ if target t is covered in schedule j, and $P_{jt} = 0$ otherwise. The set of all feasible joint schedules is denoted J, it includes all joint schedules $\boldsymbol{j}$ that respect the constraints on schedules imposed by S. We can define a mapping $M: J \rightarrow \{0, 1\}^n$, which maps a joint schedule j to its associated coverage vector $\boldsymbol{P}_j$: $M(j) = \boldsymbol{P}_j = \langle P_{jt} \rangle$, where $P_{jt} = 1$ if $t \in \cup_{s \in j} s$ (target t is covered by some schedule within the joint schedule j) and $P_{jt} = 0$ otherwise (target t is not covered by any asset within j). The defender's mixed strategies consist of the set of probability distributions over J. Such a mixed strategy looks like $\boldsymbol{x} = \langle x_j \rangle \in [0,1]^{|J|}$, where x_j is the probability of using schedule $j \in J$. Corresponding to any defender mixed strategy $\boldsymbol{x}$, we also have a vector of coverage probabilities that indicate the probability that each target in T is covered by the defender. If we let $\boldsymbol{c} = \langle c_t \rangle \in [0,1]^n$ be the vector of coverage probabilities corresponding to $\boldsymbol{x}$, then $c_t = \sum_{j \in J} P_{jt} x_j$ is the probability of covering target t. The attacker's pure strategy space is the set of targets in T that could be attacked. The attacker's mixed strategies are probability distributions $\boldsymbol{a} = \langle a_i \rangle \in [0,1]^n$ over targets T, where a_i is the probability of attacking target $t \in T$.

Given a strategy profile $\langle \boldsymbol{c}, \boldsymbol{a} \rangle$ that consists of the mixed strategies of both the defender and the attacker (where the defender's strategy is represented by the vector of coverage probabilities), the expected utilities for the defender and attacker are:

$$U_d(c,a) = \sum_{t \in T} a_t \left(c_t U_d^c(t) + (1 - c_t) U_d^u(t) \right)$$

and

$$U_a(c,a) = \sum_{t \in T} a_t \left(c_t U_a^c(t) + (1 - c_t) U_a^u(t) \right).$$

The goal in Jain and colleagues (2010) is to compute a Stackelberg equilibrium for the security game that we have just described. In particular, the equilibrium conditions of this game can be represented as a mixed-integer linear programming problem, where the integer variables correspond to the attacker strategy and the continuous variables correspond to the defender mixed strategy (we know that the attacker has an optimal pure strategy when faced with any defender mixed strategy). The resulting mixed-integer linear programming problem is large-scale and is very difficult to solve. Jain and colleagues (2010) overcome this difficulty with a two-pronged approach. First, given an attacker strategy $\boldsymbol{a}$, Jain and colleagues (2010) use a linear programming formulation and column generation to compute the optimal defender mixed strategy. Second, Jain and colleagues (2010) use branch and price to update the integer variables corresponding to the attacker strategy.

The first component of the algorithm in Jain and colleagues (2010) uses column generation to search through the exponential defender strategy space, given a fixed attacker strategy $\boldsymbol{a}$, and to compute an optimal defender mixed strategy. It is divided into a "master" and a "slave" component. The master is a restricted problem with a small subset of variables (corresponding to joint schedules). Given a candidate solution to the master problem, the slave identifies new variables to add to the master so that its solution can be improved. We introduce some additional notation:

- $\boldsymbol{P} \in \mathbb{R}^{n \times |J|}$ is a matrix which maps between targets T and joint schedules $\boldsymbol{J}$;
- $d \in \mathbb{R}$ is the defender reward;
- $k \in \mathbb{R}$ is the adversary reward;
- $\boldsymbol{d} \in \mathbb{R}^n$ is a column vector where every component is equal to d, where the dimension is the number of targets;
- $\boldsymbol{k} \in \mathbb{R}^n$ is a column vector where every component is equal to k, where the dimension is the number of targets;
- $\boldsymbol{D} \in \mathbb{R}^{n \times n}$ is the diagonal matrix of defender rewards $U_d^c(t) - U_d^u(t)$ (all off-diagonal elements are zero);
- $\boldsymbol{A} \in \mathbb{R}^{n \times n}$ is the diagonal matrix of attacker rewards $U_a^c(t) - U_a^u(t)$ (again, all off-diagonal elements are zero);
- $U_d^u \in \mathbb{R}^n$ is the vector of defender rewards for uncovered targets $U_d^u(t)$;
- $U_a^u \in \mathbb{R}^n$ is the vector of attacker rewards for uncovered targets $U_a^u(t)$;
- $M \gg 0$ is a very large positive constant.

The matrix $\boldsymbol{P}$ is a succinct representation of joint schedules; coverage probabilities can be computed from $\boldsymbol{P}$ and $\boldsymbol{x}$ via $\boldsymbol{c} = \boldsymbol{Px}$. The defender expected payoff across targets is the vector $\boldsymbol{DP}x + U_d^u \in \mathbb{R}^n$, and the payoff for target t is $(\boldsymbol{DP}x + U_d^u)_t$. Similarly, the vector of attacker expected payoffs is $\boldsymbol{AP}x + U_a^u$, and the attacker payoff for target t is $(\boldsymbol{AP}x + U_a^u)_t$. Let $e \in \mathbb{R}^n$ be the vector of all ones. We can now write the master problem that optimizes over defender mixed strategies given a fixed attacker pure strategy as:

$$\max d \tag{23.32}$$

$$\text{s.t. } d - \boldsymbol{DP}x - U_d^u \le (e - a)M, \tag{23.33}$$

$$k - \boldsymbol{AP}x - U_a^u \le (e - a)M, \tag{23.34}$$

$$\boldsymbol{AP}x - U_a^u \le k, \tag{23.35}$$

$$\sum_{j \in J} x_j = 1 \tag{23.36}$$

$$x \ge 0, \tag{23.37}$$

which is a linear programming problem in the variables ($\boldsymbol{x}$, d, k). Problems (23.32)–(23.37) maximizes the defender reward given a fixed attacker's strategy. Since we know the attacker is restricted to pure strategies, the vector $\boldsymbol{a}$ has the value 1 in one of its components, and zeroes elsewhere. Thus, the constraints (23.33) and (23.34) are only active for one target (the target t^* for which $a_{t^*} = 1$). Constraint (23.35) requires this target to be a best response to the defender's mixed strategy.

Given a candidate solution to problems (23.32)–(23.37), we want to check to see if we can improve on this solution by adding more schedules, which is equivalent to adding to the columns of $\boldsymbol{P}$. The slave problem fulfills this purpose by identifying columns to add to the matrix $\boldsymbol{P}$, which can improve on the current best candidate solution to problems (23.32)–(23.37). The columns of $\boldsymbol{P}$ correspond to the components of $\boldsymbol{x}$, so the slave checks the reduced cost of variables x_j with $j \in J$ that have not yet been included in $\boldsymbol{P}$. As we mentioned in the previous section, a negative reduced cost corresponds to a variable that can improve on the current candidate solution. The reduced cost of the variable x_j for $\boldsymbol{j} \in \boldsymbol{J}$ is

$$\bar{c}_j = w^T \left(\boldsymbol{DP}_j\right) + y^T \left(\boldsymbol{AP}_j\right) - z^T \left(\boldsymbol{AP}_j\right) - h,$$

where $\boldsymbol{w}$, $\boldsymbol{y}$, z, and h are the dual variables corresponding to constraints (23.33), (23.34), (23.35), and (23.36), respectively. The next step is to solve the problem:

$$\min_{j \in J} \left\{\bar{c}_j\right\} \tag{23.38}$$

to find the variable x_j that minimizes the reduced costs. If the optimal value of equation (23.38) is nonnegative, then the current candidate solution is optimal. Otherwise, we have identified a variable x_j with a negative reduced cost and a corresponding schedule j that should be added to the columns of $\boldsymbol{P}$.

Given the exponential space of defender strategies, it is too inefficient to enumerate $\bar{c}_j$ for all j to solve (23.38). Instead, we formulate (23.38) as a special class of linear programming problems. In particular, we write (23.38) as a minimum cost network flow (MCNF) problem. MCNF problems are well studied and can be solved efficiently, and they encompass a wide class of network optimization problems. An MCNF finds a flow through a given network that has the lowest possible cost, where a flow is a connected set of arcs from the network source to the network sink that moves through intermediate nodes. The idea is that feasible flows in the MCNF correspond to joint operating schedules for defender resources. For clarity, a flow may consist of several paths.

We can set up a network that represents possible joint schedules as follows. For each defender type $\lambda \in \Lambda$, a source node with supply r_λ is created.

Additionally, there is a single sink node with demand $\Sigma_{\lambda\in\Lambda} r_\lambda$. The targets in a schedule s are represented by pairs of nodes $(a_{s_\lambda,t}, b_{s_\lambda,t})$. For every feasible schedule $s_\lambda \in S_\lambda$, there is a path from the source to the sink of the form:

$$\langle \text{source}_\lambda, a_s, t_1, b_s, t_1, a_s, t_2, \ldots, b_s, t_L, \text{sink} \rangle$$

which visits the sequence of targets t_1, t_2, …, t_L. The capacity on every link is set to one. We also add an artificial dummy link with infinite capacity; this device is used for resources that are not assigned to targets. The total number of resources assigned to each target t is

$$\text{assigned}(t) \triangleq \sum_{s\in S} \text{flow}\left[\mathit{link}\left(a_{s,t}, b_{s,t}\right)\right].$$

Constraints are enforced so that each target is assigned at most one resource, i.e., assigned $(t) \le 1$ for all $t \in T$.

The costs of the flows in the MCNF are designed to correspond to the reduced costs $\bar{c}_j$. The cost of target t is $w_t D_t + y_t A_t - z_t A_t$, where w_t, y_t, and z_t are the tth components of $\boldsymbol{w}$, $\boldsymbol{y}$, and z, $D_t = U_d^c(t) - U_d^u(t)$, and $A_t = U_a^c(t) - U_a^u(t)$. The overall objective sums the contribution to the reduced costs from each individual flow. The resulting MCNF can be represented as

$$\min \text{flow} \left\{ \sum_{(a_{s,t}, b_{s,t})} \hat{c}_t \text{flow}\left[\left(a_{s,t}, b_{s,t}\right)\right] - h \right\}. \tag{23.39}$$

Problem (23.39) also appears in Brown and colleagues (2014) to generate schedules. The model in Brown and colleagues (2014) generalizes the one in Jain and colleagues (2010) by considering ambiguity over the attacker type, i.e., it is a robust optimization problem.

Since attacker pure strategies are integer-valued, one needs to use a combinatorial algorithm to optimize over them. It is observed in Jain and colleagues (2010) (via numerical experimentation) that the linear programming relaxation of the original mixed-integer linear programming problem performs badly, and thus a combinatorial method (such as branch and bound) is needed. Pure branch and bound can be applied; however, Jain and colleagues (2010) propose a heuristic algorithm that is observed to perform well experimentally. Consider

$$\min k \tag{23.40}$$

$$\text{s.t. } U_a(c) = Ac + U_a^u, \tag{23.41}$$

$$0 \leq k - U_a(c) \leq (e - \boldsymbol{q})M, \tag{23.42}$$

$$c_t = \sum_{s \in S} \tilde{c}_{t,s}, \quad \forall t \in T, \tag{23.43}$$

$$\sum_{s \in S_\lambda} \tilde{c}_{T_\lambda(s),s} \leq r_\lambda, \quad \forall \lambda \in \Lambda, \tag{23.44}$$

$$\sum_{t \in T} c_t \leq L \sum_{\lambda \in \Lambda} r_\lambda, \tag{23.45}$$

$$c \leq \boldsymbol{q}, \tag{23.46}$$

$$\boldsymbol{q} \in \{0,1\}^n, c \in [0,1]^n, c_{t,s} \in [0,1], \forall t \in T, s \in S. \tag{23.47}$$

Problems (23.40)–(23.47) are based on the model developed in Kiekintveld and colleagues (2009). This problem set minimizes the attacker's maximum payoff. The vector $\boldsymbol{q} \in \{0, 1\}^n$ represents the attack set; each component of $\boldsymbol{q}$ is equal to one corresponding to every target that gives the attacker a maximal expected payoff. The variables $\tilde{c}_{t,s}$ represent the coverage probability of each target $t \in T$ in schedule $s \in S$. The total coverage probability c_t is then the sum of the individual coverage probabilities, $c_t = \sum_{s \in S} \tilde{c}_{t,s}$ for all $t \in T$. The set T_λ contains one representative target from each schedule $s \in S_\lambda$, this selection is used to bound total coverage assigned to resource type λ by r_λ (equation (23.44)). Constraint (23.44) is equivalent to stating that the total flow cannot exceed the available supply. The total coverage is $\sum_{t \in T} c_t$, which must be bounded by the total amount of resources available multiplied by the maximum length L of any schedule (constraint (23.45)). Constraint (23.46) reflects the fact that it is never optimal to assign coverage to a target that will not be attacked.

In Jain and colleagues (2010), problems (23.40)–(23.47) are initially solved once. Targets in the attack set are then sorted by decreasing defender reward. The resulting maximum value of these rewards is then an upper bound on the defender's optimal reward. Next, the algorithm in Jain and colleagues (2010) sets $a_t = 1$ corresponding to the maximum valued target. A solution of problems (23.32)–(23.37) is found using column generation; the resulting solution is a lower bound on the defender payoff at equilibrium. The procedure stops if this lower bound meets the upper bound. Otherwise, the process is repeated with the next highest defender reward. By Jain and colleagues (2010, theorem 1), the defender payoff computed by this procedure is an upper bound on the defender reward in a Stackelberg equilibrium.

23.3.2 *Cut Generation*

In this section, we discuss the application of cut generation to security games. We focus on the work in Yang and colleagues (2014), which extended the literature on security games by considering bounded rationality models for attackers. More precisely, terrorists and criminals cannot be assumed to meet the common game-theoretic assumption of perfect rationality. Thus, Yang and colleagues (2014) incorporated the quantal response model, which is based on stochasticity in human behavior.

In this setting, there are M defender resources all of a single type. We do not consider schedules in this model; here, the jth defender pure strategy is captured by a vector $A_j = (A_{tj})^T \in \{0, 1\}^n$ where A_{tj} indicates whether target t is covered by A_j, and $\mathcal{A} = \{A_j\}$ is the set of all feasible resource assignments. The defender's mixed strategy over $\mathcal{A}$ is $\boldsymbol{x} = \langle x_j \rangle$ where $x_j \in [0, 1]$ is the probability of choosing A_j. In a more compact representation, the coverage probabilities on targets $c = \langle c_t \rangle$, where $c_t = \sum_{A_j \in A} x_j A_{tj}$, give the marginal probabilities of covering each target. Given c_t, the defender's expected utility if t is attacked is $c_t U_d^c(t) + (1 - c_t) U_d^u(t)$ and the attacker's expected utility is $c_t U_a^c(t) + (1 - c_t) U_a^u(t)$.

The bounded rationality model is as follows. Given the defender's marginal strategy $\boldsymbol{c}$, $q_t(\boldsymbol{c})$ reports the probability that the adversary will attack target t. Subsequently, the defender's optimization problem is

$$\max_c F(c) \triangleq \sum_{t \in T} \left[c_t U_d^c(t) + (1 - c_t) U_d^u(t) \right] q_t(c) \tag{23.48}$$

$$\text{s.t. } c \in \chi_{f1} \triangleq \left\{ c = \sum_{A_j \in A} x_j A_j, \sum_j x_j = 1, x_j \in [0,1] \right\}, \tag{23.49}$$

$$c \in \chi_{f2} \triangleq \{ Bc \le b \}. \tag{23.50}$$

The expression $q_t(\boldsymbol{x})$ depends on the particulars of the adversary model. Using the quantal response mode, the objective function is explicitly

$$F(c) = \sum_{t \in T} \frac{e^{\lambda[c_t U_a^c(t) + (1 - c_t) U_a^u(t)]}}{\sum_{i \in T} e^{\lambda[c_i U_a^c(i) + (1 - c_i) U_a^u(i)]}} \left[c_t U_d^c(t) + (1 - c_t) U_d^u(t) \right].$$

In this setting, problems (23.48)–(23.50) are a nonlinear nonconvex optimization problem – which is difficult to solve in general. Yang and colleagues

(2014) solve problems (23.48)–(23.50) by using a binary search method and approximating the resulting problem with a mixed-integer linear programming (MILP) problem. Let $f_t(c_t) = e^{\lambda\left[c_t U_a^c(t) + (1-c_t) U_a^u(t)\right]}$. This problem is based on dividing the range of each c_i into K segments and then doing a piecewise linear approximation of $f_i(c_i)$ and $c_i f_i(c_i)$ on each segment. The resulting problem is:

$$\min_{x,z} \sum_{i \in T} \left(r - P_i^d\right) \left(f_i(0) + \sum_{k=1}^{K} \gamma_{ik} c_{ik} \right) - \sum_{i \in T} \alpha_i \sum_{k=1}^{K} \mu_{ik} c_{ik} \tag{23.51}$$

$$\text{s.t. } 0 \le c_{ik} \le 1/K, \quad \forall i, k = 1, \ldots, K, \tag{23.52}$$

$$z_{ik}/K \le c_{ik}, \quad \forall i, k = 1, \ldots, K-1, \tag{23.53}$$

$$c_i(k+1) \le z_{ik} \quad \forall i, k = 1, \ldots, K-1, \tag{23.54}$$

$$z_{ik} \in \{0,1\}, \quad \forall i, k = 1, \ldots, K-1, \tag{23.55}$$

$$c \in \chi_f. \tag{23.56}$$

The objective is a piecewise linear approximation of $\sum_{i \in T} \left(r - P_i^d\right) f_i(c_i) - \sum_{i \in T} \alpha_i c_i f_i(c_i)$ where $\alpha_i = R_i^d - P_i^d$ for all $i \in T$, γ_{ik} is the slope of $f_i(c_i)$ in the kth segment and μ_{ik} is the slope of $c_i f_i(c_i)$ in the kth segment. Note that c_i is replaced by the variables $c_i = \sum_{k=1}^{K} c_{ik}$ and the integer variables z_{ik} determine the segment in which c_i lies.

There is considerable difficulty with scaling up problems (23.51)–(23.56). Instead, we use cut generation to solve our original problems (23.48)–(23.50). First, we rewrite problems (23.48)–(23.50) as

$$\max_c \left\{ F(c) : Hc \le h, Bc \le b, 0 \le c_i \le 1, \forall i \in T \right\}, \tag{23.57}$$

where the constraints $H\,\mathbf{c} \le h$ represent the convex hull of χ_{f1}. Problem (23.57) is the master problem. We solve successive iterations of the relaxed problem

$$\max_c \left\{ F(c) : \tilde{H}c \le \tilde{h}, Bc \le b, 0 \le c_i \le 1, \forall i \in T \right\}, \tag{23.58}$$

which replaces the constraints $H\,\mathbf{c} \le h$ with the approximation $\tilde{H}c \le \tilde{h}$.

Each time we solve problem (23.58), we check feasibility of the resulting solution with respect to problem (23.57), add any violated constraints to

$\tilde{H}c \leq \tilde{h}$, and then resolve problem (23.58). We let $A = \left[A_j\right]_{A_j \in \mathcal{A}}$ be a matrix whose columns represent defender pure strategies $A_j \in \mathcal{A}$. The separation problem for problem (23.57) is

$$\min\left\{\left\|Ax - \tilde{c}\right\|_1 : BAx \leq b, \sum_{A_j \in \mathcal{A}} x_j = 1, x_j \geq 0, \forall A_j \in \mathcal{A}\right\}, \tag{23.59}$$

where $\|\cdot\|_1$ denotes the 1–norm. Problem (23.59) can be written as a linear programming problem by replacing the 1-norm objective $\left\|Ax - \tilde{c}\right\|$ with $\Sigma_{i \in T} z_i$ and the constraints $z + Ax \geq \tilde{c}$ and $z - Ax \geq -\tilde{c}$ to obtain

$$\min_{a,z} \sum_{i \in T} z_i \tag{23.60}$$

$$\text{s.t. } z + Ax \geq \tilde{c}, \tag{23.61}$$

$$z - Ax \geq \tilde{c} \tag{23.62}$$

$$-BAx \geq -b, \tag{23.63}$$

$$\sum_{A_j \in \mathcal{A}} x_j = 1, x_j \geq 0, \forall A_j \in \mathcal{A}. \tag{23.64}$$

If the optimal solution to problems (23.60)–(23.64) is zero, then $\tilde{c}$ is feasible. Otherwise, we identify a violated constraint. If $\tilde{c}$ is infeasible, then the hyperplane $(\boldsymbol{y}^*)^T \boldsymbol{x} - (\boldsymbol{g}^*)^T \boldsymbol{b} + u^* = 0$ separates $\tilde{c}$ and χ_f. In fact, this cut is deep as shown in Yang and colleagues (2013, Lemma 1).

The dual of problem (23.59) is

$$\max_{y,u,g} \tilde{c}^T y - b^T g + u \tag{23.65}$$

$$\text{s.t. } A^T y - A^T B^T g + u \leq 0, \tag{23.66}$$

$$-1 \leq y \leq 1, \tag{23.67}$$

$$g \geq 0. \tag{23.68}$$

Constraint generation is also applied to solve the dual problems (23.65)–(23.68). To improve the performance of the algorithm, Yang and colleagues

(2014) propose a heuristic modification to the cut generation problem (23.60)–(23.64). In particular, they propose replacing the objective $\sum_{i\in T} z_i$ with

$$\sum_{i\in T}\left(\nabla_i F\left(\tilde{c}\right)+\xi\right)z_i,$$

where $\nabla_i F\left(\tilde{c}\right)$ is the gradient of $F\left(\tilde{c}\right)$ at $\tilde{c}$, and $\xi > 0$ is a fixed constant to ensure that $\nabla_i F\left(\tilde{c}\right)+\xi>0$ for all $i \in T$. It is argued that the expression $\sum_{i\in T}\left(\nabla_i F\left(\tilde{c}\right)+\xi\right)z_i$ better takes the original objective $F\left(\tilde{c}\right)$ of problems (23.48)–(23.50) into account. Yang and colleagues (2013, Lemma 4) confirm that $\tilde{c}$ is feasible if and only if the optimal value of problems (23.60)–(23.64) with the modified objective is zero.

23.4 Heuristic Methods to Solve Large SSGs

Very large instances of SSGs remain that are challenging to solve even with state-of-the-art computational capabilities. In particular, when the action space of the defender or the attacker (or both) consist of paths on a network, then the number of possible actions of the players increase exponentially in the number of edges and nodes. The previous work that addresses the effect of geographical location on the defender strategy either leads to computationally intensive approaches that are impractical or is so difficult to address that the developed methods can only be applied for very specific target configurations (see Alpern, Morton, & Papadaki, 2011; Jain, Conitzer, & Tambe, 2013; Jain, Korzhyk, Vanek, Conitzer, Pechoucek, & Tambe, 2011; Shieh et al., 2012). Solving games where geographical considerations are incorporated is well beyond the capabilities of current algorithms. In this section, we outline two heuristic algorithms that have been developed for SSGs on networks.

23.4.1 *Fixed Patrols against Attackers on a Path*

We begin by presenting a heuristic solution method analyzed in Spencer (2013). This solution method is proposed to address SSGs on a network where the adversary can attack some sequence of nodes and the defender security forces decide to patrol a subset of nodes from the network. A strategy profile for this problem is a pair $\boldsymbol{x}, \boldsymbol{a}$ where $\boldsymbol{x}$ corresponds to a probability distribution for the defense resources among the nodes and $\boldsymbol{a}$ indicates the likelihood that a certain route is selected to attack by the terrorists (attacker). This game could also represent community crime, where pickpockets travel across a commercial district and the police decide which areas of the city they patrol.

On real-world scales, instances for this problem have such a large number of routes that it promptly becomes computationally inoperable. If there are 100 points of interest, or nodes, then we have up to 2^{100} possible routes. Each route ends up being an additional restriction in the lineal problem (LP).

The work in Spencer (2013) proposes a heuristic algorithm based on a decomposition method that generates columns and constraints to gradually solve the complete Stackelberg integer programming formulation (as in (23.2)). The two-stage algorithm begins with only one attacker route and solves the problem for the defender and attacker iteratively. Each time, the defender allocates his resources and the attacker sees the defender's coverage and then chooses the best possible attack route. For every not previously considered route, we add it to the problem until no new route is proposed. We compare our results with the results using the standard methodology of solving linear problems and show that it can greatly reduce the required computational resources and the running time to reach the solution. This problem was the focus of Jain and colleagues (2013), which used a similar column generation method to identify optimal check points to defend critical infrastructure from attackers that would traverse the network in the zero-sum setting.

In this work, an SSG model on a network is considered, where defenders select the nodes that are covered while attackers choose a path through the network where they attempt to commit crimes. A two-stage decomposition approach for SSGs was developed in Spencer (2013). This algorithm considers an SSG with arbitrary joint schedules and routes as attack targets. Our approach takes as input the utilities for each node being attacked and computes a heuristic solution corresponding to a randomized allocation of defender recourses. We propose an algorithm that uses a column and restriction generation approach, decomposed in two separate steps or phases where two different problems are solved: the master (defender) and the slave or follower (attacker) problem. The master solves for the defender strategy C, given a restricted set of columns P (i.e., possible attacker routes). The objective function for the slave is updated based on the allocation of the master, and the slave is solved to identify the best new column to add to the master problem (explained in detail later). If no new column can improve the slave's solution, then the algorithm terminates.

This decomposition algorithm was shown in Spencer (2013) to return the optimal solution for zero sum SSGs, but it could get stuck in a non-optimal solution in the general case. Next, we present an example where this algorithm gets stuck in a local optimum, proving that in the general case this decomposition method is a heuristic.

Table 23.2. *Network patrolling problem example*

Def/Att	Covered	Uncovered
Node 1	4 / −3.2	−1 / 0.8
Node 2	2 / −2.4	−2 / 0.6

Route	Payoff
Path 1	$-4x_1 + 0.8$
Path 2	$-3x_2 + 0.6$
Path 3	$-4x_1 + -3x_2 + 1.4$

Note: Nodes are patrolled by defender, attacker decides on a path to attack. Table on the left indicates reward/penalties for attacks on nodes. These rewards are compounded for the attacker payoffs of attacking paths on the table on the right.

The example considers a network with two nodes and three paths on those nodes. Path 1 crosses node 1, path 2 crosses node 2, and path 3 crosses nodes 1 and 2. The rewards and penalties for attacks at each of the nodes are given in Table 23.2, which also shows the attacker reward for each of the possible routes for a given defender strategy $\boldsymbol{x}$.

Therefore, for each combination of defender strategies $\boldsymbol{x}$, the optimal response for the attacker is given by the maximum of these three lines, which appear in Figure 23.2. We also plot the optimal reward for the defender taking into account the optimal attacker response.

We note that the optimal reward surface for the defender is not a continuous surface because the change of strategy of the attacker can modify the non-zero-sum objective by an arbitrary amount. Given this problem, and for specific initial conditions, the decomposition method could generate paths that build the bottom surface without identifying a combination of defender coverages that would generate path 1, which is part of the upper surface. The algorithm could, depending on the initial conditions of the decomposition algorithm, terminate with a suboptimal solution.

23.4.2 *Patrolling Streets on a Graph*

Now we discuss the heuristic route patrolling strategies developed in Hochbaum and colleagues (2014). In this work, the process of building a defender strategy is tackled in two phases:

- Phase 1: Determines the optimal defender strategy as if the targets were independent. We represent the nuclear detection patrolling problem as an

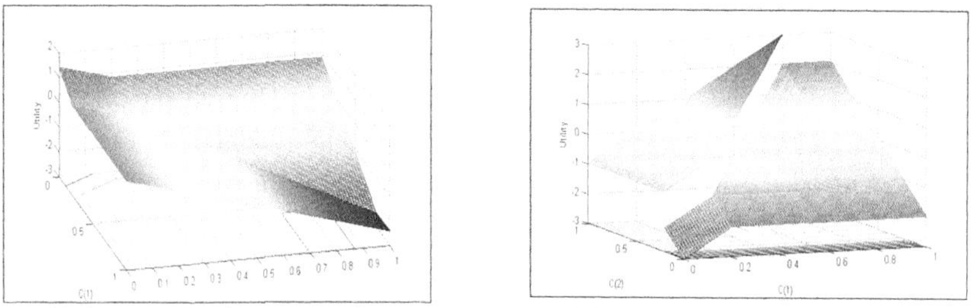

(a) Attacker optimal response surface (b) Defender optimal response surface

Figure 23.2. Optimal rewards for a network patrolling example from Spencer (2013).

SSG, where the defenders view targets independently without considering the geographical effect on defender strategies. This phase develops mixed strategies for defending the targets based only on the relative importance of their rewards. The output of this stage will be a frequency with which each target is protected, assigning a probability of protection to each target. This represents the optimal defender strategy for independent targets.

- Phase 2: Determines defender patrols that match the optimal outcome of the adversarial game (from Phase 1) given limited resources (k defenders with limited capacity) and a maximum set of targets within a certain area. We show that determining the optimal solution for this problem is the NP-hard k-vehicle Rural Postman Problem (k-RPP). We devise a $\left(2.5-\frac{1}{k}\right)$-approximation algorithm for this problem that works in any setting (including non-metric settings). This approximation algorithm is used to construct maximal reward patrolling routes for k defenders that visit the edges sampled according to the optimal defender strategy from Phase 1.

We compare the performance of the proposed strategy against simple alternate solution strategies. For this comparison, we build a realistic security instance that aims to deter and prevent theft in downtown Santiago. This domain is analogous to the SNM detection problem described earlier and the data used to construct the instances are from police crime statistics on downtown Santiago. The downtown Santiago network is modeled as an undirected geometric planar graph, where each edge represents a street and each node represents an intersection of some streets. Each node has an X-Y coordinate and the length of an edge is the Euclidean distance between the two end nodes of the edge. The graph consists of 119 nodes and 203 edges.

The attacker in this security game is a thief and we assume that he (she) steals only on nodes. The defender is the police and we assume that defenders patrol and protect edges (e.g., police vehicles travel streets monitoring activity). If an edge is protected by police, then the two end nodes of the edge are considered protected. Then, the attack on either node is deterred and prevented. The crime statistics have the number and total amount (in dollars) of thefts that have occurred at every node of the graph over three years. We use these data to compute the average amount of a theft (in dollars) on every node. We assume that the reward of the attacker for conducting an attack at an unprotected node is the average amount of theft on that node, while the penalty for the attacker when targeting a protected node is the negation of the total amount of theft on that node. Symmetrically, we assume that the reward for the defender of protecting a node from attack is the total amount of theft on that node, while the penalty for the defender of not protecting a node from attack is the negation of the average amount of theft on that node. Notice that since the defender protects edges, the game is not a zero-sum game as the defenders receive the rewards of both end nodes when protecting a certain edge.

We first compute the defender's mixed strategy on this problem instance, according to Phase 1 of our proposed solution approach. This determines the frequency of coverage on each edge (we refer to this solution as the *Stackelberg probability*). Solving this problem also determines the set of nodes that give the maximum reward to the attacker. In our experiment, we run Phase 1 with $kc = 15$, which generates a defender strategy that sets a large number of nodes (110 out of the total of 119 nodes) with the maximum reward for the attacker.

In Phase 2, we implement the 2.5-approximate k-RPP algorithm introduced in Hochbaum and colleagues (2014). In this phase, k represents the number of available police patrols (vehicles) and the capacity of each patrol is the maximum distance that the vehicle can patrol in one day, starting from and returning to the depot (police department). We simulate a time range of 100 days in the experiment. Each day, the attacker uniformly samples a node to attack from the set of nodes with maximum reward. The defender, in turn, selects edges to patrol by sampling without replacement according to the Stackelberg probability. This sampling process is continued while the edges selected can be patrolled with k patrols of limited capacity, which is checked using the k-RPP approximation algorithm. On each day we consider the patrol successful if the node attacked is incident to one edge protected by the patrols; otherwise we assume the attack was successful. The utility perceived by the defender (and attacker) depends on whether the patrol was successful: with the defender accruing either the reward or the penalty for the attacked node depending on whether it was covered. The performance of the complete two-phase strategy is measured by the *total net utility* (rewards plus penalties) of the defender accumulated over the 100 days.

For each phase, we consider a counterpart for comparison. For Phase 1, we consider another mixed strategy where the protection probability is uniform over all edges. With uniform probability, the node that gives the maximum reward to the attacker is unique. For Phase 2, we consider another simple nearest-neighbor routing heuristic: for each vehicle, starting from the depot, it always covers the nearest edge that needs protection (in E'). Then, starting from this edge, it covers next the nearest edge that needs protection. The vehicle continues this searching procedure until the remaining capacity forces it to go back to the depot.

As a summary, by doing cross combinations, we compare the following four two-phase strategies:

- Combination SK – Phase 1: Stackelberg probability. Phase 2: k-RPP algorithm.
- Combination UK – Phase 1: Uniform probability. Phase 2: k-RPP algorithm.
- Combination SN – Phase 1: Stackelberg probability. Phase 2: Nearest-neighbor algorithm.
- Combination UN – Phase 1: Uniform probability. Phase 2: Nearest-neighbor algorithm.

For each two-phase strategy, we study the change of the total net utilities as a function of k, the number of available patrols, and as a function of the patrols' capacities. To differentiate these two effects, we explore two setups:

- For a fixed number of patrolling vehicles, we study the change of the total net utilities as a function of the vehicles' capacities.
- For a fixed level of vehicles' capacities, we study the change of the total net utilities as a function of the number of available patrolling vehicles.

The results for these two cases appear in Figure 23.3 when the number of patrolling vehicles is kept constant and in Figure 23.4 for constant capacity. For every combination of the k and the capacity, we run all four two-phase strategies 10 times and report the average total net utility. The performance of the four two-phase strategies are plotted in the same figure for each testing scenario.

Results show that the Stackelberg probability gives better performance than the uniform probability. This follows because the Stackelberg probability is the optimal solution of the SSG in Phase 1, which computes the maximum net utility. This is consistent with the previous computational results showing that the Stackelberg probability outperforms the uniform probability (without considering the effect of geographical locations on the defender

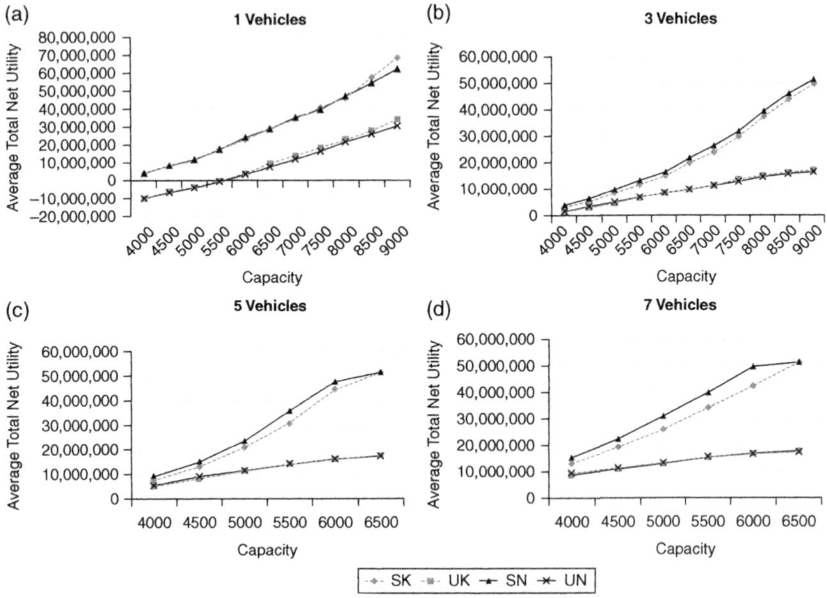

Figure 23.3. The change of the average total net utility with increasing capacities for a fixed number of vehicles.

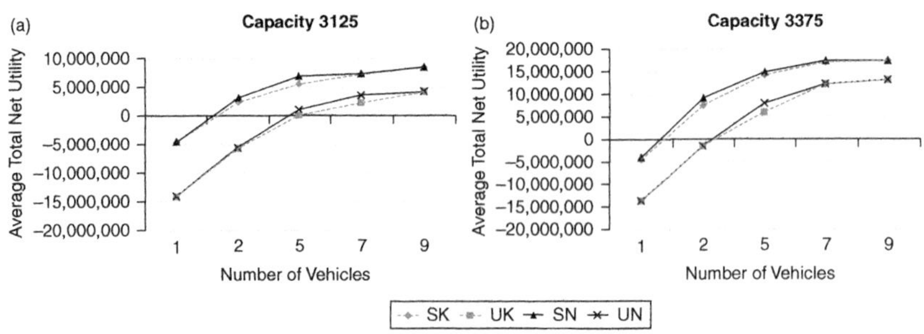

Figure 23.4. The change of the average total net utility with increasing number of vehicles and fixed vehicle capacity.

strategy) (Jain, Tsai, Pita, Kiekintveld, Rathi, Ordóñez, & Tambe, 2010; Yang, Kiekintveld, Ordóñez, Tambe, & John, 2013). On the other hand, the "worst-case" 2.5-approximate k-RPP algorithm works well for the case of 1 vehicle ($k = 1$). For the case of more vehicles, since the problem is defined in a Euclidean plane, a simple heuristic like the nearest-neighbor algorithm would do well, while the "worst-case" 2.5-approximate k-RPP algorithm does not improve the solution.

23.5 Conclusions

The success in using Stackelberg security games models to address real applications depends on the ability to solve the optimization problems that capture the salient characteristics of these real security problems. In this chapter, we describe the use of different optimization techniques to solve increasingly challenging versions of Stackelberg security games.

Standard mixed-integer programming optimization techniques are able to solve instances with a few adversaries (~10) and where the number of actions for the defender and attackers are at most in the hundreds. Larger problems than that run into memory or runtime constraints. In this chapter, we describe the use of stronger mixed-integer formulations (Subsection 23.2.1), exploiting structure in the payoff matrix (Subsection 23.2.2), and the use of column and cut generation approaches (Section 23.3) to solve exactly instances that are much larger. In particular, Section 23.3 of this chapter presents a review of column and cut generation approaches. We further show situations where column and constraint generation approaches only provide an approximate, or heuristic solution, and that for even larger problems, interesting heuristics are developed by decomposing the problem.

The work reviewed in this chapter is able to tackle specific Stackelberg security games with a large number of defender or attacker actions. Typically, the structure of these large set of actions comes from creating actions that represent combinations of defender resources over possible targets or paths on a graph. This structure is exploited specifically in the solution methods developed. It is still a challenge to develop algorithms for large instances that do not have such underlying structure in the payoff matrices. Furthermore, there are real issues of the security domain, such as the existence of multiple heterogeneous defender resources, sequential decisions, and dynamics that enable the use of on-line information, that require even larger formulations and new solution algorithms.

REFERENCES

Alpern, S., Morton, A., & Papadaki, K. (2011). Patrolling games. *Operations Research*, *59*, 1246–1257.

Barnhart, C., Johnson, E. L., Nemhauser, G. L., Savelsbergh, M. W., & Vance, P. H. (1998). Branch-and-price: Column generation for solving huge integer programs. *Operations Research*, *46*(3), 316–329.

Bertsimas, D., & Tsitsiklis, J. N. (1997). *Introduction to linear optimization*, volume 6. Belmont, MA: Athena Scientific.

Brown, M., Haskell, W. B., & Tambe, M. (2014). Addressing scalability and robustness in security games with multiple boundedly rational adversaries. In *Decision and Game Theory for Security* (pp. 23–42). Springer.

Casorrán-Amilburu, C., Fortz, B., Labbé, M., & Ordóñez, F. (2015). Novel formulations for Stackelberg security games. Technical report, Départment d'Informatique, Université Libre de Bruxelles.

de Farias, D. P., & Van Roy, B. (2014, August). On constraint sampling in the linear programming approach to approximate dynamic programming. *Mathematics of Operations Research*, *29*(3), 462–478.

Hochbaum, D. S., Lyu, C., & Ordóñez, F. (2014). Security routing games with multi-vehicle Chinese postman problem. *Networks*, *64*(3), 181–191.

Jain, M., Conitzer, V., & Tambe, M. (2013). Security scheduling for real-world networks. In *International Conference on Autonomous Agents and Multiagent Systems (AAMAS)*.

Jain, M., Kardes, E., Kiekintveld, C., Ordónez, F., & Tambe, M. (2010). Security games with arbitrary schedules: A branch and price approach. *AAAI*.

Jain, M., Korzhyk, D., Vanek, O., Conitzer, V., Pechoucek, M., & Tambe, M. (2011). A double oracle algorithm for zero-sum security games on graphs. In *International Conference on Autonomous Agents and Multiagent Systems (AAMAS)*.

Jain, M., Tsai, J., Pita, J., Kiekintveld, C., Rathi, S., Ordóñez, F., & Tambe., M. (2010). Software assistants for randomized patrol planning for the LAX airport police and the federal air marshals service. *Interfaces*, *40*(4), 267–290.

Kiekintveld, C., Jain, M., Tsai, J., Pita, J., Ordóñez, F., & Tambe, M. (2009). Computing optimal randomized resource allocations for massive security games. In *Proceedings of the 8th International Conference on Autonomous Agents and Multiagent Systems-Volume 1* (pp. 689–696). International Foundation for Autonomous Agents and Multiagent Systems.

Kortanek, K., & No, H. (1993). A central cutting plane algorithm for convex semi-infinite programming problems. *SIAM Journal on Optimization*, *3*(4), 901–918.

Lagos, F., Ordóñez, F., & Labbé, M. (2015). Column generation for a strong formulation of Stackelberg games. In *AAAI 2015 Spring Symposium Series*.

Paruchuri, P., Pearce, J., Marecki, J., Tambe, M., Ordóñez, F., & Kraus, S. (2008). Playing games for security: An efficient exact algorithm for solving Bayesian games. In *Proceedings of the 7th International Conference on Autonomous Agents and Multiagent Systems*. International Foundation for Autonomous Agents and Multiagent Systems.

Ruszczynski, A. (2006). *Nonlinear optimization*. Princeton, NJ: Princeton University Press.

Shieh, E., An, B., Yang, R., Tambe, M., Baldwin, C., DiRenzo, J., Maule, B., & Meyer, G. (2012). PROTECT: An application of computational game theory for the security of the ports of the United States. In *Conference on Artificial Intelligence (AAAI) – Spotlight Track*.

Spencer, T. (2013). A heuristic algorithm for a Stackelberg security game on a network. Master's thesis, Universidad de Chile.

Yang, R., Jiang, A. X., Tambe, M., & Ordóñez, F. (2013). Scaling-up security games with boundedly rational adversaries: A cutting-plane approach. In *Proceedings of the 23rd International Joint Conference on Artificial Intelligence* (pp. 404–410). AAAI Press.

Yang, R., Kiekintveld, C., Ordóñez, F., Tambe, M., & John, R. (2013). Improving resource allocation strategies against human adversaries in security games: An extended study. *Artificial Intelligence Journal (AIJ)*, *195*, 440–469.

Yin, Z., Jain, M., Tambe, M., & Ordóñez, F. (2011). Risk-averse strategies for security games with execution and observational uncertainty. In *Conference on Artificial Intelligence (AAAI)*.

24

Methods for Addressing the Unpredictable Real-World Element in Security

Jinshu Cui, Thanh H. Nguyen, James Pita, and Richard S. John

24.1 Introduction

Previously, we've examined how to set up and solve a Stackelberg security game (SSG) (Tambe, 2011). In doing so, we've demonstrated how game theory can be useful for reasoning about real-world security settings where security forces must protect critical assets from potential adversaries. However, an important factor in these real-world security settings is that the adversaries involved may not behave according to the standard assumptions of game-theoretic models. We will first examine three core assumptions of game theory that real-world adversaries may challenge. Then we will provide the current best known methods to address these potential deviations from the standard assumptions. Additionally, because different adversaries can challenge these assumptions in unique ways, we will discuss how to address games involving a variety of adversaries. We will demonstrate how each of these approaches can be valuable under different circumstances by providing supporting experimental data and results within the security game paradigm, and, finally, we will provide real-world examples of deployed systems that utilize some of the methods described. The summary of this chapter is described in Table 24.1.

24.2 Real-World Effects on the Standard Assumptions of Game Theory

An important factor in real-world security settings is that the adversaries involved may not behave according to the standard assumptions of game-theoretic models. Three specific assumptions that real-world adversaries challenge are: (1) that the players are all perfectly rational; (2) that the players know all the rules, payoffs, and other information in the game; and (3) that the

All coauthors have equal contributions.

Table 24.1. *A summary of this chapter*

Real-world security challenges	Methodology		Simulation	
Adversaries' bounded rationality Game payoff and action space uncertainty Execution and observation uncertainty	Robust approaches: *address worst-case scenario due to uncertainty*	Behavioral modeling: *use behavioral models, e.g., Quantal Response to capture adversaries' decision making*	Behavioral modeling: evaluation: *comparing prediction accuracy between different models*	Robust versus Modeling: *human subject experiments*

players perfectly execute their own actions and perfectly observe the actions of the other players. We will look at each of these assumptions in turn to help explain why standard solution concepts in game theory may not apply to real-world security adversaries.

24.2.1 *Rationality Assumption*

One of the strict assumptions made by standard game-theoretic approaches based on Stackelberg games is that the adversary is perfectly rational (Conitzer & Sandholm, 2006; Jain et al., 2010; Paruchuri et al., 2008). That is, the adversary is assumed to act with infallible value-maximizing rationality, maximizing his expected value based on the information available. Additionally, if the adversary faces a tie in his strategies' rewards, he will break it in favor of the leader, choosing the one that gives a higher reward to the leader (a strong Stackelberg equilibrium). This standard assumption is also shown to follow from the follower's rationality and optimal response under some conditions (von Stengal and Zamir, 2004).

Unfortunately, real-world adversaries may not be expected-value maximizers, computing optimal decisions. Instead, their decisions may be governed by their bounded rationality (Simon, 1956), which causes them to deviate from their optimal strategy. Such deviations can have arbitrarily negative impacts on the security force's expected value if not explicitly accounted for. In essence, the leader faces uncertainty over follower responses – the follower may not choose the optimal response, but from a range of possible responses – potentially significantly degrading leader rewards.

24.2.2 *Information Assumption*

Game-theoretic models require defined (i.e., fixed) potential actions on behalf of all the players involved in order to compute an optimal mixed strategy for the security force. However, real-world security settings are dynamic in nature, and real-world adversaries are highly adaptable and creative, allowing for the action space of an attacker to constantly be evolving. Furthermore, even if the action space was held static, the sheer number of potential attack methods by real-world adversaries may be too large to enumerate and solve in practice. Thus, in certain situations, intelligent methods for compactly representing the potential threat capabilities of real-world adversaries may be required.

Beyond the uncertainty about the actions available to the players involved, when dealing with real-world adversaries, there may also be uncertainty about the preferences and capabilities of those adversaries. The standard Stackelberg solution technique requires point-estimate payoffs be assigned to the outcomes in a given game (von Stengal & Zamir, 2004). However, it can be difficult to ascribe a specific point value to different potential outcomes. Information about real-world adversaries is often limited or imprecise. For example, it can be difficult to predict precisely how attackers will weight casualties, economic consequences, media exposure, and other factors when selecting targets.

24.2.3 *Execution and Observation Assumption*

A standard assumption made in game-theoretic approaches and particularly those based on Stackelberg games is that the adversary perfectly observes the defender's strategy and that the defender can perfectly execute his intended strategy (Conitzer & Sandholm, 2006; Jain et al., 2010; Paruchuri et al., 2008; von Stengal & Zamir, 2004). Unfortunately, in the real world, execution and observation may not be perfect due to unforeseen circumstances and/or real-world errors. For example, a canine unit protecting a terminal at an airport may be called off to another, more urgent assignment or, alternatively, a new unit could suddenly become available. Similarly, the attacker's observations can be noisy. The adversary may miss an officer patrolling a target, or mistake a passing car as a security patrol. Additionally, a real-world adversary will always observe the defender strategy for a limited period of time, resulting in limited knowledge of the defender strategy that may be inaccurate and lead to action choices that are not based on an accurate representation of the defender strategy. Under such circumstances, the real-world adversary may be guided by his cognitive biases and bounded rationality, causing him to deviate from the predicted optimal response, which can lead to significant losses on behalf of the security force.

24.3 Robust Approaches to Address Adversary Decision Making

A common concern in game-theoretic settings has been considering potential deviations from the game-theoretic optimal play (Camerer, 2003; McKelvey & Palfrey, 1995; Selten, 1988). The field of experimental game theory has provided a wealth of contributions and insights on deviations from equilibrium predictions and different models to explain these deviations (Camerer, 2003). Additionally, in the next section, we will present a number of behavioral solution concepts that focus on finding appropriate models of adversary decision making in security settings.

These equilibrium approaches seek stable equilibrium points where the players involved are not expected to deviate (outside of the predicted deviations) and thus do not specifically address the impact of unexpected deviations or protect against them. In contrast with these equilibrium concepts and adversary models, the approaches presented here robustly guard against potential deviations due to suboptimal play. Also, many equilibrium concepts consider that all the players involved will suffer from the same limitations (e.g., bounded rationality), whereas the robust approaches (and later the behavioral approaches) optimally compute a rational strategy for the defender by considering the limitations of the adversary.

Having a perfect model of adversary decision-making processes would lead to the optimal strategy for security forces against real-world adversaries. However, finding such perfect models is a difficult task and often requires a large amount of data to create, which can be limited in security settings. Additionally, McCubbins and colleagues (2012) have recently shown that many of the currently existing models of adversary decision making are fundamentally flawed. In the absence of such data and strong models of adversary decision making, the robust approaches presented here are a beneficial substitute and can perform significantly better than current approaches for addressing real-world adversaries.

The approaches presented here draw inspiration from robust optimization methodology, in which the decision maker optimizes against the worst outcome over some uncertainty (Aghassi & Bertsimas, 2006; Nilim & Ghaoui, 2005). In the presented Stackelberg problems, the defender will make a robust decision by considering that the adversary, who may not follow expected-value-maximizing rationality, could choose a strategy from his range of possible responses that degrades the leader expected value the most.

24.3.1 *ε-optimal Robust Response*

The first robust method introduces the idea of robust responses to ε-optimal adversaries in Stackelberg games. Specifically, assume that the adversary may select any ε-optimal response strategy. An adversary's ε-optimal response strategies are any strategies within ε-expected-value of his optimal strategy. This suboptimal choice could be made for any number of reasons, but the robust approach attempts to guard against the choices that fall within this ε-bound of the optimal response. More specifically, given multiple possible ε-optimal responses, the robust approach is to assume that the adversary could choose the one that provides the defender the worst expected value – not necessarily because the adversary attends to the defender reward, but to robustly guard against the worst-case outcome. In an adversarial setting, handling the worst-case outcome may be in the best interest of the defender. For the full mathematical model to calculate an ε-optimal robust response, the reader is referred to Pita and colleagues (2010).

24.3.2 *Robustness in Stackelberg Security Games*

The second robust method focuses on the restricted version of a Stackelberg game known as a Stackelberg security game. Whereas standard robust optimization robustly guards against a worst-case outcome within some error bound, this robust method assumes an expected-value-maximizing outcome on behalf of the attacker. However, it bounds the defender's loss for a potential deviation by the attacker based on the distance of the deviation from the adversary's expected-value-maximizing strategy. This is done in an effort to address all potential deviations while still avoiding the complex task of explicitly modeling adversary decision making.

Such an approach provides three key benefits: (i) it provides significant runtime benefits over competing approaches; (ii) it strongly couples the adversary's and defender's performance, robustly guarding against potential deviations by real-world adversaries and avoiding situations where minor deviations (i.e., deviations that result in minor losses in expected value) by the adversary may result in large losses for the defender; (iii) it avoids the task of creating an accurate opponent model. This new type of optimization is referred to as *graduated optimization*. For the full mathematical model to calculate a graduated optimization response, the reader is referred to Pita and colleagues (2012).

24.3.3 *Robustness to Execution and Observation Uncertainty*

Most existing solution approaches, including those just presented, do not model two key uncertainties of the real world. Namely, there may be noise in the defender's execution of a suggested mixed strategy and/or the observations made by an attacker can be noisy. To address this issue, a robust solution technique was developed that attempts to maximize the worse-case outcome within some execution and/or observation noise or error level from the intended strategy. Specifically, in this robust solution, the defender's execution strategy is allowed to deviate a specified amount from the intended execution strategy to a realized strategy. The attacker is then allowed to observe any potential realized strategy, and his observation may further deviate from that realized strategy by a specified amount. The robust solution is to compute a defender strategy that has the best worse-case deviation from the intended strategy given both levels of deviation (execution deviation from the intended strategy and then observational deviation from the realized strategy). While this may not be the defender's optimal strategy, it robustly defends against negative outcomes that might arise from execution and observation uncertainty. For the full mathematical model to calculate a robust execution and observation uncertainty response, the reader is referred to Yin and colleagues (2011).

24.4 Behavioral Approaches to Address Adversarial Decision Making

Adversarial decision making within a Stackelberg security game (SSG) context is an example of decision making under risk. That is, the complete payoff matrix is known for attacker (follower) rewards and penalties, and the probability of any particular play by the defender (leader) is also assumed known through (prior) surveillance. A number of well-studied heuristic decision rules deviate from expected value maximization. Several of these are enumerated next:

1. Adversary subjective value (utility) is nonlinear in the stated rewards and penalties, often due to decreasing marginal value.
2. Adversary subjective value (utility) is nonlinear in the stated rewards and penalties due to risk-aversive or risk-seeking attitudes.
3. Adversary subjective value (utility) is sensitive to framing effects around a reference point, e.g., status quo, or a goal. Positive deviations from the reference point involve risk aversion, and negative deviations from the reference point involve risk-seeking behavior.
4. Adversary decision making may involve probability neglect, in which probability differences in the defender's (leader's) strategy are attenuated (usually in the middle quartile, 0.25–0.75).
5. Adversary decision making may involve certainty effects, in which probabilities differences near 0 and 1 in the defender's (leader's) strategy are magnified.
6. Adversary decision making may involve multiple objectives that do not coincide with the objective of maximizing expected value. For example, adversaries may be more concerned with maximizing the probability of "winning" (or minimizing the probability of "not losing"), than with the expected consequences of the game play. This may be a greater concern for games played with very small or no monetary incentive.
7. Adversary decision making may involve the objective of minimizing the defender's (leader's) expected value. While the defender's payoffs are of great relevance in solving for an equilibrium solution to the game, once the defender's strategy is known to the attacker, the defender's payoffs should be irrelevant. That is, once the SSG strategy has been selected by the defender and surveilled by the attacker, the game is transformed into

a decision under risk for the attacker, and all rewards and penalties that should matter are captured in the attacker's payoff matrix. The objective of penalizing the defender, beyond that captured in the attacker's own payoff matrix, is not represented in the normative EV model for the attacker.

We address these heuristics and biases with a number of descriptive models of the attacker, defined next.

24.4.1 *Cognitive Biases*

One recent taxonomy of cognitive biases lists 53 distinct cognitive and motivational biases related to judgment and decision making (Baron, 2008). In a more recent and quite thoughtful piece, Montibeller and von Winterfeldt (2015) identify the "top 12" cognitive biases and "top 4" motivational biases relevant to modeling decisions. Four of the cognitive biases in their taxonomy are explicitly accounted for in our adversary decision models: (1) Anchoring; (2) Certainty Effect; (3) Equalizing Bias; and (4) Gain-Loss Framing Bias. There is a vacuum of research investigating the extent to which cognitive biases and heuristics in decision making under risk (against nature) apply to adversary game contexts, such as SSGs. Formally, they are equivalent, but all psychological research on heuristics and biases has been conducted in contexts that involve decision making under risk (against nature), rather than a competitive game context, against an adaptive adversarial agent.

Conservatism Bias

Conservatism is a cognitive bias in which decision makers rely too heavily on early pieces of information and attenuate the weight that should be given to subsequently observed information (Phillips & Edwards, 1966; Phillips et al., 1966; von Winterfeldt & Edwards, 1986). Conservatism is consistent with an anchoring bias (Kahneman et al., 1982; Tversky & Kahneman, 1974), in which decision makers anchor on early information and under-adjust their beliefs in light of newly acquired observations. Conservatism is also consistent with the ubiquitous confirmation bias (Nickerson, 1998), in which new information is more likely to be selected if consistent with current beliefs, and information that is selected is interpreted in a manner that confirms current beliefs. Tversky and Koehler (1994) developed a theory of subjective probability known as *support theory*, which models the conservatism phenomenon. Research based on support theory has shown that when given no information about the occurrence of a discrete set of events, decision makers will tend to assign an equal weight to the occurrence of each event (a uniform distribution) (Fox & Clemen, 2005; Fox & Rottenstreich, 2003; See et al., 2006). This is also referred to as giving full support to the ignorance prior where the ignorance prior represents a subject's baseline belief (i.e., in this case the uniform distribution over events) (Fox & Rottenstreich, 2003).

It has been shown empirically that decision makers are particularly susceptible to giving full support to this ignorance prior (uniform distribution) before they are given any information and that, once given information, they are slow to update away from this assumption (Fox & Clemen, 2005; Fox & Rottenstreich, 2003; See et al., 2006). Thus they leave some support, $\alpha \in [0 \ldots 1]$, on the ignorance prior and the rest, $1-\alpha$, on the occurrence they have actually viewed. As decision makers become more confident in what they are viewing, this bias begins to diminish, decreasing the value of α.

Models have been proposed to address this bias and predict what probability a decision maker will assign to a particular event *e* from a set of events *E* based on the evaluative assessment (i.e., assessment based on events actually viewed) they have made for the occurrence of that event. Let *e* represent a particular event, $E \setminus e$ represent the remaining events possible, let $P(e), P(E \setminus e)$ be the real probabilities and $P(e'), P((E \setminus e)')$ represent the probability a decision maker assigns to event *e* and $E \setminus e$ respectively. One model (Fox & Rottenstreich, 2003) defines the decision maker's estimated probabilities with the following equation:

$$\frac{P(e')}{P((E \setminus e)')} = \left(\frac{|e|}{|E \setminus e|}\right)^{\alpha} \circ \left(\frac{P(e)}{P(E \setminus e)}\right)^{1-\alpha}.$$

Here, $\frac{|e|}{|E \setminus e|}$ is the ratio in the case of uniform probabilities and represents the ignorance prior. The α value indicates the relative contribution of these two sources of information. Note that as α approaches 1, the estimated probability converges on the uniform assumption, while when α approaches 0, it is closer to the true probability distribution. Research suggests that as decision makers gain more relevant knowledge, they will give less support to the ignorance prior and more support to evaluative assessment, thus decreasing the value of α (Fox & Clemen, 2005; Fox & Rottenstreich, 2003; See et al., 2006).

An alternative model assumes estimated probabilities are directly calculated using a simple linear model (Fox & Clemen, 2005; Fox & Rottenstreich, 2003; Tversky & Koehler, 1994):

$$P(e') = \alpha \cdot \left(\frac{1}{|E|}\right) + (1-\alpha) \cdot P(e).$$

Anchoring bias can be applied to Stackelberg games to determine how a real-world adversary may perceive the leader strategy under different observation conditions. Although these may not be the only possible models for determining anchoring bias, they are standard in the related literature (Fox & Rottenstreich, 2003; Tversky & Koehler, 1994), and the linear model is ideal since the odds form model is not easily representable in a Mixed-Integer Linear Program (the standard solution technique in Stackelberg security games).

Bayesian Updating

As an alternative to support theory, Bayesian updating can be used to predict how adversaries will perceive the probability of each event from a set of events after obtaining some observations. However, there is more support in the literature that decision makers' beliefs are conservative, due to a combination of anchoring and confirmation bias, rather than performing Bayesian updating when evaluating evidence (Fox & Clemen, 2005; Fox & Rottenstreich, 2003; Kahneman & Tversky, 1972; See et al., 2006; Tversky & Koehler, 1994). Also, using Bayesian updates requires tracking which observations the real-world adversaries have received, while in the real world, defenders will not be aware of which days adversaries take observations and which days they do not. A descriptive conservatism model that includes both anchoring and confirmation bias is more general in that it allows the defender to work with just an estimate of the number of observations she believes an adversary will take rather than an exact representation of observations taken.

24.4.2 *Modeling the Adversary Decision-Making Process*

We introduce a number of approaches that attempt to model the adversary decision-making process. The models introduced here all measure adversaries' choices as probabilistic choices. That is, if the probability of choosing one target is higher than that of choosing an alternative, adversaries will more frequently choose that target to attack, but in some instances may choose the alternative. In decision-making research, Luce's Choice Axiom (LCA) (1959) assumes that choice behavior is probabilistic instead of deterministic. McFadden (1976, 2001) applied LCA to preferential choice in economic analysis. His model was able to exaggerate the differences between alternatives by exponentiating utilities, and the optimal choice is consistent with utility maximization. McKelvey and Palfrey (1995) later developed Quantal Response Equilibrium (QRE) in economics, which assumes that the chance of selecting a non-optimal strategy increases as the level of error increases, in which λ captures the rational level (absence of errors) of a player. Since expected utility maximizing is the baseline of a rational decision maker and it is easier to measure a parameter close to 0, the quantal response model is adjusted by reversing the parameter to let λ represent the level of error (known as the SOFTMAX function): a player chooses randomly when $\lambda \to \infty$ and maximizes expected utility when $\lambda \to 0$. Let $q_i \in [0, 1]$ represent the probability that target t_i will be attacked:

$$q_i(x_i) = \frac{e^{U_{A_i}(x_i)/\lambda}}{\sum_{t_k \in T} e^{U_{A_k}(x_k)/\lambda}}, \lambda \geq 0 \quad (24.1)$$

The following section introduces nine models varying the attacker's utility function (U_{A_i}) in (1). The nine versions of U_{A_i} can be partitioned into five categories: (1) attacker's expected value (EV); (2) attacker's EV accounting for risk attitude; (3) lens model (Brunswik, 1952; Hammond, 1955) with a weighted average of cues: p(success), attacker's reward and penalty, and defender's reward and penalty; (4) lens model accounting for risk attitude; and (5) multiattribute utility model with a weighted average of p(success), attacker's EV, and defender's EV (Keeney & Raiffa, 1976).

Attacker's Expected Value

The basic utility function of an adversary only captures the expected utility of an attacker who is risk neutral (expected value). The model was first introduced by Yang and colleagues (2012) under the name of Quantal Response. If the attacked target t_i (i = 1,2,…,8) is covered by the defender, the attacker receives penalty P_{A_i} and the defender receives reward R_{D_i}; if the attacked target is not covered by the defender, the attacker receives reward R_{A_i} and the defender receives penalty P_{D_i}. Let x_i denote the probability of a guard at target t_i; the attacker's expected utility at t_i is:

$$U_{A_i}(x_i) = x_i P_{A_i} + (1 - x_i) R_{A_i} \quad (24.2)$$

Yang and colleagues (2012) further modified the model by adding an extra weight (λ_s, $\lambda_s \geq 0$) to the target that is least protected by the defender, that is, the least defended target is given a bonus in the SOFTMAX calculation. This assumption is consistent with the "soft target" hypothesis. Let $S_i(x_i)$ denote whether a target is the least defended:

$$S_i(x_i) = \begin{cases} 1, & \textit{if } x_i \leq x_{i'} \\ 0, & \textit{otherwise} \end{cases} \quad (24.3)$$

Attacker's Expected Utility Accounting for Risk Attitude

A simple power utility function was constructed by adding a parameter α to capture risk attitude where $\alpha > 1$ indicates risk seeking and $0 < \alpha < 1$ indicates risk aversion. Assuming the same risk attitude for gain and loss, the expected utility of target t_i is:

$$U_{A_i}(x) = x_i P_{A_i}^{\alpha} + (1 - x_i) R_{A_i}^{\alpha} \tag{24.4}$$

Lens Model

The lens model suggests that attacker judgments depend on a linear combination of multiple observable cues. Therefore, the expected utility function of an attacker can be a linear combination of three attributes that are important to the decision (x_i, R_{A_i}, and P_{A_i}). The model, labeled the Subjective Utility Quantal Response (SUQR) model, was first proposed by Nguyen and colleagues (2013). The utility function was defined as:

$$U_{A_i}(x_i) = w_1 x_i + w_2 P_{A_i} + w_3 R_{A_i} \tag{24.5}$$

The utility function was then extended to a linear combination of five cues with four weighting parameters (x_i, R_{A_i}, P_{A_i},R_{D_i}, and P_{D_i}) with a common weight for the sum of the defender's penalty and reward. The model was also extended to a linear combination of all five cues with separate weighting parameters for each cue:

$$U_{A_i}(x_i) = w_1 x_i + w_2 P_{A_i} + w_3 R_{A_i} + w_4 (P_{D_i} + R_{D_i}) \tag{24.6}$$

$$U_{A_i}(x_i) = w_1 x_i + w_2 P_{A_i} + w_3 R_{A_i} + w_4 P_{D_i} + w_5 R_{D_i} \tag{24.7}$$

Lens Model accounting for Risk Attitude (lens-α)

Risk attitude can be captured by introducing the parameter α to the lens model:

$$U_{A_i}(x_i) = w_1 x_i + w_2 P_{A_i}^{\alpha} + w_3 R_{A_i}^{\alpha} \tag{24.8}$$

Risk attitude was also captured in the lens model with five cues. To reduce the number of parameters, common weights are assumed on the attacker's reward and penalty and the defender's reward and penalty. The evaluation of choosing target t_i then is:

$$U_{A_i}(x_i) = w_1 x_i + w_2 (P_{A_i}^{\alpha} + R_{A_i}^{\alpha}) + w_3 (P_{D_i} + R_{D_i}) \tag{24.9}$$

Multiattribute Utility (MAU) Model

Inspired from the lens model, which assumes utility as a linear combination of different attributes, a new model was developed of multiattribute utility considering that an adversary holds multiple objectives. The model assumes that adversaries have three objectives: (1) maximize the probability of success; (2) maximize their expected utility; and (3) minimize the defender's expected utility. The utility function of choosing target t_i can be written as:

$$U_{A_i}(x_i) = w_1 x_i + w_2 EU_{A_i} + w_3 EU_{D_i}$$

$$= w_1 x_i + w_2 [x_i P_{A_i} + (1 - x_i) R_{A_i}] + w_3 [x_i P_{D_i} + (1 - x_i) R_{D_i}] \quad (24.10)$$

A summary of the nine probabilistic choice models grouped in five categories is presented in Table 24.2. All nine models capture the inconsistency level (λ). EV is the baseline model. The five lens models and the MAU model capture an attacker's trade-offs among competing cues (or objectives). The EU–α, lens–3–α, and lens–5–α allow risk attitude to be accounted for; lens–4, lens–5, lens–5–α, and MAU model take the defender's utility into account for the attacker's utility function.

24.5 Addressing Multiple Decision Makers

One of the key challenges in real-world security problems is that the defender often has to deal with multiple types of self-interested attackers, each of whom may have different preferences, objectives, and risk attitudes. For example, the police force at Los Angeles International Airport (LAX) may face either well-funded, armed terrorists or local gangs and petty criminals (Jain et al., 2010). Similarly, in green security games (e.g., wildlife conservation domains), a recent study has shown a variety of poacher types, ranging from well-organized groups with high-technology equipment whose targets are high-valued animals for trading, such as rhinos, to local residents with low-technology equipment whose main purpose for poaching is food supply (Lemieux, 2014). Each different attacker type will follow a unique decision-making process that is guided by many factors such as capabilities, preferences, and risk attitudes. Therefore, it is important for security agencies to understand the behaviors of different types of attackers and address this multiple-attacker-type challenge when planning patrols.

This section will present a Bayesian extension to Stackelberg games that models the diversity of potential attackers with an assumption of a known distribution of attacker types. In this extension, each attacker type has a unique payoff matrix and behavioral model. We will also present an alternative robust approach that can be utilized when knowledge or available data are insufficient to precisely estimate a distribution of attacker types.

Table 24.2. *A summary of the nine models grouped in five categories*

Category	Model	Abbreviation	Equation
Attacker's expected value models	Attacker's expected value model	EV	$q_i(x_i) = \frac{e^{[x_i P_{A_i} + (1-x_i) R_{A_i}]\lambda}}{\sum_{t_k \in T} e^{[x_k P_{A_k} + (1-x_k) R_{A_k}]/\lambda}}$
	Attacker's expected utility model accounting for soft target	EU–soft target	$q_i(x_i) = \frac{e^{[x_i P_{A_i} + (1-x_i) R_{A_i}]/\lambda + \lambda_s S_i(x_i)}}{\sum_{t_k \in T} e^{[x_k P_{A_k} + (1-x_k) R_{A_k}]/\lambda + \lambda_s S_k(x_i)}}$
Attacker's expected utility model accounting for risk attitude	Attacker's expected utility model accounting for risk attitude	EU–α	$q_i(x_i) = \frac{e^{[x_i P_{A_i}^{\alpha} + (1-x_i) R_{A_i}^{\alpha}]/\lambda}}{\sum_{t_k \in T} e^{[x_i P_{A_k}^{\alpha} + (1-x_i) R_{A_k}^{\alpha}]/\lambda}}$
Lens models	Lens model – three parameters	Lens–3	$q_i(x_i) = \frac{e^{(w_1 x_i + w_2 P_{A_i} + w_3 R_{A_i})/\lambda}}{\sum_{t_k \in T} e^{(w_1 x_k + w_2 P_{A_k} + w_3 R_{A_k})/\lambda}}$
	Lens model – four parameters	Lens–4	$q_i(x_i) = \frac{e^{(w_1 x_i + w_2 P_{A_i} + w_3 R_{A_i} + w_4 (P_{D_i} + R_{D_i}))/\lambda}}{\sum_{t_k \in T} e^{(w_1 x_k + w_2 P_{A_k} + w_3 R_{A_k} + w_4 (P_{D_k} + R_{D_k}))/\lambda}}$
	Lens model – five parameters	Lens–5	$q_i(x_i) = \frac{e^{(w_1 x_i + w_2 P_{A_i} + w_3 R_{A_i} + w_4 P_{D_i} + w_5 R_{D_i})/\lambda}}{\sum_{t_k \in T} e^{(w_1 x_k + w_2 P_{A_k} + w_3 R_{A_k} + w_4 P_{D_k} + w_5 R_{D_k})/\lambda}}$

(continued)

Table 24.2 *(cont.)*

Category	Model	Abbreviation	Equation
Lens models accounting for risk attitude	Lens model – three attributes accounting for risk attitude	Lens–3–α	$q_i(x_i) = \frac{e^{(w_1 x_i + w_2 P_{A_i}^{\alpha} + w_3 R_{A_i}^{\alpha})/\lambda}}{\sum_{t_k \in T} e^{(w_1 x_k + w_2 P_{A_k}^{\alpha} + w_3 R_{A_k}^{\alpha})/\lambda}}$
	Lens model – five attributes accounting for risk attitude	Lens–5–α	$q_i(x_i) = \frac{e^{[w_1 x_i + w_2 (P_{A_i}^{\alpha} + R_{A_i}^{\alpha}) + w_3 (P_{D_i} + R_{D_i})]/\lambda}}{\sum_{t_k \in T} e^{[w_1 x_k + w_2 (P_{A_k}^{\alpha} + R_{A_k}^{\alpha}) + w_3 (P_{D_k} + R_{D_k})]/\lambda}}$
Multi-attribute utility model	Multi-attribute utility model	MAU	$q_i(x_i) = \frac{e^{w_1 x_i + w_2 [x_i P_{A_i} + (1-x_i) R_{A_i}] + w_3 [x_i P_{D_i} + (1-x_i) R_{D_i}]}}{\sum_{t_k \in T} e^{w_1 x_k + w_2 [x_i P_{A_k} + (1-x_i) R_{A_k}] + w_3 [x_i P_{D_k} + (1-x_i) R_{D_k}]}}$

24.5.1 *Bayesian Stackelberg Games*

Bayesian modeling is one of the leading approaches for addressing multiple types of attackers in security games, with the key assumption that the distribution of attacker types is known beforehand. Previous studies introduce two different solutions to capture different attacker types: (1) One represents different types of attackers by using a different payoff matrix associated with each attacker type. This results from the fact that different attackers may view targets differently in terms of the targets' values as well as the risk of being caught at these targets. Yin and colleagues (2012) introduce a Bayesian-based algorithm, HUNTER, that captures all these aspects as a distribution over attacker types; each attacker has his own payoff matrix, yet responds optimally corresponding to that payoff matrix. HUNTER then attempts to compute the optimal patrolling strategy for the defender by maximizing her expected utility over that distribution of attacker types. (2) The second solution, PAWS, is introduced by Yang and colleagues (2014), in which different attacker types are assumed to associate with different behavioral models. Unlike HUNTER, which still assumes each attacker is perfectly rational with regard to his payoff matrix, PAWS focuses on modeling the bounded rationality of every attacker based on available data such as the defender's patrols or attacking data collected by the defender (Yang et al., 2014). In particular, SUQR is the real-world behavior model used in PAWS to capture the behavior of attackers (PAWS is specifically designed for wildlife protection domains where rangers have to conduct patrols to protect animals in conservation areas against illegal poaching; yet it can be generalized to any security domain). Different attacker types will result in different weights of the SUQR model. Finally, PAWS attempts to learn the distribution of attacker types as a distribution over weights of SUQR. The optimal strategy for the defender is computed as to maximize the defender's expected utility against all SUQR-based attackers.

24.5.2 *Robust Bayesian Stackelberg Games*

One disadvantage of the standard Bayesian Stackelberg game approach is that it requires a significant amount of data in order to precisely estimate a distribution of attacker types, and thus it may be inapplicable when the data are insufficient. Robust Bayesian Stackelberg games are an alternative method that overcomes this shortcoming. Current robust-based algorithms focus on addressing the worst-case scenario for the defender by determining the attacker type whose response will affect the defender's strategy the most in terms of the utility that the defender receives.

The most recent algorithm following this approach is MIDAS (Brown et al., 2014), which is currently used to generate patrols for U.S. Coast Guard personnel in fishery protection. Similar to PAWS, MIDAS focuses on modeling different types of attackers (or illegal fishermen) using the SUQR models. However, instead of learning a distribution over attacker types, MIDAS attempts to address the worst-case SUQR-based attacker for the defender. In other words, given a particular strategy for the defender, MIDAS tries to find the attacker type whose response is the worst for that defender strategy. Then the optimal strategy for the defender is determined so as to maximize that worst-case utility for the defender. This robust idea of MIDAS is motivated by the fact that there is a lack of fishery data that can be used for predicting a distribution of attacker types.

All these modeling-based algorithms (Bayesian and Robust Bayesian) have to cope with the challenge that the resulting optimization problem for computing the optimal strategy for the defender is nonlinear and non-convex, which is hard to solve. Several novel techniques are applied for efficiently solving these problems. More algorithmic details can be found in Yin and colleagues (2012),Yang and colleagues (2014), and Brown and colleagues (2014).

24.6 Simulations

All the approaches presented are centered on addressing non-optimal and uncertain real-world responses, therefore, traditional proofs of correctness and optimality are insufficient: it is necessary to experimentally test these new approaches against existing approaches. Experimental analysis with human subjects shows how these approaches are expected to perform against real-world adversaries compared to alternative approaches.

24.6.1 *General Population as an Approximation for Threats*

A key question that needs to be addressed is if the terrorist psychiatric profile may lead to vastly different decision-making processes than the general population. An argument could be made that terrorists are completely irrational or suffer from severe psychosis, causing them to make near-arbitrary decisions. Thus they do not attempt to maximize expected value or even attempt to make boundedly rational decisions, making any game-theoretic approach, even one that considers potential deviations due to bounded rationality, irrelevant. However, research has shown that the primary shared characteristic of the psychiatric profile of the decision makers of terrorist factions is their normalcy (Abrahms, 2008; Gill & Young, 2011; Richardson, 2006). This extensive research suggests that terrorists are actually highly rational and carefully construct their attack strategies, attempting to maximize their expected value, justifying a game-theoretic approach (Abrahms, 2008; Keeney & von Winterfeldt, 2010; Richardson, 2006; Rosoff & John, 2009). As Louise Richardson, who has done significant research on the psyche of terrorists, states, "But terrorists, by and large, are not insane at all. Their primary shared characteristic is their normalcy, insofar as we understand the term. Psychological studies of *terrorists* are virtually unanimous on this point" (2006). Furthermore, Abrahms (2008) both agrees on this point and notes that these studies are based on records of dozens of terrorist organizations from the late 1960s to the present. It follows that testing against the general population can give some strategic insights on how to better defend against potential terrorists attacks given the inability to test against actual terrorists.

Based on the assumption that terrorists are rational agents, an argument could be made that it should be sufficient to calculate a game-theoretic optimal resource allocation strategy. However, modern decision theory recognizes that decision makers face cognitive and informational constraints. Additionally, terrorists sometimes face competing objectives and noisy information (Abrahms, 2008; Allison & Zelikow, 1999).

Another benefit of experimenting with the general population is that the defender strategies developed may be applied to a number of criminal scenarios beyond terrorism. Criminal activities can be broadly broken down into six categories: (i) property crimes; (ii) violent crimes; (iii) sex crimes; (iv) gangs and crime; (v) white-collar occupational crime; and (vi) drugs and crime (Pogrebin, 2012). It is the responsibility of many different security agencies to attempt to prevent these criminal activities, with each agency focusing on different categories of crime. Given the large range of criminal activities, the resulting types of criminals will span a wide variety, making the use of a general population largely applicable.

24.6.2 *Experimental Setup*

Behavioral data were collected from human subjects by designing online experiments. Specifically, a game inspired by the security domain at LAX (Pita et al., 2008), known as "The Guards and the Treasure," was developed by the TEAMCORE group at USC (Pita et al., 2009). The domain has some number of pirates – jointly acting as the defender – guarding a set of doors, and an individual subject acts as a single adversary. The subject's goal is to steal a treasure from behind a door without getting caught. Each of the doors has a different reward and penalty associated with it for the subjects as well as for the pirates. If a subject chooses a door that a pirate is guarding, the subject incurs the subject penalty for that door and the pirate receives the pirate reward for that door, or vice versa. The subjects are monetarily incentivized based on the points they earn.

Several aspects in the procedure of the experiments are challenging and merit consideration. First, the number of rounds each subject should play needs to be determined. The experimenter needs to balance the amount of data collected from an individual against the bias that will be incurred by fatigue. Second, the experimenter needs to consider how to incentivize subjects, whether to incentivize on all or a random subset of the choices subjects made, and the overall size of the rewards and penalties (monetary point values). It is undetermined in which case subjects' choices are most representative of the real world, and whether the face value of points can influence subjects' motivations. Third, the experimenter needs to determine whether to provide feedback after each play or at the end of all plays. Letting subjects know how they performed after every play enables them to better understand the game and make better decisions, but can result in learning and updating, which breaks the assumption that choices are independent. While this is not a comprehensive look at all the issues in experimental design, other aspects that need to be taken into account include the display of information, for example, whether to make all information equally salient; whether to provide experience-based or distribution-based probability information; and what training should be given to subjects before beginning the experiment.

24.6.3 *Results*

In this section, the evaluation of different behavioral models introduced in Section 24.4.2 as well as comparing between behavioral modeling–based algorithms and robust-based algorithms for handling the attacker's bounded rationality in security games is presented. While robust-based algorithms are more appropriate for security problems when available data are limited, behavioral modeling is important when sufficient data are available for learning, as experiments have shown that modeling-based algorithms outperform robust-based algorithms in terms of the expected utility the defender obtains by playing either of these algorithms.

Behavioral Modeling Evaluation

The following section discusses the comparison of the nine behavioral models introduced in 24.4.2. We utilize data from three experiments, including more than 1,000 subjects playing more than 25,000 games (Nguyen et al., 2013; Pita et al., 2012; Yang et al., 2012). The three experiments used the same game paradigm called "The Guards and the Treasure" written in PHP. Each participant was asked to play as an attacker and choose one out of eight gates to attack given x_i, R_{A_i}, P_{A_i}, R_{D_i}, and P_{D_i} for each alternative. The three experiments differed in attacker and defender payoff matrices, defenders' guarding strategies, and experimental procedures. In experiment I, 102 participants each played 40 rounds, and completed 4,080 rounds in total. Forty of the 102 participants were from the United States and 48 were from India. Thirty-six (35%) were female. In experiment II, a total of 653 U.S. participants each played 25 rounds and completed 16,325 rounds in total. Two-hundred and seventy-two (42%) were female. In experiment III, a total of 294 U.S. participants each played 25 to 33 rounds and completed 8,538 rounds in total. Eighty-nine (30%) were female.

Maximum Likelihood Estimation (MLE) (Scholz, 1985) was employed to fit the data over all the games played in each of the three experiments and estimate parameters for all nine models. The likelihood function for each model is:

$$\mathrm{L} = \prod_{i=1,2,..,N} q_i(x_i) \tag{23.11}$$

Akaike Information Criterion (AIC) (Akaike, 1974) was calculated using equation (24.12) for each model in the three experiments, where k is the number of model parameters. AIC is an estimate of the expected, relative distance between the fitted model and the unknown true mechanism that generated the observed data (Burnham & Anderson, 2003). The model with the minimum AIC is the best among the alternatives.

$$\mathrm{AIC} = -2 \ln \mathrm{L} + 2\mathrm{k} \tag{24.12}$$

The estimates of the parameters and AICs for the nine models tested in experiments I, II, and III are summarized in Table 24.3. In Experiment I, AIC results indicate that models EV and EU–α were similar in terms of fit; model EU–soft target was slightly better than EU and EU–α. The lens models fit better than model EU–soft target; among lens models, lens–5–α was the best. The MAU model did not fit as well as the linear utility models. Parameter estimates indicate that participants were consistent with maximization of the various evaluation functions ($\lambda < 0.1$) for all nine models. Both the lens models and the MAU model resulted in a negative weight on the probability of being caught, which suggests that participants tended to give a bonus to targets that are less likely to be guarded. Parameter estimates for the four models that captured the weight participants put on defender's rewards and penalties (Lens–4, lens–5, lens–5–α, and MAU) suggest that the weight on defender's side was

Table 24.3. *Estimates of parameters and AIC for experiments I, II, and III*

Model	Experiment I		Experiment II		Experiment III	
	AIC	Parameters estimation	AIC	Parameters estimation	AIC	Parameters estimation
EU	15036	$\lambda = 0.09$	60674	$\lambda = 0.08$	33334	$\lambda = 0.20$
EU–soft target	14820	$\lambda = (0.09, .59)$	50548	$\lambda = (0.07, 1.89)$	27802	$\lambda = (0.41, 1.79)$
EU–α	15012	$\lambda = 0.08$, $\alpha = 0.86$	59169	$\lambda = 0.06$, $\alpha = 0.7$	31065	$\lambda = 0.08$, $\alpha = 0.33$
Lens–3	14670	$\lambda = 0.05$, $w = (-0.32, 0.44, 0.24)$	52014	$\lambda = 0.04$, $w = (-0.42, 0.35, 0.23)$	25445	$\lambda = 0.07$, $w = (-0.16, 0.18, 0.67)$
Lens–4	14656	$\lambda = 0.02$, $w = (-0.31, 0.44, 0.23, .02)$	48218	$\lambda = 0.01$, $w = (-0.36, 0.30, 0.20, 0.14)$	22937	$\lambda = 0.02$, $w = (-0.47, 0.03, 0.19, 0.31)$
Lens–5	14658	$\lambda = 0.05$, $w = (-0.30, 0.43, 0.23, 0.02, 0.02)$	43265	$\lambda = 0.02$, $w = (-0.31, 0.26, 0.17, 0.04, 0.20)$	22592	$\lambda = 0.04$, $w = (-0.44, -0.01, 0.04, 0.30, 0.21)$
Lens–3–α	14645	$\lambda = 0.05$, $w = (-0.32, 0.35, 0.34)$, $\alpha = 1.47$	51929	$\lambda = 0.04$, $w = (-0.42, 0.30, 0.28)$, $\alpha = 1.25$	25159	$\lambda = 0.07$, $w = (-0.09, 0.11, 0.80)$, $\alpha = 1.86$
Lens–5–α	14624	$\lambda = 0.08$, $w = (-0.46, 0.5, 0.04)$, $\alpha = 1.51$	48121	$\lambda = 0.04$, $w = (-0.45, 0.32, 0.18)$, $\alpha = 1.32$	23228	$\lambda = 0.05$, $w = (-0.58, 0.07, 0.35)$, $\alpha = 0.47$
MAU	14973	$\lambda = 0.08$, $w = (-0.06, 0.84, 0.10)$	45335	$\lambda = 0.03$, $w = (-0.32, 0.39, 0.29)$	26540	$\lambda = 0.07$, $w = (-0.66, -0.01, 0.33)$

much lower than that put on the attacker's rewards and penalties (about 1/10). Finally, model EU–α indicated that participants were risk averse, while lens–3–α and lens–5–α indicated that attackers were risk seeking.

In Experiment II, the AIC fit indices were consistent with Experiment I; model EV was the worst model among the nine. Model EU–α was slightly better than EV, but was worse than EU–soft target. The lens models and MAU were again better than EU–soft target. The MAU model was not as good as the lens models. Among the five lens models, lens–4 was better than lens–3 and lens–5 was better than lens–4. Adding a parameter for risk attitude on lens–3 (lens–3–α) improved the model slightly. Adding a parameter for risk attitude on lens–5 and combining attacker's side and defender's side (lens–5–α) did not improve the model. Parameter estimates indicated that participants were rational ($\lambda < 0.1$ for EU–α, lens models, and MAU while $\lambda < 0.5$ for EV and EU–soft target). Again, lens models and the MAU model indicated that a negative weight was put on the probability of being caught. Results of the four models that capture the weight attackers place on the defender's rewards and penalties (lens–4, lens–5, lens–5–α, and MAU) suggested that the weight on defender's rewards and penalties was as high as the weight on attacker's side. Finally, EU–α and EU–5–α indicated that participants were risk averse while EU–3–α indicated that participants were risk seeking.

In Experiment III, AIC results were consistent with those from Experiments I and II in that model EV was the worst model among the nine. EU–α was slightly better than EV, but was worse than EU–soft target. EU–soft target was better than lens–3 and lens–3–α, and was worse than lens–4, lens–5, lens–5–α, and MAU (all models accounted for the defender's rewards and penalties). Among the five lens models, lens–4 was better than lens–3 and lens–5 was better than lens–4. Adding a parameter for risk attitude on lens–3 (lens–3–α) improved the model slightly. Adding a parameter for risk attitude on lens–5 and combining the attacker's rewards and penalties and the defender's rewards and penalties (lens–5–α) did not improve the model. The MAU model did not fit as well as lens–5, but was better than the other seven models. Parameter estimates indicated that participants were rational ($\lambda < 0.1$) for all models. Again, lens models and MAU indicated that a negative weight was put on the probability of being caught. Results of the four models that captured the weight participants put on the defender's rewards and penalties (lens–4, lens–5, lens–5–α, and MAU) suggested that the weight put on the defender's rewards and penalties was as high as that put on the attacker's rewards and penalties. Finally, EU–α indicated that attackers were risk averse while lens–3–α and lens–5–α indicated that attackers were risk seeking.

We expected there were individual differences in utility function parameters. For instance, some attackers may have multiple objectives of maximizing expected utility, minimizing the chance of being caught, and minimizing their enemies' (defenders) expected utility at the same time (captured in MAU). Some attackers may only maximize their own expected value (captured in EV). It is impossible to differentiate different types of "adversaries" with the nomothetic analysis alone. An ideographical analysis allows us to evaluate how each individual attacker made the decision and how that person is different from others. Again, parameters were estimated using MLE. Since the sample size (N) is small with respect to the number of parameters (k) (N/k < 40 using the k from the most complex model), AICc was calculated for comparisons over different models (Burnham & Anderson, 2002):

$$\mathrm{AICc} = -2\ln \mathrm{L} + 2\mathrm{k}\left(\frac{N}{N-k-1}\right) \tag{24.13}$$

The number of times each model has a minimum AICc is summarized in Table 24.4. Out of 102 attackers (each playing 40 games) in Experiment I, results indicated that lens–3 scored the minimum AICc most often. MAU model, EU–α, and lens–4 also scored the minimum AICc more often than the other models. In Experiment II, out of 653 attackers (playing 25 games each), results indicated that lens–5 scored the minimum AICc most often; MAU and lens–4 also scored the minimum AICc more often than other models. In Experiment III, out of 294 attackers (each playing 25–33 games), results indicated that lens–5 scored the minimum AICc most often, and lens–4 more often scored the minimum AICc compared to other models. EU and EU–α never scored the minimum AICc across all 294 attackers.

Modeling versus Robust-Based Algorithms

Game settings: In order to demonstrate the importance of behavioral modeling for addressing the attacker's bounded rationality in security games, two representative algorithms for modeling and robust-based approaches, named SU-BRQR (using the lens-3 model) and MATCH respectively, are compared. The lens-3 model is learned using data of 10 games in Pita and colleagues (2012). The experiment compares SU-BRQR against MATCH in the setting of the 8-target and 3-guard game and the 24-target and 9-guard game with the Amazon Mechanical Turk (AMT) workers. The game interface is described in Yang and colleagues (2011). In this 8-target game setting, for each game, results are averaged over at least 45 subjects. When two algorithms are compared, subjects played against both defender strategies using the same payoff structures. Participants were paid a base amount of U.S. $1.00. In addition, each participant was given a bonus based on his or her performance in the

Table 24.4. *Number of times model i has minimum AICc for Experiments I, II, and III*

	EU	EU – soft target	EU–α	Lens – 3	Lens – 4	Lens – 5	Lens–3–α	Lens–5–α	MAU	Total
Experiment I	7	3	14	28	11	3	7	8	21	102
Experiment II	7	13	6	32	105	282	14	26	168	653
Experiment III	0	29	0	34	56	96	11	33	35	294

Table 24.5. *Performance comparison, alpha = 0.05*

	SU-BRQR	Draw	MATCH
8-target case	13	8	1
24-target case	11	10	1

games for motivation. Players were not choosing targets arbitrarily by having each participant play two extra trivial games (i.e., games in which there is a target with the highest adversary reward and lowest adversary penalty and lowest defender coverage probability). Players' results were removed if they did not choose that target. A number of 22 payoff structures is generated based on covariance games in GAMUT (Nudelman et al., 2004). In covariance games, the covariance value r is adjusted in [–1, 1] to control the correlation between rewards of players. The statistical significance of experimental results is evaluated using the bootstrap-t method (Wilcox, 2002).

Experimental results: The experimental results are summarized in Table 24.5. The second column of Table 24.5 shows the number of payoffs where SU-BRQR outperforms MATCH with statistical significance (alpha = 0.05) in terms of the average defender expected reward over all the choices of the participants. The number of payoff structures where MATCH is better than SU-BRQR with statistical significance is shown in the fourth column. In the case of 8-target games, in the 22 payoff structures, SU-BRQR outperforms MATCH 13 times with statistical significance while MATCH defeats SU-BRQR only once; in the remaining eight cases, no statistical significance is obtained either way. Similarly, in the 24-target games, SU-BRQR also significantly outperforms MATCH. This result stands in stark contrast to Pita and colleagues' (2012) result and demonstrates that there is indeed value to integrating descriptive models of adversary decision making in computing defender strategies in SSGs.

24.7 Conclusions and Real-World Applications

Real-world adversaries present a critical challenge in real-world applications. That is, real-world adversaries challenge many of the key assumptions made in standard game-theoretic models. To that end, we have presented both robust and behavioral approaches for addressing real-world adversaries in security domains, as well as experimental results demonstrating the efficacy of those models against standard game-theoretic models. Due to the demonstrated success of these models against real-world adversaries, some have now been featured in real-world applications.

24.7.1 *PROTECT for U.S. Coast Guard*

The U.S. Coast Guard (USCG) focuses on the maritime security of U.S. coasts, ports, and inland waterways – a security domain that faces increased risks due to threats such as terrorism and drug trafficking. In particular, the USCG conducts patrols to protect the variety of critical infrastructure that an attacker may attack within ports. However, while the attacker can observe patrol patterns, patrols cannot be at every location all the time due to limited security resources. The Port Resilience Operational/Tactical Enforcement to Combat Terrorism (PROTECT) model assists the USCG to enhance its patrolling effectiveness in maritime security (Shieh et al., 2012). It has been in use at the port of Boston since April 2011, and at the port of New York since February 2012. PROTECT uses attacker-defender Stackelberg games to capture the maritime security problem, with the USCG as the defender against terrorists, where it uses a quantal response model of the attacker's behavior to capture terrorists' decision making.

24.7.2 *PAWS for Wildlife Security*

Recently, game-theoretic research has started focusing on resource conservation domains wherein security resources are allocated to patrol a vast geographical area (e.g., a wildlife conservation area or a maritime fishery) against environmental criminals such as poachers and illegal fishermen. Wildlife crime is a global issue where many endangered species such as tigers and rhinos are in danger of extinction from illegal poaching. In many wildlife conservation areas, e.g., Queen Elizabeth National Park in Uganda and Kruger National Park in South Africa, poachers attempt to capture animals by setting poaching traps such as wire snares at places where they think animals are likely to appear. The main objective of park rangers, the law enforcement officers within the national parks, is to prevent poaching by conducting patrols throughout the parks. The key challenge of wildlife protection is that rangers face multiple types of poachers with different rationalities and objectives. These poachers may be local villagers, whose main purpose is to hunt animals for food supplies, who do not make a poaching plan in advance. Conversely, there are groups of poachers who attempt to capture high-valued animals for trading such as tigers or rhinos, that are well equipped and spend time to observe rangers' patrols and carefully plan their poaching activity. Thus, it is critical for rangers to take into account this challenge of multiple poacher types to conduct effective patrolling strategies. Protection Assistant for Wildlife Security (PAWS) – a security game application provided by Yang and colleagues (2014) that optimizes wildlife crime patrols for stopping poachers – takes into account the Bayesian distribution of multiple poacher types. The key idea of PAWS is to adaptively improve patrolling strategies for rangers based on the rich poaching data collected during previous patrols, where behavioral models of poachers are incrementally improved with more poaching data. PAWS was tested by rangers in Uganda for protecting wildlife in Queen Elizabeth National Park in April 2014.

24.7.3 *Fish Protection for the U.S. Coast Guard*

The U.S. Coast Guard (USCG) attempts to protect the nation's fisheries against illegal, unreported, and unregulated (IUU) fishing by deploying resources to patrol over fishery areas. Since these patrolling areas are large while the patrolling resources are limited, it is impossible to protect an entire fishery at all times. Thus, the MIDAS algorithm was developed to assist the USCG to optimally randomize its limited resources to protect fisheries (Brown et al., 2014). To that end, this specific domain exhibits several challenges and opportunities. First, the defender in fishery domains has to cope with numerous adversaries whose behaviors may differ from each other. Second, due to frequent interactions between the defender and adversaries, the defender can collect data on the adversaries' activity over time. Third, the adversaries are less strategic due to the short planning windows between actions. Based on these challenges and opportunities, MIDAS utilizes available data to learn the behavior models of the adversaries using the SUQR model. Moreover, due to the limited available fishery data, instead of predicting a distribution of adversaries like PAWS, MIDAS applies the robust optimization method described in Section 24.5.2 that focuses on addressing the worst case of adversary types for the defender while considering different SUQR models that represent the bounded rationality of different adversaries. MIDAS has been successfully deployed and evaluated by the USCG in the Gulf of Mexico. It is currently under review by the USCG and is being considered for further deployment in the Gulf of Mexico as well as in other fisheries nationwide.

REFERENCES

Abrahms, M. (2008). What terrorists really want: Terrorist motives and counterterrorism strategy. *International Security*, *32*(4), 78–105.

Aghassi, M., & Bertsimas, D. (2006). Robust game theory. *Mathematical Programming*, *107*(1), 231–273.

Akaike, H. (1974). A new look at the statistical model identification. *Automatic Control, IEEE Transactions on*, *19*(6), 716–723.

Allison, G., & Zelikow, P. (1999). *Essence of decision: Explaining the Cuban Missile Crisis, 2nd Edition.* New York: Longman.

Baron, J. (2008). *Thinking and deciding*. New York: Cambridge University Press.

Brown, M., Haskell, W. B., & Tambe, M. (2014). Addressing scalability and robustness in security games with multiple boundedly rational adversaries. In *International Conference on Decision and Game Theory for Security* (pp. 23–42). New York: Springer International Publishing.

Brunswik, E. (1952). The conceptual framework of psychology. *Psychological Bulletin*, *49*(6), 654–656.

Burnham, K. P., & Anderson, D. R. (2003). *Model selection and multimodel inference:A practical information-theoretic approach.* Springer Science & Business Media.

Camerer, C. (2003). *Behavioral game theory: Experiments in strategic interaction.* Princeton, NJ: Princeton University Press.

Conitzer, V., & Sandholm, T. (2006). Computing the optimal strategy to commit to. In *ACM EC*, 82–90.

Fox, C. R., & Clemen, R. T. (2005). Subjective probability assessment in decision analysis: Partition dependence and bias toward the ignorance prior. *Management Science*, *51*(9), 1417–1432.

Fox, C. R., & Rottenstreich, Y. (2003). Partition priming in judgement under uncertainty. *Psychological Science*, *14*, 195–200.

Gill, P., & Young, J. K. (2011). Comparing role-specific terrorist profiles. http://dx.doi .org/10.2139/ssrn.1782008.

Hammond, K. R. (1955). Probabilistic functioning and the clinical method. *Psychological Review*, *62*(4), 255–262.

Jain, M., Pita, J., Tsai, J., Kiekintveld, C., Rathi, S., Ordóñez, F., & Tambe, M. (2010). Software assistants for patrol planning at LAX and Federal Air Marshals Service. In *Interfaces*, *40*(4), 267–290.

Kahneman, D., Slovic, P., & Tversky, A. (Editors) (1982). *Judgment under uncertainty: Heuristics and biases.* New York: Cambridge University Press.

Kahneman, D., & Tversky, A. (1972). Subjective probability: A judgement of representativeness. *Cognitive Psychology*, *3*, 430–454.

Keeney, R. L., & Raiffa, H. (1976). *Decisions with multiple objectives: Preferences and value trade-offs*. New York: John Wiley & Sons, Inc.

Keeney, G. L., & von Winterfeldt, D. (2010). Identifying and structuring the objectives of terrorists. *Risk Analysis*, *30*(12), 1803–1816.

Lemieux, A. (2014). *Situational prevention of poaching*. In Crime Science Series. New York: Routledge.

Luce, R. D. (1959). *Individual choice behavior*. New York: John Wiley & Sons, Inc.

McCubbins, M. D., Turner, M. B., & Weller, N. (2012). The theory of minds within the theory of games. In *Proceedings of the 2012 International Conference on Artificial Intelligence*.

McFadden, D. L. (1976). Quantal choice analysis: A survey. *Annals of Economic and Social Measurement*, *5*(4), 363–390.

McFadden, D. (2001). Economic choices. *American Economic Review*, *91*(3), 351–378.

McKelvey, R., & Palfrey, T. (1995). Quantal response equilibria for normal form games. *Games and Economic Behavior*, *10*(1), 6–38.

Montibeller, G., & von Winterfeldt, D. (2015). Cognitive and motivational biases in decision and risk analysis. *Risk Analysis*, *35*(7), 1230–1251.

Nickerson, R. S. (1998). Confirmation bias: A ubiquitous phenomenon in many guises. *Review of General Psychology*, *2*(2), 175–220.

Nilim, A., & Ghaoui, L. E. (2005). Robust control in Markov decision problems with uncertain transition matrices. *Operations Research*, *53*(5), 780–798.

Nguyen, T. H., Yang, R., Azaria, A., Kraus, S., & Tambe, M. (2013). Analyzing the effectiveness of adversary modeling in security games. In *Conference on Artificial Intelligence.*

Nudelman, E., Wortman, J., Shoham, Y., & Leyton-Brown, K. (2004). Run the gamut: A comprehensive approach to evaluating game-theoretic algorithms. *In Proceedings of the 3rd International Conference on Autonomous Agents and Multiagent Systems* (pp. 880–887).

Paruchuri, P., Pearce, J. P., Marecki, J., Tambe, M., Ordonez, F., & Kraus, S. (2008). Playing games with security: An efficient exact algorithm for Bayesian Stackelberg games. In *Proceedings of the 7th International Conference on Autonomous Agents and Multiagent Systems.*

Phillips, L. D., & Edwards, W. (1966). Conservatism in a simple probability inference task. *Journal of Experimental Psychology*, *72*(3), 346–354.

Phillips, L. D., Hays, W. L., & Edwards, W. (1966). Conservatism in complex probabilistic inference. *IEEE Transactions on Human Factors in Electronics*, *7*(1), 7–18.

Pita, J. Jain, M., Ordóñez, F., Tambe, M., Kraus, S., & Magori-Cohen, R. (2009). Effective solutions for real-world Stackelberg games: When agents must deal with real-world uncertainties. In *Proceedings of the 8th International Conference on Autonomous Agents and Multiagent Systems.*

Pita, J., Jain, M., Tambe, M., Ordóñez, F., & Kraus, S. (2010). Robust solutions to Stackelberg games: Addressing bounded rationality and limited observations in real-world cognition. *Artificial Intelligence Journal*, *174*(15), 1142–1171.

Pita, J., Jain, J., Western, C., Portway, C., Tambe, M., Ordóñez, F., Kraus, S., & Paruchuri, P. (2008). *Deployed ARMOR protection: The application of a game theoretic model for security at the Los Angeles International Airport.* In *Proceedings of the 7th International Conference on Autonomous Agents and Multiagent Systems.*

Pita, J., John, R., Maheswaran, R., Tambe, M., & Kraus, S. (2012). A robust approach to addressing real-world adversaries in security games. In *Proceedings of the 20th European Conference on Artificial Intelligence.*

Pogrebin, M. R. (2012). *About criminals: A View of the offenders' world.* Thousand Oaks, Ca.: SAGE.

Richardson, L. (2006). *What terrorist want: Understanding the enemy, containing the threat.* New York: Random House.

Rosoff, H., & John, R. (2009). Decision analysis by proxy for the rational terrorist. In *Proceedings of the Workshop on Quantitative Risk Analysis for Security Applications*, 25–32.

Scholz, F. (1985). Maximum likelihood estimation. *Encyclopedia of Statistical Sciences.*

See, K. E., Fox, C. R., & Rottenstreich, Y. (2006). Between ignorance and truth: Partition dependence and learning in judgment under uncertainty. *Journal of Experimental Psychology: Learning, Memory, and Cognition*, *32*, 1385–1402.

Selten, R. (1988). Evolutionary stability in extensive two-person fames – Correction and further development. *Mathematical Social Sciences*, *16*, 223–266.

Shieh, E., An, B., Yang, R., Tambe, M., Baldwin, C., DiRenzo, J., Maule, B., & Meyer, G. (2012). PROTECT: A deployed game theoretic system to protect the ports of the United States. In *Proceedings of the 11th International Conference on Autonomous Agents and Multiagent Systems*.

Simon, H. (1956). Rational choice and the structure of the environment. *Psychological Review*, *63*(2), 129–138.

Tambe, M. (2011). *Security and game theory: algorithms, deployed systems, lessons learned*. New York: Cambridge University Press.

Tversky, A., & Kahneman, D. (1974). Judgment under uncertainty: Heuristics and biases. *Science*, *185*, 1124–1131.

Tversky, A., & Koehler, D. (1994). Support theory: A Nonextensional representation of subjective probability. *Psychological Review*, *101*(4), 547–567.

von Stengel, B., & Zamir, S. (2004). Leadership with commitment to mixed strategies. In *CDAM Research Report LSE-CDAM-2004-01*. London School of Economics.

von Winterfeldt, D., & Edwards, W. (1986). *Decision analysis and behavioral research*. New York: Cambridge University Press.

Wilcox, R. (2002). *Applying contemporary statistical techniques*. Academic Press.

Yang, R., Ford, B., Tambe, M., & Lemieux, A. (2014). Adaptive resource allocation for wildlife protection against illegal poachers. In *Proceedings of the 13th International Conference on Autonomous Agents and Multiagent Systems*.

Yang, R., Kiekintveld, C., Ordóñez, F., Tambe, M., & John, R. (2011). Improving resource allocation strategy against real-world adversaries in security games. In *Proceedings of the 20th International Joint Conference on Artificial Intelligence* (pp. 458–464).

Yang, R., Kiekintveld, C., Ordóñez, F., Tambe, M., & John, R. (2012). Improving resource allocation strategies against human adversaries in security games: An extended study. *Artificial Intelligence*, *195*, 440–469.

Yin, Z., Jain, M., Tambe, M., & Ordóñez, F. (2011). Risk-averse strategies for security games with execution and observational uncertainty. In *Proceedings of the 25th AAAI Conference on Artificial Intelligence*.

Yin, Z., & Tambe, M. (2012). A Unified Method for Handling Discrete and Continuous Uncertainty in Bayesian Stackelberg Games. In *Autonomous Agents and Multiagent Systems*.

25

Learning to Play Stackelberg Security Games

Avrim Blum, Nika Haghtalab, and Ariel D. Procaccia

As discussed in previous chapters, algorithmic research on Stackelberg security games has had a striking real-world impact. But an algorithm that computes an optimal strategy for the defender can only be as good as the game it receives as input, and if that game is an inaccurate model of reality, then the output of the algorithm will likewise be flawed. Consequently, researchers have introduced Bayesian frameworks that capture uncertainty using a probability distribution over possible games. Others have assumed that the unknown parameters of the game lie within known intervals. These approaches are discussed in Chapter 17 of this book (Kiekintveld & Jain, 2015).

In this chapter, we present an alternative, *learning-theoretic* approach for dealing with uncertainty in Stackelberg security games. In order to paint a cohesive picture, we focus on one type of uncertainty: unknown attacker utilities (so we are assuming that the attacker is perfectly rational and has perfect surveillance). Learning will take place in a *repeated Stackelberg security game*, where the defender gathers information about the attacker purely by observing the attacker's responses to mixed strategies played by the defender.

In more detail, we wish to learn a good strategy for the defender without any initial information about the utility function of the attacker (Section 25.1), when given a distribution over attacker types (Section 25.2), and when faced with an unknown sequence of attackers (Section 25.3). In each section, we present, in some generality, the relevant learning-theoretic techniques: optimization with membership queries, Monte Carlo tree search, and no-regret learning, respectively. In Section 25.4, we briefly discuss additional work at the intersection of machine learning and Stackelberg security games.

25.1 Single Unknown Attacker

In this section, we describe a setting introduced by Letchford and colleagues (2009), and present a method designed by Blum and colleagues (2014) for using best-response queries to learn the optimal strategy against a single attacker with unknown utilities. The main result discussed in this section involves learning the optimal strategy of the defender with a number of best-response observations that is polynomial in the parameters of the game.

Consider a Stackelberg security game (SSG) denoted by a 3-tuple (N, U, M) where N is the set of n *targets*, U indicates the *utilities* associated with the defender and attacker, and M indicates all the possible subsets of targets that can be simultaneously defended in valid deployments of resources. More precisely, M is determined by the set of targets N, set of *resources* R, a set $\mathcal{D} \subseteq 2^N$ of *schedules* where each $D \in \mathcal{D}$ represents a set of targets that can be simultaneously defended by one resource, and function $A{:}R \rightarrow 2^{\mathcal{D}}$ that indicates the set of all schedules that can be defended by a given resource. Let there be m pure strategies, then M is a zero-one $n \times m$ matrix, with rows that represent targets and columns that represent pure strategies, such that $M_{ij} = 1$ if and only if target i is covered by some resource in pure strategy j.

A *mixed strategy* of the defender is a distribution over pure strategies and is represented by a probability vector $\mathbf{s} \in \mathbb{R}^m$. The *coverage probability vector* of a strategy $\mathbf{s}$ is a vector $\mathbf{p} \in [0,1]^n$ such that for all $i \in N$, p_i is the probability with which target i is defended under $\mathbf{s}$. The coverage probability vector of $\mathbf{s}$ is given by $\mathbf{p} = M\mathbf{s}$.

Let $u_d^c(i)$ and $u_d^u(i)$ indicate the defender's payoffs when target i is attacked and is, respectively, covered or not covered by the defender. Similarly, $u_a^c(i)$ and $u_a^u(i)$ are the attacker's payoffs from attacking target i when it is, respectively, covered or not covered by the defender. Then under coverage probability vector $\mathbf{p}$, the expected utility of the defender when target i is attacked is given by $U_d(i, \mathbf{p}) = u_d^c(i)p_i + u_d^u(i)(1 - p_i)$ and the expected utility of the attacker is $U_a(i, \mathbf{p}) = u_a^c(i)p_i + u_a^u(i)(1 - p_i)$. Note that $U(i, \mathbf{p})$ is a linear function in p_i. Since $U(i, \mathbf{p})$ only depends on p_i, with a slight abuse of notation, we use $U(i, p_i)$ when it is convenient.

Let $\mathbf{p}$ be the coverage probability vector of strategy $\mathbf{s}$. The attacker *best responds* to $\mathbf{s}$ by attacking target $b(\mathbf{s}) \in \operatorname{argmax}_{i \in N} U_a(i, \mathbf{p})$. Note that given a coverage probability vector and the attacker's utilities, the attacker's best response is invariant to the mixed strategy that is used to implement that coverage probability vector. Therefore, we can work directly in the space of valid coverage probability vectors.

When $U_a(\cdot)$ is unknown, one cannot find the optimal strategy of the defender directly. However, observing the attacker's response to some carefully chosen strategies can help fill in the missing information required for finding the optimal strategy. As a warmup, consider a security game with two targets and one resource that can defend only one of the targets at a time. The defender's payoff for either

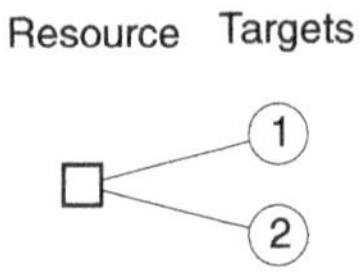

Utilities

Targets	Defender		Attacker Type θ_1		Attacker Type θ_2	
i	$u_d^c(\cdot)$	$u_d^u(\cdot)$	$u_a^c(\cdot)$	$u_a^u(\cdot)$	$u_a^c(\cdot)$	$u_a^u(\cdot)$
1	0	− 1	0	1	0	1
2	0	− 1	0	1	0	0

Figure 25.1. An example of a security game with one resource and two targets, with two possible utilities for the attacker.

of the targets is −1 when they are attacked and are left unprotected, and 0 otherwise. Consider two possible utilities for the attacker. In the first case (type θ_1), the attacker values both targets equally, with utility 1 for attacking a target that is not protected and 0 for attacking a target that is protected. In the second case (type θ_2), the attacker's utility for target 1 is the same as in case 1; however, there is 0 payoff for attacking target 2 whether or not it is protected (see Figure 25.1). The attacker's type is not known a priori. Consider the strategy where the resource covers target 1 with probability $\frac{2}{3}$ and target 2 with probability $\frac{1}{3}$; it induces the coverage probability vector $\mathbf{p} = (\frac{2}{3}, \frac{1}{3})$. An attacker of type θ_1 would respond to p by attacking target 2, whereas an attacker of type θ_2 would respond to p by attacking target 1. So, by observing the attacker's response to p, we can detect the type of the attacker in play and use the optimal strategy for that attacker type.

In the foregoing example, we demonstrated how the uncertainty about two possible discrete types of attackers was resolved by observing the best response of the attacker. However, the results discussed in this section are more general, as we can pinpoint the optimal strategy with no prior knowledge regarding the possible types of attackers. To describe these methods, we first review how the optimal strategy can be computed when $U_a(\cdot)$ is known.

One of the methods used for computing the optimal strategy in SSGs is the Multiple LPs approach (Conitzer & Sandholm, 2006). In this formulation, we solve one linear program (LP) for every target $i \in N$. This LP, as shown in equation (25.1), finds the optimal defender mixed strategy among all strategies that cause an attack on target i. The optimal defender's strategy is the best strategy among the solutions to these LPs; the one with the highest payoff to the defender.

$$\begin{aligned}
\text{maximize} \quad & U_d\Big(i, \sum_{j:M_{ij}=1} s_j\Big) \\
\text{s.t.} \quad & \forall i',\ U_a\Big(i', \sum_{j:M_{i'j}=1} s_j\Big) \le U_a\Big(i, \sum_{j:M_{ij}=1} s_j\Big) \\
& \forall j, s_j \ge 0 \\
& \sum_{j=1}^{m} s_j = 1
\end{aligned} \tag{25.1}$$

When $U_a(\cdot)$ is unknown, one cannot explicitly write down these LPs. However, the attacker's best response to $\mathbf{s}$ tells us which of the n feasible regions $\mathbf{s}$ belongs to. Therefore, in a *repeated* Stackelberg game against a single attacker, the defender can hope to construct the hidden LPs via *best-response queries*, that is, by observing the attacker's responses to a sequence of strategies. This is indeed the approach taken by Letchford and colleagues (2009), but the number of observations they require is polynomial in the number of pure leader strategies in a Stackelberg game, which, in the case of SSGs, could be exponential in the representation of the game.

Here we focus on the approach of Blum and colleagues (2014), which learns the optimal strategy using only a polynomial number of best-response queries in SSGs. This improvement in the number of queries is mainly based on the observation that the defender's utility can be optimized in the feasible region of a given LP without actually reconstructing these feasible regions. Formally:

Theorem 1 (Blum et al., 2014). *Consider an SSG with n targets. For any* $\epsilon, \delta > 0$, *with probability* $1 - \delta$, *it is possible to learn a defender strategy that is optimal up to an additive term of* ϵ, *using a number of best-response queries that is polynomial in n and logarithmic in* $1/\epsilon$ *and* $1/\delta$.

In the remainder of this section, we first provide an overview of some learning-theoretic techniques for optimizing linear programs without explicit knowledge of the constraints, and then provide an outline of the algorithm of Blum and colleagues (2014) based on these techniques.

To perform optimization over a region, we need to find at least one point in that region. Conceptually, a key challenge here is that it could be that the optimal defender strategy involves getting the attacker to attack a particular target i, and yet random allocation of resources might have a tiny chance of getting the attacker to do so; thus, some form of intelligent search through the strategy space will be needed. We first describe how to optimize over a region given an initial point in that region, and then discuss how we can perform this intelligent search.

25.1.1 *Optimization with Membership Queries*

A classic result of Khachiyan (1979) shows that one can solve linear programs in polynomial time using the *ellipsoid algorithm.* This algorithm has the additional property that it does not need to know the constraints of the linear program explicitly so long as it has access to a *separation oracle.* This is an oracle for the feasible region $\mathcal{P} \subseteq \mathbb{R}^n$ that given a proposed solution $\mathbf{x}$ will either answer "yes" if $\mathbf{x} \in \mathcal{P}$ or else will produce a violated constraint (a half-space that contains $\mathcal{P}$ but does not contain $\boldsymbol{x}$) if $\mathbf{x} \notin \mathcal{P}$. However, sometimes one has only a weaker capability: the ability to answer whether a proposed point $\mathbf{x}$ belongs to a given convex region $\mathcal{P}$, without being able to produce a violated constraint when the answer is "no." Such a capability is called a *membership oracle.* A specialized version of the ellipsoid algorithm is shown in Grötschel and colleagues (1988) to be able to solve optimization problems of this form – that is, optimizing a linear function over a convex region, where the convex region is given a membership oracle – when one is given an initial starting point $\mathbf{x}_{init} \in \mathcal{P}$ such that a ball of sufficiently large radius r about $\mathbf{x}_{init}$ is also contained in $\mathcal{P}$. However, the algorithm is quite complex and slow. More recently, Bertsimas and Vempala (2004) and Tauman Kalai and Vempala (2006) have shown how simpler and faster random-walk based algorithms can be applied to solve this problem as well, with the fastest being the algorithm of Tauman Kalai and Vempala (2006). This algorithm overall makes $\tilde{O}(n^{4.5})$ calls to the given membership oracle, where the $\tilde{O}$ notation hides polylogarithmic factors.

The algorithm of Tauman Kalai and Vempala (2006) can be viewed as a form of simulated annealing. Given an initial point $\mathbf{x}_{init}$ such that a ball of radius r about $\mathbf{x}_{init}$ is contained inside $\mathcal{P}$, it maintains a "temperature" parameter T that begins initially at a radius R such that we are guaranteed that $\mathcal{P}$ is contained within a ball of radius R. (All these algorithms will have a polylogarithmic dependence on the ratio R/r.) This temperature is then slowly lowered over time. If $\mathbf{c} \cdot \mathbf{x}$ is the objective we are aiming to minimize, the algorithm's goal on each round is to select a random point inside the convex body according to a probability distribution where the probability of some point x is proportional to $e^{-\mathbf{c}\cdot\mathbf{x}/T}$. This is an easier (though still nontrivial!) task to perform for large values of T. What Tauman Kalai and Vempala (2006) show is how to use a solution for one value of T to help solve for an answer for the next lower value of T.

The selection of a random point inside $\mathcal{P}$ from the distribution for a given value of T is done via the following random-walk procedure. Starting from the output of the previous value of T, a random vector is drawn through the current point with a distribution that depends on an estimated covariance matrix for the distribution for the previous value of T (this covariance matrix is estimated by taking several samples from the previous value of T). Then the point is updated to a random point inside $\mathcal{P}$ along that line, with probability density proportional to the objective function $e^{-\mathbf{c}\cdot\mathbf{x}/T}$. This process has a stationary distribution that equals the distribution we would like to be sampling from for this value of T, and what is proven in Tauman Kalai and Vempala (2006) is that the walk quickly approaches this distribution. Finally, once T is sufficiently small, nearly all the probability mass of the stationary distribution is close to the optimum value of the objective. Putting this together, we have the following:

Theorem 2 (Tauman Kalai & Vempala, 2006). *For any convex set $\mathcal{P} \subseteq \mathbb{R}^n$, suppose we are given a linear objective* $\mathbf{c}$, *an accuracy parameter* ϵ, *a starting point* $\mathbf{x}_{init} \in \mathcal{P}$, *and values r, R such that a ball of radius r about* $\mathbf{x}_{init}$ *is contained in* $\mathcal{P}$ *and a ball of radius R about* $\mathbf{x}_{init}$ *contains* $\mathcal{P}$. *Then, using* $\tilde{O}(n^{4.5})$ *calls to a membership oracle for* $\mathcal{P}$, *the algorithm will with high probability output a point* $\mathbf{x}^* \in \mathcal{P}$ *such that* $\mathbf{c} \cdot \mathbf{x}^* \leq \min_{\mathbf{x} \in \mathcal{P}} \mathbf{c} \cdot \mathbf{x} + \epsilon$, *where the* $\tilde{O}$ *notation hides terms logarithmic in n, R/r, and* $1/\epsilon$.

25.1.2 *Using Membership Queries to Learn SSGs*

In order to understand the connection between SSGs and optimization using membership queries, we make two changes to the LP in equation (25.1). First, since the attacker's best response is independent of the strategy that is used to implement a given coverage probability, we restate the LP to use variables that represent the coverage probability vector rather than the mixed strategy used to implement it. Second, since $U_a(\cdot)$ is unknown, we change the constraints to implicitly determine whether a given target i is attacked as a best response to a strategy.

$$\begin{aligned} &\text{maximize } U_d(i, \mathbf{p}) \\ &\text{s.t.} \quad i \text{ is attacked under } \mathbf{p} \\ &\qquad\;\; \mathbf{p} \text{ is implementable} \end{aligned} \tag{25.2}$$

To use the results from Section 25.1.1, we need to establish four properties regarding the LP given earlier:

1. The objective function is linear in **p**.
2. The optimization region is convex.
3. There is a membership oracle for the optimization region.
4. We have access to a well-centered initial feasible point in the optimization region.

Next, we briefly explain how each of these four properties is satisfied.

For the first requirement, note that by the definition of the utility function, the objective function $U_d(i, \mathbf{p})$ is linear in **p**. For the second requirement, for any i, let $\mathcal{P}_i$ denote the optimization region in equation (25.2), i.e., $\mathcal{P}_i$ is the set of all coverage probabilities that can be implemented by some mixed strategy and lead to target i being attacked. $\mathcal{P}_i$ corresponds to a linear transformation of the optimization region in LP (25.1), which is itself the intersection of

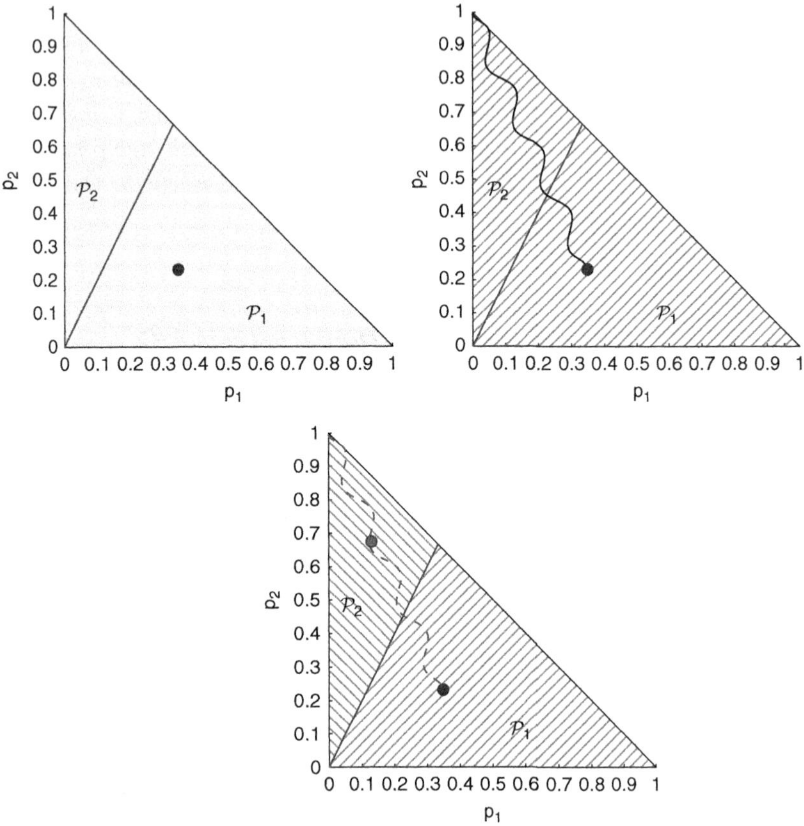

Figure 25.2. Three major steps of the algorithm for finding initial points in all regions. In the first step, one strategy is queried at random and is found to be in $\mathscr{P}_1$. In the subsequent steps, the optimization process only considers targets that are revealed so far, i.e., target 1. In the last figure, a strategy in region $\mathscr{P}_2$ is revealed, acting as an initial point in the newly discovered region $\mathscr{P}_2$.

halfspaces defined by the linear constraints and, as a result, is convex. So, $\mathscr{P}_i$ is also convex.

For the third requirement, a natural membership oracle for $\mathscr{P}_i$ involves a two-step procedure: For every $\mathbf{p}$, first solve the linear system $M\mathbf{s} = \mathbf{p}$ to find out whether $\mathbf{p}$ can be implemented by some mixed strategy $\mathbf{s}$, and then play $\mathbf{s}$ and observe the attacker's response to learn whether $\mathbf{s}$ causes an attack on target i.

The last requirement for using the results in Section 25.1.1 is having an initial feasible point in the region of optimization. A natural first attempt for finding an initial point in region $\mathscr{P}_i$ involves an algorithm that repeatedly plays a randomly chosen coverage probability vector and observes the attacker's response until one of the strategies induces an attack on target i. However, hitting a region with randomly chosen points in a high-dimensional space requires exponentially many attempts. So, this algorithm requires exponentially many best-response queries just to find initial feasible points.

Blum and colleagues (2014) take an entirely different approach for finding initial feasible points. The method is based on an intuitive observation that, if a target is left unprotected at all times and yet never attacked even as the defender increases his probability of covering other targets, then the defender cannot possibly achieve a better payoff by defending that target. This insight motivates a procedure that, at every major step, only considers the optimization tasks for the targets that have been observed so far. Moreover, we will explicitly set coverage probabilities on targets never attacked so far to 0 even if the strategies output by the optimization happen as a by-product to give them nonzero coverage. If, in this process, we observe that a strategy induces an attack on a target that has not been already seen, e.g., target i, then we have an initial point for this new region, $\mathscr{P}_i$. Otherwise, we have reached the optimal strategy (see Figure 25.2). Additional caution must be taken to ensure that these initial points are "well-centered." We leave it to the interested reader to refer to Blum and colleagues (2014) for the details of this procedure.

25.2 Single Attacker Drawn from a Distribution

In Section 25.1, we discussed the case of a single attacker who is completely unknown initially. While the method given therein eventually converges to an almost optimal strategy for the defender, the learning process itself can lead to significant inefficiencies. The purpose of the approach described in this section is to trade off *exploration* and *exploitation*, that is, take advantage of useful knowledge about the attacker *as it is obtained*, while also taking appropriate actions to learn more about the attacker.

Taking a Bayesian viewpoint, suppose there is a set of possible attacker types Θ, where each attacker type $\theta \in \Theta$ is defined by that attacker's utility function. In this section, it is assumed that a single attacker type θ is drawn from a known distribution $\mathscr{D}$ over Θ, and the defender plays a repeated Stackelberg security game with T rounds against an attacker of type θ. Importantly, the assumption that the defender has a prior over attacker types was not made in Section 25.1, that is, in this section, the defender has more information up front.

To gain some intuition about how the current setting differs from the one of Section 25.1, consider the game given in Figure 25.3, which is a slight variant

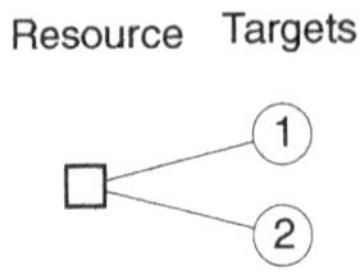

Utilities

Targets	Defender		Attacker Type θ_1		Attacker Type θ_2	
i	$u_d^c(\cdot)$	$u_d^u(\cdot)$	$u_a^c(\cdot)$	$u_a^u(\cdot)$	$u_a^c(\cdot)$	$u_a^u(\cdot)$
1	0	$-L$	0	2	0	1
2	0	-1	0	1	0	1

Figure 25.3. An example of a security game with one resource and two targets, with two possible attacker types.

of our previous example. In this game, Θ includes only two attacker types, and the attacker's identity can be determined in a single round by playing (3/5, 2/5): if target 1 is attacked, then the attacker is of type θ_1, and if target 2 is attacked, then he is of type θ_2.

Now, suppose that $\mathcal{D}$ gives probability 1/2 to each attacker type, and L is a large number such that, $T \ll L$. Then under an optimal policy, the defender would always cover target 1 with probability 2/3, because otherwise the defender would incur a huge loss in the event (which happens with probability 1/2) that the attacker is of type θ_1. Indeed, in that case, the attacker would attack target 1, causing the defender to incur a huge expected loss (at least $L/3$). In other words, an optimal policy would essentially instruct the defender to play it safe by assuming the attacker is of type θ_1, instead of using a single best-response query to learn the attacker's type and play optimally thereafter.

So how does one compute such an optimal policy, given a known distribution over attacker types? Next, we present an approach due to Marecki and colleagues (2012), which relies on Monte Carlo Tree Search.

25.2.1 *Monte Carlo Tree Search*

Monte Carlo Tree Search (MCTS) is a popular search algorithm that tries to pinpoint optimal actions in a sequential decision-making process. Its central idea is to iteratively build a search tree by using random sampling in order to evaluate each new node and update the evaluations of its ancestors. MCTS is becoming increasingly popular in artificial intelligence research, in part due to its successful applications to computer Go. Next, we briefly present some of the algorithm's most salient features; for a detailed presentation, we refer the reader to the survey by Browne and colleagues (2012).

For now, we will think of the sequential decision-making process as being represented by a tree that is gradually uncovered. In its most abstract form, MCTS has two components. First, given the currently explored tree and statistics about its nodes, a *tree policy* traverses the current tree and pinpoints a node to be expanded, and an unexplored action to take at that node. This leads to the discovery of a new child of the expanded node, which is added to the tree.

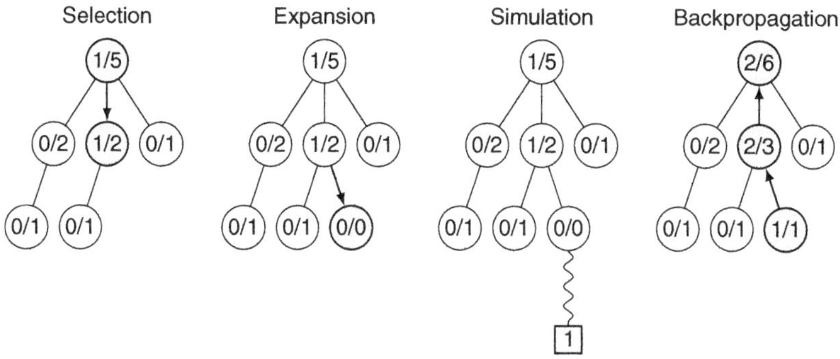

Figure 25.4. An illustration of one iteration of MCTS. In this example, the values of leaves are in $\{0, 1\}$. Each node is labeled by Q_v/N_v.

Second, a *default policy* is used to run a *simulation* – sample a path from the newly discovered node to a leaf. The value of this leaf is then used to update the evaluations of the new node as well as its ancestors.

In more detail, each node v in the search tree has a counter N_v, which counts the total number of simulations that involve v or its descendants, and a total value Q_v, which represents the sum of values of these simulations. The evaluation of v is $\bar{Q}_v = Q_v/N_v$. The main loop of MCTS has four distinct steps (illustrated in Figure 25.4):

1. *Selection*: Select a node to expand by traversing the tree according to the tree policy. This node must be an internal node that has unexplored children.
2. *Expansion*: Expand an unexplored child v, and add it to the search tree. The choice of specific child to expand is also governed by the tree policy. Set $N_v = 0$, $Q_v = 0$.
3. *Simulation*: Use the default policy to sample a path from v to a leaf node ℓ. For the purposes of this chapter, we will assume that the default policy simply selects an action uniformly at random at each node, that is, the sampled path proceeds to a uniformly random child at each step, until a leaf node is reached.
4. *Backpropagation*: Use the value of ℓ to update the evaluation of v and each of its ancestors. Specifically, for each such node v', increase N_v by 1 and Q_v by the value of ℓ.

The basic ideas behind MCTS are, therefore, quite simple. But the key to its success lies in the design of an effective tree policy. In particular, a good tree policy should strike a balance between exploration – in this case, learning more about the value of underexplored nodes – and exploitation – extracting more value from the most promising nodes.

To this end, Kocsis and Szepesvári (2006) propose using the UCB1 policy (Auer, Cesa-Bianchi, & Fischer, 2002), originally designed for the (stochastic) multi-armed bandit problem (UCB stands for "upper confidence bound"). In this problem, each arm is associated with an unknown distribution over rewards. Playing an arm gives a reward drawn from its distribution, and also contributes an observation that can be used to better estimate potential rewards from that arm. The UCB1 policy pulls arm j with probability

$$\bar{X}_j + \sqrt{\frac{2\ln r}{r_j}}, \tag{25.3}$$

where $\bar{X}_j$ is the average reward from arm j so far, r is the total number of observations of all arms, and r_j is the number of observations of arm j. The theoretical guarantees given by the UCB1 formula are beyond the scope of this chapter, but the high-level idea is clear: the first term in equation (25.3) encourages exploitation, whereas the second term drives exploration – it is larger than the less explored arm j is compared to other arms.

The variant of MCTS that uses UCB1 as the tree policy is known as the Upper Confidence Bounds for Trees (UCT) algorithm. Analogously to equation (25.3), UCT's tree policy approaches the exploration vs. exploitation dilemma by selecting the child v' of v that maximizes

$$\frac{Qv'}{Nv'} + c\sqrt{\frac{2\ln N_v}{N_{v'}}}, \tag{25.4}$$

where $C > 0$ is a parameter that determines the priority given to exploration. $N_{v'} = 0$ is taken to mean that the value of the expression is ∞, that is, unexplored children are prioritized above explored children.

UCT ultimately selects the action at the root that maximizes $\bar{Q}_v$.[1] As the number of iterations of UCT grows, the evaluations $\bar{Q}_v$ become more accurate, and approach their true values. In particular, the *failure* probability at the root – the probability that a suboptimal action seems optimal – goes to zero. Specifically, Kocsis and Szepesvári (2006, Theorem 6) show:

Theorem 3. *With an appropriate choice of C in equation (25.4), the failure probability at the root under UCT converges to* 0 *at a polynomial rate (which depends on the maximum number of actions per state and the depth of the tree) as the number of iterations grows.*

To be a bit more precise, it could be the case that the UCB1 policy guides the search toward nodes that have no unexplored children, all the way down to a leaf node. In that case, the expansion and simulation steps do not take place; values are still updated via backpropagation. This is why $\bar{Q}_v$ is gradually refined even when the entire sub-tree rooted at v has been expanded.

25.2.2 *Applying MCTS to Repeated SSGs*

In order to understand the relation between repeated SSGs and MCTS, we present a slightly different formulation of the sequential decision-making domain to which MCTS (specifically, UCT) is applied; this is, in fact, the formalism used by Kocsis and Szepesvári (2006). A *Markov Decision Process* (MDP) consists of states, available actions in each state, and a stochastic transition function T, where $T(s, a, s')$ is the probability of transitioning from state s to state s' when the action a is taken at state s. In addition, each possible transition is associated with a reward. We are interested in a *finite-horizon* MDP: the process ends after T steps, and the objective is to maximize the total accumulated reward.

An MDP with horizon T can be represented by a tree where every path from the root to a leaf is of length T. The reward at a leaf is simply the sum of rewards associated with the transitions along the path. Under this interpretation, the results of Section 25.2.1 – in particular, Theorem 3 – still hold.

Marecki and colleagues (2012) represent a repeated SSG with a prior P over types and T rounds as an MDP with horizon T. Defender mixed strategies are discretized to ensure finite state and action spaces. Each state corresponds to the history of play so far, which is a sequence of pairs – the mixed strategy of the defender, and corresponding best response of the attacker. Given such a history h of observations, let Θ_h be the set of attacker types consistent with the history. Then the posterior distribution $\mathcal{D}_h$ over attacker types is given by

$$\mathcal{D}_h(\theta) = \begin{cases} \dfrac{\mathcal{D}(\theta)}{\sum_{\theta' \in \Theta_h} \mathcal{D}(\theta')} & \theta \in \Theta_h \\ 0 & \theta \notin \Theta_h \end{cases}$$

$\mathcal{D}_h$ straightforwardly determines the transition function: for action $\mathbf{p}$ and each $\theta \in \Theta$, the pair $(\mathbf{p}, b_\theta(\mathbf{p}))$ is appended to the history h with probability $\mathcal{D}_h(\theta)$. When this transition takes place, the reward $U_d(b_\theta(\mathbf{p}), \mathbf{p})$ is obtained.

Crucially, the UCT implementation of Marecki and colleagues (2012) does not actually compute the posterior distributions at each step. Instead, in each iteration of the algorithm, they sample $\theta \sim \mathcal{D}$ upfront, and use θ to compute the attacker's best response at each step – this is equivalent to sampling the (unknown) posterior distribution at each step. In particular, Marecki and colleagues (2012) obtain the following corollary of Theorem 3:

Corollary 4. *When the Marecki and colleagues (2012) implementation of UCT is applied to a repeated SSG, the failure probability at the root converges to* 0 *at a polynomial rate (which depends on T and the number of discretized defender mixed strategies) as the number of iterations grows.*

The experiments of Marecki and colleagues (2001) show that the approach scales well with respect to number of rounds T. The experiments also suggest that trees with a large branching factor – which is the product of the number of discretized mixed strategies available to the defender, and the number of targets – may be hard to deal with, but additional heuristics provide encouraging results.

25.3 Sequence of Attackers

The methods discussed in Sections 25.1 and 25.2 are designed to use repeated interactions with a *single* attacker to learn the missing payoff information. Settings involving *multiple* attackers, however, give rise to additional, fundamentally different sources of uncertainty. For example, it might not be possible to observe which attacker carried out the latest attack, or the distribution over attackers might be unknown or changing. In this section, we address some of these issues. In particular, we describe a method introduced by Balcan and colleagues (2015) for defending against an adversarial sequence of attackers, when attackers are chosen from a known set of types but with no distributional assumptions. Additionally, this method can be adapted to work even when the identity of the attacker remains unknown after the attack.

As in the previous section, consider a set of k known attacker types $\Theta = \{\theta_1, \ldots, \theta_k\}$ and U_{θ_i} for all $i \in [k]$, such that U_{θ_i} represents the utilities of attacker type θ_i. We assume that all utilities are scaled such that the defender's payoff is in the range $[-1, 1]$. Consider playing a repeated Stackelberg game for T rounds, such that at round t, an unknown attacker $a_t \in \Theta$ is chosen to attack the targets by best responding to the mixed strategy of the defender at time t, denoted by $\mathbf{p}_t$. The defender's goal is to maximize his payoff over a period of time, even when the sequence of attackers is unknown. That is, without knowing the sequence ahead of time, at each round the defender must commit to a mixed strategy based on the history of play so far. The defender then receives some "feedback" regarding the attacker at that round (either the attacker type or merely which target was attacked), and adjusts the mixed strategy for future rounds.

In more detail, in this *online* setting, the expected payoff of the defender over T time steps is

$$\mathbb{E}\left[\sum_{t=1}^{T} U_d(b_{a_t}(\mathbf{p}_t), \mathbf{p}_t)\right],$$

where the expectation is taken only over the internal randomness of the defender's online algorithm, and $b_{a_t}(\cdot)$ indicates the best response of attacker a_t. In contrast, if the sequence of attackers, or merely the frequency of each attacker type in the sequence, is known to the defender, the defender can pre-compute a fixed mixed strategy with the best payoff against that sequence of attackers and play it at every time step. Such a strategy is called *the best mixed strategy in hindsight* and denoted by

$$\mathbf{p}^* = \arg\max_{\mathbf{p}} \sum_{t=1}^{T} U_d(b_{a_t}(\mathbf{p}), \mathbf{p}).$$

The goal is to design online algorithms for the defender with payoff that is almost as good as the payoff of the best mixed strategy in hindsight. This difference in utilities is termed *regret* and is equal to

$$R_T = \sum_{t=1}^{T} U_d(b_{a_t}(\mathbf{p}^*), \mathbf{p}^*) - \mathbb{E}\left[\sum_{t=1}^{T} U_d(b_{a_t}(\mathbf{p}_t), \mathbf{p}_t)\right].$$

The results in this section show that, even when faced with this (relatively extreme) type of uncertainty, one can compete with the best fixed mixed strategy in hindsight. In particular, the algorithms introduced here are *no-regret algorithms*: the expected *average* regret R_T/T goes to zero as T goes to infinity. This indicates that the longer the algorithm runs the better the guarantees are in terms of the regret. In addition, the regret depends only polynomially on the number of types and targets. That is

$$R_T \le o(T) \cdot \text{poly}(n, k).$$

The feedback the defender receives at each time step plays a major role in the design of the algorithm. We consider two types of feedback, *full information* and *partial information*. In the full information case, the identity of the attacker is revealed after each attack (e.g., through surveillance). An upper bound on the regret under full information feedback is as follows.

Theorem 5 (Balcan et al., 2015). *Given a repeated SSG with full information feedback, there is an algorithm that for any unknown sequence of attackers has regret*

$$R_T \leq O\left(\sqrt{Tn^2 k \log nk}\right).$$

On the other hand, in the partial information case, even after the attack occurs the identity of the attacker remains unknown – only the attacked target is observed. Note that full information is indeed strictly more informative, as knowing the attacker a_t is sufficient to compute the target attacked at that round, which is the feedback obtained in the partial information case.

Theorem 6. *Given a repeated SSG with partial information feedback, there is an algorithm that for any unknown sequence of attackers has regret*

$$R_T \leq O\left(T^{2/3} nk \log^{1/3}(nk)\right).$$

In the remainder of this section, we first provide a brief overview of basic no-regret algorithms, and then apply these techniques to repeated SSGs.

25.3.1 *Background on Regret Minimization*

Consider the following online scenario. Suppose we have a set $\mathcal{M}$ of possible actions to take each day (e.g., a collection of different possible ways to get to work). At each time step t, we probabilistically select some action j_t according to a probability distribution of our choice $\mathbf{q}_t$ over the actions. An adversary (who may observe our previous actions $j_1, \ldots, j_{t-1}$, and may even know our current distribution $\mathbf{q}_t$ but does not get to observe our selection j_t from $\mathbf{q}_t$) simultaneously chooses a loss vector $\boldsymbol{\ell}_t$ over the actions, where let us assume all losses are in the range [0,1] (i.e., $\boldsymbol{\ell}_t \in [0,1]^{|\mathcal{M}|}$). For instance, these might be the travel times of our different options for getting to work. We then incur an *expected loss* of $\mathbf{q}_t \cdot \boldsymbol{\ell}_t$.

Let $L_{\text{alg},T} = \sum_{t=1}^{T} \mathbf{q}_t \cdot \boldsymbol{\ell}_t$ denote the expected total loss of the algorithm in T time steps, and let $L_{\min,T} = \min_{j \in M} \sum_{t=1}^{T} \boldsymbol{\ell}_t(j)$ denote the total loss of the best fixed action under the sequence of loss vectors $\boldsymbol{\ell}_1, \ldots, \boldsymbol{\ell}_T$. That is, $L_{min,\,T}$ is the loss of the best fixed action in hindsight. The difference, $R_T = L_{alg,\,T} - L_{min,\,T}$, is called the *regret* of the algorithm.

As noted earlier, algorithms with regret sublinear in T are often called *no-regret algorithms* because a sublinear regret implies that the average per-day cost of the algorithm is approaching (or even less than) the average per-day cost of the best fixed action in hindsight. That is, the *per-day regret* R_T/T is approaching 0. So, one does not regret very much having not known initially what the best fixed action in hindsight is.

Hannan (1957) was the first to develop algorithms with a no-regret guarantee. In the 1990s, motivated by machine learning settings in which $\mathcal{M}$ can be quite large (e.g., corresponding to all possible prediction rules in some class), algorithms were developed whose regret furthermore has only a logarithmic dependence on the size $|\mathcal{M}|$ of the action space. The following theorem formalizes these results.

Theorem 7 (e.g., Littlestone, & Warmuth, 1994). *There is an efficient algorithm (running time per stage linear in $|\mathcal{M}|$) with the guarantee that for any adversary,*

$$R_T = O\left(\sqrt{T \log |\mathcal{M}|}\right).$$

There are by now many algorithms known to achieve the guarantees of Theorem 7. One canonical class of such algorithms is "multiplicative weights" or "randomized weighted majority" methods. Here is one such algorithm, called "polynomial weights" (Cesa-Bianchi, Mansour, & Stoltz, 2005):

Polynomial Weights algorithm (given an input $\epsilon \in (0,1)$)

1. Initialize weights $\mathbf{w}(j) = 1$ for each $j \in \mathcal{M}$.
2. Choose distribution $\mathbf{q}_t$ proportional to $\mathbf{w}$. That is, $\mathbf{q}_t(j) = \mathbf{w}(j)/W$ where $W = \Sigma_i \mathbf{w}(i)$.
3. Given $\boldsymbol{\ell}_t$, update $\mathbf{w}(j) \leftarrow \mathbf{w}(j)(1 - \epsilon \boldsymbol{\ell}_t(j))$.

The intuition behind this algorithm is that actions that are far from optimal quickly become highly penalized and so the algorithm gives them negligible probability mass. At a formal level, at each step, the total weight in the system drops by an amount that is directly related to the expected loss of the algorithm; specifically, one can verify that at each step we have $W \leftarrow W(1 - \epsilon\, \mathbf{q}_t \cdot \boldsymbol{\ell}_t)$. On the other hand, any low-loss actions will have high weight (ensuring that W remains high), and so if there are any low-loss actions, this means the algorithm's total expected loss cannot be too great. The overall guarantee is that $L_{alg,T} \leq (1+\varepsilon) L_{\min,T} + \frac{In|\mathcal{M}|}{\varepsilon}$ (in fact, the algorithm has a stronger guarantee but this suffices for our purposes here), and setting $\varepsilon = \min\left(\sqrt{\frac{\text{In}|\mathcal{M}|}{T}}, \frac{1}{2}\right)$

gives the bound of Theorem 7. For further discussion of this and related algorithms, see (Blum & Mansour, 2007).

The discussion so far assumes the algorithm has the ability to observe the entire loss vector ℓ_t; indeed, this feedback is used in Step 3 of the Polynomial Weights algorithm. However, often one receives a more impoverished form of feedback. For example, one may learn only the loss $\ell_t(j_t)$ of the action one specifically chose: this is called the (adversarial) *multi-armed bandit* problem(Auer, Cesa-Bianchi, Freund, & Schapire, 2002). Or one may receive partial information about multiple actions: e.g., in choosing a route to drive to work, one would learn about traffic congestion along the roads used in that route, which might provide partial information about other routes that also include some of those roads.

One way to tackle these partial information settings is to use the fact that algorithms for the full information setting really need only a *bounded unbiased estimate* of the loss of each action (Auer et al., 2002; Awerbuch & Mansour, 2003), that is, a bounded random variable whose expectation is equal to the loss. For example, if one breaks the overall T time steps into windows, and within each window samples each action once at a random time, then these "exploratory samples" will give an unbiased estimate of the losses of each action in the window; this approach was introduced and analyzed by Awerbuch and Mansour (2003). In the case of actions that provide partial information, an improved method proposed by Awerbuch and Kleinberg (2008) is to identify an "exploratory basis" $\mathcal{S} \subseteq \mathcal{M}$ of actions, whose feedback can be used to reconstruct unbiased estimates for all the actions, and then just perform exploration over those. For example, in the case of routes to drive to work, this might correspond to a small set of routes that together include all roads of interest. Using this approach and combining with bounds for the full information setting yields the following theorem (see Balcan et al., 2015 for details):

Theorem 8. *Suppose that $\mathcal{S} \subseteq \mathcal{M}$ is such that for any time-window τ, sampling the actions in $\mathcal{S}$ at random times in τ will produce estimates $\hat{\ell}_\tau(j)$ for each $j \in \mathcal{M}$ such that:*

1. $\mathbb{E}[\hat{\ell}_\tau(j) = \ell_\tau(j)$, *where* $\ell_\tau(j) = \frac{1}{|\tau|}\sum_{t\in\tau}\ell_t(j)$

2. $\ell_\tau(j) \in [-\kappa, \kappa]$.

Then there is an algorithm with regret guarantee $R_T = O(T^{2/3}\, |\mathcal{S}|^{1/3}\, \kappa^{1/3} \log^{1/3} |\mathcal{M}|)$ for all $T > \kappa$.

25.3.2 *Applying Regret Minimization Techniques to Repeated SSGs*

To use the results from Section 25.3.1, we first need to define what constitutes actions and loss functions in our setting. Indeed, each valid coverage probability vector p in a security game corresponds to one action, and the loss associated with this action at time t is simply the negation of its utility to the defender, $-U_d(b_{a_t}(\mathbf{p}), \mathbf{p})$. However, this creates an infinitely large set of actions, which renders the guarantees in Theorem 7 meaningless. Instead, we show that a carefully chosen subset of the strategy space that is representative of the set of all strategies can be used with the algorithms from Section 25.3.1.

To this end, we first show that the set of attacker types, Θ, partitions the space of all strategies to convex regions where the attackers' best response remains fixed. As an extension to the definition of $\mathscr{P}_i$ from Section 1.2, let $\mathscr{P}_i^j$ indicate the set of all valid coverage probabilities where an attacker of type θ_j attacks target i. Then for a given set Θ and any function $\sigma\colon \Theta \to N$, let $\mathscr{P}_\sigma$ indicate the set of all valid coverage probability vectors such that for all $\theta_j \in \Theta$, θ_j attacks $\sigma(j)$. In other words, $\mathscr{P}_\sigma = U_{\theta j \in \Theta} \mathscr{P}_{\sigma(j)}^{j}$. Since $\mathscr{P}_\sigma$ is the intersection of finitely many convex polytopes, it is itself a convex polytope. Figure 25.5 illustrates these regions.

Consider the set E of all extreme points of convex polytopes $\mathscr{P}_\sigma$ for all σ. It is easy to show that $|E| \in O((2^n + kn^2)^{2n})$. Balcan and colleagues (2015) show that a regret minimization algorithm incurs no additional loss when the set of actions is restricted to only those strategies that are in E. Together with the bound on $|E|$, the proof of Theorem 5 follows directly from Theorem 7.

Next, consider the partial information feedback model. As discussed in Section 25.3.1, we only need a mechanism to produce bounded unbiased estimators for the loss of all actions. As a naïve approach, first consider sampling each mixed strategy in E once at random in a window of time and observing its loss. The loss created this way is clearly bounded and unbiased. However, at every time step in which we sample a random strategy, we are left open to experiencing significant regret. As shown in Theorem 8, sampling each strategy individually once at random adds a regret that is polynomial in the number of mixed strategies sampled, which – based on the size of $|E|$ – is *exponential* in the number of targets. Therefore a more refined mechanism for estimating the loss is needed.

Notice that the loss incurred by different mixed strategies is not independent; rather, they all depend on the number of times each target is attacked in a window of time, which in turn depends on the frequency of attacker types. In order to estimate the type frequency, it is sufficient to distinguish between

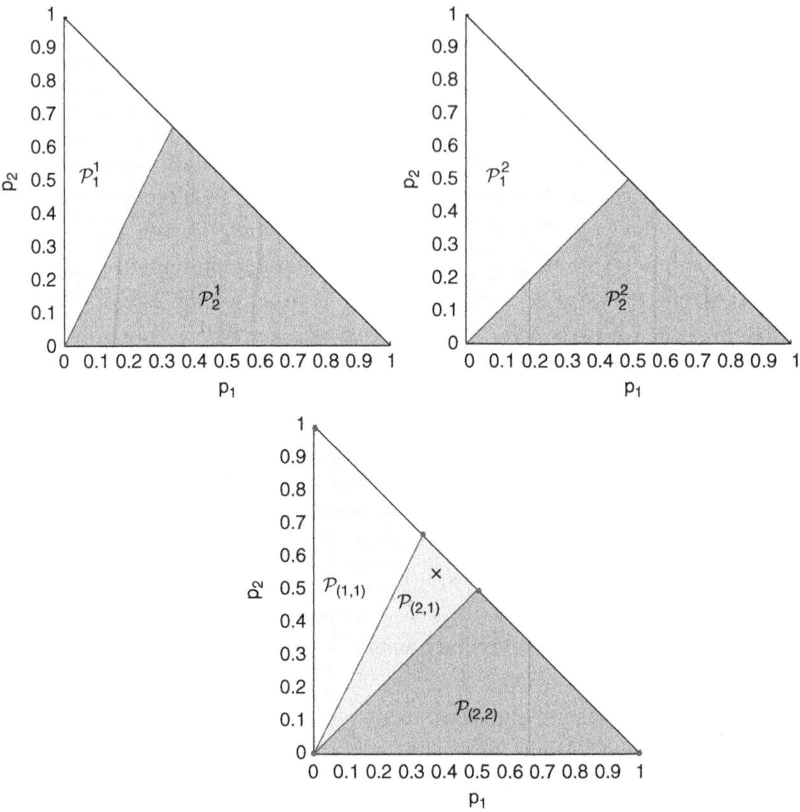

Figure 25.5. Best response regions in the space of coverage probability vectors. The first two figures define $\mathscr{P}_\sigma$ in a game where one resource can cover one of two targets, and with two attacker types. The third figure illustrates $\mathscr{P}_\sigma$ for the intersection of the best response regions of the two attackers. The red points indicate the extreme points, and the black cross is a coverage probability vector where the two attackers respond differently.

types based on their response to a number of mixed strategies. As an example, assume that there is a mixed strategy where each attacker type responds differently (see Figure 25.5).

Then sampling such a mixed strategy even once at random in a window of time allows us to construct an unbiased estimator of the frequency of all attacker types, and, as a result, to create an unbiased estimator of the loss of all actions. In general, such a mixed strategy, where each attacker responds differently, may not exist. Instead, Balcan and colleagues (2015) show how to choose k mixed strategies that construct a loss estimator for all actions, such that the loss is in the range $[-nk, nk]$. The guarantee of Theorem 6 then follows from Theorem 8. We leave it to the interested reader to refer to Balcan and colleagues (2015) for the details of this procedure.

25.4 Further Reading

In this chapter, we have focused on a single, relatively cohesive line of work on learning in SSGs. A number of other learning-based approaches have been proposed to deal with uncertainty in SSGs. Next, we briefly outline some of them.

A fundamentally different approach for handling payoff uncertainty assumes that the utilities $u_a^c(i)$ and $u_a^u(i)$ – the utility to the attacker when target i is attacked and it is covered and uncovered, respectively – are not exactly known, but lie in a known interval (Kiekintveld, Islam, & Kreinovich, 2013). Given such a set of uncertainty intervals, two per target, it is possible to reason about the *maximum regret* of the defender strategy $\mathbf{p}$: it is the maximum difference between the defender's utility when playing the optimal strategy and when playing $\mathbf{p}$, where the maximum is taken over all possible utility functions that are consistent with the given uncertainty intervals. From this viewpoint, the defender's strategy should minimize the maximum regret.

Nguyen and colleagues (2014) reduce minimax regret by eliciting information about utilities. This is done via *bound queries,* which ask an expert (say, a risk analyst) whether a given utility is above or below a given threshold. Nguyen and colleagues introduce three heuristic elicitation strategies, which aim to achieve the best possible minimax regret under a query budget.

Another type of uncertainty that can be addressed using machine learning is uncertainty about the attacker's behavioral model. We have focused on perfectly rational attackers who best respond to the defender's strategy, but studies suggest that more nuanced behavioral models give better predictions (Nguyen et al., 2013).

Yang and colleagues (2014) adopt the *subjective utility quantal response* model (Nguyen et al., 2013), which stipulates that the attacker's probability of choosing an action increases with his *subjective utility* for that action. The subjective utility is induced by three parameters that weight the relative importance of p_i, $u_a^c(i)$ and $u_a^u(i)$. Yang and colleagues learn a three-dimensional normal distribution of these parameters (i.e., they estimate its mean and covariance matrix).

Finally, we mention work by Chakraborty and colleagues (2013), which deals with learning to play against memory-bounded agents. While their framework is more general, let us describe its application to mass transit security. The goal in this domain is to deter passengers who may not buy a ticket, by deploying patrol officers. Each passenger has a fixed route, so the passengers' decision amounts to whether to buy a ticket; a risk-neutral passenger would buy a ticket if and only if the expected cost of not buying a ticket (probability of getting caught times fine) is larger than the fare. A passenger estimates his probability of being caught based on patrol strategies in the last L days, where L can be different for different passengers. Given the number of passengers on each route, but not the distribution over their memory size L (which is assumed to be in $\{2, 3, 4\}$ in the experiments), Chakraborty and colleagues (2013) use their TOMMBA algorithm to play against these forgetful passengers.

Acknowledgments

The authors thank Leandro Soriano Marcolino for helpful feedback. The authors were partially supported by the NSF under grants IIS-1350598, CCF-1215883, CCF-1116892, IIS-1065251, and CCF-1415460, and by a Sloan Research Fellowship.

NOTE

1. There are alternative approaches, e.g., selecting the action that maximizes Nv.

REFERENCES

Auer, P., Cesa-Bianchi, N., & Fischer, P. (2002). Finite-time analysis of the multiarmed bandit problem. *Machine Learning*, *47*(2), 235–256.

Auer, P., Cesa-Bianchi, N., Freund, Y., & Schapire, R. E. (2002). The nonstochastic multiarmed bandit problem. *SIAM Journal on Computing*, *32*(1), 48–77.

Awerbuch, B., & Kleinberg, R. (2008). Online linear optimization and adaptive routing. *Journal of Computer and System Sciences*, *74*(1), 97–114.

Awerbuch, B., & Mansour, Y. (2003). Adapting to a reliable network path. In *Proceedings of the 22nd ACM symposium on principles of distributed computing (PODC)* (pp. 360–367).

Balcan, M.-F., Blum, A., Haghtalab, N., & Procaccia, A. D. (2015). Commitment without regrets: Online learning in Stackelberg security games. In *Proceedings of the 16th ACM Conference on Economics and Computation (EC)* (pp. 61–78).

Bertsimas, D., & Vempala, S. (2004). Solving convex programs by random walks. *Journal of the ACM*, *51*(4), 540–556.

Blum, A., Haghtalab, N., & Procaccia, A. D. (2014). Learning optimal commitment to overcome insecurity. In *Proceedings of the 28th Annual Conference on Neural Information Processing Systems (NIPS)* (pp. 1826–1834).

Blum, A., & Mansour, Y. (2007). Learning, regret minimization, and equilibria. In N. Nisan, T. Roughgarden, É. Tardos, & V. Vazirani (Eds.), *Algorithmic game theory* (chapter 4). Cambridge University Press.

Browne, C., Powley, E. J., Whitehouse, D., Lucas, S. M., Cowling, P. I., Rohlfshagen, P., Tavener, S., Perez, D., Samothrakis, S., & Colton, S. (2012). A survey of Monte Carlo tree search methods. *IEEE Transactions on Computational Intelligence and AI in Games, 4*(1), 1–43, 2012.

Cesa-Bianchi, N., Mansour, Y., & Stoltz, G. (2005). Improved second-order bounds for prediction with expert advice. In *Proceedings of the 18th Conference on Computational Learning Theory (COLT)* (pp. 217–232).

Chakraborty, D., Agmon, N., & Stone, P. (2013). Targeted opponent modeling of memory-bounded agents. In *Proceedings of the Adaptive Learning Agents Workshop (ALA)* (pp. 211–226).

Conitzer, V., & Sandholm, T. (2006). Computing the optimal strategy to commit to. In *Proceedings of the 7th ACM Conference on Economics and Computation (EC)* (pp. 82–90).

Grötschel, M., Lovász, L., & Schrijver. A. (1988). *Geometric algorithms and combinatorial optimization.* Springer.

Hannan, J. (1957). Approximation to Bayes risk in repeated plays. In M. Dresher, A. Tucker, & P. Wolfe (Eds.), *Contributions to the theory of games* (vol. 3, pp. 97–139). Princeton University Press.

Khachiyan, L. G. (1979). A polynomial algorithm in linear programming. *Soviet Mathematics Doklady*, *20*, 191–194.

Kiekintveld, C., Islam, T., & Kreinovich, V. (2013). Security games with interval uncertainty. In *Proceedings of the 12th International Conference on Autonomous Agents and Multiagent Systems (AAMAS)* (pp. 231–238).

Kiekintveld, C., & Jain, M. (2015). Basic solution concepts and algorithms for Stackelberg security games. In A. Abbas, M. Tambe, & D. von Winterfeldt (Eds.), *Improving homeland security decisions* (chapter 17).

Kocsis, L., & Szepesvári, C. (2006). Bandit based Monte-Carlo planning. In *Proceedings of the 17th European Conference on Machine Learning (ECML)* (pp. 282–293).

Letchford, J., Conitzer, V., & Munagala, K. (2009). Learning and approximating the optimal strategy to commit to. In *Proceedings of the 2nd International Symposium on Algorithmic Game Theory (SAGT)* (pp. 250–262).

Littlestone, N., & Warmuth, M. K. (1994). The weighted majority algorithm. *Information and Computation,* pages 212–261.

Marecki, J., Tesauro, G., & Segal, R. (2012). Playing repeated Stackelberg games with unknown opponents. In *Proceedings of the 11th International Conference on Autonomous Agents and Multiagent Systems (AAMAS)* (pp. 821–828).

Nguyen, T. H., Yadav, A., An, B., Tambe, M., & Boutilier, C. (2014). Regret-based optimization and preference elicitation for Stackelberg security games with uncertainty. In *Proceedings of the 28th AAAI Conference on Artificial Intelligence (AAAI)* (pp. 756–762).

Nguyen, T. H. Yang, R. Azaria, A. Kraus, S., & Tambe, M. (2013). Analyzing the effectiveness of adversary modeling in security games. In *Proceedings of the 27th AAAI Conference on Artificial Intelligence (AAAI)* (pp. 718–724).

Tauman Kalai, A., & Vempala, S. (2006). Simulated annealing for convex optimization. *Mathematics of Operations Research, 31*(2), 253–266.

Yang, R., Ford, B. J., Tambe, M., & Lemieux, A. (2014). Adaptive resource allocation for wildlife protection against illegal poachers. In *Proceedings of the 13th International Conference on Autonomous Agents and Multiagent Systems (AAMAS)* (pp. 453–460).

26

Evaluating Deployed Decision Support Systems for Security: Challenges, Analysis, and Approaches

Matthew E. Taylor, Christopher Kiekintveld, Eric Shieh, Francesco Delle Fave, and Milind Tambe

26.1 Introduction

There are an increasing number of technically sophisticated tools to support decision making for security resource allocation in many different domains. In this chapter, we discuss the question of evaluating deployed security systems, using examples to illustrate some of the key challenges. Two of the primary difficulties are that (1) we cannot rely on adversaries to cooperate in evaluation, which makes it difficult to validate models, and (2) there is (thankfully) relatively little data available about real-world terrorist attacks.

Despite the difficulties of comprehensive evaluation in security domains, it is only by asking the question "how well does a system work?" that policy makers can adopt the best policies for allocating finite security resources. In this chapter, we discuss the goals of security systems, the elements that comprise these systems, and different approaches for evaluation. Every approach has drawbacks, so in lieu of a single ideal test we advocate comprehensive evaluations that use diverse metrics and data to perform cost-benefit analysis for the complete system. We also emphasize that the focus of the evaluation is not "is system X a perfect security system?" which is an impossible standard. The relevant question is "which of the available alternatives is best?" The goal is, therefore, to provide strong evidence that one alternative is superior to other approaches – this goal is often feasible even when providing exact quantitative comparisons is not. As a community, we must strive to perform the best possible evaluations using the tools and data available, but we cannot let the absence of an ideal evaluation prevent us from deploying effective technologies. However, we must also recognize weaknesses in the current evaluations and view these as opportunities to develop new techniques and gather new data to improve our understanding of the value of different approaches for security.

We note that the question of how well a decision support system for security resource allocation works is orthogonal to the question of what type of security resources should be deployed, or whether the security resources should be deployed at all. The focus of these decision support systems is to allocate a set number and type of security resources, and our focus is in evaluating how well these systems do this task. In particular, the decision support systems surveyed in this chapter focus on optimizing a constrained objective function: how well can a given number and type of resources be deployed to best protect a given target? Our task in evaluation is, then, to check if this decision support system does a better job than available alternatives in allocating these resources to protect the given target.

Section 26.2 gives a brief background on the security applications discussed in this chapter in the context of evaluation – more detailed descriptions of these systems are provided in other chapters of this book. Section 26.3 describes the three steps involved in formulating a decision support system for security: abstracting the model, solving the model, and then accurately deploying the solution. Section 26.4 describes the different types of evaluations that have been conducted on three systems in our case study. Section 26.5 will discuss the goals of a deployed security system and the inherent difficulties in measuring the performance of such a system. For instance, unlike many types of technical applications, a security system does not have binary behavior; no security system is able to provide 100% protection and it does not make sense to say that it "does" or "does not" work. Instead, systems must be evaluated on basis of *risk reduction*, often through indirect measures such as increasing adversary cost and uncertainty, or reducing the effectiveness of an adversary's attack.

Section 26.6 discusses the strengths and weaknesses of different evaluation techniques, tying together the discussion in Sections 26.3–26.5. Related work is discussed in Section 26.7. Finally, Section 26.8 ends this chapter with a discussion of future work, in terms of enhancing the evaluation of our specific applications, as well as the challenge of security evaluation in general.

26.2 Descriptions of Systems

The importance of security games as a research topic is driven by the recent development and deployment of several applications that assist major security organizations with making resource allocation decisions using algorithmic analysis based on game theory. These include the ARMOR system deployed at Los Angeles International Airport (LAX) (Pita et al., 2008) to assist the

Los Angeles World Airport (LAWA) police, the IRIS system deployed by the Federal Air Marshals Service (FAMS) (Tsai, Rathi, Kiekintveld, Ordóñez, & Tambe, 2009), and the PROTECT system deployed by the U.S. Coast Guard (USCG) in scheduling patrols (Shieh et al., 2012). While these systems share a similar approach in that they apply game-theoretic modeling techniques and emphasize randomized, unpredictable scheduling of security resources and activities, each system poses unique challenges in system design, modeling and solution techniques, and evaluation. ARMOR was the first system developed using this methodology, and has the longest deployment history.

The ARMOR system has been deployed since 2007 by the Los Angeles World Airport (LAWA) police at LAX, the fifth busiest airport in the United States (and largest destination), serving 70–80 million passengers per year. LAX is a primary terrorist target on the West Coast, and multiple individuals have been arrested for plotting or attempting to attack LAX (Stevens et al., 2009). Police have designed multiple "rings" of protection for LAX, including vehicular checkpoints, police patrols of roads and inside terminals (some with bomb-sniffing canine units, also known as K9 units), and passenger and baggage screening. Because of the high demands on LAX security, due in part to the large physical area and high density of passenger and cargo traffic, police do not have enough resources (e.g., officers and K9 units) to provide constant security for every area and event at the airport. This limitation leads to the question of how best to allocate the limited resources to improve security at the airport.

ARMOR addresses two specific security problems by increasing the unpredictability of security schedules and weighting the defensive strategy based on targets' importance. First, many roads serve as entry points to LAX. When and where should vehicle checkpoints be set up on these roads? Pertinent information includes typical traffic patterns on inbound roads, the areas each road accesses within LAX, and areas of LAX that may have more or less importance as terrorist targets. Second, how and when should the K9 units patrol the eight terminals at LAX? Here it is important to consider the time-dependent passenger volumes per terminal, as well as the attractiveness of different terminals. In both cases, an observant attacker can exploit a predictable pattern.

The approach ARMOR takes uses game-theoretic models to derive scheduling strategies for the police resources. This is modeled as a Bayesian Stackelberg game (Conitzer & Sandholm, 2006), in which the police (e.g., defenders) must commit to a (randomized) security policy. Each possible attacker type observes the security policy and then selects an optimal attack strategy based on its preferences (i.e., *utilities* or utility *payoff matrices*). The solution to this game is called a Strong Stackelberg Equilibrium, an optimal randomized strategy.

ARMOR uses the DOBSS (Paruchuri, Pearce, Marecki, Tambe, Ordóñez, & Kraus, 2008) algorithm to solve these game instances and to recommend schedules for checkpoints and canine patrols to the LAWA police. The schedules account for three key factors: (1) attackers are able to observe the security policy using surveillance; (2) attackers change their behavior in response to the security policy; and (3) the risk/consequence of an attack varies depending on the target.

The IRIS system was designed to address scheduling problems faced by the Federal Air Marshals Service (FAMS). FAMS is charged with law enforcement and antiterrorism on commercial airline flights, which is accomplished primarily by placing armed marshals on individual flights. U.S. airlines operate tens of thousands of flights each day, and FAMS lacks the resources to place marshals on every flight. This leads to a similar resource allocation challenge to that faced by ARMOR: how best to place the available marshals on flights to maximize security. In addition to the overall constraint on the number of air marshals available, this problem is complicated by the presence of complex physical and temporal constraints on which flight tours the air marshals can feasibly fly, as well as other idiosyncratic constraints. Flights are also diverse in the potential risk and consequence of a terrorist attack, based on many factors, including the source and destination airports, aircraft size, and flight path. The schedules IRIS produces must optimize over all of these constraints and variations in flight valuation, resulting in a significantly more complex problem space than for the ARMOR scheduler (Kiekintveld, Jain, Tsai, Pita, Ordóñez, & Tambe, 2009).

PROTECT was designed to aid the U.S. Coast Guard (USCG) in scheduling patrols and has been used at the Port of Boston since April 2011. The USCG's mission includes maritime security of U.S. coasts, ports, and inland waterways, a security domain that faces increased risks in the context of threats such as terrorism and drug trafficking. The USGS must conduct patrols to protect a variety of critical infrastructure at a given port – as with the two previous systems, the adversary can observe patrol patterns and security resources are insufficient to cover all locations continually. PROTECT is designed to use game theory to assist the USCG in maximizing its effectiveness in the Ports, Waterways, and Coastal Security (PWCS) Mission. PWCS patrols are focused on protecting critical infrastructure; limited security resources preclude covering all targets continually and optimization of security resource is critical. The task of generating patrols in the maritime domain differs from previous applications in that the distance between targets must be taken into consideration in planning the patrol routes. The patrols in PROTECT also need to model different defensive activities that could be performed in each area, creating a more complex scenario because of the activities' impact on time and utilities. Finally, the solution concepts used in PROTECT are based on models that consider human bounded rationality, in contrast with prior systems that use perfect rationality assumptions.

The TRUSTS systems is being developed in coordination with the Los Angeles County Sheriff's Department (LASD) for deployment over the Los Angeles Metro System (LA Metro). TRUSTS is a randomized scheduling system (Fave, Jiang, Yin, Zhang, Tambe, Kraus, & Sullivan, 2014). It allocates security officers (e.g., LASD sheriffs, security assistants, or metro police) to patrol the train lines of the LA Metro. A transit system is a key target for illegal activities, and the LASD must work to counter fare evasion (FE) and to run counterterrorism (CT) and crime reduction (CR) activities to protect a large number of people. As a consequence, the security of such a system poses a number of challenges. The first challenge is multi-operation patrolling. Whereas PROTECT, TRUSTS, and IRIS have focused on threats that could be represented with a single adversary model, the comprehensive security of a transit system requires different specialized security responses against the three threats (FE, CT, and CR). The second challenge is execution uncertainty. Security resources are often interrupted during their patrols (e.g., to provide assistance or arrest a suspect). Thus, traditional patrol schedules are difficult to complete in practice. The third challenge is in accounting for joint activities. In CT patrolling, security resources, such as explosive detective canine (EK9) teams, often patrol train lines in cooperation with other resources to increase their effectiveness. One important feature of this domain is that crimes like fare evasion happen frequently, providing more data for real-world evaluation.

26.3 Formulating the Problem

We now describe the basic elements of these decision-support tools for security and the process used to design them. In each case, the complete system includes a domain model based on expert elicitation, a game-theoretic solution algorithm used to analyze the model, and a software application designed around the algorithm that allows data inputs, visualization of the solution, and so on. To fully evaluate the system, we must evaluate each of these components, as well as the way in which the system is actually used in practice (e.g., how the security forces implement the recommended solutions).

We focus first on the process of formulating a model of the domain, since this is the basis for all subsequent analysis. This model must have sufficient detail that it can provide useful decision support, it must accurately represent the domain, and it must not be so complex that it is intractable for analysis.

The problem of computing a solution is discussed in the next section and is the most familiar to computer scientists. While other chapters in this volume have discussed the computational difficulties in solving complex security problems, our discussion considers more broadly the assumptions behind the solution methods. A system may be theoretically sound, but badly flawed if it is implemented poorly. The final step we discuss is implementing (and verifying) the system in a real-world setting.

26.3.1 *Abstracting the Real-World Problem*

The first step of developing a defensive measure is to determine what particular attack vector(s) should be addressed. For example, the ARMOR system was designed to counteract a perceived defensive shortcoming at LAX, as identified in a RAND study (Stevens et al., 2009). In order to construct a model that can be solved, the real problem must be formulated as a quantitative problem. In the case of ARMOR, this involved focusing on a fixed number of checkpoint positions that could be either covered or not covered. How many officers were at a checkpoint, or the skills of the particular officers, is abstracted away.

In order to decide how important the different checkpoints were, they were assigned values based on their proximity and road layout relative to the terminals. The terminals, in turn, were evaluated by experts who estimated the utilities to the defenders if an attacker successfully attacked a terminal, or was caught attempting to attack a terminal. The experts also estimated the utilities to the attacker if he successfully attacked a terminal, or was caught attacking a terminal.

Formulating a real-world problem as a solvable quantitative model requires abstraction, which by definition causes a loss of information. However, the goal is to minimize the loss of precision so that the solution to the model will be near-optimal on the physical problem. The formulation chosen will also constrain the policy space. In ARMOR, checkpoints are constrained to last at least a certain amount of time, due to the time of setting up a checkpoint. For example, ARMOR should never schedule checkpoint #1 to be set up at 9:00 A.M. and taken down at 9:05 A.M. The selection of this constrained solution space may also impact the optimality of the solution in the real world.

Modeling deterrence is also an important aspect of the model. One method is to test an attacker's optimal option with and without defense. If the attacker changes actions to a lower-valued action, s/he can be considered deterred. Another option is to consider that the attacker may choose to not attack at all, or to attack a target that is outside the scope of the security system in question. In a game-theoretic setting, one option is to add a "stay at home" action to the attacker's action space. A second option would be to add an "attack another target" option. In the case of ARMOR, this could include the action "attack another target in Los Angeles," or "attack another airport in the United States." Notice that from the point of view of LAWA, these three possibilities are equivalent since they are all successful deterrence with respect to LAX. However, for the city of Los Angeles or for the U.S. federal government, the "stay at home" action is clearly superior.

26.3.2 *Solution Concepts and Computational Considerations*

Once a model is defined, the next stage is to analyze this model to determine the best course of action (strategy) for the security forces. Given the complexity of the models, we need fast computational methods to perform this analysis.

Potential Solution Concepts

Given a precise mathematical model of a strategic interaction, the game theory literature provides many powerful tools for analyzing the model to make predictions and strategy recommendations. Unfortunately, many different solution concepts can be applied in different situations, and it is not obvious which is the "best" solution method to use. The starting point is typically Nash equilibrium, which predicts how players should act given that they are perfectly rational, and all players have exact common knowledge of the game model (including the possible actions, sequence of moves, payoffs for different outcomes, etc.). In real-world situations, these assumptions are often too strong, since players may be uncertain about different aspects of the game, or may not make mathematically optimal choices due to limits on their reasoning capabilities or other factors. Different solution concepts can be applied based on different assumptions about knowledge and rationality, but it is an important open question how well the solution concept and underlying assumptions correspond to the real-world situation, and this must be evaluated empirically.

There are alternatives to game-theoretic equilibrium solutions methods that should also be evaluated as candidates. The simplest is the uniform random strategy: given a set of possible security actions, take each of them with equal probability. This has the appeal of simplicity, and of being completely unpredictable (by definition). It also makes absolutely no assumptions about the adversary, so there is no chance that these assumptions may be incorrect or exploitable. However, it does not take into account the value of different actions, and may waste limited resources by performing less valuable actions too often. A weighted randomization can potentially account for this by selecting more valuable actions more frequently. The key question here is how to determine the weights for the randomization; in effect, a game-theoretic solution is one answer to how to find these weights in a principled way, since a game-theoretic solution is an instance of weighted randomization. There may be simpler approaches to finding a weighted randomization in some cases that could also be candidate strategies. Our evaluations of ARMOR and similar systems, discussed in subsequent sections, have consistently shown that game-theoretic solutions are superior to both uniform randomization and simple weighted randomization strategies.

A final point about solution concepts is whether the attackers are assumed to be fully rational. In classical game theory, behavior can be fully predicted if the utilities of the actors are known. In practice, humans are not perfectly rational and omniscient decision makers, and may make suboptimal choices. ARMOR's game-theoretic model makes strong assumptions about the attacker's rationality, including perfect knowledge of the game and perfect observation capabilities. However, PROTECT and other recent systems adopt more realistic assumptions about knowledge and behavior, including imperfect information and solution concepts from behavioral game theory.

Algorithmic Goals

Computer science often focuses on developing efficient algorithms with low worst-case complexity (or proving that efficient algorithms are not possible for a particular problem). For instance, a computer scientist may ask "can a solution be found in time polynomial in the number of joint actions?" or "can a solution be found with an amount of memory polynomial in the number of joint actions?" Different methods for solving games can be compared against each other by measuring their running time, memory usage, scalability, etc. However, when discussing real-world problems, a more pertinent question is "can a solution be found given our time constraints?"

In the ARMOR system, schedules for checkpoints are generated once a week on a standard desktop computer. This means that, given the ARMOR problem definition, the solution method must be able to finish within one week with a relatively pedestrian processor and less than 4GB of memory available. Therefore, speeding up a solution from nine days to three days is much more important than working to speed up the system from a run time of three days down to a single day. The standard metrics for evaluating algorithms may be less important than the practical requirements. In particular, the worst-case complexity is largely irrelevant compared to typical performance.

Optimality proofs (or approximation guarantees) are highly valuable in real-world applications. Without such guarantees, it is difficult to build confidence in the system, and there is a legitimate concern that if an adverse event occurs, it could be attributed to a poor-quality solution. While such proofs are often necessary, they are not sufficient to prove that a security system works. First, those in nontechnical fields (who are necessarily part of the system's funding, implementation, and staffing) may not value the results of a proof as highly as a mathematician. Second, just because the model has been optimally solved does not mean that the correct problem was modeled (Section 26.3.1) or that the solution was correctly implemented (Section 26.3.3).

26.3.3 *Implementing the Solution*

The final element to consider is how the tool is actually used in practice. Any decision-support system requires inputs from human users, and the recommendations it makes must be implemented by the users as well. Problems at the interface between humans and the computer system may cause the overall system to be ineffective, even if the computer system is theoretically perfect! Based on our experience with implemented systems, we believe that a tight loop between the modelers, system developers, and end users is critical to a successful system. For instance, if those building the model do not understand how the solution will be implemented, they may not account for key factors in the real-world problem. Likewise, if assumptions made when solving the model are violated, any optimality guarantees are invalidated.

For example, the ARMOR system schedules checkpoints at certain times. During our discussions with LAWA, we heard anecdotal evidence that before ARMOR, some checkpoints were typically not manned during the middle of the day because they were in direct sunlight and uncomfortably hot. This is precisely the kind of predictable behavior that an adversary could exploit, and that the ARMOR system is designed to avoid. However, if ARMOR produced a schedule calling for one of these checkpoints to be manned during the middle of the day and the schedule was violated, the assumptions made by ARMOR would be invalidated. (To the best of our knowledge, this has not happened. In fact, we have heard that officers stationed to such checkpoints previously complained to the scheduling officer, but now complaints have been reduced because it was "the computer's decision," rather than the decision of a senior officer.)

While this guideline is true of all deployed systems, security presents two unique challenges. First, those who perform the modeling and solving steps may not be able to make site visits and fully understand the proposed deployment. For instance, a game theory expert who is working on developing safe convoy routes for patrols in Iraq is unlikely to want to travel to Baghdad given the security risks. Further, in some cases, the full details of the defensive operation may not be available to the modelers or model solvers for security reasons, and they must work under assumptions that those with appropriate security clearances independently verify.

Second, data related to the deployed configuration and performance of the system may be classified or sensitive, and not available to researchers. For instance, when working with FAMS, we developed a solution method that takes the utility matrices as input and outputs a schedule. However, we did not have clearance to see the true utility matrices used by FAMS, nor are we able to view the schedules produced by the production system. This enforced disconnect means that it is impossible for all members of the design team to understand the full operation – final verification of the system falls most heavily on those with clearance to see the full picture.

Other questions related to implementation can best be answered through interviews and surveys of users. For instance:

- Do the users understand the day-to-day use of the system?
- Do users consistently enter correct inputs?
- Do users follow the recommendations of the system?
- What happens if there are unanticipated events (e.g., flight cancellations/ delays or people call in sick)? Does the system effectively handle such situations?

26.4 Evaluation Case Studies

We now explore some of the existing evaluations we have performed on these exemplar systems. Latter sections of this chapter will refer back to these examples when discussing the overall effectiveness of the systems and delving into the strengths and weaknesses of different types of evaluations and metrics.

26.4.1 *Comparison with Previous Best Practices*

The most straightforward type of evaluation is to assume that the model is correct, and to evaluate different potential strategies using the model. In particular, we can test the solutions used in ARMOR, IRIS, and PROTECT against simpler randomization approaches, such as uniform random or simple heuristic weighted randomization. We can also test against strategies designed to replicate the best practices used by security forces. For example, by comparing the schedules generated by ARMOR with the previous scheduling practices at LAX (Cruz, 2009), we can answer the question "does ARMOR improve security?" in the affirmative, with high confidence.

There are some clear differences in ARMOR schedules and the previous scheduling practices at LAX. First, in the previous scheduling system, all checkpoints remained in place for an entire day, whereas checkpoints are now moved throughout the day according to ARMOR's schedule (adding to the adversary's uncertainty). Second, before ARMOR, only a single checkpoint was manned on any given day; multiple checkpoints are now used (due in part to an increased security budget). Third, a fixed sequence of checkpoints was defined (i.e., checkpoints 2, 3, 1, etc.) to create a static mapping from date to checkpoint. This sequence was not optimized according to the importance of different targets, and the sequence would repeat (allowing the attacker to anticipate which checkpoint would be manned on any given day).

The LAX officers informed us that they previously generated checkpoint schedules based on a cyclic strategy with random perturbations. A study of past schedules showed that patterns remained evident despite these random perturbations – no checkpoint was repeated for two consecutive days. Therefore, we also compared the ARMOR strategy against two strategies: (1) a "cyclic" strategy where the checkpoints were scheduled in a cyclic order on all inbound roads; and (2) a "restricted uniform" strategy that was a uniform random strategy with the additional restriction that no checkpoint was repeated on two consecutive days.

Our first experiments attempted to replicate as closely as possible the scheduling policies in use at LAX prior to ARMOR. Police officers placed one checkpoint on any of the five inbound roads, and the rewards of the terminals were randomly chosen between 1 and 10. We varied the duration for which the adversary could make observations from 0 to 100 days, in increments of 25 days. We averaged our results over 100 trials. In these simulations, we assumed that the adversary could use simple techniques to recognize patterns in the checkpoint schedules. The adversary maintained a history of observed checkpoints and generated confidence intervals over sequences of observations.

Figure 26.1 shows our experimental results. The x-axis represents the number of observations available to the adversary, and the y-axis represents the average defender reward. The ARMOR strategy has a higher average defender reward compared to the other two strategies. The reason is that the adversary can better predict the defender action in case of the cyclic and restricted uniform strategies, compared to the ARMOR strategy. Therefore, simple pattern recognition techniques are sufficient for the adversary to exploit the patterns in the cyclic and restricted uniform strategies. These patterns can be avoided by the use of uniform random strategies, but a uniform random strategy does not take into account the different preferences over targets of the defender. ARMOR provides weights for the different targets such that the average defender reward is the highest when compared against both cyclic and restricted uniform strategies. Additionally, ARMOR strategies are not just weighted random since they also account for the fact that the adversary observes the defender's strategy and then makes an informed, rational choice.

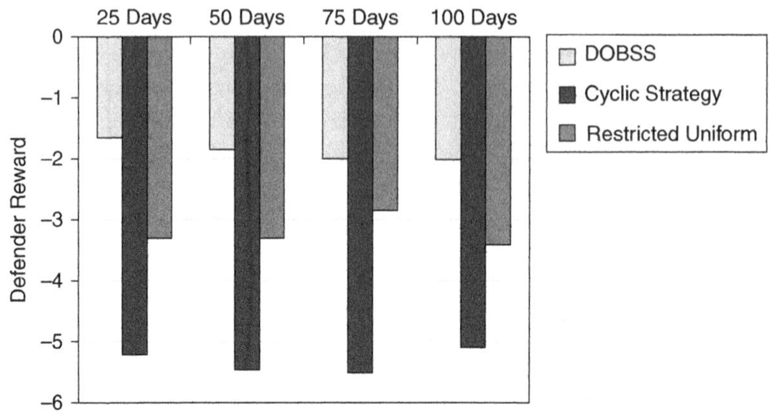

Figure 26.1. This figure compares ARMOR strategies (which use the DOBBS algorithm) and policies representative of previous methods where adversaries have the option to attack once every 25 days.

Comparing schedules generated by ARMOR against a benchmark uniform random schedule shows that ARMOR's schedule is much more efficient. For example, Figure 26.2 shows the expected reward for the police using ARMOR's schedule (calculated using DOBSS) compared with a uniform random benchmark strategy when scheduling canine patrols in the airport. ARMOR is able to make such effective use of resources that using three canines scheduled with DOBSS yields higher utility than using six canines with uniform random scheduling!

Scheduling for the FAMS domain has undergone multiple tests using real-world schedules and has withstood expert scrutiny, but the exact results of these tests are not public. Instead, publicly released results rely on mathematical evaluations. For instance, Figure 26.3(a) shows the results of an evaluation done with hypothetical payoffs to better measure the success of the system. Results show the expected payoff when scheduling a single air marshal in the FAMS domain over 20, 100, and 200 schedules. In these experiments, this equated to 10, 50, and 100 flights, with half as "departures" and half as "return" flights for the air marshals. The x-axis in these graphs shows the expected payoff for IRIS, six different methods of weighting the schedules, uniform random, and no coverage (i.e., the single air marshal protects no planes).

In the full FAMS problem, there are too many schedules to enumerate and it is difficult to decide how to weight the flights, even if enumerated. This weighting is a simple way of determining which schedules to cover, and may improve the performance of the defender relative to uniform random. In these simple tests, the problem space was small enough that all schedules could be easily enumerated.

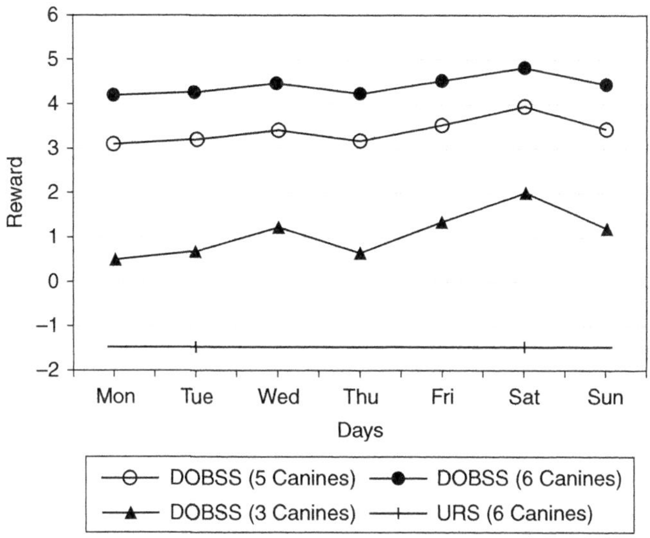

Figure 26.2. This figure shows that ARMOR (powered by the DOBBS algorithm) can outperform the baseline solution method, even using many fewer K9 units, which may achieve substantial cost savings. The x-axis shows the results from seven different days, and the y-axis shows the expected utility for the different scheduling methods.

Utilities are generated as random numbers drawn uniformly from the ranges of numbers as follows:

- Defender payoff for covering a flight that is attacked: [+5,000, +10,000]
- Defender payoff for not covering a flight that is attacked: [–10,000, –5,000]
- Attacker payoff for attacking a flight that is not covered: [+5,000, +10,000]
- Attacker payoff for attacking a flight that is covered: [–10,000, –5,000]

The six baseline weighted random strategies are specified next. In all cases, a mixed probability distribution is obtained by normalizing these weights. *Max. of attacker rewards*: the weight of a schedule is the maximum of the attacker rewards for a successful attack over all flights covered in the schedule. *Min. of attacker penalty*: the weight of a schedule is the minimum of the attacker penalty for a failed attack over all flights covered in the schedule. *Min. of defender penalty*: the weight of a schedule is the minimum of defender penalties for a successful attack over all flights covered in the schedule. *Max. of defender reward*: the weight of a schedule is the maximum of defender rewards for capturing an attacker over all flights covered in the schedule. *Sum of defender penalties*: the weight of a schedule is the sum of the defender penalties for

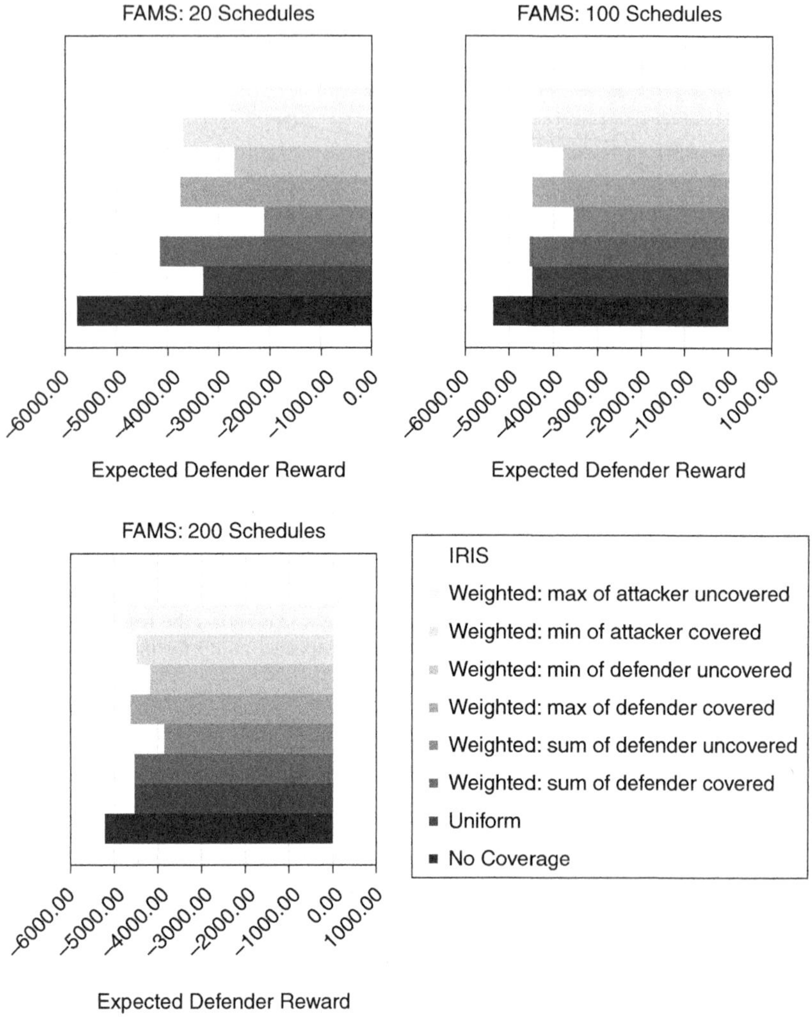

Figure 26.3. These graphs show the results of scheduling with the IRIS algorithm (which is called ASPEN in this version) vs. other possible scheduling methods. In the case of (a) 20, (b) 100, and (c) 200 schedules, the IRIS scheduling method is superior. Results are averaged over 30 independent trials each.

a successful attack over all flights covered in the schedule. *Sum of defender rewards*: the weight of a schedule is sum of defender rewards for capture over all flights covered in the schedule.

In contrast to ARMOR and IRIS, the PROTECT system is based on a quantal response (QR) model of an adversary, which has shown to be a promising model of human decision making (McKelvey & Palfrey, 1995; Rogers, Palfrey, & Camerer, 2009; Yang, Kiekintveld, Ordóñez, Tambe, & John, 2011). A relevant study of the QR model was conducted by Yang and colleagues (2011) in the context of security games, where this model was shown to outperform competitors in modeling human subjects.

The QR model adapts ideas from the literature, which presumes that humans will choose better actions at a higher frequency, but with noise added to the decision-making process following a logit distribution as defined next:

$$q_i = \frac{e^{\lambda G_i^a(x_i)}}{\sum_{j=1}^{T} e^{\lambda G_j^a(x_i)}}$$

The parameter λ represents the amount of noise in the attacker's strategy. λ can range from 0 to ∞ with a value of 0 representing a uniform random probability over attacker strategies while a value of ∞ represents a perfectly rational attacker. qi corresponds to the probability that the attacker chooses a target i; G^a_i (x_i) corresponds to the attacker's expected utility of attacking target *i* given x_i, the probability that the defender covers target i; and T is the total number of targets. To apply the QR model in a Stackelberg framework, PROTECT employs an algorithm named PASAQ (Yang, Tambe, & Ordóñez, 2012). PASAQ computes the optimal defender strategy (within a guaranteed error bound) given a QR model of the adversary by solving a nonlinear and non-convex optimization problem.

Figure 26.4 shows the results of the defender's expected utility of PASAQ with λ = 1.5 in comparison with three baselines: (i) a uniform random defender's strategy; (ii) a mixed strategy assuming the attacker chooses any target uniformly at random (λ = 0); and (iii) a mixed strategy assuming a fully rational attacker (which is equivalent to DOBSS, where λ = ∞). The expected utilities of the four strategies are compared against different attacker strategies based on the λ value on the x-axis. Both uniform random strategies perform well when the attacker's strategy is based on λ = 0. However, as λ increases, both strategies quickly drop to a very low defender-expected utility. In contrast, the PASAQ strategy with λ = 1.5 provides a higher expected utility than that assuming a fully rational attacker over a range of attacker λ values (and indeed over the range of interest), not just at λ = 1.5.

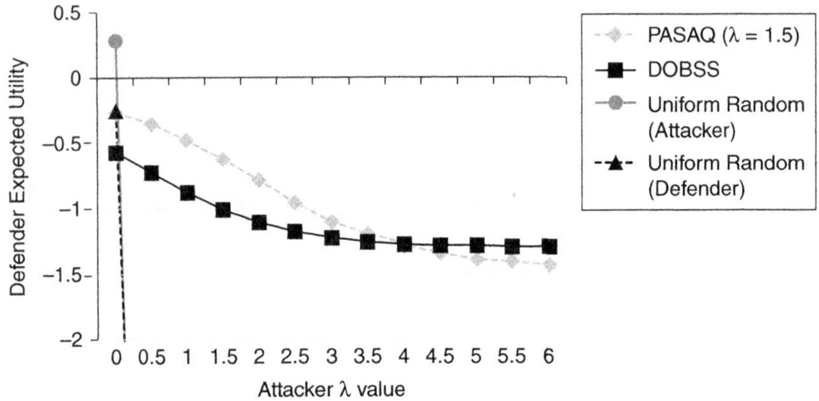

Figure 26.4. This figure compares strategies when varying λ for the attacker's strategy.

26.4.2 *Mathematical Sensitivity Analysis*

The analyses in the previous section compare the effectiveness of different scheduling policies, but do so based on all of the assumptions in the model (e.g., that the actions and payoffs are correctly specified for both players, and that the attacker is perfectly rational). To build more confidence in this approach, we must also validate the model itself. A first step in this is to better understand the impact of the different assumptions using sensitivity analysis Figure 26.4.

In this type of evaluation, important parameters of the model are varied to test how the output of the model changes to different inputs. One important input to the ARMOR model is the distribution of different types of attackers. For example, some attackers may be highly organized and motivated, while others are amateurish and more likely to surrender. Different types of attackers can be modeled as having different payoff matrices. Changing the percentages of each attacker can help show the system's sensitivity to assumptions regarding the composition of likely attackers, and (indirectly) the system's dependence on precise utility elicitation. Figure 26.5(a)–5(c) depicts two adversary types with different reward matrices. Figure 26.5(a) demonstrates that DOBSS has a higher expected utility than that of a uniform random strategy on a single checkpoint, regardless of the percentage of "type one" and "type two" adversaries. Figures 26.5(b) and (c) shows that DOBSS again dominates uniform random for two and three checkpoints, respectively.

Further sensitivity analysis can be applied to other parameters of the model. The payoffs that describe the preferences of the two players for different outcomes

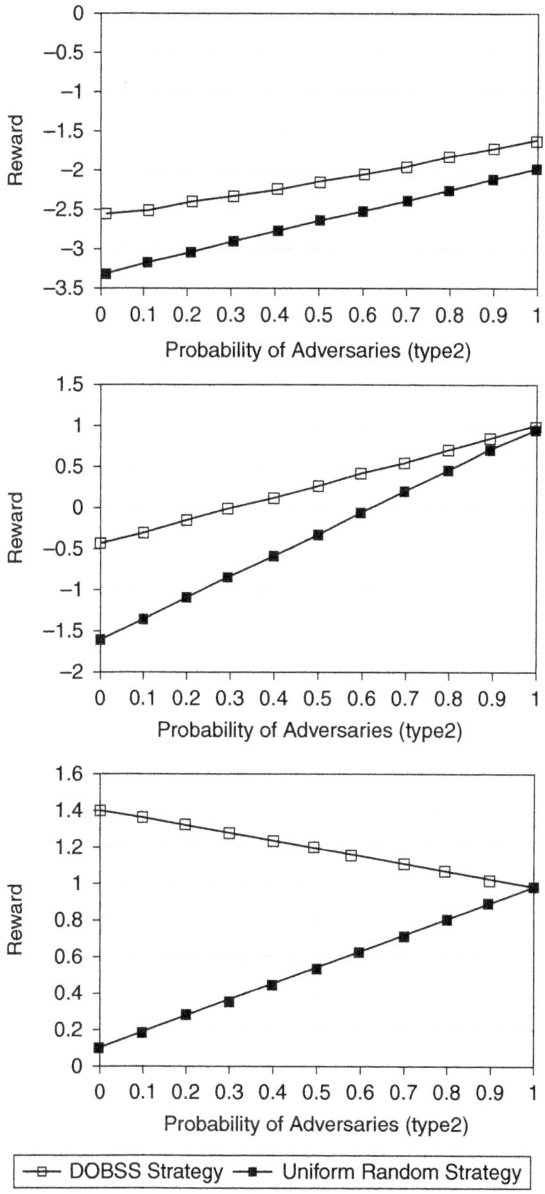

Figure 26.5. This figure compares ARMOR's schedules (generated by the algorithm named DOBBS) with a uniform random baseline schedule. Figures a–c show the utility of schedules for one to three vehicle checkpoints varying the relative probability of two different attacker types. The x-axes show the probability of the two attacker types (where 0 corresponds to 0% attack type 2, and 100% attack type 1), and the y-axes show the expected utility of ARMOR and a uniform random defense strategy.

are a very important set of parameters. These parameters are estimates of the true utilities determined through elicitation sessions with experts. Unfortunately, it is known that game-theoretic models can be quite sensitive to payoff noise (Kiekintveld & Wellman, 2008), and arbitrary changes in payoffs can lead to arbitrary changes in the optimal schedule. There is some evidence that ARMOR is robust to certain types of variations. In one experiment, we multiplied all of the defender's negative payoffs for successful attacks by a factor of four, essentially increasing the impact of a successful attack. We found that in the one- and three-checkpoint cases, the strategies were unchanged. In the two-checkpoint case, the actions were slightly different, but the overall strategy and utility were unchanged. Unfortunately, there is also evidence that this does not generalize to all payoffs in security games. Kiekintveld and colleagues (2011) show that in general, adding small amounts of noise to the attacker's payoffs in security games can cause large deviations in the defender's payoffs (though the changes in the optimal strategy are less drastic).

We tested the PROTECT system to show the robustness of the PASAQ solution concept versus a defender strategy that assumes a perfectly rational attacker. The hypothesis is that PASAQ's expected defender utility will not degrade as much as DOBSS over the range of attacker λ values, when we add observation, execution, and payoff noise. In the real world, observation, execution, and payoffs are not always perfect due to noise in the attacker's surveillance of the defender's patrols, the many tasks and responsibilities of the USCG where the crew may be pulled off a patrol, and limited knowledge of the attacker's payoff values. A comparison of the uniform random strategies was not included due to its poor performance (as shown in Figure 26.4).

For Figures 26.6–26.8, the y-axis represents the defender's expected utility and the x-axis is the attacker's λ value, with error bars depicting the standard error. The first experiment considers observational noise, which means that the attacker has noise associated with observing the defender's patrol strategy as shown in Figure 26.6. Contrary to expectation, observation error leads to an increase in defender expected utility in PASAQ, but a potential decrease (or no change) in DOBSS – thus PASAQ ends up dominating DOBSS by a larger margin over wider ranges of λ. For the second experiment, noise is added to the execution phase of the defender in Figure 26.7, showing that execution error leads to PASAQ dominating DOBSS over all tested values of λ. In the third experiment (Figure 26.8), payoff noise is added to the attacker's original payoff values (similar to Kiekintveld et al., 2011), with DOBSS being dominated by PASAQ, indicating the robustness of the PASAQ model.

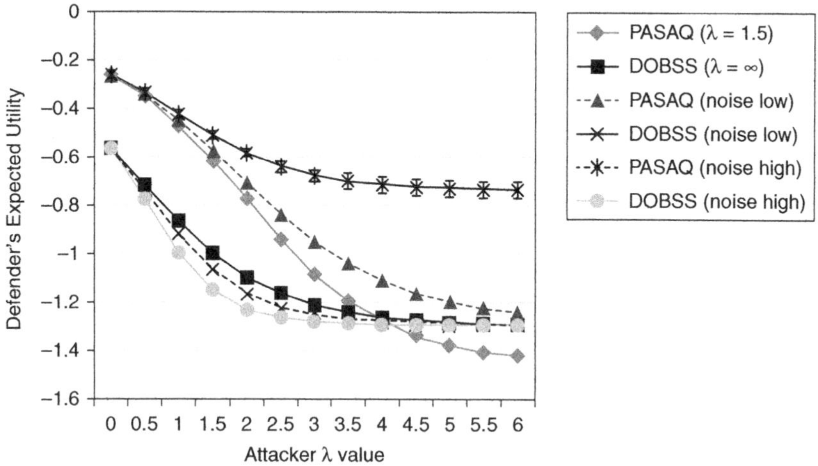

Figure 26.6. Defender's expected utility: observation noise.

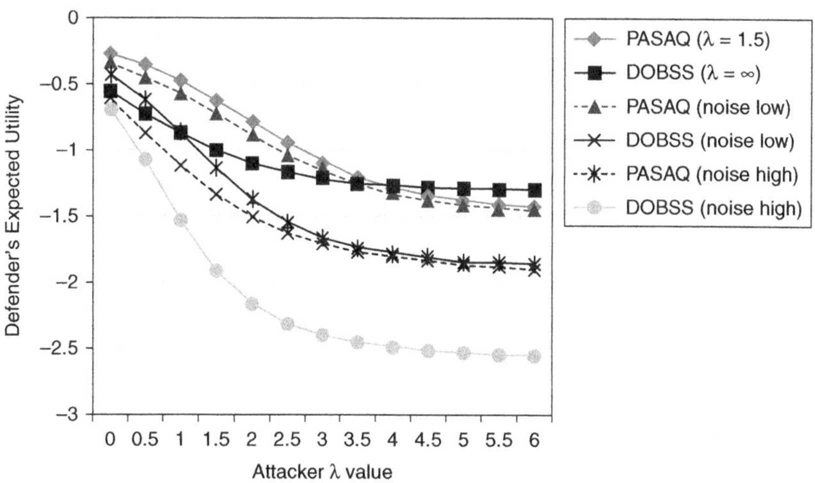

Figure 26.7. Defender's expected utility: execution noise.

26.4.3 *Human Trials*

Another set of assumptions in the game models used for ARMOR and IRIS is that the attackers are perfectly rational, and will always choose the mathematically optimal attack strategy. To test the sensitivity of the solutions to variations in human behavior, we have run a series of controlled laboratory experiments with human subjects (Pita, Jain, Tambe, Ordóñez, Kraus, &

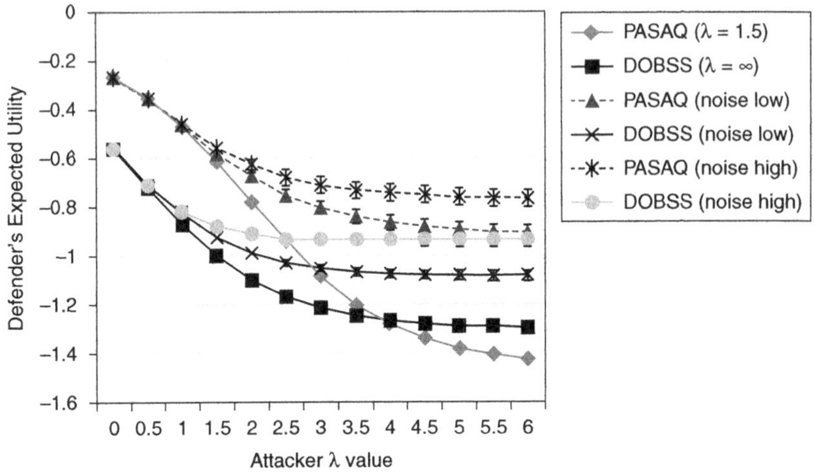

Figure 26.8. Defender's expected utility: payoff noise.

Magori-Cohen, 2009). In these experiments, subjects play a "pirates and treasure" game designed to simulate an adversary planning an attack on an LAX terminal. Subjects are given information about the payoffs for different actions and the pirates' strategy for defending their gold (analogous to the security policy for defending airport terminals). Subjects receive cash payments at the conclusion of the experiment, based on their performance in the game.

These experiments have provided additional support for the quality of ARMOR's schedules against human opponents. First, they suggest that the assumptions the game-theoretic model imposes are reasonable. Second, we have tested many conditions, varying both the payoff structure and the observation ability, ranging from no observation of the defense strategy to perfect observation. The results show that ARMOR's schedules achieve higher payoffs than the uniform random benchmark across all of the experimental conditions tested, often by a large margin. These results demonstrate that ARMOR schedules outperform competing methods when playing against human adversaries.

26.4.4 *Quantitative Data*

Ideally, we would be able to use data from the operation of a deployed system to provide further validation of the modeling assumptions. For example, in the case of ARMOR, we might be interested in the number of attacks the system prevents. Unfortunately, such data are very limited in the case of ARMOR; there have been no major attacks on the airport, but it is impossible to say how many attacks would have occurred *without* ARMOR.

We were able to obtain some data on the arrest record at ARMOR checkpoints. Although the use of such data has multiple caveats (see Section 26.5 for more discussion), it can still provide useful information. We received summarized and actual reports from the LAX police regarding the number of violations that they detected at checkpoints in 2007, 2008, and January 2009. For example, we received the following report for January 2009:

1. January 3, 2009: Loaded 9mm pistol discovered
2. January 3, 2009: Loaded 9mm handgun discovered (no arrest)
3. January 9, 2009: 16 handguns, 4 rifles, 1 pistol, and 1 assault rifle discovered – some fully loaded
4. January 10, 2009: 2 unloaded shotguns discovered (no arrest)
5. January 12, 2009: Loaded 22cal rifle discovered
6. January 17, 2009: Loaded 9mm pistol discovered
7. January 22, 2009: Unloaded 9mm pistol discovered (no arrest)

Figure 26.9 tabulates the number of violations for the year prior to the deployment of ARMOR and during 2008 when ARMOR was in use. The x-axis breaks down the violations into different types and the y-axis represents the number of violations. The number of violations is substantially higher at LAX after ARMOR was deployed than in the preceding period. For example, only four drug-related offenses were detected before the deployment of ARMOR, while 30 such offenses were detected after the deployment. While we must be careful about drawing too many conclusions from these data due to the large number of uncontrolled variables (for example, the number of checkpoints was not consistent during this period), the ARMOR checkpoints do appear to be effective on this measure.

PROTECT was evaluated by comparing data from the USCG patrols before and after deployment at the Port of Boston. Figures 26.10 and 26.11 show the frequency of visits by the USCG to different patrol areas over a number of weeks. The x-axis is the day of the week, and the y-axis is the number of times a patrol area is visited for a given day of the week. The y-axis is intentionally unlabeled for security reasons. There are more lines in Figure 26.10 than in Figure 26.11 because during the implementation of PROTECT, new patrol areas were formed that contained more targets and thus fewer patrol areas in the post-PROTECT figure. Figure 26.10 depicts a definite pattern in the patrols. While there is a spike in patrols executed on Day 5, there is a dearth of patrols on Day 2. Besides this pattern, the lines in Figure 26.10 intersect, indicating that during some days, a higher-value target was visited more often while on other days, it was visited less often, even though the value of a target does not change from day to day. This means that there was not a consistently high

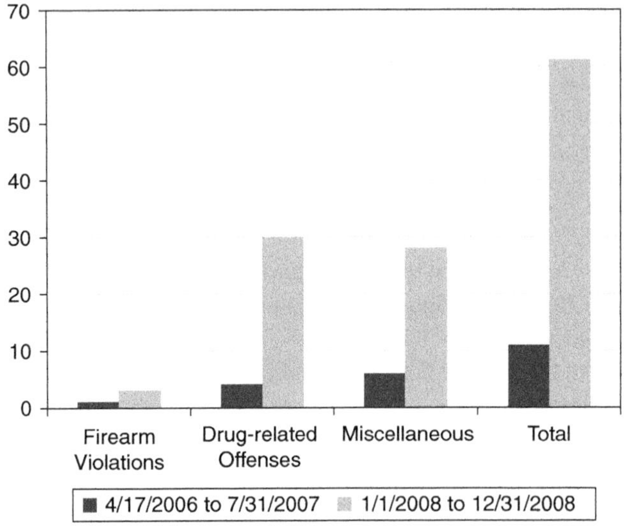

Figure 26.9. This figure shows how the number of arrests in three categories changed before and after the ARMOR deployment at LAX.

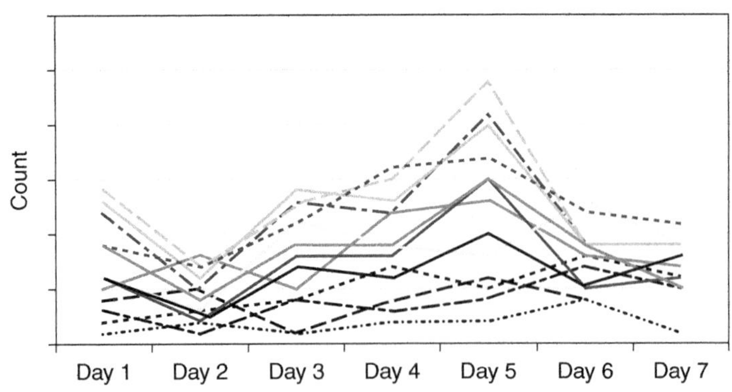

Figure 26.10. Patrol visits per day by area pre-PROTECT.

frequency of coverage of higher-value targets before PROTECT. These figures show a decrease in predictable patrolling patterns when using PROTECT.

26.4.5 *Adversarial Perspective Team*

Another type of evaluation that was done for PROTECT was the use of an Adversarial Perspective Team (APT). The APT was tasked with measuring the level of discernible tactical deterrence, if any, of the PROTECT model's

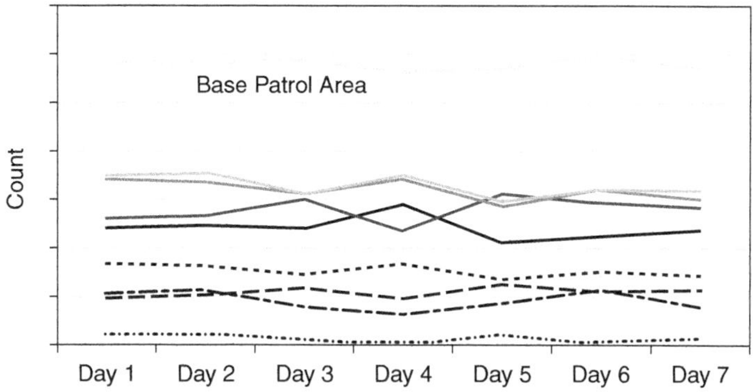

Figure 26.11. Patrol visits per day by area post-PROTECT.

randomization of patrol schedules against a known adversary. The secondary goal was to identify, through the adversary's perspective, the tactical deterrence value of purposeful activities during a maritime security patrol at specific maritime critical infrastructure.

The APT incorporated the adversary's known intent, capabilities, skills, commitment, resources, and cultural influences. The APT included a spectrum of threats characterized by both measurable capabilities (knowledge, access, and resources) and intangibles such as risk tolerance and motivation. The Port of Boston was used as a proof-of-concept in which to implement and evaluate the PROTECT model scheduling algorithm for Coast Guard small boat patrols. As part of that evaluation, the APT conducted a pre- and post-PROTECT assessment of the algorithm's impact on an adversary's deterrence.

The adversary was defined as an individual(s) with ties to al-Qaida or al-Qaida affiliates. Incorporating the adversary's known intent, capabilities, and resources, the APT identified the potential value of each maritime critical infrastructure in Boston to an al-Qaida cell, using a waterborne incendiary explosive device (WBIED) as the only attack method. In the abstract, the APT provides a framework that could be applied and expanded to other international and domestic terrorism teams.

Recall the primary goal of the APT is to calculate a deterrence value by observing law enforcement assets. If multiple law enforcement assets were observed during one APT observation, the deterrence values of those individual assets were summed to obtain an overall, daily deterrence score. The daily score is highly variable and depends on the random chance of multiple law enforcement assets, including the Coast Guard, Boston Police Department, and possibly others, being observed in a 15–60-minute window by the APT.

The analysis by the APT uncovered a positive trend where the effectiveness of deterrence increased from the pre- to post-PROTECT observations.

26.4.6 *Qualitative Expert Evaluations*

Given the sparseness and limitations of the available data from the field, evaluations of the system by security experts become a very important source of information. Although they are typically qualitative in nature, they are one of the few ways to gather evidence on the quality of the modeling assumptions and the effectiveness of the holistic system, as it is actually deployed and used on a day-to-day basis. Security procedures at LAX are subject to numerous internal and external security reviews (not all of which are public). The available qualitative reviews indicate ARMOR is both effective and highly visible.

Director James Butts of the LAWA police reported that ARMOR "makes travelers safer" and even gives them "a greater feeling of police presence" (Murr, 2007). Erroll Southers, assistant chief of the LAWA police, told a congressional hearing that "LAX is safer today than it was eighteen months ago," due in part to ARMOR. A recent external study by Israeli transportation security experts concluded that ARMOR was a key component of the LAX defensive setup. The ARMOR team has also been awarded letters of commendation from the City of Los Angeles in recognition of the efforts toward securing LAX. Thus, the domain experts have been highly supportive of ARMOR, and it would be very hard to deploy the system without their support. They are also likely to identify potential problems with the system quickly. While such studies are not very useful for quantifying ARMOR's benefit, they all suggest that the domain experts believe that ARMOR generates better schedules as compared to their previous approaches.

ARMOR was designed as a mixed-initiative system so that police could override the recommended policies. In practice, users have not chosen to modify the recommended schedules, suggesting that users are confident in the outputs. While such studies are not very useful for directly quantifying ARMOR's benefit, it would be very hard to deploy the system without the support of such experts. Furthermore, if there were an "obvious" problem with the system, such experts would likely identify it quickly.

Expert opinions have emphasized that an important benefit of the system is its transparency and visibility, which contribute to deterrence. ARMOR assumes that adversaries are intelligent and have the ability to observe the security policy: knowing about the system does not reduce its effectiveness. The deployment of ARMOR has been quite visible: ARMOR has been covered on local TV stations (including FOX and NBC), in newspapers (including the

LA Times and the *International Herald Tribune*), and in a national magazine (*Newsweek*).

The IRIS system has been tested both qualitatively and quantitatively, but the results of these tests are restricted. However, James B. Curren, special assistant in the Office of Flight Operations at the Federal Air Marshals Service, has affirmed the effectiveness of IRIS:

> We have tested IRIS and found it to be applicable to our needs in creating uncertainty as to FAM presence on selected city pairs of flights. After extensive testing we have implemented IRIS on a trial run of flight selections and have continued to expand the number of flights scheduled using IRIS. Our exact use of IRIS is sensitive information and we can only state that we are satisfied with IRIS and confident in using this scheduling approach.

Furthermore, internal governmental studies have recommended that random scheduling be implemented, which is precisely what IRIS accomplishes (T. U. S. G. A. O. GAO, 2009)[1]:

> Because the number of air marshals is less than the number of daily flights, FAMS's operational approach is to assign air marshals to selected flights it deems high risk – such as the nonstop, long-distance flights targeted on September 11, 2001. In assigning air marshals, FAMS seeks to maximize coverage of flights in 10 targeted high-risk categories, which are based on consideration of threats, vulnerabilities, and consequences. In July 2006, the Homeland Security Institute, a federally funded research and development center, independently assessed FAMS's operational approach and found it to be reasonable. However, the institute noted that certain types of flights were covered less often than others. The institute recommended that FAMS increase randomness or unpredictability in selecting flights and otherwise diversify the coverage of flights within the various risk categories. As of October 2008, FAMS had taken actions (or had ongoing efforts) to implement the Homeland Security Institute's recommendations. GAO found the institute's evaluation methodology to be reasonable.

The use of PROTECT and APT's improved guidance given to boat crews on how to conduct the patrol jointly provided a noticeable increase in the quality and effectiveness of the patrols. Prior to implementing PROTECT, there were no documented reports of illicit activity. After implementation, USCG crews reported more illicit activities within the port and provided a noticeable "on the water" presence with industry port partners commenting, "the Coast Guard seems to be everywhere, all the time." Coast Guard Station Boston leadership and boat crews reported that the schedule was easily implemented and improved their ability to plan and schedule crew rest intervals, required training evolutions, and boat maintenance. Success was measured not only by the elimination of patrol patterns previously identified, but by exceeding all service visit count requirements without increasing the number of boat hours or crews, and higher revisit rates on critical infrastructure that presented the greatest risk as measured by the Coast Guard's Maritime Security Risk Assessment Model.

Rear Admiral Daniel Neptun, commander, 1st Coast Guard District, which includes the ports of Boston and New York, noted:

> The partnership across the DHS Centers of Excellence, and with the Coast Guard, added synergy and excitement into an important, but otherwise routine process of conducting harbor patrols to improve maritime domain awareness and enhance harbor safety and security. The result was invigorated boatcrews more focused upon providing effective presence, reducing predictability, and enhancing the safety and security within our ports. PROTECT-guided patrols became a source of pride for my boatcrews, and it doesn't get much better than that! The results have been exceptional, and we continue to refine the product for what I hope will be the eventual servicewide use (Doane & DiRezno, 2012).

Vice Admiral Robert C. Parker, commander, LantArea, explained the value of the project when he noted:

> The example of the collaboration on the PROTECT Model is a wonderful partnership of academia and the Coast Guard because it empowers us to be more random, and less predictable in our Ports, Waterways, and Coastal Security mission. Our maritime safety and security presence enables American prosperity and synergies, and partnerships like PROTECT help make America safer. (Doane & DiRezno, 2012)

26.4.7 *Real-World Evaluations*

In collaboration with the Los Angeles Sheriff's Department (LASD), we designed two types of real-world tests, one for fare evasion and one for counterterrorism (a preliminary evaluation of TRUSTS's crime component can be found in Fave et al., 2014). The fare evasion experiment constitutes the first real-world deployment of game-theoretic schedules and analyzes their performance against real adversaries (fare evaders). Similarly, the counterterrorism experiment is the first real-world, head-to-head comparison between game-theoretic and human-generated schedules.

Fare Evasion Experiment

This experiment took place over 21 days during the months of July and August 2013. The organization of the experiment (e.g., train the security officers, design and organize the experiment in collaboration with the LASD) required approximately two weeks. This experiment had two key purposes. The first was to validate the robustness of the Fare Evasion module against execution uncertainty (e.g., schedules could not be completed due to the officers being delayed). The second was to run a head-to-head comparison between TRUSTS and a standard random resource allocation, i.e., the canonical type of randomized allocation system used by several security agencies. This section discusses the setup of the experiment and the results that we obtained.

Experiment Setup: The fare evasion experiment took place on the Blue line of the LA Metro system (see Figure 26.12 for the map of the metro line). This line consists of 22 different stations and is one of the longest lines in the LA Metro system. It was selected by the LASD, which helped to organize the experiment (e.g., assign security officers and patrol times).

Each day, a team of two security officers (see Figure 26.13), was randomly selected by the LASD to patrol the line for a duration of up to two hours. Patrols were run during both the morning and the afternoon. Some days, the tests ended early due to the officers being reassigned. One of the two officers acted as the leader of the team: s/he was given the smartphone, had to read the schedule to the other officers, collected the data, and eventually updated it whenever a delay occurred. An update could be made either during a station check, i.e., when checking riders at a station, or during a train check, i.e., when checking riders on the coach of a train. In the latter case, the officers were required to leave the train at the next station to request an update. This was required because of how the defender strategy was defined. Every week the team was provided with one of two types of schedules:

Game-theoretic schedules (GT): This type of schedule was generated using TRUSTS's fare evasion component.

UR schedules (UR): This type of schedule was generated by modeling the problem as a Markov Decision Process (see Fave et al., 2014 for the technical details) and by setting the corresponding Markov strategy as a uniform probability distribution. The officers were not told which schedule they were using as not to bias their performance. Before the experiment, we anticipated that the officers might view some of the schedules as leading to low performance in terms of catching very few fare evaders. In such situations, the officers, in order to avoid poor performance, might end up voluntarily deviating from their given schedules to reach a better location because they were unsatisfied with the current one. In anticipation of such voluntary deviations, we augmented both the game-theoretic and UR schedules with the ability to perform updates. More specifically, we allowed the officers to request *voluntary* or *involuntary* updates. Voluntary updates consisted of the officers updating the current schedule because, in their opinion, the current specified action was not fruitful as a venue to check fares. Officers were allowed to choose a new location that they considered more fruitful for catching fare evaders and to request a new schedule from there. Involuntary occurred officers requested a new schedule because they were delayed (e.g., from issuing citations or arresting a suspect) and were unable to perform the next action on their schedule. This type of update could be requested every time an officer was delayed. As we will see, the

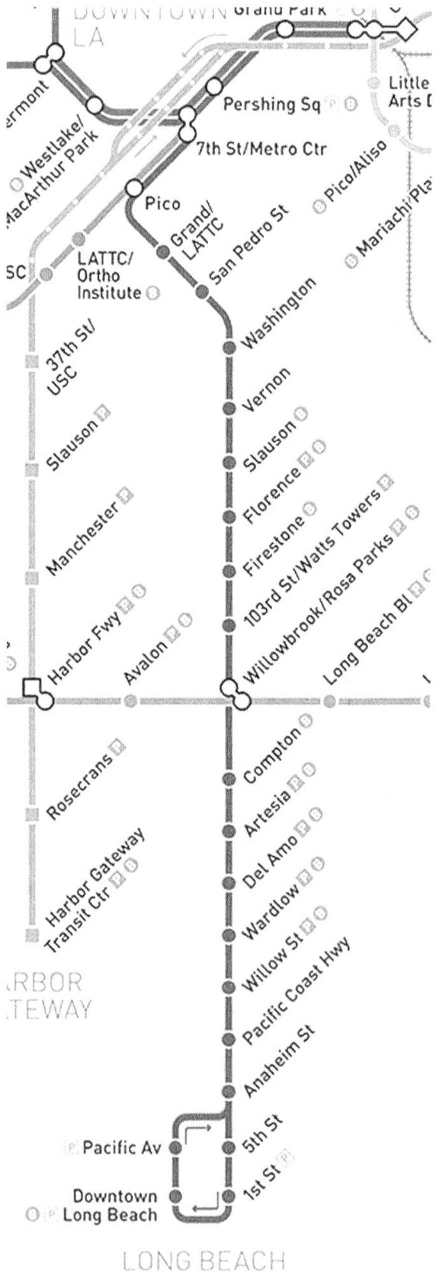

Figure 26.12. Map of the Blue line of the LA Metro 44.

Figure 26.13. Two security officers performing fare checks on a train.

officers used voluntary updates almost every day with the UR schedules, but never with the GT schedules.

Finally, note that, given the duration of our experiment, the game-theoretic schedules are essentially testing a maximin strategy. The fare-evasion component computes a Stackelberg strategy, a strategy based on the assumption that the riders will conduct surveillance and observe the defender's mixed strategy. However, because there were only 21 days of patrol and the officers could patrol for only a few hours per day, we cannot assume that the riders had sufficient time to conduct accurate surveillance, observe the mixed strategy, and best respond to it. Nonetheless, the FE component solves a zero-sum game for which a Stackelberg equilibrium and the maximin strategy are known to be equivalent (Yin & Tambe, 2012). Thus, since the maximin strategy provides a guaranteed level of defender utility without making any assumption about the adversary's surveillance of the defender's mixed strategy, these experiments compare the benefit of using a maximin strategy against other (non-game-theoretic) approaches for generating patrol schedules.

Results: During the 21 weekdays of the experiment, we were able to run GT schedules for 11 days of testing while UR schedules were deployed for 10 days, resulting in 855 and 765 patrol minutes, respectively. The schedules were compared using two different metrics. First, we counted the number of passengers checked and the number of captures at the end of each patrol. The captures were defined as the sum of the number of warnings, citations, and arrests. Passengers without a valid ticket could be given a warning or cited for a violation on the discretion of the officer. This metric was chosen because it

Table 26.1. *Patrol duration over each of the 21 days*

	D1	D2	D3	D4	D5	D6	D7	D8	D9	D10	D11	total
GT	60	60	90	60	90	10	90	110	90	90	105	855
UR	60	60	60	60	60	75	100	100	100	90		765

would allow us to measure the performance of each schedule in the real world. Second, we counted the number of times that the update function was used voluntarily and involuntarily. While involuntary updates helped determine the value of using MDPs (discussed later in this chapter), voluntary updates measured the human (officer) perception of the quality of the schedules – the more such voluntary updates, the more the officers were dissatisfied with their given action. Table 26.1 shows the duration of each day of patrol for both GT and UR schedules.[2]

As shown in the table, the actual duration of a daily patrol was often different over the 21 days of the experiment for GT and UR schedules. For this reason, providing a comparison normalized over the days of the experiment was impossible. However, on most of the days, we were able to collect data for multiples of 30 minutes (e.g., 60 or 90 minutes). To best compare our results, we divided our data into 30-minute segments. More specifically, we considered all the train and station checks within a time window of 30 minutes and collected the data resulting from these actions. In so doing, whereas the measurements can be considered independent one from the other, the locations that were checked will actually have some small dependency. Indeed, measurements are independent because within each segment the officers will check different people who are unable to affect each other. Each measurement corresponds to a sample of different train riders, taken at different times and locations. Not only did the officers never check the same rider twice, but, most important, during 30 minutes, they were able to visit different locations by riding the trains (roughly, one train every six minutes in the Blue line) and inspecting the stations (on-station operations lasted no longer than 20 minutes). In contrast, the locations that were checked will have some dependency since they were calculated using game theory. More specifically, locations calculated using UR schedules can be considered independent because they were sampled uniformly randomly. In contrast, there might be some small dependency for locations corresponding to GT schedules because these are calculated from an MDP policy whereby a specific action is based on the previous action. By introducing this bias, however, the performance of GT schedules is damaged as UR schedules could pick all locations whereas GT schedules could not.

Table 26.2. *Number of involuntary (delays) deviations for each day of patrol*

	D1	D2	D3	D4	D5	D6	D7	D8	D9	D10	D11	total
GT	0	1	3	1	1	0	2	2	4	2	1	18
UR	0	2	1	1	1	2	2	2	3	2		16

Table 26.3. *Number of voluntary (updates) deviations for each day of patrol*

	D1	D2	D3	D4	D5	D6	D7	D8	D9	D10	D11	total
GT	0	0	0	0	0	0	0	0	0	0	0	0
UR	1	0	1	1	1	0	1	1	1	1		8

Validation of the MDP model: As discussed at the beginning of this section, both GT and UR schedules were calculated by solving an MDP. For this reason, both schedules could be updated to request a new schedule. Tables 26.2 and 26.3 then show, for each day of patrol, the number of voluntary and involuntary deviations requested by the officers. In total, GT schedules were updated 18 times, all of which were *involuntary* deviations, i.e., delays. All these update requests confirm that the MDP model was able to provide schedules that could be updated whenever necessary.

All involuntary deviations were due to the officers writing citations or helping people. The average delay length was 12 minutes and the longest delay was 20 minutes. In each case, a new schedule was provided starting at the officers' current location and closest time. Table 26.3 shows that voluntary deviations were used only with UR schedules. This result strongly suggests that the officers were mostly satisfied with GT schedules. In addition, it means that GT schedules did not really compete against UR schedules only. Rather, the comparison was between UR schedules which were augmented with real-time human intelligence for most of the time (8 out of 10 days). We discuss the results of such comparison next.

Game Theory vs. Uniform Random: The results that we obtained are shown in Figure 26.14 and in Table 26.5. Figure 26.14 shows eight boxplots depicting the data that we collected during each patrol, using both GT and UR schedules. Respectively, the four figures present data collected on captures, warnings, violations, and passengers checked per 30 minutes of patrolling.[3] For each boxplot, the top and bottom of the box represent the 75th and 25th percentiles, respectively, while the middle line indicates the median of the collected data. The "+" data points indicate statistical outliers, while the whiskers show the

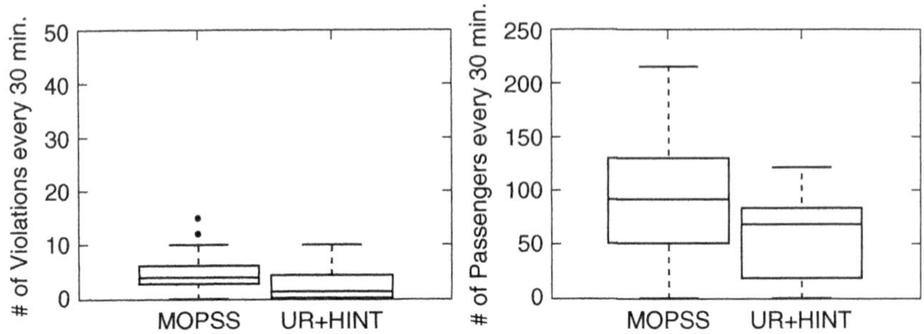

Figure 26.14. Results of the fare evasion tests.

Table 26.4. *Average captures (C), warnings (W), violations (V), and passengers (P) based on the results obtained in Figure 26.14.*

	Days	avg. C	avg. W	avg. V	avg. P
GT	11	15.52	10.42	5.03	96.77
UR	10	9.55	6.48	3.07	60.85

Table 26.5. *Count of individual activities*

	S1	S2	S3	S4	S5	S6	S7	S8	S9	S10
Manual	3	3	3	2	3	2	2	2	2	2
TRUSTS	2	2	3	3	2	2	2	3	3	2

most extreme non-outlier data points. Each of the four figures (captures, warnings, violations, and passengers checked) shows that the data collected using GT schedules had higher values than the data collected using UR schedules. As shown in Table 26.5, on average, GT schedules led to, respectively, 15.5 captures, 10.4 warnings, and 5.0 violations issued every 30 minutes compared to 9.6 captures, 6.5 warnings, and 3.1 violations obtained using UR schedules. To confirm the statistical significance of these results, we ran a number of weighted unpaired student t-tests ($p = 0.05$) (Bland & Kerry, 1998; Goldberg, Kercheval, & Lee, 2005) and verified, for each metric, that the difference in the results was statistically significant. We used a weighted t-test because some data segments had a duration shorter than 30 minutes and we wanted to use all the available data for our analysis. As shown in Table 26.1, not all the patrol durations could be properly divided into a finite number of 30-minute segments (e.g., UR: D6, D7, D8, D9, and GT: D6, D8, and D11). Therefore, we calculated a weighted average for each of the metrics defined earlier, whereby each segment was given a weight that was defined based on the segment's duration (longer segments corresponded to higher weights).

From a practical perspective, the magnitude of the difference between the two approaches is significant: over a period of 21 days, GT would capture many more fare evaders. This result can be emphasized even further if we correlate it with the results shown in Tables 26.3 and 26.2. While running UR schedules, the officers were requesting involuntary deviations daily, whereas no such deviations were requested while running GT schedules. In other words, they were using real-time situation awareness to augment the quality of the schedules, thus making the UR schedule more compelling.

The results in Table 26.5 also indicate that GT schedules led to 96.8 passengers checked every 30 minutes against 60.9 passengers checked by using UR schedules. As discussed elsewhere (Jiang, Yin, Zhang, Tambe, & Kraus, 2013), GT schedules are generated by leveraging all the possible sequences of train and station checks and by taking into account key dimensions such as the train schedules, the officers' effectiveness, and, most important, the daily ridership statistics. This means that stations or trains with a higher presence of riders will be given a higher coverage probability since they are more likely to contain fare evaders. Hence, these results confirm the accuracy of the model as both Figure 26.14(d) and Table 26.5 show that GT schedules led the officers to check more passengers than UR schedules.

This raises the question of whether a static type of schedule, which deploys the officers only at the most crowded locations, would lead to similar or even better results than those obtained with GT. Given the limited amount of time that we had to conduct our experiments, we were unable to compare GT schedules against a static deployment – where the key weakness is predictability in the longer term. Indeed, effective randomization was one of the main reasons for LASD to collaborate on these experiments – security agencies know that static schedules become predictable in the long term. After a certain amount of time, the passengers would know where the officers are located and could exploit this information to avoid paying the fare.

Counterterrorism Experiment

The purpose of this experiment was to run a head-to-head comparison between TRUSTS and a human allocation, a second standard methodology adopted by security agencies. Security agencies refer to this type of experiment as a mass transit full-scale exercise (FSE). An FSE is a training exercise where multiple security agencies analyze the way in which their resources cooperate to secure a specific area while simulating a critical scenario. This scenario typically describes a "high-level" threat, e.g., intelligence reports confirming that a terrorist attack might take place in the LA Metro system. The FSE consists of simulating the response to this threat, such as increasing the number of resources patrolling a train line on a daily basis to improve the quality of the security. Setup: The FSE consisted of patrolling 10 stations of one train line of the LA Metro system for 12 hours. Each station on the train line is composed of three levels (street level, platform level, and mezzanine), except station 1, which is composed of five levels (two more platform levels). The exercise involved multiple security agencies, each participating with a number of resources. Overall, 80 security personnel were involved. These resources were divided into 14 teams, each with different abilities. Five different types of teams were defined, which we will refer to as type 1 to type 5.[4] The first two sorties were used to run the head-to-head comparison. The sorties were run under the same settings: the same number of officers had to cover the 10 stations for a cumulative time of 450 minutes. The first two sorties were run during off-peak times (9h00 to 12h00 and 13h00 to 16h00, respectively), hence the type and the number of riders of the train lines are approximately constant. The purpose of Sortie 3 was to test whether the officers were capable of following TRUSTS schedules for a longer period (900 minutes instead of 450) during peak time. We found out that the officers were able to follow the schedules successfully. Because the purpose of this sortie was unrelated to our comparison, we will focus on the first two sorties in the remainder of this section. Each type of schedule was generated as follows:

TRUSTS schedules: The schedules were generated by (i) instantiating a CT game using the specifics of the FSE discussed earlier; (ii) solving this problem instance; and (iii) sampling a pure strategy to generate the patrol schedule for each of the different resources involved. This pure strategy contained a schedule for each resource.

Manual schedules: The schedules were generated by human expert schedulers of the LASD using a two-step process. First, each station was assigned a coverage duration of 45 minutes (i.e., one-tenth of the time). The idea was to have the officers perform three "observe" actions at each station. Second, the human expert schedulers assigned teams to each station so that each station was covered for exactly 45 minutes. Joint team activities were used six times in six different stations. This simple two-step process was adopted to avoid the cognitive burden involved with leveraging the effectiveness of each team to cover the different stations individually or while coordinating with other teams. In practice, this process was difficult for the human expert schedulers because it involved several discussions and required one entire day of work.

Table 26.6. *Count of joint activities*

	S1	S2	S3	S4	S5	S6	S7	S8	S9	S10
Manual	0	0	0	1	0	1	1	1	1	1
TRUSTS	1	0	0	0	0	2	0	1	1	1

Table 26.7. *This table summarizes the types of experiments we have conducted on our security systems and shows how they help verify the problem formulation and/or that the system is accomplishing its goals*

Evaluation	Problem Formulation Abstraction Solution Implementation	Goal Accomplishments Direct Benefit Indirect Benefit
Algorithmic Cost/Benefit Relative Benefit Human Experiments Operational Record High-Level Evaluations	... .	... ...

Results: We first analyze the type of schedules generated as a result of using either TRUSTS or manual scheduling. Then, we evaluate the results obtained by deploying the schedules during Sorties 1 and 2 and measuring their performance in the real world.

The numbers of individual and joint activities for both the schedules generated during the FSE are shown in Tables 26.6 and 26.7. In both tables, we can see that the number of individual (IA) and joint (JA) activities for both approaches are the same. All joint activities in the TRUSTS schedules are performed by CRM and EK9 teams, i.e., the teams with a positive joint effectiveness. This is similar to the behavior of the manual generated schedules, where joint activities are mostly performed by resources of types 4 and 5 (and once by a team of resources of type 3). The remaining individual activities are performed by resources of types 1–3.

There are two important differences between the two types of schedules. First, TRUSTS sent the most effective type, type 3, to the most important stations because its individual effectiveness is greater than the effectiveness of other teams. This was not seen in the human schedule. Second, the schedules generated using TRUSTS assigned the different teams to cover all the different levels of the different stations, whereas manual schedules did not specify such levels. The reason for this is that human schedulers were not able to reach this level of detail, and thus they preferred to leave the decision of which level to patrol to the teams once they were deployed. In addition, the effort required to generate the schedules using TRUSTS was much lower than the effort required to generate manual schedules, which required one day of work due to its significant cognitive burden. Since typically such patrols would be conducted day in and day out for several days in high-threat periods, the savings of human effort achieved by game-theoretic schedulers are thus very significant.

Each type of security allocation (either manual or game-theoretic based on TRUSTS) was evaluated by security experts. A team of security experts (SEs) was placed at each station for the entire length of the exercise. Their task was to observe and evaluate the officers' patrolling activity during each sortie, and to determine how their behavior was affecting the quality of the security within each station. The SEs did not know what type of schedules (to avoid bias). To translate the observers' observations into a comparable value, each observer was asked to fill out a questionnaire every 30 minutes. The objective was to define a number of key sentences that could help to qualify the way in which the security officers had been patrolling the station in the past 30 minutes. Each questionnaire contained 11 assertions about the level of security within the station. The assertions were defined in collaboration with a team of SEs from the LASD and with social scientists. Each SE had to determine his level of agreement with each assertion, which was defined in the integer interval

$$0, 6,$$

where 0 meant a strong disagreement and 6 meant a strong agreement.

Figures 26.15(a) and 26.15(b) show the results that we obtained. Figure 26.15(a) shows the weighted average agreement obtained for each assertion calculated over all the stations (the average was calculated considering each station's corresponding weight). Figure 26.15(b) shows the average agreement obtained for each station calculated over all the assertions. The error bars in both figures show the standard error of the mean calculated for each specific assertion (in Figure 26.15(a)) and station (in Figure 26.15(b)). As we can see, the differences between some data points of the two approaches do not seem to be statistically significant. A student's t-test confirmed this trend. This is expected, since we were able to collect data for only a few hours of a single day. Nonetheless, we can still acquire some interesting information about the performance of game-theoretic schedules in the field, by analyzing the results that are statistically significant.

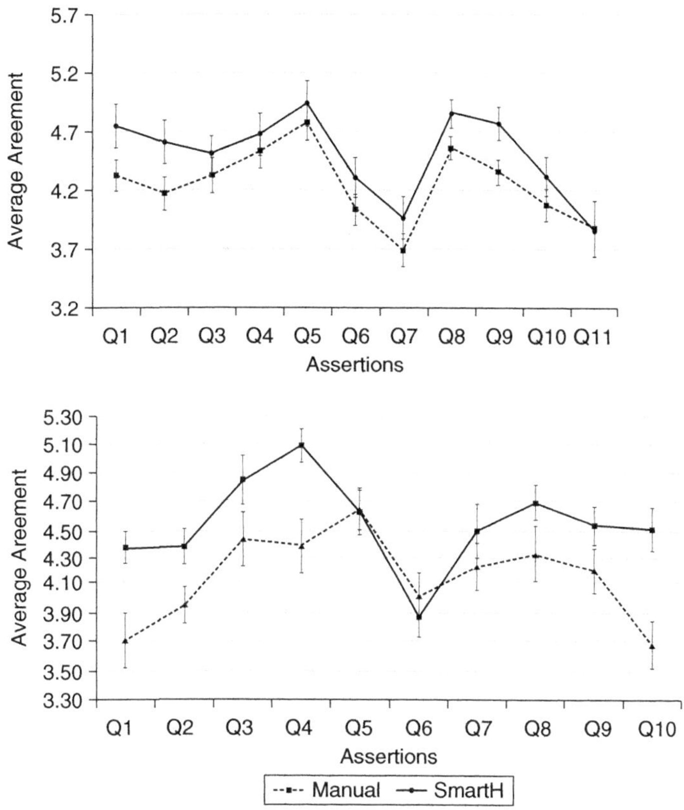

Figure 26.15. Evaluation of the FSE: average agreement over the different questions and stations.

In Figure 26.15(a), we can see that TRUSTS schedules yield a higher level of agreement than manual schedules over all questions. As shown in the figure, the difference is significant only for assertions Q_1, Q_2, Q_8, and Q_9. These four assertions correspond to very general statements about the security at each station that address the efficiency of the schedules, their ability to provide a strong feeling of safety and to allow the officers to patrol each area as much as needed.

Similarly, in Figure 26.15(b), we can see that the average agreement is higher for TRUSTS schedules over manual schedules for stations S_1, S_2, S_3, S_4, S_8, S_9, and S_{10}. Some of these stations (S_1, S_8, and S_9) are the ones assigned a higher set of payoffs, as discussed earlier. Hence, they correspond to the ones given a higher coverage by TRUSTS.

These results indicate that game-theoretic schedules were more effective than manual schedules. By analyzing the differences between the schedules, we can infer two key causes. First, as discussed earlier, manual schedules were generated by leaving the decision of which level of a station to patrol to each deployed team.

The officers were not able to coordinate over the different levels, and therefore they ended up patrolling the same levels. Second, TRUSTS produced a schedule that more effectively allocated resources of type 3, i.e., the team with the highest effectiveness for covering each target. More specifically, the resources of type 3 patrolled *all* the most important stations at key levels. In contrast, manual schedules assigned the same type of resources, without accounting for their effectiveness. This made an impact on the security evaluators, which considered the game-theoretic allocation more effective than the manual allocation because it leveraged the abilities of the resources in a way that human experts did not.

26.5 Goals for Security Decision Support Systems

Any security system can be evaluated using a wide variety of metrics, including both costs and benefits. For the purposes of analysis, we divide benefits into two categories: *direct* and *indirect*. Direct benefits speak to benefits that may be measured, such as

reduced security costs,
attacks prevented or mitigated during execution,
increased numbers of attackers caught,
reduced damage from successful attacks.
In contrast, indirect benefits include
attacks prevented through deterrence,
increased attacker planning time,
increased requirements for a successful attack,
improved public perceptions of security, or
improved (qualitative) assessments of security by experts.

Regardless of how such benefits are partitioned, some are easier to directly measure than others. However, all of them speak to the idea of defender utility: Given a finite amount of resources, how can the defender maximize security per dollar spent? As discussed previously, no security system can be 100% effective – there is always a chance that, given enough resources and planning, an adversary may compromise a system. A system may, however, require higher amounts of equipment, manpower, and/or planning in order to compromise it, relative to having the system disabled. Thus the question becomes not "is the system effective?" but "is system A more effective than system B?" or "how much does the system improve defensive capabilities per dollar spent?" We next elaborate on two of these themes: utility and deterrence.

26.5.1 *Security per Dollar*

Evaluating security systems requires a full cost-benefit analysis that takes a comprehensive view of both the costs and the benefits of the system. There is always additional defensive capability that could be purchased – another guard could be hired, one more piece of technology could be installed or upgraded, etc. Of course, additional security typically comes with diminishing returns, and there is a policy decision to make about how many resources to devote to security. Improving the efficiency of security can be used to either (1) increase defensive abilities on a fixed budget, and/or (2) decrease expenditure for a fixed defensive capability. Such increases in efficiency are the primary goal of the decision support systems discussed previously.

Recall that Figure 26.2 showed how the defender's utility changed with different numbers of K9 units. Such a graph of the trade-off between defensive capability and cost can allow policy makers to more easily see the trade-off between money and security when using the ARMOR system. More important, one can easily see how much ARMOR helps increase security relative to uniform random (or other randomization strategies), which is a critical factor when deciding whether to implement a given security technique. It directly follows that countermeasure investments with the highest cost-effectiveness ratio should be implemented first, assuming that there is no interdependence between proposed measures (Edmunds & Wheeler, 2009). Determining exactly how much safer ARMOR makes the airport is a very difficult question, but as we will discuss in the following sections, arguing that ARMOR improves safety on a per-office basis (and thus "safety per dollar" spent) is not difficult.

26.5.2 *Threat Deterrence*

A key goal of many security systems is *deterrence*: an effective system will not only identify and prevent successful attacks, but will also dissuade potential attackers. Unfortunately, it is typically impossible to directly measure the deterrence effect. To measure deterrence directly, one needs to know how many attacks did not occur due to security, a generally unmeasurable counterfactual.

In all cases relevant to this chapter, we assume that the attacker is adaptive and does not act blindly, but takes the defender's actions into account. It is only through this adaptability that defensive measures could deter an attacker. For instance, one could model the uncertainty due to chance or attacker choices via an *event tree model* or a *decision tree model* (Edmunds & Wheeler, 2009). We choose to focus on game-theoretic methods as they have strong mathematical underpinnings and are ideally suited to reasoning about adaptable, rational actors. Game models naturally factor in deterrence, in that an intelligent attacker is assumed to switch to a different strategy if the defender uses security resources to make the original strategy undesirable.

How to best understand and accurately model deterrence is a topic of current research in the threat assessment community (Taquechel, 2010). For instance, for a threat (i.e., a defensive measure) to be an effective deterrent, it must be *credible* in the eyes of the attacker (Berejikian, 2002), which may involve complex *signaling* effects between the parties (Gerson & Boyars, 2007). As mentioned previously, ARMOR has been well publicized in the popular press. While many other less-public security measures are employed at LAX, ARMOR may lead a potential attacker to decide not to attack LAX. Another possible outcome of ARMOR's publicity is that an attacker selects a more vulnerable target. While this is a win from the standpoint of LAX, it may or may not be an improvement for Los Angeles, the State of California, or the United States, depending on the target selected. However, if LAX is not targeted due to deterrence measures, the eventual target is presumably less attractive to the attackers, resulting in a net increase in the defender's utility.

26.6 Summary of Evaluation Types

This section of this chapter ties together the discussion of the previous sections, showing how the different evaluations performed on ARMOR, IRIS, and PROTECT help to verify that the goals of the security system are achieved.

26.6.1 *Model-Based/Algorithmic*

The first type of evaluations (and the most natural for computer scientists) is based on analysis of the model and the underlying algorithms. Given assumptions about the attacker (e.g., the payoff matrix is known), game-theoretic tools can be used to determine the attacker's expected payoff. Additionally, deterrence can be measured by including a "stay home" action, returning neutral reward. These types of analyses speak to such concerns.

First, the sensitivity of the solution method can be examined. No model will be perfect, but a solution to that model would ideally be robust to small imperfections in the model. This type of analysis is able to speak to the potential impact of different kinds of abstraction and modeling error on the final solution.

Second, the utility of the solution method can be directly estimated from algorithmic evaluations. For instance, the attacker resources vs. expected attacker utility and the defender resources vs. expected defender utility can both be estimated. To the extent that the model is correct, these utilities can be determined exactly, providing an important tool to both policy makers and security professionals when deciding what defensive measures to implement.

Finally, the underlying algorithms can be evaluated, both to prove correctness (or a close approximation), and to evaluate the computational effort required to compute a solution for problems with different sizes and properties.

26.6.2 *Cost-Benefit Analysis*

Another type of analysis, which can combine algorithmic results and those from the implementation, is a full cost-benefit analysis of the deployed system. For instance, the cost of implementation and maintenance on the system can be directly measured. The benefit of the system can be measured both by changes in the defender's utility (e.g., Figure 26.2) and in terms of less-tangible factors (such as such as quantifying increases in travel time or a decrease in civil liberties).

A system-wide cost-benefit analysis can help security professionals decide where to allocate finite resources in order to best protect the entire area by changing staffing levels and/or implementing new security measures. Additionally, such studies can provide important information to other sites that may consider deploying a new security measure, such as ARMOR. The ultimate goal of this analysis is to answer the questions "how useful is this defensive measure?" and "how does utility change for different levels of defender/attacker resources?"

26.6.3 *Relative Benefit*

Measuring the *relative* benefit of defensive measures is similar to measuring the absolute utility, as described in the previous section. However, an important difference is that such measurements may be less brittle to model inaccuracies. For instance, when deciding whether to implement ARMOR, a site may consider the absolute utility of ARMOR using a cost-benefit analysis, which depends on the assumptions about the attacker. A relatively simple analysis could instead look at how utility will be changed by a proposed security method.

For instance, Figure 26.1 shows how the utility of ARMOR compares to different scheduling strategies. A relative benefit analysis makes it easier to answer the question "should this measure be implemented?" and may be particularly useful when discussing security with professionals who are leery of implementing decision-theoretic methods.

26.6.4 *Human Behavioral Experiments*

Human psychological studies can help to better simulate attackers in the real world (e.g., Figure 26.9). Evaluations on an abstract version of the game may test base assumptions, or a detailed rendition of the target in a virtual reality setting with physiological stress factors could test situated behavior. Human subjects may allow researchers to better simulate the actions of attackers, who may not be fully rational. Human tests suffer from the fact that participants are not drawn from the same population as the actual attackers (i.e., undergraduate college students).

In some situations, it may be possible to conduct realistic human studies in the true setting, i.e., employ a "red team." Such tests can use qualified security personnel to probe security defenses and provide realistic information in life-like situations using the true defenses (including those that are not visible). However, such a test is very difficult to conduct as some security must be alerted (so that the team is not endangered) while remaining realistic, the tests are often not repeatable, and a single test is likely unrepresentative. To the extent that such tests are feasible, they would speak to both the implementation of the system and how well it accomplishes its direct goals.

26.6.5 *Operational Record*

Analysis of the operational record of a system provides confirmation that the system works, but is not as useful as one may initially think. Ideally, the system would be enabled and disabled for randomized periods of time, allowing for a careful study of the system. Unfortunately, this is typically infeasible.

Consider the ARMOR system: that arrests result from the system shows that the system does allow rule breakers to be caught. However, comparing the number of arrests before and after ARMOR may or may not be useful. For instance, if the number of arrests increases, this could be an indication that ARMOR is more effective than the previous system. However, it could also mean that more people are violating the rules during this time period, and in fact a lower number of criminals is caught. In the opposite case, a lower number of arrests could mean that deterrence from ARMOR is convincing more criminals to stay away. Or, a lower number of arrests could mean that more people are circumventing the system. Without knowing the numbers of criminals that are not caught, it is impossible to tell how well a particular piece of security is performing.

An additional complication is that, thankfully, most security threats are very low-frequency events. This means that data are collected relatively infrequently, making analysis of the operational record more sparse than in other (non-security) deployed applications.

26.6.6 *High-Level Evaluations*

While computer scientists traditionally prioritize precise, repeatable studies, this is not always possible in the security community; computer scientists are used to quantitative evaluations in controlled studies, whereas security specialists are more accepting of qualitative metrics on deployed systems. For instance, Lazarick (1999) summarized a multiyear airport security initiative by the FAA where the highest-ranked evaluation methodology (of seven) relied on averaging *qualitative* expert evaluations.

Assuming the high-level evaluation is appropriately done, it may address the abstraction and implementation questions arising from problem formulation, as well as direct and indirect goals. While these studies generally do not produce quantitative numbers that can be used to determine the utility of a security measure, they may uncover flaws in that measure, or in the security of the entire system as a whole. If no such flaws are found, the study may support the hypothesis that the security measure in question does improve overall security. As with the other types of evaluation, the goal is not to prove that the system works, but to provide evidence toward this effect.

26.7 Related Work

Security is a complex research area, spanning many disciplines, and policy evaluation is a persistent challenge. Many security applications are evaluated primarily on the basis of theoretical results; situated evaluations and even laboratory experiments with human subjects are relatively rare. In addition, existing general methodologies for risk and security evaluation often rely heavily on expert opinions and qualitative evaluations.

Lazarick (1999) is a representative example that relies heavily on expert opinions. In the study, seven tools/approaches used to evaluate airport security were compared as part of a competitive bidding process. At the end of the multiyear security initiative, the highest-ranked evaluation methodology relied on averaging qualitative expert evaluations.

A second example of a high-level methodology for per-facility and regional risk assessment is described by Baker (2005). The methodology relies heavily on expert opinions and evaluations from local technical staff/experts, similar to Lazarick (1999). The three key questions in the methodology are: (1) Based on the vulnerabilities identified, what is the likelihood that the system will fail? (2) What are the consequences of such failure (e.g., cost or lives)? (3) Are these consequences acceptable? Such an approach enumerates all vulnerabilities and threats in an attempt to determine what should (or must) be improved. There is no quantitative framework for evaluating risk.

Many in the risk-analysis community have recently argued for game theory as a paradigm for security evaluation, with the major advantage that it explicitly models the adaptive behavior of an intelligent adversary. Cox (2008) provides a detailed discussion of the common "Risk = Threat × Vulnerability × Consequence" model, including analysis of an example use of the model. Several arguments are raised as weaknesses of this approach, including (1) the values are fundamentally subjective; (2) rankings of risk are often used, but are insufficient; (3) there are mathematical difficulties with the equation, including dependencies between the multiplied terms; and (4) the model does not account for adaptive, intelligent attackers. One of the main recommendations of the paper is to adopt more intelligent models of attacker behavior, instead of more simple, static risk estimates.

Even in situations where the adversary may not be fully strategic, Shan and Zhuang (2013) demonstrate the usefulness of game-theoretic methods. They compare game models to non-game models in relation to the defender's belief of both a nonstrategic and fully strategic terrorist, and find that the game-theoretic model is preferred even when the probability of a nonstrategic terrorist is greater than 50%. In addition, game-theoretic methods for security have explored investment in technology as an option for both the attacker and the defender (Jose & Zhuang, 2013). These models have been tested in simulation, but have yet to be evaluated in deployed systems.

Bier and colleagues (2009) provide a high-level discussion of game-theoretic analysis in security applications and their limitations. The main argument is that the *adaptive* nature of the terrorist threat leads to many problems with static models – such models may overstate the protective value of a policy by not anticipating an attacker's options to circumvent the policy. They explicitly propose using quantitative risk analysis to provide probability/consequence numbers for game-theoretic analysis.

Bier (2007)performs a theoretical analysis of the implications of a Bayesian Stackelberg security game very similar to the one solved by ARMOR, although most of the analysis assumes that the defender does *not* know the attacker's payoffs. The primary goal is to examine intuitive implications of the model, such as the need to leave targets uncovered in some cases so as not to drive attackers toward more valuable targets. There are no "real-world" evaluations of the model. Other work (Bier, Haphuriwat, Menoyo, Zimmerman, & Culpen,

2008) considers high-level budget allocation (e.g., to large metropolitan areas). While the study uses real data, its focus is not model evaluation, but the implications resulting from the model.

Game theory does have much to offer in our view, but should not be considered a panacea for security evaluation. One difficulty is that human behavior often does not correspond exactly to game-theoretic predictions in controlled studies. Weibull(2004) describes many of the complex issues associated with testing game-theoretic predictions in a laboratory setting, including a discussion of the ongoing argument regarding whether people typically play the Nash equilibrium (a point discussed at length in the literature, such as in Erev, Roth, Slonim, & Barron, 2002). This is one reason we believe behavioral studies with humans are an important element for security system evaluation.

Many of the issues we describe in acquiring useful real-world data for evaluation purposes are mirrored in other types of domains. Blundell and Costa-Dias (2009) describe approaches for experimental design and analysis of policy proposals in microeconomics, where data are limited in many of the same ways: it is often not possible to run controlled experiments and many desired data cannot be observed. They describe several classes of statistical methods for these cases, some of which may be valuable in the security setting (though data sensitivity and sparse observations pose significant additional challenges). In truth, it is often hard to evaluate complex deployed systems in general – in our field a test of the prototype often suffices (c.f., Scerri, Goten, Fudge, Owens, & Sycara, 2008).

Jacobson and colleagues (2005) describe a deployed model for screening airline passenger baggage. The model includes detailed information regarding estimated costs of many aspects of the screening process, including variables for probability of attack and cost of a failed detection, but these are difficult to estimate and left to other security experts to determine. One particularly interesting aspect of the approach is that they perform sensitivity analysis on the model in order to assess the effect of different values on the overall decisions. Unfortunately, the authors have little to say about actually setting the input values to their model; in fact, no empirical data validate their screening approach.

Kearns and Ortiz (2003) introduce algorithms for a class of "interdependent" security games, where the security investment of one player has a positive externality and increases the security of other players. They run the algorithms on data from the airline domain but do not directly evaluate their approach, instead looking at properties of the equilibrium solution and considering the broad insight that this solution yields regarding the benefits of subsidizing security in such games.

Last, the field of *fraud detection* (Kou & Lu, 2004), encompassing credit card fraud, computer intrusion, and telecommunications fraud, is also related. Similar to the physical security problem, data are difficult to access, researchers often do not share techniques, and deterrence is difficult (or impossible) to measure. Significant differences include:

1. Humans can often classify (in retrospect) false positives and false negatives, allowing researchers to accurately evaluate strategies.
2. Companies have significant amounts of data regarding known attacks, even if they do not typically share the data outside the company. Some datasets do exist for common comparisons (c.f., the 1998 DARPA Intrusion Detection Evaluation data[5]).
3. The frequency of such attacks is much higher than physical terrorist attacks, providing significant training/evaluation data.
4. Defenders can evaluate multiple strategies (e.g., classifiers) on real-time data, whereas physical security may employ and evaluate only one strategy at a time.

26.8 Conclusions

This chapter has discussed existing evaluations of the deployed ARMOR, IRIS, PROTECT, and TRUSTS systems. These results show how such systems can be reasonably evaluated, and in particular show that these systems work well in theory and that security experts agree they are beneficial. In many ways, this level of evaluation goes beyond what is typical of applications, even those deployed in real-world settings. Overall, we find strong evidence to support the use of ARMOR and PROTECT over previous methods.

Another point worth stressing is that ARMOR, IRIS, and PROTECT are relatively easy to use. ARMOR in particular has been instrumental in aiding police forces to efficiently and more conveniently generate schedules to deploy more and more units. For example, consider a situation when only two canines need to be scheduled for two hours each over any of the seven terminals. Each canine could be assigned to any of the seven terminals each hour, making the search space as large as 74 = 2,401 combinations. This search space grows exponentially with the number of canines and the number of hours for which the schedule needs to be generated, making it impractical for human schedulers. Thus, ARMOR has played a significant role in reducing, if not completely eliminating, the work of officers who manually construct patrolling schedules. Additionally, the use of ARMOR has also made it possible for the security officers to update the generated schedules, in case more resources become available or new constraints need to be incorporated. Furthermore, even though ARMOR was designed as a mixed-initiative system, users have chosen not to modify ARMOR schedules in practice, which suggests that the output schedules are indeed high quality, and that domain experts have not chosen to "tweak" the system's decisions. These added benefits have themselves been a contributing factor toward the continued use of schedules generated by ARMOR. Most important, when considering the cost of implementing a decision support system, it is important to consider ways in which the system may actually reduce security costs.

While none of the evaluation tests presented can calculate a measure's utility with absolute accuracy, understanding what each test *can* provide will help evaluators better understand what tests *should* be run on deployed systems. The goal of such tests will always be to provide better understanding to the "customer," be it researchers, users, or policy makers. By running multiple types of tests, utility (the primary quantity) can be approximated with increasing reliability.

At a higher level, thorough cost-benefit analyses can provide information to policy makers at the inter-domain level. For instance, consider the following example from Tengs and Graham (1996):

> To regulate the flammability of children's clothing we spend $1.5 million per year of life saved, while some 30% of those children live in homes without smoke alarms, an investment that costs about $200,000 per year of life saved.

While such a comparative cost-benefit analysis is beyond the scope of the current study, these statistics show how such an analysis can be used to compare how effective measures are across very different domains, and could be used to compare different proposed security measures.

In the future we plan to use this framework to help decide which evaluation tests are most important to determining the utility of a deployed, security-focused decision support system. Additionally, we intend to continue collaborating with security experts to determine if our evaluations are sufficiently general to cover all existing types of security tests.

Currently, the majority of our evaluations has been conducted on both the ARMOR and PROTECT systems. However, we intend to continue testing IRIS and other newly developed domains, to both better evaluate the domains, and to attempt to discover or improve evaluation techniques in the context of deployed systems.

Finally, we expect that additional data from TRUSTS, in which we have much more data on attacker data, will further improve our ability to measure the ground truth and determine how deterrence effects the number and determination of attackers.

Acknowledgments

The authors would like to thank Manish Jain, James Pita, and Vicki Bier for helpful comments and suggestions, Sergeant Jim Holcomb for supplying ARMOR-related arrest data, and First Sargent Cruz for discussing pre-ARMOR scheduling procedures. This research was supported in part by the U.S. Department of Homeland Security through the National Center for Risk and Economic Analysis of Terrorism Events (CREATE) under grant number 2007-ST-061-000001. However, any opinions, findings, and conclusions or recommendations in this document are those of the authors and do not necessarily reflect views of the U.S. Department of Homeland Security.

This research has taken place in part at the Intelligent Robot Learning (IRL) Lab, Washington State University. IRL research is supported by grants from AFRL FA8750-14-1-0069, AFRL FA8750-14-1-0070, NSF IIS-1149917, NSF IIS-1319412, and USDA 2014-67021-22174.

NOTES

1. This report is a public version of the restricted report GAO-09-53SU.
2. As shown in Table 26.1, each day of patrol corresponds to a two-day test where GT schedules were tested on the first day and UR schedules were tested on the second, both at identical times.
3. GT schedules also led to two arrests on Day 6. This is why the patrol lasted only 10 minutes. The exercise was divided into three different "sorties," each consisting of three hours of patrolling and one hour of debriefing. Human-generated schedules were used during the first sortie while TRUSTS schedules were used during the second.
4. The real name of each type is omitted as requested by the agencies participating in the FSE.
5. For data and program details, see www.ll.mit.edu/mission/ communications/ist/ index.html.

REFERENCES

Baker, G. H. (2005). A vulnerability assessment methodology for critical infrastructure sites. In *DHS symposium: R and D partnerships in homeland security*.

Berejikian, J. (2002). A cognitive theory of deterrence. *Journal of Peace Research*, *39*, 165–183.

Bier, V. M. (2007). Choosing what to protect. *Risk Analysis*, *27*(3), 607–620.

Bier, V. M., Cox, J. L. A., & Azaiez, M. N. (2009). Why both game theory and reliability theory are important in defending infrastructure against intelligent attacks. In *Game theoretic risk analysis and security threats*, vol. 128. Springer US.

Bier, V. M., Haphuriwat, N., Menoyo, J., Zimmerman, R., & Culpen. A. M. (2008). Optimal resource allocation for defense of targets based on differing measures of attractiveness. *Risk Analysis*, *28*(3), 763–770.

Bland, M. J., & Kerry, S. M. (1998). Weighted comparison of means. *BMJ: British Medical Journal*, *316*, 125–129.

Blundell, R., & Costa-Dias, M. (2009). Alternative approaches to evaluation in empirical microeconomics. *Journal of Human Resources.*

Conitzer, V., & Sandholm, T. (2006). Computing the optimal strategy to commit to. In *Proceedings of EC.*

Cox, J. L. A. (2008). Some limitations of "risk = threat x vulnerability x consequence" for risk analysis of terrorist attacks. *Risk Analysis, 28*(6), 1749–1761.

Cruz, F. S. (2009, August 20). Personal communication.

Doane, C., & DiRezno, III, J. (2012). The port resilience operational/tactical enforcement to combat terrorism model. *Coast Guard Outlook 2012 Summer Edition.*

Edmunds, T., & Wheeler, R. (2009). Setting priorities. In S. M. Maurer (Ed.), *WMD terrorism* (chapter 7, pp. 191–209). Cambridge, MA: MIT Press.

Erev, I., Roth, A. E., Slonim, R. L., & Barron, G. (2002). Predictive value and usefulness of game theoretic models. *International Journal of Forecasting, 18*(3), 359–368.

Fave, F. M. D., Brown, M., Shieh, E., Zhang, C., Jiang, A. X., Rosoff, H., Tambe, M., & Sullivan, J. P. (2014). Security games in the field: Deployments on a transit system. *Lecture Notes in Artificial Intelligence (LNAI).*

Fave, F. M. D., Jiang, A. X., Yin, Z., Zhang, C., Tambe, M., Kraus, S., & Sullivan. J. (2014). Game-theoretic security patrolling with dynamic execution uncertainty and a case study on a real transit system. *Journal of Artificial Intelligence Research, 50,* 321–367.

Gerson, M., & Boyars, J. (2007). The future of U.S. deterrence: Constructing effective strategies to deter state and non-state actors. www.cna.org/ documents/D0017171 .A2.pdf.

Goldberg, L. R., Kercheval, A. N., & Lee, K. (2005). T-statistics for weighted means in credit risk modeling. *Journal of Risk Finance, 6*(4), 349–365.

Jacobson, S. H., Karnai, T., & Kobza, J. E. (2005). Assessing the impact of deterrence on aviation checked baggage screening strategies. *International Journal of Risk Assessment and Management, 5*(1), 1–15.

Jiang, A. X., Yin, Z., Zhang, C., Tambe, M., & Kraus, S. (2013). Game-theoretic randomization for security patrolling with dynamic execution uncertainty. In *Proceedings of the International Conference on Autonomous Agents and Multiagent Systems (AAMAS).*

Jose, V. R. R., & Zhuang, J. (2013). Technology adoption, accumulation, and competition in multi-period attacker-defender games. *Military Operations Research, 18*(2), 33–47.

Kearns, M., & Ortiz, L. E. (2003). Algorithms for interdependent security games. In *Neural information processing systems (NIPS).*

Kiekintveld, C., Jain, M., Tsai, J., Pita, J., Ordóñez, F., & Tambe, M. (2009). Computing optimal randomized resource allocations for massive security games. In *Proceedings of the International Conference on Autonomous Agents and Multiagent Systems (AAMAS).*

Kiekintveld, C., Marecki, J., & Tambe, M. (2011). Approximation methods for infinite Bayesian Stackelberg games: Modeling distributional payoff uncertainty. In *Proceedings of the International Conference on Autonomous Agents and Multiagent Systems (AAMAS).*

Kiekintveld, C., & Wellman, M. (2008). Selecting strategies using empirical game models: An experimental analysis of meta-strategies. In *Proceedings of the International Conference on Autonomous Agents and Multiagent Systems (AAMAS)*.

Kou, Y., Lu, C., Sinvongwattana, S., &Huang, Y. (2004). Survey of fraud detection techniques. In *Proceedings of IEEE Networking*.

Lazarick, R. (1999). Airport vulnerability assessment – a methodology evaluation. In *Proceedings of 33rd IEEE International Carnahan Conference on Security Technology*.

McKelvey, R. D., & Palfrey, T. R. (1995). Quantal response equilibria for normal form games. *Games and Economic Behavior, 10*(1), 6–38.

Murr, A. (2007, September). Random checks. In *Newsweek National News*. www.newsweek.com/id/43401.

Paruchuri, P., Pearce, J. P., Marecki, J., Tambe, M., Ordóñez, F., & Kraus, S. (2008). Playing games with security: An efficient exact algorithm for Bayesian Stackelberg games. In *Proceedings of the International Conference on Autonomous Agents and Multiagent Systems (AAMAS)*.

Pita, J., Jain, M., Tambe, M., Ordóñez, F., Kraus, S., & Magori-Cohen, R. (2009). Effective solutions for real-world Stackelberg games: When agents must deal with human uncertainties. In *Proceedings of the International Conference on Autonomous Agents and Multiagent Systems (AAMAS)*.

Pita, J., Jain, M., Western, C., Portway, C., Tambe, M., Ordóñez, F., Kraus, S., & Paruchuri, P. (2008). Deployed ARMOR protection: The application of a game theoretic model for security at the Los Angeles International Airport. In *Proceedings of the International Conference on Autonomous Agents and Multiagent Systems (AAMAS)*.

Rogers, B. W., Palfrey, T. R., & Camerer., C. F. (2009). Heterogeneous quantal response equilibrium and cognitive hierarchies. *Journal of Economic Theory, 144*, 1440–1467.

Scerri, P., Goten, T. V., Fudge, J., Owens, S., & Sycara, K. (2008). Transitioning multiagent technology to UAV applications. In *Proceedings of the International Conference on Autonomous Agents and Multiagent Systems (AAMAS) (Industry Track)*.

Shan, X., & Zhuang, J. (2013). Hybrid defensive resource allocations in the face of partially strategic attackers in a sequential defender-attacker game. *European Journal of Operational Research, 228*(1), 262–272.

Shieh, E., An, B., Yang, R., Tambe, M., Baldwin, C., DiRenzo, J., Maule, B., & Meyer, G. (2012). PROTECT: A deployed game theoretic system to protect the ports of the United States. In *Proceedings of the International Conference on Autonomous Agents and Multiagent Systems (AAMAS)*.

Stevens, D., Hamilton, T., Schaffer, M., Dunham-Scott, D. Medby, J. J., Chan, E. W., Gibson, J., Eisman, M., Mesic, R., Kelly, C. T., Kim, J., LaTourrette, T., & Riley, K. J. (2009). Implementing security improvement options at Los Angeles International Airport. www.rand.org/pubs/documented_ briefings/2006/RAND_ DB499-1.pdf.

Taquechel, E. F. (2010, March). Validation of rational deterrence theory: Analysis of U.S. government and adversary risk propensity and relative emphasis on gain or loss. Master's thesis, Naval Postgraduate School.

Tengs, T. O., & Graham, J. D. (1996). Risks, costs, and lives saved: Getting better results from regulation. In *The opportunity costs of haphazard social investments in lifesaving*. Washington, DC: American Enterprise Institute.

Tsai, J., Rathi, S., Kiekintveld, C., Ordóñez, F., & Tambe, M. (2009). IRIS: Tools for strategic security allocation in transportation networks. In *Proceedings of the International Conference on Autonomous Agents and Multiagent Systems (AAMAS) (Industry Track)*.

T. U. S. G. A. O. GAO. (2009, January). Aviation security: Federal air marshal service has taken actions to fulfill its core mission and address workforce issues, but additional actions are needed to improve workforce survey. GAO-09-273.

Weibull, J. (2004). Testing game theory. In S. Huck (Ed.), *Advances in understanding strategic behavior: Game theory, experiments and bounded rationality* (pp. 85–104). Palgrave MacMillan.

Yang, R., Kiekintveld, C., Ordóñez, F., Tambe, M., & John, R. (2011). Improving resource allocation strategy against human adversaries in security games. In *Proceedings of the 22nd International Joint Conference on Artificial Intelligence (IJCAI)* (pp. 458–464).

Yang, R., Tambe, M., & Ordóñez, F. (2012). Computing optimal strategy against quantal response in security games. In *Proceedings of the International Conference on Autonomous Agents and Multiagent Systems (AAMAS)*.

Yin, Z., & Tambe, M. (2012). A unified method for handling discrete and continuous uncertainty in Bayesian Stackelberg games. In *Proceedings of the International Conference on Autonomous Agents and Multiagent Systems (AAMAS)*.

27

Homeland Security Resource Allocation Games: Considering Partially Strategic Attackers and Equity

Xiaojun Shan and Jun Zhuang

27.1 Introduction

The terrorist attacks of September 11, 2001, have drawn huge amounts of money into homeland security in the United States. For instance, Figure 27.1(a) shows the annual budget requests for the U.S. Department of Homeland Security (DHS) since it was formed in 2002. However, the effectiveness of such large expenditures is unknown. "Pork-barrel politics," where political factors are driving resources toward low-risk targets, is used by critics of how homeland security resources are utilized (e.g., de Rugy, 2005). Toward risk-based optimization of resource allocation, DHS has employed a three-step method, which is based on perceived risk, in directing grant distribution since 2006 (U.S. Government Accountability Office, 2008). Construction of a comprehensive risk management framework would mandate more theoretical and practical work.

Operations research (OR) techniques have proven valuable in developing counterterrorism strategies. Harris (2004) highlighted how probabilistic risk analysis, randomization, and game theory could play a role in defending against attacks initiated by terrorists. Brown and colleagues (2005) built a two-sided optimization model whose goal is to optimally pre-localize defensive missile platforms in the face of adaptive attackers. By modeling and data analysis, Kaplan and Kress (2005) studied the effectiveness of a suicide-bomber-detector scheme in decreasing fatalities resulted from suicide bombing. Brown

This research was partially supported by the U.S. Department of Homeland Security (DHS) through the National Center for Risk and Economic Analysis of Terrorism Events (CREATE) under award number 2010-ST-061-RE0001. This research was also partially supported by the U.S. National Science Foundation (NSF) under award numbers 1200899 and 1334930. However, any opinions, findings, and conclusions or recommendations in this document are those of the authors and do not necessarily reflect views of the DHS, CREATE, or NSF.

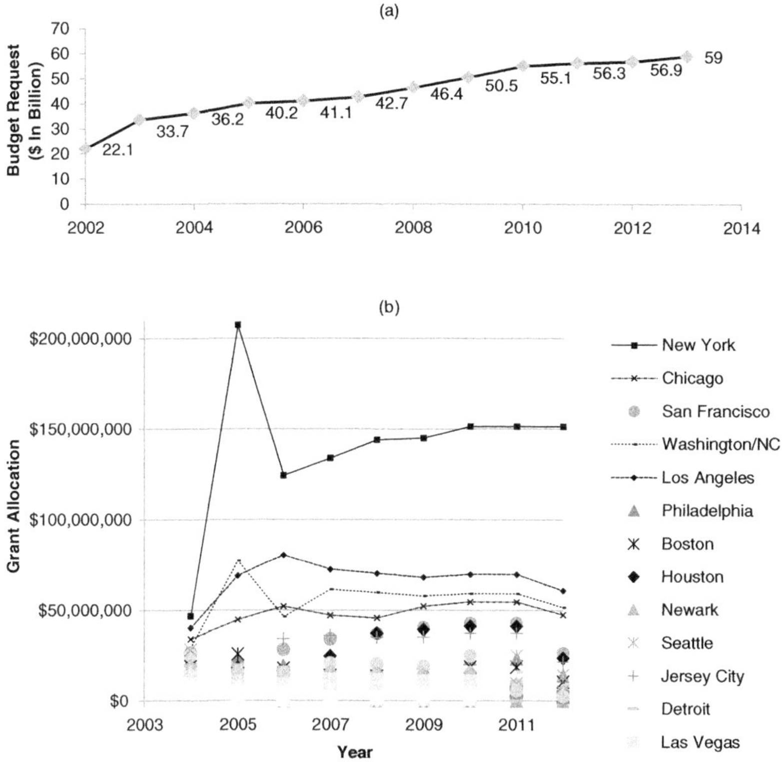

Figure 27.1. (a) Budget requests for the U.S. Department of Homeland Security from FY 2002 to FY 2013 and (b) UASI grant allocations to U.S. urban areas from 2004 to 2012.
Source: U.S. Deparwtment of Homeland Security (2013).

and colleagues (2006) constructed innovative bi-level and tri-level models to explore optimal strategies to defending critical infrastructure. Lin and colleagues (2009) studied a M/G/1 queue to develop optimal scheduling policies to be effectively used by an antiterrorist surveillance system. Wein (2009) discussed how mathematical modeling can assist with policy recommendations. With queuing theory and Markov population processes, Kaplan (2010) investigated the optimal strategies to be employed by undercover intelligence agents in order to infiltrate and stop terror plots. Jose and Zhuang (2013) considered technologies adoption of an attacker, which is repeatedly interacting with the same defender using sequential games with uncertainty. One main contribution of OR methods to tackling homeland-security problems is to understand the attacker's objectives. This could be achieved by building "value models" (Keeney, 2007). For instance, Wang and Bier (2010, 2011) built a multiattribute utility function to model terrorist's preference behaviors. Different approaches of weighting multiple attributes have been compared by Stillwell and colleagues (1987).

With OR methodologies, especially game theory and optimization techniques, we systematically study defense allocation. To our knowledge, we are one of the first to study defense allocations against a partially strategic attacker and to incorporate equity consideration in defense allocations.

The following notation has been used throughout this chapter. First, N represents the number of top valuable targets attracting a non-strategic attacker. Second, L^*, L^-, and $L^\wedge$ are expected losses for the following three scenarios: (1) the defender knows the attacker characteristic (probability of being strategic); (2) the defender believes the attacker is strategic while the attacker could be non-strategic or partially strategic; and (3) the defender believes that the attacker is non-strategic while the attacker could be strategic or partially strategic, respectively.

The rest of this chapter is organized as follows: Section 27.2 reviews the literature on strategic and non-strategic threats and equity in resource allocation. Section 27.3 introduces a hybrid model originally developed by Shan and Zhuang (2013b) where a centralized defender distributes limited resources to protect many targets in order to minimize total expected loss facing an attacker, who could be fully strategic, partially strategic or fully non-strategic. Section 27.4 introduces an optimization model originally developed by Shan and Zhuang (2013a), where a centralized defender uses both risk-based optimization and predetermined equity allocation schemes in distributing defense resources to protect multiple targets. Section 27.5 draws conclusions and provides recommendations on the basis of the findings reported in this chapter.

27.2 Literature Review

Homeland security problems have inspired many mathematical and economic models (Atkinson et al., 2008; Harris, 2004; Rosoff & von Winterfeldt, 2007). In particular, to deal with the adaptiveness of strategic attackers, several intelligent risk analysis models (Ezell et al., 2010; Parnell et al., 2010; Rio Insua et al., 2009) and game-theoretic models (Bier et al., 2008; Golalikhani & Zhuang, 2011; Hausken, 2002; Jiang et al., 2013; Nikoofal & Zhuang, 2012; Powell, 2009; Zhuang & Bier, 2007) have been built to investigate problems related to defense allocations. One important issue in resource allocation is the large amount of uncertainty surrounding cost, which Dillon and colleagues (2002) have studied in the context of technology investment. Besides, game theory has been used in studying other aspects of counterterrorism. For instance, Shan and Zhuang (2014b) constructed a complex four-player game to study the effectiveness of subsidization in mitigating risks posed by weapons of mass destruction (WMD). Shan and Zhuang (2014a) built a three-stage game to model credibility of retaliation in deterring the smuggling of nuclear weapons. Xu and colleagues (2015) modeled and studied mitigation of supply chain disruption in a defender-attacker game.

27.2.1 *Strategic and Non-strategic Threats*

Two main types of threats plague human society: natural disasters or non-intentional threats and terrorism or intentional threats. One main distinction is that terrorists are adaptive and strategic whereas natural disasters are not. As a consequence, the government's optimal protective strategies against natural disasters could have a dramatic difference with those tackling terrorism (Levitin & Hausken, 2009; Zhuang & Bier, 2007). The government's understanding of whether and how much the terrorist is strategic usually comes from intelligence (Kress & Szechtman, 2009). Specifically, Kaplan and colleagues (2010) showed that if the government does not have adequate intelligence information, strategic insurgents could relatively easily escape from attacks by the government.

Players' rationality is one typical assumption of game-theoretic models. In practice, humans are known not to behave fully rationally due to facts such as cognitive limitations of decision makers. Simon (1957, 1982) proposed bounded rationality to account for it. Gigerenzer and Selten (2002) further discuss a wide spectrum of methods to model human decision making and a number of important topics, including judgment heuristics, risky choice, and framing effects (Kahneman, 2003). The concept of bounded rationality has been applied to understand many human behaviors. For example, Crawford (2003) developed cases of intention misrepresentation, in which he finds deception as an equilibrium strategy. While in general terrorists are considered adaptive and strategic, their decisions of target selection could be influenced by other things. As a result, they might not manifest fully rational or strategic from a game-theoretic point of view. Therefore, to modify fully endogenous models by incorporating the possibility that the attacker is partially strategic or fully non-strategic is imperative. Further, not only threats brought by terrorists could be non-strategic, other types of threats such as natural disasters are non-strategic as well. Using past data regarding the magnitude and frequency of natural disasters, non-strategic attack probabilities toward targets of value to the defender could be estimated. It is not surprising that human rationality or cognition affects risk-related decision making. Von Winterfeldt and colleagues (1981) showed that cognitive factors play important roles in risk ranking. Human practical reasoning has been explored in Georgeff and colleagues (1999). To the best of our knowledge, limited research has studied strategic defense allocations with the scenario that the attacker could be partially strategic or non-strategic with some probability. Hao and colleagues (2009) pioneered some research on defense allocation against a possibly non-strategic attacker whereas Section 27.3 continues this research effort by analyzing the robustness of game-theoretic and non-game-theoretic models. Nguyen and colleagues (2013) evaluated two approaches to study boundedly rational human adversaries and found that using techniques from robust optimization that circumvent adversary modeling outperformed an integrated model of human decision making and a game-theoretic algorithm.

27.2.2 *Equity Considerations in Risk Analysis*

Homeland security has justifiably attracted huge amounts of money. However, its effectiveness is frequently unclear (U.S. Government Accountability Office, 2008) and criticized with the phrase "pork-barrel politics," where political factors play a large role in fund distribution (e.g., de Rugy, 2005). Considering a recent budget cut experienced by homeland security (Associated Press, 2011), it is imperative to distribute limited defense resources in an optimal way, taking into account critical factors, including possibly strategic and/or non-strategic adversary behavior and equity or fairness.

With respect to equity, although "pork-barrel politics" is not equivalent to equity, it does result in homeland security expenditures being more equally distributed. For instance, political factors have driven somehow equitable resource allocation to different geographical areas without considering the risk or threat level. The allocation guideline mandates at least 0.75% of the State Homeland Security Grant to be allocated to every state, which translates to that about 60% of the total budget is determined without optimization on the basis of risk or threats. In the year 2008, this percentage was reduced to 0.375% (U.S. Government Accountability Office, 2008). Overall, equity is an important topic that has not been extensively studied in homeland security research, with only a few exceptions. Specifically, Yetman (2004) studied how to incorporate equity when screening passengers in airports. Wang and Zhuang (2011) compared discriminatory and nondiscriminatory screening policies with strategic applicants, who attempt to enter an organization but have private information.

With respect to types of equity, different equity types include: individual vs. group vs. society equity (Fishburn, 1984; Fishburn & Sarin, 1991), vertical vs. horizontal equity (Davidson et al., 2005), ex ante vs. ex post equity

(Keller & Sarin, 1988, 1995; Sounderpandian, 1989), maximum equity vs. minimum equity (Chanta et al., 2011), and proportional vs. max-min equity (Bertsimas et al., 2011). Moreover, Gini coefficient is another standard metric for equity commonly used in economics (Lambert & Aronson, 1993).

With respect to application areas of equity, public resource allocations considering equity have been studied in siting hazardous waste facilities (Keller & Sarin, 1995), global warming (Rose et al., 1998), and regional earthquake risk mitigation (Davidson et al., 2005). Other areas of equity study include economics (Young, 1994), transportation (U.S. Congress, 1998), health care (Chanta et al., 2011), education (Gillborn & Youdell, 1999), and energy (Illich, 2000).

To our knowledge, no previous work has studied the trade-off between equity and efficiency in homeland security resource allocation, especially considering an adaptive adversary. Section 27.4 fills this gap by developing a novel model in which a government allocates defensive resources among multiple potential targets, while reserving a portion of defensive resources (represented by the equity coefficient) for equal distribution (according to geographical areas, population, density, etc.).

27.3 Hybrid Defensive Resource Allocation in the Face of a Partially Strategic Attacker

Section 27.3 is structured as follows. Section 27.3.1 formulates the model and discusses data sources. Section 27.3.2 provides a numerical example, investigating one particular type of non-strategic attack probabilities: evenly distributed to top N targets. Section 27.3.3 introduces two types of false beliefs, defines robustness, and conducts both one-way and two-way sensitivity analyses to investigate the robustness of game-theoretic models. The research reported in this section was originally introduced in Shan and Zhuang (2013b).

27.3.1 *Model Formulation and Data Sources*

In this research effort, we build a hybrid model in which a single defender distributes defense resources of limited quantities among many targets to minimize the expected loss. We make the assumption that the attacker could be fully strategic, fully non-strategic, or partially strategic, whereas the defender is fully strategic. Similar to Powell (2007) and Bier and colleagues (2008), we use a sequential game to model the interaction between the attacker and defender. In this game, the defender is the first mover and allocates from a total budget to many targets; after observing defenders' allocations, a strategic attacker chooses to attack a target with maximal expected loss. If this maximal expected loss is equal for more than one target, those targets are assumed to be attacked with the same probabilities. On the other hand, we assume that a non-strategic attacker, who is bounded-rational or irrational, uses exogenously determined probabilities to decide on where to attack. The equilibria are calculated based on minimizing expected loss by the defender after taking into account the best response of the attacker. Readers are referred to Shan and Zhuang (2013b) for details on the computation of equilibrium. There could be a couple of reasons for the attacker behaving non-strategically besides its intrinsic nature. For instance, erroneous choices (Zhuang, 2010) could lead to this non-strategic behavior.

27.3.2 *Numerical Illustrations*

Similar to Bier and colleagues (2008), Hao and colleagues (2009), and Hu and colleagues (2011), we employ expected property loss data for 47 U.S. urban areas as reported in Willis and colleagues (2005) to be target valuations. This dataset is replicated in column 3 of Table 27.1. For ease of reading, this dataset is sorted in a descending manner in expected property loss. Column 7 of Table 27.1 shows the defense resource allocation to those 47 urban areas in the fiscal year of 2004 by the Office of Grants and Training. We employ the total of FY 2004 Urban Area Security Initiative (UASI) Grant Allocations ($675 M) to be the entire defense budget since expected property losses data in Willis and colleagues (2005) are based on estimation in 2004. However, FY 2005–2012 data are available and are demonstrated in Figure 27.1(b).

Similar to Bier and colleagues (2008), Hao and colleagues (2009), and Golalikhani and Zhuang (2011), we assume that the success probability of attacks is affected by the cost-effectiveness of defense and amount of defense resources. That is, if the allocation is high or cost-effectiveness of defense is high for a particular target, that target is less likely to suffer from an attack.

We focus on the following type of non-strategic attack behavior: attack probabilities are the same for N most-valued urban areas, and zero for the other urban areas. With the dataset in Table 27.1, we find the equilibrium (optimal) defense allocations as a function of the probability of the attacking being non-strategic $(1 - q)$ when cost-effectiveness coefficient of defense (λ) equals 0.01, 0.05, and 1, respectively, and N equals 1, 5, and 47, respectively.

Figure 27.2 demonstrates the defender's optimal decision regarding defense allocations (in percentage of the allocations out of the entire budget) as a function of the probability that the attacker is non-strategic $(1 - q)$ given that the number of top valuable targets attracting the strategic attacker (N) equals 1, 5, and 47 and cost-effectiveness coefficient of defense (λ) equals 0.01, 0.05, and 1,

Table 27.1. *Expected property losses, populations, population densities, density-weighted populations, and FY2004 UASI budget allocations for 47 urban areas in the United States*

#	Urban Area	Expected Property Loss ($Million)†	Population†	Density (per Square Mile)†	Weighted Population‡	FY2004 UASI Allocation ($Million)‡
1	New York City	413.0	9,314,235	8,159	75,991,762,554	47
2	Chicago	115.0	8,272,768	1,634	13,519,096,414	34
3	San Francisco	57.0	1,731,183	1,705	2,951,064,038	26
4	Washington, DC	36.0	4,923,153	756	3,723,526,125	29
5	Los Angeles	34.0	9,519,338	2,344	22,314,867,674	40
6	Philadelphia, PA-NJ	21.0	5,100,931	1,323	6,749,136,215	23
7	Boston, MA-NH	18.0	3,406,829	1,685	5,740,709,241	19
8	Houston	11.0	4,177,646	706	2,948,039,040	20
9	Newark	7.3	2,032,989	1,289	2,619,713,383	15
10	Seattle-Bellevue	6.7	2,414,616	546	1,318,032,823	17
11	Jersey City	4.4	608,975	13,044	7,943,237,618	17
12	Detroit	4.2	4,441,551	1,140	5,062,484,593	134
13	Las Vegas	4.1	1,563,282	40	62,076,079	11
14	Oakland	4.0	2,392,557	1,642	3,927,449,645	8
15	Orange County	3.7	2,846,289	3,606	10,262,626,470	25
16	Cleveland	3.0	2,250,871	832	1,871,707,337	10
17	San Diego	2.8	2,813,833	670	1,885,205,299	10
18	Miami	2.7	2,253,362	1,158	2,609,185,020	20

(*continued*)

Table 27.1 *(cont.)*

#	Urban Area	Expected Property Loss ($Million)†	Population†	Density (per Square Mile)†	Weighted Population‡	FY2004 UASI Allocation ($Million)‡
19	Minneapolis-St. Paul	2.7	2,968,806	490	1,453,687,745	19
20	Denver	2.5	2,109,282	561	1,183,064,989	9
21	Baltimore	2.4	2,552,994	979	2,498,144,264	16
22	Atlanta	2.3	4,112,198	672	2,761,386,037	11
23	Dallas	2.1	3,519,176	569	2,002,093,120	12
24	St. Louis	2.1	2,603,607	407	1,060,496,877	11
25	Portland	2.0	1,918,009	381	731,703,925	8
26	Phoenix	1.9	3,251,876	223	725,649,640	12
27	San Jose	1.7	1,682,585	1,304	2,193,476,169	10
28	Charlotte	1.1	1,499,293	444	665,682,378	7
29	Kansas City	1.1	1,776,062	329	583,476,273	13
30	Milwaukee	1.1	1,500,741	1,028	1,542,728,464	10
31	New Haven	1.1	542,149	1,261	683,670,545	9
32	Buffalo	1.0	1,170,111	747	873,657,856	10
33	Pittsburgh	1.0	2,358,695	510	1,202,742,683	12
34	Cincinnati	0.9	1,646,395	493	811,141,960	13
35	Tampa	0.9	2,395,997	938	2,247,784,596	9

36	New Orleans	0.8	1,337,726	394	526,405,217	7
37	Columbus	0.7	1,540,157	490	755,141,752	9
38	Indianapolis	0.7	1,607,486	456	733,470,541	10
39	Sacramento	0.7	1,628,197	399	649,623,296	8
40	Louisville	0.6	1,025,598	495	507,651,616	9
41	Orlando	0.6	1,644,561	471	774,794,778	9
42	Memphis	0.5	1,135,614	378	428,953,952	10
43	Albany	0.4	875,583	272	237,926,588	7
44	Richmond	0.4	996,512	338	337,254,906	7
45	San Antonio	0.4	1,592,383	479	762,291,362	6
46	Baton Rouge	0.2	602,894	380	229,154,762	7
47	Fresno	0.2	922,516	114	105,084,482	7
	Total	788.7	122,581,611	58,281	200,768,260,341	675

Sources: † Willis et al. (2005); ‡ U.S. Department of Homeland Security (2004).
Adapted from table 2 in Shan and Zhuang (2013a).

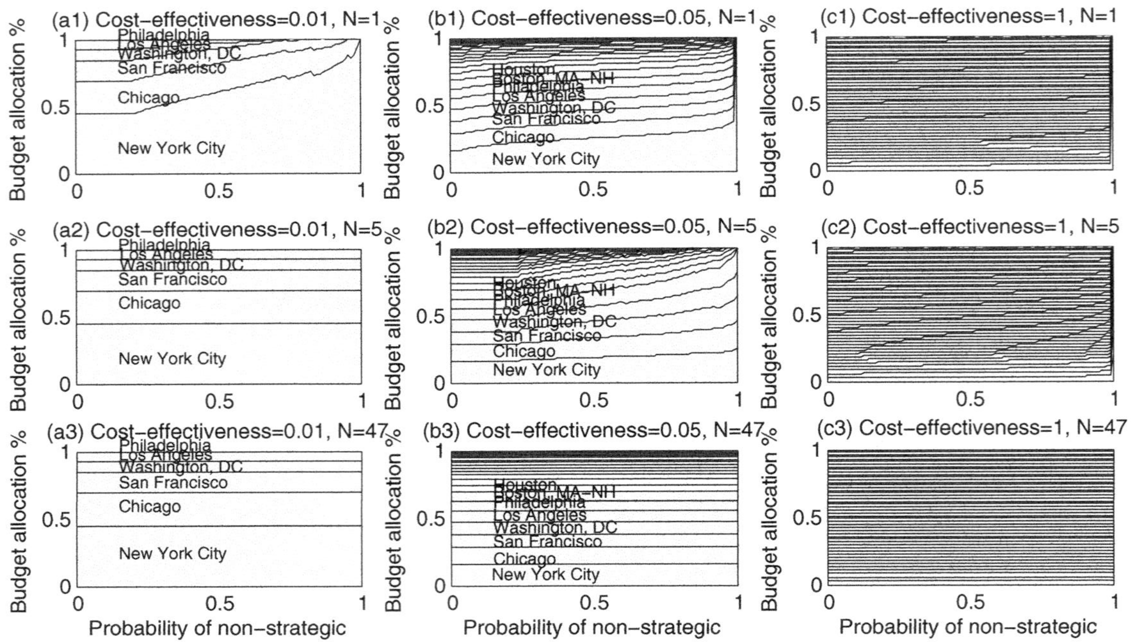

Figure 27.2. Optimal defense allocations as a function of the probability of the attacker being non-strategic if cost-effectiveness (coefficient of defense) equals 0.01, 0.05, or 1 and the number of top valuable targets (N) equals 1, 5, or 47, respectively. This figure is adapted from figure 5 in Shan and Zhuang (2013b).

respectively. Three columns in Figure 27.2 show that if the attacker is fully strategic ($1 - q = 0$), the defender will allocate the defense budget to the top 6, 25, and 47 urban areas regardless of the number of top valuable targets attracting the strategic attacker (N), for cost-effectiveness coefficient of defense (λ) equals 0.01, 0.05, and 1, respectively. (Note that Philadelphia's allocation may be too small and its labels might be difficult to read.) When the cost-effectiveness of defense is high ($\lambda = 1$), the defender can protect more urban areas. Further, from all subfigures in Figure 27.2, we find that the defense allocations for the valued urban areas to the strategic attacker weakly increase in the probability of the attacker being non-strategic ($1 - q$), whereas the allocations for other urban areas weakly decrease. If the attacker is fully non-strategic, other urban areas than the top N valued targets will not be defended since the non-strategic attacker will not attack them. When the strategic attacker targets all 47 urban areas with equal probabilities, the defender decides only to protect the top 6, 25, and 47 valued urban areas no matter what the probability that the attacker being non-strategic is given that cost-effectiveness (coefficient of defense) equals 0.01, 0.05, and 1, respectively.

Generally speaking, Figure 27.2 illustrates that the optimal defender's decisions regarding defense allocation are not affected by the probability of the attacker being non-strategic, especially if the attacker is unlikely to be non-strategic, the defense is highly cost-effective, or the number of top-valued urban areas attracting the strategic attacker is large.

27.3.3 *Robustness Analyses*

Two False Beliefs and Definitions of Robustness

In this subsection, we compare optimal defensive resource allocations and the corresponding optimal total expected loss with two other alternative allocation schemes. The reason behind this investigation is that, in reality, the defender might (possibly falsely) hold the belief that the attacker is fully strategic (or non-strategic) and make decisions on defense allocations with game-theoretic (or non-game-theoretic) models. It would be instrumental to evaluate the costs of these two false beliefs along with the case where the defender has complete information with regard to the rationality of the attacker assuming that these beliefs are false.

Figure 27.3 illustrates the three types of defender's expected property losses after using the three defense allocation strategies, as a function of the probability of the attacker being non-strategic given cost-effectiveness (coefficient of defense) equals 0.01, 0.05, and 1, and the number of top valued targets attracting the strategic attacker (N) equals 1, 5, and 47, respectively. One robustness

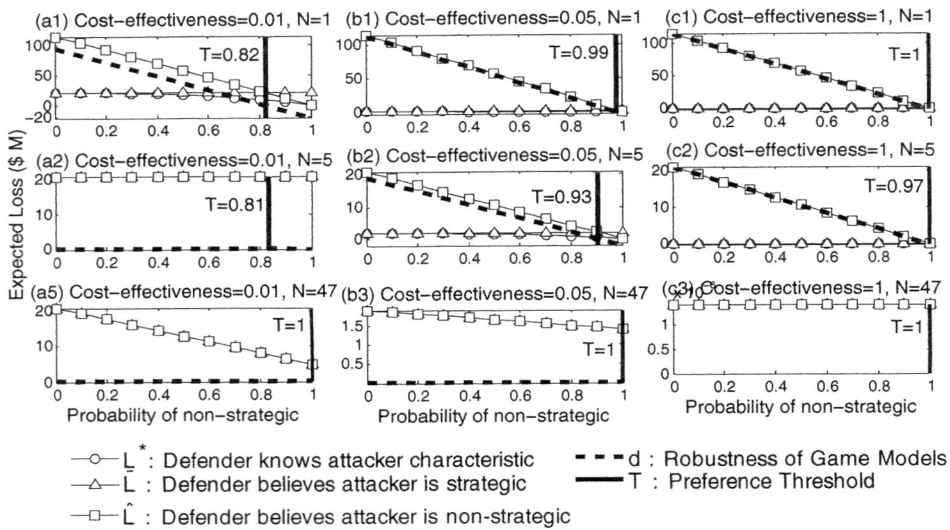

Figure 27.3. Total expected loss of the defender resulting from three strategies of defense allocations and corresponding preference threshold (T) and robustness measure (d) as a function of the probability of the attacker being non-strategic if cost-effectiveness (coefficient of defense) equals 0.01, 0.05, and 1 and the number of top valuable targets attracting the strategic attacker (N) equals 1, 5, and 47, respectively. This figure is adapted from figure 4 in Shan and Zhuang (2013b).

measure – preference threshold (T) – represents the probability of the attacker being non-strategic, where expected losses are equal ($\hat{L} = \bar{L}$) under the two scenarios that the defender falsely believes that the attacker to be strategic or non-strategic, respectively. If the probability of the attacker being non-strategic is less than T, the expected loss (resulting from the scenario where the defender falsely believes that the attacker is non-strategic whereas the attacker is strategic or partially strategic) is higher than that (resulting from the scenario where the defender believes that the attacker is strategic while the attacker is non-strategic or partially strategic). Further, Figure 27.3 also presents another robustness measure (difference between expected losses under the two scenarios where the defender potentially falsely believes the attacker to be strategic or non-strategic; i.e., $d = \hat{L} - \bar{L}$) to compare the costs of these two false beliefs. The scenario $d = 0$ happens when the probability of the attacker being non-strategic equals T, which implies that the game-theoretic model and the non-game-theoretic model have the same performance in total expected loss. The scenarios $d > 0$ (and $d < 0$) denote the situations, where the game-theoretic model is superior (and inferior) to the non-game-theoretic model. We have the following observations: First, from all subfigures of Figure 27.3, we have several observations. (a) When the probability of the attacker being strategic is 1, expected loss if the defender has complete information about the rationality of the attacker equals expected loss if the defender holds the belief that the attacker is fully strategic while the belief could be wrong ($L^* = \bar{L}$); and when the probability of the attacker being non-strategic is 1, expected loss if the defender has complete information about the rationality of the attacker equals expected loss if the defender holds the belief that the attacker is fully non-strategic while the belief could be wrong ($L^* = \hat{L}$); (b) all three expected losses ($L^*, \bar{L}, \hat{L}$) weakly decrease in the probability of the attacker being non-strategic; and (c) expected loss when the defender knows the probability of the attacker being strategic is less than or equal to expected loss when the defender might falsely believe the attacker is strategic ($L^* \le \bar{L}$) and expected loss when the defender knows the probability of the attacker being strategic is less than or equal to expected loss when the defender might falsely believe the attacker is non-strategic ($L^* \le \hat{L}$) for all probabilities of the attacker being non-strategic.

Second, when the probability of the attacker being non-strategic is less than the preference threshold ($1 - q \le T$), expected loss when the defender might falsely believe the attacker is strategic is less than or equal to expected loss when the defender might falsely believe the attacker is non-strategic ($\bar{L} \le \hat{L}$) for T = 0.82, 0.81, 1, 0.99, 0.93, 1, 1, 0.97, and 1 in Figure 27.3 (a1–a3, b1–b3, and c1–c3), respectively. All scenarios that we examined have large values of preference threshold ($T \ge 0.81$), which implies that unless the probability of the attacker being non-strategic is greater than 0.81, the defender always prefers game-theoretic models over non-game-theoretic models. Note these numbers are much greater than 0.5. Further, this implies that if the defender has non-informative beliefs (i.e., the attacker could be strategic or non-strategic with the same probabilities), it is beneficial to the defender by making decisions about defense allocations with game-theoretic models.

Third, from Figure 27.3 (a3, b3, and c3), we find that if the strategic attackers target all 47 urban areas with the same probabilities, the three expected losses ($L*$, $\bar{L}$, and $\hat{L}$) are the same regardless of the probability of the attacker being non-strategic. This is because, regardless of the defender's belief about the attacker's rationality, the defender will protect the top 6 (if cost-effectiveness coefficient of defense equals 0.01), 25 (if cost-effectiveness coefficient of defense equals 0.05), and 47 (if cost-effectiveness coefficient of defense equals 1) urban areas, respectively. Since the optimal defender's decisions regarding defense allocation are the same for the three scenarios, the resulting total expected losses also equal each other. Figure 27.2 (a3, b3, c3) supports this argument.

Finally, from Figure 27.3, we find that if the defender has no information about the rationality of the attacker, it is a better strategy to allocate defensive resources on the basis of the belief of the attacker being fully strategic than the belief of the attacker being fully non-strategic. The reason is that expected loss if the defender holds the false belief of the attacker being fully strategic ($\bar{L}$) is merely marginally higher than that if the defender has complete information about the attacker being strategic ($L*$), whereas expected loss if the defender holds the false belief of the attacker being non-strategic ($\hat{L}$) is dramatically higher than that if the defender has complete information about the attacker being strategic ($L*$). This difference marginally increases in cost-effectiveness (coefficient of defense), decreases in the number of targets attracting the non-strategic attacker (N), and decreases in the probability of the attacker being non-strategic.

Sensitivity Analyses

The following sensitivity analyses were conducted with regard to preference threshold (T) and difference measure of robustness (d). The goal is to investigate how budget variation impacts the defense system. Figure 27.4 demonstrates how the preference threshold (T) for game-theoretic models changes with changes in defense budget if the number of top-valued targets attracting the non-strategic attacker (N) equals 1, 5, and 47 and cost-effectiveness (coefficient of defense) equals 0.01, 0.05, and 1, respectively. We find that preference threshold (T) is always greater than 0.58 for all combinations of defense

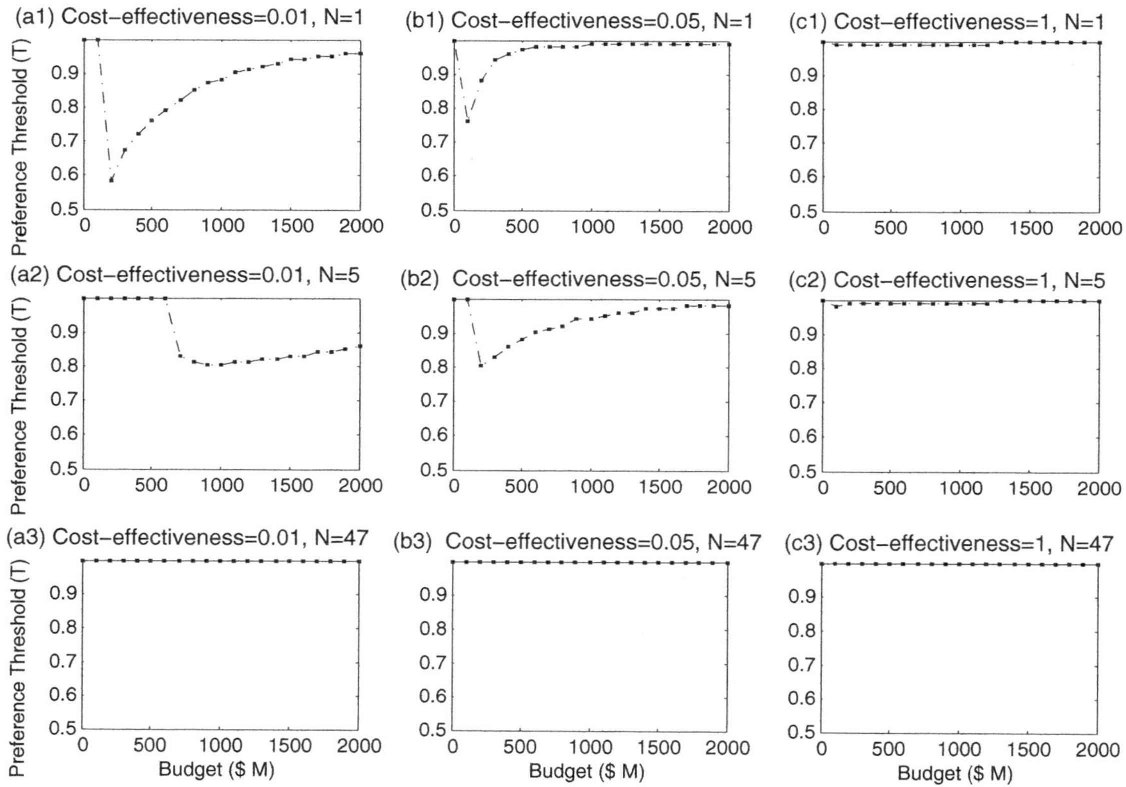

Figure 27.4. Preference threshold (T) as a function of budget (C) when N = 1, 2, 5, and 47, and λ = 0.01, 0.05, and 1, respectively. This figure is adapted from figure 6 in Shan and Zhuang (2013b).

budget, cost-effectiveness coefficient of defense, and the number of top valuable targets attracting the non-strategic attacker (N), echoing the findings from Figure 27.3. From Figure 27.4 (a3, b3, and c3), we find that when the number of top-valued targets attracting the strategic attacker is 47, the preference threshold is always 1 for all values of defense budget and cost-effectiveness (coefficient of defense), which reinforces the findings in Figure 27.3 (a3, b3, c3). We also find that in general, the preference threshold (T) is high when cost-effectiveness (coefficient of defense) or the number of top-valued targets attracting the non-strategic attacker (N) is high, or when the defense budget is either high or low.

Then, we conduct a two-way sensitivity analysis to investigate how the robustness measures varies with changes in probability of the attacker being non-strategic, defense budget, cost-effectiveness (coefficient of defense), and the number of top-valued targets attracting the non-strategic attacker (N). (Note that we consider the probability of the attacker being non-strategic in investigating robustness measured, whereas in investigating preference threshold T, that probability is irrelevant.) That is, Figure 27.5 shows contours of difference measure of robustness of game-theoretic models (d), where the first dimension is budget, and the second dimension is the probability of the attacker being non-strategic.

Note that if the non-strategic attacker is equally likely to attack all 47 targets (Figure 27.5 (a3, b3, c3)), the difference measure of robustness d is always 0 for all values of cost-effectiveness coefficient of defense and defense budget. This finding is reinforced by the results in Figures 27.2 (a3, b3, c3) and 27.3 (a3, b3, c3).

Generally, Figure 27.5 illustrates that the difference measure of robustness d is positive most of the time as indicated by areas in the figure, which implies that game-theoretic models perform better than non-game-theoretic models in terms of lower consequent total expected loss. Especially, this is true if the probability of the attacker being non-strategic is low or medium, defense budget is either high or low, or cost-effectiveness (coefficient of defense) is high.

27.4 Cost of Equity in Defensive Resource Allocation in the Face of a Strategic Attacker

Equity is one of the major factors affecting resource allocations and has been extensively studied in public resource allocation. However, to the best of the authors' knowledge, little research has been devoted to studying equity issues in homeland security resource allocation. In this section, we present a rigorous quantitative study to investigate the trade-off between equity and efficiency

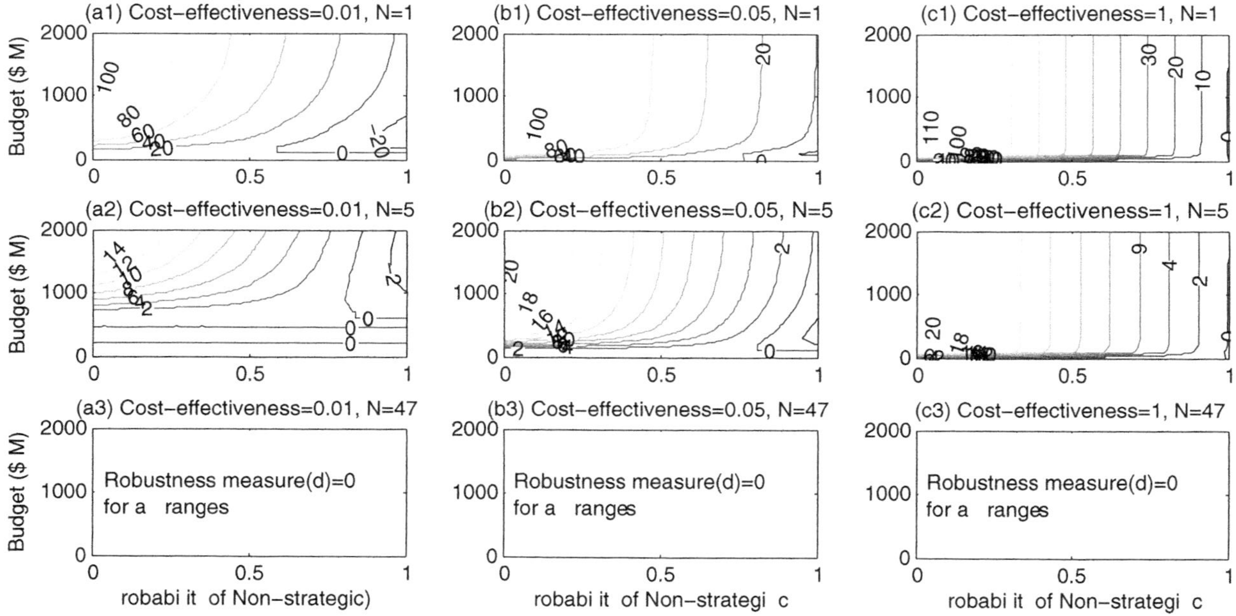

Figure 27.5. Robustness measured for game-theoretic models as a function of defense budget and the probability of the attacker being non-strategic given cost-effectiveness coefficient of defense equals 0.01, 0.05, and 1, and the number of top-valued targets attracting the non-strategic attacker equals 1, 5, and 47, respectively. This figure is adapted from figure 7 in Shan and Zhuang (2013b).

in home security resource allocations and compare different types of equity. Section 27.4 is organized in the following way. Section 27.4.1 lists assumptions and model formulation. Section 27.4.2 presents data sources. As the numerical illustration, Section 27.4.3 reports an extensive sensitivity analysis of optimal defense allocations and the consequent total expected property losses for three system parameters, which are type of equity, cost-effectiveness (coefficient of defense), and total defense budget. The research reported in this section was originally introduced in Shan and Zhuang (2013a).

27.4.1 *Model Formulation*

Due to the fact that the defender and/or attacker has some information about the other player at the time of decision making, the interaction between a defender and an attacker can be modeled as a sequential game (Nikoofal & Zhuang, 2012). In this formulation, we adopt the sequential game, where the defender is the first mover, and allocates a total defense budget to many targets. After observing the defender's decision, the attacker chooses to attack targets with conditional probabilities, which also depends on the exogenously determined total probability of attack (e.g., due to funding and technological constraints). The attacker is assumed to select the target with the highest expected loss as his best response. If more than one target has highest expected loss, we assume that the attacker is equally likely to select those targets.

In practice, other factors could also influence target selection by the attacker, including likelihood of successful attacks, physical proximity, and funding availability. However, those are minor considerations. For instance, although terrorists in San Francisco (SF) might be more likely to attack SF than Chicago, the cost of traveling from SF to Chicago is much smaller than known terrorists' financial resources (e.g., Osama Bin Laden was known to have control over approximately $300 million worth of assets; see Ackman, 2001) and terrorists' operation costs (e.g., the price tag on the 9/11/2001 terrorist attacks could be almost $500 thousands; see Gómez & Miguel, 2010). In fact, the model does consider likelihood of successful attacks in selecting a target by introducing the probability function of a successful attack. Further, total probability of attack is likely to be constrained by the attacker's funding availability.

We assume that the defender aims to minimize the expected loss by distributing defensive resources with two mechanisms. Particularly, a fraction of the total budget is devoted to equity distributions, whereas the remainder is dedicated to risk-based minimization incorporating the best response of the attacker. The following five types of equity allocations are investigated: (1) the total budget is evenly distributed to all targets (per-target equity); (2) target distribution is made to be proportional to target valuation (per-valuation equity); (3) target distribution is made to be proportional to the population size of each target (per-capita equity); (4) target distribution is made to be proportional to the population density of each target (per-density equity), and (5) target distribution is made to be proportional to the population weighted population density of each target (per-weighted-capita equity). There are mainly three factors in the defender's objective function, which represent threat, vulnerability, and consequences, respectively. These three factors are also considered by the U.S. Department of Homeland Security in its risk analysis for terrorist attacks (Cox, 2009). Note that unless full equity is considered, expected loss for each target in the group of targets to be defended with non-reserved part of defense resources is the same as a result of the defender's equilibrium strategy. Unless full equity is considered and the optimization model is no longer applied, the equilibria are calculated based on minimizing expected loss subject to the equity constraints by the defender after considering the best response of the attacker. Readers are referred to Shan and Zhuang (2013a) for details on the computation of equilibrium.

27.4.2 *Data Sources*

Whereas the practice in homeland security resource allocation where equity is achieved at the state level motivates the model, and the model is generic in nature and any level (e.g., state, county, and city) of equity could be investigated, the dataset used for the numerical illustration is at the urban area level. Particularly, Willis and colleagues (2005) report a dataset of 47 urban areas in terms of homeland security target valuations, which has found applications in a wide range of model illustrations (Nikoofal & Zhuang, 2012). Similar to those studies, we also employ expected property losses for the 47 U.S. urban areas in Willis and colleagues (2005) to represent the target valuations (one key input), which are shown in column 3 in Table 27.1, sorted in descending order. The model is not constrained by the dataset in any way and could be used with other data as inputs as well, to obtain new results. Note also that Table 27.1 reports the property losses from potential damages if an attack is successful, which is determined probabilistically. If the attack fails, it could not lead to the losses in column 3 of Table 27.1. As the basis for the equity distribution, we consider population sizes, population densities, and density-weighted population sizes for the 47 urban areas in Willis and colleagues (2005), which are also replicated in columns 4–6 of Table 27.1. The defense allocations to those 47 urban areas from the Office of Grants and Training in FY2004 are listed in column 7 of Table 27.1. The reason the total FY2004 UASI Grant Allocations

($675 million) serves as the entire defense budget in the model is because the dataset of expected property losses in Willis and colleagues (2005) is based on the year 2004.

27.4.3 *Sensitivity Analyses*

In this section, we conducted sensitivity analyses to compare five types of equity funding distribution and to study the effects of various budget levels and cost-effectiveness of defense. The following parameter values are used to define the baseline: per-target equity and defense budget is $675 million, which equals the entire defense budget for UASI in FY 2004 as shown in Table 27.1.

Equity Type

With the dataset discussed in Section 27.4.2, we find the optimal defense allocations considering different degrees of equity distribution. Five types of equity are considered: (1) per-target equity (where equity distribution is equal for each target); (2) per-valuation equity (where expected property losses serve as proxies for target valuation and equity distribution is proportional to target valuations); (3) per-capita equity (where equity distribution is proportional of population size of each target); (4) per-density equity (where equity distribution is proportional to population density of each target); and (V) per-weighted-capita equity (where equity distribution is proportional to population-density-weighted population size of each target). Note that we select 0.005 to be the value for cost-effectiveness coefficient of defense in order to fully illustrate the differences between the five types of equity in terms of optimal defense allocations and consequent expected loss.

From Figure 27.6(I_1), where the defender considers per-target equity, we find that if no equity distribution is considered, the defender protects the top three most-valued targets. As the defender allows more equity distribution, the number of defended targets increases. Finally, if the entire defense distribution is based on equity principle, the defender protects all 47 targets equally. If there is no consideration of equity, the expected property loss is $45.25 million and increases convexly as more equity distribution is implemented and turns $384.38 million if the entire defense distribution is based on equity principle (Figure 27.6(I_2)). From Figure 27.6(II_1–II_2, III_1–III_2, IV_1–IV_2, V_1–V_2), where per-valuation, per-capita, per-density, and per-weighted-capita equities are considered, respectively, we observe that defense allocations and the consequent expected property loss change similarly.

By comparing the five types of equity in Figure 27.6(I–V), we find: (1) considering different equity types leads to different optimal defense allocations,

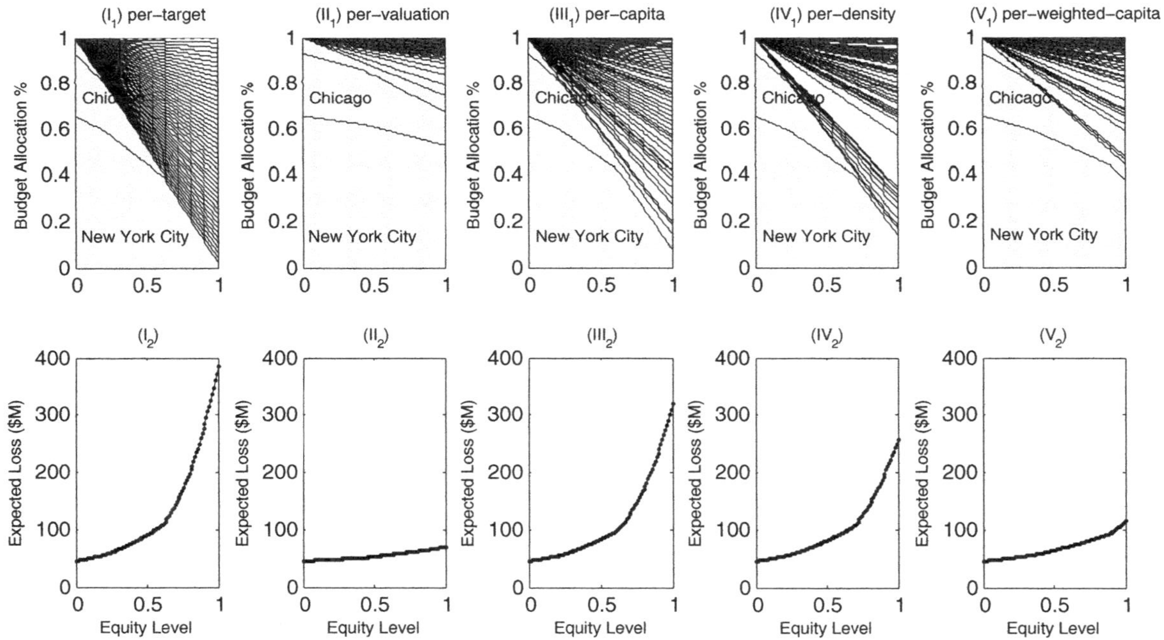

Figure 27.6. Optimal defensive allocations and the consequent expected loss as more equity distribution is considered for five types of equity. This figure is adapted from figure 2 in Shan and Zhuang (2013a).

and the cost of equity increases convexly as more equity distribution is allowed regardless of the equity type; (2) per-valuation equity results in the smallest expected property loss regardless of the degree of equity distribution; and (3) per-target equity leads to the largest expected property loss.

Cost-effectiveness of Defense

In order to understand the effects of cost-effectiveness of defense, we conduct sensitivity analysis regarding the cost-effectiveness coefficient of defense. We consider three levels of cost-effectiveness coefficient of defense, 0.005, 0.05, and 0.5, to represent hypothetical scenarios where the defense is not cost-effective, moderately cost-effective, or highly cost-effective. We find that when the cost-effectiveness (coefficient of defense) increases from 0.005 in Figure 27.7(a) to 0.05 in Figure 27.7(b) and to 0.5 in Figure 27.7(c), more top-valued urban areas are defended (e.g., from 3 to 25 and to 47, respectively, if no equity allocation is considered). In the meantime, the expected property loss is reduced from \$45.25 million to \$1.92 million and to \$0.0019 million, respectively, if no equity allocation is considered, whereas it is reduced from \$357.75 million to \$98.23 million and to \$0.0019 million, respectively, if the entire defense budget is distributed based on equity principle. This finding suggests that if defense is more cost-effective, cost of equity is lower and the defender can protect more targets for any degree of equity distribution.

Total Defense Budget

In order to understand the effects of amount of defense budget, sensitivity analysis is conducted regarding the entire defense budget and a moderate value of cost-effectiveness of defense is selected (0.01). We consider three budget levels: 1) \$ 675 million as the baseline, which is the defense budget in FY 2004 for the Urban Areas Security Initiative; 2) \$67.5 million, which is 10 times less than the baseline, and 3) \$6,750 million, which is 10 times more than the baseline. We find that when the defense budget increases from \$67.5 million in Figure 27.8(a) to \$675 million in Figure 27.8(b) and to \$6,750 million in Figure 27.8(c), the number of defended urban areas increases from 1 to 6 and to 45, respectively, if no equity allocation is considered. Meanwhile, the consequent expected property loss decreases from \$210.28 million to \$20.82 million and to \$0.64 million, respectively, if no equity allocation is considered, or from \$401.30 million to \$309.89 million and to \$23.36 million, respectively, if all allocations are based on equity principle. This finding suggests that if the amount of defense budget is large, cost of equity would be small, and the defender will be able to protect more targets for any degree of equity allocation.

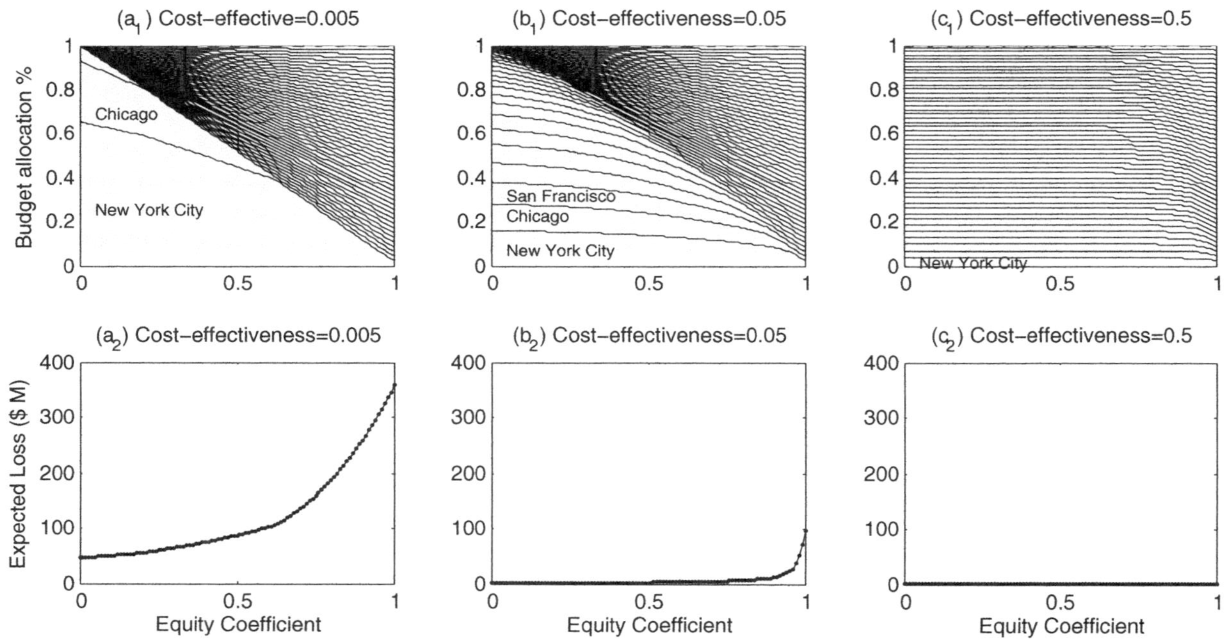

Figure 27.7. Optimal defense allocations and the consequent expected property loss as the degree of equity distribution increase from 0% to 100% with cost-effectiveness (coefficient of defense) equal 0.005, 0.05, and 0.5, respectively, when the defender employs per-target equity and defense budget equals $675 million. This figure is adapted from figure 3 in Shan and Zhuang (2013a).

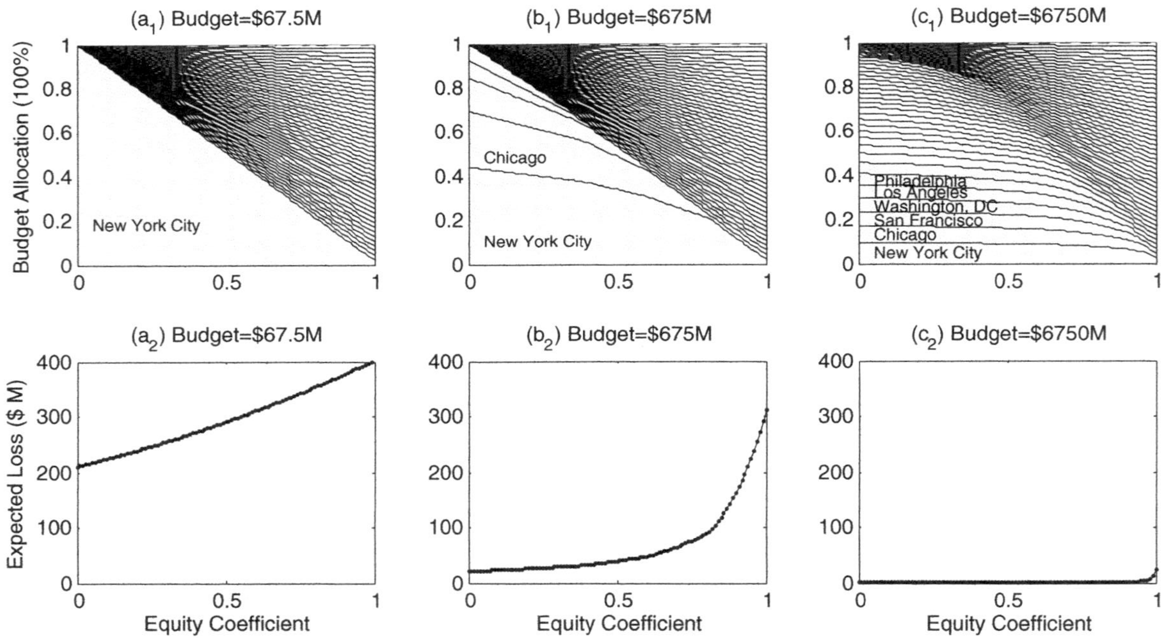

Figure 27.8. Optimal defense allocations and the consequent expected property loss as the degree of equity distribution increases from 0% to 100% given the defense budget equals \$67.5 million, \$675 million, and \$6,750 million, respectively, when the defender uses per-target equity and defense is moderately cost-effective. This figure is adapted from figure 4 in Shan and Zhuang (2013a).

27.5 Conclusion

In this chapter, we formulate and solve an optimization problem based on game theory, where the defender allocates defense resources among many targets facing an unknown attacker who could be fully strategic, partially strategic, or fully non-strategic. Further, we study the effects of the equity principle in defense allocation by incorporating equity principles as constraints to the optimization problem. First, we introduce a hybrid model, where a single defender allocates defense resources to protect many targets in order to minimize expected loss caused by an adversary of varying degree of rationality. That is, the attacker could be fully strategic, partially strategic, or fully non-strategic. Generally speaking, the defense strategies facing such an unknown attacker differ from those against a fully strategic or a fully non-strategic attacker. A smaller expected loss can be reached if the attacker is highly likely to be non-strategic, no matter what the defender believes about the degree of the attacker's rationality. The numerical example suggests that the optimal defense allocations are not much affected by the probability of the attacker being non-strategic. This is especially true if that probability is low, defense is highly cost-effective, or the number of top valuable urban areas attracting the non-strategic terrorist is large.

Since both fully exogenous and fully endogenous models are special cases of the hybrid model, the hybrid model allows us to compare these three models with a numerical example and sensitivity analyses. Results suggest that game-theoretic models perform better even if the probability of the attacker being non-strategic is much larger than pure chance. Specifically, defense allocations on the basis of game-theoretic models are preferred over those on the basis of non-game-theoretic models in terms of smaller total expected loss, especially if the budget is not moderate, defense is highly cost-effective, or the probability of the attacker being strategic is high.

Given the debate regarding application of game theory to terrorism analysis, the finding that optimal defense allocations on the basis of game theory is "robust" to the probability that the terrorist is non-strategic. That is, believing facing a strategic terrorist is "conservative" because in general it results in smaller expected loss than believing facing a non-strategic terrorist. This research yields valuable insights on this issue. In particular, the findings suggest the necessity of more research on applying game theory to terrorism analysis. Results show that game-theoretic models are useful even if the terrorist's behavior is at odds with a fundamental assumption of game theory.

Even though we find that game-theoretic models perform more robustly than non-game-theoretic models, we note that in reality there might be challenges in game-theoretic models, and they might have difficulty in scaling up to solve complicated real problems.

Second, based on the aforementioned model, we develop another optimization model where a certain fraction of the total budget is reserved for equity distribution. We study how optimal defensive resource allocations change as the degree of equity distribution varies. We find that the cost of equity in terms of expected loss increases convexly as more equity distribution is implemented for all five possible equity types that we consider, and the per-valuation and per-target equities lead to smallest and largest expected property losses, respectively.

The findings suggest that high cost-effectiveness of or large amount of defense resources could alleviate the effects of equity distribution. In other words, if the defensive system is highly cost-effective or the defense budget is high, the defender would be able to implement a high level of equity by reserving more resources for equity allocations. By contrast, extra attention is needed in reserving defensive resources for equity allocation if the budget is low or the defensive investments are not effective.

Potential policy implications follow. Specifically, the model presents a tool where the degree of equity can be customized, and illustrates a complete Pareto frontier of the equity-efficiency trade-off when facing a strategic attacker. The defender could select an appropriate (equity, efficiency) pair, based on her own individual preferences. Since the results show that the cost of equity increases convexly in the degree of equity distribution, moderate degree of equity would likely lead to a reasonable balance between cost and equity in practice. The exact "optimal level" of equity would be determined by several factors including budget, cost-effectiveness of defense, and tolerance of expected loss.

REFERENCES

Ackman, D. (2001). The cost of being Osama bin Laden. www.forbes.com/2001/09/14/0914ladenmoney.html. Accessed in February 2015.

Associated Press. (2011). Cities affected by anti-terrorism grant cuts. *Washington Post*. www.foxnews.com/us/2011/05/19/cities-affected-anti-terrorism-grant- cuts/. Accessed in August 2012.

Atkinson, M. P., Cao, Z., & Wein, L. M. (2008). Optimal stopping analysis of a radiation detection system to protect cities from a nuclear terrorist attack. *Risk Analysis, 28*(2), 353–371.

Bertsimas, D., Farias, V. F., & Trichakis, N. (2011). The price of fairness. *Operations Research, 59*(1), 17–31.

Bier, V. M., Haphuriwat, N., Menoyo, J., Zimmerman, R., & Culpen, A. M. (2008). Optimal resource allocation for defense of targets based on differing measures of attractiveness. Risk *Analysis, 28*(3), 763–770.

Brown, G., Carlyle, M., Diehl, D., Kline, J., & Wood, K. (2005). A two-sided optimization for theater ballistic missile defense. *Operations Research, 53*(5), 263–275.

Brown, G., Carlyle, M., Salmeron, J., & Wood, K. (2006). Defending critical infrastructure. *Interfaces, 36*(6), 530–544.

Chanta, S., Mayorga, M. E., Kurz, M. E., & McLay, L. A. (2011). The minimum p-envy location problem: A new model for equitable distribution of emergency resources. *IIE Transactions on Healthcare Systems Engineering, 1*(2), 101–115.

Cox, L. A. (2009). Game theory and risk analysis. *Risk Analysis, 29*(8), 1062–1068.

Crawford, V. P. (2003). Lying for strategic advantages: Rational and boundedly rational misrepresentation of intentions. *American Economic Review, 93*(1), 133–149.

Davidson, R. A., Nozick, L. K., Xu, N., & Dodo, A. (2005, June 19–23). Equity in regional earthquake risk mitigation investment decisions. Safety and reliability of engineering systems and structures. *Proceedings of the Ninth International Conference on Structural Safety and Reliability ICOSSAR '05, Rome, Italy* (pp. 2721–2728).

de Rugy, V. (2005). Homeland security pork. *Washington Times*. www.aei.org/publication/homeland-security-pork/. Accessed in February 2015.

Dillon, R. L., John, R., & von Winterfeldt, D. (2002). Assessment of cost uncertainties for large technology projects: A methodology and an application. *Interfaces, 32*(4), 52–66.

Ezell, B. C., Bennett, S. P., von Winterfeldt, D., Sokolowski, J., & Collins, A. J. (2010). Probabilistic risk analysis and terrorism risk. *Risk Analysis, 30*(4), 575–589.

Fishburn, P. C. (1984). Equity axioms for public risks. *Operations Research, 32*(5), 901–908.

Fishburn, P. C., & Sarin, R. K. (1991). Dispersive equity and social risk. *Management Science, 37*(7), 751–769.

Georgeff, M., Pell, P., Pollack, M., Tambe, M., & Wooldridge, M. (1999). The belief-desire-intention model of agency. In *Intelligent agents V: Agents theories, architectures, and languages* (pp. 1–10). Berlin Heidelberg: Springer.

Gigerenzer, G., & Selten. R. (2002). *Bounded rationality: The adaptive toolbox.* Cambridge, MA: MIT Press.

Gillborn, D., & Youdell, D. (1999). *Rationing education: Policy, practice, reform and equity*. Philadelphia, PA: Open University Press.

Golalikhani, M., & Zhuang, J. (2011). Modeling arbitrary layers of continuous-level defenses in facing with strategic attackers. *Risk Analysis, 31*(4), 533–547.

Gomez, D. C., & Miguel, J. (2010). A financial profile of the terrorism of Al-Qaeda and its affiliates. *Perspectives on Terror, 4*(4), 3–27.

Hao, M., Jin, S., & Zhuang, J. (2009). Robustness of optimal defensive resource allocations in the face of less fully rational attacker. *Proceedings of the 2009 Industrial Engineering Research Conference* (pp. 886–891).

Harris, B. (2004). Mathematical methods in combating terrorism. *Risk Analysis, 24*(4), 985–988.

Hausken, K. (2002). Probabilistic risk analysis and game theory. *Risk Analysis, 22*(1), 17–27.

Hu, J., Homem-de Mello, T., & Mehrotra, S. (2011). Risk adjusted budget allocation models with application in homeland security. *IIE Transactions, 43*(12), 819–839.

Illich, I. (2000). *Energy and equity*. London: Marion Boyars Publishers Ltd.
Jiang, A. X., Yin, Z., Zhang, C., Tambe, M., & Kraus, S. (2013). Game-theoretic randomization for security patrolling with dynamic execution uncertainty. *Proceedings of the 2013 International Conference on Autonomous Agents and Multiagent Systems* (pp. 207–214).
Jose, V. R. R., & Zhuang, J. (2013). Technology adoption, accumulation, and competition in multi-period attacker-defender games. *Military Operations Research, 18*(2), 33–47.
Kahneman, D. (2003). A perspective on judgment and choice: Mapping bounded rationality. *American Psychologist, 58*(9), 697–720.
Kaplan, E. H. (2010). Terror queues. *Operations Research, 58*(4), 773–784.
Kaplan, E. H., & Kress, M. (2005). Operational effectiveness of suicide-bomber-detector schemes: A best-case analysis. *Proceedings of the National Academy of Science, 102*(29), 10399–10404.
Kaplan, E. H., Kress, M., & Szechtman, R. (2010). Confronting entrenched insurgents. *Operations Research, 58*(2), 329–341.
Keeney, R. L. (2007). Modeling values for anti-terrorism analysis. *Risk Analysis, 27*(3), 585–596.
Keller, L. R., & Sarin, R. K. (1988). Equity in social risk: Some empirical observations. *Risk Analysis, 8*(1), 135–146.
Keller, L. R., & Sarin, R. K. (1995). Fair processes for societal decisions involving distributional inequalities. *Risk Analysis, 15*(1), 49–59.
Kress, M., & Szechtman, R. (2009). Why defeating insurgencies is hard: The effect of intelligence in counterinsurgency operations – a best case scenario. *Operations Research, 57*(3), 578–585.
Lambert, P. J., & Aronson, J. R. (1993). Inequality decomposition analysis and the Gini coefficient revisited. *The Economic Journal, 103*(420), 1221–1227.
Levitin, G., & Hausken, K. (2009). Redundancy vs. protection in defending parallel systems against unintentional and intentional impacts. *IEEE Transactions on Reliability, 58*(4), 679–690.
Lin, K. Y., Kress, M., & Szechtman, R. (2009). Scheduling policies for an antiterrorist surveillance system. *Naval Research Logistics, 56*(2). 113–126.
Nguyen, T. H., Yang, R., Azaria, A., Kraus, S., & Tambe. M. (2013). Analyzing the effectiveness of adversary modeling in security games.
Nikoofal, M. E., & Zhuang, J. (2012). Robust allocation of a defensive budget considering an attacker's private information. *Risk Analysis, 32*(5), 930–943.
Parnell, G. S., Smith, C. M., & Moxley, F. I. (2010). Intelligent adversary risk analysis: a bioterrorism risk management model. *Risk Analysis, 30*(1), 32–48.
Powell, R. (2007). Defending against terrorist attacks with limited resources. *American Political Science Review, 101*(3), 527–541.
Powell, R. (2009). Sequential, nonzero-sum "Blotto": Allocating defensive resources prior to attack. *Games and Economic Behavior, 67*(2), 611–615.
Rio Insua, D., Jesus, J., & Banks, D. (2009). Adversarial risk analysis. *Journal of the American Statistical Association, 104*(486), 841–854.
Rose, A., Stevens, B., Edmonds, J., & Wise, M. (1998). International equity and differentiation in global warming policy. *Environmental and Resource Economics, 12*(1), 25–51.

Rosoff, H., & von Winterfeldt, D. (2007). A risk and economic analysis of dirty bomb attacks on the ports of Los Angeles and Long Beach. *Risk Analysis, 27*(3), 533–546.

Shan, X., & Zhuang, J. (2013a). Cost of equity in homeland security resource allocations in the face of a strategic attacker. *Risk Analysis, 33*(6), 1083–1099.

Shan, X., & Zhuang, J. (2013b). Hybrid defensive resource allocations in the face of a partially strategic attacker. *European Journal of Operational Research, 228*(1), 262–272.

Shan, X., & Zhuang, J. (2014a). Modeling credible retaliation in deterring smuggling of nuclear weapons – A three-stage game. *Decision Analysis, 11*(1), 43–62.

Shan, X., & Zhuang, J. (2014b). Subsidizing to disrupt a terrorism supply chain – a four-player game. *Journal of the Operational Research Society, 65*(7), 1108–1119.

Simon, H. A. (1957). *Model of man: Social and rational*. New York: Wiley.

Simon, H. A. (1982). *Models of bounded rationality: Empirically grounded economic reason*, vol. 3. Cambridge, MA: MIT Press.

Sounderpandian, J. (1989). Ex ante equity in public risk. *Operations Research, 37*(4), 528–530.

Stillwell, W. G., Von Winterfeldt, D., & John, R. S. (1987). Comparing hierarchical and nonhierarchical weighting methods for eliciting multiattribute value models. *Management Science, 33*(4), 442–450.

U.S. Congress. (1998). Transportation Equity Act for the 21st Century. www.fhwa .dot.gov/tea21/h2400enr.htm. Accessed in February 2015.

U.S. Department of Homeland Security. (2004). Fiscal Year 2004 Urban Areas Security Initiative Grant Program. http://web.archive.org/web/*/http://www.ojp .usdoj.gov/odp/docs/fy04uasi.pdf. Accessed in February 2015.

U. S. Department of Homeland Security. (2013). Budget and performance. www.dhs .gov/xabout/budget/dhs-budget.shtm. Accessed in February 2015.

U.S. Government Accountability Office. (2008). DHS risk-based grant methodology is reasonable, but current version's measure of vulnerability is limited. www.gao.gov/ new.items/d08852.pdf. Accessed in February 2015.

Wang, C., & Bier, V. M. (2010). Impact of intelligence on target-hardening decisions based on multiattribute terrorist utility. HST'09. *IEEE Conference on Technologies for Homeland Security* (pp. 373–380).

Wang, C., & Bier, V. M. (2011). Target-hardening decisions based on uncertain multiattribute terrorist utility. *Decision Analysis, 8*(4), 286–302.

Wang, X., & Zhuang, J. (2011). Balancing congestion and security in the presence of strategic applicants with private information. *European Journal of Operational Research, 212*(1), 100–111.

Wein, L. M. (2009). Homeland security: From mathematical models to policy implementation: The 2008 Philip McCord Morse lecture. *Operations Research, 57*(4), 801–811.

Willis, H. H., Morral, A. R., Kelly, T. K., & Medby. J. J. (2005). Estimating terrorism risk. www.rand.org/pubs/monographs/MG388.html. Accessed in February 2015.

Winterfeldt, D., John, R. S., & Borcherding, K. (1981). Cognitive components of risk ratings. *Risk Analysis, 1*(4), 277–287.

Xu, J., Zhuang, J., & Liu, Z. (2015). Modeling and mitigating the effects of supply chain disruption in a defender-attacker game. *Annals of Operations Research, 236*(1), 255–270.

Yetman, J. (2004). Suicidal terrorism and discriminatory screening: An efficiency-equity trade-off. *Defence and Peace Economics, 15*(3), 221–230.

Young, P. (1994). *Equity: In theory and practice*. Princeton, NJ: Princeton University Press.

Zhuang, J. (2010). Impacts of subsidized security on stability and total social costs of equilibrium solutions in an n-player game with error. *The Engineering Economist, 55*(2), 131–149.

Zhuang, J., & Bier, V. M. (2007). Balancing terrorism and natural disasters – Defensive strategy with endogenous attacker effort. *Operations Research, 55*(5), 976–991.

28

Decision Analysis by Proxy for Adaptive Adversaries

Heather Rosoff and Richard S. John

Understanding motivations that drive terrorist group behavior is critical. Current methods for terrorism risk assessment focus on target vulnerability, terrorist capability and resources, and attack consequence. However, threat assessment requires consideration of the influence of terrorist group values and beliefs in shaping militant behavior. The understanding of terrorist values and beliefs has the potential to contribute to probabilistic estimates of terrorist threats.

Since we cannot collect information directly from terrorists, individuals who have studied the general topic of terrorism as well as Islamic terrorist groups (such as Al Qaeda) are asked to act as experts regarding the values of terrorist leaders. These proxies include individuals holding positions as former intelligence specialists, policy analysts, researchers familiar with the study of terrorism and associated events, and former residents of Middle Eastern Islamic countries familiar with the perspective and motivations of Islamic terrorist organizations. Proxies are expected to possess overlapping knowledge, but are also expected to provide potentially conflicting assessments. A diversity across proxies is critical for obtaining various perspectives on how terrorist leader values and beliefs motivate attack decisions.

Using a value-focused decision framework (Keeney, 1992), which we refer to as "proxy utility modeling," we assess how the values and beliefs of terrorist leaders might influence the selection of an attack strategy. We then use a random utility modeling approach to compare the risk profiles of alternative attack strategies and estimate the relative likelihood of a terrorist leader selecting a particular attack strategy.

In this chapter, we first describe the use of risk assessment in the evaluation of terrorist threats. Next, we summarize an analysis of terrorist attack alternatives and describe the multiattribute utility (MAU) elicitation procedures employed. We then present results from an example demonstration of the MAU model and the implications of the findings for the overall probability of attack. Finally, we include overall observations and general conclusions regarding the proposed decision modeling by proxy approach.

28.1 Terrorism Context

Several applied case studies exist assessing specific terrorism risk to U.S. infrastructure, e.g., disruption of the electrical power grid and oil and gas pipelines (Simonoff et al., 2007); terrorist attacks on U.S. commercial aviation (Gordon et al., 2007; Shafieezadeh et al., 2015; von Winterfeldt & O'Sullivan, 2006); radiological (dirty bomb) attack on a major U.S. port (Rosoff & von Winterfeldt, 2007); terror attack on the electrical power system of a major U.S. city (Rose et al., 2007); an urban chlorine truck attack (Barrett & Casman, 2013); an urban terrorist nuclear detonation (Wein et al., 2010); terrorist attacks on bridges (Leung et al., 2004); and terrorist attacks on the U.S. food supply (Hartnett et al., 2009). These reports offer tremendous insight into the technical and resource (manpower) capabilities of terrorist organizations, the relative feasibility of carrying out and defending against specific attacks, and the economic, health, and psychological consequences that might ensue. However, to fully assess the threat of terrorism, it is important to incorporate the probability of an attack being selected for execution into the analysis.

A terrorist organization's commitment to execution of an attack involves a complex decision process. Just as the U.S. Department of Homeland Security makes decisions on national counterterrorism policies (Keeney, 2007b; Keeney & von Winterfeldt, 2011), terrorist organizational leadership must decide on the best attack strategy given its current values and beliefs. If particular values weigh heavily in a terrorist leader's decision-making process, then certain attack types are expected to have an increased likelihood of occurring. This chapter describes an approach to modeling the decision problem of a terrorist leader. By modeling terrorist values and beliefs, additional information is acquired about the likelihood that particular terrorist organizations will attack a given target.

Estimating the likelihood of terrorism consequences is complicated by uncertainty related to the nature of the attack type in terms of the time, location, and mode of attack. Assessing other non-terror extreme events, whether they are technological accidents or natural disasters, involves similar predictive challenges. Yet, researchers still attempt to characterize the probability of these events. They conduct geological studies to evaluate earthquakes, oceanographic studies to understand hurricanes, and risk analysis to assess the threat of industrial accidents.

The study of terrorism is further complicated by the fact that it is difficult to identify terrorist leaders preferences. Collecting data on terrorist leaders' values and beliefs is a formidable task, given the sensitive nature of the information and limited number of public resources. For example, a suicide truck bombing might be the most feasible attack alternative, but there is uncertainty about whether the attack's outcome meets the objectives of the terrorist leader. Alternatively, a dirty bomb attack might be desirable to a terrorist, but the attack may not be feasible if the radioactive material cannot be acquired. Any model of terrorist preferences should account for terrorist uncertainties about events related to resource acquisition, interdiction, attack execution success, and the consequences of various attack alternatives. In addition, the model should also capture uncertainty related to the modeler's understanding of terrorist values, including trade-offs among conflicting objectives and risk attitude.

Terrorist attacks are created and caused by human agents and thus are dynamic in nature. Knowledge about the functionality of terrorist leaders, their organization, and their capabilities is perpetually evolving and difficult to acquire. There is a need for a systematic approach to assessing the uncertainty associated with the decision-making threat posed by terrorist leaders. A detailed representation of the adversary's objective function may give some direction as to the probabilities associated with different attack types. This chapter describes the construction of a value model used to represent terrorist leader preferences. To accomplish this, we utilize and expand the multiattribute utility modeling approach (Keeney, 2007a; Keeney & Raiffa, 1976).

28.2 Proxy Multiple Objectives Value Modeling

The MAU modeling approach utilizes proxies' understanding of terrorist leaders' beliefs and motivations to determine preferences among attack alternatives, including time, location, and mode of attack. The construction of the model involves seven primary steps:

(1) Identify and structure fundamental objectives for terrorist organizations,
(2) Define attribute scales for the fundamental objectives,
(3) Assess risk attitude for each attribute in the form of a utility function,
(4) Define value trade-offs (weights) among the conflicting objectives and attributes,
(5) Define beliefs about consequences of each attack alternative on each scale in the form of a probability distribution,

(6) Specify uncertainties in model parameters for both single-attribute utility functions and trade-off parameters (weights),
(7) Use Monte Carlo simulation to obtain risk profiles (CDFs) for each alternative, and probabilities that the utility of each alternative attack results in maximum utility.

As previously noted, since we cannot directly collect information from terrorist leaders, subject matter experts (SMEs) are utilized as proxies. SME's assessments are used to provide both a multi-objective framework for the problem (Keeney & von Winterfeldt, 2011), in the form of an objectives hierarchy for the terrorist group, as well as quantitative estimates of MAU model parameters. Probability distributions are assessed over attribute scale scores for each alternative, single-attribute utility (risk attitude) parameters, and trade-off (weight) parameters to capture proxies' uncertainty about terrorist beliefs and values. In addition to capturing uncertainty about proxy terrorists' values and beliefs, there is also considerable uncertainty associated with events relevant to whether the attack is executed and the consequences associated with the attack. An event tree is incorporated into the model to account for uncertainty about attack execution, which should impact preferences for attack alternatives varying in feasibility. Overall, this model is designed to represent terrorist leaders' beliefs and values, providing insight into terrorist preferences for attack alternatives and the threat associated with each alternative.

28.2.1 *Methodology*

This section describes a demonstration of the proxy utility modeling methodology for quantifying Al Qaeda organizational preferences among generic attack modes on the United States. The demonstration includes identification and structuring of Al Qaeda objectives, definition of attribute scales, elicitation of trade-offs among conflicting objectives, elicitation of single-attribute utility functions to capture risk attitude, and assessment of uncertainties related to attack execution and consequences.

28.3 Decision Maker and Context

To identify fundamental objectives of terrorist organizations, it is important to gather information about terrorist leaders' values. In the context of this study, the decision maker is a hypothetical leader of Al Qaeda. While the actions terrorist organizations carry out are interpreted by many to indicate that terrorist leaders are irrational and their decision processes are rooted in psychological pathologies (Lankford, 2014), other researchers argue for a strategic model, consistent with a collective rationality assumption (Sprinak, 2000; Victoroff, 2005). A rational justification for terrorism has been based on an assertion that violence is a means to an end (Crenshaw, 1995). Terrorists are thought to pursue goals while recognizing that the consequences of violence have a practical purpose. Thus, terrorists are assumed to operate as a collective unit that requires a high level of organization and careful planning to succeed (Rosoff & von Winterfeldt, 2007).

The idea of a rational terrorist group generalizes to the process of how terrorist leaders make strategic decisions. In this framework, terrorism is perceived as an instrumental activity designed to achieve a set of goals. The terrorist leader is thought to make decisions by looking ahead and evaluating consequences, which in this case refers to the decision to commit violent acts and the target and mode of attack selected. Much like any other major business or government decision, a terrorist leader attempts to maximize expected consequences, including minimizing expenditures of both human and financial resources.

28.4 Objectives Hierarchy

Terrorists' fundamental objectives are identified by asking proxy terrorist leaders (experts) about what matters to terrorist organizations and what they want to accomplish by executing acts of terror. To elicit this information, the proxies are interviewed individually, and then from the individual assessments a comprehensive set of fundamental objectives is developed. An example of such a set of fundamental objectives (Rosoff, 2009), organized into a hierarchy, is illustrated in Figure 28.1.

Proxies indicated that an Al Qaeda leader's fundamental objectives fall into three categories: (1) maintaining Al Qaeda's organizational strength; (2) managing Al Qaeda's operational expenditures; and (3) ensuring Al Qaeda has an impact on the United States. Further investigation into the primary objectives resulted in a compilation of attributes that are used to quantify the likely consequences of attack options. For example, one objective of a terrorist leader is to continue the formation of training camps. This is a means to the larger objective of maintaining internal organizational strength. This is also a means to contributing to the fear of terrorism in the United States, as the suggested existence of training camps implies the Al Qaeda threat is in fact a reality. Overall, the attributes identified were of a health, economic, or psychological nature.

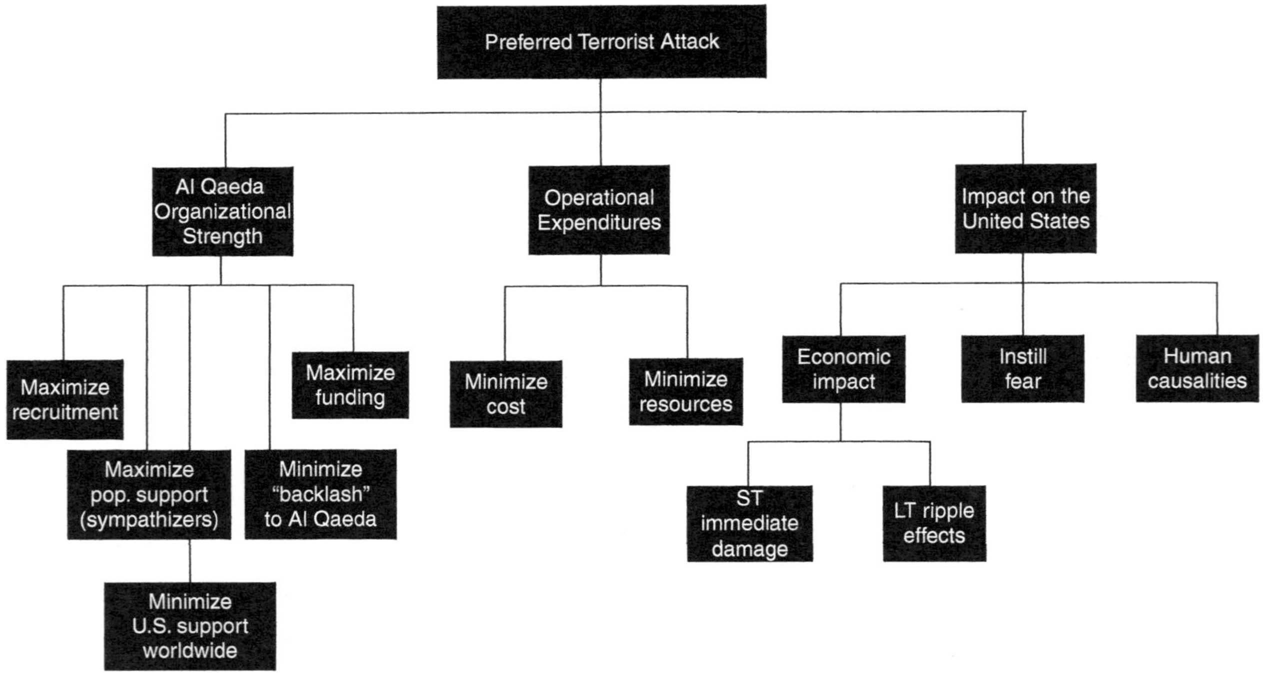

Figure 28.1. SME proxy assessment of terrorist leader objectives hierarchy.

Table 28.1. *Nine identified attack types considered*

Attack Alternatives
No attack (baseline)
IED* in the engine room of naval vessel
Explosion resulting in dam failure
MANPADS** attack on an airplane
Portable nuclear bomb in a major city
Explosion on mass transport(s)
Release of anthrax (movie or sports arena)
Detonation of a dirty bomb
Smallpox release in a major city

* Improvised explosive device
** Man Portable Air Defense Systems

28.5 Attack Alternatives

Proxies were asked to identify a range of attack modes (strategies) that Al Qaeda leaders might contemplate. These suggestions were collected during open dialogue about the terrorist organization's operational capabilities and the feasibility of various attack options.

Table 28.1 is the compiled list of attack strategies. The strategy list resulted in a range of attack modes including explosive, nuclear, and biological alternatives. The "no attack" alternative was included as a baseline, or status quo alternative for comparison. Some proxies also indicated doubt that Al Qaeda leaders were actively preparing to execute attacks within the United States. They posited that the organization might be directing its resources to internal development or to the execution of attacks outside of the United States.

The attack strategies were deliberately broadly defined as strategic initiatives that allowed each proxy to flexibly interpret the attacks in terms of size, frequency, and location. The intent of using attack strategies rather than specific attack scenarios was to fully capture the range of possible attack alternatives Al Qaeda leadership might be considering.

28.6 Decision Tree Probability Estimates

A critical component to modeling preferences for terror attack strategies is the uncertainty that Al Qaeda leaders might have regarding their success in executing each of the attack types. We postulate that a terrorist attack operates much like any other complex management project, starting with an attack-planning phase, followed by the actual preparations for the attack, and culminating with

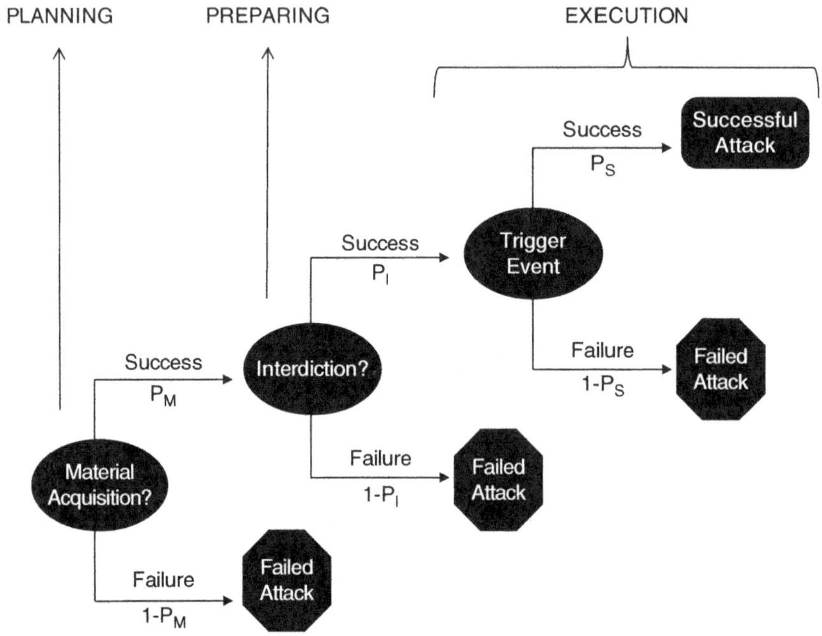

Figure 28.2. Attacker event tree with 3 key uncertainties.

attack execution (Rosoff & von Winterfeldt, 2007). Figure 28.2 is a simple event tree illustrating these three critical phases; event nodes (circles) are conditional on the preceding event outcomes to the left.

In the planning phase, the acquisition of material is instrumental to ensuring that the attack strategy is viable. Avoiding interdiction by anti-terror forces is an ongoing concern for the terrorist leader during the planning phase, including such tasks as collection of restricted materials and surveillance of the target, both essential to attack success. Lastly, the execution phase refers to the critical final steps required to complete the attack. Emphasis is on whether the triggering device is effective and whether the executioner successfully carries out the attack. If at any point within the planning, preparing, or execution phases a task is not successfully completed, the attack is assumed terminated.

Each of the planning, preparing, and execution phases is associated with a probability of detection and disruption of the attack. To determine how these probability estimates affect the overall attack success probability, the proxies estimated the probability of success of each phase for each attack type. Each probability estimate varied depending on the difficulty of the task as perceived by the proxy.

The elicitation process included two parts (Hora, 2007). First, proxies provided a preliminary estimate of the probability of success for a given attack strategy of obtaining necessary resources (including materials), successfully avoiding interdiction, and execution of attack. Next, they considered the uncertainty in their estimates. To account for proxy uncertainty, general Beta distributions were specified for each probability estimate. Proxies specified an interquartile range (25th and 75th percentiles on the cumulative distribution) for each probability estimate. Beta parameters were estimated using the @Risk software (Best Fit module) for each distribution using these two points on the cumulative Beta distribution ($F = 0.25$ and $F = 0.75$).

The resulting probability distribution was then compared to the probability originally assessed and proxies were allowed to adjust their estimates to resolve inconsistencies. For example, one expert cited that the probability of a terrorist successfully acquiring radioactive material might vary depending on the number of persons involved and the type of material acquired. Another suggested that the probability of successful execution might vary depending on whether the attack was carried out using a triggering device or a suicide bomber.

28.7 Attribute Definition and Measurement

An important part of multiattribute utility assessment is specifying attribute scales that allow for comparison of the alternative attack strategies relative to the fundamental objectives defined earlier. Proxies provided both measures, operationalizing how the attributes are defined, as well as scales indicating the units and range over which the measures are defined. Table 28.2 shows how the attribute units and scales were defined by one proxy. This particular proxy found it convenient to base scales on a comparison with outcomes following the attacks of September 11 (9/11).

In developing the model, proxies were encouraged to define the attribute scales as they desired (as opposed to forcing a consistent set of units and scales across proxies). This enabled the proxies to consider uncertainty about consequences in a form that is most natural and familiar to each, presumably resulting in more valid estimates for each.

Given the limited number of international terrorist attacks on the United States and the inability to interview Al Qaeda leaders directly, a high degree of uncertainty surrounding proxies' understanding of terrorist leaders is not unusual. Consequently, proxies provided probability distributions over each attribute scale for each attack type. Distributions were defined by each proxy

Table 28.2. *Attribute units and scales defined for one proxy*

Attribute	Measure (Unit)	Scale
Short-term economic	% of 9/11	0–400%
Long-term economic	% of 9/11	0–500%
Recruitment	% of 9/11 militaristic recruitment	0–400%
Popular support (for Al Qaeda)	% of 9/11 popular support	0–170%
Damage to Al Qaeda	% of 9/11 organization damaged	0–60%
Instill fear in the United States	% life change (relative to 9/11)	0–80%
Worldwide U.S. support (post attack)	% of 9/11 worldwide support	0–100%
Kill Americans	# killed	0–100,000
Funding	% of 9/11 annual funding	0–400%
Cost	Dollars (relative to 9/11)	0–$450,000
Resources	# of people required (relative to 9/11)	0–25

using general Beta distributions on the entire matrix of nine attack strategies by 11 attributes. For each of the 99 cells of the matrix, each proxy provided an estimate of the consequence of each of the nine attack strategies relative to each of the 11 individual attribute scales.

The initial assessment was conducted assuming a successful attack, but modified assessments were also obtained for the three unsuccessful end nodes of the event tree described previously in Figure 28.2. As with the attack event tree in Section 28.2.4, the uncertainty distribution elicitation process included two steps. First, the estimate was assessed as a median, such that the proxy indicated a 50/50 chance that the true score was above or below the estimate. Then, the Beta distribution was obtained by eliciting both a range (minimum and maximum possible scores) and an interquartile range (25th and 75th percentiles on the cumulative distribution). The Best Fit module of @Risk was again used to calculate parameters of the Beta distribution consistent with these four estimates. The median of the Beta distribution fit to these estimates was then compared to the median obtained in the original estimate, and the proxy was allowed to make adjustments to resolve inconsistencies. The proxies were provided with several fractiles of the resulting distributions, including more extreme percentiles (5th and 95th), in addition to the values corresponding to the elicited fractiles (25th, 50th, and 75th).

Table 28.3. *Impact of interdicted or failed attack on attributes by alternative score matrix*

	Failed Material Acquisition	Attack Interdiction During Preparing	No Impact (utility = 1)
Minimize blowback to Al Qaeda	No impact (utility = 1)	% of execution utility	No impact (utility = 1)
Minimize U.S. worldwide support	No impact (utility = 1)	% of execution utility	No impact (utility = 1)
Attack cost	% of execution utility	full execution utility	full execution utility
Attack resources	% of execution utility	full execution utility	full execution utility

Execution utility refers to the utility value assigned to the attribute assuming a successful attack.

The resulting Beta distributions captured the degree of overall uncertainty in the proxy estimates, as well as any skew in the direction of the uncertainty. The resulting distribution is intended to quantify uncertainty about how well each attack alternative satisfies the attributes defining each fundamental objective.

By having proxies characterize attribute performance using different scales and uncertainty distributions, the uniqueness of the model as a tool for capturing perceived Al Qaeda leader motivations and beliefs is preserved. In addition, it is still possible to compare relative expected utilities and probability estimates that each alternative is the "best," as calculated in the Monte Carlo simulation.

As noted previously, the initial proxy scale scores (and distributions) are conditional on a successful attack. However, there are instances in which the attack fails – whether it is during resource acquisition, discovery and interdiction during adversary surveillance, or unsuccessful execution. For unsuccessful attack outcomes, many attributes scores will be quite different. If, for instance, the terrorist is unable to acquire the radioactive material for a dirty bomb, certain costs and resources still are expended in the process. However, only a fraction of the originally estimated costs and resources for the full attack would be expended. For many attributes, a failed attack would result in "no impact," leading to a very different scale score than that resulting from a successful attack outcome. Table 28.3 displays how the utility values of these select attributes might be configured for the current example in the event of a failed attack.

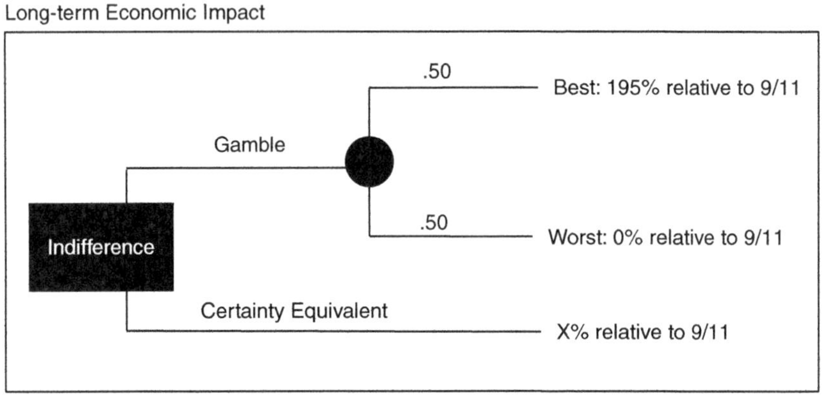

Figure 28.3. Certainty equivalent assessment for long-term economic impact single attribute utility function.

28.8 Risk Attitudes across Attributes

The first instance of variation in proxy perceptions of Al Qaeda leaders' motivations and beliefs was expressed through attribute definition and uncertainty about impacts of attack outcomes on the scores for attack strategies on each attribute. A second variation over proxies is the single-attribute utility functions that capture the terrorist leaders' attitude toward risk. While the proxies may share similar attack attributes, their perspectives relative to risk attitude may not be in agreement. Each proxy can have different views regarding terrorist leader attitudes toward risk and in turn, may be willing to accept different levels of risk.

Utility functions are traditionally used to capture an individual's attitude toward risk. The concavity of the utility function indicates whether that individual is risk averse, neutral, or seeking. Through the acquisition of certainty equivalents from the proxies, we are able to estimate the nature of proxy perceptions of terrorist leader risk attitude with respect to each attribute. To accomplish this, an exponential utility function is estimated for each proxy for each of the 11 attributes by assessing a certainty equivalent on each attribute for a 50/50 gamble between the worst and best outcomes for each attribute. The proxy was asked to estimate a sure outcome that would make the terrorist leader indifferent between playing the gamble and taking the sure thing. Figure 28.3 displays two proxies' decision trees capturing the certainty equivalent for the long-term economic impact attribute.

Proxy 1 evaluated long-term economic impact in terms of damage relative to 9/11. Proxy 1 perceives 55% of the economic impact following 9/11 to be of equal value to the gamble between his best (195%) and worst (0%) estimates relative to 9/11. (Note: The range of the attribute, 0% to 195% of the 9/11 economic impact, was provided by Proxy 1.) Proxy 2 evaluated long-term economic impact in terms of a percentage of $2 trillion. Proxy 2 perceived a terrorist leader would value 30% of $2 trillion as much as the best-worst gamble. Interestingly, while the proxies used different units for attribute definition, their certainty equivalents represented a similar percentage of the total measure, roughly 30%.

As with the definition of attribute uncertainties across proxies, generalized Beta distributions were assessed for the proxies' uncertainty about the terrorist leaders' certainty equivalent for the gamble (Hora, 2007). (The same four-point elicitation procedure of obtaining a minimum, a maximum, and an interquartile range was used as for the score matrix. An iterative process for comparing the median of the obtained distribution to the original estimate was also used to resolve inconsistencies.)

28.9 Value Trade-offs across Attributes

Another essential component of the MAU model is the proxies' perception of how terrorist leaders assess the trade-offs among the attributes representing attacker objectives. In general, the model form is verified by questioning the proxies regarding attribute independence. In the present example, an additive model is assumed. Each proxy rank orders the attributes based on a "swing weight" assessment procedure (Keeney et al., 1990). Proxies identify the attribute for which the change (swing) from worst to best represents the largest impact for the terrorist leader. All remaining attributes are assigned a percentage between 0 and 100% to reflect the relative desirability of changing (swinging) a score from worst to best on that attribute.

Generalized Beta distributions are assessed over each assigned swing weight to reflect uncertainty about terrorist leaders' relative value of the swing from worst to best (Hora, 2007). The same four-point elicitation procedure of obtaining a minimum, maximum, and an interquartile range is used as for the trade-off uncertainty matrix.

Lastly, the swing weights are normalized to sum to 1.0. Table 28.4 shows the mean normalized swing weight assessments for all four proxies. While two of the proxies prioritized logistical feasibility, such as cost of the attack, one other emphasized economic impacts, and the fourth stressed the psychological implications of an attack – the ability to instill fear.

Table 28.4. *Normalized swing weights by attributes and proxies*

	ST Economic Impact	LT Economic Impact	Recruitment	Popular Support	Retaliation	Instill Fear
Proxy 1	0.15	0.16	0.08	0.10	0.04	0.12
Proxy 2	0.10	0.14	0.07	0.10	0.05	0.10
Proxy 3	0.05	0.05	0.07	0.08	0.08	0.07
Proxy 4	0.15	0.14	0.04	0.07	0.08	0.16

	U.S. Support	Kill Americans	Funding	Cost	Resources
Proxy 1	0.08	0.12	0.08	0.05	0.05
Proxy 2	0.09	0.04	0.00	0.24	0.07
Proxy 3	0.09	0.08	0.07	0.29	0.06
Proxy 4	0.11	0.15	0.02	0.03	0.05

28.10 Results

28.10.1 Attack Alternative Ranking

The model includes full implementation of the event tree, with four possible outcomes (three failures and one success), as well as the additive multiattribute utility function, utilizing exponential single attribute utility functions and trade-offs defined by swing weights. Uncertainty is captured in all proxy assessments, including event tree probabilities, the attributes by alternatives score matrix, swing weights, and certainty equivalents defining single-attribute utility functions. A risk profile for each attack strategy is generated using Monte Carlo simulation. Expected utilities are calculated as the means of these distributions. In the present demonstration, distributions are estimated from 10,000 iterations using the @Risk software (Palisades, Inc.).

Figure 28.4 displays one proxy's multiattribute utility distribution, representing a risk profile for the attack alternative: "explosions on mass transportation." As illustrated by the distribution range and shape, there was some uncertainty concerning the proxy's perception of the desirability of this attack alternative for the terrorist leader. For this particular proxy, the Al Qaeda terrorist leader's desirability for an attack involving explosions on mass transportation was calculated to be between 0.21 and 0.41, and the distribution is right skewed, with a mean EU = 0.34.

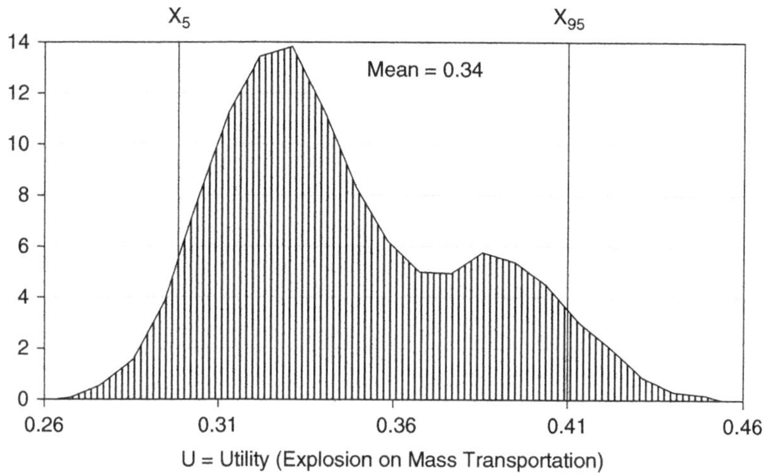

Figure 28.4. PDF from Monte Carlo simulation of attacker utility for Mass transportation attack alternative.

Table 28.5. *Expected utilities for 9 attack alternatives and 4 proxies*

Attack Alternative	Proxy 1	Proxy 2	Proxy 3	Proxy 4
No attack	0.18	0.49	0.58	0.41
IED	0.15	0.45	0.34	0.21
Dam explosion	0.14	0.47	0.29	0.15
MANPADS	0.14	0.46	0.38	0.33
Portable nuclear device	0.06	0.44	0.10	0.14
Transportation system	0.14	0.48	0.35	0.36
Anthrax	0.16	0.45	0.32	0.37
Dirty bomb	0.14	0.42	0.22	0.27
Smallpox	0.09	0.51	0.17	0.14

Results presented in Table 28.5 provide a relative ordering of the 9 attack alternatives in terms of expected utility for each proxy. Findings indicate that when accounting for the possibility of attack failures, the attack with the highest mean expected utility was no attack for three of the four proxies (all except Proxy 2). Proxy 2 indicated terrorist leader's preference for a smallpox attack, and this utility was only 0.02 greater than that for no attack. Interestingly, the utility outputs for Proxy 2 are all very similar. This result suggests that Proxy 2 does not believe Al Qaeda leaders' preferences for attack type vary, or, alternatively, that there is considerable uncertainty about terrorist leaders' preferences.

28.10.2 *Utilities by Proxy*

Cumulative utility distributions (risk profiles) are presented for one proxy for all nine attack strategies in Figure 28.5. Each curve represents a different attack strategy. Utility increases from left to right, so curves on the right generally reflect more desired attack strategies than curves on the left. The "no attack" alternative is the right-most distribution.

The cumulative risk profiles depict considerable variability in utility distributions across attack alternatives. Several of the distributions intersect on the right, indicating little or no stochastic dominance among the more attractive attack alternatives. In addition, there is some variations in curve shapes, indicating more uncertainty associated with some attacks compared to others.

For example, the greater "S" formation in the no attack distribution compared to that of the smallpox attack implies greater uncertainty in the terror leader's utility for the smallpox attack compared to the no attack alternative.

Ultimately, the model produces estimated attack probabilities. These estimates are derived by sampling from the expected utility distributions. We use the risk profiles to calculate the probability that the utility for each alternative attack strategy is the maximum. Table 28.6 displays each proxy's estimated subjective probabilities and specifies the most attractive attack strategy for each.

Given the possibility of not acquiring necessary material, getting caught, and failing to successfully execute an attack, the most probable scenario for two of the proxies is "no attack." Of the remaining two proxy models, one indicated a smallpox attack is most likely, and one suggested that a MANPADS attack is most likely. The probability estimates for Proxies 2 and 4 are considerable lower than those for Proxies 1 and 3. This suggests that Proxies 2 and 4 believe the attack strategies in this study are not strongly favored by terrorist leaders. However, since an attack strategy was preferred to the "no attack" alternative, it is possible the proxies felt that other attack strategies (not included in the list) might better capture a terrorist leader's preferences. Lastly, the probability estimates across some of the attack strategies are extremely low for all four proxies, such as an IED attack and dam failure.

28.11 Conclusions

The proxy utility modeling approach introduced in this chapter represents an alternative to game theory models for adversary decision making, e.g., Shan and Zhuang (2013). In the example demonstration, analysts worked closely with proxies to formulate an understanding of how terrorist leaders' values and

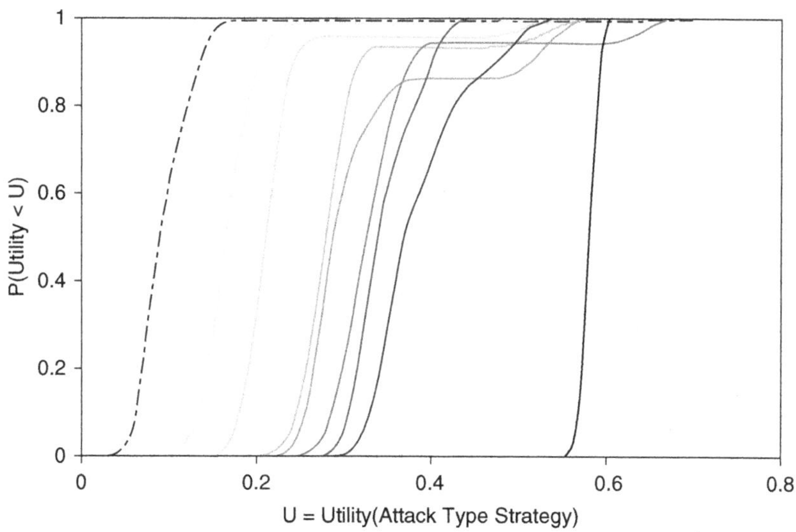

Figure 28.5. Attacker's risk profiles (CDF for overall utility) for 9 attack options; from left to right at the median, (1) Portable nuclear device, (2) Smallpox, (3) Dirty bomb, (4) Dam explosion, (5) Anthrax, (6) IED, (7) Transportation system, (8) MANPADS, (9) No attack.

Table 28.6. *Probabilities of 9 attack alternatives for 4 proxies*

Attack Alternative	Proxy 1	Proxy 2	Proxy 3	Proxy 4
No attack	0.69	0.13	0.94	0.10
IED	0.01	0.01	0.05	0.00
Dam explosion	0.00	0.03	0.00	0.01
MANPADS	0.00	0.09	0.00	0.24
Portable nuclear device	0.03	0.12	0.01	0.09
Transportation system	0.00	0.16	0.00	0.12
Anthrax	0.08	0.05	0.00	0.22
Dirty bomb	0.14	0.07	0.00	0.17
Smallpox	0.05	0.34	0.00	0.05

beliefs influence their attack strategy preferences. Data were collected from proxies on attack strategy attributes, their risk preferences for the attributes, their value trade-offs across attributes and objectives, and their estimates of attack strategy feasibility. Probability distributions also were elicited over all model parameters to account for the uncertainty that terrorists have about events leading up to attack execution and consequences following an executed attack, as well as proxy uncertainties about terrorist leaders' values. Analysts then used the information collected from proxies to estimate the threat of the assessed attack strategies, in terms of a random utility model and probability estimates that each attack mode would be judged "best" by the terrorist leaders. The resulting model produces an overall probability of attack strategy selection by the terrorist leadership.

Results indicate that after taking into account the possibility of not acquiring the necessary material for an attack, getting caught, or not successfully executing an attack, the scenario with the highest mean expected utility was "no attack" for three of the four proxies. The remaining proxy indicated that a smallpox attack had the highest expected utility. When probability estimates were calculated, two of the proxies indicated that the terrorist leadership was most likely to select the "no attack" alternative. The third proxy indicated that a MANPADS attack would be most likely selected by the Al Qaeda leadership, even though "no attack" had the highest expected utility. The fourth proxy indicated that a smallpox attack was the most likely to be selected by the terrorist leader, consistent with the attack alternative with the highest expected utility.

28.12 Model Challenges

By design, the proxy utility model does not capture changes in the objective and attribute inputs over time. Some of the variability in proxy terrorist leader values is captured by the probability distributions over model parameters. However, significant political, economic, and social changes will likely occur, resulting in the need to restructure and redefine the fundamental objectives hierarchy. For instance, a terror organization's leadership may change. Whether it is because its leadership structure has evolved, or a leader is killed or captured, new leaders rise in the ranks. The new leaders' beliefs and motivations might diverge from those of their predecessors. Alternatively, as DHS counterterrorism efforts are introduced, the terrorist leaders' attack preferences and strategies might be altered to adjust to the new environment. Whatever the reason, both terrorist organizations' and DHS's values and preferences are bound to change, and the associated adversary utility model should be adjusted accordingly.

One way to deal with the dynamic nature of an adaptive adversary is to reevaluate the proxy utility model on a regular basis. However, ongoing assessments would require resources and commitment on the part of the proxies and trained analysts. To avoid some of the complications with frequent data elicitation, the development of a dynamic utility/feedback model of a terrorist organization that accounts for the fluctuations in terrorist leader beliefs and motivations might be a viable alternative.

28.13 Model Applications

While the proxy utility model's outputs for the example demonstration are mostly illustrative of the methodology used, they do show how the model might help the Department of Homeland Security (DHS) make better decisions when responding to terrorist threats. In any decision context, it is difficult to determine how to allocate security resources when one does not understand the nature of the threat, vulnerability, and consequences. The proxy value model presents a formalized approach for understanding the influence of various terrorist motivations and capabilities on the selection of potential attack strategies.

The comparative nature of the decision model output might furthermore help DHS address some of the challenges and complexity associated with the allocation of the department's annual budget. Taking time to consider the uncertainties regarding terrorist leader objectives and the resulting impact this might have on alternative selection (e.g., target, mode of attack, etc.) could help the department in allocating constrained resources. As such, by investing in modeling the values and beliefs of adaptive adversaries, DHS resources can be more wisely directed.

Last, DHS also might explore the contributions of the proxy value model to counter-intelligence analyses. Intelligence analysts present completely different sources of data input for the model. Intelligence analysts have access to proprietary and recent information about terrorist activities, communications, and capabilities. Adjusting the model to account for multiple data sources could provide valuable insights regarding the evolving nature of the threat from an adaptive adversary.

REFERENCES

Barrett, A., & Casman, E. (2013). Should cities invest in sheltering-in-place measures against chlorine truck attacks by terrorists? *Risk Analysis*, *33*(5), 931–944.

Crenshaw, M. (Ed.) (1995) *Terrorism in context*. State College: Pennsylvania State University Press.

Gordon, P., Moore, J., Park, J., & Richardson, H. (2007). The economic impacts of a terrorist attack on the U.S. commercial aviation system. *Risk Analysis*, *27*(3), 505–512.

Hartnett, E., Paoli, G., & Schaffner, D. (2009). Modeling the public health system response to a terrorist event in the food supply. *Risk Analysis*, *29*(11), 1506–1520.

Hora, S. (2007). Eliciting probabilities from experts. In W. Edwards, R. F. Miles, & D. von Winterfeldt (Eds.), *Advances in decision analysis: From foundations to applications* (pp. 129–153). New York: Cambridge University Press.

Keeney, G., & von Winterfeldt, D. (2010). Identifying and structuring the objectives of terrorists. *Risk Analysis*, *30*(12), 1803–1816.

Keeney, R. L. (1992). *Value focused thinking*. Cambridge, MA: Harvard University Press.

Keeney, R. L. (2007a). Developing objectives and attributes. In W. Edwards, R. F. Miles, & D. von Winterfeldt (Eds.), *Advances in decision analysis: From foundations to applications* (pp. 104–128). New York: Cambridge University Press.

Keeney, R. L. (2007b). Modeling values for anti-terrorism analysis. *Risk Analysis*, *27*(3), 585–596.

Keeney, R. L., & Raiffa, H. (1976). *Decisions with multiple objectives: Preferences and value trade-offs*. New York: John Wiley & Sons, Inc.

Keeney, R. L., & von Winterfeldt, D. (2011). A value model for evaluating homeland security decisions. *Risk Analysis*, *31*(9), 1470–1487.

Keeney, R. L., von Winterfeldt, D., & Eppel, T. (1990). Eliciting public values for complex policy decisions. *Management Science*, *36*(9), 1011–1030.

Lankford, A. (2014). Précis of "The myth of martyrdom: What really drives suicide bombers, rampage shooters, and other self-destructive killers." *Behavioral and Brain Sciences*, *37*, 351–393.

Leung, M., Lambert, J., & Mosenthal, A. (2004). A risk-based approach to setting priorities in protecting bridges against terrorist attacks. *Risk Analysis*, *24*(4), 963–984.

Rose, A., Oladosu, G., & Liao, S. (2007). Business interruption impacts of a terrorist attack on the electric power system of Los Angeles: Customer resilience to a total blackout. *Risk Analysis*, *27*(3), 513–531.

Rosoff, H. (2009). *Using decision and risk analysis to assist in policy making about terrorism.* Doctoral Dissertation. University of Southern California, Sol Price School of Public Policy.

Rosoff, H., & von Winterfeldt, D. (2007). A risk and economic analysis of dirty bomb attacks on the ports of Los Angeles and Long Beach. *Risk Analysis*, *27*(3), 533–546.

Shafieezadeh, E., Cha, J., & Ellingwood, B. (2015). A decision framework for managing risk to airports from terrorist attack. *Risk Analysis*, *35*(2), 292–306.

Shan, X., & Zhuang, J. (2013). Hybrid defensive resource allocations in the face of partially strategic attackers in a sequential defender-attacker game. *European Journal of Operational Research*, *228*(1), 262–272.

Simonoff, J., Restrepo, C., Zimmerman, R., & Naphtali, Z. (2007). Analysis of electrical power and oil and gas pipeline failures. In E. Goetz & S. Shenoi (Eds.), *Critical infrastructure protection* (pp. 381–394). Boston, MA: Springer.

Sprinzak, E. (2000). Rational fanatics. *Foreign Policy*, *120*, 66–73.

Victoroff, J. (2005). The mind of the terrorist: A review and critique of psychological approaches. *Journal of Conflict Resolution, 49*(1), 3–42.

Von Winterfeldt, D., & O'Sullivan, T. (2006). Should we protect commercial airplanes against surface-to-air missile attacks by terrorists? *Decision Analysis, 3*(2), 63–75.

Wein, L., Choi, Y., & Denuit, S. (2010). Analyzing evacuation versus shelter-in-place strategies after a terrorist nuclear detonation. *Risk Analysis, 30*(9), 1315–1327.

29

Asymmetric Prescriptive/Descriptive Game Theory for Counterterrorism

Jason R. W. Merrick and Philip Leclerc

29.1 Introduction

Raiffa (1985) describes three types of game theory (GT) research. The first is symmetrically descriptive research where one aims to describe the behavior of all decision makers in a competitive situation, now known as behavioral game theory (Camerer, 2003). The second is symmetrically prescriptive research where one aims to prescribe the normatively best actions of each decision maker. The third is asymmetrically prescriptive/descriptive research where one aims to prescribe the action one decision maker should take, while describing the actions that competing decision makers will take.

Expected-utility theory (EUT) (Ramsey, 1931; Savage, 1954) is the accepted model for prescriptive research. Von Neumann and Morgenstern (1944) use EUT to model human preferences over risk in their theory of games, an assumption carried forward in the concept of a Nash equilibrium (Nash, 1950). This makes classical game theory symmetrically prescriptive. However, the descriptive validity of EUT has been questioned, and many experimental anomalies have been identified that EUT cannot descriptively match. In an effort to explain these anomalies, a number of alternative models of preferences under risk have been developed; Starmer (2000) provides a survey of work in this area. Of particular importance for descriptive research is the original prospect theory (PT) of Kahneman and Tversky (1979) and its more modern version, prospect theory (PT), described in Tversky and Kahneman (1992). PT is a generalization of EUT aimed at accommodating three commonly observed, systematic deviations of human behavior from EUT: reference dependence, probabilistic insensitivity, and rank dependence. PT replaces EUT's assumption of linearity in probabilities with a probability weighting function and replaces the utility function with a reference-dependent value function.

A small but growing literature seeks to address how descriptive preferences should impact game theory. Colman (2003) criticizes the dominance of traditional, EUT rationality in GT, while Kadane and Larkey (1982a, 1982b) discuss asymmetrically prescriptive/descriptive analysis and propose the use of an "empirically supported psychological theory making at least probabilistic predictions about the strategies people are likely to use" for descriptive analyses. Shalev (2000) developed "loss-aversion equilibrium" (LAE) to address reference dependence, but maintained EUT's linearity in probabilities. Metzger and Rieger (2010) developed an equilibrium concept that featured probability weighting but neither rank nor reference dependence, while Ritzberger (1994) defined an equilibrium concept that incorporated both probability weighting and rank dependence, but did not feature reference dependence. Berejikian (2002a, 2002b) and Butler (2007) considered PT's impacts on game theory without explicitly defining new equilibrium concepts. Butler applied PT to an ultimatum game, but evaluated only the impact of reference dependence, assuming linearity in probabilities.

This approach to asymmetrically prescriptive/descriptive decisions has been fully accepted at the Center for Risk and Economic Analysis of Terrorist Events at the University of Southern California. Yang and colleagues (2013) introduce PT preferences in a Stackelberg game for security decisions, an approach extended in Kar and colleagues (2015) to include experiments with human subjects. Nguyen and colleagues (2013) follow a similar approach, but assume that the attacker chooses his strategy using quantal response equilibria. Merrick and Leclerc (2016) develop a general asymmetric/descriptive approach to counterterrorism decisions assuming PT preferences for the attacker.

In this chapter, we make the case for asymmetric prescriptive/descriptive game theory in counterterrorism research and describe the experimental observations that led to prospect theory. We review the work on reference dependence and loss aversion in Shalev (2000) and the work on probability weighting in Ritzberger (1994). Both authors demonstrate that equilibrium can fail to exist in certain conditions, so we prove existence in the types of games of interest in counterterrorism. We finish by considering asymmetric prescriptive/descriptive game theory with EUT preferences for one player and PT preferences for other players, reviewing the work in Merrick and Leclerc (2016) demonstrating the impacts of prospect theory on a simple container screening decision.

29.2 Asymmetrically Prescriptive/Descriptive Applications in Counterterrorism

Merrick and Parnell (2011) reviewed a number of approaches for making counterterrorism decisions, including decision analysis, game theory, and hybrid approaches such as intelligent adversary risk analysis (Parnell et al., 2010) and adversarial risk analysis (Rios Insua et al., 2009). Figure 29.1 shows the main difference between decision analysis and game theory approaches to counterterrorism decisions. In each case in Figure 29.1, the defender's decision is whether to invest in screening containers entering the United States for radioactive material. Subsequently, the attacker attempts to obtain radioactive material to use in a radioactive dispersal device (RDD or dirty bomb), and then makes the choice whether to use the RDD or attack with conventional methods. The chance of success of an attack is reduced if the defender is screening, while conventional attacks are not affected by screening. In the decision analysis view (left of Figure 29.1), the defender decides whether to screen and the attacker's decision to use an RDD is an uncertainty for the defender. In the game theory view (right of Figure 29.1), the defender still decides whether to screen, but the attacker's decision to use an RDD is a decision for the attacker that is made by maximizing the attacker's preferences. One can view the right side of Figure 29.1 as a sequential game with perfect information, and if we assume that the defender is minimizing the expected cost of an attack and the attacker is maximizing the expected cost, then it is a zero-sum game. These are the assumptions made in Parnell and colleagues (2010).

When solving for the defender decision in Figures 29.1a and 1b, we are working in a prescriptive mode to help the defender make a better decision. When estimating the probability of the attacker choosing either alternative in Figure 29.1a, we are eliciting probabilities from intelligence experts (see Merrick et al., 2015, for a review). However, when we are solving for the attacker decision in Figure 29.1b, we are not attempting to help the attacker make a better decision; we are attempting to predict his behavior and so are working in a descriptive mode. The standard assumption of EUT preferences for the attacker (see Merrick & Parnell, 2011) has been shown to be a poor descriptive model, and PT preferences are more suitable based on experimental evidence (Tversky & Kahneman, 1992).

29.3 Prospect Theory

Kahneman (2011) provides simple thought experiments to demonstrate reference dependence, loss aversion, and likelihood insensitivity, the descriptive decision-making behavior described by prospect theory. In Table 29.1, we consider in each case (row) a choice between lottery A and lottery B. For Case 1a, would you prefer to win $90 for sure (lottery A) or have a 90% chance of winning $100 and a 10% chance of winning nothing? Most people choose the sure thing in lottery A, demonstrating risk aversion for gains (winning additional

(a)

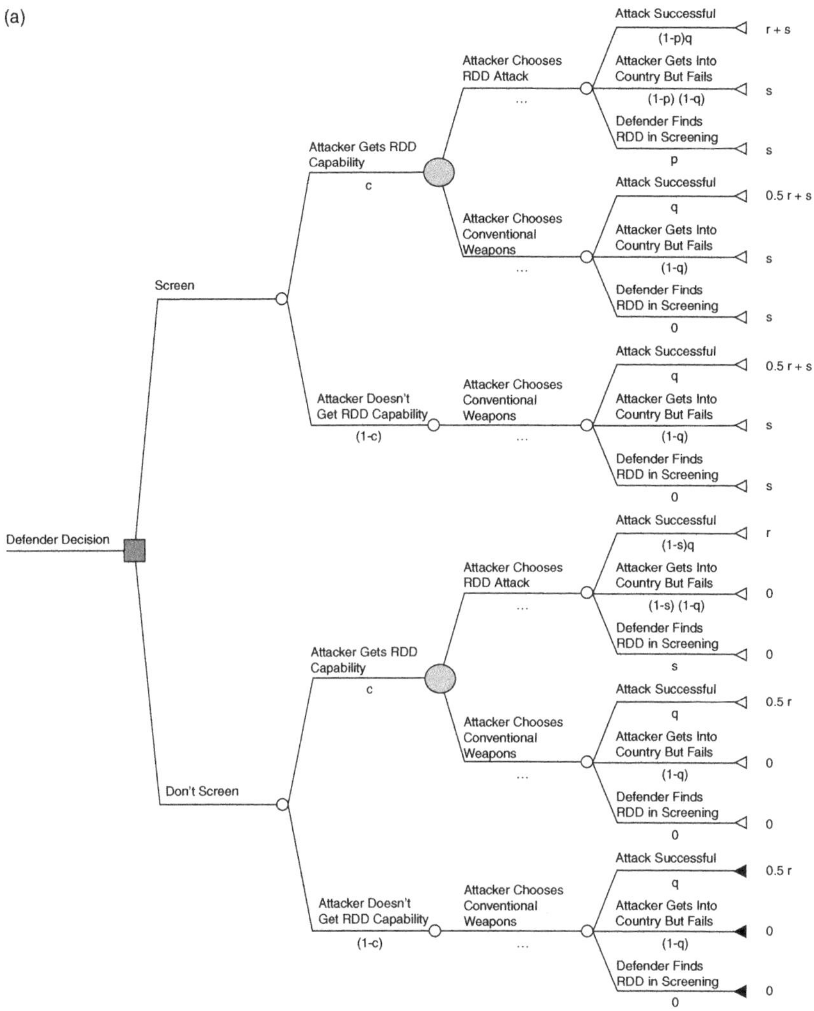

Figure 29.1a. A simple container screening decision from the perspective of decision analysis.

money). However, in Case 1b where the outcomes become losses, most people choose lottery B, the gamble that gives them a chance of not losing. Thus, they are risk-seeking for losses. This also demonstrates reference dependence, as their risk behavior changes above and below the reference point (no win and no loss). Case 2 asks you to consider the status quo in lottery A and a lottery that gives you a 50-50 chance of losing \$100 or winning an amount \$X.

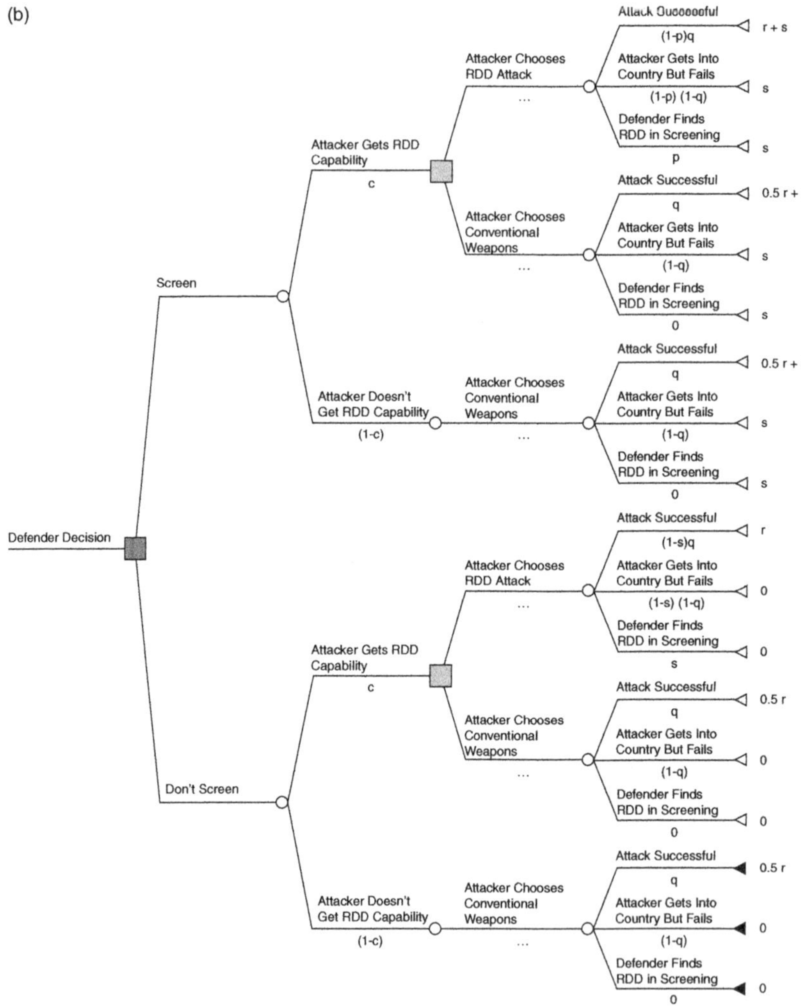

Figure 29.1b. A simple container screening decision from the perspective of game theory.

If X = \$100 most people choose against the gamble. If X = \$10,000, almost everyone would choose to gamble. At what point does this switch happen? For most people, the switch happens between X = \$150 and X = \$300. This demonstrates loss aversion as we weight the loss of \$100 higher than the gain of \$100 and most people weight the loss 1.5 to 3 times as high as the gain.

Table 29.1. *Example lotteries that demonstrate reference dependence, Loss Aversion, and Likelihood Insensitivity*

	Lottery A	Lottery B
Case 1a	Win \$90 for sure	Win \$100 with 90% chance
Case 1b	Lose \$90 for sure	Lose \$100 with 90% chance
Case 2	Win nothing Lose nothing	Win \$X with 50% chance Lose \$100 with 50% chance
Case 3a	61% chance to win \$520,000	63% chance to win \$500,000
Case 3b	98% chance to win \$520,000	100% chance to win \$500,000

Cases 1 and 2 demonstrate the way decision makers consider the outcome of a decision. Case 3 demonstrates the way we consider the probabilities. Case 3a gives us lotteries that are close in probability and close in outcome. In considering lotteries A and B, most people consider the difference between 61% and 63% inconsequential and so choose lottery A with the higher outcome of \$520,000. However, the probabilities in Case 3b also only differ by 2%, but the impact on the decision maker is more pronounced. In lottery B, the decision maker is now offered a sure thing and even though the winnings are lower, he chooses against the 2% probability of winning nothing in lottery A, despite the higher possible outcome. This demonstrates that we are overly drawn to a sure thing (the certainty effect) and that we are insensitive to probabilities in the middle range (likelihood insensitivity).

Tversky and Kahneman (1992) codify these effects mathematically. Suppose we have n possible outcomes that are denoted $x_1 \geq \cdots \geq x_k \geq 0 \geq x_{k+1} \geq \cdots \geq x_n$. Thus, the outcomes are ordered from highest to lowest and there are k that are considered gains and $n-k$ that are considered losses. The probabilities of these outcomes are $p_1, \ldots, p_n$. We compute the cumulative probabilities $P(X \geq x_j) = \sum_{i=1}^{j} p_i$ and $P(X \leq x_j) = \sum_{i=j}^{n} p_i$. The expected utility of the gamble is

$$\sum_{j=1}^{n} p_j u(x_j)$$

or equivalently

$$\sum_{j=1}^{n} \left(P(X \leq x_j) - P(X < x_j) \right) u(x_j).$$

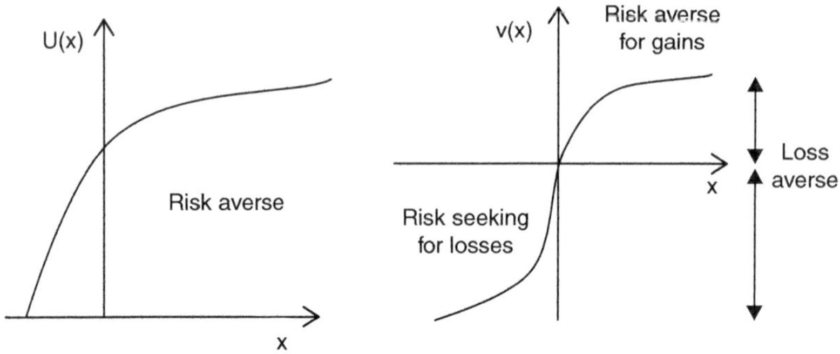

Figure 29.2. A typical utility function from EUT (left) and a typical value function from PT (right).

The prospect theoretic value of the gamble is then given by

$$\sum_{j=1}^{k}\left(w\left(P\left(X \geq x_j\right)\right)-w\left(P\left(X > x_j\right)\right)\right) v(\mathrm{x}_j)$$
$$+\sum_{j=k+1}^{n}\left(w\left(P\left(X \leq x_j\right)\right)-w\left(P\left(X < x_j\right)\right)\right) v(\mathrm{x}_j).$$

One can immediately observe that the prospect theoretic value separates the expression into a sum for gains and a sum for losses. In each sum, the utility function $u(\mathrm{x})$ is replaced by a value function $v(x)$. Figure 29.2 shows the difference between the typical utility function and the typical value function. Utility functions do not consider a reference point and are typically concave to represent risk aversion. The value function has a different shape above and below the reference point. Above the reference point, it looks like the utility function and is risk averse for gains.

However, below the reference point it is convex to represent risk-seeking behavior for losses. Also, the magnitude of $v(+x)$ is lower than for $v(-x)$ which represents loss aversion. The cumulative probabilities are weighted by the weighting function $w(p)$. However, that weighting is applied to $P\left(X \geq x_j\right)$ and $P\left(X > x_j\right)$ before taking their difference in the gains sum and to $P\left(X \leq x_j\right)$ and $P\left(X < x_j\right)$ before taking their difference in the losses sum. This maintains stochastic dominance. Figure 29.3 shows a typical weighting function. The function is flat in the middle range of probabilities to represent likelihood insensitivity. The function is then steeper near probability one to represent the certainty effect and near probability zero to represent its counterpart,

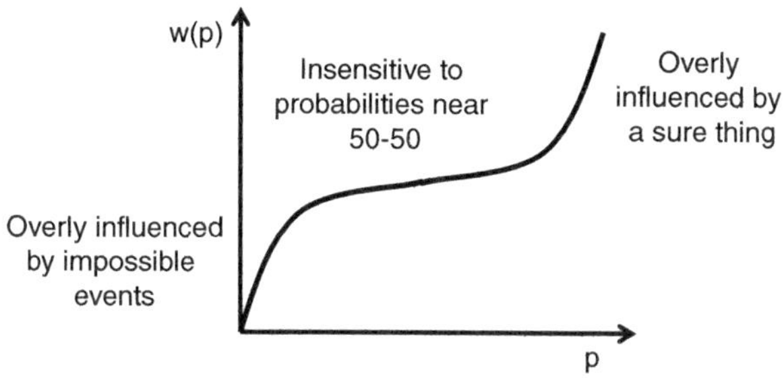

Figure 29.3. A typical probability weighting function from PT.

the possibility effect. Thus, comparisons of lotteries using the prospect theory form can more closely resemble those experimentally observed for human decision makers.

29.4 Existing Work on PT in Game Theory

Ritzberger (1994) developed an equilibrium concept where player's preferences are described by rank-dependent utility (RDU) (Quiggin 1982, 1991). RDU is a special case of PT that incorporates rank-dependence and probability weighting (Figure 29.3), but maintains the utility function from EUT (Figure 29.2 left) rather than the loss-averse value function from PT (Figure 29.2 right). Ritzberger showed that the following game

	L	R
T	0,2	2,0
B	3,0	1,1

does not have an equilibrium when player 1 (choosing T or B) has preferences $w_1(p) = p^2$ and $v_1(x) = x$ (i.e., RDU preferences) and player 2 (choosing L or R) has preferences $w_2(p) = p$ and $v_2(x) = x$ (i.e., EUT preferences). However, the RDU player's probability weighting function is not as commonly observed in experimental studies as inverse-S shapes (Wu & Gonzalez, 1996, 1999). Leclerc (2014) shows that when player 1 has the more commonly observed inverse-S shaped preferences, one can find values of A and B so that no equilibrium exists for the following game

	L	R
T	0,A	2,0
B	3,0	1,B

Thus, PT preferences can cause lack of existence of equilibrium in simultaneous games. This is problematic as this approach would then provide no prediction of either player's actions in such games. While lack of existence only occurs for more pathological examples, the fact that a pathological example can be found in a 2 by 2 game is concerning.

Shalev (2000) developed "loss-aversion equilibrium" (LAE) to address reference dependence, but maintained EUT's linearity in probabilities. He introduced three approaches for the treatment of the reference point: in the first, the reference point is exogenous and fixed throughout a game (eLAE). In the remaining two approaches, reference points are endogenously determined by the game's structure and equilibrium conditions; the first of these is "myopic" in that it remains constant throughout any given game (mLAE), while the latter is "non-myopic" in that it may change throughout the play of a game (nLAE) as players' payoffs change. Shalev (2000) shows that eLAE and mLAE exist for all finite games, but also demonstrates an extensive-form game where nLAE may fail to exist as a result of dynamic inconsistency.

There is no formal theory of how reference points are determined. However, one need only consider a case where you think you have won a prize only to later learn you will not receive the prize due to a technicality to see that we change our reference points. In this case, you have never received the prize, so you should be indifferent. However, when you think you have gained the prize, you have moved your reference point and then when it is "taken away," you see it as a loss and weight the loss more than the initial gain. This would indicate that the non-myopic version of Shalev's equilibrium is of interest and again the lack of existence is worrying. Fortuitously for our purposes, Leclerc (2014) shows that the lack of existence in Shalev's example is primarily due to his choice of value function and if one restricts the $v(0) = 0$, then existence is returned. There are still pathological examples, but if one assumes sub-game perfection, then existence can be guaranteed in sequential games like Figure 29.1.

29.5 Impacts of Modeling the Adversary with PT Preferences

To follow the asymmetric prescriptive/descriptive approach, we assume that the attacker will make his decision to maximize his prospect theoretic value and the defender can make his decision assuming that the attacker will follow these predictions (including assuming sub-game perfection). In this decision, there is no need to consider changes in either player's reference point as each player makes just one decision. Thus, we can solve the decision in Figure 29.1

right using prospect theory preferences for the attacker and EUT preferences for the defender without concern for existence of the equilibrium.

Merrick and Leclerc (2014) assume a value function from Tversky and Kahneman (1992)

$$v(x) = \begin{cases} x^{\alpha} & x \geq 0 \\ -\lambda(-x)^{\beta} & x < 0 \end{cases},$$

where $\alpha, \beta \geq 0$ and $\lambda \geq 1$, and the probability weighting function from Chateauneuf and colleagues (2007)

$$\pi(p) = \begin{cases} 1 & p = 1 \\ \kappa p + \frac{1}{2}(1-\kappa) & 0 < p < 1, \\ 0 & p = 0 \end{cases}$$

where $0 < \kappa \leq 1$. Thus, α and β represent risk attitude, λ represents the level of loss aversion, and κ represents the level of likelihood insensitivity, with $\lambda = 1$ representing no loss aversion and $\kappa = 1$ representing no probability weighting.

Merrick and Leclerc (2014) observe that typical values of each parameter are α and β between 1 and 0.2, λ between 2 and 3, and κ between 0.6 and 0.8. The graphs in Figure 29.4 vary λ and κ in these ranges. There are four graphs showing increasing risk aversion by setting α and β to 0.8, 0.7, 0.6, and 0.5. The gray area in each graph represents the attacker choosing an RDD if the defender doesn't screen and a conventional attack if the defender does screen, i.e., the attacker is deterred from an RDD because of screening, making screening worthwhile. The white area in each graph represents the attacker choosing a conventional weapons attack irrespective of the defender's decision. In this case, screening is not worthwhile as it does not influence the attacker's decision. One can see in Figure 29.4 that the attacker is deterred by screening when he has lower levels of loss aversion. Higher levels of loss aversion lead the attacker to not choose an RDD whether the defender is screening or not as there is still a lower chance of detection. Likelihood insensitivity has a smaller effect in this range, although more likelihood insensitivity does slightly increase the level of loss aversion where the attacker switches from deterrence to no deterrence. The level of risk aversion has a larger effect as increasing risk aversion reduces the size of the area where the attacker is deterred.

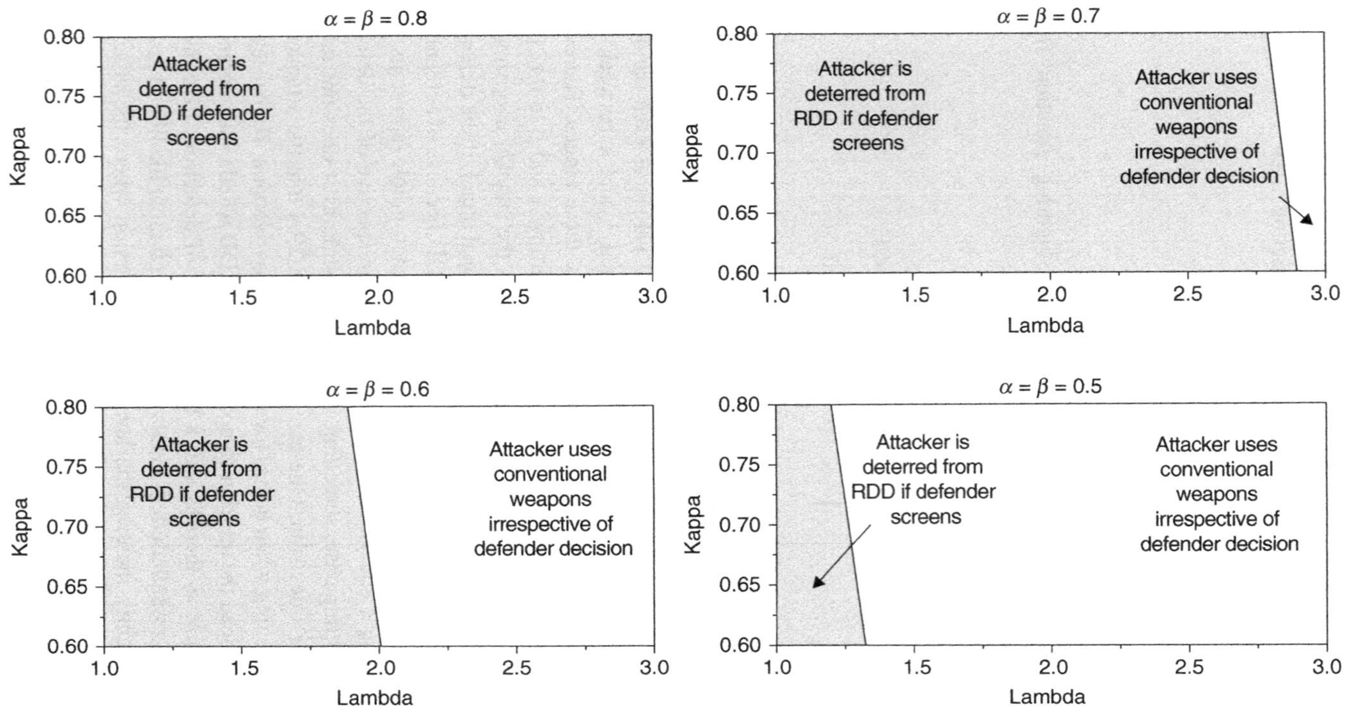

Figure 29.4. The effect of loss aversion (λ), likelihood insensitivity (κ), and risk aversion (α, β) on the attacker decision in Figure 29.1 right.

In summary, we can see from Figure 29.4 that modeling an attacker's preferences using prospect theory has a significant effect on the prediction of the attacker's decision and hence changes the defender's decision (see Merrick & Leclerc, 2015) for additional details and examples). It is also clear from these results that it is important to understand the level of loss aversion, likelihood insensitivity, and risk aversion exhibited by the attacker to accurately predict his behavior.

29.6 Conclusions and Further Research

In this chapter, we have advocated for an asymmetrically prescriptive/descriptive approach for counterterrorism decision making. The analyst is seeking to help the defender make better decisions, so defender decisions are handled in a prescriptive mode using expected utility preferences. In contrast, the analyst is modeling the behavior of the attacker, so the attacker decision should be handled in a descriptive mode. We reviewed the leading descriptive model for decisions under uncertainty, namely, prospect theory (Tversky & Kahneman, 1992) and the decision-making behavior it represents. We then reviewed the existing literature on problems that are introduced by modeling preferences in game theory with prospect theory, arguing that many counterterrorism decisions do not fall prey to these problems. We finished by demonstrating the impact of the attacker having prospect theory preferences on a container screening decision.

The asymmetrically prescriptive/descriptive approach can improve counterterrorism decision making. However, there is an important area for future research. Keeney (2007) and Keeney and von Winterfeldt (2010) argue that terrorists are not simply trying to maximize the economic impact on their target and so we should model the multiple objectives they are trying to achieve if we are to better understand and model their decision making. This view is expanded in Rosoff and John (2009) arguing that the analyst is modeling the attacker's multiple objective decisions by proxy. However, Rosoff and John represent the attacker as rational, meaning that they assume expected utility preferences. If we follow the asymmetric prescriptive/descriptive approach, then we require a descriptive approach to multiple objective decisions. It is not clear that decision makers make explicit trade-offs as implied by multiple objective utility theory without considerable help (Simon, 1955). Several descriptive models for multiple-objective decision making have been proposed with quite different implications for descriptive decision behavior. Tverksy's (1972) elimination by aspects and Gigerenzer and colleagues' (1999) take-the-best heuristic each assume that the decision maker eliminates alternatives using one objective at a time. In elimination by aspects, the objectives are used in order of importance and alternatives are eliminated if they do not reach a given aspiration level. In take-the-best, the objectives are used in order of validity or lack of ambiguity. Rather than using such non-compensatory approaches, Zank (2001) and Bleichrodt and colleagues (2009) extend prospect theory to the multiple objective case, with Zank assuming an overall reference point for the multiple objective value and Bleichrodt and colleagues assuming a reference point for each objective. Einhorn (1970, 1971) and Tversky and colleagues (1988) find that decision makers exhibit noncompensatory effects in their choice patterns, so further developments of the asymmetric prescriptive/descriptive approach to include noncompensatory models of multiple objective preferences are necessary.

Acknowledgments

This work was developed partially under grants from the National Science Foundation (SES-1227324). The research was done at Virginia Commonwealth University. The views and conclusions contained in this document are those of the authors and should not be interpreted as necessarily representing the official policies, either expressed or implied, of the National Science Foundation.

REFERENCES

Berejikian, J. (2002a). A cognitive theory of deterrence. *Journal of Peace Research, 39*(2), 165–183.

Berejikian, J. (2002b). Model building with prospect theory: A cognitive approach to international relations. *Political Psychology, 23*(4), 759–786.

Bleichrodt, H., Schmidt, U., & Zank, H. (2009). Additive utility in prospect theory. *Management Science, 55*(5), 863–873.

Butler, C. (2007). Prospect theory and coercive bargaining. *Journal of Conflict Resolution, 51*(2), 227–250.

Camerer, C. (2003). *Behavioral game theory*. Princeton, NJ: Princeton University Press.

Chateauneuf, A., Eichberger, J., & Grant, S. (2007). Choice under uncertainty with the best and worst in mind: NEO-additive capacities. *Journal of Economic Theory, 137*, 538–567.

Colman, A. (2003). Cooperation, psychological game theory, and limitations of rationality in social interaction. *Behavioral and Brain Sciences, 26*, 139–198.

Einhorn, H. J. (1970). The use of nonlinear, noncompensatory models in decision making. *Psychological Bulletin, 73*(3), 221.

Einhorn, H. J. (1971). Use of nonlinear, noncompensatory models as a function of task and amount of information. *Organizational Behavior and Human Performance, 6*(1), 1–27.

Gigerenzer, G., & Todd, P. M. (1999). *Simple heuristics that make us smart*. New York: Oxford University Press.

Kadane, J. B., & Larkey, P. D. (1982a). Subjective probability and the theory of games. *Management Science, 28*(2), 1365–1379.
Kadane, J. B., & Larkey, P. D. (1982b). The confusion of is and ought in game theoretic contexts. *Management Science, 29*(12), 113–120.
Kahneman, D. (2011). *Thinking fast and slow*. New York: Farrar, Straus and Girou.
Kahneman, D., & Tversky, A. (1979). Prospect theory: An analysis of decision under risk. *Econometrica, 47*, 263–292.
Kar, D., Fang, F., Fave, F. D., Sintov, N., & Tambe, M. (2015). "A game of thrones": When human behavior models compete in repeated Stackelberg security games. In *Proceedings of the International Conference on Autonomous Agents and Multiagent Systems (AAMAS).*
Keeney, G. L., & von Winterfeldt, D. (2010). Identifying and structuring the objectives of terrorists. *Risk Analysis, 30*(12), 1803–1816.
Keeney, R. L. (2007). Modeling values for anti-terrorism analysis. *Risk Analysis, 27*(3), 585–596.
Leclerc, P. (2014). *Prospect theory preferences in noncooperative game theory*. PhD Dissertation. Virginia Commonwealth University.
McKelvey, R., & Palfrey, T. (1995). Quantal response equilibria for normal form games. *Games and Economic Behavior, 10*(1), 6–38.
Merrick, J. R. W., & LeClerc, P. (2016). Modeling adversaries in counterterrorism decisions using prospect theory. *Risk Analysis, 36*(4), 681–693.
Merrick, J. R. W., LeClerc, P., Trenkov, H., & Olsen, R. (2015). Outthinking the terrorists. In Anthony Cox (Ed.), *Breakthroughs in risk analysis* (pp. 287–312). Hoboken, NJ: Wiley.
Merrick, J. R. W., & Parnell, G. S. (2011). A comparative analysis of PRA and intelligent adversary methods for counterterrorism risk management. *Risk Analysis, 31*(9), 1488–1510.
Metzger, L., & Rieger, M. (2010). Equilibria in games with prospect theory preferences. Unpublished manuscript.
Nash, J. (1950). Equilibrium points in N-person games. *Proceedings of the National Academy of Sciences, 36*(36), 48–49.
Nguyen, T. H., Yang. R., Azaria, A., Kraus, S., & Tambe, M. (2013, July). Analyzing the effectiveness of adversary modeling in security games. In *Proceedings of the 27th AAAI Conference on Artificial Intelligence (AAAI).*
Parnell, G. S., Smith, C. M., & Moxley, F. I. (2010). Intelligent adversary risk analysis: A bioterrorism risk management model. *Risk Analysis, 30*(1), 32–48.
Quiggin, J. (1982). A theory of anticipated utility. *Journal of Economic Behavior and Organization, 3*(4), 323–343.
Quiggin, J. (1991). *Generalized expected utility theory: The rank-dependent model.* Boston, MA: Kluwer Academic Publishers.
Raiffa, H. (1985). *The art and science of negotiation*. Cambridge, MA: Harvard University Press.
Ramsey, F. (1931). *Foundations essays in philosophy, logic, mathematics and economics*. London: Humanities Press, Routledge.
Rios Insua, D., Rios, J., & Banks, D. (2009). Adversarial risk analysis. *Journal of the American Statistical Association, 104*(486), 841–854.

Ritzberger K. (1994). On games under expected utility with rank dependent probabilities. *Theory and Decision, 40*(1), 1–27.

Rosoff, H., & John, R. (2009). Decision analysis by proxy for the rational terrorist. Proc. Internat. Joint Conf. Artificial Intelligence (IJCAI-09), Quant. Risk Anal. Security Applications. Accessed January 4, 2013, http://teamcore.usc.edu/QRASA-09/SubmissionsFinal/Paper_4.pdf.

Savage, L. J. (1954). *Foundations of statistics*. New York: John Wiley.

Shalev, J. (2000). Loss aversion equilibrium. *International Journal of Game Theory, 29*(2), 269–287.

Simon, H. A. (1955). A behavioral model of rational choice. *The Quarterly Journal of Economics, 69*(1), 99–118.

Starmer, C. (2000). Developments in non-expected utility theory: The hunt for a descriptive theory of choice under risk. *Journal of Economic Literature, 38*(2), 332–382.

Tversky, A. (1972). Elimination by aspects: A theory of choice. *Psychological Review, 79*(4), 281–299.

Tversky, A., & Kahneman, D. (1992). Advances in prospect theory: Cumulative representation of uncertainty. *Journal of Risk and Uncertainty, 5*(4), 297–323.

Tversky, A., Sattath, S., & Slavic, P. (1988). Contingent weighting in judgment and choice. *Psychological Review, 95*(37), 1–84.

von Neumann J, & Morgenstern O. (1944). *The theory of games and economic behaviour*. Princeton, NJ: Princeton University Press.

Wu, G., & Gonzalez, R. (1996). Curvature of the probability weighting function. *Management Science, 42*(12), 1676–1690.

Wu, G., & Gonzalez, R. (1999). On the shape of the probability weighting function. *Cognitive Psychology, 38*, 129–166.

Yang, R., Kiekintveld, C., Ordonez, F., Tambe, M., & John, R. (2013). Improving resource allocation strategy against human adversaries in security games: An extended study. *Artificial Intelligence Journal, 195*, 440–469.

Zank, H. (2001). Cumulative prospect theory for parametric and multiattribute utilities. *Mathematics of Operations Research, 26*, 67–81.

30

Near-Misses and Decision Making Under Uncertainty in the Context of Cybersecurity

Robin L. Dillon and Catherine H. Tinsley

30.1 Introduction

Most people think of near-misses as dreadful close calls that could have been a lot worse, but recent research has focused on another class of near-misses, ones that are far more common and overlooked. These more common events are often misinterpreted or ignored as signs that a system is resilient rather than signals that a negative event nearly happened. Dillon and Tinsley (2008) examine near-misses in the context of space systems motivated by NASA's loss of the space shuttle *Columbia* after prior problems with foam debris on nearly every previous mission but without catastrophe. Dillon, Tinsley, and Cronin (2011) and Tinsley, Dillon and, Cronin (2012) examine near-misses in the context of extreme weather events in particular hurricanes motivated by poor evacuation decision making during Hurricane Katrina where prior near-miss experience and risk taking lead to successful outcomes. Dillon, Tinsley, and Burns (2014a and 2014b) examine terrorism threats in the United States where events since September 11 have largely been near-misses. They describe how near-miss terrorist attacks can produce increased vigilance among the general population, but find that this awareness decreases with time as behavior gradually returns to some prior baseline. In this research, we continue to examine the near-miss event, but explore a new context, the cyber-attack.

On December 19, 2013, the Target retailer announced that it had experienced a data breach that resulted in the loss of 40 million credit card numbers. As described by Riley and colleagues (2014), six months *before* the actual credit card theft, Target began installing a $1.6 million malware detection tool made by the computer security firm FireEye (whose customers also included the CIA and the Pentagon). Target had a team of security specialists in Bangalore that monitored its computers around the clock, and notified Target's security operations center in Minneapolis if anything suspicious was detected with the network. On Saturday, November 30, 2013, as hackers uploaded exfiltration malware to move stolen credit card numbers, FireEye did spot them. Bangalore specialists were alerted, who passed along the alert to the security team in Minneapolis. The question is, why did the security team in Minneapolis not respond further? After the breach, the company did a thorough audit of its security processes. While poring over computer logs, Target found FireEye's alerts from November 30, 2013, and additional ones from December 2, 2013, when hackers installed yet another version of the malware. Not only should those alarms have been impossible to miss, they went off early enough that the hackers had not begun transmitting the stolen card data out of Target's network. In other words, had the Minneapolis team responded to the alarms, the theft of the credit card numbers from this malware attempt could have been prevented. With multiple lawsuits pending, Target management did not speculate on why no one responded to the alarms.

In this chapter, however, we *do* speculate on why Target's security team, and why people in general, might not respond to such alarms. We hypothesize that it could have been past experience with warnings and how these events were mentally categorized as successes rather than near-misses (or, more precisely, near-failures) that might have influenced how Target's security team in Minneapolis viewed the warnings in November and December 2013.

30.2 Background – Warning Systems and Near-Miss Events

Warning systems are commonly categorized regarding the quality of their warning signal in terms of Type I errors (false positives) and Type II errors (false negatives). But the binary categorizations of true/false and success/failure oversimplify the complexity in many security systems. These systems rely on a subjective interpretation of warnings by people who depend on their memory of the past events and outcomes (Paté-Cornell, 1985). For example, warnings of a stream of unusual traffic against a system's firewall could be triggered by a misconfigured application somewhere in the world or it could be a probe of the firewall's defenses (Singer & Friedman, 2014). The warning correctly issues an alert of unusual traffic (true positive), but people must judge the cause of the traffic (i.e., benign or harmful) that created the warning. In a similar context, consider the security systems at our nation's airports that correctly detect a passenger who packed toothpaste or sunscreen larger than three ounces. The system is correct in detecting the contraband item (true positive), but the item detected was not a bomb (or dangerous or hazardous in any way). In both of these examples, the system is correctly alerting officials of a problem (i.e., unusual firewall traffic and a prohibited item), but there is a human response over time that depends on the officials interpreting these warnings and maintaining a level of vigilance that will distinguish between general Internet traffic noise and a probe of a network and between toothpaste and an explosive device. This is also the challenge in signal detection theory (Canfield, Fischhoff, & Davis, 2015).

In addition to interpreting whether a signal is a warning of a hazard, there is a related challenge of interpreting whether events are false alarms. In the toothpaste example, most people could agree that since the toothpaste is a prohibited item, it is not a false alarm. But when the toothpaste is innocently packed by a 72-year-old woman or a 13-year-old boy, who then has the tube confiscated, other passengers might feel like this is a false alarm. A muddier case could be when a tornado warning is issued for a county in Iowa. If one city in the county (e.g., Newton) does not see a tornado, but a different city in the county does (e.g., Baxter), was the tornado warning a false alarm for Newton? People might differ in their interpretation.

Rather than a categorization that considers events binary (success or failure), we define additional events, specifically different types of near-misses, that reside on a continuum between true success and true failure where their location on the continuum is largely determined by subjective interpretation as to how close the event came to a recognizable failure.

Understanding how people interpret events and respond to warnings is important because a too-common response to systems that are interpreted as "over-warning" is to turn them off. Warning alarms were disabled on the Deepwater Horizon oil rig prior to the April 20, 2010, disaster because "they did not want people woken up at three o'clock in the morning due to false alarms" (Pilkington, 2010), and an alarm box near the front entrance of the White House designed to alert guards to intruders "had been muted at what officers believed was a request of the usher's office" prior to the White House intruder incident in September 2014 (Leonning, 2014). We believe that Target personnel's experience with past warnings and their interpretation of those events as evidence of the network's resilience could have contributed to missing the warnings of the malware intrusions in 2013.

30.2.1 *Resilient Near-Misses*

A near-miss is some event when a negative outcome could have happened because of hazardous conditions but did not (Dillon & Tinsley, 2008). Tinsley and colleagues (2012) demonstrate that some near-misses have cues that highlight resilience (a disaster was successfully avoided). People experiencing these "resilient near-misses" can observe that the negative outcome did not impact them and then illegitimately underestimate the danger of subsequent hazardous situations. False alarms are one type of resilient near-miss when events forecasted to happen do not occur, but other resilient near-misses could include events that do occur but with no obvious damage, such as a forecasted tornado that occurs far from any populated area or a terrorist attack when the bomb is a decoy provided by the Federal Bureau of Investigation (FBI). Tinsley and colleagues (2012) further demonstrate that individuals who experience resilient near-misses are then likely to ignore or discount future hazard warnings. A cybersecurity example of a resilient near-miss would be when a person downloads malware from a fraudulent e-mail but fortunately, the computer's anti-virus software (that can only detect a limited set of malware programs) did happen to immediately detect *this* malware and delete it.

30.2.2 *Vulnerable Near-Misses*

At the other end of the continuum, vulnerable near-misses have cues that highlight danger (a disaster almost happened). When the near-miss experience also highlights the harm the event could have caused, it adds information to counteract the basic resilient near-miss effect. Reconsidering the cybersecurity example, the same person downloaded malware from a fraudulent e-mail and the computer's anti-virus software was able to detect this malware and delete it, but only after the computer crashed and needed to be restarted a half-dozen times. Another alternative type of vulnerable near-miss could be a failure that happens to someone else. For example, the same person received the fraudulent e-mail and deleted it, but the person's friend did not. The friend downloaded the malware from a fraudulent e-mail and his anti-virus software was not able to detect the malware before his entire hard drive was lost.

Other scenarios could be imagined that would fall elsewhere on the continuum, closer to success (delete the fraudulent e-mail without clicking on it) or closer to failure (hard drive is corrupted, but data can be salvaged by technology support staff). In a series of studies, we explore several hypotheses associated with near-misses and risky decision making in the cybersecurity context.

30.3 Study 1

As discussed previously, near-misses offer evidence of both a system's resilience (failure averted) and its vulnerability (failure narrowly averted) (March et al., 1991). Therefore, a near-miss resembles a success in terms of ultimate outcome, but resembles a failure in other details, such as problems arising before the outcome. However, consistent with outcome bias (Baron & Hershey, 1988), decision makers should evaluate a near-miss with no obviously hazardous salient features like a success based on the ultimate outcome. Thus, we expect near-misses to be judged in a similar manner as successes and in a significantly different manner from failures despite the fact that such near-misses would have been failures except for an element of chance.

Hypothesis 1: Both resilient near-miss outcomes and success outcomes will be judged significantly more successful than failure outcomes.

30.3.1 *Materials and Methods*

The premise in this first study that everyone reads is as follows:

Chris owns a small business that sells custom-made table protectors for dining room tables. He is not getting rich selling table protectors, but enjoys the freedoms and challenges of running a small business. As the new year starts, he faces two challenges. He only has the resources to resolve one this year and one will have to wait until next year.

1) *His chief supplier is located in a country recently identified as "high risk" on a U.S. government ranking of how corrupt a country's public sectors are. While he has not had any problems with his supplier to date, he is concerned about the stability of the supplier given the country's corruption score. Thus he is considering a different materials supplier in a country with a low corruption score. But switching will create an upfront cost for him because most suppliers need a guaranteed purchase amount and require prepayment for initial purchases.*
2) *A few months ago, Microsoft stopped issuing security updates for the XP operating system, which his computers use to receive online orders and store customer data. Although he also uses Blast antivirus software, Microsoft's security updates were important to defend against malware and viruses that target Microsoft XP vulnerabilities. Microsoft of course recommends upgrading to Windows 8, which is Microsoft supported, but for him that means also purchasing all new computers.*

Half are told that Chris changes the supplier and half are told that Chris upgrades the computers. For the half told that Chris changes the supplier, there are three conditions (bolded here to highlight differences):

- ***Success***: *Chris decides the supplier is a greater concern and changes suppliers. Since the switch there have been no problems, but it has been a short period of time. Chris closely monitors computers and orders are processed smoothly. Did Chris make a good decision to invest in a new supplier?*
- ***Near-Miss***: *Chris decides the supplier is a greater concern and changes suppliers. Since the switch there have been no problems, but it has been a short period of time. At one point, two of the computers intermittently*

respond very slowly and have in fact been infected with some malware. Chris suspects a virus attack and runs the latest version of the Blast antivirus software. ***By chance, Chris' antivirus software is one that is able to isolate this malware program***. *No damage is done.*

Despite the risk of future XP problems, all orders are processed smoothly. Did Chris make a good decision to invest in new a supplier?

- ***Failure***: *Chris decides the supplier is a greater concern and changes suppliers. Since the switch there have been no problems, but it has been a short period of time. For the next few months, Chris closely watches the computers for signs of malware or viruses. At one point, two of the computers intermittently respond very slowly and have in fact been infected with some malware. Chris suspects a virus attack and runs the latest version of the Blast antivirus software.* ***By chance, Chris' antivirus software is not one that is able to isolate this malware program.***

  For several weeks, the malware program secretly captures customer data until Blast is again updated and can detect the problem. Chris' small business incurs significant costs associated with notifying customers and supplying them with fraud detection services, and the risks of future XP problems still exist. Did Chris make a good decision to invest in a new supplier?

For the half told that Chris upgrades the computers, there are three conditions (bolded here to highlight differences):

- ***Success***: *Chris decides computers are a greater concern and upgrades the computers. Since the switch there have been no problems, but it has been a short period of time. Chris closely monitors supplier and orders are processed smoothly. Did Chris make a good decision to invest in new computers?*
- ***Near-Miss***: *Chris decides computers are a greater concern and upgrades the computers. Since the switch there have been no problems, but it has been a short period of time. Chris closely monitors his supplier. Two senior government officials are arrested on fraud charges, but* ***"by chance" these officials are not associated with foreign commerce and trade, so no damage is done and orders are processed smoothly.*** *Did Chris make a good decision to invest in new computers?*
- ***Failure***: *Chris decides computers are a greater concern and upgrades the computers. Since the switch there have been no problems, but it has been a short period of time. Chris closely monitors his supplier. Two senior government officials are arrested on fraud charges, and* ***"by chance" these officials are associated with foreign commerce and trade, so Chris' supplier has difficulty shipping product*** *and Chris' small business incurs significant*

costs and still has risks of future problems with suppliers. Did Chris make a good decision to invest in new computers?

Participants were recruited from Amazon Mechanical Turk and 1,078 passed an attention filter and completed the exercise randomly distributed across the six conditions. The sample was 54% male with an average age of 33.5. Of the participants, 87% had at least some college education, and 79% were white.

30.3.2 *Results*

After reading their assigned condition, participants were asked whether Chris made a good decision. The results are shown in Table 30.1. Condition has a significant effect on the dependent variables ($\chi(5) = 166.34$, $p < 0.001$), where the failure outcome is perceived as significantly different than either the success or the near-miss even though the difference between the near-miss and the failure case was chance, supporting hypothesis 1.

The perception that near-misses were equivalent to successes and different from failures was the same regardless of whether the near-misses occurred to Chris's computers or to Chris's supply chain. At the same time, if we compare only the participants in the failure conditions, participants who read about a failure because of a computer virus were significantly more likely to agree with Chris on the decision to use resources to buy new computers (despite the supplier failure) than were participants who read about a failure because of a computer virus (because Chris had spent resources replacing the supplier ($\chi(1) = 7.66$, $p < 0.01$). Thus the overall pattern of results (comparing across all three conditions) suggests participants react to cyber threats the way they do to non-cyber threats. Results for just the failure conditions, though, additionally suggest that the cyber threat was seen as more hazardous than the supplier threat since more participants agreed that it was still wise of Chris to use resources to protect against a cyber threat even when the supplier risk was realized.

Additionally, we wanted to examine if participants might differ in their perception of the cyber threat depending on their knowledge of computer security–related concerns. Thus we asked them to rate their knowledge on a 0 to 10 scale on various computer security–related items (phishing, malware, spyware, botnet, computer virus, SQL injection, denial of service (DOS), encryption, virtual private network (VPN), anti-virus software, intrusion detection system, and firewall). We used the 10 items as a scale to assess a participant's perception of his technical knowledge (scale Cronbach's alpha: 0.94). The mean for the participants was 5.3 (st. dev. = 2.3).

Table 30.1. *Study 1 Results for "Did Chris make a good decision?"*

	Outcome: Success	Outcome: Near-Miss	Outcome: Failure
Condition: Changes suppliers	Yes: 82.5%	Yes: 83.9%	Yes: 38.0%
Condition: Buys new computers	Yes: 79.9%	Yes: 81.3%	Yes: 52.6%

Table 30.2. *Study 1 Binary logistic regression model*

Variables in the Equation	B	S.E.	Wald	Df	Sig.	Exp(B)
Technical Knowledge	–0.058	0.041	1.988	1	0.159	0.944
Problem (Supplier/ Computers)	0.570	0.343	2.760	1	0.097	1.768
Tech Know x Problem	–0.069	0.059	1.375	1	0.241	0.933
Outcome			173.559	2	0.000	
Outcome (1)	1.710	0.163	110.078	1	0.000	6.616
Outcome (2)	1.889	0.168	126.174	1	0.000	6.616
Constant	–0.004	0.243	0.000	1	0.985	0.996
Variable entered in step 1: Outcome						

We used a binary logistic regression where the dependent variable was whether participants agreed with the decision Chris made in their condition. The independent variables were: level of knowledge about computer security, the type of problem Chris had decided to fix, the interaction of these two variables, and the outcome that was realized (a failure, success, or near-miss). The type of outcome failure was the reference category. Results, shown in Table 30.2, indicate that the technical knowledge had no main effect on whether participants agreed with Chris's protective actions, nor was there any interaction between participants' technical knowledge and the type of problem Chris decided to fix on their agreement with Chris's actions. As in Table 30.1, the only variable that influenced whether participants agreed with Chris's actions was whether they read that Chris subsequently experienced a success or a near-miss rather than a failure.

30.3.3 *Discussion*

Our data show that a decision maker whose decision resulted in a near-miss, like those whose decisions resulted in success, is evaluated more favorably than those whose decisions resulted in failure. These evaluations demonstrate that participants are judging near-miss outcomes as successes rather than failures. Because prior research shows that past successes contribute to overconfidence and over-optimism in future decision making, interpreting near-misses as successes can be antithetical to learning (Kahneman & Lovallo, 1993) and thus jeopardize future decision making.

This near-miss bias reinforces the challenge of correctly interpreting warnings that the system is being challenged if the system regularly passes the test even if by chance. If Target had previous network probes that did not seem to be near failures, its staff could have learned the wrong lesson about its system's resiliency rather than vulnerability. Yet, since no conditions in Study 1 had any salient cues about danger, the near-misses examined all would be "resilient near-misses." We test the impact of a vulnerable near-miss in the experiment that follows.

30.4 Study 2

Identifying near-misses requires a subjective interpretation of events that in most cases had no obviously bad outcome, but instead only a slight deviation from normal that the observer has to interpret as a sign of a potentially larger problem. As Study 1 showed, observers are poorly equipped to interpret near-misses as such – typically interpreting them as successes as long as the outcome was not an overt failure, and consequently failing to extract valuable information from them. If people experience a near-miss as a disaster that almost happened rather than as a disaster that was avoided, then their hazard category should be associated with vulnerability. We distinguish these "vulnerable" near-misses (wherein a disaster almost happened and results in the perceived vulnerability of the system) from "resilient" near-misses (wherein a disaster could have but did not happen and results in the perceived resilience of the system).

In this study, we examine the influence of the type of near-miss on one's decision to mitigate risks.

> *Hypothesis 2*: People with resilient near-miss information (that highlights how a negative event did not happen) will be less likely to take mitigating action for an impending hazard than people without this information.

> *Hypothesis 3*: People with vulnerable near-miss information (that highlights how a negative event almost happened) will be more likely to take mitigating action for an impending hazard than people with resilient near-miss information.

30.4.1 *Materials and Methods*

Participants read a similar version to Study 1, but instead of Chris, a third-party decision maker, participants are told they are the decision maker. In addition to being presented with the two challenges (their chief supplier and the end of support for Windows XP), they read additional text based on random assignment to one of four conditions:

- *Control*: *You must now decide which investment to make for your business (switch suppliers or upgrade computers). You know the business can't afford both investments.*
- *Resilient Near-Miss*: *Since Microsoft stopped the security updates for Windows XP, you have been monitoring your computers for signs of malware or viruses. Last month, two of the computers began intermittently responding very slowly. You ran Blast antivirus, and the issue seems to be resolved. You have had no trouble since then. You must now decide …*
- *Vulnerable Near-Miss*: *Since Microsoft stopped the security updates, you have been monitoring your computers for signs of malware or viruses. Last month, two of the computers began intermittently responding very slowly. Then your system froze as you tried to download a Blast antivirus upgrade. You were able to download the upgrade from another computer, reboot the two infected computers, and upload the antivirus update, and the issue seems to be resolved. You have had no trouble since then. You must now decide …*
- *Extremely Vulnerable Near-Miss*: *Since Microsoft stopped the security updates, you have been monitoring your computers for signs of malware or viruses. Last month, two of the computers began intermittently responding very slowly. Then your system froze as you tried to download a Blast antivirus upgrade. You were able to download the upgrade from another computer, reboot the two infected computers, and upload the antivirus update, and the issue seems to be resolved. You have had no trouble since then, but you thought about how small businesses can go bankrupt if their customer data is breached as the costs of notification/fraud monitoring protection drain the company's assets. You must now decide …*

Participants were recruited from Amazon Mechanical Turk and 640 passed an attention filter and completed the exercise randomly distributed across the conditions. The sample was 66% male with an average age of 30.0. Of the participants, 88% had at least some college education, and 76% were white.

30.4.2 *Results*

After reading their assigned condition, participants were asked to choose whether they would invest in a new supplier or invest in new computers. The results are shown in Table 30.3. Condition has a significant effect on the dependent variables ($\chi(3) = 42.4$, $p < 0.001$), where the resilient near-miss outcome makes people less willing to buy new computers instead of choosing to change suppliers, but the vulnerable near-miss condition and the extremely vulnerable near-miss condition encourage participants to purchase the new computers instead of choosing to change suppliers, supporting hypotheses 2 and 3.

Similar to the previous study, participants were also asked to rate on a 0 to 10 scale how knowledgeable they were about 10 computer security–related items (phishing, malware, spyware, botnet, computer virus, SQL injection, denial of service (DOS), encryption, virtual private network (VPN), anti-virus software, intrusion detection system, and firewall). We used the 10 items as a scale to assess a participant's perception of his technical knowledge (scale Cronbach's alpha: 0.94). The mean for the participants was 5.7 (st. dev. = 2.3).

We used a binary logistic regression with decision (to invest in the computer problems = 1 and invest in the supply chain problem = 0) as the dependent variable and participants' self-reported technical knowledge and condition (control, resilient near-miss, vulnerable near-miss, extremely vulnerable near-miss) as the independent variables. Because our condition variable was a nested variable, we used Helmert contrast tests. This required making contrast variables that are orthogonal to each other. Our first variable (Contrast 1) contrasts the control condition with the three near miss conditions and thus was coded as −3 for the control participants and 1 for any of the three types of near-miss participants. The second variable (Contrast 2) contrasts the resilient near-miss condition from the two vulnerable near-miss conditions and thus has a value of −2 if a participant is in the resilient near-miss condition and 1 if the participant is in either of the vulnerable near-miss conditions. The third variable (Contrast 3) contrasts the vulnerable near-miss from the extremely vulnerable near-miss and therefore has a value of 1 if the participant is in the vulnerable near-miss condition and has a value of −1 if the participant is in the extremely vulnerable near-miss condition.

Results in Table 30.4 show there was no main effect for participants' technical knowledge on whether they decided to invest in protective measures against the computer versus supply chain hazard. This is similar to Study 1, where technical knowledge does not seem to influence participants' assessments of cyber versus non-cyber hazards. The first contrast, Contrast 1, is not significant, meaning there is no overall effect for having read a near-miss scenario

Table 30.3. *Study 2 Results for "What investment do you make?"*

	Outcome Condition: Resilient Near-Miss	Outcome Condition: Control	Outcome Condition: Vulnerable Near-Miss	Outcome Condition: Extremely Vulnerable Near-Miss
Decision: Change suppliers	57.3%	50.3%	43.2%	22.8%
Decision: Buy new computers	42.7%	49.7%	56.8%	77.2%

Table 30.4. *Study 2 Binary logistic regression model*

Variables in the Equation	B	S.E.	Wald	Df	Sig.	Exp(B)
Technical Knowledge	−0.047	0.036	1.720	1	0.190	0.954
Contrast1(1)	−0.294	0.228	1.663	1	0.197	0.745
Contrast2			6.342	1	0.012	
Contrast2(2)	0.572	0.227	6.342	1	0.012	1.771
Contrast3			14.669	1	0.000	
Contrast3(2)	0.940	0.245	14.669	1	0.000	2.560
Constant	0.266	0.264	1.015	1	0.314	1.304

Variable entered in step 1: Contrast1, Contrast2, Contrast3

rather than no information about a computer virus. This is not surprising since we expect the influence of the vulnerable near-misses and the resilient near-misses to pull participants in opposite directions, effectively canceling out an overall near-miss effect (when compared to a control condition). Consistent with the table of percentages reported earlier, the second contrast is significant (Contrast 2), and shows that those who experience a resilient cyber-related near-miss are less likely to take protective action against the cyber hazard than those who experience a vulnerable cyber-related near-miss. The final contrast (Contrast 3) is significant and shows that those who experience only the vulnerable near-miss are significantly less likely to take protective action against the cyber hazard than those who experience the extremely vulnerable cyber near-miss. These results are fully consistent with Table 30.3 and the hypotheses.

30.4.3 *Discussion*

If a participant is told that he or she had previously experienced a problem (computers responding slowly, freezing), he or she is more willing to use available resources to fix the computers rather than the supplier. Moreover, if the participant is reminded of the potentially severe negative consequences (small businesses can go bankrupt if their customer data is breached as the costs of notification/fraud monitoring protection drain the company's assets), they are even more likely to use available resource to fix the computers. But if a problem arises without any associated signs of vulnerability as in the resilient case, the participant is less willing to fix the computers than those in the vulnerable near-miss conditions, but also is even less likely than those in the control condition to invest in protective action. In other words, people are less willing to invest in a protective action for a hazard once they've experienced it than they are if they never experienced a problem in the first place. This demonstrates the potential dilemma that can occur at an organization like Target; if the systems seem to be defending against attacks, this can increase the decision maker's confidence that the systems will defend against future attacks even if many of the events are the result of good fortune. Good fortune would occur in most cases when the attacker's tools align with the defender's capabilities and thus the defender is able to successfully defend against the attack. But of course this alignment cannot be guaranteed to exist in future events.

30.5 General Discussion

People know they cannot protect against every hazard – it is not economically or practically feasible. Thus, understandably, people set priorities and protect against those that feel most risky. The only problem with this approach is that perceptions of risk are sometimes not based on the objective likelihoods and consequences, but rather on how "close" they feel a hazard is to becoming a bad outcome, and this assessment of closeness of the hazard can be over- or underestimated. Our results show, on the one hand, that if people experience a near-miss without any associated cues about danger, they tend to underestimate the risk of the hazard and are less likely to take protective action – even compared to those who had experienced no problems. On the other hand, if people experience a near-miss associated with system vulnerability, the hazard feels quite close to producing a bad outcome, and people may overestimate the hazard risk. Both of these conditions can occur repeatedly in systems exposed to threats and hazards.

Researchers have been studying these topics in non-cyber contexts such as natural disasters for some time (Yi & Bier, 1998). The added challenge in the cybersecurity context is that understanding the cybersecurity environment and the corresponding threats and defenses is challenging to many non-computer scientists. This makes research in the decision-making component additionally challenging. For example, in a February 2014 congressional hearing following the Target attack, Representative Joe Barton (R-Texas) asked the Target executives about what "occurred when criminals came into stores and used credit cards that infected the systems at the point of purchase" (Henneberger, 2014). The agent in charge of the criminal investigation division of the Secret Service's cyber-operations had to explain to the congressman during the hearing that no one came into the store, but instead people infiltrated the computer networks and inserted malware code. This story exemplifies how hard it is for some to grasp the technical details of the cyber threat. So while decision making under risk is a challenging topic that warrants further study in all contexts, the added challenge of many people's aversion to the technical details has limited the study in this context. This chapter is a start to contributing to this area of study.

30.6 Conclusions

Based on our own research (Tinsley et al., 2011), we remind decision makers in situations where near-misses are common to beware of high-pressure situations, to beware of normalization of deviance (Vaughan, 1996), and to search for root causes of weak signals. Too many catastrophes (e.g., *Challenger* space shuttle, Deep Water Horizon oil drilling rig, etc.) have happened in high-pressure situations. Decision makers need to question if, given more time and more resources, the same risky decision would be made. Additionally, as organizations experience more resilient near-misses, their behaviors toward risky decisions can change as described by the normalization of deviance paradigm. Decision makers need to be concerned if their policies toward risk change over time to be continually accepting more risk. Finally, in most resilient near-misses, there is only a weak signal of the problem. Even when the outcome seems to have resolved itself satisfactorily, decision makers need to question: Why did this effect happen? What was required to produce this? What do we need to do to address the root cause?

Significant investments have been made to improve the technologies associated with cybersecurity. As Singer and Friedman (2014) point out, though, we have to study the human component of the decision making also:

> The real battle in cybersecurity is not just about high technology. It is also driven by the human factor, the fight over our behavior. To use a metaphor from the wars of old, the walls protecting a network don't always have to be knocked down or tunneled under by an attacker. Sometimes, as in ancient Troy, oblivious defenders simply open up the gates without recognizing the attacker's deception. (Singer & Friedman, 2014, p. 65)

Again, as reported by Riley and colleagues (2014), Target had strong technological system in place, but it failed in the human interpretation of the data. Thus, the attack was successful. Research is required to further help decision makers to remain vigilant over time even when their systems may appear to be resilient.

REFERENCES

Baron, J., & Hershey, J. C. (1988). Outcome bias in decision evaluation. *Journal of Personality and Social Psychology*, *54*, 569–579.

Canfield, C., Fischhoff, B., & Davis, A. (2015). *Using signal detection theory to measure phishing detection ability and behavior*. Working paper, Carnegie Mellon University, Pittsburgh, PA.

Dillon, R. L., & Tinsley, C. H. (2008). How near-misses influence decision making under risk: A missed opportunity for learning. *Management Science*, *54*(8), 1425–1440.

Dillon, R. L., Tinsley, C. H., & Burns, W. J. (2014a). Evolving risk perceptions about near-miss terrorist events. *Decision Analysis*, *11*(1), 27–42.

Dillon, R. L., Tinsley, C. H., & Burns, W. J. (2014b). Near-misses and future disaster preparedness. *Risk Analysis*, *34*(10), 1907–1922.

Dillon, R. L., Tinsley, C. H., & Cronin, M A. (2011). Why near-miss events can decrease an individual's protective response to hurricanes. *Risk Analysis, 31*(3), 440–449.

Henneberger, M. (2014, February 5). Can credit card data breaches be prevented? *The Washington Post*. Accessed at: www.stltoday.com/business/local/can-credit-card-data-breaches-be-prevented/article_67246ae9-1325-581b-978d-9aeff021279c.html.

Kahneman, D., & Lovallo, D. (1993). Timid choices and bold forecasts: A cognitive perspective on risk taking. *Management Science*, *37*(1), 17–31.

Leonning, C. (2014, September 29). White House fence-jumper made it far deeper into building than previously known. *The Washington Post*. Accessed at: www.washingtonpost.com/politics/white-house-fence-jumper-made-it-far-deeper-into-building-than-previously-known/2014/09/29/02efd53e-47ea-11e4-a046-120a8a855cca_story.html.

March, J. G., Sproul, L., & Tamuz, M. (1991). Learning from samples of one or fewer. *Organization Science*, *2*, 1–13.

Paté-Cornell, M. E. (1985). Warning systems and risk reduction. *Risk analysis in the private sector. Advances in Risk Analysis*, *220*, 469–482.

Pilkinton, E. (2010, July 23). Deepwater Horizon alarms were switched off "to help workers sleep." *The Guardian*. Accessed at: www.theguardian.com/environment/2010/jul/23/deepwater-horizon-oil-rig-alarms.

Riley, M., Elgin, B., Lawrence, D., & Matlack, C. (2014, March 13). Missed alarms and 40 million stolen credit card numbers: How Target blew it. *Business Week.* Accessed at: www.businessweek.com/printer/articles/189573-missed-alarms-and-40-million-stolen-credit-card-numbers-how-target-blew-it.

Singer, P. W., & Friedman, A. (2014). *Cybersecurity and cyberwar: What everyone needs to know*. Oxford: Oxford University Press.

Tinsley, C. H., Dillon, R. L., & Cronin, M. A. (2012). How near-miss events amplify or attenuate risky decision making. *Management Science*, *58*(9), 1596–1613.

Tinsley, C. H., Dillon, R. L., & Madsen, P. M. (2011, April). How to avoid catastrophe. *Harvard Business Review*, 90–97.

Vaughan, D. (1996). *The Challenger launch decision: Risky technology, culture, and deviance at NASA*. Chicago, IL: University of Chicago Press.

Yi, W., & Bier, V. (1998). An application of copulas to accident precursor analysis. *Management Science, 44*(12), S257–S270.

Index

Page numbers in *italics* and **bold** refer to figures and tables, respectively